Martingales and Stochastic Analysis

SERIES ON MULTIVARIATE ANALYSIS

Editor: M M Rao

Forthcoming

Vol. 2: Mathematical Methods on Sample Surveys
       H. F. Tucker

Martingales and Stochastic Analysis

J. Yeh

Department of Mathematics
University of California, Irvine

World Scientific
Singapore • New Jersey • London • Hong Kong

Published by

World Scientific Publishing Co. Pte. Ltd.
P O Box 128, Farrer Road, Singapore 912805
USA office: Suite 1B, 1060 Main Street, River Edge, NJ 07661
UK office: 57 Shelton Street, Covent Garden, London WC2H 9HE

Library of Congress Cataloging-in-Publication Data
Yeh, James.
    Martingales and stochastic analysis / James Yeh.
       p.   cm. -- (Series on multivariate analysis ; vol. 1)
    Includes bibliographical references and index.
    ISBN 981022477X
    1. Martingales (Mathematics)   2. Stochastic analysis.   I. Title.
II. Series.
QA274.5.Y44   1995
519.2'87--dc20   95-42874
                                                                         CIP

British Library Cataloguing-in-Publication Data
A catalogue record for this book is available from the British Library.

Copyright © 1995 by World Scientific Publishing Co. Pte. Ltd.

All rights reserved. This book, or parts thereof, may not be reproduced in any form or by any means, electronic or mechanical, including photocopying, recording or any information storage and retrieval system now known or to be invented, without written permission from the Publisher.

For photocopying of material in this volume, please pay a copying fee through the Copyright Clearance Center, Inc., 222 Rosewood Drive, Danvers, MA 01923, USA.

This book is printed on acid-free paper.

Printed in Singapore by Uto-Print

Preface

This monograph is an introduction to martingales and stochastic analysis assuming only real analysis and some basic concepts in probability theory. Stochastic independence, conditional expectation, and regular conditional probability are included in the appendix. In writing this monograph I strove for clarity and precision at the cost of concision. Details of proofs are worked out for readability and for reference.

Chapter 1 begins with a collection of theorems concerning σ-algebras that will be need in the sequel. Measurable processes, progressively measurable processes, right-continuous and left-continuous processes, predictable processes, and approximation by sequences of simple processes are discussed Stopping times are introduced as a method of truncation for stochastic processes. Uniform integrability, which plays an important role in martingale theory, is included.

Chapter 2 introduces martingale and submartingale properties as monotonicity conditions in terms of conditional expectation. The chapter includes the fundamental martingale inequalities, optional stopping theorem, optional sampling theorem, martingale convergence theorem, closing a martingale by a final element, roles of uniform integrability in convergence questions for martingales, increasing processes as a class of submartingales, increasing processes as integrators in stochastic integration, and the Doob-Meyer decomposition theorem.

Chapter 3 treats Itô's stochastic integral in the language of martingale theory. The integrand is a left-continuous, and more generally a predictable, process satisfying some integrability condition on the sample functions and the integrator process is a right-continuous L_2-martingale. The stochastic integral is then extended to local L_2-martingales as integrators. The chapter also includes Itô's formula for quasimartingales and Itô's stochastic calculus.

Chapter 4 starts with a study of the space of continuous functions on which solutions of stochastic differential equations will be constructed. Function space representation of solutions, initial value problems, uniqueness of solution in the sense of probability law, pathwise uniqueness of solution, existence and uniqueness of strong solutions are the topics included in this chapter.

The main part of the text consists of four chapters in twenty sections: §1 to §20. One counter for all the twenty sections runs through the chapters. Within each section, one counter is used for definitions, lemmas, propositions, theorems, corollaries, remarks, observations, and examples. Thus for instance in §5, Definition 5.1 is followed by Definition

5.2, Theorem 5.3, Proposition 5.4, Observation 5.5, and so on. Most sections are divided in to subsections and these are numbered with upper case roman numerals. For instance §5 is divided into five subsections: [I] to [V]. Numbering of definitions, lemmas, propositions, etc., in a section is independent of the subsections. The appendix has four parts: A to D. Within each one of the four parts in the appendix, one counter is used for definitions, lemmas, propositions, etc. Thus for instance in Appendix A, Definition A.1 is followed by Proposition A.2, Observation A.3, Theorem A.4, and so on.

In writing this monograph I am indebted to [21] P. A. Meyer and [7] N. Ikeda and S. Watanabe as valuable guides. It is hoped that this monograph will also serve as an introduction as well as a supplement to the latter. I wish to express my gratitude to Dr. G. Engl and N. Malik for their comments on the manuscript. I thank Professor M. M. Rao for inviting me to write this monograph. Finally I express my appreciation to the editorial staff of World Scientific Publishing Co. for their cooperation in the production of the manuscript.

August, 1995   J. Yeh

Contents

Preface		v
Notations		xi

1 Stochastic Processes   1
- §1   Generated σ-algebras 1
- §2   Stochastic Processes 11
  - [I] Measurable and Progressively Measurable Processes 11
  - [II] Predictable Processes 19
- §3   Stopping Times 25
  - [I] Stopping Times for Stochastic Processes with Continuous Time 25
  - [II] Stopping Times for Stochastic Processes with Discrete Time 39
  - [III] Random Variables at Stopping Times 41
  - [IV] Stopped Processes and Truncated Processes 45
- §4   Convergence in L_p and Uniform Integrability 49
  - [I] Convergence in L_p 49
  - [II] Uniformly Integrable Systems of Random Variables 53

2 Martingales   71
- §5   Martingale, Submartingale and Supermartingale 71
  - [I] Martingale, Submartingale and Supermartingale Properties 71
  - [II] Convexity Theorems 76
  - [III] Discrete Time Increasing Processes and Doob Decomposition 78
  - [IV] Martingale Transform 81
  - [V] Some Examples 84
- §6   Fundamental Submartingale Inequalities 86
  - [I] Optional Stopping and Optional Sampling 87
  - [II] Maximal and Minimal Inequalities 93
  - [III] Upcrossing and Downcrossing Inequalities 101

§7 Convergence of Submartingales 110
  [I] Convergence of Submartingales with Discrete Time 110
  [II] Convergence of Submartingales with Continuous Time 114
  [III] Closing a Submartingale with a Final Element 116
  [IV] Discrete Time L_2-Martingales 119
§8 Uniformly Integrable Submartingales 122
  [I] Convergence of Uniformly Integrable Submartingales 122
  [II] Submartingales with Reversed Time 124
  [III] Optional Sampling by Unbounded Stopping Times 131
  [IV] Uniform Integrability of Random Variables at Stopping Times 141
§9 Regularity of Sample Functions of Submartingales 145
  [I] Sample Functions of Right-Continuous Submartingales 145
  [II] Right-Continuous Modification of a Submartingale 149
§10 Increasing Processes 155
  [I] The Lebesgue Stieltjes Integral 155
  [II] Integration with Respect to Increasing Processes 158
  [III] Doob-Meyer Decomposition 171
  [IV] Regular Submartingales 187

3 Stochastic Integrals 197
§11 L_2-Martingales and Quadratic Variation Processes 197
  [I] The Space of Right-Continuous L_2-Martingales 197
  [II] Signed Lebesgue-Stieltjes Measures 205
  [III] Locally Bounded Variation Processes 207
  [IV] Quadratic Variation Processes 212
§12 Stochastic Integrals with Respect to Martingales 221
  [I] Stochastic Integral of Bounded Left-Continuous Adapted
      Simple Processes w. r. t. L_2-Martingales 222
  [II] Stochastic Integral of Predictable Processes w. r. t. L_2-Martingales .. 235
§13 Adapted Brownian Motions 262
  [I] Processes with Independent Increments 262
  [II] Brownian Motions in $\mathbb{R}^d$ 270
  [III] 1-Dimensional Brownian Motions 286
  [IV] Stochastic Integrals with Respect to a Brownian Motion 290
§14 Extensions of the Stochastic Integral 297
  [I] Local L_2-Martingales and Their Quadratic Variation Processes 297
  [II] Extensions of the Stochastic Integral to Local Martingales 306

	§15	Itô's Formula	318
		[I] Continuous Local Semimartingales and Itô's Formula	318
		[II] Stochastic Integrals with Respect to Quasimartigales	333
		[III] Exponential Quasimartingales	335
		[IV] Multidimensional Itô's Formula	340
	§16	Itô's Stochastic Calculus	344
		[I] The Space of Stochastic Differentials	344
		[II] Fisk-Stratonovich Integrals	352
4	**Stochastic Differential Equations**		**357**
	§17	The Space of Continuous Functions	357
		[I] Function Space Representation of Continuous Processes	357
		[II] Metrization of the Space of Continuous Functions	364
	§18	Definition and Function Space Representation of Solution	368
		[I] Definition of Solutions	368
		[II] Function Space Representation of Solutions	372
		[III] Initial Value Problems	385
	§19	Existence and Uniqueness of Solutions	395
		[I] Uniqueness in Probability Law and Pathwise Uniqueness of Solutions	395
		[II] Simultaneous Representation of Two Solutions on a Function Space	405
	§20	Strong Solutions	419
		[I] Existence of Strong Solutions	419
		[II] Uniqueness of Strong Solutions	432
A	**Stochastic Independence**		**443**
B	**Conditional Expectations**		**453**
C	**Regular Conditional Probabilities**		**475**
D	**Multidimensional Normal Distributions**		**487**
	Bibliography		495
	Index		498

Notations

$\mathbb{N}$	the natural numbers	1
$\mathbb{Z}$	the integers	1
$\mathbb{Z}_+$	the nonnegative integers	1
$\mathbb{R}$	the real numbers	1
$\mathbb{R}_+$	$[0, \infty)$	1
$\mathbb{T}$	$\mathbb{R}_+$ or $\mathbb{Z}_+$	72
$\overline{\mathbb{N}}$	$\mathbb{N} \cup \{\infty\}$	1
$\overline{\mathbb{Z}}_+$	$\mathbb{Z}_+ \cup \{\infty\}$	1
$\overline{\mathbb{R}}_+$	$\mathbb{R}_+ \cup \{\infty\}$	1
$\overline{\mathbb{T}}$	$\mathbb{T} \cup \{\infty\}$	72
$\overline{\mathbb{R}}$	$\{-\infty\} \cup \mathbb{R} \cup \{\infty\}$	1
$\sigma(\mathfrak{E})$	σ-algebra generated by $\mathfrak{E}$	1
$\alpha(\mathfrak{E})$	algebra generated by $\mathfrak{E}$	1
$d(\mathfrak{E})$	d-class generated by $\mathfrak{E}$	5
$\sigma(f)$	σ-algebra generated by f	10
$(\Omega, \mathfrak{F}, \{\mathfrak{F}_t\}, P)$	a filtered space	14
$X^{T \wedge}$	stopped process of X by stopping time T	45
$X^{[T]}$	truncated process of X by stopping time T	47
$P \cdot \lim_{n \to \infty} X_n$	limit of convergence in probability measure P	52
$C \bullet X$	martingale transform	81
$\mathbf{M}_2(\Omega, \mathfrak{F}, \{\mathfrak{F}_t\}, P)$	space of null at 0 right-continuous L_2 martingales on a standard filtered space	198
$\mathbf{M}_2^c(\Omega, \mathfrak{F}, \{\mathfrak{F}_t\}, P)$	space of null at 0 a. s. continuous L_2 martingales on a standard filtered space	198
$|\cdot|_t$	a seminorm on $\mathbf{M}_2(\Omega, \mathfrak{F}, \{\mathfrak{F}_t\}, P)$	198
$|\cdot|_\infty$	a quasinorm on $\mathbf{M}_2(\Omega, \mathfrak{F}, \{\mathfrak{F}_t\}, P)$	198
$\mathbf{M}_2^{loc}(\Omega, \mathfrak{F}, \{\mathfrak{F}_t\}, P)$	space of null at 0 right-continuous local L_2 martingales on a standard filtered space	300
$\mathbf{M}_2^{c,loc}(\Omega, \mathfrak{F}, \{\mathfrak{F}_t\}, P)$	space of null at 0 a. s. continuous local L_2 martingales on a standard filtered space	300
$[M]$	quadratic variation process of M	214
$[M, N]$	quadratic variation process of M and N	214

$\mathbf{A}(\Omega, \mathfrak{F}, \{\mathfrak{F}_t\}, P)$	space of a.s. increasing processes on a standard filtered space	210
$\mathbf{A}^c(\Omega, \mathfrak{F}, \{\mathfrak{F}_t\}, P)$	space of a.s. continuous a.s. increasing processes on a standard filtered space	210
$\mathbf{A}^{loc}(\Omega, \mathfrak{F}, \{\mathfrak{F}_t\}, P)$	see Definition 14.6	300
$\mathbf{A}^{c,loc}(\Omega, \mathfrak{F}, \{\mathfrak{F}_t\}, P)$	see Definition 14.6	300
$\mathbf{V}(\Omega, \mathfrak{F}, \{\mathfrak{F}_t\}, P)$	space of a.s. locally bounded variation processes on a standard filtered space	210
$\mathbf{V}^c(\Omega, \mathfrak{F}, \{\mathfrak{F}_t\}, P)$	space of a.s. continuous a.s. locally bounded variation processes on a standard filtered space	210
$\mathbf{V}^{loc}(\Omega, \mathfrak{F}, \{\mathfrak{F}_t\}, P)$	see Definition 14.6	300
$\mathbf{V}^{c,loc}(\Omega, \mathfrak{F}, \{\mathfrak{F}_t\}, P)$	see Definition 14.6	300
$\lvert V \rvert$	total variation process of $V \in \mathbf{V}(\Omega, \mathfrak{F}, \{\mathfrak{F}_t\}, P)$	208
m_L	the Lebesgue measure	59
m_W	the Wiener measure	362
μ_A	family of Lebesgue-Stieltjes measures determined by an a.s. increasing process A	159
$\int_{[0,t]} X(s)\, dA(s)$	alternate for $\int_{[0,t]} X(s) \mu_A(ds)$	159
$\mathbf{L}_{p,\infty}(\mathbb{R}_+ \times \Omega, \mu_A, P)$	see Definition 11.17	212
$\lVert \cdot \rVert_{p,t}^{A,P}$	a seminorm on $\mathbf{L}_{p,\infty}(\mathbb{R}_+ \times \Omega, \mu_A, P)$	212
$\lVert \cdot \rVert_{p,\infty}^{A,P}$	a quasinorm on $\mathbf{L}_{p,\infty}(\mathbb{R}_+ \times \Omega, \mu_A, P)$	212
$\mathbf{L}_{2,\infty}^{loc}(\mathbb{R}_+ \times \Omega, \mu_A, P)$	see Definition 14.14	307
$\mathbf{L}_{1,\infty}^{loc}(\mathbb{R}_+ \times \Omega, \mu_A, P)$	see Definition 15.8	333
$\mathbf{L}_{p,\infty}(\mathbb{R}_+ \times \Omega, m_L \times P)$	see Definition 13.35	292
$\lVert \cdot \rVert_{p,t}^{m_L \times P}$	a seminorm on $\mathbf{L}_{p,\infty}(\mathbb{R}_+ \times \Omega, m_L \times P)$	292
$\lVert \cdot \rVert_{p,\infty}^{m_L \times P}$	a quasinorm on $\mathbf{L}_{p,\infty}(\mathbb{R}_+ \times \Omega, m_L \times P)$	292
$\mathbf{L}_0(\Omega, \mathfrak{F}, \{\mathfrak{F}_t\}, P)$	the collection of all bounded adapted left-continuous processes on a standard filtered space $(\Omega, \mathfrak{F}, \{\mathfrak{F}_t\}, P)$	222
$X \bullet M$	stochastic integral of X with respect to M	223, 313
$\int_{[0,t]} X(s)\, dM(s)$	alternate for $(X \bullet M)_t$	223, 314

NOTATIONS

$\mathbf{C}(\mathbb{R}_+ \times \Omega)$	space of measurable processes with a.s. continuous sample functions	333
$\mathbf{B}(\mathbb{R}_+ \times \Omega)$	space of measurable processes with a.s. locally bounded sample functions	333
$\mathbf{Q}(\Omega, \mathfrak{F}, \{\mathfrak{F}_t\}, P)$	collection of all quasimartingales on a standard filtered space	345
$\mathbf{dQ}$	collection of the equivalence classes in $\mathbf{Q}(\Omega, \mathfrak{F}, \{\mathfrak{F}_t\}, P)$	346
$\mathbf{dM}_2^{c,loc}, \mathbf{dV}^{c,loc}$	subcollections of $\mathbf{dQ}$	347
$\mathbf{W}^d$	space of d-dimensional continuous functions on $\mathbb{R}_+$	357
$(\mathbf{W}^d, \mathfrak{W}^d, m_W^d)$	d-dimensional Wiener space	362
$(\mathbf{W}^d, \mathfrak{W}^d, \{\mathfrak{W}_t^d\})$	see Definition 17.3	359
$(\mathbf{W}^r, \mathfrak{W}^{r,w}, \{\mathfrak{W}_t^{r,w}\}, m_W^r)$	see Definition 19.10	406
$\mathbf{M}^{d \times r}(\mathbf{W}^d, \mathfrak{W}^d, \{\mathfrak{W}_t^d\})$	see Definition 18.2	369
$(\mathbf{W}^{r+d}, \mathfrak{W}^{r+d,*}, \{\mathfrak{W}_t^{r+d,*}\}, P_{(B,X)})$	see Definition 18.10	374
$(\mathbf{W}^{r+d}, \mathfrak{W}^{r+d,*,x}, \{\mathfrak{W}_t^{r+d,*,x}\}, P_{(B,X)}^x)$	see Definition 18.17	387
$\mathbf{L}_{2,\infty}^{d,c}(\mathbb{R}_+ \times \Omega, \{\mathfrak{F}_t\}, m_L \times P)$	see Definition 20.2	419

Chapter 1

Stochastic Processes

§1 Generated σ-algebras

For a collection $\mathfrak{E}$ of subsets of a set S, we write $\sigma_S(\mathfrak{E})$ for the σ-algebra of subsets of S generated by $\mathfrak{E}$, that is, the smallest σ-algebra of subsets of S containing $\mathfrak{E}$. Similarly, we write $\alpha_S(\mathfrak{E})$ for the algebra of subsets of S generated by $\mathfrak{E}$. The subscript S will be dropped from σ_S and α_S when there is no ambiguity. For a topological space S we write $\mathfrak{B}_S$ or $\mathfrak{B}(S)$ for the Borel σ-algebra of subsets of S, that is, the σ-algebra of subsets of S generated by the collection of the open sets in S. In particular, $\mathfrak{B}_\mathbb{R}$ is the smallest σ-algebra containing all the open sets in the space of the real numbers $\mathbb{R}$ with the usual topology, and $\mathfrak{B}_{\overline{\mathbb{R}}}$ is the smallest σ-algebra containing all the open sets in the space of the extended real numbers $\overline{\mathbb{R}} = \{-\infty\} \cup \mathbb{R} \cup \{\infty\}$ with the topology of the two-point compactification of $\mathbb{R}$, that is, $\mathfrak{B}_{\overline{\mathbb{R}}}$ is the smallest σ-algebra of subsets of $\overline{\mathbb{R}}$ containing $\mathfrak{B}_\mathbb{R}$, $\{-\infty\}$, and $\{\infty\}$.

For a collection $\mathfrak{E}$ of subsets of a set S and for a subset A of S we write $\mathfrak{E} \cap A$ for the collection $\{E \cap A : E \in \mathfrak{E}\}$ of subsets of S. Similarly for a subset A of a set S and for a collection $\mathfrak{E}$ of subsets of a set T we write $A \times \mathfrak{E}$ for the collection $\{A \times E : E \in \mathfrak{E}\}$ of subsets of $S \times T$.

We use the symbols $\mathbb{N}, \mathbb{Z}$ and $\mathbb{Z}_+$ respectively for the collections of the positive integers, all integers and nonnegative integers. Let $\overline{\mathbb{N}} = \mathbb{N} \cup \{\infty\}$, $\overline{\mathbb{Z}}_+ = \mathbb{Z}_+ \cup \{\infty\}$, $\mathbb{R}_+ = [0, \infty)$ and $\overline{\mathbb{R}}_+ = [0, \infty]$.

The following theorems concerning generation of a σ-algebra will be used frequently.

Theorem 1.1. *Let f be a mapping of a set S into a set T. Then for an arbitrary collection $\mathfrak{E}$ of subsets of T, we have $\sigma(f^{-1}(\mathfrak{E})) = f^{-1}(\sigma(\mathfrak{E}))$.*

Proof. Since $\mathfrak{E} \subset \sigma(\mathfrak{E})$ we have $f^{-1}(\mathfrak{E}) \subset f^{-1}(\sigma(\mathfrak{E}))$. Since $\sigma(\mathfrak{E})$ is a σ-algebra of subsets of T, $f^{-1}(\sigma(\mathfrak{E}))$ is a σ-algebra of subsets of S. Then $\sigma(f^{-1}(\mathfrak{E})) \subset f^{-1}(\sigma(\mathfrak{E}))$.

To prove the reverse inclusion, let $\mathfrak{A}_1$ be an arbitrary σ-algebra of subsets of S and let $\mathfrak{A}_2 = \{A \subset T : f^{-1}(A) \in \mathfrak{A}_1\}$. To show that $\mathfrak{A}_2$ is a σ-algebra of subsets of T, note first of all that $f^{-1}(T) = S \in \mathfrak{A}_1$ so that $T \in \mathfrak{A}_2$. Secondly, since f is defined on the entire S, for every $A \in \mathfrak{A}_2$ we have $f^{-1}(A^c) = f^{-1}(T - A) = f^{-1}(T) - f^{-1}(A) \in \mathfrak{A}_1$ so that $A^c \in \mathfrak{A}_2$. Finally for any $\{E_n : n \in \mathbb{N}\} \subset \mathfrak{A}_2$ we have $f^{-1}(\cup_{n \in \mathbb{N}} A_n) = \cup_{n \in \mathbb{N}} f^{-1}(A_n) \in \mathfrak{A}_1$ so that $\cup_{n \in \mathbb{N}} A_n \in \mathfrak{A}_2$ and thus $\mathfrak{A}_2$ is a σ-algebra of subsets of T. In particular, if we let $\mathfrak{A} = \{A \subset T : f^{-1}(A) \in \sigma(f^{-1}(\mathfrak{E}))\}$, then $\mathfrak{A}$ is a σ-algebra of subsets of T and furthermore $\mathfrak{A} \supset \mathfrak{E}$ so that $\mathfrak{A} \supset \sigma(\mathfrak{E})$ and thus $\sigma(f^{-1}(\mathfrak{E})) \supset f^{-1}(\sigma(\mathfrak{E}))$. Therefore $\sigma(f^{-1}(\mathfrak{E})) = f^{-1}(\sigma(\mathfrak{E}))$. ∎

Corollary 1.2. *Let S be a set and $\mathfrak{E}$ be a collection of subsets of a set T. Then $S \times \sigma(\mathfrak{E}) = \sigma(S \times \mathfrak{E})$.*

Proof. Let π be the projection of $S \times T$ onto T. Then $\pi^{-1}(\sigma(\mathfrak{E})) = \sigma(\pi^{-1}(\mathfrak{E})) = \sigma(S \times \mathfrak{E})$ by Theorem 1.1. On the other hand $\pi^{-1}(\sigma(\mathfrak{E})) = S \times \sigma(\mathfrak{E})$. Therefore $S \times \sigma(\mathfrak{E}) = \sigma(S \times \mathfrak{E})$. ∎

Lemma 1.3. *Let $\mathfrak{E}_i$ be an arbitrary collection of subsets of a set S_i for $i = 1, \ldots, d$ where $d \in \mathbb{N}$. Then*

(1) $$\sigma(\mathfrak{E}_1) \times \cdots \times \sigma(\mathfrak{E}_d) \subset \sigma(\mathfrak{E}_1 \times \cdots \times \mathfrak{E}_d).$$

Proof. Let $E_2 \in \mathfrak{E}_2, \ldots, E_d \in \mathfrak{E}_d$ be arbitrarily fixed and let T_1 be the projection of $S_1 \times E_2 \times \cdots \times E_d$ onto S_1. Then by Theorem 1.1 we have $T_1^{-1}(\sigma(\mathfrak{E}_1)) = \sigma(T_1^{-1}(\mathfrak{E}_1))$, that is,

$$\sigma(\mathfrak{E}_1) \times E_2 \times \cdots \times E_d = \sigma(\mathfrak{E}_1 \times E_2 \times \cdots \times E_d) \subset \sigma(\mathfrak{E}_1 \times \mathfrak{E}_2 \times \cdots \times \mathfrak{E}_d).$$

By the arbitrariness of $E_2 \in \mathfrak{E}_2, \ldots, E_d \in \mathfrak{E}_d$, we have

(2) $$\sigma(\mathfrak{E}_1) \times \mathfrak{E}_2 \times \cdots \times \mathfrak{E}_d \subset \sigma(\mathfrak{E}_1 \times \mathfrak{E}_2 \times \cdots \times \mathfrak{E}_d).$$

Next, let $F_1 \in \sigma(\mathfrak{E}_1)$, $E_3 \in \mathfrak{E}_3, \ldots, E_d \in \mathfrak{E}_d$ be arbitrarily fixed. By (2) we have

(3) $$F_1 \times \mathfrak{E}_2 \times E_3 \times \cdots \times E_d \subset \sigma(\mathfrak{E}_1 \times \mathfrak{E}_2 \times \mathfrak{E}_3 \times \cdots \times \mathfrak{E}_d).$$

§1. GENERATED σ-ALGEBRAS

Let T_2 be the projection of $F_1 \times S_2 \times E_3 \times \cdots \times E_d$ onto S_2. Then by Theorem 1.1 we have $T_2^{-1}(\sigma(\mathfrak{E}_2)) = \sigma(T_2^{-1}(\mathfrak{E}_2))$, that is,

$$F_1 \times \sigma(\mathfrak{E}_2) \times E_3 \times \cdots \times E_d = \sigma(F_1 \times \mathfrak{E}_2 \times E_3 \times \cdots \times E_d)$$
$$\subset \sigma(\mathfrak{E}_1 \times \mathfrak{E}_2 \times \mathfrak{E}_3 \times \cdots \times \mathfrak{E}_d),$$

where the last set inclusion is by (3). From the arbitrariness of $F_1 \in \sigma(\mathfrak{E}_1)$, $E_3 \in \mathfrak{E}_3, \ldots$, $E_d \in \mathfrak{E}_d$, we have

$$\sigma(\mathfrak{E}_1) \times \sigma(\mathfrak{E}_2) \times \mathfrak{E}_3 \times \cdots \times \mathfrak{E}_d \subset \sigma(\mathfrak{E}_1 \times \mathfrak{E}_2 \times \mathfrak{E}_3 \times \cdots \times \mathfrak{E}_d).$$

Thus proceeding we have (1). ∎

Theorem 1.4. *Let S_i be a topological space satisfying the second axiom of countability, let $\mathfrak{O}_{S_i}$ be the collection of the open sets in S_i, and let $\mathfrak{B}_{S_i} = \sigma(\mathfrak{O}_{S_i})$, that is, the Borel-σ-algebra of subsets of S_i, for $i = 1, \ldots, d$. Then for the Borel-σ-algebra $\mathfrak{B}_{S_1 \times \cdots \times S_d}$ of subsets of $S_1 \times \cdots \times S_d$ in its product topology we have*

$$(1) \qquad \mathfrak{B}_{S_1 \times \cdots \times S_d} = \sigma(\mathfrak{B}_{S_1} \times \cdots \times \mathfrak{B}_{S_d}).$$

Proof. Let $\mathfrak{V}_i$ be a countable base for $\mathfrak{O}_{S_i}$ for $i = 1, \ldots, d$. Let $\mathfrak{O}_{S_1 \times \cdots \times S_d}$ be the collection of the open sets in the product topology of $S_1 \times \cdots \times S_d$. Then $\mathfrak{V}_1 \times \cdots \times \mathfrak{V}_d$ is a countable base for $\mathfrak{O}_{S_1 \times \cdots \times S_d}$. Since every member of $\mathfrak{O}_{S_1 \times \cdots \times S_d}$ is a countable union of members of $\mathfrak{V}_1 \times \cdots \times \mathfrak{V}_d$, we have

$$\mathfrak{O}_{S_1 \times \cdots \times S_d} \subset \sigma(\mathfrak{V}_1 \times \cdots \times \mathfrak{V}_d) \subset \sigma(\mathfrak{O}_{S_1} \times \cdots \times \mathfrak{O}_{S_d})$$
$$\subset \sigma(\sigma(\mathfrak{O}_{S_1}) \times \cdots \times \sigma(\mathfrak{O}_{S_d})) = \sigma(\mathfrak{B}_{S_1} \times \cdots \times \mathfrak{B}_{S_d})$$

and hence

$$(2) \qquad \mathfrak{B}_{S_1 \times \cdots \times S_d} = \sigma(\mathfrak{O}_{S_1 \times \cdots \times S_d}) \subset \sigma(\mathfrak{B}_{S_1} \times \cdots \times \mathfrak{B}_{S_d}).$$

On the other hand by Lemma 1.3 we have

$$\mathfrak{B}_{S_1} \times \cdots \times \mathfrak{B}_{S_d} = \sigma(\mathfrak{O}_{S_1}) \times \cdots \times \sigma(\mathfrak{O}_{S_d})$$
$$\subset \sigma(\mathfrak{O}_{S_1} \times \cdots \times \mathfrak{O}_{S_d}) \subset \sigma(\mathfrak{O}_{S_1 \times \cdots \times S_d}) = \mathfrak{B}_{S_1 \times \cdots \times S_d}$$

so that

$$(3) \qquad \sigma(\mathfrak{B}_{S_1} \times \cdots \times \mathfrak{B}_{S_d}) \subset \mathfrak{B}_{S_1 \times \cdots \times S_d}.$$

With (2) and (3) we have (1). ∎

Theorem 1.5. *Let $\mathfrak{E}$ be a collection of subsets of a set S and let $A \subset S$. Then $\sigma_A(\mathfrak{E} \cap A) = \sigma(\mathfrak{E}) \cap A$.*

Proof. Note that the subscript A in σ_A indicates that it is a σ-algebra of subsets of A, not a σ-algebra of subsets of S.

Since $\mathfrak{E} \subset \sigma(\mathfrak{E})$ we have $\mathfrak{E} \cap A \subset \sigma(\mathfrak{E}) \cap A$. From the fact that $\sigma(\mathfrak{E})$ is a σ-algebra of subsets of S and $A \subset S$ it follows that $\sigma(\mathfrak{E}) \cap A$ is a σ-algebra of subsets of A. Thus

(1) $$\sigma_A(\mathfrak{E} \cap A) \subset \sigma(\mathfrak{E}) \cap A.$$

Therefore, to prove the theorem it remains to show

(2) $$\sigma(\mathfrak{E}) \cap A \subset \sigma_A(\mathfrak{E} \cap A).$$

Let $\mathfrak{K}$ be the collection of subsets K of S of the type

(3) $$K = (C \cap A^c) \cup B \quad \text{where } C \in \sigma(\mathfrak{E}) \text{ and } B \in \sigma_A(\mathfrak{E} \cap A).$$

Observe that since $B \subset A$, the union in (3) is a disjoint union. By (3), $S \in \mathfrak{K}$ and $\mathfrak{K}$ is closed under countable unions. To show that $\mathfrak{K}$ is also closed under complementation, let $K \in \mathfrak{K}$ be as given by (3). Then

$$K^c = S - K = [(S \cap A^c) \cup A] - [(C \cap A^c) \cup B] = [(S \cap A^c) - (C \cap A^c)] \cup (A - B)$$

since $S \cap A^c \supset C \cap A^c$ and $A \supset B$. But $(S \cap A^c) - (C \cap A^c) = C^c \cap A^c$. Therefore

$$K^c = (C^c \cap A^c) \cup (A - B) \in \mathfrak{K}.$$

Thus $\mathfrak{K}$ is closed under complementation and is therefore a σ-algebra of subsets of S. Next, observe that for any $K \in \mathfrak{K}$ as given by (3) we have $K \cap A \in \sigma_A(\mathfrak{E} \cap A)$ so that $\mathfrak{K} \cap A \subset \sigma_A(\mathfrak{E} \cap A)$. Thus to show (2) it suffices to show that $\sigma(\mathfrak{E}) \cap A \subset \mathfrak{K} \cap A$. Since $\mathfrak{K}$ is a σ-algebra of subsets of S, it remains only to show that $\mathfrak{E} \subset \mathfrak{K}$. Let $E \in \mathfrak{E}$ and write $E = (E \cap A^c) \cup (E \cap A)$. Since $E \cap A \in \sigma_A(\mathfrak{E} \cap A)$, E is a subset of S of the type (3). Thus $E \in \mathfrak{K}$ and therefore $\mathfrak{E} \subset \mathfrak{K}$. This completes the proof. ∎

Definition 1.6. *A collection $\mathfrak{E}$ of sets is called a π-class if it is closed under finite intersection, that is, $A, B \in \mathfrak{E} \Rightarrow A \cap B \in \mathfrak{E}$. A collection $\mathfrak{G}$ of subsets of a set S is called a d-class of subsets of S if*

§1. GENERATED σ-ALGEBRAS

$1°$. $S \in \mathfrak{G}$,

$2°$. $A, B \in \mathfrak{G}, A \subset B \Rightarrow B - A \in \mathfrak{G}$,

$3°$. $A_n \in \mathfrak{G}, n \in \mathbb{N}, A_n \uparrow \Rightarrow \lim_{n \to \infty} A_n \in \mathfrak{G}$.

For a collection $\mathfrak{E}$ of subsets of a set S, the d-class of subsets of S generated by $\mathfrak{E}$, that is, the smallest d-class of subsets of S containing $\mathfrak{E}$, is denoted by $d(\mathfrak{E})$.

Note that for an arbitrary collection $\mathfrak{E}$ of subsets of a set S, the collection of all subsets of S is a d-class of subsets of S containing $\mathfrak{E}$. If we take the intersection of all d-classes of subsets of S containing $\mathfrak{E}$, then it is again a d-class of subsets of S containing $\mathfrak{E}$ and this is $d(\mathfrak{E})$.

Clearly a σ-algebra of subsets of a set S is both a π-class and a d-class of subsets of S. Conversely, a collection $\mathfrak{G}$ of subsets of a set S which is both a π-class and a d-class of subsets of S is a σ-algebra of subsets of S. Indeed, if $A, B \in \mathfrak{G}$, then $A^c \cap B^c \in \mathfrak{G}$ by $1°$ and $2°$ of Definition 1.6 and by the fact that $\mathfrak{G}$ is also a π-class. Thus $A \cup B \in \mathfrak{G}$ and therefore $\mathfrak{G}$ is an algebra of subsets of S. Then $\mathfrak{G}$ is a σ-algebra of subsets of S by $3°$ of Definition 1.6.

A d-class may be called a monotone class of sets since it satisfies condition $3°$ of Definition 1.6. The following monotone class theorem for sets by E. B. Dynkin shows that the d-class and the σ-algebra generated by a π-class are equal.

Theorem 1.7. *Let $\mathfrak{E}$ be a π-class of subsets of a set S. Then $d(\mathfrak{E}) = \sigma(\mathfrak{E})$.*

Proof. Since a σ-algebra of subsets of S is also a d-class of subsets of S, the collection of all σ-algebras of subsets of S containing $\mathfrak{E}$ is a subcollection of all d-classes of subsets of S containing $\mathfrak{E}$. Therefore $d(\mathfrak{E}) \subset \sigma(\mathfrak{E})$.

To prove the reverse inclusion, recall that a d-class of subsets of a set which is also a π-class is a σ-algebra of subsets of the set. Thus, if we show that $d(\mathfrak{E})$ is a π-class, then it is a σ-algebra of subsets of S containing $\mathfrak{E}$ so that $\sigma(\mathfrak{E}) \subset d(\mathfrak{E})$. To show that $d(\mathfrak{E})$ is a π-class, let $\mathfrak{G}_1$ be the collection of all subsets A of S such that

(1) $\qquad\qquad A \cap B \in d(\mathfrak{E}) \qquad$ for every $B \in \mathfrak{E}$.

It is easily verified that $\mathfrak{G}_1$ satisfies $1°$, $2°$ and $3°$ of Definition 1.6 and is thus a d-class of subsets of S. Since $\mathfrak{E}$ is a π-class, every $A \in \mathfrak{E}$ satisfies (1) and therefore $\mathfrak{E} \subset \mathfrak{G}_1$. Then

$d(\mathfrak{E}) \subset \mathfrak{G}_1$. This implies by (1) that

(2) $$A \in d(\mathfrak{E}), B \in \mathfrak{E} \Rightarrow A \cap B \in d(\mathfrak{E}).$$

Next, let $\mathfrak{G}_2$ be the collection of all subsets B of S such that

(3) $$B \cap A \in d(\mathfrak{E}) \quad \text{for every } A \in d(\mathfrak{E}).$$

It is easily verified that $\mathfrak{G}_2$ is a d-class of subsets of S. By (2), $\mathfrak{E} \subset \mathfrak{G}_2$. Thus $d(\mathfrak{E}) \subset \mathfrak{G}_2$. Therefore by (3),
$$A, B \in d(\mathfrak{E}) \Rightarrow A \cap B \in d(\mathfrak{E}).$$
This shows that $d(\mathfrak{E})$ is a π-class. Therefore, $\sigma(\mathfrak{E}) \subset d(\mathfrak{E})$. ∎

Corollary 1.8. *Let $\mathfrak{E}$ be a π-class of subsets of a set S and let μ and ν be two finite measures on the measurable space $(S, \sigma(\mathfrak{E}))$ such that $\mu(S) = \nu(S)$. If $\mu = \nu$ on $\mathfrak{E}$, then $\mu = \nu$ on $\sigma(\mathfrak{E})$.*

Proof. Let us show first that the collection $\mathfrak{G}$ of all members A of $\sigma(\mathfrak{E})$ such that $\mu(A) = \nu(A)$ is a d-class. Now since $\mu(S) = \nu(S)$, S is in $\mathfrak{G}$. Suppose $A, B \in \mathfrak{G}$ and $A \subset B$. Then since $\mu(S) = \nu(S) < \infty$, we have

$$\mu(B - A) = \mu(B) - \mu(A) = \nu(B) - \nu(A) = \nu(B - A),$$

and thus $B - A \in \mathfrak{G}$. Let $A_n \in \mathfrak{G}$, $n \in \mathbb{N}$, and $A_n \uparrow$. Then

$$\mu(\lim_{n \to \infty} A_n) = \lim_{n \to \infty} \mu(A_n) = \lim_{n \to \infty} \nu(A_n) = \nu(\lim_{n \to \infty} A_n),$$

so that $\lim_{n \to \infty} A_n$ is in $\mathfrak{G}$. Thus $\mathfrak{G}$ is a d-class of subsets of S by Definition 1.6.

Now since $\mu = \nu$ on $\mathfrak{E}$, we have $\mathfrak{E} \subset \mathfrak{G}$ and thus $d(\mathfrak{E}) \subset \mathfrak{G}$. According to Theorem 1.7, $\sigma(\mathfrak{E}) = d(\mathfrak{E})$. Thus $\sigma(\mathfrak{E}) \subset \mathfrak{G}$ and therefore $\sigma(\mathfrak{E}) = \mathfrak{G}$. This shows that $\mu(A) = \nu(A)$ for every $A \in \sigma(\mathfrak{E})$. ∎

Corollary 1.9. *Let $\mathfrak{E}$ be a π-class of subsets of a set S and let λ be a measure on the measurable space $(S, \sigma(\mathfrak{E}))$. Let f and g be two nonnegative integrable functions on $(S, \sigma(\mathfrak{E}))$. If $\int_S f \, d\lambda = \int_S g \, d\lambda$ and $\int_A f \, d\lambda = \int_A g \, d\lambda$ for every $A \in \mathfrak{E}$, then $\int_A f \, d\lambda = \int_A g \, d\lambda$ for every $A \in \sigma(\mathfrak{E})$, and therefore $f = g$ a.e. on $(S, \sigma(\mathfrak{E}), \lambda)$.*

Proof. Define two measures μ and ν on $(S, \sigma(\mathfrak{E}))$ by setting $\mu(A) = \int_A f \, d\lambda$ and $\nu(A) = \int_A g \, d\lambda$ for $A \in \sigma(\mathfrak{E})$. Since $\int_S f \, d\lambda = \int_S g \, d\lambda < \infty$, μ and ν are two finite measures on

§1. GENERATED σ-ALGEBRAS

$(S, \sigma(\mathfrak{E}))$ such that $\mu(S) = \nu(S)$ and $\mu(A) = \nu(A)$ for every $A \in \mathfrak{E}$. Thus by Corollary 1.8, we have $\mu(A) = \nu(A)$ for every $A \in \sigma(\mathfrak{E})$, that is, $\int_A f \, d\lambda = \int_A g \, d\lambda$ for every $A \in \sigma(\mathfrak{E})$. This then implies that $f = g$ a.e. on $(S, \sigma(\mathfrak{E}), \lambda)$. ∎

The following monotone class theorem for functions is useful in verifying that all $\sigma(\mathfrak{E})$-measurable functions belong to a linear space of functions characterized by a certain property so that they have that property.

Theorem 1.10. *Let $\mathfrak{E}$ be a π-class of subsets of a set S. Let $\mathbf{V}$ be a collection of real valued functions on S such that*

1°. $\mathbf{1}_A \in \mathbf{V}$ *for every* $A \in \mathfrak{E}$ *and* $\mathbf{1}_S \in \mathbf{V}$,

2°. $\mathbf{V}$ *is a linear space,*

3°. *if* $\{f_n : n \in \mathbb{N}\}$ *is an increasing sequence of nonnegative functions in $\mathbf{V}$ such that* $f \equiv \lim_{n \to \infty} f_n$ *is real valued (resp. bounded), then* $f \in \mathbf{V}$.

Then $\mathbf{V}$ contains all real valued (resp. bounded) $\sigma(\mathfrak{E})$-measurable functions on S.

Proof. Let us show first that $\mathbf{1}_A \in \mathbf{V}$ for every $A \in \sigma(\mathfrak{E})$. Let $\mathfrak{G}$ be the collection of all subsets A of S such that $\mathbf{1}_A \in \mathbf{V}$. By 1°, $\mathfrak{E} \subset \mathfrak{G}$. If we show that $\mathfrak{G}$ is a d-class of subsets of S then $\sigma(\mathfrak{E}) = d(\mathfrak{E}) \subset \mathfrak{G}$ by Theorem 1.7 so that $\mathbf{1}_A \in \mathbf{V}$ for every $A \in \sigma(\mathfrak{E})$.

To verify that $\mathfrak{G}$ is a d-class of subsets of S, note first that by 1°, $S \in \mathfrak{G}$. Let $A, B \in \mathfrak{G}$ and $A \subset B$. Then $\mathbf{1}_A, \mathbf{1}_B \in \mathbf{V}$ and $\mathbf{1}_{B-A} = \mathbf{1}_B - \mathbf{1}_A \in \mathbf{V}$ by 2°. Thus $B - A \in \mathfrak{G}$. Finally let $\{A_n : n \in \mathbb{N}\}$ be an increasing sequence in $\mathfrak{G}$. Then for $A_\infty \equiv \lim_{n \to \infty} A_n$ we have

$$\mathbf{1}_{A_\infty} = \lim_{n \to \infty} \mathbf{1}_{A_n} \in \mathbf{V}$$

by 3° so that $\lim_{n \to \infty} A_n \in \mathfrak{G}$. Thus $\mathfrak{G}$ is a d-class of subsets of S. This shows that $\mathbf{1}_A \in \mathbf{V}$ for every $A \in \sigma(\mathfrak{E})$.

Let f be a simple function on the measurable space $(S, \sigma(\mathfrak{E}))$, that is,

$$f = \sum_{i=1}^{k} c_i \mathbf{1}_{A_i} \quad \text{where } A_i \in \sigma(\mathfrak{E}), \ c_i \in \mathbb{R}, \text{ for } i = 1, 2, \ldots, k.$$

By 2°, we have $f \in \mathbf{V}$.

Let f be a real valued (resp. bounded) $\sigma(\mathfrak{E})$-measurable function on S and let $f = f^+ - f^-$. Then there exists an increasing sequence of nonnegative simple functions $\{f_n : n \in \mathbb{N}\}$

on $(S, \sigma(\mathfrak{E}))$ such that $f_n \uparrow f^+$ on S. Thus by 3°, $f^+ \in \mathbf{V}$. Similarly $f^- \in \mathbf{V}$. By 2°, we have $f \in \mathbf{V}$. Therefore $\mathbf{V}$ contains all real valued (resp. bounded) $\sigma(\mathfrak{E})$-measurable functions on S. ∎

As a variant of Theorem 1.10 we have the following corollary for non-negative functions measurable with respect to a generated σ-algebra.

Corollary 1.11. *Let $\mathfrak{E}$ be a π-class of subsets of a set S. Let $\mathbf{V}$ be a collection of real nonnegative valued functions on S such that*

1°. $\mathbf{1}_A \in \mathbf{V}$ *for every $A \in \mathfrak{E}$ and $\mathbf{1}_S \in \mathbf{V}$,*

2°. *if $f, g \in \mathbf{V}$ and $\alpha, \beta \geq 0$ then $\alpha f + \beta g \in \mathbf{V}$; if $f, g \in \mathbf{V}$ and $f \leq g$ then $g - f \in \mathbf{V}$,*

3°. *if $\{f_n : n \in \mathbb{N}\}$ is an increasing sequence of nonnegative functions in $\mathbf{V}$ such that $f \equiv \lim_{n \to \infty} f_n$ is real valued (resp. bounded), then $f \in \mathbf{V}$.*

Then $\mathbf{V}$ contains all real nonnegative valued (resp. bounded and nonnegative valued) $\sigma(\mathfrak{E})$-measurable functions on S.

Proof. The Corollary is proved exactly in the same way as Theorem 1.10. ∎

Let $\mathfrak{A}$ be an algebra of subsets of a set S. By a simple function based on $\mathfrak{A}$ on the measurable space $(S, \sigma(\mathfrak{A}))$, we mean a function of the form

$$f = \sum_{i=1}^{k} c_i \mathbf{1}_{A_i} \quad \text{with } A_i \in \mathfrak{A},\ c_i \in \mathbb{R} \text{ for } i = 1, \ldots, k.$$

Proposition 1.12. *Let $\mathfrak{A}$ be an algebra of subsets of a set S. Then for every real valued $\sigma(\mathfrak{A})$-measurable function f on S there exists a sequence of simple functions based on $\mathfrak{A}$, $\{f_n : n \in \mathbb{N}\}$, such that $f = \lim_{n \to \infty} f_n$.*

Proof. It suffices to show that for every real nonnegative valued $\sigma(\mathfrak{A})$-measurable function f on S there exists an increasing sequence $\{f_n : n \in \mathbb{N}\}$ of simple functions based on $\mathfrak{A}$ such that $f_n \uparrow f$ on S as $n \to \infty$. To show this, let $\mathbf{V}$ be the collection of all real nonnegative valued functions f on S for which such a sequence exists. Let $\mathfrak{A}$ be the π-class $\mathfrak{E}$ in Proposition 1.11. If we show that our $\mathbf{V}$ satisfies conditions 1°, 2°, and 3° of Proposition 1.11, then $\mathbf{V}$ contains all real nonnegative $\sigma(\mathfrak{A})$-measurable function f on S

§1. GENERATED σ-ALGEBRAS

and we are done. Clearly **V** satisfies 1° and 2°. To verify 3°, let $\{f_n : n \in \mathbb{N}\}$ be an increasing sequence in **V** such that $f \equiv \lim_{n \to \infty} f_n$ is real valued. Let us show that f is in **V**. Now for every $n \in \mathbb{N}$, since f_n is in **V** there exists an increasing sequence $\{f_{n,k} : k \in \mathbb{N}\}$ of simple functions based on $\mathfrak{A}$ such that $f_{n,k} \uparrow f_n$ on S as $k \to \infty$. Consider the collection $\{f_{n,k} : n, k \in \mathbb{N}\}$. For every $m \in \mathbb{N}$, $m \geq 2$, let g_m be a real valued function on S defined by $g_m = \max_{\{n,k \in \mathbb{N} : n+k=m\}} f_{n,k}$. Then g_m is a simple function based on $\mathfrak{A}$, $g_m \uparrow$ as $m \to \infty$, and $g_m \leq f$ on S. Let us show that $g_m \uparrow f$ on S. Let $x_0 \in S$ be fixed and let $\varepsilon > 0$ be arbitrarily given. Since $f_n(x_0) \uparrow f(x_0)$ as $n \to \infty$, there exists $n_0 \in \mathbb{N}$ such that $f_{n_0}(x_0) \geq f(x_0) - \varepsilon/2$. Since $f_{n_0,k}(x_0) \uparrow f_{n_0}(x_0)$ as $k \to \infty$, there exists $k_0 \in \mathbb{N}$ such that $f_{n_0,k_0}(x_0) \geq f_{n_0}(x_0) - \varepsilon/2$. Then $f_{n_0,k_0}(x_0) \geq f(x_0) - \varepsilon$. Let $m_0 = n_0 + k_0$. Then since $g_{m_0} \geq f_{n_0,k_0}$ on S, we have $g_{m_0}(x_0) \geq f(x_0) - \varepsilon$ and therefore $g_m(x_0) \geq f(x_0) - \varepsilon$ for $m \geq m_0$. This shows that $g_m(x_0) \uparrow f(x_0)$ as $m \to \infty$. From the arbitrariness of $x_0 \in S$ we have $g_m \uparrow f$ on S as $m \to \infty$. This verifies 3° of Proposition 1.11 for our **V** and completes the proof. ∎

For a finite measure defined on the σ-algebra generated by an algebra of subsets of a set we have the following approximation theorem in terms of symmetric difference.

Theorem 1.13. *Let $\mathfrak{A}$ be an algebra of subsets of a set S and let μ be a finite measure on the σ-algebra $\sigma(\mathfrak{A})$ generated by $\mathfrak{A}$. Then for every $E \in \sigma(\mathfrak{A})$ and $\varepsilon > 0$ there exists $A \in \mathfrak{A}$ such that $\mu(E \triangle A) < \varepsilon$.*

Proof. Let $\mathfrak{K}$ be the collection of all $E \in \sigma(\mathfrak{A})$ such that for every $\varepsilon > 0$ there exists $A \in \mathfrak{A}$ with $\mu(E \triangle A) < \varepsilon$. To prove the theorem it suffices to show that $\sigma(\mathfrak{A}) \subset \mathfrak{K}$. Clearly $\mathfrak{A} \subset \mathfrak{K}$. Thus, if we show that $\mathfrak{K}$ is a σ-algebra of subsets of S then we have $\sigma(\mathfrak{A}) \subset \mathfrak{K}$. To show that $\mathfrak{K}$ is a σ-algebra of subsets of S note first that $S \in \mathfrak{A} \subset \mathfrak{K}$. To show that $\mathfrak{K}$ is closed under complementation, note that for any two sets E and A we have $E^c \triangle A^c = E \triangle A$ so that if $E \in \mathfrak{K}$ then $E^c \in \mathfrak{K}$ also. To show that $\mathfrak{K}$ is closed under countable unions, let $\{E_n : n \in \mathbb{N}\} \subset \mathfrak{K}$ and let $E = \cup_{n \in \mathbb{N}} E_n \in \sigma(\mathfrak{A})$. Let $\varepsilon > 0$ be arbitrarily given. If we let $F_n = \cup_{k=1}^n E_k$ then $F_n \in \sigma(\mathfrak{A})$ for $n \in \mathbb{N}$ and $F_n \uparrow E$ so that $\mu(F_n) \uparrow \mu(E)$. Since $\mu(E) < \infty$ there exists $N \in \mathbb{N}$ such that

$$(1) \qquad \mu(E) - \frac{\varepsilon}{2} < \mu(F_N).$$

Now since $E_k \in \mathfrak{K}$ there exists $A_k \in \mathfrak{A}$ such that $\mu(E_k \triangle A_k) < \varepsilon/2^{k+1}$ for $k \in \mathbb{N}$. Since

$$\left(\cup_{k=1}^N E_k\right) \triangle \left(\cup_{k=1}^N A_k\right) \subset \cup_{k=1}^N (E_k \triangle A_k),$$

we have

(2) $$\mu\left[\left(\cup_{k=1}^N E_k\right) \triangle \left(\cup_{k=1}^N A_k\right)\right] \leq \sum_{k=1}^N \mu(E_k \triangle A_k) \leq \frac{\varepsilon}{2}.$$

By the triangle inequality for symmetric difference, $B_1 \triangle B_2 \subset (B_1 \triangle B_3) \cup (B_3 \triangle B_2)$ for any three sets B_1, B_2 and B_3, we have by (1) and (2)

$$\mu\left[E \triangle \left(\cup_{k=1}^N A_k\right)\right] \leq \mu(E \triangle F_N) + \mu\left[\left(\cup_{k=1}^N E_k\right) \triangle \left(\cup_{k=1}^N A_k\right)\right] < \varepsilon.$$

Since $\cup_{k=1}^N A_k \in \mathfrak{A}$, this shows that $E \in \mathfrak{K}$. Therefore $\mathfrak{K}$ is a σ-algebra of subsets of S. ∎

Let $(S, \mathfrak{F})$ and $(T, \mathfrak{G})$ be two measurable spaces. A mapping f with its domain of definition $\mathfrak{D}(f) \subset S$ and range $\mathfrak{R}(f) \subset T$ is said to be $\mathfrak{F}/\mathfrak{G}$-measurable if $f^{-1}(\mathfrak{G}) \subset \mathfrak{F}$, that is, $f^{-1}(G) \in \mathfrak{F}$ for every $G \in \mathfrak{G}$. This condition implies in particular that $\mathfrak{D}(f) = f^{-1}(T) \in \mathfrak{F}$. When $(T, \mathfrak{G}) = (\mathbb{R}, \mathfrak{B}_{\mathbb{R}})$ or $(\overline{\mathbb{R}}, \mathfrak{B}_{\overline{\mathbb{R}}})$ and f is $\mathfrak{F}/\mathfrak{B}_{\mathbb{R}}$- or $\mathfrak{F}/\mathfrak{B}_{\overline{\mathbb{R}}}$-measurable, we say simply that f is $\mathfrak{F}$-measurable.

For two measurable spaces $(S, \mathfrak{F})$ and $(T, \mathfrak{G})$ and an $\mathfrak{F}/\mathfrak{G}$-measurable mapping of S into T, we write $\sigma(f)$ for the σ-algebra of subsets of S generated by f, that is, the smallest sub-σ-algebra $\mathfrak{F}_0$ of $\mathfrak{F}$ such that f is $\mathfrak{F}_0/\mathfrak{G}$-measurable. Clearly

$$\sigma(f) = f^{-1}(\mathfrak{G}) \subset \mathfrak{F}.$$

For a collection $\{f_\lambda : \lambda \in \Lambda\}$ of $\mathfrak{F}/\mathfrak{G}$-measurable mappings of S into T we write $\sigma\{f_\lambda : \lambda \in \Lambda\}$ for the σ-algebra of subsets of S generated by the collection of mappings, that is, the smallest σ-algebra $\mathfrak{F}_0$ of subsets of S such that f_λ is $\mathfrak{F}_0/\mathfrak{G}$-measurable for every $\lambda \in \Lambda$. Thus

$$\sigma\{f_\lambda : \lambda \in \Lambda\} = \sigma(\cup_{\lambda \in \Lambda} \sigma(f_\lambda)) \subset \mathfrak{F}.$$

When S is a set, $(T, \mathfrak{G})$ is a measurable space and f is a mapping of S into T, we write $\sigma(f)$ for the smallest σ-algebra $\mathfrak{F}$ of subsets of S such that f is $\mathfrak{F}/\mathfrak{G}$-measurable. In this case we have $\mathfrak{F} = f^{-1}(\mathfrak{G})$. Similarly for a collection $\{f_\lambda : \lambda \in \Lambda\}$ of mappings of S into T we write $\sigma\{f_\lambda : \lambda \in \Lambda\}$ for the σ-algebra $\mathfrak{F}$ of subsets of S such that f_λ is $\mathfrak{F}/\mathfrak{G}$-measurable for every $\lambda \in \Lambda$. Thus $\mathfrak{F} = \sigma(\cup_{\lambda \in \Lambda} \sigma(f_\lambda))$.

Definition 1.14. *A σ-algebra $\mathfrak{F}$ of subsets of a set S is said to be separable or countably generated if there exists a countable collection $\mathfrak{E}$ of subsets of S such that $\mathfrak{F} = \sigma(\mathfrak{E})$.*

§2. STOCHASTIC PROCESSES

Theorem 1.15. *For an arbitrary sequence $\{f_n : n \in \mathbb{N}\}$ of extended real valued functions on a set S, the σ-algebra $\sigma\{f_n : n \in \mathbb{N}\}$ is separable. Indeed, if we let $\{r_m : m \in \mathbb{N}\}$ be the collection of all rational numbers and let $\mathfrak{E} = \{E_{n,m} : n, m \in \mathbb{N}\}$ where*

$$E_{n,m} = f_n^{-1}([-\infty, r_m)) = \{s \in S : f_n(s) < r_m\},$$

then $\alpha(\mathfrak{E})$, the algebra of subsets of S generated by $\mathfrak{E}$, is a countable collection and $\sigma(\alpha(\mathfrak{E})) = \sigma\{f_n : n \in \mathbb{N}\}$.

Proof. With $\mathfrak{E}^* = \{E_{n,m}, E_{n,m}^c : n, m \in \mathbb{N}\}$, let $\mathfrak{D}$ be the collection of all finite intersections of members of $\mathfrak{E}^*$ and $\mathfrak{A}$ be the collection of all finite unions of members of $\mathfrak{D}$. Clearly, $\mathfrak{E}, \mathfrak{E}^*, \mathfrak{D}$ and $\mathfrak{A}$ are all countable collections. It is easy to verify that $\mathfrak{A}$ is an algebra of subsets of S. Since no member of $\mathfrak{A}$ can be absent from an algebra of subsets of S containing $\mathfrak{E}$, our $\mathfrak{A}$ is indeed the smallest algebra containing $\mathfrak{E}$. Hence, $\mathfrak{A} = \alpha(\mathfrak{E})$.

It remains to show that $\sigma(\mathfrak{A}) = \sigma\{f_n : n \in \mathbb{N}\}$. Now

$$E_{n,m} \in f_n^{-1}(\mathfrak{B}_{\overline{\mathbb{R}}}) = \sigma\{f_n\} \subset \sigma\{f_n : n \in \mathbb{N}\}.$$

Then, since $\mathfrak{A} = \alpha(\mathfrak{E})$, we have $\mathfrak{A} \subset \sigma\{f_n : n \in \mathbb{N}\}$ and consequently $\sigma(\mathfrak{A}) \subset \sigma\{f_n : n \in \mathbb{N}\}$. On the other hand, since $\mathfrak{B}_{\overline{\mathbb{R}}} = \sigma\{[-\infty, r_m) : m \in \mathbb{N}\}$, Theorem 1.1 implies that

$$f_n^{-1}(\mathfrak{B}_{\overline{\mathbb{R}}}) = \sigma\{f_n^{-1}([-\infty, r_m)) : m \in \mathbb{N}\} = \sigma\{E_{n,m} : m \in \mathbb{N}\} \subset \sigma(\mathfrak{A}).$$

Hence

$$\sigma\{f_n : n \in \mathbb{N}\} = \sigma\left(\cup_{n \in \mathbb{N}} f_n^{-1}(\mathfrak{B}_{\overline{\mathbb{R}}})\right) \subset \sigma(\mathfrak{A}).$$

This completes the proof of the equality of the two σ-algebras. ∎

§2 Stochastic Processes

[I] Measurable and Progressively Measurable Processes

A probability space is a measure space $(\Omega, \mathfrak{F}, P)$ with $P(\Omega) = 1$. A member A of $\mathfrak{F}$ is called an event. A null set in $(\Omega, \mathfrak{F}, P)$ is a member A of $\mathfrak{F}$ with $P(A) = 0$. A probability space $(\Omega, \mathfrak{F}, P)$ is called complete if it is a complete measure space, that is, every subset A_0 of a null set A in $(\Omega, \mathfrak{F}, P)$ is a member of $\mathfrak{F}$ so that $P(A_0)$ is defined and is equal to 0. A sub-σ-algebra $\mathfrak{G}$ of $\mathfrak{F}$ is said to be augmented if it contains all the null sets in $(\Omega, \mathfrak{F}, P)$.

Let $(\Omega, \mathfrak{F}, P)$ be a probability space and let $(S, \mathfrak{A})$ be an arbitrary measurable space. An S-valued random variable on the probability space is an $\mathfrak{F}/\mathfrak{A}$-measurable mapping of

Ω into S. Unless otherwise specified, by a random variable we mean an extended real valued random variable. Thus a random variable X, without any qualification, is an $\mathfrak{F}/\mathfrak{B}_{\overline{\mathbb{R}}}$-measurable mapping of Ω into $\overline{\mathbb{R}}$. We say that X is real valued only when it assumes neither ∞ nor $-\infty$ as its value. For a random variable X and a member E of $\mathfrak{B}_{\overline{\mathbb{R}}}$ we write $\{X \in E\}$ for the event $\{\omega \in \Omega : X(\omega) \in E\} \in \mathfrak{F}$.

The expectation $\mathbf{E}(X)$ of a random variable X on a probability space $(\Omega, \mathfrak{F}, P)$ is defined by
$$\mathbf{E}(X) = \int_\Omega X(\omega) P(d\omega),$$
provided the integral exists in the extended real number system $\overline{\mathbb{R}}$. When $\mathbf{E}(X)$ is finite, that is, $\mathbf{E}(X) \in \mathbb{R}$ or equivalently $\mathbf{E}(|X|) < \infty$, we say that X is integrable.

For an integrable random variable X on a probability space $(\Omega, \mathfrak{F}, P)$ and for a sub-σ-algebra $\mathfrak{G}$ of $\mathfrak{F}$, we write $\mathbf{E}(X | \mathfrak{G})$ for the conditional expectation of X with respect to $\mathfrak{G}$. We use the same notation both for the conditional expectation as an equivalence class of random variables and for an arbitrary representative of the equivalence class.

Let $(S, \mathfrak{A})$ be a measurable space. An S-valued stochastic process on a probability space $(\Omega, \mathfrak{F}, P)$ is a collection of S-valued random variables on the probability space, that is, $\mathfrak{F}/\mathfrak{A}$-measurable mappings of Ω into S, which are indexed by $t \in D$ where D is a subset of $\mathbb{R}$. In most cases D will be equal to $\mathbb{R}_+$. When we speak of a stochastic process without a qualification we mean a real valued stochastic process. Thus a stochastic process is a collection of real valued random variables $X = \{X_t : t \in D\}$ on $(\Omega, \mathfrak{F}, P)$. It is then a mapping X of $\Omega \times D$ into $\mathbb{R}$ such that $X(t, \cdot) \equiv X_t(\cdot)$ is a real valued random variable on $(\Omega, \mathfrak{F}, P)$ for every $t \in D$. The notation $X(t)$ is also used for the random variable $X(t, \cdot)$. While a random variable is in general extended real valued, the random variables X_t in a stochastic process $X = \{X_t : t \in D\}$ are real valued. The advantage of this restriction is that for any pair $s, t \in D$, $X_s + X_t$ and $X_s - X_t$ are always defined on the entire space Ω, that is, indeterminates such as $\infty - \infty$ and $-\infty + \infty$ never arise. We refer to t as the time variable or the time parameter. For each $\omega \in \Omega$, the real valued function $X(\cdot, \omega)$ on D is called the sample function or the sample path corresponding to ω.

Let $\{t_n : n \in \mathbb{Z}_+\}$ be a strictly increasing sequence in $\mathbb{R}_+$. A sequence of real valued random variables $\{X_{t_n} : n \in \mathbb{Z}_+\}$ will be called a stochastic process with discrete time. A particular case is when $t_n = n$ for $n \in \mathbb{Z}_+$. In a stochastic process with discrete time $\{X_{t_n} : n \in \mathbb{Z}_+\}$, the sample path corresponding to a sample point $\omega \in \Omega$ is a sequence of real numbers $\{X_{t_n}(\omega) : n \in \mathbb{Z}_+\}$.

Definition 2.1. *Two stochastic processes* $X = \{X_t : t \in \mathbb{R}_+\}$ *and* $Y = \{Y_t : t \in \mathbb{R}_+\}$ *on a*

§2. STOCHASTIC PROCESSES

probability space $(\Omega, \mathfrak{F}, P)$ *are said to be equivalent if for a.e.* $\omega \in \Omega$ *the sample functions* $X(\cdot, \omega)$ *and* $Y(\cdot, \omega)$ *on* $\mathbb{R}_+$ *are identical, that is, there exists a null set* Λ *in* $(\Omega, \mathfrak{F}, P)$ *such that* $X(t, \omega) = Y(t, \omega)$ *for all* $t \in \mathbb{R}_+$ *when* $\omega \in \Lambda^c$.

Definition 2.2. *A stochastic process* $X = \{X_t : t \in \mathbb{R}_+\}$ *on a probability space* $(\Omega, \mathfrak{F}, P)$ *is said to be continuous, left-continuous, or right-continuous if* $X(\cdot, \omega)$ *is a continuous, left-continuous, or right-continuous function on* $\mathbb{R}_+$ *for every* $\omega \in \Omega$. X *is said to be a.s. continuous, a.s. left-continuous, or a.s. right-continuous if* $X(\cdot, \omega)$ *is continuous, left-continuous, or right-continuous on* $\mathbb{R}_+$ *for a.e.* $\omega \in \Omega$.

Note that by left-continuity of $X(\cdot, \omega)$ on $\mathbb{R}_+$ we mean left-continuity on $(0, \infty)$.

Theorem 2.3. *Let* $X = \{X_t : t \in \mathbb{R}_+\}$ *and* $Y = \{Y_t : t \in \mathbb{R}_+\}$ *be two a.s. left- (or a.s. right-) continuous processes on a probability space* $(\Omega, \mathfrak{F}, P)$. *If* $X_t = Y_t$ *a.e. on* $(\Omega, \mathfrak{F}, P)$ *for every* $t \in \mathbb{R}_+$ *then* X *and* Y *are equivalent processes.*

Proof. Suppose X and Y are a.s. left-continuous. Then there exist null sets Λ_X and Λ_Y in $(\Omega, \mathfrak{F}, P)$ such that $X(\cdot, \omega)$ is left-continuous on $\mathbb{R}_+$ when $\omega \in \Lambda_X^c$ and $Y(\cdot, \omega)$ is left-continuous on $\mathbb{R}_+$ when $\omega \in \Lambda_Y^c$.

Let $\{r_n : n \in \mathbb{Z}_+\}$ be an arbitrary enumeration of the collection Q of rational numbers in $\mathbb{R}_+$ with $r_0 = 0$. Then for every $n \in \mathbb{Z}_+$ there exists a null set Λ_n in $(\Omega, \mathfrak{F}, P)$ such that $X(r_n, \omega) = Y(r_n, \omega)$ for $\omega \in \Lambda_n^c$. Consider the null set $\Lambda = \Lambda_X \cup \Lambda_Y \cup (\cup_{n \in \mathbb{Z}_+} \Lambda_n)$. Then

$$X(r_n, \omega) = Y(r_n, \omega) \quad \text{for all } n \in \mathbb{Z}_+ \text{ when } \omega \in \Lambda^c.$$

For any $t \in (0, \infty)$, let $\{s_k : k \in \mathbb{N}\}$ be a sequence in Q such that $s_k \uparrow t$. Then $X(s_k, \omega) = Y(s_k, \omega)$ for all $k \in \mathbb{N}$ when $w \in \Lambda^c$ so that by the left-continuity of $X(\cdot, \omega)$ and $Y(\cdot, \omega)$ we have $X(t, \omega) = Y(t, \omega)$ when $\omega \in \Lambda^c$. Therefore $X(t, \omega) = Y(t, \omega)$ for all $t \in (0, \infty)$ when $\omega \in \Lambda^c$. Since $\Lambda^c \subset \Lambda_0^c$ we have $X(0, \omega) = Y(0, \omega)$ also when $\omega \in \Lambda^c$. Therefore $X(t, \omega) = Y(t, \omega)$ for all $t \in \mathbb{R}_+$ when $\omega \in \Lambda^c$. This proves the equivalence of the two processes X and Y for this case.

The case where X and Y are a.s. right-continuous can be treated in the same way by approaching an arbitrary $t \in \mathbb{R}_+$ by a sequence $\{s_k : k \in \mathbb{N}\}$ in Q such that $s_k \downarrow t$. ∎

Definition 2.4. *A stochastic process* $X = \{X_t : t \in \mathbb{R}_+\}$ *on a probability space* $(\Omega, \mathfrak{F}, P)$ *is called a measurable process if* X *is* $\sigma(\mathfrak{B}_{\mathbb{R}_+} \times \mathfrak{F})/\mathfrak{B}_{\mathbb{R}}$-*measurable as a mapping of* $\mathbb{R}_+ \times \Omega$ *into* $\mathbb{R}$.

A measurable process X is a mapping of $\mathbb{R}_+ \times \Omega$ into $\mathbb{R}$ which is measurable with respect to the product σ-algebra $\sigma(\mathfrak{B}_{\mathbb{R}_+} \times \mathfrak{F})$ on $\mathbb{R}_+ \times \Omega$. Then for every $\omega \in \Omega$ the section of the mapping X at ω, namely $X(\cdot, \omega)$, is measurable with respect to the factor σ-algebra $\mathfrak{B}_{\mathbb{R}_+}$ of the product σ-algebra. This shows that every sample function of a measurable process is a Borel-measurable function on $\mathbb{R}_+$.

In a probability space $(\Omega, \mathfrak{F}, P)$, the σ-algebra $\mathfrak{F}$ of subsets of the sample space Ω consists of all possible events A with probabilities $P(A) \in [0, 1]$. If $\mathfrak{F}_t$ is a sub-σ-algebra of $\mathfrak{F}$ consisting of all possible events at the time t and if $\mathfrak{F}_t$ increases with t, then $\mathfrak{F}_t$ consists of all possible events up to the time t. If a real valued random variable Y on the probability space is $\mathfrak{F}_t$-measurable, then the σ-algebra of all events of the type $\{Y \in E\}$ for $E \in \mathfrak{B}_{\mathbb{R}}$, namely $Y^{-1}(\mathfrak{B}_{\mathbb{R}})$, is contained in $\mathfrak{F}_t$. If $X = \{X_t : t \in \mathbb{R}_+\}$ is a stochastic process on the probability space such that X_t is $\mathfrak{F}_t$-measurable for every $t \in \mathbb{R}_+$, then $\mathfrak{F}_t$ contains all events of the type $\{X_s \in E\}$ for $E \in \mathfrak{B}_{\mathbb{R}}$ and $0 \leq s \leq t$, that is, $\cup_{s \in [0,t]} X_s^{-1}(\mathfrak{B}_{\mathbb{R}}) \subset \mathfrak{F}_t$.

Definition 2.5. *Let $(\Omega, \mathfrak{F}, P)$ be a probability space. A system of sub-σ-algebras of $\mathfrak{F}$, $\{\mathfrak{F}_t : t \in \mathbb{R}_+\}$, is called a filtration of the probability space if it is an increasing system in the sense that $\mathfrak{F}_s \subset \mathfrak{F}_t$ for $s, t \in \mathbb{R}_+$, $s < t$. The quadruple $(\Omega, \mathfrak{F}, \{\mathfrak{F}_t : t \in \mathbb{R}_+\}, P)$, or more briefly $(\Omega, \mathfrak{F}, \{\mathfrak{F}_t\}, P)$, is called a filtered space. We call $\mathfrak{F}_\infty \equiv \sigma(\cup_{t \in \mathbb{R}_+} \mathfrak{F}_t)$ the σ-algebra at infinity of the filtration. A stochastic process $X = \{X_t : t \in \mathbb{R}_+\}$ on a probability space $(\Omega, \mathfrak{F}, P)$ for which a filtration $\{\mathfrak{F}_t : t \in \mathbb{R}_+\}$ is designated is said to be $\{\mathfrak{F}_t\}$-adapted, or simply adapted when $\{\mathfrak{F}_t : t \in \mathbb{R}_+\}$ is understood, if X_t is an $\mathfrak{F}_t/\mathfrak{B}_{\mathbb{R}}$-measurable mapping of Ω into $\mathbb{R}$ for every $t \in \mathbb{R}_+$.*

Note that if $X = \{X_t : t \in \mathbb{R}_+\}$ is a stochastic process on a probability space $(\Omega, \mathfrak{F}, P)$ and if we let $\mathfrak{F}_t^X = \sigma\{X_s : s \in [0, t]\}$, then $\{\mathfrak{F}_t^X : t \in \mathbb{R}\}$ is a filtration on the probability space and X is an $\{\mathfrak{F}_t^X\}$-adapted process. We call $\{\mathfrak{F}_t^X : t \in \mathbb{R}\}$ the filtration generated by X.

Definition 2.6. *For a filtration $\{\mathfrak{F}_t : t \in \mathbb{R}_+\}$ on a probability space $(\Omega, \mathfrak{F}, P)$ we define $\mathfrak{F}_{t+} = \cap_{u > t} \mathfrak{F}_u$ for $t \in \mathbb{R}_+$. We say that the filtration $\{\mathfrak{F}_t : t \in \mathbb{R}_+\}$ is right-continuous if $\mathfrak{F}_{t+} = \mathfrak{F}_t$ for every $t \in \mathbb{R}_+$. The filtered space $(\Omega, \mathfrak{F}, \{\mathfrak{F}_t\}, P)$ is then called a right-continuous filtered space.*

Note that if $\{\mathfrak{F}_t : t \in \mathbb{R}_+\}$ is an arbitrary filtration on a probability space $(\Omega, \mathfrak{F}, P)$, then the system $\{\mathfrak{G}_t : t \in \mathbb{R}_+\}$ of sub-σ-algebras of $\mathfrak{F}$ where $\mathfrak{G}_t$ is defined by $\mathfrak{G}_t = \mathfrak{F}_{t+}$ for

§2. STOCHASTIC PROCESSES

$t \in \mathbb{R}_+$ is a right-continuous filtration of $(\Omega, \mathfrak{F}, P)$ since

$$\mathfrak{G}_{t+} = \bigcap_{u>t} \mathfrak{G}_u = \bigcap_{u>t} \mathfrak{F}_{u+} = \bigcap_{u>t} \bigcap_{v>u} \mathfrak{F}_v = \bigcap_{u>t} \mathfrak{F}_u = \mathfrak{F}_{t+} = \mathfrak{G}_t.$$

Definition 2.7. *A stochastic process $X = \{X_t : t \in \mathbb{R}_+\}$ on a probability space $(\Omega, \mathfrak{F}, P)$ is called a right-continuous simple process if there exist a strictly increasing sequence $\{t_k : k \in \mathbb{Z}_+\}$ in $\mathbb{R}_+$ with $t_0 = 0$ and $\lim_{k \to \infty} t_k = \infty$ and a sequence $\{\xi_k : k \in \mathbb{N}\}$ of real valued random variables on $(\Omega, \mathfrak{F}, P)$ such that*

(1) $\qquad X(t, \omega) = \xi_k(\omega) \quad \text{for } t \in [t_{k-1}, t_k), \ k \in \mathbb{N}, \text{ and } \omega \in \Omega.$

Similarly, X is called a left-continuous simple process if condition (1) is replaced by

(2) $\qquad X(t, \omega) = \xi_k(\omega) \quad \text{for } t \in (t_{k-1}, t_k], \ k \in \mathbb{N}, \text{ and } \omega \in \Omega.$

Thus, a right-continuous simple process X is completely determined by the two sequences $\{t_k : k \in \mathbb{Z}_+\}$ and $\{\xi_k : k \in \mathbb{N}\}$, but to determine a left-continuous simple process we need also the random variable $X(0, \cdot)$. Every sample function of a right-continuous simple process is a right-continuous step function on $\mathbb{R}_+$. On the other hand, every sample function of a left-continuous simple process is a left continuous step function when restricted to the domain $(0, \infty)$. Note that if there exists $N \in \mathbb{N}$ such that $\xi_k = \xi_N$ for all $k \geq N$ then the sample functions are step functions with finitely many steps.

Lemma 2.8. *Let $X = \{X_t : t \in \mathbb{R}_+\}$ be a stochastic process on a probability space $(\Omega, \mathfrak{F}, P)$. If X is a left-continuous process, then let $I_{n,k} = ((k-1)2^{-n}, k2^{-n}]$ for $k, n \in \mathbb{N}$ and define a sequence of left-continuous simple processes $\{X^{(n)} : n \in \mathbb{N}\}$ by*

(1) $\qquad \begin{cases} X^{(n)}(t, \omega) = X((k-1)2^{-n}, \omega) & \text{for } t \in I_{n,k}, \ k \in \mathbb{N}, \text{ and } \omega \in \Omega \\ X^{(n)}(0, \omega) = X(0, \omega) & \text{for } \omega \in \Omega. \end{cases}$

If X is a right-continuous process, then let $I_{n,k} = [(k-1)2^{-n}, k2^{-n})$ for $k, n \in \mathbb{N}$ and define a sequence of right-continuous simple processes $\{X^{(n)} : n \in \mathbb{N}\}$ by

(2) $\qquad X^{(n)}(t, \omega) = X(k2^{-n}, \omega) \quad \text{for } t \in I_{n,k}, \ k \in \mathbb{N}, \text{ and } \omega \in \Omega.$

Then for both cases we have

(3) $\qquad \lim_{n \to \infty} X^{(n)}(t, \omega) = X(t, \omega) \quad \text{for } (t, \omega) \in \mathbb{R}_+ \times \Omega.$

Also, if $\{\mathfrak{F}_t : t \in \mathbb{R}_+\}$ is a filtration on $(\Omega, \mathfrak{F}, P)$ and X is an arbitrary $\{\mathfrak{F}_t\}$-adapted process on $(\Omega, \mathfrak{F}, P)$, then $X^{(n)}$ defined by (1) is $\{\mathfrak{F}_t\}$-adapted.

Proof. Consider the case where X is left-continuous. Since $X^{(n)}(0, \cdot) = X(0, \cdot)$ for every $n \in \mathbb{N}$, to prove (3) we need only consider $(t, \omega) \in (0, \infty) \times \Omega$. For each $n \in \mathbb{N}$, consider the decomposition of $(0, \infty)$ into $I_{n,k} = ((k-1)2^{-n}, k2^{-n}]$, $k \in \mathbb{N}$. For each $n \in \mathbb{N}$, there exists a unique $k_n \in \mathbb{N}$ such that $t \in I_{n,k_n}$. Then by (1)

$$X^{(n)}(t, \omega) = X((k_n - 1)2^{-n}, \omega) \quad \text{for } n \in \mathbb{N}.$$

Since $(k_n - 1)2^{-n} < t$ and $t - (k_n - 1)2^{-n} < 2^{-n}$ for every $n \in \mathbb{N}$, we have $(k_n - 1)2^{-n} \uparrow t$ as $n \to \infty$. Thus by the left-continuity of $X(\cdot, \omega)$ at t we have

$$X(t, \omega) = \lim_{n \to \infty} X((k_n - 1)2^{-n}, \omega) = \lim_{n \to \infty} X^{(n)}(t, \omega),$$

proving (3). For a right-continuous process X, (3) is proved in the same manner.

When X is $\{\mathfrak{F}_t\}$-adapted, $X((k-1)2^{-n}, \cdot)$ is $\mathfrak{F}_{(k-1)2^{-n}}$-measurable and $X^{(n)}(t, \cdot)$ defined by (1) is $\mathfrak{F}_{(k-1)2^{-n}}$-measurable for t in the interval $I_{n,k}$. Then $X^{(n)}(t, \cdot)$ is $\mathfrak{F}_t$-measurable since $\mathfrak{F}_{(k-1)2^{-n}} \subset \mathfrak{F}_t$. Therefore $X^{(n)}$ is $\{\mathfrak{F}_t\}$-adapted. ∎

We have shown that a left- (resp. right-) continuous process is the limit of a sequence of left- (resp. right-) continuous simple processes. Actually a left-continuous process is also the limit of a sequence of right-continuous simple processes and similarly a right-continuous process is also the limit of a sequence of left-continuous simple processes. This is proved in the next lemma. The proof parallels closely that of Lemma 2.8.

Lemma 2.9. *Let X be a stochastic process on a probability space $(\Omega, \mathfrak{F}, P)$. If X is left-continuous, then let $I_{n,k} = [(k-1)2^{-n}, k2^{-n})$ for $k, n \in \mathbb{N}$ and define a sequence of right-continuous simple processes $\{Y^{(n)} : n \in \mathbb{N}\}$ by*

(1) $\qquad Y^{(n)}(t, \omega) = X((k-1)2^{-n}, \omega) \quad \text{for } t \in I_{n,k}, \ k \in \mathbb{N}, \text{ and } \omega \in \Omega.$

If X is right-continuous, then let $I_{n,k} = ((k-1)2^{-n}, k2^{-n}]$ for $k, n \in \mathbb{N}$ and define a sequence of left-continuous simple processes $\{Y^{(n)} : n \in \mathbb{N}\}$ by

(2) $\qquad \begin{cases} Y^{(n)}(t, \omega) = X(k2^{-n}, \omega) & \text{for } t \in I_{n,k}, \ k \in \mathbb{N}, \text{ and } \omega \in \Omega \\ Y^{(n)}(0, \omega) = X(0, \omega) & \text{for } \omega \in \Omega. \end{cases}$

For both cases,

(3) $\qquad \lim_{n \to \infty} Y^{(n)}(t, \omega) = X(t, \omega) \quad \text{for } (t, \omega) \in \mathbb{R}_+ \times \Omega.$

§2. STOCHASTIC PROCESSES

If X is an adapted process with respect to a filtration $\{\mathfrak{F}_t : t \in \mathbb{R}_+\}$ on the probability space, then $Y^{(n)}$ defined by (1) is $\{\mathfrak{F}_t\}$-adapted.

Proof. Consider the case where X is left-continuous. To prove (3), let $(t, \omega) \in \mathbb{R}_+ \times \Omega$. For each $n \in \mathbb{N}$, consider the decomposition of $\mathbb{R}_+$ into $I_{n,k} = [(k-1)2^{-n}, k2^{-n}), k \in \mathbb{N}$. For each $n \in \mathbb{N}$ there exists a unique $k_n \in \mathbb{N}$ such that $t \in [(k_n - 1)2^{-n}, k_n 2^{-n})$. Then by (1)

$$Y^{(n)}(t, \omega) = X((k_n - 1)2^{-n}, \omega) \quad \text{for } n \in \mathbb{N}.$$

Now $(k_n - 1)2^{-n} \leq t$ and $t - (k_n - 1)2^{-n} < 2^{-n}$ for every $n \in \mathbb{N}$. Then $(k_n - 1)2^{-n} \uparrow t$ as $n \to \infty$. By the left-continuity of $X(\cdot, \omega)$ at t we have

$$X(t, \omega) = \lim_{n \to \infty} X((k_n - 1)2^{-n}, \omega) = \lim_{n \to \infty} Y^{(n)}(t, \omega),$$

proving (3). Similarly for a right-continuous process X.

When X is $\{\mathfrak{F}_t\}$-adapted, $Y^{(n)}$ defined by (1) is $\{\mathfrak{F}_t\}$-adapted by the same argument as in the proof of Lemma 2.8. ∎

Theorem 2.10. *A stochastic process $X = \{X_t : t \in \mathbb{R}_+\}$ on a probability space $(\Omega, \mathfrak{F}, P)$ which is either left-continuous or right-continuous is always a measurable process and in particular every sample function of such a process is a Borel-measurable real valued function on $\mathbb{R}_+$. If in addition X is adapted to a filtration $\{\mathfrak{F}_t : t \in \mathbb{R}_+\}$ on the probability space, then the left- or right- continuity of X implies its $\sigma(\mathfrak{B}_{\mathbb{R}_+} \times \mathfrak{F}_\infty)/\mathfrak{B}_\mathbb{R}$-measurability.*

Proof. Let X be a left-continuous process. Consider the sequence $\{X^{(n)} : n \in \mathbb{N}\}$ of left-continuous simple processes defined by (1) of Lemma 2.8. Let $n \in \mathbb{N}$ be fixed. Then for every $E \in \mathfrak{B}_\mathbb{R}$ we have

$$(X^{(n)})^{-1}(E) = (\{0\} \times X_0^{-1}(E)) \cup [\bigcup_{k \in \mathbb{N}} (I_{n,k} \times X_{(k-1)2^{-n}}^{-1}(E))].$$

By the $\mathfrak{F}$-measurability of X_t for every $t \in \mathbb{R}_+$, $X_0^{-1}(E)$ and $X_{(k-1)2^{-n}}^{-1}(E)$ are in $\mathfrak{F}$ for all $k \in \mathbb{N}$. Since $\{0\}$ and $I_{n,k}$ are in $\mathfrak{B}_{\mathbb{R}_+}$ for all $k \in \mathbb{N}$, $(X^{(n)})^{-1}(E)$ is a countable union of members of $\mathfrak{B}_{\mathbb{R}_+} \times \mathfrak{F}$ and therefore $(X^{(n)})^{-1}(E) \in \sigma(\mathfrak{B}_{\mathbb{R}_+} \times \mathfrak{F})$. This shows the $\sigma(\mathfrak{B}_{\mathbb{R}_+} \times \mathfrak{F})/\mathfrak{B}_\mathbb{R}$-measurability of $X^{(n)}$ as a mapping of $\mathbb{R}_+ \times \Omega$ into $\mathbb{R}$. Then by (3) of Lemma 2.8, X is a $\sigma(\mathfrak{B}_{\mathbb{R}_+} \times \mathfrak{F})/\mathfrak{B}_\mathbb{R}$-measurable mapping of $\mathbb{R}_+ \times \Omega$ into $\mathbb{R}$, that is, X is a measurable process.

The measurability of a right-continuous process X is proved similarly by using the sequence $\{X^{(n)} : n \in \mathbb{N}\}$ of right-continuous simple functions defined by (2) of Lemma 2.8.

If X is adapted to a filtration $\{\mathfrak{F}_t : t \in \mathbb{R}_+\}$, then X is a stochastic process on the probability space $(\Omega, \mathfrak{F}_\infty, P)$ so that by replacing $\mathfrak{F}$ in the argument above by $\mathfrak{F}_\infty$ we have the $\sigma(\mathfrak{B}_{\mathbb{R}_+} \times \mathfrak{F}_\infty)/\mathfrak{B}_\mathbb{R}$-measurability of X. ∎

Definition 2.11. *A stochastic process $X = \{X_t : t \in \mathbb{R}_+\}$ on a filtered space $(\Omega, \mathfrak{F}, \{\mathfrak{F}_t\}, P)$ is said to be $\{\mathfrak{F}_t\}$-progressively measurable, or simply progressively measurable when the filtration is understood, if for every $t \in \mathbb{R}_+$ the restriction of X to $[0, t] \times \Omega$ is a $\sigma(\mathfrak{B}_{[0,t]} \times \mathfrak{F}_t)/\mathfrak{B}_\mathbb{R}$-measurable mapping of $[0, t] \times \Omega$ into $\mathbb{R}$.*

Observation 2.12. An $\{\mathfrak{F}_t\}$-progressively measurable process $X = \{X_t : t \in \mathbb{R}_+\}$ on a filtered space $(\Omega, \mathfrak{F}, \{\mathfrak{F}_t\}, P)$ is always an $\{\mathfrak{F}_t\}$-adapted measurable process.

Proof. If X is an $\{\mathfrak{F}_t\}$-progressively measurable process then for every $t \in \mathbb{R}_+$ the restriction of X to $[0, t] \times \Omega$ is a $\sigma(\mathfrak{B}_{[0,t]} \times \mathfrak{F}_t)/\mathfrak{B}_\mathbb{R}$-measurable mapping of $[0, t] \times \Omega$ into $\mathbb{R}$. Then $X(s, \cdot)$ is an $\mathfrak{F}_t$-measurable mapping of Ω into $\mathbb{R}$ for every $s \in [0, t]$ and in particular $X(t, \cdot)$ is $\mathfrak{F}_t$-measurable. Thus X is an $\{\mathfrak{F}_t\}$-adapted process.

If X is an $\{\mathfrak{F}_t\}$-progressively measurable process, then for each $n \in \mathbb{N}$ the restriction of X to $[0, n] \times \Omega$ is a $\sigma(\mathfrak{B}_{[0,n]} \times \mathfrak{F}_n)$-measurable and hence $\sigma(\mathfrak{B}_{[0,n]} \times \mathfrak{F}_\infty)$-measurable mapping of $[0, n] \times \Omega$ into $\mathbb{R}$. Thus $X \cdot \mathbf{1}_{[0,n] \times \Omega}$ is a $\sigma(\mathfrak{B}_{\mathbb{R}_+} \times \mathfrak{F}_\infty)$-measurable mapping of $\mathbb{R}_+ \times \Omega$ into $\mathbb{R}$. Then $X = \lim_{n \to \infty} X \cdot \mathbf{1}_{[0,n] \times \Omega}$ is a $\sigma(\mathfrak{B}_{\mathbb{R}_+} \times \mathfrak{F}_\infty)$-measurable mapping of $\mathbb{R}_+ \times \Omega$ into $\mathbb{R}$, that is, X is a measurable process. ∎

Proposition 2.13. *If $X = \{X_t : t \in \mathbb{R}_+\}$ is a left- or right-continuous $\{\mathfrak{F}_t\}$-adapted process on a filtered space $(\Omega, \mathfrak{F}, \{\mathfrak{F}_t\}, P)$, then X is an $\{\mathfrak{F}_t\}$-progressively measurable process.*

Proof. Let X be a left-continuous $\{\mathfrak{F}_t\}$-adapted process. For $t \in \mathbb{R}_+$ fixed, let $I_{n,k} = ((k-1)2^{-n}t, k2^{-n}t]$ for $k = 1, \ldots, 2^n, n \in \mathbb{N}$, and define a sequence of left-continuous simple processes $\{X^{(n)} : n \in \mathbb{N}\}$ on $[0, t] \times \Omega$ by

$$\begin{cases} X^{(n)}(s, \omega) = X((k-1)2^{-n}t, \omega) & \text{for } s \in I_{n,k},\ k = 1, \ldots, 2^n,\ \text{and } \omega \in \Omega \\ X^{(n)}(0, \omega) = X(0, \omega) & \text{for } \omega \in \Omega. \end{cases}$$

We then have $\lim_{n \to \infty} X^{(n)}(s, \omega) = X(s, \omega)$ for $(s, \omega) \in [0, t] \times \Omega$. By the same argument as in the proof of Theorem 2.10 but replacing $\mathfrak{F}$ there with $\mathfrak{F}_t$ now, we have the $\sigma(\mathfrak{B}_{[0,t]} \times \mathfrak{F}_t)$-measurability of $X^{(n)}$. Then the convergence of $X^{(n)}$ to X on $[0, t] \times \Omega$ implies the $\sigma(\mathfrak{B}_{[0,t]} \times \mathfrak{F}_t)$-measurability of the restriction of X to $[0, t] \times \Omega$. This proves the $\{\mathfrak{F}_t\}$-progressive measurability of X.

§2. STOCHASTIC PROCESSES

Similarly, if X is a right-continuous $\{\mathfrak{F}_t\}$-adapted process then we let $I_{n,k} = [(k-1)2^{-n}t, k2^{-n}t)$ for $k = 1, \ldots, 2^n, n \in \mathbb{N}$, and define a sequence of right-continuous simple processes $\{X^{(n)} : n \in \mathbb{N}\}$ on $[0, t] \times \Omega$ by

$$X^{(n)}(s, \omega) = X(k2^{-n}t, \omega) \quad \text{for } s \in I_{n,k}, \ k = 1, \ldots, 2^{-n}, \text{ and } \omega \in \Omega.$$

In this case too we have $\lim_{n \to \infty} X^{(n)}(s, \omega) = X(s, \omega)$ for $(s, \omega) \in [0, t] \times \Omega$ and the proof of the $\{\mathfrak{F}_t\}$-progressive measurability of X follows in the same way as above. ■

[II] Predictable Processes

If $X = \{X_t : t \in \mathbb{R}_+\}$ is a left-continuous process on a probability space $(\Omega, \mathfrak{F}, P)$ then at every $t \in \mathbb{R}_+$ and for every $\omega \in \Omega$ we have $\lim_{s \uparrow t} X(s, \omega) = X(t, \omega)$. Thus, if we know the behavior of the sample function $X(\cdot, \omega)$ in a time interval $(t - \delta, t)$ for some $\delta > 0$, then we know $X(t, \omega)$. In this sense, a left-continuous process is predictable. The definition of the class of predictable adapted processes on a filtered space will be given in Definition 2.19.

Let $X = \{X_t : t \in \mathbb{R}_+\}$ be a left-continuous adapted process on a filtered space $(\Omega, \mathfrak{F}, \{\mathfrak{F}_t\}, P)$. As we saw in Theorem 2.10, X is a $\sigma(\mathfrak{B}_{\mathbb{R}_+} \times \mathfrak{F})/\mathfrak{B}_{\mathbb{R}}$-measurable transformation of $\mathbb{R}_+ \times \Omega$ into $\mathbb{R}$. Now since $\mathfrak{B}_{\mathbb{R}}$ is a σ-algebra of subsets of $\mathbb{R}$, $X^{-1}(\mathfrak{B}_{\mathbb{R}})$ is a σ-algebra of subsets of $\mathbb{R}_+ \times \Omega$ and in fact it is the smallest of all sub-σ-algebras $\mathfrak{A}$ of $\sigma(\mathfrak{B}_{\mathbb{R}_+} \times \mathfrak{F})$ with respect to which X is $\mathfrak{A}/\mathfrak{B}_{\mathbb{R}}$-measurable.

Let $\mathfrak{X}$ be the collection of all left-continuous adapted processes on $(\Omega, \mathfrak{F}, \{\mathfrak{F}_t\}, P)$. Then $\sigma(\cup_{X \in \mathfrak{X}} X^{-1}(\mathfrak{B}_{\mathbb{R}}))$ is the smallest of all sub-σ-algebras $\mathfrak{A}$ of $\sigma(\mathfrak{B}_{\mathbb{R}_+} \times \mathfrak{F})$ with respect to which every $X \in \mathfrak{X}$ is $\mathfrak{A}/\mathfrak{B}_{\mathbb{R}}$-measurable.

Definition 2.14. *Let $(\Omega, \mathfrak{F}, \{\mathfrak{F}_t\}, P)$ be a filtered space. The smallest σ-algebra $\mathfrak{S}$ of subsets of $\mathbb{R}_+ \times \Omega$ with respect to which all left-continuous $\{\mathfrak{F}_t\}$-adapted processes are $\mathfrak{S}/\mathfrak{B}_{\mathbb{R}}$-measurable is called the $\{\mathfrak{F}_t\}$-predictable σ-algebra, or simply the predictable σ-algebra when the filtration is understood. The smallest σ-algebra $\mathfrak{T}$ of subsets of $\mathbb{R}_+ \times \Omega$ with respect to which all right-continuous $\{\mathfrak{F}_t\}$-adapted processes are $\mathfrak{T}/\mathfrak{B}_{\mathbb{R}}$-measurable is called the $\{\mathfrak{F}_t\}$-well measurable σ-algebra, or simply the well-measurable σ-algebra.*

Theorem 2.15. *For the predictable σ-algebra $\mathfrak{S}$ and the well-measurable σ-algebra $\mathfrak{T}$ for a filtered space $(\Omega, \mathfrak{F}, \{\mathfrak{F}_t\}, P)$, we have $\mathfrak{S} \subset \mathfrak{T}$.*

Proof. Let X be an arbitrary left-continuous $\{\mathfrak{F}_t\}$-adapted process. If we show that X is a $\mathfrak{T}/\mathfrak{B}_{\mathbb{R}}$-measurable mapping of $\mathbb{R}_+ \times \Omega$ into $\mathbb{R}$, then we have $\mathfrak{S} \subset \mathfrak{T}$ by the definition of

$\mathfrak{S}$ as the smallest σ-algebra of subsets of $\mathbb{R}_+ \times \Omega$ with respect to which all left-continuous $\{\mathfrak{F}_t\}$-adapted processes are $\mathfrak{S}/\mathfrak{B}_\mathbb{R}$-measurable.

Let $\{Y^{(n)} : n \in \mathbb{N}\}$ be the sequence of right-continuous simple processes defined by (1) of Lemma 2.9. Then $Y^{(n)}$ is $\{\mathfrak{F}_t\}$-adapted and therefore $Y^{(n)}$ is a $\mathfrak{T}/\mathfrak{B}_\mathbb{R}$-measurable mapping of $\mathbb{R}_+ \times \Omega$ into $\mathbb{R}$ by the definition of $\mathfrak{T}$. Then by (3) of Lemma 2.9, X is a $\mathfrak{T}/\mathfrak{B}_\mathbb{R}$-measurable transformation of $\mathbb{R}_+ \times \Omega$ into $\mathbb{R}$. ∎

Note that an argument parallel to that of the proof above does not lead to the conclusion $\mathfrak{T} \subset \mathfrak{S}$. If we start with an arbitrary right-continuous $\{\mathfrak{F}_t\}$-adapted process X, then the sequence $\{Y^{(n)} : n \in \mathbb{N}\}$ of left-continuous simple processes defined by (2) of Lemma 2.9 does satisfy $\lim_{n \to \infty} Y^{(n)}(t, \omega) = X(t, \omega)$ for $(t, \omega) \in \mathbb{R}_+ \times \Omega$. However in this case $Y^{(n)}$ may not be $\{\mathfrak{F}_t\}$-adapted and thus it may not be $\mathfrak{S}/\mathfrak{B}_\mathbb{R}$-measurable.

The predictable σ-algebra $\mathfrak{S}$ has another characterization based on the predictable rectangles defined below.

Definition 2.16. *Let $(\Omega, \mathfrak{F}, \{\mathfrak{F}_t\}, P)$ be a filtered space. By an $\{\mathfrak{F}_t\}$-predictable rectangle we mean a subset of $\mathbb{R}_+ \times \Omega$ of one of the following three types:*
(a) $(t', t''] \times A$, where $t', t'' \in \mathbb{R}_+$, $t' < t''$, and $A \in \mathfrak{F}_{t'}$,
(b) $\{0\} \times A$, where $A \in \mathfrak{F}_0$,
(c) $\emptyset$.
We write $\mathfrak{R}$ for the collection of all $\{\mathfrak{F}_t\}$-predictable rectangles. $\{\mathfrak{F}_t\}$-predictable rectangles will be referred to as predictable rectangles when the filtration is understood.

Lemma 2.17. *The collection $\mathfrak{R}$ of all the predictable rectangles in a filtered space $(\Omega, \mathfrak{F}, \{\mathfrak{F}_t\}, P)$ is a π-class.*

Proof. Let us take two arbitrary members of $\mathfrak{R}$. If they are not both of the type (a) in Definition 2.16, then their intersection is of the type (b) or $\emptyset$ so that it is in $\mathfrak{R}$. If they are both of the type (a) and are given as $(t'_1, t''_1] \times A_1$ and $(t'_2, t''_2] \times A_2$, then their intersection is given by

$$(t', t''] \times (A_1 \cap A_2) \text{ where } t' = \max\{t'_1, t'_2\} \text{ and } t'' = \min\{t''_1, t''_2\}.$$

Since $A_1 \in \mathfrak{F}_{t'_1}$ and $A_2 \in \mathfrak{F}_{t'_2}$ and since $\mathfrak{F}_t \uparrow$ as $t \to \infty$ we have $A_1 \cap A_2 \in \mathfrak{F}_{t'}$. Thus the intersection is of the type (a) if it is not $\emptyset$. ∎

Theorem 2.18. *For the predictable σ-algebra $\mathfrak{S}$ and the π-class $\mathfrak{R}$ of predictable rectangles in a filtered space $(\Omega, \mathfrak{F}, \{\mathfrak{F}_t\}, P)$, we have $\mathfrak{S} = \sigma(\mathfrak{R}) = d(\mathfrak{R})$.*

§2. STOCHASTIC PROCESSES

Proof. 1) Let $R \in \mathfrak{R}$ and consider the mapping X of $\mathbb{R}_+ \times \Omega$ into $\mathbb{R}$ defined by $X = 1_R$. If $R = (t', t''] \times A$ where $t', t'' \in \mathbb{R}_+$, $t' < t''$, and $A \in \mathfrak{F}_{t'}$, then X is a left-continuous $\{\mathfrak{F}_t\}$-adapted process since $X(t, \omega) = \mathbf{1}_{(t', t''] \times A}(t, \omega)$ for $(t, \omega) \in \mathbb{R}_+ \times \Omega$. Thus by the definition of $\mathfrak{S}$, we have $(t', t''] \times A = X^{-1}(\{1\}) \in \mathfrak{S}$. Similarly when $R = \{0\} \times A$ with $A \in \mathfrak{F}_0$. Therefore $\mathfrak{R} \subset \mathfrak{S}$ and thus $\sigma(\mathfrak{R}) \subset \mathfrak{S}$.

2) To show that $\mathfrak{S} \subset \sigma(\mathfrak{R})$ we show that every left-continuous $\{\mathfrak{F}_t\}$-adapted process X is $\sigma(\mathfrak{R})/\mathfrak{B}_{\mathbb{R}}$-measurable. This will imply $\mathfrak{S} \subset \sigma(\mathfrak{R})$ since $\mathfrak{S}$ is the smallest σ-algebra of subsets of $\mathbb{R}_+ \times \Omega$ with respect to which all left-continuous $\{\mathfrak{F}_t\}$-adapted processes are $\mathfrak{S}/\mathfrak{B}_{\mathbb{R}}$-measurable. Now let $\{X^{(n)} : n \in \mathbb{N}\}$ be the sequence of left-continuous $\{\mathfrak{F}_t\}$-adapted simple processes defined by (1) of Lemma 2.8. As we showed in the proof of Theorem 2.10 for any $E \in \mathfrak{B}_{\mathbb{R}}$ we have

$$(X^{(n)})^{-1}(E) = (\{0\} \times X_0^{-1}(E)) \cup [\bigcup_{k \in \mathbb{N}} (I_{n,k} \times X_{(k-1)2^{-n}}^{-1}(E))],$$

where $I_{n,k} = ((k-1)2^{-n}, k2^{-n}]$ for $k \in \mathbb{N}$. Since X is $\{\mathfrak{F}_t\}$-adapted, we have $X_0^{-1}(E) \in \mathfrak{F}_0$ and $X_{(k-1)2^{-n}}^{-1}(E) \in \mathfrak{F}_{(k-1)2^{-n}}$ for $k \in \mathbb{N}$. Thus $(X^{(n)})^{-1}(E)$ is a countable union of members of $\mathfrak{R}$ and is therefore in $\sigma(\mathfrak{R})$. This shows that $X^{(n)}$ is $\sigma(\mathfrak{R})/\mathfrak{B}_{\mathbb{R}}$-measurable. Then by (3) of Lemma 2.8, X is $\sigma(\mathfrak{R})/\mathfrak{B}_{\mathbb{R}}$-measurable.

3) Finally, since $\mathfrak{R}$ is a π-class, we have $\sigma(\mathfrak{R}) = d(\mathfrak{R})$ by Theorem 1.7. ∎

Definition 2.19. *Let $\mathfrak{S}$ and $\mathfrak{T}$ be the $\{\mathfrak{F}_t\}$-predictable σ-algebra and the $\{\mathfrak{F}_t\}$-well measurable σ-algebra in a filtered space $(\Omega, \mathfrak{F}, \{\mathfrak{F}_t\}, P)$. A stochastic process $X = \{X_t : t \in \mathbb{R}_+\}$ on the filtered space is called an $\{\mathfrak{F}_t\}$-predictable process if it is an $\mathfrak{S}/\mathfrak{B}_{\mathbb{R}}$-measurable mapping of $\mathbb{R}_+ \times \Omega$ into $\mathbb{R}$; it is called an $\{\mathfrak{F}_t\}$-well measurable process if it is a $\mathfrak{T}/\mathfrak{B}_{\mathbb{R}}$-measurable mapping.*

If X is a left-continuous $\{\mathfrak{F}_t\}$-adapted process, then X is $\mathfrak{S}/\mathfrak{B}_{\mathbb{R}}$-measurable by the definition of the $\{\mathfrak{F}_t\}$-predictable σ-algebra $\mathfrak{S}$ so that it is an $\{\mathfrak{F}_t\}$-predictable process.

Note that since $\mathfrak{S} \subset \mathfrak{T}$, a predictable process is always a well-measurable process. We show next that every $\{\mathfrak{F}_t\}$-well measurable process, and in particular, every $\{\mathfrak{F}_t\}$-predictable process is an $\{\mathfrak{F}_t\}$-adapted process. This is a consequence of the following lemma.

Lemma 2.20. *Let $\mathfrak{Y}$ be an arbitrary nonempty collection of $\{\mathfrak{F}_t\}$-adapted processes Y on a filtered space $(\Omega, \mathfrak{F}, \{\mathfrak{F}_t\}, P)$ and let $\mathfrak{G}$ be the σ-algebra of subsets of $\mathbb{R}_+ \times \Omega$ generated by $\mathfrak{Y}$. Then every $\mathfrak{G}/\mathfrak{B}_{\mathbb{R}}$-measurable mapping X of $\mathbb{R}_+ \times \Omega$ into $\mathbb{R}$ is an $\{\mathfrak{F}_t\}$-adapted process.*

Proof. To show that an arbitrary $\mathfrak{G}/\mathfrak{B}_\mathbb{R}$-measurable mapping X of $\mathbb{R}_+ \times \Omega$ into $\mathbb{R}$ is an $\{\mathfrak{F}_t\}$-adapted process, we show that for every $t \in \mathbb{R}_+$, X_t is $\mathfrak{F}_t/\mathfrak{B}_\mathbb{R}$-measurable. Thus, we are to show that $X_t^{-1}(E) \in \mathfrak{F}_t$ for every $E \in \mathfrak{B}_\mathbb{R}$. Note first that

(1) $$\{t\} \times X_t^{-1}(E) = X^{-1}(E) \cap (\{t\} \times \Omega).$$

Now

$$\mathfrak{G} = \sigma_{\mathbb{R}_+ \times \Omega}(\bigcup_{Y \in \mathfrak{Y}} Y^{-1}(\mathfrak{B}_\mathbb{R})) = \sigma_{\mathbb{R}_+ \times \Omega}\{Y^{-1}(A) : A \in \mathfrak{B}_\mathbb{R}, Y \in \mathfrak{Y}\}$$

and thus by Theorem 1.5, the expression (1) above for Y, and Corollary 1.2,

$$\begin{aligned}
\mathfrak{G} \cap (\{t\} \times \Omega) &= \sigma_{\{t\} \times \Omega}\{Y^{-1}(A) \cap (\{t\} \times \Omega) : A \in \mathfrak{B}_\mathbb{R}, Y \in \mathfrak{Y}\} \\
&= \sigma_{\{t\} \times \Omega}\{\{t\} \times Y_t^{-1}(A) : A \in \mathfrak{B}_\mathbb{R}, Y \in \mathfrak{Y}\} \\
&= \{t\} \times \sigma_\Omega\{Y_t^{-1}(A) : A \in \mathfrak{B}_\mathbb{R}, Y \in \mathfrak{Y}\}.
\end{aligned}$$

Since $Y \in \mathfrak{Y}$ is $\{\mathfrak{F}_t\}$-adapted, $Y_t^{-1}(A) \in \mathfrak{F}_t$ for $A \in \mathfrak{B}_\mathbb{R}$. Thus

(2) $$\mathfrak{G} \cap (\{t\} \times \Omega) \subset \{t\} \times \mathfrak{F}_t.$$

Now since X is $\mathfrak{G}/\mathfrak{B}_\mathbb{R}$-measurable, we have $X^{-1}(E) \in \mathfrak{G}$ for $E \in \mathfrak{B}_\mathbb{R}$. Then we have

(3) $$X^{-1}(E) \cap (\{t\} \times \Omega) \in \mathfrak{G} \cap (\{t\} \times \Omega).$$

Using (1) and (2) in (3) we have $\{t\} \times X_t^{-1}(E) \in \{t\} \times \mathfrak{F}_t$ and therefore $X_t^{-1}(E) \in \mathfrak{F}_t$. This shows that X is an $\{\mathfrak{F}_t\}$-adapted process. ∎

Proposition 2.21. *Every $\{\mathfrak{F}_t\}$-well measurable process $X = \{X_t : t \in \mathbb{R}_+\}$ and in particular every $\{\mathfrak{F}_t\}$-predictable process on a filtered space $(\Omega, \mathfrak{F}, \{\mathfrak{F}_t\}, P)$ is an $\{\mathfrak{F}_t\}$-adapted process.*

Proof. Let $\mathfrak{Y}$ be the collection of all right-continuous $\{\mathfrak{F}_t\}$-adapted processes on the filtered space. Since the $\{\mathfrak{F}_t\}$-well measurable σ-algebra $\mathfrak{T}$ is generated by $\mathfrak{Y}$, every $\mathfrak{T}/\mathfrak{B}_\mathbb{R}$-measurable mapping X of $\mathbb{R}_+ \times \Omega$ into $\mathbb{R}$ is an $\{\mathfrak{F}_t\}$-adapted process by Lemma 2.20. ∎

The following monotone class theorem for stochastic processes will be used later in proving that all $\{\mathfrak{F}_t\}$-predictable processes have a certain property.

Theorem 2.22. *Let **V** be a collection of stochastic processes on a filtered space $(\Omega, \mathfrak{F}, \{\mathfrak{F}_t\}, P)$ satisfying the following conditions:*

§2. STOCHASTIC PROCESSES

1°. **V** *contains all bounded left-continuous $\{\mathfrak{F}_t\}$-adapted processes,*

2°. **V** *is a linear space,*

3°. *if $\{X_n : n \in \mathbb{N}\}$ is an increasing sequence of nonnegative processes in **V** such that $X \equiv \lim_{n \to \infty} X_n$ is real valued (resp. bounded), then $X \in$ **V**.*

*Then **V** contains all real valued (resp. bounded) $\{\mathfrak{F}_t\}$-predictable processes.*

Proof. Let $\mathfrak{R}$ be the π-class of $\{\mathfrak{F}_t\}$-predictable rectangles in $\mathbb{R}_+ \times \Omega$. Then for every $A \in \mathfrak{R}$, $\mathbf{1}_A$ is a bounded left-continuous $\{\mathfrak{F}_t\}$-adapted process so that $\mathbf{1}_A \in$ **V** by 1°. For the same reason $\mathbf{1}_{\mathbb{R}_+ \times \Omega} \in$ **V**. Therefore condition 1° in Theorem 1.10 is satisfied by **V**. Conditions 2° and 3° of the same theorem are also satisfied on account of our 2° and 3°. Thus **V** contains all real valued (resp. bounded) $\sigma(\mathfrak{R})$-measurable transformations of $\mathbb{R}_+ \times \Omega$ into $\mathbb{R}$. Since $\sigma(\mathfrak{R}) = \mathfrak{S}$, **V** contains all real valued (resp. bounded) $\mathfrak{S}$-measurable, that is, $\{\mathfrak{F}_t\}$-predictable, processes. ∎

Proposition 2.23. *An arbitrary $\{\mathfrak{F}_t\}$-well measurable process, in particular an arbitrary $\{\mathfrak{F}_t\}$-predictable process, $X = \{X_t : t \in \mathbb{R}_+\}$ on a filtered space $(\Omega, \mathfrak{F}, \{\mathfrak{F}_t\}, P)$ is an $\{\mathfrak{F}_t\}$-progressively measurable process. In particular it is an $\{\mathfrak{F}_t\}$-adapted measurable process and therefore every sample function is a Borel-measurable real valued function on $\mathbb{R}_+$.*

Proof. Let $\mathfrak{Y}$ be the collection of all right-continuous $\{\mathfrak{F}_t\}$-adapted process on the filtered space. By definition the $\{\mathfrak{F}_t\}$-well measurable σ-algebra $\mathfrak{T}$ is the σ-algebra of subsets of $\mathbb{R}_+ \times \Omega$ generated by $\mathfrak{Y}$, that is,

(1) $$\mathfrak{T} = \sigma\{Y^{-1}(E) : E \in \mathfrak{B}_{\mathbb{R}}, Y \in \mathfrak{Y}\}.$$

For $t \in \mathbb{R}_+$, let $\mathfrak{T}_t$ be the σ-algebra of subsets of $[0,t] \times \Omega$ generated by the restrictions to $[0,t] \times \Omega$ of all members of $\mathfrak{Y}$, that is,

(2) $$\mathfrak{T}_t = \sigma\{Y^{-1}(E) \cap [0,t] \times \Omega : E \in \mathfrak{B}_{\mathbb{R}}, Y \in \mathfrak{Y}\}.$$

By Theorem 1.5, we have

(3) $$\mathfrak{T}_t = \mathfrak{T} \cap [0,t] \times \Omega.$$

Now according to Proposition 2.13, every $Y \in \mathfrak{Y}$ is an $\{\mathfrak{F}_t\}$-progressively measurable process and thus its restriction to $[0,t] \times \Omega$ is a $\sigma(\mathfrak{B}_{[0,t]} \times \mathfrak{F}_t)/\mathfrak{B}_{\mathbb{R}}$-measurable mapping of

$[0, t] \times \Omega$ into $\mathbb{R}$. Since $\mathfrak{T}_t$ is the smallest σ-algebra of subsets of $[0, t] \times \Omega$ with respect to which the restrictions to $[0, t] \times \Omega$ of all members of $\mathfrak{Y}$ are measurable, we have

(4) $$\mathfrak{T}_t \subset \sigma(\mathfrak{B}_{[0,t]} \times \mathfrak{F}_t).$$

Now our $\{\mathfrak{F}_t\}$-well measurable process X is a $\mathfrak{T}/\mathfrak{B}_\mathbb{R}$-measurable mapping of $\mathbb{R}_+ \times \Omega$ into $\mathbb{R}$ and thus its restriction to $[0, t] \times \Omega$ is a $\mathfrak{T} \cap [0, t] \times \Omega/\mathfrak{B}_\mathbb{R}$-measurable mapping of $[0, t] \times \Omega$ into $\mathbb{R}$. Therefore the restriction of X is $\mathfrak{T}_t/\mathfrak{B}_\mathbb{R}$-measurable by (3) and thus $\sigma(\mathfrak{B}_{[0,t]} \times \mathfrak{F}_t)/\mathfrak{B}_\mathbb{R}$-measurable by (4). This shows that X is an $\{\mathfrak{F}_t\}$-progressively measurable process. Then by Observation 2.12, X is an $\{\mathfrak{F}_t\}$-adapted measurable process. ∎

Theorem 2.24. *For every $\{\mathfrak{F}_t\}$-predictable process $X = \{X_t : t \in \mathbb{R}_+\}$ on a filtered space $(\Omega, \mathfrak{F}, \{\mathfrak{F}_t\}, P)$, there exists a sequence $\{X^{(n)} : n \in \mathbb{N}\}$ of left-continuous $\{\mathfrak{F}_t\}$-adapted processes on the filtered space such that $\lim_{n \to \infty} X^{(n)} = X$ on $\mathbb{R}_+ \times \Omega$. Thus every sample function of a predictable process is the limit of a sequence of left-continuous functions on $\mathbb{R}_+$.*

Proof. Consider the collection $\mathfrak{R}$ of all $\{\mathfrak{F}_t\}$-predictable rectangles on $\mathbb{R}_+ \times \Omega$. Let $\mathfrak{D}$ be the collection of subsets of $\mathbb{R}_+ \times \Omega$ of the type $F_1 \cap \cdots \cap F_n$ where F_i is either a member of $\mathfrak{R}$ or the complement of a member of $\mathfrak{R}$ for $i = 1, \ldots, n$. Let $\mathfrak{A}$ be the collection of all finite unions of members of $\mathfrak{D}$. It is easily verified that $\mathfrak{A}$ is the algebra of subsets of $\mathbb{R}_+ \times \Omega$ generated by $\mathfrak{R}$. Thus by Theorem 2.18, we have $\mathfrak{S} = \sigma(\mathfrak{R}) = \sigma(\mathfrak{A})$.

Let us show that for every $A \in \mathfrak{A}$, $\mathbf{1}_A$ is a left-continuous $\{\mathfrak{F}_t\}$-adapted process. Note first that if $F \in \mathfrak{R}$, then by Definition 2.16, $\mathbf{1}_F$ is a left-continuous $\{\mathfrak{F}_t\}$-adapted process. If $F^c \in \mathfrak{R}$, then $\mathbf{1}_{F^c}$ is a left-continuous $\{\mathfrak{F}_t\}$-adapted process so that $\mathbf{1}_F = \mathbf{1}_{\mathbb{R}_+ \times \Omega} - \mathbf{1}_{F^c}$ is a left-continuous $\{\mathfrak{F}_t\}$-adapted process. Next, let us show that $\mathbf{1}_D$ is a left-continuous $\{\mathfrak{F}_t\}$-adapted process for every $D \in \mathfrak{D}$. Let $D = F_1 \cap \cdots \cap F_n$ where either $F_i \in \mathfrak{R}$ or $F_i^c \in \mathfrak{R}$ for $i = 1, \ldots, n$. Note that in general for any two sets E_1 and E_2 we have

$$\mathbf{1}_{E_1 \cap E_2} = \{(\mathbf{1}_{E_1} + \mathbf{1}_{E_2}) - 1\} \vee 0.$$

Now $\mathbf{1}_{F_1}$ is a left-continuous $\{\mathfrak{F}_t\}$-adapted process. Suppose $\mathbf{1}_{F_1 \cap \cdots \cap F_k}$ is a left-continuous $\{\mathfrak{F}_t\}$-adapted process. Then since

$$\mathbf{1}_{F_1 \cap \cdots \cap F_{k+1}} = \{(\mathbf{1}_{F_1 \cap \cdots \cap F_k} + \mathbf{1}_{F_k}) - 1\} \vee 0,$$

$\mathbf{1}_{F_1 \cap \cdots \cap F_{k+1}}$ is a left-continuous $\{\mathfrak{F}_t\}$-adapted process. Thus by induction $\mathbf{1}_{F_1 \cap \cdots \cap F_n}$ is a left-continuous $\{\mathfrak{F}_t\}$-adapted process. Now let $A \in \mathfrak{A}$. Then $A = D_1 \cap \cdots \cap D_m$ where $D_i \in \mathfrak{D}$

for $i = 0, \ldots, m$. In general for any two sets E_1 and E_2, we have

$$\mathbf{1}_{E_1 \cup E_2} = \mathbf{1}_{E_1} + \mathbf{1}_{E_2} - \mathbf{1}_{E_1 \cap E_2}.$$

Using this equality we show by induction that $\mathbf{1}_A$ is a left-continuous $\{\mathfrak{F}_t\}$-adapted process.

Since $\mathfrak{S} = \sigma(\mathfrak{A})$, according to Proposition 1.12 for every real valued $\mathfrak{S}$-measurable function X on $\mathbb{R}_+ \times \Omega$ there exists a sequence $\{X^{(n)} : n \in \mathbb{N}\}$ of simple functions based on $\mathfrak{A}$ such that $X = \lim_{n \to \infty} X^{(n)}$ on $\mathbb{R}_+ \times \Omega$. Now

$$X^{(n)} = \sum_{j=1}^{n_p} c_{n,j} \mathbf{1}_{A_{n,j}} \quad \text{where } c_{n,j} \in \mathbb{R}, A_{n,j} \in \mathfrak{A} \text{ for } j = 1, \ldots, n_p.$$

Since $\mathbf{1}_{A_{n,j}}$ is a left-continuous $\{\mathfrak{F}_t\}$-adapted process for each $j = 1, \ldots, n_p$, so is $X^{(n)}$. Thus we have a sequence $\{X^{(n)} : n \in \mathbb{N}\}$ of left-continuous $\{\mathfrak{F}_t\}$-adapted processes on the filtered space such that $\lim_{n \to \infty} X^{(n)} = X$ on $\mathbb{R}_+ \times \Omega$. ∎

§3 Stopping Times

[I] Stopping Times for Stochastic Processes with Continuous Time

Stopping times are a method of truncating the sample functions of a stochastic process. Let $X = \{X_t : t \in \mathbb{R}_+\}$ be an adapted process on a filtered space $(\Omega, \mathfrak{F}, \{\mathfrak{F}_t\}, P)$. Suppose for instance that the sample functions $X(\cdot, \omega), \omega \in \Omega$, are to be stopped from varying with $t \in \mathbb{R}_+$ after some time point and suppose whether to stop a sample function or not at any given time depends on the behavior of the sample function up to the time point. Since the sub-σ-algebra $\mathfrak{F}_t$ of $\mathfrak{F}$ is a store of all data on the behavior of all the sample functions up to the time point t, we require that a stopping time be a random variable T satisfying the condition $\{\omega \in \Omega : T(\omega) \leq t\} \in \mathfrak{F}_t$ for every $t \in \mathbb{R}_+$. Since some sample functions may not be stopped at any time, we permit a stopping time to assume the value ∞. Corresponding to a stopping time T, the stopped process $X^{T\wedge}$ of the process X then is defined by $X^{T\wedge}(t, \omega) = X(T(\omega) \wedge t, \omega)$ for $(t, \omega) \in \mathbb{R}_+ \times \Omega$. The formal definition of a stopping time is given as follows.

Definition 3.1. *Let $(\Omega, \mathfrak{F}, \{\mathfrak{F}_t\}, P)$ be a filtered space. An $\{\mathfrak{F}_t\}$-stopping time is an $\overline{\mathbb{R}}_+$-valued function T on Ω satisfying the condition $\{\omega \in \Omega : T(\omega) \leq t\} \in \mathfrak{F}_t$ for every $t \in \mathbb{R}_+$. An $\{\mathfrak{F}_t\}$-stopping time will be referred to simply as a stopping time when the filtration $\{\mathfrak{F}_t : t \in \mathbb{R}_+\}$ is understood.*

Since $\mathfrak{F}_\infty = \sigma(\cup_{t\in\mathbb{R}_+}\mathfrak{F}_t)$, a stopping time T is $\mathfrak{F}_\infty$-measurable and in particular $\{\omega \in \Omega : T(\omega) = \infty\} \in \mathfrak{F}_\infty$.

A function T on Ω which is identically equal to a value $t_0 \in \mathbb{R}_+$ satisfies the condition in Definition 3.1 in a trivial way and is thus an $\{\mathfrak{F}_t\}$-stopping time. A constant stopping time is also called a deterministic time. Just as we write c for a function f which is identically equal to the constant c, let us write t for a stopping time that is identically equal t. Let us regard the σ-algebra $\mathfrak{F}_t$ in the filtration as a sub-σ-algebra of $\mathfrak{F}$ corresponding to the constant stopping time t. A sub-σ-algebra of $\mathfrak{F}$ corresponding to an arbitrary $\{\mathfrak{F}_t\}$-stopping time T is defined as follows.

Definition 3.2. *Let T be an $\{\mathfrak{F}_t\}$-stopping time on a filtered space $(\Omega, \mathfrak{F}, \{\mathfrak{F}_t\}, P)$. The sub-$\sigma$-algebra of $\mathfrak{F}$ at stopping time T is a subcollection of $\mathfrak{F}_\infty$ defined by*

$$\mathfrak{F}_T = \{A \in \mathfrak{F}_\infty : A \cap \{T \leq t\} \in \mathfrak{F}_t \text{ for every } t \in \mathbb{R}_+\}.$$

The fact that the collection $\mathfrak{F}_T$ is indeed a σ-algebra of subsets of Ω will be shown next. Note that if T is a deterministic time, say $T(\omega) = t_0 \in \mathbb{R}_+$ for all $\omega \in \Omega$, then $\mathfrak{F}_T = \mathfrak{F}_{t_0}$. This follows from the fact that when $T = t_0$ on Ω, then we have $\{T \leq t\} = \emptyset$ for $t < t_0$ and $\{T \leq t\} = \Omega$ for $t \geq t_0$ so that the condition $A \cap \{T \leq t\} \in \mathfrak{F}_t$ for every $t \in \mathbb{R}_+$ imposed on $A \in \mathfrak{F}_\infty$ is equivalent to the condition $A \in \mathfrak{F}_t$ for every $t \geq t_0$, or equivalently, $A \in \mathfrak{F}_{t_0}$, and therefore $\mathfrak{F}_T = \{A \in \mathfrak{F}_\infty : A \in \mathfrak{F}_{t_0}\} = \mathfrak{F}_{t_0}$.

Theorem 3.3. *Let T be a stopping time on a filtered space $(\Omega, \mathfrak{F}, \{\mathfrak{F}_t\}, P)$. Then*
1) $\mathfrak{F}_T$ is a σ-algebra of subsets of Ω.
2) T is $\mathfrak{F}_T$-measurable.
3) $\sigma(T) \subset \mathfrak{F}_T \subset \mathfrak{F}_\infty \subset \mathfrak{F}$.

Proof. 1) To show that $\mathfrak{F}_T$ is a σ-algebra of subsets of Ω, note first that $\Omega \in \mathfrak{F}_T$ since $\Omega \cap \{T \leq t\} = \{T \leq t\} \in \mathfrak{F}_t$ for every $t \in \mathbb{R}_+$. To show that $\mathfrak{F}_T$ is closed under complementation, let $A \in \mathfrak{F}_T$. Since $A^c = \Omega - A$, we have for every $t \in \mathbb{R}_+$

$$A^c \cap \{T \leq t\} = [\Omega \cap \{T \leq t\}] - [A \cap \{T \leq t\}].$$

Since $\Omega, A \in \mathfrak{F}_T$, the two sets on the right side are in $\mathfrak{F}_t$ and then so is their difference. To show that $\mathfrak{F}_T$ is closed under countable unions, let $A_n \in \mathfrak{F}_T$ for $n \in \mathbb{N}$. Then for every

§3. STOPPING TIMES

$t \in \mathbb{R}_+$ we have

$$(\cup_{n \in \mathbb{N}} A_n) \cap \{T \leq t\} = \bigcup_{n \in \mathbb{N}} [A_n \cap \{T \leq t\}] \in \mathfrak{F}_t,$$

and therefore $\cup_{n \in \mathbb{N}} A_n \in \mathfrak{F}_T$. This completes the verification that $\mathfrak{F}_T$ is a σ-algebra of subsets of Ω.

2) Since T is $\overline{\mathbb{R}}_+$-valued, to show that it is $\mathfrak{F}_T$-measurable, it suffices to show that $\{T \leq t_0\} \in \mathfrak{F}_T$ for every $t_0 \in \mathbb{R}_+$. Now since T is $\mathfrak{F}_\infty$-measurable by Definition 3.1, we have $\{T \leq t_0\} \in \mathfrak{F}_\infty$. Furthermore

$$\{T \leq t_0\} \cap \{T \leq t\} = \begin{cases} \{T \leq t_0\} \in \mathfrak{F}_{t_0} \subset \mathfrak{F}_t & \text{for } t \in \mathbb{R}_+,\ t \geq t_0 \\ \{T \leq t\} \in \mathfrak{F}_t & \text{for } t \in \mathbb{R}_+,\ t < t_0. \end{cases}$$

Thus, by Definition 3.2, $\{T \leq t_0\} \in \mathfrak{F}_T$.

3) Since T is $\mathfrak{F}_T$-measurable, we have $\sigma(T) \subset \mathfrak{F}_T$. By Definition 3.2, $\mathfrak{F}_T \subset \mathfrak{F}_\infty \subset \mathfrak{F}$. ∎

When the filtration is right-continuous, a stopping time has the following characterization:

Theorem 3.4. *Let $(\Omega, \mathfrak{F}, \{\mathfrak{F}_t\}, P)$ be a right-continuous filtered space. An $\overline{\mathbb{R}}_+$-valued function T on Ω is a stopping time if and only if*

(1) $\qquad\qquad\qquad \{\omega \in \Omega : T(\omega) < t\} \in \mathfrak{F}_t \quad \text{for every } t \in \mathbb{R}_+.$

For a stopping time T on the right-continuous filtered space we have

(2) $\qquad\qquad\qquad \mathfrak{F}_T = \{A \in \mathfrak{F}_\infty : A \cap \{T < t\} \in \mathfrak{F}_t \quad \text{for every } t \in \mathbb{R}_+\}.$

Proof. If T is a stopping time then for every $t \in \mathbb{R}_+$ we have by Definition 3.1

$$\{T < t\} = \bigcup_{n \in \mathbb{N}} \left\{T \leq t - \frac{1}{n}\right\} \in \mathfrak{F}_t$$

so that (1) holds.

Conversely suppose T is a $\overline{\mathbb{R}}_+$-valued function on Ω satisfying (1). Assume the right-continuity of the filtration. Then for every $t \in \mathbb{R}_+$

$$\{T \leq t\} = \bigcap_{k \in \mathbb{N}} \left\{T < t + \frac{1}{k}\right\} = \bigcap_{k \geq n} \left\{T < t + \frac{1}{k}\right\} \in \mathfrak{F}_{t + \frac{1}{n}} \quad \text{for every } n \in \mathbb{N}$$

by the fact that $\{T < t + 1/k\} \downarrow$ as $k \to \infty$, then by (1) and the fact that $\{\mathfrak{F}_t : t \in \mathbb{R}_+\}$ is an increasing system. Therefore

$$\{T \leq t\} \in \bigcap_{n \in \mathbb{N}} \mathfrak{F}_{t+\frac{1}{n}} = \mathfrak{F}_{t+} = \mathfrak{F}_t$$

by the right-continuity of the filtration. Thus T is a stopping time by Definition 3.1.

The equality (2) is proved by the same sort of argument as above using Definition 3.2 and the right-continuity of the filtration. ∎

Proposition 3.5. *Let $(\Omega, \mathfrak{F}, \{\mathfrak{F}_t\}, P)$ be a right-continuous filtered space and let $X = \{X_t : t \in \mathbb{R}_+\}$ be a right-continuous $\{\mathfrak{F}_t\}$-adapted process. The first passage time, or the hitting time, of a subset E of $\mathbb{R}$ by X is an $\overline{\mathbb{R}}_+$ valued function on Ω defined by*

$$T(\omega) = \begin{cases} \inf\{s \in \mathbb{R}_+ : X_s(\omega) \in E\} \\ \infty \text{ if the set above is } \emptyset. \end{cases}$$

If E is an open set then T is a stopping time.

Proof. Since the filtration is right-continuous, according to Theorem 3.4 it suffices to show that

$$\{\omega \in \Omega : T(\omega) < t\} \in \mathfrak{F}_t \quad \text{for every } t \in \mathbb{R}_+.$$

Actually it suffices to verify this for $t > 0$ since $\{\omega \in \Omega : T(\omega) < 0\} = \emptyset \in \mathfrak{F}_0$. Let $\omega_0 \in \Omega$ and $t > 0$ be fixed. Then $T(\omega_0) < t$ if and only if there exists $s \in (0, t)$ such that $X_s(\omega_0) \in E$. Let Q be the collection of all positive rational numbers. Then the right-continuity of the sample function $X(\cdot, \omega_0)$ and the openness of E imply that $T(\omega_0) < t$ if and only if there exists $r \in Q$ such that $r \in (0, t)$ and $X_r(\omega_0) \in E$. Thus $T(\omega_0) < t$ if and only if $\omega_0 \in \{\omega \in \Omega : X_r(\omega) \in E\}$ for some $r \in Q$, $r < t$. Therefore

$$\{\omega \in \Omega : T(\omega) < t\} = \bigcup_{r \in Q, r < t} \{\omega \in \Omega : X_r(\omega) \in E\} = \bigcup_{r \in Q, r < t} X_r^{-1}(E).$$

Since $X_r^{-1}(E) \in \mathfrak{F}_r \subset \mathfrak{F}_t$ for $r < t$, we have $\{\omega \in \Omega : T(\omega) < t\} \in \mathfrak{F}_t$ for every $t > 0$. Therefore T is a stopping time. ∎

It is known that the first passage time of every $E \in \mathfrak{B}_\mathbb{R}$ is a stopping time. See [21] P. A. Meyer.

Theorem 3.6. *Let T be a stopping time on a filtered space $(\Omega, \mathfrak{F}, \{\mathfrak{F}_t\}, P)$. Then every $\overline{\mathbb{R}}_+$-valued $\mathfrak{F}_T$-measurable function S on Ω satisfying the condition $S \geq T$ on Ω is also a stopping time.*

§3. STOPPING TIMES

Proof. To show that S is a stopping time, we show that $\{S \leq t\} \in \mathfrak{F}_t$ for every $t \in \mathbb{R}_+$. Now since S is $\mathfrak{F}_T$-measurable we have $\{S \leq t\} \in \mathfrak{F}_T$. This implies according to Definition 3.2

$$\{S \leq t\} \cap \{T \leq t\} \in \mathfrak{F}_t.$$

Since $S \geq T$ we have $\{S \leq t\} \subset \{T \leq t\}$. Thus the last expression reduces to $\{S \leq t\} \in \mathfrak{F}_t$. ∎

Theorem 3.7. *If S and T are stopping times on a filtered space $(\Omega, \mathfrak{F}, \{\mathfrak{F}_t\}, P)$, then so are $S \vee T$, $S \wedge T$, and $S + c$ for every $c \in \mathbb{R}_+$. If the filtration is right-continuous, then $S + T$ is a stopping time.*

Proof. For every $t \in \mathbb{R}_+$ we have

$$\{S \vee T \leq t\} = \{S \leq t\} \cap \{T \leq t\} \in \mathfrak{F}_t$$

so that $S \vee T$ is a stopping time. On the other hand

$$\{S \wedge T > t\} = \{S > t\} \cap \{T > t\}$$

so that

$$\{S \wedge T \leq t\} = \{S \leq t\}^c \cup \{T \leq t\}^c \in \mathfrak{F}_t,$$

showing that $S \wedge T$ is a stopping time.

For every $c \in \mathbb{R}_+$, $S + c$ is a stopping time by Theorem 3.6.

When the filtration is right-continuous, let $t \in \mathbb{R}_+$ and let Q_t be the collection of all rational numbers in $[0, t]$. Then by Theorem 3.4,

$$\{S + T < t\} = \bigcup_{u, v \in Q_t, u+v < t} (\{S < u\} \cap \{T < v\}) \in \mathfrak{F}_t$$

so that $S + T$ is a stopping time. ∎

Lemma 3.8. *Let T be a stopping time on a filtered space $(\Omega, \mathfrak{F}, \{\mathfrak{F}_t\}, P)$. Then for every $t \in \mathbb{R}_+$, $T \wedge t$ is $\mathfrak{F}_t$-measurable.*

Proof. Since $T \wedge t$ is an $\mathbb{R}_+$-valued function on Ω, to show that it is $\mathfrak{F}_t$-measurable we show that for every $\alpha \in \mathbb{R}_+$ we have $\{T \wedge t \leq \alpha\} \in \mathfrak{F}_t$. Since $T \wedge t$ is a stopping time according to Theorem 3.7, we have for every $\alpha \in \mathbb{R}_+$

$$\{T \wedge t \leq \alpha\} \in \mathfrak{F}_\alpha.$$

If $\alpha \leq t$ then $\mathfrak{F}_\alpha \subset \mathfrak{F}_t$ so that $\{T \wedge t \leq \alpha\} \in \mathfrak{F}_t$. If on the other hand $\alpha > t$ then $\{T \wedge t \leq \alpha\} = \Omega \in \mathfrak{F}_t$. In any case $\{T \wedge t \leq \alpha\} \in \mathfrak{F}_t$. ∎

According to Definition 3.2, the σ-algebra $\mathfrak{F}_S$ at a stopping time S on a filtered space $(\Omega, \mathfrak{F}, \{\mathfrak{F}_t\}, P)$ consists of all members A of $\mathfrak{F}_\infty$ satisfying the condition $A \cap \{S \leq t\} \in \mathfrak{F}_t$ for every $t \in \mathbb{R}_+$. The following theorem shows that actually every member A of $\mathfrak{F}_S$ satisfies the condition above not only with all deterministic times but also with all stopping times T.

Theorem 3.9. *Let S be a stopping time on a filtered space $(\Omega, \mathfrak{F}, \{\mathfrak{F}_t\}, P)$. If $A \in \mathfrak{F}_S$ then $A \cap \{S \leq T\} \in \mathfrak{F}_T$ for every stopping time T on the filtered space. If the filtered space is right-continuous, then we have $A \cap \{S < T\} \in \mathfrak{F}_T$ also.*

Proof. Let $A \in \mathfrak{F}_S$ and T be a stopping time on the filtered space. To show that $A \cap \{S \leq T\} \in \mathfrak{F}_T$, we verify according to Definition 3.2 that for every $t \in \mathbb{R}_+$ we have

(1) $$A \cap \{S \leq T\} \cap \{T \leq t\} \in \mathfrak{F}_t.$$

For this, let us show first that

(2) $$\{S \leq T\} \cap \{T \leq t\} = \{S \leq t\} \cap \{T \leq t\} \cap \{S \wedge t \leq T \wedge t\}.$$

Now if $S(\omega) \leq T(\omega)$ and $T(\omega) \leq t$, then $S(\omega) \leq t$ and $S(\omega) \wedge t \leq T(\omega) \wedge t$. Conversely, if $S(\omega) \leq t, T(\omega) \leq t$ and $S(\omega) \wedge t \leq T(\omega) \wedge t$, then $S(\omega) \wedge t = S(\omega)$ and $T(\omega) \wedge t = T(\omega)$ so that $S(\omega) \leq T(\omega)$. This proves (2). With (2) we have

$$A \cap \{S \leq T\} \cap \{T \leq t\} = A \cap \{S \leq t\} \cap \{T \leq t\} \cap \{S \wedge t \leq T \wedge t\}.$$

Since $A \in \mathfrak{F}_S$, we have $A \cap \{S \leq t\} \in \mathfrak{F}_t$. Since T is a stopping time, $\{T \leq t\} \in \mathfrak{F}_t$. Since both $S \wedge t$ and $T \wedge t$ are $\mathfrak{F}_t$-measurable according to Lemma 3.8 we have $\{S \wedge t \leq T \wedge t\} \in \mathfrak{F}_t$. Thus (1) holds.

When the filtered space is right-continuous, to show that $A \cap \{S < T\} \in \mathfrak{F}_T$, we verify that $A \cap \{S < T\} \cap \{T < t\} \in \mathfrak{F}_t$ for every $t \in \mathbb{R}_+$ according to Theorem 3.4. The verification is done by the same method as above. ∎

Theorem 3.10. *Let S and T be stopping times on a filtered space $(\Omega, \mathfrak{F}, \{\mathfrak{F}_t\}, P)$. If $S \leq T$ on Ω then $\mathfrak{F}_S \subset \mathfrak{F}_T$.*

Proof. If $A \in \mathfrak{F}_S$ then by Theorem 3.9, $A \cap \{S \leq T\} \in \mathfrak{F}_T$. If $S \leq T$ on Ω then $\{S \leq T\} = \Omega$ so that $A \in \mathfrak{F}_T$. Thus $\mathfrak{F}_S \subset \mathfrak{F}_T$. ∎

§3. STOPPING TIMES

Lemma 3.11. *Let S and T be stopping times on a filtered space $(\Omega, \mathfrak{F}, \{\mathfrak{F}_t\}, P)$. Then $\{S \leq T\}, \{S \geq T\} \in \mathfrak{F}_S \cap \mathfrak{F}_T$.*

Proof. Let $A \in \mathfrak{F}_S$. By Theorem 3.9, $A \cap \{S \leq T\} \in \mathfrak{F}_T$ and in particular with $A = \Omega$ we have

(1) $$\{S \leq T\} \in \mathfrak{F}_T.$$

Now

$$\{S < T\} = \{T \leq S\}^c = \{S \wedge T = T\}^c.$$

Since $S \wedge T$ is a stopping time by Theorem 3.7, $S \wedge T$ is $\mathfrak{F}_{S \wedge T}$-measurable by Theorem 3.3. Since $S \wedge T \leq T$, we have $\mathfrak{F}_{S \wedge T} \subset \mathfrak{F}_T$ by Theorem 3.10. Thus $S \wedge T$ is $\mathfrak{F}_T$-measurable. Since both $S \wedge T$ and T are $\mathfrak{F}_T$-measurable, we have $\{S \wedge T = T\} \in \mathfrak{F}_T$. Therefore

(2) $$\{S < T\} \in \mathfrak{F}_T.$$

From (1) and (2) we have

(3) $$\{S = T\} = \{S \leq T\} - \{S < T\} \in \mathfrak{F}_T.$$

From (2)

(4) $$\{S \geq T\} \in \mathfrak{F}_T.$$

By (3) and (4)

(5) $$\{S > T\} \in \mathfrak{F}_T.$$

Summarizing (2), (3), and (5), we have

$$\{S < T\}, \{S = T\}, \{S > T\} \in \mathfrak{F}_T.$$

Interchanging the roles of S and T, we have

$$\{S < T\}, \{S = T\}, \{S > T\} \in \mathfrak{F}_S.$$

Therefore

(6) $$\{S < T\}, \{S = T\}, \{S > T\} \in \mathfrak{F}_S \cap \mathfrak{F}_T.$$

In particular we have $\{S \leq T\}, \{S \geq T\} \in \mathfrak{F}_S \cap \mathfrak{F}_T$. ∎

Theorem 3.12. *Let S and T be stopping times on a filtered space $(\Omega, \mathfrak{F}, \{\mathfrak{F}_t\}, P)$. Then for the stopping time $S \wedge T$ we have $\mathfrak{F}_{S \wedge T} = \mathfrak{F}_S \cap \mathfrak{F}_T$.*

Proof. Since $S \wedge T \leq S, T$ we have by Theorem 3.10, $\mathfrak{F}_{S \wedge T} \subset \mathfrak{F}_S, \mathfrak{F}_T$ and thus $\mathfrak{F}_{S \wedge T} \subset \mathfrak{F}_S \cap \mathfrak{F}_T$.

To prove the reverse inclusion, let $A \in \mathfrak{F}_S \cap \mathfrak{F}_T$. To show that $A \in \mathfrak{F}_{S \wedge T}$ we verify according to Definition 3.2 that for every $t \in \mathbb{R}_+$

(1) $$A \cap \{S \wedge T \leq t\} \in \mathfrak{F}_t.$$

Now since $\Omega = \{S \leq T\} \cup \{S > T\}$, we have

$$\begin{aligned} \{S \wedge T \leq t\} &= \{S \leq T, S \wedge T \leq t\} \cup \{S > T, S \wedge T \leq t\} \\ &= \{S \leq T, S \leq t\} \cup \{S > T, T \leq t\}, \end{aligned}$$

and therefore

(2) $\quad A \cap \{S \wedge T \leq t\} = [A \cap \{S \leq T\} \cap \{S \leq t\}] \cup [A \cap \{S > T\} \cap \{T \leq t\}].$

Since $A \in \mathfrak{F}_S$ and $\{S \leq T\} \in \mathfrak{F}_S$ by Lemma 3.11, we have $A \cap \{S \leq T\} \in \mathfrak{F}_S$ and therefore $A \cap \{S \leq T\} \cap \{S \leq t\} \in \mathfrak{F}_t$ according to Definition 3.2. Similarly, since $A \in \mathfrak{F}_T$ and since $\{S > T\} \in \mathfrak{F}_T$ by (6) in the proof of Lemma 3.11, we have $A \cap \{S > T\} \in \mathfrak{F}_T$ and therefore $A \cap \{S > T\} \cap \{T \leq t\} \in \mathfrak{F}_t$ by Definition 3.2. Using these results in (2) we have (1). ∎

Corollary 3.13. *If S and T are stopping times on a filtered space $(\Omega, \mathfrak{F}, \{\mathfrak{F}_t\}, P)$ then*

$$\{S < T\}, \{S \leq T\}, \{S > T\}, \{S \geq T\} \in \mathfrak{F}_{S \wedge T}.$$

Proof. By (6) in the proof of Lemma 3.11 and Theorem 3.12 we have the corollary. ∎

Let us consider the conditional expectation of an integrable random variable with respect to the σ-algebra $\mathfrak{F}_{S \wedge T}$. To fix notation, let us recall that the conditional expectation $\mathbf{E}(X|\mathfrak{G})$ of an integrable random variable X with respect to a sub-σ-algebra $\mathfrak{G}$ of $\mathfrak{F}$ in a probability space $(\Omega, \mathfrak{F}, P)$ is by definition the equivalence class of $\mathfrak{G}$-measurable functions Y on Ω such that $\int_G Y\, dP = \int_G X\, dP$ for every $G \in \mathfrak{G}$, the equivalence relation being that of a.e. equality on $(\Omega, \mathfrak{G}, P)$. An arbitrary member in the equivalence class $\mathbf{E}(X|\mathfrak{G})$ is called a

§3. STOPPING TIMES

version of the conditional expectation. For two integrable random variables X_1 and X_2 and two sub-σ-algebras $\mathfrak{G}_1$ and $\mathfrak{G}_2$ of $\mathfrak{F}$, we write

$$E(X_1|\mathfrak{G}_1) \supset E(X_2|\mathfrak{G}_2)$$

to indicate that every version of $E(X_2|\mathfrak{G}_2)$ is also a version of $E(X_1|\mathfrak{G}_1)$. We write

$$E(X_1|\mathfrak{G}_1) = E(X_2|\mathfrak{G}_2)$$

if and only if the two equivalence classes are identical. At times it is convenient to write $E(X|\mathfrak{G})$ for an arbitrary member of the equivalence class. Whether the notation $E(X|\mathfrak{G})$ stands for the entire equivalence class or an arbitrary member of it should be clear from context. For instance, the expression $Y \in E(X|\mathfrak{G})$ indicates that the function Y is a member of the equivalence class $E(X|\mathfrak{G})$, and in an expression such as $\int_G E(X|\mathfrak{G})\,dP$, $E(X|\mathfrak{G})$ is an arbitrary version of the conditional expectation. We also use the notation $E(X|\mathfrak{G}_1|\mathfrak{G}_2)$ as an abbreviation for $E[E(X|\mathfrak{G}_1)|\mathfrak{G}_2]$. Regarding the conditional expectations of a random variable with respect to two different σ-algebras we have the following lemma.

Lemma 3.14. *Let X be an integrable random variable and $\mathfrak{G}_1$ and $\mathfrak{G}_2$ satisfying $\mathfrak{G}_1 \supset \mathfrak{G}_2$ be two sub-σ-algebras of $\mathfrak{F}$ in a probability space $(\Omega, \mathfrak{F}, P)$. Then*

(1) $$E(X|\mathfrak{G}_1|\mathfrak{G}_2) = E(X|\mathfrak{G}_2),$$

and

(2) $$E(X|\mathfrak{G}_2|\mathfrak{G}_1) \supset E(X|\mathfrak{G}_2).$$

If every version of $E(X|\mathfrak{G}_1)$ is $\mathfrak{G}_2$-measurable, then we have

(3) $$E(X|\mathfrak{G}_2) \supset E(X|\mathfrak{G}_1).$$

Note that under the condition $\mathfrak{G}_1 \supset \mathfrak{G}_2$ alone, there is in general no comparison between $E(X|\mathfrak{G}_1)$ and $E(X|\mathfrak{G}_2)$. Under additional conditions either one may contain the other. Also the set inclusion in (2) can be strict. For the proof of Lemma 3.14 we refer to Theorem B.12 and Theorem B.13. The following theorem regarding the conditional expectation of a random variable with respect to the σ-algebra $\mathfrak{F}_{S \wedge T}$ where S and T are stopping times is attributed to H. Asano in [7] Ikeda and Watanabe.

Theorem 3.15. *Let S and T be stopping times and let X be an integrable random variable on a right-continuous filtered space $(\Omega, \mathfrak{F}, \{\mathfrak{F}_t\}, P)$. Then*

(1) $$\mathbf{E}(\mathbf{1}_{\{S<T\}} X | \mathfrak{F}_S) = \mathbf{E}(\mathbf{1}_{\{S<T\}} X | \mathfrak{F}_{S \wedge T}),$$

(2) $$\mathbf{E}(\mathbf{1}_{\{S \leq T\}} X | \mathfrak{F}_S) = \mathbf{E}(\mathbf{1}_{\{S \leq T\}} X | \mathfrak{F}_{S \wedge T}),$$

(3) $$\mathbf{E}[X | \mathfrak{F}_S | \mathfrak{F}_T] \supset \mathbf{E}(X | \mathfrak{F}_{S \wedge T}).$$

Proof. Note that by Corollary 3.13, $\{S < T\}, \{S \leq T\} \in \mathfrak{F}_{S \wedge T}$ so that the random variables $\mathbf{1}_{\{S<T\}}$ and $\mathbf{1}_{\{S \leq T\}}$ are $\mathfrak{F}_{S \wedge T}$-measurable and a fortiori $\mathfrak{F}_S$- and $\mathfrak{F}_T$-measurable.

To prove (1), let us show first that $\mathbf{E}(\mathbf{1}_{\{S<T\}} X | \mathfrak{F}_S)$ is not only $\mathfrak{F}_S$-measurable but in fact $\mathfrak{F}_{S \wedge T}$-measurable. Note that since $\mathbf{1}_{\{S<T\}}$ is $\mathfrak{F}_S$-measurable we have

$$\mathbf{E}(\mathbf{1}_{\{S<T\}} X | \mathfrak{F}_S) = \mathbf{1}_{\{S<T\}} \mathbf{E}(X | \mathfrak{F}_S).$$

Since
$$\mathbf{1}_{\{S<T\}} \mathbf{E}(X | \mathfrak{F}_S) = \begin{cases} \mathbf{E}(X | \mathfrak{F}_S) & \text{on } \{S < T\} \\ 0 & \text{on } \{S < T\}^c, \end{cases}$$

and since $\{S < T\} \in \mathfrak{F}_{S \wedge T}$, to show the $\mathfrak{F}_{S \wedge T}$-measurability of $\mathbf{1}_{\{S<T\}} \mathbf{E}(X | \mathfrak{F}_S)$, it suffices to show that for every $\alpha \in \mathbb{R}$ we have

$$\{\mathbf{E}(X | \mathfrak{F}_S) \leq \alpha\} \cap \{S < T\} \in \mathfrak{F}_{S \wedge T}.$$

To prove this, we verify according to Definition 3.2 that for every $t \in \mathbb{R}_+$

(4) $$\{\mathbf{E}(X | \mathfrak{F}_S) \leq \alpha\} \cap \{S < T\} \cap \{S \wedge T \leq t\} \in \mathfrak{F}_t.$$

Now since
$$\{S < T\} \cap \{S \wedge T \leq t\} = \{S < T\} \cap \{S \leq t\}$$

we have
$$\{\mathbf{E}(X | \mathfrak{F}_S) \leq \alpha\} \cap \{S < T\} \cap \{S \wedge T \leq t\}$$
$$= [\{\mathbf{E}(X | \mathfrak{F}_S) \leq \alpha\} \cap \{S \leq t\}] \cap [\{S < T\} \cap \{S \leq t\}].$$

Since $\mathbf{E}(X | \mathfrak{F}_S)$ is $\mathfrak{F}_S$-measurable we have $\{\mathbf{E}(X | \mathfrak{F}_S) \leq \alpha\} \in \mathfrak{F}_S$. Then recalling Definition 3.2 for $\mathfrak{F}_S$ we have

$$\{\mathbf{E}(X | \mathfrak{F}_S) \leq \alpha\} \cap \{S \leq t\} \in \mathfrak{F}_t.$$

§3. STOPPING TIMES

Similarly, since $\{S < T\} \in \mathfrak{F}_S$ we have

$$\{S < T\} \cap \{S \leq t\} \in \mathfrak{F}_t.$$

Therefore (4) holds. This proves the $\mathfrak{F}_{S \wedge T}$-measurability of $\mathbf{E}(\mathbf{1}_{\{S<T\}} X | \mathfrak{F}_S)$.

Now since $\mathbf{E}(\mathbf{1}_{\{S<T\}} X | \mathfrak{F}_S)$ is $\mathfrak{F}_{S \wedge T}$-measurable and $\mathfrak{F}_S \supset \mathfrak{F}_{S \wedge T}$, we have according to (3) of Lemma 3.14,

$$\mathbf{E}(\mathbf{1}_{\{S<T\}} X | \mathfrak{F}_S) \subset \mathbf{E}(\mathbf{1}_{\{S<T\}} X | \mathfrak{F}_{S \wedge T}).$$

To show that actually the two conditional expectations are equal, we show that an arbitrary version of $\mathbf{E}(\mathbf{1}_{\{S<T\}} X | \mathfrak{F}_{S \wedge T})$ is also a version of $\mathbf{E}(\mathbf{1}_{\{S<T\}} X | \mathfrak{F}_S)$. Now since $\mathbf{E}(\mathbf{1}_{\{S<T\}} X | \mathfrak{F}_{S \wedge T})$ is $\mathfrak{F}_{S \wedge T}$-measurable and hence $\mathfrak{F}_S$-measurable, it remains only to verify that

(5) $$\int_A \mathbf{E}(\mathbf{1}_{\{S<T\}} X | \mathfrak{F}_{S \wedge T}) \, dP = \int_A \mathbf{1}_{\{S<T\}} X \, dP$$

not only for every $A \in \mathfrak{F}_{S \wedge T}$ but furthermore for every $A \in \mathfrak{F}_S$. Let $A \in \mathfrak{F}_S$. Since the filtered space is right-continuous, we have $A \cap \{S < T\} \in \mathfrak{F}_T$ according to Theorem 3.9. On the other hand according to Corollary 3.13 we have $\{S < T\} \in \mathfrak{F}_{S \wedge T} \subset \mathfrak{F}_S$ so that $A \cap \{S < T\} \in \mathfrak{F}_S$. Therefore $A \cap \{S < T\} \in \mathfrak{F}_S \cap \mathfrak{F}_T = \mathfrak{F}_{S \wedge T}$ by Theorem 3.12. Now

$$\int_A \mathbf{E}(\mathbf{1}_{\{S<T\}} X | \mathfrak{F}_{S \wedge T}) \, dP = \int_A \mathbf{1}_{\{S<T\}} \mathbf{E}(X | \mathfrak{F}_{S \wedge T}) \, dP = \int_{A \cap \{S<T\}} \mathbf{E}(X | \mathfrak{F}_{S \wedge T}) \, dP.$$

Since the domain of integration in the last integral above is in $\mathfrak{F}_{S \wedge T}$, the integral is equal to $\int_{A \cap \{S<T\}} X \, dP$ which in turn is equal to $\int_A \mathbf{1}_{\{S<T\}} X \, dP$. This verifies (5) and completes the proof of (1). The equality (2) is proved in the same way.

To prove (3), let us take the conditional expectation of both sides of (1) with respect to $\mathfrak{F}_T$ and then apply (2) of Lemma 3.14 to the right side to obtain

(6) $$\mathbf{E}[\mathbf{1}_{\{S<T\}} X | \mathfrak{F}_S | \mathfrak{F}_T] \supset \mathbf{E}[\mathbf{1}_{\{S<T\}} X | \mathfrak{F}_{S \wedge T}].$$

Interchanging the roles of S and T in (2), we have

(7) $$\mathbf{E}(\mathbf{1}_{\{T \leq S\}} X | \mathfrak{F}_T) = \mathbf{E}(\mathbf{1}_{\{T \leq S\}} X | \mathfrak{F}_{S \wedge T}).$$

Applying (7) to the integrable random variable $\mathbf{E}(X | \mathfrak{F}_S)$, we have

$$\mathbf{E}[\mathbf{1}_{\{T \leq S\}} \mathbf{E}(X | \mathfrak{F}_S) | \mathfrak{F}_T] = \mathbf{E}[\mathbf{1}_{\{T \leq S\}} \mathbf{E}(X | \mathfrak{F}_S) | \mathfrak{F}_{S \wedge T}].$$

By the $\mathfrak{F}_S$-measurability of $\{T \leq S\}$ and by applying (1) of Lemma 3.14 to the right side, we have

(8) $$\mathbf{E}[\mathbf{1}_{\{T \leq S\}} X | \mathfrak{F}_S | \mathfrak{F}_T] = \mathbf{E}[\mathbf{1}_{\{T \leq S\}} X | \mathfrak{F}_{S \wedge T}].$$

Adding (6) and (8) we have (3). ∎

Let us observe that an equality in (3) of Theorem 3.15 does not hold in general. Consider for instance the trivial case in which $S = s$, $T = t$ where $s, t \in \mathbb{R}_+$, $s < t$, and X is identically equal to 1 on Ω. In this case we have $\mathbf{E}[X | \mathfrak{F}_S | \mathfrak{F}_T] = \mathbf{E}[1 | \mathfrak{F}_s | \mathfrak{F}_t]$ and $\mathbf{E}(X | \mathfrak{F}_{S \wedge T}) = \mathbf{E}(1 | \mathfrak{F}_s)$ and in general $\mathbf{E}[1 | \mathfrak{F}_s | \mathfrak{F}_t]$ and $\mathbf{E}(1 | \mathfrak{F}_s)$ are not equal.

Now let us consider limits of sequences of stopping times.

Theorem 3.16. *If $\{T_n : n \in \mathbb{N}\}$ is a sequence of stopping times on a filtered space $(\Omega, \mathfrak{F}, \{\mathfrak{F}_t\}, P)$, then $\sup_{n \in \mathbb{N}} T_n$ is a stopping time on the filtered space. If the filtration is right-continuous, then $\inf_{n \in \mathbb{N}} T_n$, $\liminf_{n \to \infty} T_n$ and $\limsup_{n \to \infty} T_n$ are stopping times, and if $\lim_{n \to \infty} T_n$ exists everywhere on Ω then it is a stopping time.*

Proof. Note that $\sup_{n \in \mathbb{N}} T_n$ is a stopping time since for every $t \in \mathbb{R}_+$ we have

$$\left\{\sup_{n \in \mathbb{N}} T_n \leq t\right\} = \bigcap_{n \in \mathbb{N}} \{T_n \leq t\} \in \mathfrak{F}_t.$$

For $\inf_{n \in \mathbb{N}} T_n$ we have

$$\left\{\inf_{n \in \mathbb{N}} T_n \geq t\right\} = \bigcap_{n \in \mathbb{N}} \{T_n \geq t\} = \bigcap_{n \in \mathbb{N}} \{T_n < t\}^c.$$

Assume that the filtration is right-continuous. Then $\{T_n < t\} \in \mathfrak{F}_t$ for every $n \in \mathbb{N}$ by Theorem 3.4 and thus $\left\{\inf_{n \in \mathbb{N}} T_n \geq t\right\} \in \mathfrak{F}_t$. Again by Theorem 3.4, $\inf_{n \in \mathbb{N}} T_n$ is a stopping time. Since $\liminf_{n \to \infty} T_n = \sup_{n \in \mathbb{N}} \left\{\inf_{k \geq n} T_k\right\}$ and $\limsup_{n \to \infty} T_n = \inf_{n \in \mathbb{N}} \left\{\sup_{k \geq n} T_k\right\}$, they are stopping times.

If $\lim_{n \to \infty} T_n$ exists on Ω, then $\lim_{n \to \infty} T_n = \liminf_{n \to \infty} T_n$ so that it is a stopping time. ∎

§3. STOPPING TIMES

Theorem 3.17. *Let $\{T_n : n \in \mathbb{N}\}$ be a decreasing sequence of stopping times on a right-continuous filtered space $(\Omega, \mathfrak{F}, \{\mathfrak{F}_t\}, P)$. Then $T = \lim_{n \to \infty} T_n$ is a stopping time and $\mathfrak{F}_T = \bigcap_{n \in \mathbb{N}} \mathfrak{F}_{T_n}$.*

Proof. By Theorem 3.16, T is a stopping time. Since $T \leq T_n$ we have $\mathfrak{F}_T \subset \mathfrak{F}_{T_n}$ for every $n \in \mathbb{N}$ by Theorem 3.10 and thus $\mathfrak{F}_T \subset \bigcap_{n \in \mathbb{N}} \mathfrak{F}_{T_n}$. To show the reverse inclusion, let $A \in \bigcap_{n \in \mathbb{N}} \mathfrak{F}_{T_n}$. Now for any $t \in \mathbb{R}_+$ we have

$$\{T < t\} = \bigcup_{n \in \mathbb{N}} \{T_n < t\}$$

so that

$$A \cap \{T < t\} = \bigcup_{n \in \mathbb{N}} [A \cap \{T_n < t\}] \in \mathfrak{F}_t$$

by Theorem 3.4 applied to T_n. Thus $A \in \mathfrak{F}_T$ by Theorem 3.4 again. Therefore $\bigcap_{n \in \mathbb{N}} \mathfrak{F}_{T_n} \subset \mathfrak{F}_T$. ∎

Let T be a stopping time on a filtered space $(\Omega, \mathfrak{F}, \{\mathfrak{F}_t\}, P)$ and let T' be an $\overline{\mathbb{R}}_+$-valued function on Ω such that $T' = T$ a.e. on Ω. If we assume that $(\Omega, \mathfrak{F}, P)$ is a complete measure space, then T' is $\mathfrak{F}$-measurable but still may not be a stopping time since it may not satisfy the condition that $\{T' \leq t\} \in \mathfrak{F}_t$ for every $t \in \mathbb{R}_+$. The difficulty can be overcome by assuming that $\mathfrak{F}_0$ is augmented, that is, $\mathfrak{F}_0$ contains all the null sets of the measure space $(\Omega, \mathfrak{F}, P)$. This is proved in the next lemma. Note that if $\mathfrak{F}_0$ is augmented then so is $\mathfrak{F}_t$ for every $t \in \mathbb{R}_+$ since $\{\mathfrak{F}_t : t \in \mathbb{R}_+\}$ is an increasing system.

Lemma 3.18. *Let $(\Omega, \mathfrak{F}, \{\mathfrak{F}_t\}, P)$ be a filtered space and assume that the probability space $(\Omega, \mathfrak{F}, P)$ is complete and $\mathfrak{F}_0$ is augmented. If S and T are $\overline{\mathbb{R}}_+$-valued functions on Ω, $S = T$ a.e. on $(\Omega, \mathfrak{F}, P)$, and S is a stopping time, then T is a stopping time.*

Proof. Let Λ be a null set in $(\Omega, \mathfrak{F}, P)$ such that $S(\omega) = T(\omega)$ for $\omega \in \Lambda^c$. For every $t \in \mathbb{R}_+$ we have

$$\{T \leq t\} = [\{T \leq t\} \cap \Lambda^c] \cup [\{T \leq t\} \cap \Lambda].$$

Now

$$\{T \leq t\} \cap \Lambda^c = \{S \leq t\} \cap \Lambda^c \in \mathfrak{F}_t$$

since $\{S \leq t\} \in \mathfrak{F}_t$ and $\Lambda^c \in \mathfrak{F}_0 \subset \mathfrak{F}_t$. On the other hand, as a subset of the null set Λ, $\{T \leq t\} \cap \Lambda$ is a null set by the completeness of $(\Omega, \mathfrak{F}, P)$ and thus it is in $\mathfrak{F}_t$ since $\mathfrak{F}_t$ is augmented. Therefore, $\{T \leq t\} \in \mathfrak{F}_t$ for every $t \in \mathbb{R}_+$ so that T is a stopping time. ∎

Theorem 3.19. *Let $\{T_n : n \in \mathbb{N}\}$ be a sequence of stopping times on a filtered space $(\Omega, \mathfrak{F}, \{\mathfrak{F}_t\}, P)$. Suppose $\lim_{n\to\infty} T_n$ exists a.e. on $(\Omega, \mathfrak{F}, P)$ and T is a $\overline{\mathbb{R}}_+$-valued function on Ω such that $\lim_{n\to\infty} T_n = T$ a.e. on $(\Omega, \mathfrak{F}, P)$. Then T is a stopping time under the assumptions that $(\Omega, \mathfrak{F}, P)$ is a complete measure space, $\mathfrak{F}_0$ is augmented and $\{\mathfrak{F}_t : t \in \mathbb{R}_+\}$ is right-continuous.*

Proof. Let Λ be a null set in $(\Omega, \mathfrak{F}, P)$ such that $\lim_{n\to\infty} T_n(\omega) = T(\omega)$ for $\omega \in \Lambda^c$. For $n \in \mathbb{N}$, let T'_n be defined by

$$T'_n(\omega) = \begin{cases} T_n(\omega) & \text{for } \omega \in \Lambda^c \\ 0 & \text{for } \omega \in \Lambda, \end{cases}$$

and let T' be defined by

$$T'(\omega) = \begin{cases} T(\omega) & \text{for } \omega \in \Lambda^c \\ 0 & \text{for } \omega \in \Lambda. \end{cases}$$

Since T_n is a stopping time and $T'_n = T_n$ a.e. on $(\Omega, \mathfrak{F}, P)$, T'_n is a stopping time by Lemma 3.18. Since $\lim_{n\to\infty} T'_n = \lim_{n\to\infty} T_n = T = T'$ on Λ^c and $T'_n = T' = 0$ on Λ, we have $\lim_{n\to\infty} T'_n = T'$ everywhere on Ω. Thus T' is a stopping time by Theorem 3.16. Then since $T = T'$ a.e. on Ω, T is a stopping time by Lemma 3.18. ∎

The next theorem shows that a stopping time can always be approached from above by a sequence of discrete valued stopping times

Theorem 3.20. *Let T be a stopping time on a filtered space $(\Omega, \mathfrak{F}, \{\mathfrak{F}_t\}, P)$. For each $n \in \mathbb{N}$, let ϑ_n be an $\overline{\mathbb{R}}_+$-valued function on $\overline{\mathbb{R}}_+$ defined by*

(1) $$\vartheta_n(t) = \begin{cases} k2^{-n} & \text{for } t \in [(k-1)2^{-n}, k2^{-n}) \text{ for } k \in \mathbb{N} \\ \infty & \text{for } t = \infty \end{cases}$$

and let

(2) $$T_n = \vartheta_n \circ T.$$

Then $\{T_n : n \in \mathbb{N}\}$ is a decreasing sequence of stopping times with T_n assuming values in $\{k2^{-n} : k \in \mathbb{N}\} \cup \{\infty\}$ and $T_n \downarrow T$ uniformly on Ω as $n \to \infty$.

Proof. By (1), ϑ_n defined on $\overline{\mathbb{R}}_+$ is a right-continuous step function when restricted to $\mathbb{R}_+$ and satisfies $t < \vartheta_n(t) \leq t + 2^{-n}$ for $t \in \mathbb{R}_+$. Thus $\vartheta_n(t) \downarrow t$ uniformly on $\overline{\mathbb{R}}_+$ as $n \to \infty$. If X is an $\overline{\mathbb{R}}_+$-valued random variable on $(\Omega, \mathfrak{F}, P)$ and if we define X_n by $X_n = \vartheta_n \circ X$, then since X is an $\mathfrak{F}/\mathfrak{B}_{\overline{\mathbb{R}}_+}$-measurable transformation of Ω into $\overline{\mathbb{R}}_+$ and ϑ_n is a $\mathfrak{B}_{\overline{\mathbb{R}}_+}/\mathfrak{B}_{\overline{\mathbb{R}}_+}$-measurable

§3. STOPPING TIMES

transformation of $\overline{\mathbb{R}}_+$ into $\overline{\mathbb{R}}_+$, X_n is an $\mathfrak{F}/\mathfrak{B}_{\overline{\mathbb{R}}_+}$-measurable transformation of Ω into $\overline{\mathbb{R}}_+$. Thus X_n is a random variable on $(\Omega, \mathfrak{F}, P)$ assuming values in $\{k2^{-n} : k \in \mathbb{N}\} \cup \{\infty\}$. Also $X(\omega) < X_n(\omega) \leq X(\omega) + 2^{-n}$ at $\omega \in \Omega$ for which $X(\omega) \in \mathbb{R}_+$ and $X_n(\omega) = \infty$ at $\omega \in \Omega$ for which $X(\omega) = \infty$. Thus $X_n(\omega) \downarrow X(\omega)$ uniformly for $\omega \in \Omega$ as $n \to \infty$.

All the above holds when X is a stopping time T in particular. Now if T is a stopping time, then T is an $\mathfrak{F}_T/\mathfrak{B}_{\overline{\mathbb{R}}_+}$-measurable mapping of Ω into $\overline{\mathbb{R}}_+$. Since ϑ_n is an $\mathfrak{B}_{\overline{\mathbb{R}}_+}/\mathfrak{B}_{\overline{\mathbb{R}}_+}$-measurable mapping of $\overline{\mathbb{R}}_+$ into $\overline{\mathbb{R}}_+$, T_n is an $\mathfrak{F}_T/\mathfrak{B}_{\overline{\mathbb{R}}_+}$-measurable mapping of Ω into $\overline{\mathbb{R}}_+$. We also have $T_n \geq T$ on Ω. Thus T_n is a stopping time by Theorem 3.6. ∎

[II] Stopping Times for Stochastic Processes with Discrete Time

The concepts of filtration and stopping time for discrete time are defined in similar ways to the continuous case. Since a stopping time is a random variable assuming values in the range of the time parameter, not only the order type of the range of the time parameter but also the values of the time parameter matter. Therefore in considering stopping times for stochastic processes with discrete time parameters it is not enough to consider only $\mathbb{Z}_+$ as the time parameter. Thus, we consider an arbitrary strictly increasing sequence in $\mathbb{R}_+$ as time parameter.

Definition 3.21. *Let $(\Omega, \mathfrak{F}, P)$ be a probability space and let $\{t_n : n \in \mathbb{Z}_+\}$ be a strictly increasing sequence in $\mathbb{R}_+$. An increasing sequence $\{\mathfrak{F}_{t_n} : n \in \mathbb{Z}_+\}$ of sub-σ-algebras of $\mathfrak{F}$ is called a filtration with discrete time of the probability space and $(\Omega, \mathfrak{F}, \{\mathfrak{F}_{t_n} : n \in \mathbb{Z}_+\}, P)$, or more briefly $(\Omega, \mathfrak{F}, \{\mathfrak{F}_{t_n}\}, P)$, is called a filtered space with discrete time. Let $t_\infty = \lim\limits_{n\to\infty} t_n \in \overline{\mathbb{R}}_+$ and $\mathfrak{F}_{t_\infty} = \sigma(\cup_{n\in\mathbb{Z}_+}\mathfrak{F}_{t_n})$. A sequence of random variables $X = \{X_{t_n} : n \in \mathbb{Z}_+\}$ on the filtered space with discrete time is said to be $\{\mathfrak{F}_{t_n}\}$-adapted if X_{t_n} is $\mathfrak{F}_{t_n}$-measurable for every $n \in \mathbb{Z}_+$. X is said to be $\{\mathfrak{F}_{t_n}\}$-predictable if X_{t_n} is $\mathfrak{F}_{t_{n-1}}$-measurable for every $n \in \mathbb{N}$.*

Note that predictability imposes no condition on the random variable X_{t_0} in a process with discrete time $X = \{X_{t_n} : n \in \mathbb{Z}_+\}$.

Definition 3.22. *Given a filtered space $(\Omega, \mathfrak{F}, \{\mathfrak{F}_{t_n} : n \in \mathbb{Z}_+\}, P)$ with discrete time. A function T on Ω with values in $\{t_n : n \in \overline{\mathbb{Z}}_+\}$ is called an $\{\mathfrak{F}_{t_n}\}$-stopping time if*

(1) $$\{T \leq t_n\} \in \mathfrak{F}_{t_n} \quad \text{for every } n \in \mathbb{Z}_+.$$

The σ-algebra at the stopping time T is defined by

(2) $\quad \mathfrak{F}_T = \{A \in \mathfrak{F}_{t_\infty} : A \cap \{T \leq t_n\} \in \mathfrak{F}_{t_n} \text{ for every } n \in \mathbb{Z}_+\}.$

Note that a stopping time T is an $\mathfrak{F}_{t_\infty}$-measurable random variable, and by the same argument as in the case of continuous time, $\mathfrak{F}_T$ is a sub-σ-algebra of $\mathfrak{F}$, T is $\mathfrak{F}_T$-measurable and $\sigma(T) \subset \mathfrak{F}_T \subset \mathfrak{F}_{t_\infty} \subset \mathfrak{F}$.

Remark 3.23. 1) Unlike the case of continuous time, the defining condition (1) of a stopping time in discrete time in Definition 3.22 is equivalent to the condition

(3) $\quad \{T = t_n\} \in \mathfrak{F}_{t_n} \quad \text{for every } n \in \mathbb{Z}_+.$

This follows from the fact that on one hand $\{T = t_0\} = \{T \leq t_0\}$ and $\{T = t_n\} = \{T \leq t_n\} - \{T \leq t_{n-1}\}$ for $n \in \mathbb{N}$ and on the other hand $\{T \leq t_n\} = \cup_{k=0}^n \{T = t_k\}$ for $n \in \mathbb{Z}_+$. By the same reason we have

(4) $\quad \mathfrak{F}_T = \{A \in \mathfrak{F}_{t_\infty} : A \cap \{T = t_n\} \in \mathfrak{F}_{t_n}, \text{ for every } n \in \mathbb{Z}_+\}.$

Furthermore, (1) implies

(5) $\quad \{T < t_n\} \in \mathfrak{F}_{t_n} \quad \text{for every } n \in \mathbb{Z}_+.$

without additional conditions on the filtration such as the right-continuity in the case of continuous time.

2) $\{\mathfrak{F}_{t_n}\}$-stopping times arise for instance in the following way. Let T be an $\{\mathfrak{F}_t\}$-stopping time on a right-continuous filtered space $(\Omega, \{\mathfrak{F}\}, \{\mathfrak{F}_t : t \in \mathbb{R}_+\}, P)$. For fixed $n \in \mathbb{N}$, let T_n be defined by (2) of Theorem 3.20 so that T_n is an $\{\mathfrak{F}_t\}$-stopping time assuming values in $\{k2^{-n} : k \in \mathbb{N}\} \cup \{\infty\}$. Now $\{\mathfrak{F}_{k2^{-n}} : k \in \mathbb{Z}_+\}$ is a filtration with discrete time. Also $\{T_n = k2^{-n}\} = \{T_n \leq k2^{-n}\} - \{T_n \leq (k-1)2^{-n}\} \in \mathfrak{F}_{k2^{-n}}$ for $k \in \mathbb{Z}_+$. Therefore T_n is a stopping time with respect to the filtration $\{\mathfrak{F}_{k2^{-n}} : k \in \mathbb{N}\}$.

Such properties of $\{\mathfrak{F}_t\}$-stopping times as those in Theorem 3.6 through Theorem 3.15 are valid for $\{\mathfrak{F}_{t_n}\}$-stopping times. Regarding sequences of $\{\mathfrak{F}_{t_n}\}$-stopping times we have the following theorem.

Theorem 3.24. *Let $\{T_m : m \in \mathbb{N}\}$ be a sequence of stopping times on a filtered space with discrete time $(\Omega, \mathfrak{F}, \{\mathfrak{F}_{t_n}\}, P)$. Then $\sup_{m \in \mathbb{N}} T_m$, $\inf_{m \in \mathbb{N}} T_m$, $\liminf_{m \to \infty} T_m$ and $\limsup_{m \to \infty} T_m$ are all*

§3. STOPPING TIMES

stopping times. If $\lim_{n\to\infty} T_m$ exists everywhere on Ω then it is a stopping time. If $\{T_m : m \in \mathbb{N}\}$ is a decreasing sequence then $T = \lim_{m\to\infty} T_m$ is a stopping time and $\mathfrak{F}_T = \bigcap_{m\in\mathbb{N}} \mathfrak{F}_{T_m}$.

Proof. These statements can be proved by the same arguments as in Theorem 3.16 and Theorem 3.17 with the exception that due to the property (5) in Remark 3.23 we now have for every $n \in \mathbb{Z}_+$,

$$\left\{\inf_{m\in\mathbb{N}} T_m \geq t_n\right\} = \bigcap_{m\in\mathbb{N}} \{T_m \geq t_n\} = \bigcap_{m\in\mathbb{N}} \{T_m < t_n\}^c \in \mathfrak{F}_{t_n}$$

without additional assumption on the filtration such as the right-continuity in the case of continuous time. ∎

[III] Random Variables at Stopping Times

Let $X = \{X_t : t \in \mathbb{R}_+\}$ be an adapted process and T be a stopping time on a filtered space $(\Omega, \mathfrak{F}, \{\mathfrak{F}_t\}, P)$. For any $\omega \in \Omega$ such that $T(\omega) < \infty$, the function $X_T(\omega) \equiv X(T(\omega), \omega)$ is defined. However, if $T(\omega) = \infty$, then $X_T(\omega)$ as given above is meaningless since the range $\mathbb{R}_+$ of the time parameter does not include ∞. If X_∞ is an arbitrary extended real valued $\mathfrak{F}_\infty$-measurable random variable and if we let $\overline{X}(\omega) = \{X_t : t \in \overline{\mathbb{R}}_+\}$, then $X_T \equiv \overline{X}(T(\omega), \omega)$ is defined for every $\omega \in \Omega$. We shall show that if X is a $\sigma(\mathfrak{B}_{\mathbb{R}_+} \times \mathfrak{F}_\infty)/\mathfrak{B}_\mathbb{R}$-measurable mapping of $\mathbb{R}_+ \times \Omega$ into $\mathbb{R}$, then the function X_T is an $\mathfrak{F}_\infty$-measurable random variable. This random variable at the stopping time T is called an optional stopping. The random variable X_∞ used in defining X_T is called a random variable at infinity. We permit X_∞ to be equal to ∞ and $-\infty$ so that we may let $X_\infty = \lim_{t\to\infty} X_t$ in case $\lim_{t\to\infty} X_t$ exists in $\overline{\mathbb{R}}$.

Definition 3.25. *Let $X = \{X_t : t \in \mathbb{R}_+\}$ be an adapted process and T be a stopping time on a filtered space $(\Omega, \mathfrak{F}, \{\mathfrak{F}_t\}, P)$. Let X_∞ be an arbitrary extended real valued $\mathfrak{F}_\infty$-measurable random variable on $(\Omega, \mathfrak{F}, P)$ and let $\overline{X} = \{X_t : t \in \overline{\mathbb{R}}_+\}$. We define X_T by setting*

(1) $$X_T(\omega) = X_{T(\omega)}(\omega) = \overline{X}(T(\omega), \omega) \quad \text{for } \omega \in \Omega.$$

Similarly, for an adapted process with discrete time $X = \{X_{t_n} : n \in \mathbb{Z}_+\}$ and a stopping time T on a filtered space $(\Omega, \mathfrak{F}, \{\mathfrak{F}_{t_n}\}, P)$, let X_{t_∞} be an arbitrary extended real valued $\mathfrak{F}_{t_\infty}$- measurable random variable on $(\Omega, \mathfrak{F}, P)$ and let $\overline{X} = \{X_{t_n} : n \in \overline{\mathbb{Z}}_+\}$. We define X_T by the same expression (1) above.

Thus defined, X_T depends on the choice of X_∞ (or X_{t_∞}). In what follows whenever we speak of X_T we understand that an extended real valued $\mathfrak{F}_\infty$-measurable random variable X_∞ (or an extended real valued $\mathfrak{F}_{t_\infty}$-measurable random variable X_{t_∞}) has been selected in defining X_T. Recall however that when T does not assume the value ∞, X_∞ is not needed in defining X_T. In particular when T is an arbitrary stopping time, $T \wedge t$ for a fixed $t \in \mathbb{R}_+$ is a bounded stopping time and $X_{T \wedge t}$ is defined without X_∞.

Next, let us consider the measurability of X_T as a function on the probability space $(\Omega, \mathfrak{F}, P)$. The next lemma shows that the measurability of X_T is unrelated to the filtration in terms of which the stopping time T is defined.

Lemma 3.26. *Let $X = \{X_t : t \in \mathbb{R}_+\}$ be a stochastic process, S be an $\overline{\mathbb{R}}_+$-valued random variable, and X_∞ be an extended real valued random variable on a probability space $(\Omega, \mathfrak{F}, P)$. With $\overline{X} = \{X_t : t \in \overline{\mathbb{R}}_+\}$, let X_S be an extended real valued function on Ω defined by $X_S(\omega) = \overline{X}(S(\omega), \omega)$ for $\omega \in \Omega$. If X is a $\sigma(\mathfrak{B}_{\mathbb{R}_+} \times \mathfrak{F})/\mathfrak{B}_{\overline{\mathbb{R}}}$-measurable mapping of $\mathbb{R}_+ \times \Omega$ into $\overline{\mathbb{R}}$, then X_S is an extended real valued random variable on $(\Omega, \mathfrak{F}, P)$.*

Proof. Consider the mapping τ of Ω into $\overline{\mathbb{R}}_+ \times \Omega$ defined by $\tau(\omega) = (S(\omega), \omega)$ and the mapping $\overline{X}$ of $\overline{\mathbb{R}}_+ \times \Omega$ into $\overline{\mathbb{R}}$. Then

$$(\overline{X} \circ \tau)(\omega) = \overline{X}(S(\omega), \omega) = X_S(\omega) \quad \text{for } \omega \in \Omega,$$

that is, X_S is the composition of τ and $\overline{X}$. If we show that τ is $\mathfrak{F}/\sigma(\mathfrak{B}_{\overline{\mathbb{R}}_+} \times \mathfrak{F})$-measurable and $\overline{X}$ is $\sigma(\mathfrak{B}_{\overline{\mathbb{R}}_+} \times \mathfrak{F})/\mathfrak{B}_{\overline{\mathbb{R}}}$-measurable, then the composite mapping X_S is $\mathfrak{F}/\mathfrak{B}_{\overline{\mathbb{R}}}$-measurable, that is, X_S is an extended real valued random variable on $(\Omega, \mathfrak{F}, P)$.

Now since S is an $\mathfrak{F}/\mathfrak{B}_{\overline{\mathbb{R}}_+}$-measurable mapping of Ω into $\overline{\mathbb{R}}_+$, τ is an $\mathfrak{F}/\sigma(\mathfrak{B}_{\overline{\mathbb{R}}_+} \times \mathfrak{F})$-measurable mapping of Ω into $\overline{\mathbb{R}}_+ \times \Omega$. This can be shown as follows. If we let ι be the identity mapping of Ω into Ω then for an arbitrary $B \in \mathfrak{B}_{\overline{\mathbb{R}}_+}$ and $A \in \mathfrak{F}$ we have

$$\tau^{-1}(B \times A) = S^{-1}(B) \cap \iota^{-1}(A) = S^{-1}(B) \cap A \in \mathfrak{F},$$

and thus $\tau^{-1}(\mathfrak{B}_{\overline{\mathbb{R}}_+} \times \mathfrak{F}) \subset \mathfrak{F}$, and then by Theorem 1.1, $\tau^{-1}(\sigma(\mathfrak{B}_{\overline{\mathbb{R}}_+} \times \mathfrak{F})) = \sigma(\tau^{-1}(\mathfrak{B}_{\overline{\mathbb{R}}_+} \times \mathfrak{F})) \subset \mathfrak{F}$.

Regarding $\overline{X} = \{X_t : t \in \overline{\mathbb{R}}_+\}$, let us decompose its domain of definition $\overline{\mathbb{R}}_+ \times \Omega$ as $(\mathbb{R}_+ \times \Omega) \cup (\{\infty\} \times \Omega)$. The restrictions of $\overline{X}$ to these two subdomains are X and X_∞ respectively. Hence for every $E \in \mathfrak{B}_{\overline{\mathbb{R}}}$ we have

$$\overline{X}^{-1}(E) = X^{-1}(E) \cup (\{\infty\} \times X_\infty^{-1}(E)) \subset (\mathbb{R}_+ \times \Omega) \cup (\{\infty\} \times \Omega).$$

§3. STOPPING TIMES

Since X is a $\sigma(\mathfrak{B}_{\mathbb{R}_+} \times \mathfrak{F})/\mathfrak{B}_{\mathbb{R}}$-measurable mapping of $\mathbb{R}_+ \times \Omega$ into $\mathbb{R}$, we have

$$X^{-1}(E) \in \sigma_{\mathbb{R}_+ \times \Omega}\left(\mathfrak{B}_{\mathbb{R}_+} \times \mathfrak{F}\right) \subset \sigma_{\overline{\mathbb{R}}_+ \times \Omega}\left(\mathfrak{B}_{\overline{\mathbb{R}}_+} \times \mathfrak{F}\right).$$

Also, since $X_\infty^{-1}(E) \in \mathfrak{F}$,

$$\{\infty\} \times X_\infty^{-1}(E) \in \{\infty\} \times \mathfrak{F} \subset \mathfrak{B}_{\overline{\mathbb{R}}_+} \times \mathfrak{F} \subset \sigma_{\overline{\mathbb{R}}_+ \times \Omega}\left(\mathfrak{B}_{\overline{\mathbb{R}}_+} \times \mathfrak{F}\right).$$

Therefore

$$\overline{X}^{-1}(E) \in \sigma_{\overline{\mathbb{R}}_+ \times \Omega}\left(\mathfrak{B}_{\overline{\mathbb{R}}_+} \times \mathfrak{F}\right),$$

that is, $\overline{X}$ is $\sigma(\mathfrak{B}_{\overline{\mathbb{R}}_+} \times \mathfrak{F})/\mathfrak{B}_{\overline{\mathbb{R}}}$-measurable. This completes the proof. ∎

Theorem 3.27. *1) Let $X = \{X_t : t \in \mathbb{R}_+\}$ be an adapted process and T be a stopping time on a filtered space $(\Omega, \mathfrak{F}, \{\mathfrak{F}_t\}, P)$. Let X_T be defined with an arbitrary extended real valued $\mathfrak{F}_\infty$-measurable random variable X_∞ on $(\Omega, \mathfrak{F}, P)$. If X is a $\sigma(\mathfrak{B}_{\mathbb{R}_+} \times \mathfrak{F}_\infty)/\mathfrak{B}_{\mathbb{R}}$-measurable mapping of $\mathbb{R}_+ \times \Omega$ into $\mathbb{R}$, and in particular if X is an $\{\mathfrak{F}_t\}$-well measurable process, then X_T is an extended real valued $\mathfrak{F}_\infty$-measurable random variable.*
2) If $X = \{X_{t_n} : n \in \mathbb{Z}_+\}$ is an adapted process and T is a stopping time on a filtered space $(\Omega, \mathfrak{F}, \{\mathfrak{F}_{t_n}\}, P)$, then X_T defined with an arbitrary extended real valued $\mathfrak{F}_{t_\infty}$-measurable random variable X_{t_∞} on $(\Omega, \mathfrak{F}, P)$ is always an extended real valued $\mathfrak{F}_{t_\infty}$-measurable random variable on $(\Omega, \mathfrak{F}, P)$.

Proof. 1) Recall that an $\{\mathfrak{F}_t\}$-well measurable process is always a $\sigma(\mathfrak{B}_{\mathbb{R}_+} \times \mathfrak{F}_\infty)/\mathfrak{B}_{\mathbb{R}}$-measurable mapping of $\mathbb{R}_+ \times \Omega$ into $\mathbb{R}$ according to Proposition 2.23 and Observation 2.12. Now if $X = \{X_t : t \in \mathbb{R}_+\}$ is a $\sigma(\mathfrak{B}_{\mathbb{R}_+} \times \mathfrak{F}_\infty)/\mathfrak{B}_{\mathbb{R}}$-measurable mapping of $\mathbb{R}_+ \times \Omega$ into $\mathbb{R}$, then by regarding $(\Omega, \mathfrak{F}_\infty, P)$ as the probability space on which the stochastic process X is defined, and by applying Lemma 3.26, we conclude that X_T is an extended real valued random variable on $(\Omega, \mathfrak{F}_\infty, P)$.
    2) For $X = \{X_{t_n} : n \in \mathbb{Z}_+\}$ and $\overline{X} = \{X_{t_n} : n \in \overline{\mathbb{Z}}_+\}$, we have for every $\omega \in \Omega$,

$$X_T(\omega) = \overline{X}(T(\omega), \omega) = \sum_{n \in \mathbb{Z}_+} \mathbf{1}_{\{T=t_n\}} X(t_n, \omega) + \mathbf{1}_{\{T=t_\infty\}} X_{t_\infty}(\omega).$$

Since T and X_{t_n} for $n \in \overline{\mathbb{Z}}_+$ are all $\mathfrak{F}_{t_\infty}$-measurable random variables, each summand in the equation above is an $\mathfrak{F}_{t_\infty}$-measurable random variable. The countable sum X_T is then an $\mathfrak{F}_{t_\infty}$-measurable random variable. ∎

In the next theorem we show that actually X_T is $\mathfrak{F}_T$-measurable for a discrete time process. The same is true for a continuous time process but under the assumption of right-continuity of both the stochastic process and the filtration.

Theorem 3.28. *1) For an adapted process with discrete time $X = \{X_{t_n} : n \in \mathbb{Z}_+\}$ and a stopping time T on a filtered space $(\Omega, \mathfrak{F}, \{\mathfrak{F}_{t_n}\}, P)$, the random variable X_T is $\mathfrak{F}_T$-measurable.*
2) For a right-continuous adapted process $X = \{X_t : t \in \mathbb{R}_+\}$ and a stopping time T on a right-continuous filtered space $(\Omega, \mathfrak{F}, \{\mathfrak{F}_t\}, P)$, X_T is an $\mathfrak{F}_T$-measurable random variable.

Proof. 1) Consider a discrete time adapted process. Since $\sigma(T) \subset \mathfrak{F}_T$ we have $\{T = t_k\} \in \mathfrak{F}_T$ for $k \in \overline{\mathbb{Z}}_+$. Therefore to show the $\mathfrak{F}_T$-measurability of X_T on Ω it suffices to show that X_T is $\mathfrak{F}_T$-measurable on $\{T = t_k\}$ for each $k \in \overline{\mathbb{Z}}_+$. For this we show that for every $\alpha \in \mathbb{R}$ we have

(1) $$\{X_T \leq \alpha\} \cap \{T = t_k\} \in \mathfrak{F}_T.$$

Now $\{X_T \leq \alpha\} \cap \{T = t_k\} \in \mathfrak{F}_{t_k} \subset \mathfrak{F}_{t_\infty}$. Furthermore for every $j \in \mathbb{Z}_+$ we have

$$\{X_T \leq \alpha\} \cap \{T = t_k\} \cap \{T \leq t_j\} = \begin{cases} \{X_T \leq \alpha\} \cap \{T = t_k\} \in \mathfrak{F}_{t_k} \subset \mathfrak{F}_{t_j} & \text{for } k \leq j \\ \emptyset \in \mathfrak{F}_{t_j} & \text{for } k > j. \end{cases}$$

This proves (1) according to (2) of Definition 3.22.

2) For a continuous time adapted process, assume that both the filtration and the process are right-continuous. For each $n \in \mathbb{N}$, let ϑ_n be defined on $\overline{\mathbb{R}}_+$ by setting

$$\vartheta_n(t) = \begin{cases} k2^{-n} & \text{for } t \in [(k-1)2^{-n}, k2^{-n}) \text{ for } k \in \mathbb{N} \\ \infty & \text{for } t = \infty \end{cases}$$

and let $T_n = \vartheta_n \circ T$. As we showed in Theorem 3.20, T_n is a stopping time whose values are contained in the countable set $\{k2^{-n} : k \in \mathbb{N}\} \cup \{\infty\}$ and $\{T_n : n \in \mathbb{N}\}$ is a decreasing sequence such that $T_n \downarrow T$ uniformly on Ω as $n \to \infty$. For each fixed $n \in \mathbb{N}$, if we let $t_k = k2^{-n}$ for $k \in \mathbb{Z}_+$ then $\{X_{t_k} : k \in \mathbb{Z}_+\}$ is an adapted process with discrete time and T_n is a stopping time on the filtered space $(\Omega, \mathfrak{F}, \{\mathfrak{F}_{t_k}\}, P)$ so that by our result in 1) X_{T_n} is $\mathfrak{F}_{T_n}$-measurable. Since $T_n \downarrow T$ on Ω and X is right-continuous we have

$$\lim_{n \to \infty} X_{T_n}(\omega) = \lim_{n \to \infty} X(T_n(\omega), \omega) = X(T(\omega), \omega) = X_T(\omega) \quad \text{for } \omega \in \Omega.$$

Since X_{T_n} is $\mathfrak{F}_{T_n}$-measurable for every $n \in \mathbb{N}$ and since $\mathfrak{F}_{T_n}$, $n \in \mathbb{N}$, is a decreasing sequence by Theorem 3.10, $\lim_{k \to \infty} X_{T_k}$ is $\mathfrak{F}_{T_n}$-measurable for every $n \in \mathbb{N}$. Therefore $\lim_{k \to \infty} X_{T_k}$

§3. STOPPING TIMES

is $\bigcap_{n \in \mathbb{N}} \mathfrak{F}_{T_n}$-measurable. But $\bigcap_{n \in \mathbb{N}} \mathfrak{F}_{T_n} = \mathfrak{F}_T$ by Theorem 3.17. Thus $\lim_{n \to \infty} X_{T_n}$ is $\mathfrak{F}_T$-measurable, that is, X_T is $\mathfrak{F}_T$-measurable. ∎

Remark 3.29. Let $X = \{X_{t_n} : n \in \mathbb{Z}_+\}$ be an adapted process and T be a stopping time on a filtered space $(\Omega, \mathfrak{F}, \{\mathfrak{F}_{t_n}\}, P)$. Regarding the integrability of X_T, let us note that even when X is an L_1-process and $X_{t_\infty} \in L_1(\Omega, \mathfrak{F}_{t_\infty}, P)$, X_T may not be integrable since X_T in general consists of patches of X_{t_n} for infinitely many $n \in \overline{\mathbb{Z}}_+$. Doob's Optional Stopping Theorem gives sufficient conditions for the integrability of X_T when X is a submartingale. (See Theorem 6.1).

[IV] Stopped Processes and Truncated Processes

Definition 3.30. Let $X = \{X_t : t \in \mathbb{R}_+\}$ be an adapted process and T be a stopping time on a filtered space $(\Omega, \mathfrak{F}, \{\mathfrak{F}_t\}, P)$. The stopped process of X by T is the process $X^{T \wedge} = \{X_t^{T \wedge} : t \in \mathbb{R}_+\}$ where $X_t^{T \wedge}$ is defined by

(1) $$X_t^{T \wedge}(\omega) = X_{T(\omega) \wedge t}(\omega) = X(T(\omega) \wedge t, \omega) \quad \text{for } \omega \in \Omega,$$

that is, for $\omega \in \Omega$ with $T(\omega) < \infty$ we have

(2) $$X_t^{T \wedge}(\omega) = \begin{cases} X(t, \omega) & \text{for } t \in [0, T(\omega)] \\ X(T(\omega), \omega) & \text{for } t \in (T(\omega), \infty) \end{cases}$$

and for $\omega \in \Omega$ with $T(\omega) = \infty$ we have

(2′) $$X_t^{T \wedge}(\omega) = X(t, \omega) \quad \text{for } t \in \mathbb{R}_+.$$

For an adapted process with discrete time $X = \{X_{t_n} : n \in \mathbb{Z}_+\}$ and a stopping time on a filtered space $(\Omega, \mathfrak{F}, \{\mathfrak{F}_{t_n}\}, P)$, the stopped process of X by T is the process $X^{T \wedge} = \{X_{t_n}^{T \wedge} : n \in \mathbb{Z}_+\}$ where $X_{t_n}^{T \wedge}$ is defined by

(3) $$X_{t_n}^{T \wedge}(\omega) = X_{T(\omega) \wedge t_n}(\omega) = X(T(\omega) \wedge t_n, \omega) \quad \text{for } \omega \in \Omega.$$

Regarding the stopped process $X^{T \wedge}$ and the increasing system of σ-algebras $\{\mathfrak{F}_{T \wedge t} : t \in \mathbb{R}_+\}$ we have the following theorem.

Theorem 3.31. *1) The stopped process $X^{T \wedge}$ of an adapted process with discrete time $X = \{X_{t_n} : n \in \mathbb{Z}_+\}$ on a filtered space $(\Omega, \mathfrak{F}, \{\mathfrak{F}_{t_n}\}, P)$ is a stochastic process adapted to the*

filtration $\{\mathfrak{F}_{T\wedge t_n} : n \in \mathbb{Z}_+\}$ of the probability space $(\Omega, \mathfrak{F}, P)$.
2) *The stopped process* $X^{T\wedge}$ *of a right-continuous adapted process* $X = \{X_t : t \in \mathbb{R}_+\}$ *on a right-continuous filtered space* $(\Omega, \mathfrak{F}, \{\mathfrak{F}_t\}, P)$ *is a right-continuous process adapted to the right-continuous filtration* $\{\mathfrak{F}_{T\wedge t} : t \in \mathbb{R}_+\}$ *of the probability space* $(\Omega, \mathfrak{F}, P)$.

Proof. 1) For a discrete time adapted process X, since $\{T \wedge t_n : n \in \mathbb{Z}_+\}$ is an increasing sequence of stopping times, $\{\mathfrak{F}_{T\wedge t_n} : n \in \mathbb{Z}_+\}$ is an increasing sequence of sub-σ-algebras of $\mathfrak{F}$, that is, a filtration of the probability space $(\Omega, \mathfrak{F}, P)$. Since $X_{T\wedge t_n}$ is $\mathfrak{F}_{T\wedge t_n}$-measurable by Theorem 3.28, $X^{T\wedge}$ is an $\{\mathfrak{F}_{T\wedge t_n}\}$-adapted process.

2) Similarly for a continuous time adapted process X, $\{T \wedge t : t \in \mathbb{R}_+\}$ is an increasing system of stopping times so that $\{\mathfrak{F}_{T\wedge t} : t \in \mathbb{R}_+\}$ is an increasing system of sub-σ-algebras of $\mathfrak{F}$, that is, a filtration of the probability space $(\Omega, \mathfrak{F}, P)$.

To show that the right-continuity of $\{\mathfrak{F}_t : t \in \mathbb{R}_+\}$ implies that of $\{\mathfrak{F}_{T\wedge t} : t \in \mathbb{R}_+\}$, let $t_0 \in \mathbb{R}_+$ be fixed and let $\{t_n : n \in \mathbb{N}\}$ be a sequence in $\mathbb{R}_+$ such that $t_n \downarrow t_0$. Then $T \wedge t_n \downarrow T \wedge t_0$ on Ω so that by Theorem 3.17 we have

$$\mathfrak{F}_{T\wedge t_0} = \bigcap_{n\in\mathbb{N}} \mathfrak{F}_{T\wedge t_n} = \bigcap_{t>t_0} \mathfrak{F}_{T\wedge t}.$$

From the arbitrariness of $t_0 \in \mathbb{R}_+$ we have the right-continuity of the filtration $\{\mathfrak{F}_{T\wedge t} : t \in \mathbb{R}_+\}$. By Theorem 3.28, $X_{T\wedge t}$ is $\mathfrak{F}_{T\wedge t}$-measurable. Thus $X^{T\wedge}$ is an $\{\mathfrak{F}_{T\wedge t}\}$-adapted process. By the right-continuity of X and by (2) of Definition 3.30, $X^{T\wedge}$ is right-continuous. ∎

Remark 3.32. Let $X = \{X_{t_n} : n \in \mathbb{Z}_+\}$ be an adapted process and T be a stopping time on a filtered space $(\Omega, \mathfrak{F}, \{\mathfrak{F}_{t_n}\}, P)$. Consider the stopped process $X^{T\wedge} = \{X_{T\wedge t_n} : n \in \mathbb{Z}_+\}$. When X is an L_p-process for some $p \in (0, \infty)$, that is, $|X_{t_n}|^p$ is integrable for every $n \in \mathbb{Z}_+$, then $X^{T\wedge} = \{X_{T\wedge t_n} : n \in \mathbb{Z}_+\}$ is an L_p-process. This follows from the fact that for every $n \in \mathbb{Z}_+$ we have

$$X_{T\wedge t_n} = \mathbf{1}_{\{T=t_0\}}X_{t_0} + \mathbf{1}_{\{T=t_1\}}X_{t_1} + \cdots + \mathbf{1}_{\{T=t_{n-1}\}}X_{t_{n-1}} + \mathbf{1}_{\{T\geq t_n\}}X_{t_n}$$

so that

$$\int_\Omega |X_{T\wedge t_n}|^p dP = \sum_{k=0}^{n-1} \int_{\{T=t_k\}} |X_{t_k}|^p dP + \int_{\{T\geq t_n\}} |X_{t_n}|^p dP \leq \sum_{k=0}^{n} \int_\Omega |X_{t_k}|^p dP < \infty.$$

For an L_p-process with continuous time $X = \{X_t : t \in \mathbb{R}_+\}$ and a stopping time T on a filtered space $(\Omega, \mathfrak{F}, \{\mathfrak{F}_t\}, P)$, if we assume that $X_T \in L_p(\Omega, \mathfrak{F}_T, P)$ then $X^{T\wedge} = \{X_{T\wedge t} : t \in \mathbb{R}_+\}$ is an L_p-process since for every $t \in \mathbb{R}_+$ we have

$$X_{T\wedge t} = \mathbf{1}_{\{T\leq t\}}X_T + \mathbf{1}_{\{T>t\}}X_t$$

§3. STOPPING TIMES

and then
$$\int_\Omega |X_{T \wedge t}|^p dP \leq \int_\Omega |X_T|^p dP + \int_\Omega |X_t|^p P < \infty.$$

Definition 3.33. *Let $X = \{X_t : t \in \mathbb{R}_+\}$ be a stochastic process and T be a stopping time on a filtered space $(\Omega, \mathfrak{F}, \{\mathfrak{F}_t\}, P)$. Consider the subset $\{(\cdot) \leq T\}$ of $\mathbb{R}_+ \times \Omega$ defined by*

(1) $$\{(\cdot) \leq T\} = \{(t, \omega) \in \mathbb{R}_+ \times \Omega : t \leq T(\omega)\}.$$

The truncated process $X^{[T]}$ is a real valued function on $\mathbb{R}_+ \times \Omega$ defined by

(2) $$X^{[T]}(t, \omega) = \mathbf{1}_{\{(\cdot) \leq T\}}(t, \omega) X(t, \omega) \quad \text{for } (t, \omega) \in \mathbb{R}_+ \times \Omega.$$

Observation 3.34. Regarding $\mathbf{1}_{\{(\cdot) \leq T\}}$, observe that for $\omega \in \Omega$ with $T(\omega) < \infty$ we have

(1) $$\mathbf{1}_{\{(\cdot) \leq T\}}(t, \omega) = \begin{cases} 1 & \text{for } t \in [0, T(\omega)] \\ 0 & \text{for } t \in (T(\omega), \infty), \end{cases}$$

and for $\omega \in \Omega$ with $T(\omega) = \infty$ we have

(1') $$\mathbf{1}_{\{(\cdot) \leq T\}}(t, \omega) = 1 \quad \text{for } t \in \mathbb{R}_+.$$

Thus for $\omega \in \Omega$ with $T(\omega) < \infty$ we have

(2) $$X^{[T]}(t, \omega) = \begin{cases} X(t, \omega) & \text{for } t \in [0, T(\omega)] \\ 0 & \text{for } t \in (T(\omega), \infty). \end{cases}$$

and for $\omega \in \Omega$ with $T(\omega) = \infty$ we have

(2') $$X^{[T]}(t, \omega) = X(t, \omega) \quad \text{for } t \in \mathbb{R}_+.$$

Similarly for every $t \in \mathbb{R}_+$, we have

(3) $$\mathbf{1}_{\{(\cdot) \leq T\}}(t, \omega) = \begin{cases} 1 & \text{on } \{T \geq t\} \\ 0 & \text{on } \{T < t\}, \end{cases}$$

and therefore for every $t \in \mathbb{R}_+$, we have

(4) $$X^{[T]}(t, \omega) = \begin{cases} X(t, \omega) & \text{on } \{T \geq t\} \\ 0 & \text{on } \{T < t\}. \end{cases}$$

If the filtered space is right-continuous, then by Theorem 3.4 we have $\{T < t\} \in \mathfrak{F}_t$ for every $t \in \mathbb{R}_+$ so that $\mathbf{1}_{\{(\cdot)\leq T\}}(t, \cdot)$ as given by (3) is $\mathfrak{F}_t$-measurable. Therefore $\mathbf{1}_{\{(\cdot)\leq T\}}$ is an adapted process on the filtered space. By (1), every sample function of $\mathbf{1}_{\{(\cdot)\leq T\}}$ is left-continuous. Therefore $\mathbf{1}_{\{(\cdot)\leq T\}}$ is a predictable process.

Observation 3.35. 1) $X^{[T]}$ is always a stochastic process.
2) If the filtered space $(\Omega, \mathfrak{F}, \{\mathfrak{F}_t\}, P)$ is right-continuous and X is an adapted process, then so is $X^{[T]}$.
3) If the filtered space $(\Omega, \mathfrak{F}, \{\mathfrak{F}_t\}, P)$ is right-continuous and X is a predictable process, then so is $X^{[T]}$.

Proof. 1) Since $\mathbf{1}_{\{(\cdot)\leq T\}}(t, \cdot)$ and $X(t, \cdot)$ are both $\mathfrak{F}$-measurable, so is their product $X^{[T]}(t, \cdot)$ for every $t \in \mathbb{R}_+$. Thus $X^{[T]}$ is a stochastic process.

2) If X is an adapted process, then the $\mathfrak{F}_t$-measurability of both $\mathbf{1}_{\{(\cdot)\leq T\}}(t, \cdot)$ and $X(t, \cdot)$ implies that of their product $X^{[T]}(t, \cdot)$ for every $t \in \mathbb{R}_+$. Thus $X^{[T]}$ is an adapted process.

3) If X is a predictable process, then both X and $\mathbf{1}_{\{(\cdot)\leq T\}}$ are $\mathfrak{S}/\mathfrak{B}_{\mathbb{R}}$-measurable mappings of $\mathbb{R}_+ \times \Omega$ into $\mathbb{R}$. Then the product $X^{[T]} = \mathbf{1}_{\{(\cdot)\leq T\}} \cdot X$ is an $\mathfrak{S}/\mathfrak{B}_{\mathbb{R}}$-measurable mapping of $\mathbb{R}_+ \times \Omega$ into $\mathbb{R}$, that is, it is a predictable process. ∎

Observation 3.36. If S and T are two stopping times, then by (1) of Observation 3.34 we have
$$\mathbf{1}_{\{(\cdot)\leq S\}} \cdot \mathbf{1}_{\{(\cdot)\leq T\}} = \mathbf{1}_{\{(\cdot)\leq S\wedge T\}},$$
and hence by (2) and (2') of Definition 3.33 we have
$$(X^{[S]})^{[T]} = (X^{[T]})^{[S]} = X^{[S\wedge T]}.$$
In particular when $S \leq T$, we have $\mathbf{1}_{\{(\cdot)\leq S\}} \cdot \mathbf{1}_{\{(\cdot)\leq T\}} = \mathbf{1}_{\{(\cdot)\leq S\}}$ and consequently $(X^{[T]})^{[S]} = X^{[S]}$.

Lemma 3.37. *Let T be a stopping time on a filtered space $(\Omega, \mathfrak{F}, \{\mathfrak{F}_t : t \in \mathbb{R}_+\}, P)$. For each $n \in \mathbb{N}$, let $\{t_{n,k} : k \in \mathbb{Z}_+\}$ be a strictly increasing sequence in $\mathbb{R}_+$ with $t_{n,0} = 0$ and $\lim_{k\to\infty} t_{n,k} = \infty$ such that $\{t_{n,k} : k \in \mathbb{Z}_+\} \subset \{t_{n+1,k} : k \in \mathbb{Z}_+\}$ for $n \in \mathbb{N}$. Assume further that $\Delta_n \equiv \sup_{k\in\mathbb{N}}(t_{n,k} - t_{n,k-1}) < \infty$ for every $n \in \mathbb{N}$ and $\lim_{n\to\infty} \Delta_n = 0$. For each $n \in \mathbb{N}$, let ϑ_n be an $\overline{\mathbb{R}}_+$-valued function on $\overline{\mathbb{R}}_+$ defined by*

(1) $$\vartheta_n(t) = \begin{cases} t_{n,1} & \text{for } t \in [t_{n,0}, t_{n,1}] \\ t_{n,k} & \text{for } t \in (t_{n,k-1}, t_{n,k}] \text{ and } k \geq 2 \\ \infty & \text{for } t = \infty, \end{cases}$$

and let

(2) $$T_n = \vartheta_n \circ T.$$

Then T_n is a stopping time on the filtered space assuming values in $\{t_{n,k} : k \in \mathbb{N}\} \cup \{\infty\}$ and $T_n \downarrow T$ on Ω as $n \to \infty$.

Proof. Note that ϑ_n is a left-continuous monotone increasing step function and $\vartheta_n(t) \downarrow t$ uniformly on $\overline{\mathbb{R}}_+$ as $n \to \infty$. Since T is $\mathfrak{F}_T/\mathfrak{B}_{\overline{\mathbb{R}}_+}$-measurable and ϑ_n is $\mathfrak{B}_{\overline{\mathbb{R}}_+}/\mathfrak{B}_{\overline{\mathbb{R}}_+}$-measurable, T_n is $\mathfrak{F}_T/\mathfrak{B}_{\overline{\mathbb{R}}_+}$-measurable. Since $\vartheta_n(t) \geq t$ for $t \in \overline{\mathbb{R}}_+$, we have $T_n(\omega) \geq T(\omega)$ for every $\omega \in \Omega$. Thus by Theorem 3.6, T_n is a stopping time. Also, $\vartheta_n(t) \downarrow t$ on $\overline{\mathbb{R}}_+$ implies $T_n(\omega) \downarrow T(\omega)$ for every $\omega \in \Omega$. ∎

Theorem 3.38. *Let $X = \{X_t : t \in \mathbb{R}_+\}$ be a stochastic process and T be a stopping time on a filtered space $(\Omega, \mathfrak{F}, \{\mathfrak{F}_t\}, P)$. Let $T_n, n \in \mathbb{N}$, be as in Lemma 3.37. Then*

$$\lim_{n \to \infty} X^{[T_n]}(t, \omega) = X^{[T]}(t, \omega) \quad \text{for } (t, \omega) \in \mathbb{R}_+ \times \Omega.$$

Proof. By Definition 3.33 we have $X^{[T_n]}(t, \omega) = \mathbf{1}_{\{(\cdot) \leq T_n\}}(t, \omega) X(t, \omega)$ and similarly $X^{[T]}(t, \omega) = \mathbf{1}_{\{(\cdot) \leq T\}}(t, \omega) X(t, \omega)$ for $(t, \omega) \in \mathbb{R}_+ \times \Omega$. Thus to prove the theorem it suffices to show that

(1) $$\lim_{n \to \infty} \mathbf{1}_{\{(\cdot) \leq T_n\}}(t, \omega) = \mathbf{1}_{\{(\cdot) \leq T\}}(t, \omega) \quad \text{for } (t, \omega) \in \mathbb{R}_+ \times \Omega.$$

Let $(t_0, \omega_0) \in \mathbb{R}_+ \times \Omega$ be arbitrarily fixed. If $\mathbf{1}_{\{(\cdot) \leq T\}}(t_0, \omega_0) = 1$, then $(t_0, \omega_0) \in \{(\cdot) \leq T\}$ so that $t_0 \leq T(\omega_0)$. Then $t_0 \leq T_n(\omega_0)$ for every $n \in \mathbb{N}$ and hence $(t_0, \omega_0) \in \{(\cdot) \leq T_n\}$ for every $n \in \mathbb{N}$, that is, $\mathbf{1}\{(\cdot) \leq T_n\}(t_0, \omega_0) = 1$ for every $n \in \mathbb{N}$. Thus (1) holds for our (t_0, ω_0) in this case. If on the other hand $\mathbf{1}_{\{(\cdot) \leq T\}}(t_0, \omega_0) = 0$, then $(t_0, \omega_0) \notin \{(\cdot) \leq T\}$ so that $t_0 > T(\omega_0)$. Since $\lim_{n \to \infty} T_n(\omega_0) = T(\omega_0)$, we have $t_0 > T_n(\omega_0)$ for sufficiently large n. Thus $(t_0, \omega_0) \notin \{(\cdot) \leq T_n\}$, that is, $\mathbf{1}_{\{(\cdot) \leq T_n\}}(t_0, \omega_0) = 0$ for sufficiently large n and therefore (1) holds for our (t_0, ω_0) in this case too. This proves (1). ∎

§4 Convergence in L_p and Uniform Integrability

[I] Convergence in L_p

Let $(\Omega, \mathfrak{F}, P)$ be a probability space. For each $p \in (0, \infty)$ consider the linear space $L_p(\Omega, \mathfrak{F}, P)$ of equivalence classes of extended real valued random variables X on $(\Omega, \mathfrak{F}, P)$

satisfying the condition $\|X\|_p \equiv \{\int_\Omega |X|^p \, dP\}^{1/p} < \infty$, the equivalence relation being that of a.e. equality on $(\Omega, \mathfrak{F}, P)$.

We say that an extended real valued random variable X on $(\Omega, \mathfrak{F}, P)$ is essentially bounded if there exists $B > 0$ such that $|X(\omega)| \leq B$ for all $\omega \in \Lambda^c$ where Λ is a null set in $(\Omega, \mathfrak{F}, P)$. Such a B is called an essential bound of X. Let $L_\infty(\Omega, \mathfrak{F}, P)$ be the linear space of the equivalence classes of the essentially bounded random variables on $(\Omega, \mathfrak{F}, P)$. For $X \in L_\infty(\Omega, \mathfrak{F}, P)$ we define $\|X\|_\infty$ as the infimum of all the essential bounds of X.

When $p \in [1, \infty]$, $\|\cdot\|_p$ is a norm and the binary function ρ on $L_p(\Omega, \mathfrak{F}, P)$ defined by $\rho(X,Y) = \|X - Y\|_p$ for $X, Y \in L_p(\Omega, \mathfrak{F}, P)$ is a metric and $L_p(\Omega, \mathfrak{F}, P)$ is complete with respect to this metric. However, when $p \in (0,1)$, $\|\cdot\|_p$ is not a norm. We shall show nevertheless that $\|\cdot\|_p^p$ is a *quasinorm*, that is, it satisfies the following conditions:

1°. $\|X\|_p^p \in [0,\infty)$ for $X_p \in L_p(\Omega, \mathfrak{F}, P)$ and $\|X\|_p^p = 0$ if and only if $X = 0$.

2°. $\|-X\|_p^p = \|X\|_p^p$ for $X \in L_p(\Omega, \mathfrak{F}, P)$.

3°. $\|X+Y\|_p^p \leq \|X\|_p^p + \|Y\|_p^p$ for $X, Y \in L_p(\Omega, \mathfrak{F}, P)$.

If we define $\rho(X,Y) = \|X - Y\|_p^p$ for $X, Y \in L_p(\Omega, \mathfrak{F}, P)$, then ρ is a metric on $L_p(\Omega, \mathfrak{F}, P)$.

For $a, b \geq 0$ and $p \in (0, \infty)$ we have $(a+b)^p \leq 2^p(a^p + b^p)$. To prove the triangle inequality for $\|X\|_p^p$ we need the following inequalities.

Lemma 4.1. *For $a, b > 0$, we have*

(1) $$(a+b)^p < a^p + b^p \quad \text{when } p \in (0,1),$$

and

(2) $$(a+b)^p > a^p + b^p \quad \text{when } p \in (1, \infty).$$

Proof. With $a > 0$ and $p \in (0, \infty)$ fixed, consider the function

(3) $$\varphi_p(x) = a^p + x^p - (a+x)^p \quad \text{for } x \in [0, \infty).$$

Then $\varphi_p(0) = 0$ and for $b > 0$ we have by the Mean Value Theorem,

(4) $$\varphi_p(b) = \varphi'(\xi)b \quad \text{with some } \xi \in (0, b).$$

§4. CONVERGENCE IN L_P AND UNIFORM INTEGRABILITY

Now

(5) $$\varphi'_p(x) = p\{x^{p-1} - (a+x)^{p-1}\} \quad \text{for } x \in (0, \infty).$$

Note that since $a > 0$ we have $x < a + x$ for $x \in (0, \infty)$ and thus $\varphi'_p(x) > 0$ for $x \in [0, \infty)$ when $p \in (0, 1)$ and $\varphi'_p(x) < 0$ for $x \in (0, \infty)$ when $p \in (1, \infty)$. Therefore by (4), $\varphi_p(b) > 0$ when $p \in (0, 1)$ and $\varphi_p(b) < 0$ when $p \in (1, \infty)$. This proves (1) and (2). ∎

Theorem 4.2. $\|\cdot\|_p^p$ *is a quasinorm on the linear space* $L_p(\Omega, \mathfrak{F}, P)$ *when* $p \in (0, 1)$, *and the binary function* ρ *defined by* $\rho(X, Y) = \|X - Y\|_p^p$, *for* $X, Y \in L_p(\Omega, \mathfrak{F}, P)$, *is a metric. With respect to this metric,* $L_p(\Omega, \mathfrak{F}, P)$ *is a complete metric space.*

Proof. It is obvious that $\|X\|_p^p \in [0, \infty)$ for $X \in L_p(\Omega, \mathfrak{F}, P)$, $\|X\|_p^p = 0$ if and only if $X = 0 \in L_p(\Omega, \mathfrak{F}, P)$ and $\|-X\|_p^p = \|X\|_p^p$ for $X \in L_p(\Omega, \mathfrak{F}, P)$. By (1) of Lemma 4.1,

$$|X + Y|^p \leq (|X| + |Y|)^p \leq |X|^p + |Y|^p.$$

Integrating both sides of this inequality we have the triangle inequality,

$$\|X + Y\|_p^p \leq \|X\|_p^p + \|Y\|_p^p \quad \text{for } X, Y \in L_p(\Omega, \mathfrak{F}, P).$$

This completes the proof that $\|\cdot\|_p^p$ is a quasinorm.

The fact that ρ is a metric follows immediately from the fact that $\|\cdot\|_p^p$ is a quasinorm. The completeness of $L_p(\Omega, \mathfrak{F}, P)$ with respect to ρ is proved in the same way as for $p \geq 1$. ∎

Lemma 4.3. *Let* $X_n \in L_p(\Omega, \mathfrak{F}, P), n \in \mathbb{N}$, *and* $X \in L_p(\Omega, \mathfrak{F}, P)$ *where* $p \in (0, \infty)$. *If* $\lim_{n \to \infty} \|X_n\|_p = \|X\|_p$ *and* $\lim_{n \to \infty} X_n = X$ *a.s., then* $\lim_{n \to \infty} \|X_n - X\|_p = 0$.

Proof. Note that from $|\alpha + \beta|^p \leq 2^p\{|\alpha|^p + |\beta|^p\}$ for $\alpha, \beta \in \mathbb{R}$ and $p \in (0, \infty)$, we have

$$2^p\{|X_n|^p + |X|^p\} - |X_n - X|^p \geq 0.$$

By the a.s. convergence of X_n to X we have

$$\lim_{n \to \infty} \{2^p(|X_n|^p + |X|^p) - |X_n - X|^p\} = 2^{p+1}|X|^p \quad a.s.$$

By Fatou's Lemma and the convergence of $\|X_n\|_p$ to $\|X\|_p$, we have

$$\int_\Omega 2^{p+1}|X|^p \, dP \leq \liminf_{n \to \infty} \int_\Omega \{2^p(|X_n|^p + |X|^p) - |X_n - X|^p\} \, dP$$

$$= 2^{p+1} \int_{\Omega} |X|^p \, dP + \liminf_{n \to \infty} \int_{\Omega} (-|X_n - X|^p) \, dP$$
$$= 2^{p+1} \int_{\Omega} |X|^p \, dP - \limsup_{n \to \infty} \int_{\Omega} |X_n - X|^p \, dP$$

and thus $\limsup_{n \to \infty} \int_{\Omega} |X_n - X|^p \, dP \leq 0$. On the other hand from the nonnegativity of the integrands the limit inferior is nonnegative. Hence $\lim_{n \to \infty} \int_{\Omega} |X_n - X|^p \, dP = 0$. ∎

Notation. For extended real valued random variables $X_n, n \in \mathbb{N}$, and X on a probability space $(\Omega, \mathfrak{F}, P)$ we write $P \cdot \lim_{n \to \infty} X_n = X$ if X_n converges to X in probability, that is, $\lim_{n \to \infty} P\{|X_n - X| \geq \varepsilon\} = 0$ for every $\varepsilon > 0$.

Theorem 4.4. *For $p \in (0, \infty)$, let $X_n \in L_p(\Omega, \mathfrak{F}, P), n \in \mathbb{N}$, and $X \in L_p(\Omega, \mathfrak{F}, P)$ also. Then $\lim_{n \to \infty} \|X_n - X\|_p = 0$ if and only if we have both $\lim_{n \to \infty} \|X_n\|_p = \|X\|_p$ and $P \cdot \lim_{n \to \infty} X_n = X$.*

Proof. When $p \in [1, \infty)$, let $\rho_p(X, Y) = \|X - Y\|_p$ and when $p \in (0, 1)$ let $\rho_p(X, Y) = \|X - Y\|_p^p$ for $X, Y \in L_p(\Omega, \mathfrak{F}, P)$. In any case, ρ_p is a metric on $L_p(\Omega, \mathfrak{F}, P)$ and $\lim_{n \to \infty} \|X_n - X\|_p = 0$ if and only if $\lim_{n \to \infty} \rho_p(X_n, X) = 0$.

1) Suppose $\lim_{n \to \infty} \rho_p(X_n, X) = 0$. By the triangle inequality for the metric ρ_p we have $\rho_p(X_n, 0) \leq \rho_p(X_n, X) + \rho_p(X, 0)$ so that $\rho_p(X_n, 0) - \rho_p(X, 0) \leq \rho_p(X_n, X)$. Interchanging the roles of X_n and X we have $\rho_p(X, 0) - \rho_p(X_n, 0) \leq \rho_p(X_n, X)$ and therefore $|\rho_p(X_n, 0) - \rho_p(X, 0)| \leq \rho_p(X_n, X)$. Thus $\lim_{n \to \infty} \rho_p(X_n, X) = 0$ implies $\lim_{n \to \infty} \rho_p(X_n, 0) = \rho_p(X, 0)$, that is, $\lim_{n \to \infty} \|X_n\|_p = \|X\|_p$.

To prove $P \cdot \lim_{n \to \infty} X_n = X$, note that for any $\varepsilon > 0$ we have by the Markov Inequality

$$P\{|X_n - X| \geq \varepsilon\} \leq \frac{\mathbf{E}(|X_n - X|^p)}{\varepsilon^p} = \frac{\|X_n - X\|_p^p}{\varepsilon^p}.$$

Since $\lim_{n \to \infty} \rho_p(X_n, X) = 0$ we have $\lim_{n \to \infty} \|X_n - X\|_p^p = 0$ and therefore we have $\lim_{n \to \infty} P\{|X_n - X| \geq \varepsilon\} = 0$, that is, $P \cdot \lim_{n \to \infty} X_n = X$.

2) Conversely assume $\lim_{n \to \infty} \|X_n\|_p = \|X\|_p$ and $P \cdot \lim_{n \to \infty} X_n = X$. Consider an arbitrary subsequence $\{X_{n_k} : k \in \mathbb{N}\}$ of $\{X_n : n \in \mathbb{N}\}$. Since $P \cdot \lim_{n \to \infty} X_n = X$ we have $P \cdot \lim_{k \to \infty} X_{n_k} = X$ and therefore there exists a subsequence $\{X_{n_{k_\ell}} : \ell \in \mathbb{N}\}$ such that

§4. CONVERGENCE IN L_P AND UNIFORM INTEGRABILITY

$\lim_{\ell \to \infty} X_{n_{k_\ell}} = X$ a.s. Then by Lemma 4.3 we have $\lim_{\ell \to \infty} \|X_{n_{k_\ell}} - X\|_p = 0$, that is, $\lim_{\ell \to \infty} \rho_p(X_{n_{k_\ell}}, X) = 0$. Thus we have shown that an arbitrary subsequence $\{X_{n_k} : k \in \mathbb{N}\}$ of $\{X_n : n \in \mathbb{N}\}$ always has a subsequence $\{X_{n_{k_\ell}} : \ell \in \mathbb{N}\}$ such that $\lim_{\ell \to \infty} \rho_p(X_{n_{k_\ell}}, X) = 0$. This implies that $\lim_{n \to \infty} \rho_p(X_n, X) = 0$. To show this, suppose that $\lim_{n \to \infty} \rho_p(X_n, X) = 0$ does not hold. Then there exists $\varepsilon_0 > 0$ such that $\rho_p(X_n, X) \geq \varepsilon_0$ for infinitely many $n \in \mathbb{N}$ and therefore we can select a subsequence $\{n_k\}$ of $\{n\}$ such that $\rho_p(X_{n_k}, X) \geq \varepsilon_0$ for all $k \in \mathbb{N}$. Then no subsequence $\{X_{n_{k_\ell}} : \ell \in \mathbb{N}\}$ of $\{X_{n_k} : k \in \mathbb{N}\}$ has the property that $\lim_{\ell \to \infty} \rho_p(X_{n_{k_\ell}}, X) = 0$. This is a contradiction. ∎

[II] Uniformly Integrable Systems of Random Variables

In order to relate the uniform integrability of a system of random variables to the integrability of a single random variable X, let us give a condition equivalent to the latter in terms of $P\{|X| > \lambda\}$ as $\lambda \to \infty$.

Lemma 4.5. *An extended real valued random variable X on a probability space $(\Omega, \mathfrak{F}, P)$ is integrable if and only if $\int_{\{|X|>\lambda\}} |X| \, dP \downarrow 0$ as $\lambda \to \infty$.*

Proof. 1) Suppose X is integrable. Then $|X| < \infty$ a.e. on $(\Omega, \mathfrak{F}, P)$. If we let $A_n = \{|X| > n\}$ for $n \in \mathbb{N}$, then $A_n \downarrow$ and $\lim_{n \to \infty} A_n = \cap_{n \in \mathbb{N}} A_n = \{|X| = \infty\}$ so that $\mathbf{1}_{A_n} \downarrow \mathbf{1}_{\{|X|=\infty\}}$. Since $|\mathbf{1}_{A_n} X| \leq |X|$ and $|X|$ is integrable we have by the Dominated Convergence Theorem

$$\lim_{n \to \infty} \int_{A_n} |X| \, dP = \lim_{n \to \infty} \int_\Omega \mathbf{1}_{A_n} |X| \, dP = \int_\Omega \lim_{n \to \infty} \mathbf{1}_{A_n} |X| \, dP = \int_\Omega \mathbf{1}_{\{|X|=\infty\}} |X| \, dP = 0.$$

Since $P\{|X| > \lambda\} \leq P\{|X| > [\lambda]\}$ where $[\lambda]$ is the greatest nonnegative integer not exceeding λ, the last equality implies $\int_{\{|X|>\lambda\}} |X| \, dP \downarrow 0$ as $\lambda \to \infty$.

2) Conversely suppose $\int_{\{|X|>\lambda\}} |X| \, dP \downarrow 0$ as $\lambda \to \infty$. Then for every $\varepsilon > 0$ there exists $\lambda > 0$ such that $\int_{\{|X|>\lambda\}} |X| \, dP < \varepsilon$. With such $\lambda > 0$, we have

$$\int_\Omega |X| \, dP = \int_{\{|X| \leq \lambda\}} |X| \, dP + \int_{\{|X|>\lambda\}} |X| \, dP \leq \lambda + \varepsilon < \infty,$$

that is, X is integrable. ∎

Definition 4.6. *A system of extended real valued random variables $\{X_\alpha : \alpha \in A\}$ on a* probability space $(\Omega, \mathfrak{F}, P)$ is said to be uniformly integrable if

$$(1) \qquad \sup_{\alpha \in A} \int_{\{|X_\alpha|>\lambda\}} |X_\alpha|\, dP \downarrow 0 \quad as \; \lambda \to \infty,$$

or equivalently, for every $\varepsilon > 0$ there exists $\lambda > 0$ such that

$$(2) \qquad \sup_{\alpha \in A} \int_{\{|X_\alpha|>\lambda\}} |X_\alpha|\, dP < \varepsilon,$$

or equivalently, for every $\varepsilon > 0$ there exists $\lambda > 0$ such that

$$(3) \qquad \int_{\{|X_\alpha|>\lambda\}} |X_\alpha|\, dP < \varepsilon \quad for\; all \; \alpha \in A.$$

We say that $\{X_\alpha : \alpha \in A\}$ is pth-order uniformly integrable if $\{|X_\alpha|^p : \alpha \in A\}$ is uniformly integrable for some $p \in (0, \infty)$.

Thus, the uniform integrability of $\{X_\alpha : \alpha \in A\}$ implies by Lemma 4.5 the integrability of X_α for every $\alpha \in A$. The integrability of X_α then implies that for every $\varepsilon > 0$ there exists $\delta > 0$ such that $\int_E |X_\alpha|\, dP < \varepsilon$ for every $E \in \mathfrak{F}$ with $P(E) < \delta$.

Theorem 4.7. *A system of extended real valued random variables $\{X_\alpha : \alpha \in A\}$ on a probability space $(\Omega, \mathfrak{F}, P)$ is uniformly integrable if and only if*

1°. $\sup_{\alpha \in A} \mathbf{E}(|X_\alpha|) < \infty$,

2°. *for every $\varepsilon > 0$ there exists $\delta > 0$ such that $\int_E |X_\alpha|\, dP < \varepsilon$ for all $\alpha \in A$ whenever $E \in \mathfrak{F}$ and $P(E) < \delta$.*

Proof. 1) Suppose $\{X_\alpha : \alpha \in A\}$ is uniformly integrable. Then by (3) of Definition 4.6 for every $\varepsilon > 0$ there exists $\lambda > 0$ such that

$$\mathbf{E}(|X_\alpha|) = \int_{\{|X_\alpha|\leq \lambda\}} |X_\alpha|\, dP + \int_{\{|X_\alpha|>\lambda\}} |X_\alpha|\, dP \leq \lambda + \varepsilon \quad for \; all \; \alpha \in A$$

and therefore 1° is satisfied.

To verify 2°, note that by (3) of Definition 4.6 again for every $\varepsilon > 0$ there exists $\lambda > 0$ such that $\int_{\{|X_\alpha|>\lambda\}} |X_\alpha|\, dP < \varepsilon/2$ for all $\alpha \in A$. Then for every $E \in \mathfrak{F}$ we have

$$\begin{aligned}
\int_E |X_\alpha|\, dP &= \int_{E\cap\{|X_\alpha|>\lambda\}} |X_\alpha|\, dP + \int_{E\cap\{|X_\alpha|\leq\lambda\}} |X_\alpha|\, dP \\
&\leq \int_{\{|X_\alpha|>\lambda\}} |X_\alpha|\, dP + \lambda P(E) < \frac{\varepsilon}{2} + \lambda P(E).
\end{aligned}$$

§4. CONVERGENCE IN L_P AND UNIFORM INTEGRABILITY

Let $\delta = \varepsilon/2\lambda$. Then for $E \in \mathfrak{F}$ with $P(E) < \delta$ we have $\int_E |X_\alpha| \, dP < \varepsilon$, verifying 2°.
2) Conversely suppose $\{X_\alpha : \alpha \in A\}$ satisfies 1° and 2°. Now for any $\lambda > 0$

$$P\{|X_\alpha| > \lambda\} \leq \frac{\mathbf{E}(|X_\alpha|)}{\lambda} \leq \frac{M}{\lambda} \quad \text{for all } \alpha \in A$$

where $M = \sup_{\alpha \in A} \mathbf{E}(|X_\alpha|) < \infty$ by 1°. Let $\varepsilon > 0$ be arbitrarily given. Let $\lambda > 0$ be so large that $M\lambda^{-1} < \delta$ as in 2°. Then

$$\int_{\{|X_\alpha| > \lambda\}} |X_\alpha| \, dP < \varepsilon \quad \text{for all } \alpha \in A,$$

verifying (3) of Definition 4.6. Thus $\{X_\alpha : \alpha \in A\}$ is uniformly integrable. ∎

The following is an example of a system of random variables which satisfies 2° but not 1° of Theorem 4.7.

Example 4.8. Consider the probability space $([0,1], \mathfrak{B}_{[0,1]}, P)$ where P is the unit mass concentrated at $\{0\}$, that is, P is a probability measure satisfying the condition $P(\{0\}) = 1$. Consider the system of random variables $\{X_n : n \in \mathbb{N}\}$ where X_n is defined by $X_n(\omega) = n$ for all $\omega \in \Omega$, $n \in \mathbb{N}$. Let $\varepsilon > 0$ be arbitrarily given. For any $\delta \in (0,1)$, if $E \in \mathfrak{B}_{[0,1]}$ and $P(E) < \delta$ then $0 \notin E$ so that $E \subset (0,1]$ and thus $P(E) = 0$. Therefore $\int_E |X_n| \, dP = 0 < \varepsilon$ for all $n \in \mathbb{N}$ whenever $E \in \mathfrak{B}_{[0,1]}$ and $P(E) < \delta$ and 2° of Theorem 4.7 is satisfied. But $\sup_{n \in \mathbb{N}} \mathbf{E}(|X_n|) = \sup\{n : n \in \mathbb{N}\} = \infty$.

Proposition 4.9. *1) If $\{X_\alpha : \alpha \in A\}$ is a uniformly integrable system of extended real valued random variables on a probability space $(\Omega, \mathfrak{F}, P)$ and $\{c_\alpha : \alpha \in A\}$ is a bounded system of real numbers, then $\{c_\alpha X_\alpha : \alpha \in A\}$ is a uniformly integrable system.*
2) If $\{X_\alpha : \alpha \in A\}$ is a uniformly integrable system and $Y \in L_\infty(\Omega, \mathfrak{F}, P)$, then $\{X_\alpha Y : \alpha \in A\}$ is uniformly integrable.
3) If $\{X_\alpha : \alpha \in A\}$ and $\{Y_\alpha : \alpha \in A\}$ are uniformly integrable systems then so is $\{X_\alpha + Y_\alpha : \alpha \in A\}$.

Proof. 1) Suppose $|c_\alpha| \leq M$ for $\alpha \in A$ with some $M > 0$. Then

$$\int_{\{|c_\alpha X_\alpha| > \lambda\}} |c_\alpha X_\alpha| \, dP \leq \int_{\{|X_\alpha| > \lambda M^{-1}\}} M|X_\alpha| \, dP$$

so that

$$\lim_{\lambda \to \infty} \sup_{\alpha \in A} \int_{\{|c_\alpha X_\alpha| > \lambda\}} |c_\alpha X_\alpha| \, dP \leq M \lim_{\lambda \to \infty} \sup_{\alpha \in A} \int_{\{|X_\alpha| > \lambda M^{-1}\}} |X_\alpha| \, dP = 0,$$

that is, $\{c_\alpha X_\alpha : \alpha \in A\}$ satisfies (1) of Definition 4.6.

2) The uniform integrability of $\{X_\alpha : \alpha \in A\}$ implies according to Theorem 4.7

$$\sup_{\alpha \in A} \mathbf{E}[|X_\alpha Y|] \leq \|Y\|_\infty \sup_{\alpha \in A} \mathbf{E}[|X_\alpha|] < \infty.$$

When $\|Y\|_\infty = 0$, the assertion is trivially true. Thus assume $\|Y\|_\infty > 0$. Then for every $\varepsilon > 0$ there exists $\delta > 0$ such that $\int_E |X_\alpha|\, dP < \varepsilon/\|Y\|_\infty$ for all $\alpha \in A$ whenever $E \in \mathfrak{F}$ and $P(E) < \delta$. Then for $E \in \mathfrak{F}$ with $P(E) < \delta$ we have

$$\int_E |X_\alpha Y|\, dP \leq \|Y\|_\infty \int_E |X_\alpha|\, dP < \varepsilon.$$

Thus by Theorem 4.7 the system $\{X_\alpha Y : \alpha \in A\}$ is uniformly integrable.

3) If $\{X_\alpha : \alpha \in A\}$ and $\{Y_\alpha : \alpha \in A\}$ are uniformly integrable systems, then each one of the two satisfies 1° and 2° of Theorem 4.7. From this follows immediately that $\{X_\alpha + Y_\alpha : \alpha \in A\}$ satisfies the same conditions 1° and 2° and is therefore uniformly integrable. ∎

Proposition 4.10. *A finite system $\{X_n : n = 1, \ldots, N\}$ of extended real valued integrable random variables on a probability space is always uniformly integrable.*

Proof. If X_n is integrable then it satisfies the condition in Lemma 4.5, that is,

$$\lim_{\lambda \to \infty} \int_{\{|X_n| > \lambda\}} |X_n|\, dP = 0.$$

Then

$$\lim_{\lambda \to \infty} \sup_{n=1,\ldots,N} \int_{\{|X_n| > \lambda\}} |X_n|\, dP \leq \lim_{\lambda \to \infty} \sum_{n=1}^{N} \int_{\{|X_n| > \lambda\}} |X_n|\, dP = 0.$$

Thus $\{X_n : n = 1, \ldots, N\}$ satisfies (1) of Definition 4.6 and is therefore uniformly integrable. ∎

As immediate consequences of Definition 4.6 we have the following statements.

Proposition 4.11. 1) *If $\{X_\alpha : \alpha \in A\}$ is uniformly integrable then so is any subsystem of $\{X_\alpha : \alpha \in A\}$.*
2) *$\{X_\alpha : \alpha \in A\}$ is uniformly integrable if and only if $\{|X_\alpha| : \alpha \in A\}$ is.*
3) *Let $Y_\alpha = X_\alpha$ or $-X_\alpha$ for each $\alpha \in A$. Then $\{X_\alpha : \alpha \in A\}$ is uniformly integrable if*

§4. CONVERGENCE IN L_P AND UNIFORM INTEGRABILITY

and only if $\{Y_\alpha : \alpha \in A\}$ is.

4) If $|X_\alpha| \leq |Y_\alpha|$ for every $\alpha \in A$ and $\{Y_\alpha : \alpha \in A\}$ is uniformly integrable, then so is $\{X_\alpha : \alpha \in A\}$. In particular if $|X_\alpha| \leq Y$ for every $\alpha \in A$ for some integrable random variable Y then $\{X_\alpha : \alpha \in A\}$ is uniformly integrable.

5) $\{X_\alpha : \alpha \in A\}$ is uniformly integrable if and only if both $\{X_\alpha^+ : \alpha \in A\}$ and $\{X_\alpha^- : \alpha \in A\}$ are.

6) If $\{X_\alpha : \alpha \in A\}$ and $\{Y_\beta : \beta \in B\}$ are uniformly integrable systems then so is $\{X_\alpha, Y_\beta : \alpha \in A, \beta \in B\}$.

The following theorem compares different orders of uniform integrability.

Theorem 4.12. *Let $\{X_\alpha : \alpha \in A\}$ be a system of extended real valued random variables on a probability space $(\Omega, \mathfrak{F}, P)$. If $\sup_{\alpha \in A} \|X_\alpha\|_{p_0} < \infty$ for some $p_0 \in (0, \infty)$, then $\{X_\alpha : \alpha \in A\}$ is pth-order uniformly integrable, that is, $\{|X_\alpha|^p : \alpha \in A\}$ is uniformly integrable for every $p \in (0, p_0)$.*

Proof. Let $p \in (0, p_0)$. Then for $0 < \eta < \xi$ we have $\xi^p = \xi^{p-p_0} \xi^{p_0} < \eta^{p-p_0} \xi^{p_0}$ so that

$$\int_{\{|X_\alpha|^p > \eta^p\}} |X_\alpha|^p \, dP \leq \int_{\{|X_\alpha|^p > \eta^p\}} \eta^{p-p_0} |X_\alpha|^{p_0} \, dP \leq \eta^{p-p_0} \|X_\alpha\|_{p_0}^{p_0}$$

and therefore

$$\lim_{\eta \to \infty} \sup_{\alpha \in A} \int_{\{|X_\alpha|^p > \eta^p\}} |X_\alpha|^p \, dP \leq \lim_{\eta \to \infty} \eta^{p-p_0} \sup_{\alpha \in A} \|X_\alpha\|_{p_0}^{p_0} = 0$$

by the fact that $\sup_{\alpha \in A} \|X_\alpha\|_{p_0}^{p_0} < \infty$ and $p - p_0 < 0$. Therefore writing λ for η^p, we have

$$\lim_{\lambda \to \infty} \sup_{\alpha \in A} \int_{\{|X_\alpha|^p > \lambda\}} |X_\alpha|^p \, dP = 0.$$

This verifies (1) of Definition 4.6 and proves the uniform integrability of $\{|X_\alpha|^p : \alpha \in A\}$ for $p \in (0, p_0)$. ∎

Note that on account of Theorem 4.7, Theorem 4.12 implies that if $\{X_\alpha : \alpha \in A\}$ is p_0th-order uniformly integrable, then it is pth-order uniformly integrable for every $p \in (0, p_0)$.

Theorem 4.13. *Let $\{X_\alpha : \alpha \in A\}$ and $\{Y_\alpha : \alpha \in A\}$ be systems of extended real valued random variables on a probability space $(\Omega, \mathfrak{F}, P)$. If $\{X_\alpha : \alpha \in A\}$ and $\{Y_\alpha : \alpha \in A\}$*

are respectively pth-order and qth-order uniformly integrable for some $p, q \in (1, \infty)$ such that $1/p + 1/q = 1$, then $\{X_\alpha Y_\alpha : \alpha \in A\}$ is uniformly integrable.

Proof. If $\{|X_\alpha|^p : \alpha \in A\}$ and $\{|Y_\alpha|^q : \alpha \in A\}$ are uniformly integrable, then by Theorem 4.7

(1) $$\sup_{\alpha \in A} \|X_\alpha\|_p < \infty \quad \text{and} \quad \sup_{\alpha \in A} \|Y_\alpha\|_q < \infty$$

and for every $\varepsilon > 0$ there exists $\delta > 0$ such that

(2) $$\int_E |X_\alpha|^p \, dP < \varepsilon \quad \text{and} \quad \int_E |Y_\alpha|^q \, dP < \varepsilon \quad \text{for all } \alpha \in A \text{ whenever } E \in \mathfrak{F} \text{ and } P(E) < \delta.$$

Now for every $E \in \mathfrak{F}$ we have by the Hölder Inequality

(3) $$\int_E |X_\alpha Y_\alpha| \, dP \leq \{\int_E |X_\alpha|^p \, dP\}^{\frac{1}{p}} \{\int_E |Y_\alpha|^q \, dP\}^{\frac{1}{q}}.$$

With $E = \Omega$ in (3) we have by (1)

$$\sup_{\alpha \in A} \mathbf{E}(|X_\alpha Y_\alpha|) \leq \sup_{\alpha \in A} \{\|X_\alpha\|_p \|Y_\alpha\|_q\} \leq \{\sup_{\alpha \in A} \|X_\alpha\|_p\}\{\sup_{\alpha \in A} \|Y_\alpha\|_q\} < \infty.$$

Also, by (3) and (2)

$$\int_E |X_\alpha Y_\alpha| \, dP \leq \varepsilon^{\frac{1}{p}} \varepsilon^{\frac{1}{q}} = \varepsilon \quad \text{for all } \alpha \in A \text{ whenever } E \in \mathfrak{F} \text{ and } P(E) < \delta.$$

Thus $\{X_\alpha Y_\alpha : \alpha \in A\}$ satisfies 1° and 2° of Theorem 4.7 and is therefore uniformly integrable. ∎

Let us turn to the role played by uniform integrability in convergence in L_p. We shall show that a sequence of random variables converges in L_p if and only if it is pth-order uniformly integrable and converges in probability. Toward this end we prepare the following lemma.

Lemma 4.14. Let $X_n \in L_p(\Omega, \mathfrak{F}, P)$, $n \in \mathbb{N}$, where $p \in (0, \infty)$. If $\lim_{n \to \infty} \|X_n\|_p = 0$ then $\{X_n : n \in \mathbb{N}\}$ is pth-order uniformly integrable.

Proof. If $\lim_{n \to \infty} \|X_n\|_p = 0$ then for every $\varepsilon > 0$ there exists $N \in \mathbb{N}$ such that

(1) $$\int_\Omega |X_n|^p \, dP < \varepsilon \quad \text{when } n \geq N + 1.$$

§4. CONVERGENCE IN L_P AND UNIFORM INTEGRABILITY

Since $\{|X_1|^p, \ldots, |X_N|^p\}$ is a finite system of integrable random variables it is uniformly integrable by Proposition 4.10. Therefore for our $\varepsilon > 0$ there exists $\lambda > 0$ such that

$$(2) \qquad \int_{\{|X_n|^p > \lambda\}} |X_n|^p \, dP < \varepsilon \quad \text{for } n = 1, \ldots, N.$$

By (1) and (2) we have $\int_{\{|X_n|^p > \lambda\}} |X_n|^p \, dP < \varepsilon$ for all $n \in \mathbb{N}$ for our $\lambda > 0$. This proves the uniform integrability of $\{|X_n|^p : n \in \mathbb{N}\}$ by (3) of Definition 4.6. ∎

In the last lemma we showed that if $\lim_{n \to \infty} \|X_n\|_p = 0$, then the sequence $\{X_n : n \in \mathbb{N}\}$ is pth-order uniformly integrable. We observe that if $\lim_{n \to \infty} \|X_n\|_p = c$ where c is an arbitrary real number and $c \neq 0$ then the sequence need not be pth-order uniformly integrable. See Example 4.15 below. The difference is that while the convergence $\lim_{n \to \infty} \|X_n\|_p = 0$ is not only the convergence of the numerical sequence $\{\|X_n\|_p : n \in \mathbb{N}\}$ to 0 but also the convergence of the sequence of the random variables in L_p to the identically vanishing random variable, the convergence $\lim_{n \to \infty} \|X_n\|_p = c$ does not imply $\lim_{n \to \infty} \|X_n - c\|_p = 0$. The next theorem shows that if $\lim_{n \to \infty} \|X_n\|_p = \|X\|_p$ for some $X \in L_p(\Omega, \mathfrak{F}, P)$ and if $P \cdot \lim_{n \to \infty} X_n = X$ then the sequence is pth-order uniformly integrable.

Example 4.15. Consider the probability space $((0, 1], \mathfrak{B}_{(0,1]}, m_L)$ where $\mathfrak{B}_{(0,1]}$ is the Borel σ-algebra on $(0, 1]$ and m_L is the Lebesgue measure. For each $n \in \mathbb{N}$, let X_n be defined by $X_n(\omega) = n$ for $\omega \in (0, 1/n]$ and $X_n(\omega) = 0$ for $\omega \in (1/n, 1]$. Let X be defined by $X(\omega) = 1$ for $\omega \in (0, 1]$. Then $\mathbf{E}(|X_n|) = 1$ for $n \in \mathbb{N}$ and $\mathbf{E}(|X|) = 1$ so that X_n and X are all in $L_1(\Omega, \mathfrak{F}, P)$ with $\lim_{n \to \infty} \mathbf{E}(|X_n|) = \mathbf{E}(|X|)$.

To show that $\{|X_n| : n \in \mathbb{N}\}$ is not uniformly integrable, we show that if $\varepsilon \in (0, 1)$ then for any $\delta > 0$ we can always find some $n \in \mathbb{N}$ and some $E \in \mathfrak{F}$ with $P(E) < \delta$ such that $\int_E |X_n| \, dP > \varepsilon$. Indeed, if $n \in \mathbb{N}$ is so large that $1/n < \delta$, then $P((0, 1/n]) = 1/n < \delta$ but $\int_{(0,1/n]} |X_n| \, dP = 1 > \varepsilon$.

Theorem 4.16. *Let $X_n \in L_p(\Omega, \mathfrak{F}, P)$, $n \in \mathbb{N}$, where $p \in (0, \infty)$. Let X be an extended real valued random variable on $(\Omega, \mathfrak{F}, p)$ and assume that $P \cdot \lim_{n \to \infty} X_n = X$. Then the following three conditions are equivalent:*

(1) $\qquad \{X_n : n \in \mathbb{N}\}$ *is pth-order uniformly integrable.*

(2) $\quad X \in L_p(\Omega, \mathfrak{F}, P) \quad and \quad \lim_{n \to \infty} \|X_n - X\|_p = 0.$

(3) $\quad X \in L_p(\Omega, \mathfrak{F}, P) \quad and \quad \lim_{n \to \infty} \|X_n\|_p = \|X\|_p.$

Proof. 1) (1) $\Rightarrow$ (2). Assume (1), that is, $\{|X_n|^p : n \in \mathbb{N}\}$ is uniformly integrable. Let us show first that $X \in L_p(\Omega, \mathfrak{F}, P)$. Now since $P \cdot \lim_{n \to \infty} X_n = X$ there exists a subsequence $\{X_{n_k} : k \in \mathbb{N}\}$ such that $\lim_{k \to \infty} X_{n_k} = X$ a.s. and thus $\lim_{k \to \infty} |X_{n_k}|^p = |X|^p$ a.s. Then by Fatou's Lemma
$$\mathbf{E}(|X|^p) \leq \liminf_{n \to \infty} \mathbf{E}(|X_{n_k}|^p).$$
Since the uniform integrability of $\{|X_n|^p : n \in \mathbb{N}\}$ implies $\sup_{n \in \mathbb{N}} \mathbf{E}(|X_n|^p) < \infty$ according to 1° of Theorem 4.7, we have $\mathbf{E}(|X|^p) < \infty$ and thus $X \in L_p(\Omega, \mathfrak{F}, P)$.
To prove $\lim_{n \to \infty} \|X_n - X\|_p = 0$, note that
$$|X_n - X|^p \leq 2^p \{|X_n|^p + |X|^p\}.$$
Now the uniform integrability of $\{|X_n|^p : n \in \mathbb{N}\}$ and the integrability of $|X|^p$ imply according to Proposition 4.9 the uniform integrability of $\{2^p(|X_n|^p + |X|^p) : n \in \mathbb{N}\}$. Thus by 4) of Proposition 4.11, $\{|X_n - X|^p : n \in \mathbb{N}\}$ is uniformly integrable. Then by 2° of Theorem 4.7 for every $\varepsilon > 0$ there exists $\delta > 0$ such that
$$\int_E |X_n - X|^p \, dP < \varepsilon \quad \text{for all } n \in \mathbb{N} \text{ whenever } E \in \mathfrak{F} \text{ and } P(E) < \delta.$$
Now $P \cdot \lim_{n \to \infty} X_n = X$ implies that for our $\varepsilon > 0$ and $\delta > 0$ there exists $N \in \mathbb{N}$ such that $P\{|X_n - X| \geq \varepsilon\} < \delta$ for $n \geq N$ and thus $\int_{\{|X_n - X| \geq \varepsilon\}} |X_n - X|^p \, dP < \varepsilon$ for $n \geq N$ and consequently
$$\|X_n - X\|_p^p = \int_{\{|X_n - X| \geq \varepsilon\}} |X_n - X|^p \, dP + \int_{\{|X_n - X| < \varepsilon\}} |X_n - X|^p \, dP < \varepsilon + \varepsilon^p$$
for $n \geq N$. This implies $\limsup_{n \to \infty} \|X_n - X\|_p^p \leq \varepsilon + \varepsilon^p$. By the arbitrariness of $\varepsilon > 0$, the limit superior is equal to 0 and therefore $\lim_{n \to \infty} \|X_n - X\|_p = 0$.

2) (2) $\Rightarrow$ (1). Assume (2). According to Lemma 4.14, $\lim_{n \to \infty} \|X_n - X\|_p = 0$ implies the uniform integrability of $\{|X_n - X|^p : n \in \mathbb{N}\}$. Note that
$$|X_n|^p = |X_n - X + X|^p \leq 2^p \{|X_n - X|^p + |X|^p\}.$$

§4. CONVERGENCE IN L_P AND UNIFORM INTEGRABILITY

Since $|X|^p$ is integrable we have the uniform integrability of $\{2^p(|X_n-X|^p+|X|^p) : n \in \mathbb{N}\}$ by Proposition 4.9 and from this follows the uniform integrability of $\{|X_n|^p : n \in \mathbb{N}\}$ by 4) of Proposition 4.11.

3) The equivalence of (2) and (3) is from Theorem 4.4. ∎

Corollary 4.17. *Let $X_n \in L_1(\Omega, \mathfrak{F}, P)$ and $X_n \geq 0$ a.e. on $(\Omega, \mathfrak{F}, P)$ for $n \in \mathbb{N}$. Let X be an extended real valued random variable on $(\Omega, \mathfrak{F}, P)$ and assume that $P \cdot \lim_{n \to \infty} X_n = X$. Then the following two conditions are equivalent:*

(1) $\qquad\qquad\qquad \{X_n : n \in \mathbb{N}\}$ *is uniformly integrable.*

(2) $\qquad\qquad\qquad \mathbf{E}(X) < \infty$ *and* $\lim_{n \to \infty} \mathbf{E}(X_n) = \mathbf{E}(X)$.

Proof. Since $P \cdot \lim_{n \to \infty} X_n = X$ implies the existence of a subsequence $\{X_{n_k} : k \in \mathbb{N}\}$ such that $\lim_{k \to \infty} X_{n_k} = X$ a.e. on $(\Omega, \mathfrak{F}, P)$ we have $X \geq 0$ a.e. on $(\Omega, \mathfrak{F}, P)$. Thus $\mathbf{E}(X) < \infty$ is equivalent to $\mathbf{E}(|X|) < \infty$, that is, $X \in L_1(\Omega, \mathfrak{F}, P)$. Then the equivalence of (1) and (2) is implied by Theorem 4.16. ∎

Although the substance of the next theorem is contained in Theorem 4.16 and Theorem 4.4, we state it separately because of its simplicity.

Theorem 4.18. *Let $X_n, n \in \mathbb{N}$, and X be in $L_p(\Omega, \mathfrak{F}, P)$ where $p \in (0, \infty)$. Then $\lim_{n \to \infty} \|X_n - X\|_p = 0$ if and only if $P \cdot \lim_{n \to \infty} X_n = X$ and $\{X_n : n \in \mathbb{N}\}$ is pth-order uniformly integrable.*

Proof. If $P \cdot \lim_{n \to \infty} X_n = X$ and $\{X_n : n \in \mathbb{N}\}$ is pth-order uniformly integrable then $\lim_{n \to \infty} \|X_n - X\|_p = 0$ by Theorem 4.16. Conversely, if $\lim_{n \to \infty} \|X_n - X\|_p = 0$ then by Theorem 4.4 we have $P \cdot \lim_{n \to \infty} X_n = X$ and thus $\{X_n : n \in \mathbb{N}\}$ is pth-order uniformly integrable by Theorem 4.16. ∎

For $p \in (0, \infty)$, $L_p(\Omega, \mathfrak{F}, P)$ is a metric space with the metric derived from the norm $\|\cdot\|_p$ when $p \in [1, \infty)$ and with the metric derived from the quasinorm $\|\cdot\|_p^p$ when $p \in (0, 1)$. We show next that the closure of a pth-order uniformly integrable set in $L_p(\Omega, \mathfrak{F}, P)$ with respect to the metric topology is pth-order uniformly integrable.

Theorem 4.19. *For $p \in (0, \infty)$, let $\mathfrak{H}$ be a subcollection of $L_p(\Omega, \mathfrak{F}, P)$. If $\mathfrak{H}$ is pth-order uniformly integrable, then so is its closure $\overline{\mathfrak{H}}$ in the metric topology of $L_p(\Omega, \mathfrak{F}, P)$.*

Proof. According to Theorem 4.7, the pth-order uniform integrability of $\mathfrak{H}$ implies that

(1) $$M \equiv \sup_{X \in \mathfrak{H}} \mathbf{E}(|X|^p) < \infty$$

and for every $\varepsilon > 0$ there exists $\delta > 0$ such that

(2) $$\int_E |X|^p \, dP < \varepsilon \quad \text{for all } X \in \mathfrak{H} \text{ whenever } E \in \mathfrak{F} \text{ and } P(E) < \delta.$$

If $Y \in \overline{\mathfrak{H}}$ then there exists a sequence $\{X_n : n \in \mathbb{N}\}$ in $\mathfrak{H}$ such that $\lim_{n \to \infty} \|X_n - Y\|_p^p = 0$. For $p \in [1, \infty)$ the triangle inequality of the norm $\|\cdot\|_p$ implies $\big|\|\xi\|_p - \|\eta\|_p\big| \leq \|\xi - \eta\|_p$. Similarly for $p \in (0, 1)$ the triangle inequality of the quasinorm $\|\cdot\|_p^p$ implies the inequality $\big|\|\xi\|_p^p - \|\eta\|_p^p\big| \leq \|\xi - \eta\|_p^p$. Thus for every $p \in (0, \infty)$, $\lim_{n \to \infty} \|X_n - Y\|_p^p = 0$ implies

(3) $$\lim_{n \to \infty} \mathbf{E}(|X_n|^p) = \mathbf{E}(|Y|^p)$$

and thus for every $E \in \mathfrak{F}$ we have

(4) $$\lim_{n \to \infty} \int_E |X_n|^p \, dP = \int_E |Y|^p \, dP.$$

From (3) and (1) we have

(5) $$\mathbf{E}(|Y|^p) \leq M < \infty \quad \text{for all } Y \in \overline{\mathfrak{H}}.$$

For $\varepsilon > 0$, let $\delta > 0$ be as specified in (2). Then by (4) and (2) we have

(6) $$\int_E |Y|^p \, dP \leq \varepsilon \quad \text{for all } Y \in \overline{\mathfrak{H}} \text{ whenever } E \in \mathfrak{F} \text{ and } P(E) < \delta.$$

According to Theorem 4.7, the inequalities (5) and (6) above establish the uniform integrability of $\{|Y|^p : Y \in \overline{\mathfrak{H}}\}$. ∎

Definition 4.20. *A sequence $\{X_n : n \in \mathbb{N}\}$ in $L_1(\Omega, \mathfrak{F}, P)$ is said to converge weakly to $X \in L_1(\Omega, \mathfrak{F}, P)$ if*

$$\lim_{n \to \infty} \int_\Omega X_n Y \, dP = \int_\Omega XY \, dP \quad \text{for every } Y \in L_\infty(\Omega, \mathfrak{F}, P).$$

§4. CONVERGENCE IN L_P AND UNIFORM INTEGRABILITY

X is called the weak limit of the sequence.

Note that the weak limit of a sequence, if it exists, is uniquely determined up to a null set in $(\Omega, \mathfrak{F}, P)$. Indeed if $\{X_n : n \in \mathbb{N}\}$ converges weakly to X and X', then by using $Y = 1_E$ for an arbitrary $E \in \mathfrak{F}$ we have $\int_E X\, dP = \int_E X'\, dP$ and therefore $X = X'$ a.e. on $(\Omega, \mathfrak{F}, P)$. Note also that if a sequence $\{X_n : n \in \mathbb{N}\}$ in $L_1(\Omega, \mathfrak{F}, P)$ converges to $X \in L_1(\Omega, \mathfrak{F}, P)$ in L_1, then $\{X_n : n \in \mathbb{N}\}$ converges to X weakly. This follows from the fact that for every $Y \in L_\infty(\Omega, \mathfrak{F}, P)$ we have

$$\left| \int_\Omega X_n Y\, dP - \int_\Omega XY\, dP \right| \le \int_\Omega |X_n - X||Y|\, dP \le \|Y\|_\infty \|X_n - X\|_1.$$

In the next theorem we show that for an arbitrary probability space $(\Omega, \mathfrak{F}, P)$ any uniformly integrable subset of $L_1(\Omega, \mathfrak{F}, P)$ is relatively weakly compact. The converse of this theorem is also true. See [21] P. A. Meyer.

Theorem 4.21. *Let* $\mathfrak{H} \subset L_1(\Omega, \mathfrak{F}, P)$ *be uniformly integrable. Then for every sequence* $\{X_n : n \in \mathbb{N}\}$ *in* $\mathfrak{H}$ *there exists a subsequence* $\{X_{n_k} : k \in \mathbb{N}\}$ *and some* $X \in L_1(\Omega, \mathfrak{F}, P)$ *such that*

$$\lim_{k \to \infty} \int_\Omega X_{n_k} Y\, dP = \int_\Omega XY\, dP \quad \text{for every } Y \in L_\infty(\Omega, \mathfrak{F}, P).$$

Proof. 1) For an arbitrary sequence $\{X_n : n \in \mathbb{N}\}$ in $\mathfrak{H}$, let $\mathfrak{G} = \sigma\{X_n : n \in \mathbb{N}\}$. According to Theorem 1.15, the σ-algebra is separable and in fact there exists a countable sub-algebra $\mathfrak{A}$ of $\mathfrak{F}$ such that $\sigma(\mathfrak{A}) = \mathfrak{G}$. Let $\mathfrak{A} = \{A_m : m \in \mathbb{N}\}$. For every A_m the uniform integrability of $\{X_n : n \in \mathbb{N}\}$ implies that $|\int_{A_m} X_n\, dP| \le \int_\Omega |X_n|\, dP \le \sup_{n \in \mathbb{N}} \mathbf{E}(|X_n|) < \infty$ for all $n \in \mathbb{N}$. Then $\{\int_{A_m} X_n\, dP : n \in \mathbb{N}\}$ is a bounded sequence in $\mathbb{R}$ and thus has a convergent subsequence $\{\int_{A_m} X_{n_k}\, dP : k \in \mathbb{N}\}$. Since this is true for every A_m, there exists by the diagonal procedure a subsequence of $\{X_n : n \in \mathbb{N}\}$ which we denote by $\{X_k : k \in \mathbb{N}\}$ such that $\{\int_{A_m} X_k\, dP : k \in \mathbb{N}\}$ is a convergent sequence in $\mathbb{R}$ for every A_m.

Let us show next that the sequence $\{\int_G X_k\, dP : k \in \mathbb{N}\}$ converges for every $G \in \mathfrak{G}$. For an arbitrary $\varepsilon > 0$ the uniform integrability of $\{X_k : k \in \mathbb{N}\}$ implies the existence of some $\delta > 0$ such that

(1) $$\int_E |X_k|\, dP < \varepsilon \quad \text{for all } k \in \mathbb{N} \text{ whenever } E \in \mathfrak{F} \text{ and } P(E) < \delta.$$

Since $G \in \mathfrak{G}$ and $\mathfrak{G} = \sigma(\mathfrak{A})$, there exists some $A \in \mathfrak{A}$ such that $P(G \triangle A) < \delta$. Then

$$\left| \int_G X_k\, dP - \int_A X_k\, dP \right| \le \int_{G \triangle A} |X_k|\, dP < \varepsilon \quad \text{for all } k \in \mathbb{N}.$$

From this we have

$$\limsup_{k\to\infty} \int_G X_k\, dP \leq \lim_{k\to\infty} \int_A X_k\, dP + \varepsilon$$

and

$$\liminf_{k\to\infty} \int_G X_k\, dP \geq \lim_{k\to\infty} \int_A X_k\, dP - \varepsilon$$

Then from the arbitrariness of $\varepsilon > 0$ we have

$$\lim_{k\to\infty} \int_G X_k\, dP = \lim_{k\to\infty} \int_A X_k\, dP \in \mathbb{R}.$$

This shows the convergence of the sequence $\{\int_G X_k\, dP : k \in \mathbb{N}\}$ in $\mathbb{R}$ for every $G \in \mathfrak{G}$.

2) Let us define a real valued set function Q on $\mathfrak{G}$ by setting

(2) $$Q(G) = \lim_{k\to\infty} \int_G X_k\, dP \quad \text{for } G \in \mathfrak{G}.$$

Clearly $Q(\emptyset) = 0$. To show that Q is a signed measure it remains only to show that Q is countably additive on $\mathfrak{G}$, that is, for an arbitrary disjoint collection $\{G_m; m \in \mathbb{N}\}$ in $\mathfrak{G}$ we have $Q(\cup_{m\in\mathbb{N}} G_m) = \sum_{m\in\mathbb{N}} Q(G_m)$. Since Q is clearly finitely additive on $\mathfrak{G}$ from its definition by (2), it suffices to show that if $\{G_m : m \in \mathbb{N}\}$ is an increasing sequence in $\mathfrak{G}$ and if we let $G = \cup_{m\in\mathbb{N}} G_m$ then $Q(G) = \lim_{m\to\infty} Q(G_m)$. Let $\varepsilon > 0$ be arbitrarily given. By the uniform integrability of $\{X_k : k \in \mathbb{N}\}$ there exists $\delta > 0$ such that (1) holds. Since $G_m \uparrow G$ there exists $N \in \mathbb{N}$ such that $P(G - G_m) < \delta$ for $m \geq N$. Then by (1) we have

$$|Q(G - G_m)| = \left|\lim_{k\to\infty} \int_{G-G_m} X_k\, dP\right| \leq \varepsilon \quad \text{for } m \geq N.$$

This shows that $\lim_{m\to\infty} Q(G - G_m) = 0$ and therefore $Q(G) = \lim_{m\to\infty} Q(G_m)$. This completes the proof that Q is a signed measure on $\mathfrak{G}$.

Since $P(G) = 0$ implies $Q(G) = 0$ by (2) for $G \in \mathfrak{G}$, the signed measure Q is absolutely continuous with respect to P on the measurable space $(\Omega, \mathfrak{G})$. By the Radon-Nikodym Theorem there exists an extended real valued $\mathfrak{G}$-measurable function X on Ω such that $Q(G) = \int_G X\, dP$ for every $G \in \mathfrak{G}$, in other words,

(3) $$\lim_{k\to\infty} \int_G X_k\, dP = \int_G X\, dP \quad \text{for } G \in \mathfrak{G}.$$

Since Q assumes values in $\mathbb{R}$ only it is a finite signed measure. This implies that its Radon-Nikodym derivative X is integrable on $(\Omega, \mathfrak{F}, P)$.

§4. CONVERGENCE IN L_P AND UNIFORM INTEGRABILITY

3) We show next that for every $Y \in L_\infty(\Omega, \mathfrak{G}, P)$ we have

(4) $$\lim_{k \to \infty} \int_\Omega X_k Y \, dP = \int_\Omega XY \, dP.$$

To apply Theorem 1.10, let us use a real valued representative for $Y \in L_\infty(\Omega, \mathfrak{G}, P)$. Let **V** be the collection of all $Y \in L_\infty(\Omega, \mathfrak{G}, P)$ which satisfy (4). By (3), $\mathbf{1}_G \in \mathbf{V}$ for every $G \in \mathfrak{G}$. Also **V** is a linear space. Thus **V** satisfies conditions 1° and 2° of Theorem 1.10. To show that condition 3° of the same theorem is also satisfied, we show that if $\{Y_m : m \in \mathbb{N}\}$ is an increasing sequence of nonnegative functions in **V** and if $Y \equiv \lim_{m \to \infty} Y_m$ is in $L_\infty(\Omega, \mathfrak{G}, P)$ then $Y \in \mathbf{V}$. For convenience let us write X_0 for X and Y_0 for Y. Since $\{X_k : k \in \mathbb{N}\}$ is uniformly integrable and X_0 is integrable, $\{X_k : k \in \mathbb{Z}_+\}$ is uniformly integrable. Since $0 \leq Y_m \leq Y_0$ on Ω for all $m \in \mathbb{N}$ and since $Y_0 \in L_\infty(\Omega, \mathfrak{G}, P)$, the system $\{X_k Y_m : k, m \in \mathbb{Z}_+\}$ is uniformly integrable by 2) of Proposition 4.9 and 4) of Proposition 4.11. Thus according to Theorem 4.7, for an arbitrary $\varepsilon > 0$ there exists some $\delta > 0$ such that

(5). $$\int_E |X_k Y_m| \, dP < \varepsilon \quad \text{for all } k, m \in \mathbb{Z}_+ \text{ when } E \in \mathfrak{F} \text{ and } P(E) < \delta.$$

Since $\lim_{m \to \infty} Y_m = Y_0$ on Ω we have $P \cdot \lim_{m \to \infty} Y_m = Y_0$. Thus $\lim_{m \to \infty} P\{|Y_m - Y_0| \geq \varepsilon\} = 0$ and hence there exists $N \in \mathbb{N}$ such that

(6) $$P\{|Y_m - Y_0| \geq \varepsilon\} < \delta \quad \text{for } m \geq N.$$

Now with an arbitrary $m \geq N$ we have

(7) $$\begin{aligned} &\left| \int_\Omega X_k Y_0 \, dP - \int_\Omega X_0 Y_0 \, dP \right| \\ &\leq \left| \int_\Omega X_k Y_0 \, dP - \int_\Omega X_k Y_m \, dP \right| + \left| \int_\Omega X_k Y_m \, dP - \int_\Omega X_0 Y_m \, dP \right| \\ &\quad + \left| \int_\Omega X_0 Y_m \, dP - \int_\Omega X_0 Y_0 \, dP \right|. \end{aligned}$$

With $a \equiv \sup_{k \in \mathbb{Z}_+} \mathbf{E}(|X_k|) < \infty$, applying (5) and (6) to the first member on the right side of the inequality (7) we have

$$\begin{aligned} &\left| \int_\Omega X_k Y_0 \, dP - \int_\Omega X_k Y_m \, dP \right| \\ &\leq \int_{\{|Y_m - Y_0| \geq \varepsilon\}} |X_k Y_0 - X_k Y_m| \, dP + \int_{\{|Y_m - Y_0| < \varepsilon\}} |X_k||Y_0 - Y_m| \, dP \\ &\leq 2\varepsilon + a\varepsilon. \end{aligned}$$

By exactly the same argument we have the same estimate for the third member on the right side of (7). Using these estimates in (7) we have

(8) $\quad |\int_\Omega X_k Y_0 \, dP - \int_\Omega X_0 Y_0 \, dP| \leq 2(2+a)\varepsilon + |\int_\Omega X_k Y_m \, dP - \int_\Omega X_0 Y_m \, dP|.$

Since $Y_m \in \mathbf{V}$ the second member on the right side of (8) converges to 0 as $k \to \infty$. Therefore

$$\limsup_{k \to \infty} |\int_\Omega X_k Y_0 \, dP - \int_\Omega X_0 Y_0 \, dP| \leq 2(2+a)\varepsilon.$$

Since this holds for every $\varepsilon > 0$ the limit superior above is equal to 0. Hence we have the equality $\lim_{k \to \infty} \int_\Omega X_k Y_0 \, dP = \int_\Omega X_0 Y_0 \, dP$. Thus $Y_0 \in \mathbf{V}$. This completes the verification that $\mathbf{V}$ satisfies all the conditions in Theorem 1.10. Therefore (4) holds for every $Y \in L_\infty(\Omega, \mathfrak{G}, P)$.

Finally, to complete the proof of the theorem we show that (4) holds for every $Y \in L_\infty(\Omega, \mathfrak{F}, P)$. Now if $Y \in L_\infty(\Omega, \mathfrak{F}, P)$ then $\mathbf{E}(Y|\mathfrak{G}) \in L_\infty(\Omega, \mathfrak{G}, P)$ and therefore

$$\lim_{k \to \infty} \int_\Omega X_k \mathbf{E}(Y|\mathfrak{G}) \, dP = \int_\Omega X \mathbf{E}(Y|\mathfrak{G}) \, dP.$$

By the $\mathfrak{G}$-measurability of X_k and X we have $X_k \mathbf{E}(Y|\mathfrak{G}) = \mathbf{E}(X_k Y | \mathfrak{G})$ and $X \mathbf{E}(Y|\mathfrak{G}) = \mathbf{E}(XY|\mathfrak{G})$ a.e. on $(\Omega, \mathfrak{G}, P)$. Therefore

$$\lim_{k \to \infty} \int_\Omega \mathbf{E}(X_k Y | \mathfrak{G}) \, dP = \int_\Omega \mathbf{E}(XY|\mathfrak{G}) \, dP.$$

Since $\Omega \in \mathfrak{G}$, by the definition of conditional expectation the last equality implies

$$\lim_{k \to \infty} \int_\Omega X_k Y \, dP = \int_\Omega XY \, dP.$$

This completes the proof. ∎

Let us consider the family of conditional expectations with respect to a fixed σ-algebra of all random variables in a uniformly integrable family. In Theorem 4.22 below we show that such a family maintains the uniform integrability. In Theorem 4.24 we show that the family of conditional expectations of an integrable random variable with respect an arbitrary family of sub-σ-algebras in the probability space is always uniformly integrable.

Theorem 4.22. *Let $\{X_\alpha : \alpha \in A\}$ be a uniformly integrable system of extended real valued random variables on a probability space $(\Omega, \mathfrak{F}, P)$. Let $\mathfrak{G}$ be a sub-σ-algebra of $\mathfrak{F}$ and let*

§4. CONVERGENCE IN L_p AND UNIFORM INTEGRABILITY

Y_α be an arbitrary version of $\mathbf{E}(X_\alpha|\mathfrak{G})$ for $\alpha \in A$. Then $\{Y_\alpha : \alpha \in A\}$ is a uniformly integrable system.

Proof. Since $\{Y_\alpha : \alpha \in A\}$ is a system of extended real valued random variables on the probability space $(\Omega, \mathfrak{G}, P)$, to show its uniform integrability it is sufficient to verify according to Theorem 4.7 that

(1) $$\sup_{\alpha \in A} \mathbf{E}(|Y_\alpha|) < \infty$$

and for every $\varepsilon > 0$ there exists $\delta > 0$ such that

(2) $$\int_E |Y_\alpha|\,dP < \varepsilon \quad \text{for all } \alpha \in A \text{ whenever } E \in \mathfrak{G} \text{ and } P(E) < \delta.$$

Now

(3) $$|Y_\alpha| = |\mathbf{E}(X_\alpha|\mathfrak{G})| \leq \mathbf{E}(|X_\alpha|\,|\mathfrak{G}) \quad \text{a.e. on } (\Omega, \mathfrak{G}, P)$$

so that

$$\mathbf{E}(|Y_\alpha|) \leq \mathbf{E}[\mathbf{E}(|X_\alpha|\,|\mathfrak{G})] = \mathbf{E}(|X_\alpha|)$$

and then since $\sup_{\alpha \in A} \mathbf{E}(|X_\alpha|) < \infty$ by 1° of Theorem 4.7 we have (1).

By 2° of Theorem 4.7, for every $\varepsilon > 0$ there exists $\delta > 0$ such that $\int_E |X_\alpha|\,dP < \varepsilon$ for all $\alpha \in A$ whenever $E \in \mathfrak{F}$ and $P(E) < \delta$. Thus for any $E \in \mathfrak{G} \subset \mathfrak{F}$ with $P(E) < \delta$ we have by (3)

$$\int_E |Y_\alpha|\,dP \leq \int_E \mathbf{E}(|X_\alpha|\,|\mathfrak{G})\,dP = \int_E |X_\alpha|\,dP < \varepsilon \quad \text{for all } \alpha \in A,$$

proving (2). ∎

Theorem 4.23. Let $\{X_n : n \in \mathbb{N}\} \subset L_1(\Omega, \mathfrak{F}, P)$ be uniformly integrable so that there exists a subsequence $\{n_k\}$ of $\{n\}$ and $X \in L_1(\Omega, \mathfrak{F}, P)$ such that

(1) $$\lim_{k \to \infty} \int_\Omega X_{n_k} \xi\,dP = \int_\Omega X \xi\,dP \quad \text{for } \xi \in L_\infty(\Omega, \mathfrak{F}, P).$$

Let $\mathfrak{G}$ be a sub-σ-algebra of $\mathfrak{F}$. Then there exists a subsequence $\{n_\ell\}$ of $\{n_k\}$ and $Y \in L_1(\Omega, \mathfrak{G}, P)$ such that

(2) $$Y = \mathbf{E}(X|\mathfrak{G}) \quad \text{a.e. on } (\Omega, \mathfrak{G}, P),$$

and

(3) $$\lim_{\ell\to\infty}\int_\Omega \mathbf{E}(X_{n_\ell}|\mathfrak{G})\xi\, dP = \int_\Omega Y\xi\, dP \quad \text{for } \xi \in L_\infty(\Omega,\mathfrak{F},P).$$

Proof. The uniform integrability of $\{X_{n_k} : k \in \mathbb{N}\}$ implies that of $\{\mathbf{E}(X_{n_k}|\mathfrak{G}) : k \in \mathbb{N}\} \subset L_1(\Omega,\mathfrak{G},P)$ by Theorem 4.21. Thus by Theorem 4.22 there exists a subsequence $\{n_\ell\}$ of $\{n_k\}$ and $Y \in L_1(\Omega,\mathfrak{G},P)$ such that

(4) $$\lim_{\ell\to\infty}\int_\Omega \mathbf{E}(X_{n_\ell}|\mathfrak{G})\eta\, dP = \int_\Omega Y\eta\, dP \quad \text{for } \eta \in L_\infty(\Omega,\mathfrak{G},P).$$

Since

$$\int_\Omega \mathbf{E}(X_{n_\ell}|\mathfrak{G})\eta\, dP = \int_\Omega \mathbf{E}(X_{n_\ell}\eta|\mathfrak{G})\, dP = \int_\Omega X_{n_\ell}\eta\, dP \quad \text{for } \eta \in L_\infty(\Omega,\mathfrak{G},P),$$

(1) implies that

(5) $$\lim_{\ell\to\infty}\int_\Omega \mathbf{E}(X_{n_\ell}|\mathfrak{G})\eta\, dP = \lim_{\ell\to\infty}\int_\Omega X_{n_\ell}\eta\, dP = \int_\Omega X\eta\, dP \quad \text{for } \eta \in L_\infty(\Omega,\mathfrak{G},P).$$

By (4) and (5) we have

(6) $$\int_\Omega X\eta\, dP = \int_\Omega Y\eta\, dP \quad \text{for } \eta \in L_\infty(\Omega,\mathfrak{G},P).$$

In particular, with $\eta = \mathbf{1}_E$ where $E \in \mathfrak{G}$, we have by (6)

(7) $$\int_E \mathbf{E}(X|\mathfrak{G})\, dP = \int_E X\, dP = \int_\Omega X\mathbf{1}_E\, dP = \int_\Omega Y\mathbf{1}_E\, dP = \int_E Y\, dP.$$

Since $\mathbf{E}(X|\mathfrak{G})$ and Y are both $\mathfrak{G}$-measurable and since (7) holds for every $E \in \mathfrak{G}$, we have (2).

To prove (3), note that for every $\xi \in L_\infty(\Omega,\mathfrak{F},P)$, we have

$$\int_\Omega \mathbf{E}(X_{n_\ell}|\mathfrak{G})\xi\, dP = \int_\Omega \mathbf{E}[\mathbf{E}(X_{n_\ell}|\mathfrak{G})\xi|\mathfrak{G}]\, dP = \int_\Omega \mathbf{E}(X_{n_\ell}|\mathfrak{G})\mathbf{E}(\xi|\mathfrak{G})\, dP.$$

Since $\mathbf{E}(\xi|\mathfrak{G}) \in L_\infty(\Omega,\mathfrak{G},P)$, we have by (4)

$$\lim_{\ell\to\infty}\int_\Omega \mathbf{E}(X_{n_\ell}|\mathfrak{G})\xi\, dP = \int_\Omega Y\mathbf{E}(\xi|\mathfrak{G})\, dP = \int_\Omega \mathbf{E}(Y\xi|\mathfrak{G})\, dP = \int_\Omega Y\xi\, dP,$$

proving (3). ∎

§4. CONVERGENCE IN L_P AND UNIFORM INTEGRABILITY

Theorem 4.24. *Let X be an integrable extended real valued random variable on a probability space $(\Omega, \mathfrak{F}, P)$. Let $\{\mathfrak{G}_\alpha : \alpha \in A\}$ be an arbitrary system of sub-σ-algebras of $\mathfrak{F}$ and let Y_α be an arbitrary version of $\mathbf{E}(X | \mathfrak{G}_\alpha)$ for $\alpha \in A$. Then $\{Y_\alpha : \alpha \in A\}$ is a uniformly integrable system of random variables on $(\Omega, \mathfrak{F}, P)$.*

Proof. According to (3) of Definition 4.6, to show the uniform integrability of $\{Y_\alpha : \alpha \in A\}$ we show that for every $\varepsilon > 0$ there exists $\lambda > 0$ such that

$$\tag{1} \int_{\{|Y_\alpha| > \lambda\}} |Y_\alpha| \, dP < \varepsilon \quad \text{for all } \alpha \in A.$$

Now for each $\alpha \in A$ we have

$$\tag{2} |Y_\alpha| = |\mathbf{E}(X|\mathfrak{G}_\alpha)| \leq \mathbf{E}(|X| \, | \, \mathfrak{G}_\alpha) \quad \text{a.e. on } (\Omega, \mathfrak{G}_\alpha, P)$$

and then

$$\mathbf{E}(|Y_\alpha|) \leq \mathbf{E}[\mathbf{E}(|X| \, | \, \mathfrak{G}_\alpha)] = \mathbf{E}(|X|).$$

Thus for every $\lambda > 0$ we have

$$\tag{3} \lambda P\{|Y_\alpha| > \lambda\} \leq \int_{\{|Y_\alpha| > \lambda\}} |Y_\alpha| \, dP \leq \mathbf{E}(|Y_\alpha|) \leq \mathbf{E}(|X|) \quad \text{for all } \alpha \in A.$$

Let $\varepsilon > 0$ be arbitrarily given. The integrability of X implies that there exists $\delta > 0$ such that

$$\tag{4} \int_E |X| \, dP < \varepsilon \quad \text{whenever } E \in \mathfrak{F} \text{ and } P(E) < \delta.$$

Let $\lambda > 0$ be so large that

$$\tag{5} \frac{1}{\lambda} \int_\Omega |X| \, dP < \delta.$$

Then for such $\lambda > 0$ we have by (3) and (5)

$$\tag{6} P\{|Y_\alpha| > \lambda\} \leq \frac{1}{\lambda} \mathbf{E}(|X|) < \delta \quad \text{for all } \alpha \in A$$

and thus for every $\alpha \in A$ we have

$$\int_{\{|Y_\alpha| > \lambda\}} |Y_\alpha| \, dP \leq \int_{\{|Y_\alpha| > \lambda\}} \mathbf{E}(|X| \, | \, \mathfrak{G}_\alpha) \, dP = \int_{\{|Y_\alpha| > \lambda\}} |X| \, dP < \varepsilon$$

by (2), the fact that $\{|Y_\alpha| > \lambda\} \in \mathfrak{G}_\alpha$, (4) and (6). This proves (1). ∎

Chapter 2

Martingales

§5 Martingale, Submartingale and Supermartingale

[I] Martingale, Submartingale and Supermartingale Properties

Definition 5.1. *Let us fix some terminologies for stochastic processes $X = \{X_t : t \in D\}$ on a probability space $(\Omega, \mathfrak{F}, P)$ where D is a subset of $\mathbb{R}$.*
1) X is null at 0 if $0 \in D$ and $X_0 = 0$ a.e. on $(\Omega, \mathfrak{F}, P)$.
2) X is nonnegative if $X_t \geq 0$ a.e. on $(\Omega, \mathfrak{F}, P)$ for every $t \in D$.
3) X is bounded if there exists $M > 0$ such that $|X(t, w)| \leq M$ for all $(t, w) \in D \times \Omega$.
4) X is an L_p-process for some $p \in (0, \infty)$ if $X_t \in L_p(\Omega, \mathfrak{F}, P)$ for all $t \in D$.
5) X is L_p-bounded, or bounded in L_p, if $\sup_{t \in D} \mathbf{E}(|X_t|^p) < \infty$.
6) X is uniformly integrable if the system of random variables $\{X_t : t \in D\}$ is uniformly integrable. X is pth-order uniformly integrable if for some $p \in (0, \infty)$ the system of random variables $\{|X_t|^p : t \in D\}$ is uniformly integrable for some $p \in (0, \infty)$.

Recall that a system of sub-σ-algebras of $\mathfrak{F}$, $\{\mathfrak{F}_t : t \in D\}$, is a filtration if it is an increasing system, that is, $\mathfrak{F}_s \subset \mathfrak{F}_t$ for $s, t \in D$ such that $s < t$. $(\Omega, \mathfrak{F}, \{\mathfrak{F}_t : t \in D\}, P)$, or briefly $(\Omega, \mathfrak{F}, \{\mathfrak{F}_t\}, P)$, is called a filtered space. X is $\{\mathfrak{F}_t\}$-adapted if X_t is $\mathfrak{F}_t$-measurable for every $t \in D$.

Martingales, submartingales and supermartingales are stochastic processes defined in terms of the conditional expectation $\mathbf{E}(X_t | \mathfrak{F}_s)$ for $s, t \in D$, $s < t$, as follows.

Definition 5.2. *A stochastic process $X = \{X_t : t \in D\}$ on a filtered space $(\Omega, \mathfrak{F}, \{\mathfrak{F}_t : t \in D\}, P)$ is called a martingale with respect to $(\{\mathfrak{F}_t\}, P)$ if it satisfies the following*

conditions:
(i) X is $\{\mathfrak{F}_t\}$-adapted.
(ii) X is an L_1-process.
(iii) $\mathbf{E}(X_t|\mathfrak{F}_s) = X_s$ a.e. on $(\Omega, \mathfrak{F}_s, P)$ for $s, t \in D$, $s < t$.
X is called a submartingale if instead of (iii) it satisfies the condition
(iv) $\mathbf{E}(X_t|\mathfrak{F}_s) \geq X_s$ a.e. on $(\Omega, \mathfrak{F}_s P)$ for $s, t \in D$, $s < t$.
X is called a supermartingale if instead of (iii) it satisfies the condition
(v) $\mathbf{E}(X_t|\mathfrak{F}_s) \leq X_s$ a.e. on $(\Omega, \mathfrak{F}_s, P)$ for $s, t \in D$, $s < t$.

The monotonicity conditions (iii), (iv) and (v) in Definition 5.2 will be referred to as the martingale, submartingale and supermartingale property of the process. According to these properties, a submartingale increases on average, a supermartingale decreases on average and a martingale remains constant on average as t increases.

Two prominent examples of the index set D are $\mathbb{R}_+$ and $\mathbb{Z}_+$. Let $\mathbb{T}$ represent both $\mathbb{R}_+$ and $\mathbb{Z}_+$ and let $\overline{\mathbb{T}} = \mathbb{T} \cup \{\infty\}$. For a filtered space $(\Omega, \mathfrak{F}, \{\mathfrak{F}_t : t \in \mathbb{T}\}, P)$ we define $\mathfrak{F}_\infty = \sigma(\cup_{t \in \mathbb{T}} \mathfrak{F}_t)$. When D is a finite set we call X a finite martingale, submartingale or supermartingale. When D is a strictly increasing sequence $\{t_n : n \in \mathbb{Z}_+\}$ in $\mathbb{R}$, we let $t_\infty = \lim_{n \to \infty} t_n \in \overline{\mathbb{R}}_+$ and let $\mathfrak{F}_{t_\infty} = \sigma(\cup_{n \in \mathbb{Z}_+} \mathfrak{F}_{t_n})$.

As an example of submartingales consider an $\{\mathfrak{F}_t\}$-adapted L_1-process $X = \{X_t : t \in \mathbb{T}\}$ on a filtered space $(\Omega, \mathfrak{F}, \{\mathfrak{F}_t : t \in \mathbb{T}\}, P)$ whose sample functions are increasing functions on $\mathbb{T}$. For such a process we have $X_t \geq X_s$ for $s, t \in \mathbb{T}$, $s < t$, so that $\mathbf{E}(X_t|\mathfrak{F}_s) \geq X_s$ a.e. on $(\Omega, \mathfrak{F}_s, P)$.

To construct a martingale, let $(\Omega, \mathfrak{F}, \{\mathfrak{F}_t : t \in \mathbb{T}\}, P)$ be a filtered space, let ξ be an integrable random variable on the probability space $(\Omega, \mathfrak{F}, P)$ and let X_t be an arbitrary real valued version of $\mathbf{E}(\xi|\mathfrak{F}_t)$ for each $t \in \mathbb{T}$. Then $X = \{X_t : t \in \mathbb{T}\}$ is an $\{\mathfrak{F}_t\}$-adapted L_1-process on the filtered space. It has the martingale property since for $s, t \in \mathbb{T}$, $s < t$ we have

$$\mathbf{E}(X_t|\mathfrak{F}_s) = \mathbf{E}[\mathbf{E}(\xi|\mathfrak{F}_t)|\mathfrak{F}_s] = \mathbf{E}(\xi|\mathfrak{F}_s) = X_s \quad \text{a.e. on } (\Omega, \mathfrak{F}_s, P).$$

This martingale is a uniformly integrable stochastic process by Theorem 4.24 and in particular it is L_1-bounded by Theorem 4.7. Now for an arbitrary version η of $\mathbf{E}(\xi|\mathfrak{F}_\infty)$ where $\mathfrak{F}_\infty = \sigma(\cup_{t \in \mathbb{T}} \mathfrak{F}_t)$, we have $\mathfrak{F}_t \subset \mathfrak{F}_\infty$ for every $t \in \mathbb{T}$ and therefore

$$\mathbf{E}(\eta|\mathfrak{F}_t) = \mathbf{E}[\mathbf{E}(\xi|\mathfrak{F}_\infty)|\mathfrak{F}_t] = \mathbf{E}(\xi|\mathfrak{F}_t) = X_t \quad \text{a.e. on } (\Omega, \mathfrak{F}_t, P).$$

For a martingale $X = \{X_t : t \in \mathbb{T}\}$ on a filtered space $(\Omega, \mathfrak{F}, \{\mathfrak{F}_t : t \in \mathbb{T}\}, P)$, if there exists

§5. MARTINGALE, SUBMARTINGALE AND SUPERMARTINGALE

an extended real valued integrable $\mathfrak{F}_\infty$-measurable random variable η such that $\mathbf{E}(\eta|\mathfrak{F}_t) = X_t$ a.e. on $(\Omega, \mathfrak{F}_t, P)$ for every $t \in \mathbb{T}$ then we call η a final element for X. In Theorem 8.2 we show that a final element exists for a martingale X if and only if X is uniformly integrable. In Theorem 5.3 next, we show that if a final element exists for a martingale then it is uniquely determined up to a null set in $(\Omega, \mathfrak{F}_\infty, P)$.

Theorem 5.3. (Uniqueness of the Final Element of a Martingale) *Let $X = \{X_t : t \in \mathbb{T}\}$ be a martingale on a filtered space $(\Omega, \mathfrak{F}, \{\mathfrak{F}_t : t \in \mathbb{T}\}, P)$ and let $\mathfrak{F}_\infty = \sigma(\cup_{t \in \mathbb{T}} \mathfrak{F}_t)$. If there exists an integrable $\mathfrak{F}_\infty$-measurable random variable ξ on $(\Omega, \mathfrak{F}, P)$ such that $\mathbf{E}(\xi|\mathfrak{F}_t) = X_t$ a.e. on $(\Omega, \mathfrak{F}_t, P)$ for every $t \in \mathbb{T}$, then ξ is uniquely determined up to a null set in $(\Omega, \mathfrak{F}_\infty, P)$.*

Proof. Suppose there exist two final elements ξ and η of the martingale which are not a.e. equal on $(\Omega, \mathfrak{F}_\infty, P)$. We may assume without loss of generality that $\xi > \eta$ on some $A \in \mathfrak{F}_\infty$ with $P(A) > 0$. Then there exists some $k \in \mathbb{N}$ such that $\xi - \eta \geq 1/k$ on some $A_1 \subset A$, $A_1 \in \mathfrak{F}_\infty$ with $P(A_1) > 0$. Since $|\xi| + |\eta|$ is integrable, for an arbitrary $\varepsilon > 0$ and in particular for $(2k)^{-1}P(A_1) > 0$ there exists $\delta > 0$ such that for every $E \in \mathfrak{F}_\infty$ with $P(E) < \delta$ we have

$$\int_E \{|\xi| + |\eta|\}\, dP < \frac{1}{2k}P(A_1).$$

Now since $\cup_{t \in \mathbb{T}} \mathfrak{F}_t$ is an algebra of subsets of Ω and since $\mathfrak{F}_\infty = \sigma(\cup_{t \in \mathbb{T}} \mathfrak{F}_t)$, there exists according to Theorem 1.13 a set $A_2 \in \mathfrak{F}_t$ for some $t \in \mathbb{T}$ such that $P(A_1 \triangle A_2) < \delta$. Then

$$\left|\int_{A_1}\{\xi - \eta\}\, dP - \int_{A_2}\{\xi - \eta\}\, dP\right| = \left|\int_{A_1 - A_2}\{\xi - \eta\}\, dP - \int_{A_2 - A_1}\{\xi - \eta\}\, dP\right|$$

$$\leq \int_{A_1 \triangle A_2}\{|\xi| + |\eta|\}\, dP < \frac{1}{2k}P(A_1).$$

From this inequality and from $\int_{A_1}\{\xi - \eta\}\, dP \geq 1/k\, P(A_1)$ we have

(1) $$\int_{A_2}\{\xi - \eta\}\, dP \geq \int_{A_1}\{\xi - \eta\}\, dP - \frac{1}{2k}P(A_1) \geq \frac{1}{2k}P(A_1) > 0.$$

From the equality $\mathbf{E}(\xi|\mathfrak{F}_t) = X_t = \mathbf{E}(\eta|\mathfrak{F}_t)$ a.e. on $(\Omega, \mathfrak{F}_t, P)$ we have $\mathbf{E}[\{\xi - \eta\}|\mathfrak{F}_t] = 0$ a.e. on $(\Omega, \mathfrak{F}_t, P)$. Then since $A_2 \in \mathfrak{F}_t$ we have

$$\int_{A_2}\{\xi - \eta\}\, dP = \int_{A_2}\mathbf{E}[\{\xi - \eta\}|\mathfrak{F}_t]\, dP = 0,$$

which contradicts (1). Therefore $\xi = \eta$ a.e. on $(\Omega, \mathfrak{F}_\infty, P)$. ∎

Proposition 5.4. *Let $X = \{X_t : t \in \mathbb{T}\}$ be a stochastic process on a filtered space $(\Omega, \mathfrak{F}, \{\mathfrak{F}_t : t \in \mathbb{T}\}, P)$.*
1) If X is a martingale then $\mathbf{E}(X_t) = const$ for $t \in \mathbb{T}$.
2) If X is a submartingale then $\mathbf{E}(X_t) \uparrow$ as $t \to \infty$.
3) If X is a supermartingale then $\mathbf{E}(X_t) \downarrow$ as $t \to \infty$.
4) A submartingale X is a martingale if and only if $\mathbf{E}(X_t) = const$ for $t \in \mathbb{T}$. Similarly for a supermartingale.

Proof. 1), 2) and 3) are immediate consequences of (iii), (iv) and (v) respectively of Definition 5.2. To prove for 2), note that by (iv) of Definition 5.2, we have $\mathbf{E}[\mathbf{E}(X_t|\mathfrak{F}_s)] \geq \mathbf{E}(X_s)$ for $s < t$. But $\mathbf{E}[\mathbf{E}(X_t|\mathfrak{F}_s)] = \mathbf{E}(X_t)$. Thus $\mathbf{E}(X_t) \geq \mathbf{E}(X_s)$. Therefore $\mathbf{E}(X_t) \uparrow$ as $t \to \infty$. Similarly we have 1) and 3).

To prove 4), suppose X is a submartingale and $\mathbf{E}(X_t) = const$ for $t \in \mathbb{T}$. Then for every $s, t \in \mathbb{T}$, $s < t$, we have $\mathbf{E}(X_t|\mathfrak{F}_s) \geq X_s$ a.e. on $(\Omega, \mathfrak{F}_s, P)$. On the other hand, $\mathbf{E}[\mathbf{E}(X_t|\mathfrak{F}_s)] = \mathbf{E}(X_t) = \mathbf{E}(X_s)$, namely, $\int_\Omega \mathbf{E}(X_t|\mathfrak{F}_s) dP = \int_\Omega X_s dP$. Therefore, $\mathbf{E}(X_t|\mathfrak{F}_s) = X_s$ a.e. on $(\Omega, \mathfrak{F}_s, P)$. This shows that X is a martingale. Similarly for a supermartingale. ∎

Observation 5.5. The monotonicity conditions (iii), (iv) and (v) in Definition 5.2 for a martingale, submartingale and supermartingale are respectively equivalent to the following conditions.

$(iii)'$ $$\int_E X_t\, dP = \int_E X_s\, dP \quad \text{for } E \in \mathfrak{F}_s,\, s, t \in \mathbb{T},\, s < t.$$

$(iv)'$ $$\int_E X_t\, dP \geq \int_E X_s\, dP \quad \text{for } E \in \mathfrak{F}_s,\, s, t \in \mathbb{T},\, s < t.$$

$(v)'$ $$\int_E X_t\, dP \leq \int_E X_s\, dP \quad \text{for } E \in \mathfrak{F}_s,\, s.t. \in \mathbb{T},\, s < t.$$

Let us show the equivalence of (iv) and (iv)$'$ for instance. Recall that by the definition of $\mathbf{E}(X_t|\mathfrak{F}_s)$ we have

$$\int_E X_t dP = \int_E \mathbf{E}(X_t|\mathfrak{F}_s) dP \quad \text{for every } E \in \mathfrak{F}_s.$$

Thus if (iv) holds, then (iv)$'$ holds. Conversely if (iv)$'$ holds then by the last equality we have

$$\int_E \mathbf{E}(X_t|\mathfrak{F}_s) dP \geq \int_E X_s dP \quad \text{for every } E \in \mathfrak{F}_s.$$

§5. MARTINGALE, SUBMARTINGALE AND SUPERMARTINGALE

Then the $\mathfrak{F}_s$-measurability of both $\mathbf{E}(X_t|\mathfrak{F}_s)$ and X_s implies (iv).

Proposition 5.6. *When* $\mathbb{T} = \mathbb{Z}_+$, *conditions (iii), (iv) and (v) in Definition 5.2 are respectively equivalent to the following conditions.*

$(iii)''$ $\qquad\qquad \mathbf{E}(X_{n+1}|\mathfrak{F}_n) = X_n \quad a.e. \text{ on } (\Omega, \mathfrak{F}_n, P) \text{ for } n \in \mathbb{Z}_+.$

$(iv)''$ $\qquad\qquad \mathbf{E}(X_{n+1}|\mathfrak{F}_n) \geq X_n \quad a.e. \text{ on } (\Omega, \mathfrak{F}_n, P) \text{ for } n \in \mathbb{Z}_+.$

$(v)''$ $\qquad\qquad \mathbf{E}(X_{n+1}|\mathfrak{F}_n) \leq X_n \quad a.e. \text{ on } (\Omega, \mathfrak{F}_n, P) \text{ for } n \in \mathbb{Z}_+.$

Proof. Let us show the equivalence of (iv) and $(iv)''$. Clearly (iv) implies $(iv)''$. To prove the converse assume $(iv)''$. Let $n, m \in \mathbb{Z}_+$ and $n < m$. Then by iterated application of $(iv)''$ we have

$$\begin{aligned} \mathbf{E}(X_m|\mathfrak{F}_n) &= \mathbf{E}[\mathbf{E}(X_m|\mathfrak{F}_{m-1})|\mathfrak{F}_n] \geq \mathbf{E}(X_{m-1}|\mathfrak{F}_n) \\ &= \mathbf{E}[\mathbf{E}(X_{m-1}|\mathfrak{F}_{m-2})|\mathfrak{F}_n] \geq \mathbf{E}(X_{m-2}|\mathfrak{F}_n) \\ &\vdots \\ &= \mathbf{E}(X_{n+1}|\mathfrak{F}_n) \geq X_n \qquad \text{a.e. on } (\Omega, \mathfrak{F}_n, P), \end{aligned}$$

proving (iv). ∎

Proposition 5.7. *Let* $X = \{X_t : t \in \mathbb{T}\}$ *be a stochastic process on a filtered space* $(\Omega, \mathfrak{F}, \{\mathfrak{F}_t : t \in \mathbb{T}\}, P)$. *Let* $\{\mathfrak{F}_t^X : t \in \mathbb{T}\}$ *be the filtration of* $(\Omega, \mathfrak{F}, P)$ *generated by* X, *namely,*

$$\mathfrak{F}_t^X = \sigma\{X_s : s \in \mathbb{T}, s \leq t\} \qquad \text{for } t \in \mathbb{T}.$$

If X is a martingale, submartingale or supermartingale with respect to $(\{\mathfrak{F}_t\}, P)$, then it is also a martingale, submartingale or supermartingale with respect to $(\{\mathfrak{F}_t^X\}, P)$.

Proof. Let X be a submartingale with respect to $(\{\mathfrak{F}_t\}, P)$. Then X_t is $\{\mathfrak{F}_t\}$-measurable and therefore $\mathfrak{F}_t^X \subset \mathfrak{F}_t$ for every $t \in \mathbb{T}$. By the definition of $\mathfrak{F}_t^X$, the process X is $\{\mathfrak{F}_t^X\}$-adapted. Therefore according to Observation 5.5, to show that X is in submartingale with respect to $(\{\mathfrak{F}_t^X\}, P)$, it remains only to verify

$$\int_E X_t dP \geq \int_E X_s dP \quad \text{for } E \in \mathfrak{F}_s^X, s, t \in \mathbb{T}, s < t.$$

Since X is a submartingale with respect to $(\{\mathfrak{F}_t\}, P)$, the last inequality holds for every $E \in \mathfrak{F}_s$. Since $\mathfrak{F}_s^X \subset \mathfrak{F}_s$, the inequality holds for every $E \in \mathfrak{F}_s^X$. Thus X is a submartingale with respect to $(\{\mathfrak{F}_t^X\}, P)$. Similarly for a martingale and a supermartingale. ∎

Proposition 5.8. *For an adapted L_1-process $X = \{X_t : t \in \mathbb{T}\}$ on a filtered space $(\Omega, \mathfrak{F}, \{\mathfrak{F}_t\}, P)$ the following statements hold.*
1) X is a martingale if and only if it is both a submartingale and a supermartingale.
2) X is a submartingale (resp. supermartingale) if and only if $-X = \{-X_t : t \in \mathbb{T}\}$ is a supermartingale (resp. submartingale).
3) If X is a martingale then so is $cX = \{cX_t : t \in \mathbb{T}\}$ for $c \in \mathbb{R}$.
4) If X is a submartingale (resp. supermartingale), then so is cX for $c \geq 0$.
5) If X and Y are both martingales (resp. submartingales or super martingales) then so is $X + Y = \{X_t + Y_t : t \in \mathbb{T}\}$.
6) If X is a martingale and Y is a submartingale (resp. supermartingale) then $X + Y$ is a submartingale (resp. supermartingale).

Proof. These statements are immediate consequences of Definition 5.2. ∎

[II] Convexity Theorems

Let $X = \{X_t : t \in \mathbb{T}\}$ be a stochastic process on a probability space. For a real valued function f on $\mathbb{R}$, let us write $f(X)$ or $f \circ X$ for $\{f \circ X_t : t \in \mathbb{T}\}$. For instance $X^+ = \{X_t \vee 0, t \in \mathbb{T}\}$ and $X^2 = \{X_t^2 : t \in \mathbb{T}\}$. Similarly we write $X \vee Y = \{X_t \vee Y_t : t \in \mathbb{T}\}$.

Theorem 5.9. *If $X = \{X_t : t \in \mathbb{T}\}$ and $Y = \{Y_t : t \in \mathbb{T}\}$ are submartingales (or supermartingales) on a filtered space $(\Omega, \mathfrak{F}, \{\mathfrak{F}_t\}, P)$, then so is $X \vee Y$ (or $X \wedge Y$).*

Proof. Suppose X and Y are submartingales. Then for any $t \in \mathbb{T}$, X_t and Y_t are $\mathfrak{F}_t$-measurable so that $X_t \vee Y_t$ is $\mathfrak{F}_t$-measurable and therefore $X \vee Y$ is an adapted process. Since $X_t, Y_t \in L_1(\Omega, \mathfrak{F}_t, P)$, and $|X_t \vee Y_t| \leq |X_t| + |Y_t|$, we have $X_t \vee Y_t \in L_1(\Omega, \mathfrak{F}_t, P)$. Thus $X \vee Y$ is an L_1-process. Now for $s, t \in \mathbb{T}$, $s < t$, we have $X_t \vee Y_t \geq X_t, Y_t$ so that

$$\mathbf{E}(X_t \vee Y_t | \mathfrak{F}_s) \geq \mathbf{E}(X_t | \mathfrak{F}_s) \geq X_s \quad \text{a.e. on } (\Omega, \mathfrak{F}_s, P)$$

and similarly

$$\mathbf{E}(X_t \vee Y_t) | \mathfrak{F}_s) \geq Y_s \quad \text{a.e. on } (\Omega, \mathfrak{F}, P).$$

Therefore

$$\mathbf{E}(X_t \vee Y_t | \mathfrak{F}_s) \geq X_s \vee Y_s \quad \text{a.e. on } (\Omega, \mathfrak{F}_s, P).$$

§5. MARTINGALE, SUBMARTINGALE AND SUPERMARTINGALE

This shows that $X \vee Y$ is a submartingale. Similarly $X \wedge Y$ is a supermartingale when X and Y are supermartingales. ∎

Corollary 5.10. *Let* $X = \{X_t : t \in \mathbb{T}\}$ *be a stochastic process on a filtered space* $(\Omega, \mathfrak{F}, \{\mathfrak{F}_t\}, P)$.
1) If X is a submartingale then so is X^+.
2) If X is a supermartingale then X^- is a submartingale.
3) If X is a martingale then $X = X' - X''$ where X' and X'' are nonnegative submartingales.

Proof. 1) If X is a submartingale then since $Y = 0$ is a martingale, $X^+ = X \vee 0$ is a submartingale by Theorem 5.9.

2) If X is a supermartingale then since $Y = 0$ is a martingale, $X \wedge 0$ is a supermartingale by Theorem 5.9. Then $X^- = -(X \wedge 0)$ is a submartingale.

3) If X is a martingale, then it is a submartingale so that X^+ is a submartingale by 1). But X is also a supermartingale so that X^- is a submartingale by 2). Since $X = X^+ - X^-$, the process X is the difference of two nonnegative submartingales. ∎

Theorem 5.11. *Let* $X = \{X_t : t \in \mathbb{T}\}$ *be an adapted L_1-process on a filtered space* $(\Omega, \mathfrak{F}, \{\mathfrak{F}_t\}, P)$. *Let f be a real valued increasing function on $\mathbb{R}$ such that $f \circ X_t \in L_1(\Omega, \mathfrak{F}_t, P)$ for every $t \in \mathbb{T}$.*
1) If X is a submartingale and f is a convex function then $f \circ X$ is a submartingale.
2) If X is a supermartingale and f is concave function then $f \circ X$ is a supermartingale.
3) When X is a martingale, the conclusions of 1) and 2) still hold when the monotonicity condition on f is removed.

Proof. 1) Suppose X is a submartingale. Since f is a real valued increasing function on $\mathbb{R}$, f is $\mathfrak{B}(\mathbb{R})/\mathfrak{B}(\mathbb{R})$-measurable and therefore $f \circ X_t$ is $\mathfrak{F}_t/\mathfrak{B}(\mathbb{R})$-measurable, that is, $f \circ X$ is an adapted process. Since X is a submartingale, for $s, t \in \mathbb{T}$, $s < t$, we have

(1) $\qquad \mathrm{E}(X_t | \mathfrak{F}_s) \geq X_s \qquad$ a.e. on $(\Omega, \mathfrak{F}_s, P)$.

Then since f is an increasing function

(2) $\qquad f(\mathrm{E}(X_t | \mathfrak{F}_s)) \geq f(X_s) \qquad$ a.e. on $(\Omega, \mathfrak{F}_s, P)$.

By the Conditional Jensen Inequality

(3) $\qquad \mathrm{E}(f(X_t) | \mathfrak{F}_s) \geq f(\mathrm{E}(X_t | \mathfrak{F}_s)) \qquad$ a.e. on $(\Omega, \mathfrak{F}_s, P)$.

From (2) and (3), we have $\mathbf{E}(f(X_t)|\mathfrak{F}_s) \geq f(X_s)$ a.e. on $(\Omega, \mathfrak{F}_s, P)$. Therefore $f(X)$ is a submartingale.

When X is a martingale then the equality in (1) holds and therefore the equality in (2) holds without assuming that f is an increasing function. Thus, in this case $f(X)$ is a submartingale when the monotonicity condition on f is removed.

2) Suppose X is a supermartingale. Then for $s, t \in \mathbb{T}$, $s < t$, we have $\mathbf{E}(X_t|\mathfrak{F}_s) \leq X_s$ a.e. on $(\Omega, \mathfrak{F}_s, P)$. Since f is an increasing function, we have $f(\mathbf{E}(X_t|\mathfrak{F}_s)) \leq f(X_s)$ a.e. on $(\Omega, \mathfrak{F}_s, P)$. If f is concave then $-f$ is convex. Thus by the Conditional Jensen Inequality $\mathbf{E}(-f(X_t)|\mathfrak{F}_s) \geq -f(\mathbf{E}(X_t|\mathfrak{F}_s))$. Therefore $\mathbf{E}(f(X_t)|\mathfrak{F}_s) \leq f(X_s)$ a.e. on $(\Omega, \mathfrak{F}_s, P)$. This shows that $f(X)$ is a supermartingale. ∎

Corollary 5.12. *1) Let $X = \{X_t : t \in \mathbb{T}\}$ be an L_p-martingale, that is, an L_p-process and a martingale, on a filtered space $(\Omega, \mathfrak{F}, \{\mathfrak{F}_t\}, P)$ where $p \in [1, \infty)$. Then $|X|^p = \{|X_t|^p : t \in \mathbb{T}\}$ is a submartingale.*

2) If X is a nonnegative L_p-submartingale then X^p is a submartingale for $p \in [1, \infty)$.

Proof. 1) The function f on $\mathbb{R}$ defined by $f(\xi) = |\xi|^p$ for $\xi \in \mathbb{R}$ is a convex function when $p \in [1, \infty)$. If X is an L_p-martingale then $f \circ X_t = |X_t|^p \in L_1(\Omega, \mathfrak{F}_t, P)$ so that $f \circ X$ is an L_1-process. Thus by 3) of Theorem 5.11, $|X|^p$ is a submartingale.

2) For $p \in [1, \infty)$, define a real valued increasing convex function f on $\mathbb{R}$ by setting $f(\xi) = 0$ for $\xi \in (-\infty, 0)$ and $f(\xi) = \xi^p$ for $\xi \in [0, \infty)$. If X is a nonnegative L_p-submartingale, then $X_t^p \in L_1(\Omega, \mathfrak{F}, P)$ and $X_t^p = f \circ X_t$ for $t \in \mathbb{T}$. Then by Theorem 5.11, X^p is a submartingale. ∎

[III] Discrete Time Increasing Processes and Doob Decomposition

Let us give a formal definition for discrete time increasing processes.

Definition 5.13. *By an increasing process we mean a stochastic process $A = \{A_n : n \in \mathbb{Z}_+\}$ on a filtered space $(\Omega, \mathfrak{F}, \{\mathfrak{F}_n\}, P)$ satisfying the following conditions.*
1°. *A is $\{\mathfrak{F}_n\}$-adapted.*
2°. *A is an L_1-process.*
3°. *$\{A_n(\omega) : n \in \mathbb{Z}_+\}$ is an increasing sequence with $A_0(\omega) = 0$ for every $\omega \in \Omega$.*
A is called an almost surely increasing process if it satisfies conditions 1°, 2° and the following condition.
4°. *There exists a null set Λ in $(\Omega, \mathfrak{F}_\infty, P)$ such that $\{A_n(\omega) : n \in \mathbb{Z}_+\}$ is an increasing sequence with $A_0(\omega) = 0$ for every $\omega \in \Lambda^c$.*

§5. MARTINGALE, SUBMARTINGALE AND SUPERMARTINGALE

An almost surely increasing process $A = \{A_n : n \in \mathbb{Z}_+\}$ is always a submartingale since for every $n \in \mathbb{Z}_+$ we have $A_{n+1} \geq A_n$ a.e. on $(\Omega, \mathfrak{F}_\infty, P)$ and consequently $\mathbf{E}(A_{n+1}|\mathfrak{F}_n) \geq \mathbf{E}(A_n|\mathfrak{F}_n) = A_n$ a.e. on $(\Omega, \mathfrak{F}_n, P)$. On the other hand there are submartingales which are not almost surely increasing processes. In fact, if A is an almost surely increasing process and thus a submartingale and if M is a martingale, then $X = A + M$ is a submartingale by Proposition 5.8, but it may not be an almost surely increasing process since it may not satisfy condition 4° in Definition 5.13. The Doob Decomposition Theorem shows that a submartingale in discrete time is always the sum of a martingale and an almost surely increasing process. Let us define the predictability of discrete time processes. This condition ensures the uniqueness in the Doob decomposition.

Definition 5.14 *An adapted discrete time process $X = \{X_n : n \in \mathbb{Z}_+\}$ on a filtered space $(\Omega, \mathfrak{F}, \{\mathfrak{F}_n\}, P)$ is called an $\{\mathfrak{F}_n\}$-predictable process if X_n is $\mathfrak{F}_{n-1}$-measurable for every $n \in \mathbb{N}$.*

Let us observe that if $X = \{X_t : t \in \mathbb{T}\}$ is an adapted L_1-process on a filtered space $(\Omega, \mathfrak{F}, \{\mathfrak{F}_t\}, P)$ and if we define a null at 0 adapted L_1-process $Y = \{Y_t : t \in \mathbb{T}\}$ by setting $Y_t = X_t - X_0$ for $t \in \mathbb{T}$. Then Y is respectively a martingale, submartingale or supermartingale if and only if X is.

Theorem 5.15. (Doob Decomposition) *Let $X = \{X_n : n \in \mathbb{Z}_+\}$ be an adapted L_1-process on a filtered space $(\Omega, \mathfrak{F}, \{\mathfrak{F}_n\}, P)$. Then X has the Doob decomposition*

$$X = X_0 + M + A$$

where M is a null at 0 martingale and A is a null at 0 predictable L_1-process. Moreover the decomposition is unique in the sense that if $X = X_0 + M' + A'$ is another such decomposition then $M(\cdot, \omega) = M'(\cdot, \omega)$ and $A(\cdot, \omega) = A'(\cdot, \omega)$ for a.e. ω in $(\Omega, \mathfrak{F}_\infty, P)$. Furthermore an adapted L_1-process X on the filtered space is a submartingale if and only the null at 0 predictable L_1-process A in the decomposition is an almost surely increasing process.

Proof. 1) Let $X = \{X_n : n \in \mathbb{Z}_+\}$ be an adapted L_1-process. Define a null at 0 process $A = \{A_n : n \in \mathbb{Z}_+\}$ by setting

(1) $\quad \begin{cases} A_0 = 0 \\ A_n = A_{n-1} + \mathbf{E}(X_n - X_{n-1}|\mathfrak{F}_{n-1}) \end{cases}$ for $n \in \mathbb{N}$.

Then A_n is $\mathfrak{F}_{n-1}$-measurable for every $n \in \mathbb{N}$ so that A is a predictable process. Also A_n is integrable for every $n \in \mathbb{Z}_+$ so that A is an L_1-process.

For the process $X - A$ we have for every $n \in \mathbb{N}$

$$\begin{aligned}\mathbf{E}(X_n - A_n | \mathfrak{F}_{n-1}) &= \mathbf{E}(X_n | \mathfrak{F}_{n-1}) - \mathbf{E}(A_{n-1} | \mathfrak{F}_{n-1}) - \mathbf{E}(X_n - X_{n-1} | \mathfrak{F}_{n-1}) \\ &= X_{n-1} - A_{n-1} \quad \text{a.e. on } (\Omega, \mathfrak{F}_{n-1}, P).\end{aligned}$$

This shows that $X - A$ is a martingale. If we set $M = (X - A) - X_0$ then since A is null at 0, so is M. Thus we have a Doob decomposition $X = X_0 + M + A$.

2) To prove the uniqueness of the decomposition, suppose an adopted L_1-process X is given as

$$X = X_0 + M + A = X_0 + M' + A',$$

where M and M' are null at 0 martingales and A and A' are predictable L_1-processes. By $X = X_0 + M + A$ we have for every $n \in \mathbb{N}$

$$\mathbf{E}(X_n - X_{n-1} | \mathfrak{F}_{n-1}) = \mathbf{E}[(M_n + A_n) - (M_{n-1} + A_{n-1}) | \mathfrak{F}_{n-1}] = A_n - A_{n-1}$$

a.e. on $(\Omega, \mathfrak{F}_{n-1}, P)$ by the martingale property of M and by the predictability of A. Similarly by $X = X_0 + M' + A'$ we have $\mathbf{E}(X_n - X_{n-1} | \mathfrak{F}_{n-1}) = A'_n - A'_{n-1}$ a.e. on $(\Omega, \mathfrak{F}_{n-1}, P)$. Thus for every $n \in \mathbb{N}$ we have

(2) $\qquad A_n - A_{n-1} = A'_n - A'_{n-1} \quad \text{a.e. on } (\Omega, \mathfrak{F}_{n-1}, P).$

Since A and A' are both null at 0, $A_0 = A'_0$ a.e on $(\Omega, \mathfrak{F}_0, P)$. Then (2) implies that $A_1 = A'_1$ a.e. on $(\Omega, \mathfrak{F}_0, P)$. By iterated application of (2) we have $A_n = A'_n$ a.e. on $(\Omega, \mathfrak{F}_{n-1}, P)$ for every $n \in \mathbb{N}$. By the countability of $\mathbb{Z}_+$ there exists a null set Λ in $(\Omega, \mathfrak{F}_\infty, P)$ such that $A(\cdot, w) = A'(\cdot, w)$ for $\omega \in \Lambda^c$. Since $M = X - X_0 - A$ and $M' = X - X_0 - A'$, the last equality implies $M(\cdot, w) = M'(\cdot, w)$ for $\omega \in \Lambda^c$.

3) Suppose X is a submartingale and $X = X_0 + M + A$ is its Doob decomposition. Since M is a martingale, $A = X - M - X_0$ is a submartingale and thus for every $n \in \mathbb{N}$ we have $A_n = \mathbf{E}(A_n | \mathfrak{F}_{n-1}) \geq A_{n-1}$ a.e. on $(\Omega, \mathfrak{F}_{n-1}, P)$ by the predictability of A. Since $\mathbb{N}$ is a countable set, there exists a null set Λ in $(\Omega, \mathfrak{F}_\infty, P)$ such that $A_{n-1}(\omega) \leq A_n(\omega)$ for all $n \in \mathbb{N}$ when $\omega \in \Lambda^c$. This shows that A is an almost surely increasing process. Conversely suppose A is an almost surely increasing process. Then A is a submartingale. Since M is a martingale, $X = X_0 + M + A$ is a submartingale. ∎

An extension of the Doob decomposition to continuous time submartingales, the Doob-Meyer decomposition, will be treated in §10.

§5. MARTINGALE, SUBMARTINGALE AND SUPERMARTINGALE

[IV] Martingale Transform

The martingale transform is a prototype of the stochastic integral with respect to martingales. It is the discrete time analog of the stochastic integral in continuous time.

Definition 5.16. *Let* $X = \{X_n : n \in \mathbb{Z}_+\}$ *be a martingale, a submartingale or a supermartingale and let* $C = \{C_n : n \in \mathbb{Z}_+\}$ *be a predictable process on a filtered space* $(\Omega, \mathfrak{F}, \{\mathfrak{F}_n\}, P)$. *By the martingale transform of X by C we mean the stochastic process $C \bullet X$ defined by*

$$\begin{cases} (C \bullet X)_0 = 0 \\ (C \bullet X)_n = \sum_{k=1}^n C_k \cdot \{X_k - X_{k-1}\} & \text{for } n \in \mathbb{N}. \end{cases}$$

Observation 5.17. Note that C_0 plays no role in the definition of $C \bullet X$. Note also that C_k, X_k and X_{k-1} for $k = 1, \ldots, n$, are all $\mathfrak{F}_n$-measurable so that $C \bullet X$ is an adapted process. If C is also a bounded process then $\sum_{k=1}^{n} C_k \cdot \{X_k - X_{k-1}\} \in L_1(\Omega, \mathfrak{F}_n, P)$ since X is an L_1-process so that $C \bullet X$ is an L_1-process. If both C and X are L_2-processes then $\sum_{i=1}^{n} C_k \cdot \{X_k - X_{k-1}\} \in L_1(\Omega, \mathfrak{F}_n, P)$ by Hölder's Inequality so that $C \bullet X$ is an L_1-process.

Theorem 5.18. (Martingale Transform) *Let* $(\Omega, \mathfrak{F}, \{\mathfrak{F}_n : n \in \mathbb{Z}_+\}, P)$ *be a filtered space.*
1) If C is a bounded nonnegative predictable process and X is a martingale, a submartingale or a supermartingale on the filtered space then the martingale transform $C \bullet X$ is respectively a null at 0 martingale, submartingale or supermartingale.
2) If C is a bounded predictable process and X is a martingale then $C \bullet X$ is a null at 0 martingale.
3) In 1) and 2) the boundedness condition on C can be replaced by the condition that both C and X are L_2-processes.

Proof. As we noted in Observation 5.17, under the hypothesis in any one of 1), 2) and 3), $C \bullet X$ is an adapted L_1-process. Note also that for any $n \in \mathbb{N}$ we have

(1) $$\begin{aligned} E[(C \bullet X)_n | \mathfrak{F}_{n-1}] &= E[\sum_{k=1}^{n-1} C_k \cdot \{X_k - X_{k-1}\} | \mathfrak{F}_{n-1}] \\ &\quad + E[C_n \cdot \{X_n - X_{n-1}\} | \mathfrak{F}_{n-1}] \end{aligned}$$

$$= \sum_{k=1}^{n-1} C_k \cdot \{X_k - X_{k-1}\} + C_n \cdot \{\mathbf{E}(X_n|\mathfrak{F}_{n-1}) - X_{n-1}\}$$
$$= (C \bullet X)_{n-1} + C_n\{\mathbf{E}(X_n|\mathfrak{F}_{n-1}) - X_{n-1}\}$$

a.e. on $(\Omega, \mathfrak{F}_{n-1}, P)$ by the $\mathfrak{F}_{n-1}$-measurability of C_n.

1) Now suppose C is bounded and nonnegative. According as X is a submartingale, a martingale or a supermartingale, we have $\mathbf{E}(X_n|\mathfrak{F}_{n-1}) \geq, =, \leq X_{n-1}$ a.e. on $(\Omega, \mathfrak{F}_{n-1}, P)$. Then since $C_n \geq 0$ a.e. on $(\Omega, \mathfrak{F}_{n-1}, P)$, we have $C_n \cdot \{\mathbf{E}(X_n|\mathfrak{F}_{n-1}) - X_{n-1}\} \geq, =, \leq 0$ a.e. on $(\Omega, \mathfrak{F}_{n-1}, P)$. Using this in (1), we have $\mathbf{E}[(C \bullet X)_n|\mathfrak{F}_{n-1}] \geq, =, \leq (C \bullet X)_{n-1}$ a.e. on $(\Omega, \mathfrak{F}_{n-1}, P)$, that is, $C \bullet X$ is a submartingale, a martingale or a supermartingale respectively.

2) Suppose C is bounded and X is a martingale. Then $\mathbf{E}(X_n|\mathfrak{F}_{n-1}) = X_{n-1}$ a.e. on $(\Omega, \mathfrak{F}_{n-1}, P)$ and thus $C_n \cdot \{\mathbf{E}(X_n|\mathfrak{F}_{n-1}) - X_{n-1}\} = 0$ a.e. on $(\Omega, \mathfrak{F}_{n-1}, P)$. Using this in (1) we have $\mathbf{E}[(C \bullet X)_n|\mathfrak{F}_{n-1}] = (C \bullet X)_{n-1}$ a.e. on $(\Omega, \mathfrak{F}_{n-1}, P)$. This shows that $C \bullet X$ is a martingale. ∎

Definition 5.19. *Let T be a stopping time on a filtered space $(\Omega, \mathfrak{F}, \{\mathfrak{F}_n : n \in \mathbb{Z}_+\}, P)$. By the stopping process derived from T we mean the predictable process $C^{(T)} = \{C_n^{(T)} : n \in \mathbb{Z}_+\}$ where $C_n^{(T)}$ is defined by*

(1) $$C_n^{(T)} = \mathbf{1}_{\{n \leq T\}} \quad \text{for } n \in \mathbb{Z}_+,$$

in other words,

$$C_n^{(T)}(\omega) = \mathbf{1}_{\{n \leq T(\omega)\}}(\omega) = \begin{cases} 1 & \text{for } n \leq T(\omega) \\ 0 & \text{for } n > T(\omega). \end{cases}$$

The fact that $C^{(T)}$ is a predictable process follows from the fact that $\mathbf{1}_{\{n \leq T\}} = \mathbf{1}_{\{T < n\}^c}$ and $\{T < n\} = \{T \leq n - 1\} \in \mathfrak{F}_{n-1}$ so that $C_n^{(T)}$ is $\mathfrak{F}_{n-1}$-measurable for $n \in \mathbb{N}$. The following theorem relates a stopped martingale to the martingale transform by the stopping process.

Theorem 5.20. *Let $X = \{X_n : n \in \mathbb{Z}_+\}$ be a submartingale, a martingale or a supermartingale and let T be a stopping time on a filtered space $(\Omega, \mathfrak{F}, \{\mathfrak{F}_n : n \in \mathbb{Z}_+\}, P)$. Then $X^{T \wedge} = X_0 + C^{(T)} \bullet X$.*

Proof. By Definition 5.16 and Definition 5.19, $(C^{(T)} \bullet X)_0 = 0$ and for $n \in \mathbb{N}$ we have

§5. MARTINGALE, SUBMARTINGALE AND SUPERMARTINGALE

$$(C^{(T)} \bullet X)_n = \sum_{k=1}^{n} \mathbf{1}_{\{k \leq T\}} \cdot \{X_k - X_{k-1}\}$$
$$= \mathbf{1}_{\{1 \leq T\}} \cdot \{X_1 - X_0\} + \mathbf{1}_{\{2 \leq T\}} \cdot \{X_2 - X_1\} + \cdots + \mathbf{1}_{\{n \leq T\}} \cdot \{X_n - X_{n-1}\}$$
$$= -X_0 \cdot \mathbf{1}_{\{1 \leq T\}} + \sum_{k=1}^{n-1} X_k \cdot \{\mathbf{1}_{\{k \leq T\}} - \mathbf{1}_{\{k+1 \leq T\}}\} + X_n \cdot \mathbf{1}_{\{n \leq T\}}.$$

Since

$$X_0 \cdot \mathbf{1}_{\{1 \leq T\}} = X_0 \cdot \{\mathbf{1}_{\{0 \leq T\}} - \mathbf{1}_{\{T=0\}}\} = X_0 \cdot \{\mathbf{1}_{\Omega} - \mathbf{1}_{\{T=0\}}\} = X_0 - X_0 \cdot \mathbf{1}_{\{T=0\}}$$

and

$$\mathbf{1}_{\{k \leq T\}} - \mathbf{1}_{\{k+1 \leq T\}} = \mathbf{1}_{\{T=k\}} \quad \text{for } k = 1, \ldots, n-1,$$

we have

$$(C^{(T)} \bullet X)_n = -X_0 + \sum_{k=0}^{n-1} X_k \cdot \mathbf{1}_{\{T=k\}} + X_n \cdot \mathbf{1}_{\{T \geq n\}}$$
$$= -X_0 + X_{T \wedge n} = -X_0 + X_n^{T \wedge},$$

that is,

$$X_n^{T \wedge} = X_0 + (C^{(T)} \bullet X)_n \quad \text{for } n \in \mathbb{N}.$$

For $n = 0$ we have

$$X_0^{T \wedge} = X_{T \wedge 0} = X_0 = X_0 + (C^{(T)} \bullet X)_0.$$

This completes the proof. ∎

Theorem 5.21. *Let $X = \{X_n : n \in \mathbb{Z}_+\}$ be a submartingale, a martingale or a supermartingale and let T be a stopping time on a filtered spare $(\Omega, \mathfrak{F}, \{\mathfrak{F}_n : \in \mathbb{Z}_+\}, P)$. Then the stopped process $X^{T \wedge} = \{X_{T \wedge n} : n \in \mathbb{Z}_+\}$ is a submartingale, martingale or supermartingale respectively and in particular $\mathbf{E}(X_{T \wedge n}) \geq, =, \leq \mathbf{E}(X_0)$ respectively for $n \in \mathbb{Z}_+$.*

Proof. Since our $X = \{X_n : n \in \mathbb{Z}_+\}$ is an L_1-process, so is $X^{T \wedge}$ by Remark 3.31. By Theorem 5.20 we have $X^{T \wedge} = X_0 + (C^{(T)} \bullet X)$. Since $C^{(T)}$ is a bounded nonnegative predictable process, according to Theorem 5.18, $C^{(T)} \bullet X$ is a null at 0 submartingale, martingale or supermartingale according as X is a submartingale, a martingale or a supermartingale. Therefore $X^{T \wedge}$ is a submartingale, a martingale or a supermartingale respectively. ∎

[V] Some Examples

Example 5.22. Consider the probability space $(\Omega, \mathfrak{F}, P) = ((0,1], \mathfrak{B}_{(0,1]}, m_L)$ where m_L is the Lebesgue measure on the Borel σ-algebra $\mathfrak{B}_{(0,1]}$ of subsets of $(0,1]$. For each $n \in \mathbb{Z}_+$, let $\mathfrak{I}_n = \{I_{n,k} : k = 1, 2, \ldots, 2^n\}$ where $I_{n,k} = ((k-1)2^{-n}, k2^{-n}]$ for $k = 1, 2, \ldots, 2^n$. Let $\mathfrak{F}_n = \sigma(\mathfrak{I}_n)$. Now $\mathfrak{I}_n$ consist of 2^n disjoint intervals and $\mathfrak{F}_n$ has $2^{2^n} = 4^n$ members. Also, $\mathfrak{F}_n \uparrow$ as $n \to \infty$ and $\mathfrak{F}_\infty \equiv \sigma(\cup_{n \in \mathbb{Z}_+} \mathfrak{F}_n) = \mathfrak{B}_{(0,1]}$. Thus we have filtered space $((0,1], \mathfrak{B}_{(0,1]}, \{\mathfrak{F}_n\}, m_L)$. Let f be an arbitrary extended real valued integrable function on the measure space $((0,1], \mathfrak{B}_{(0,1]}, m_L)$. For each $n \in \mathbb{Z}_+$ define a real valued function M_n on $(0,1]$ by

$$(1) \qquad M_n(\omega) = \frac{1}{m_L(I_{n,k})} \int_{I_{n,k}} f(\omega) m_L(d\omega) \quad \text{for } \omega \in I_{n,k} \text{ for } k = 1, 2, \ldots, 2^n,$$

that is, M_n is an averaging of f on the members of $\mathfrak{I}_n$. Now since M_n is constant on each member of $\mathfrak{I}_n$, M_n is $\mathfrak{F}_n$-measurable. From (1) we also have

$$(2) \qquad \int_{I_{n,k}} M_n \, dm_L = \int_{I_{n,k}} f \, dm_L \quad \text{for } k = 1, 2, \ldots, 2^n.$$

Since $(0,1]$ is the union of the finitely many disjoint members $I_{n,k}$ of $\mathfrak{I}_n$, (2) implies

$$\int_{(0,1]} M_n \, dm_L = \int_{(0,1]} f \, dm_L \in \mathbb{R},$$

that is, M_n is integrable. Therefore $M = \{M_n : n \in \mathbb{Z}_+\}$ is an adapted L_1-process on the filtered space $((0,1], \mathfrak{B}_{(0,1]}, \{\mathfrak{F}_n\}, m_L)$.

Let us show that for every $n \in \mathbb{Z}_+$, M_n is a version of $\mathbf{E}(f|\mathfrak{F}_n)$, that is, $\mathbf{E}(f|\mathfrak{F}_n) = M_n$ a.e. on $((0,1], \mathfrak{F}_n, m_L)$. Now since M_n is $\mathfrak{F}_n$-measurable we need only verify that

$$(3) \qquad \int_E M_n \, dm_L = \int_E f \, dm_L \quad \text{for every } E \in \mathfrak{F}_n.$$

But every $E \in \mathfrak{F}_n$ is the union of finitely many disjoint members of $\mathfrak{I}_n$. Thus (3) follows from (2). Therefore we have shown that $\mathbf{E}(f|\mathfrak{F}_n) = M_n$ a.e. on $((0,1], \mathfrak{F}_n, m_L)$ for every $n \in \mathbb{Z}_+$. This implies that $M = \{M_n : n \in \mathbb{Z}_+\}$ is a uniformly integrable martingale.

We show in §8 that if ξ is an integrable random variable on a probability space $(\Omega, \mathfrak{F}, P)$ and if we define a martingale $X = \{X_t : t \in \mathbb{T}\}$ on a filtered space $(\Omega, \mathfrak{F}, \{\mathfrak{F}_t : t \in \mathbb{T}\}, P)$ by letting X_t be an arbitrary real valued version of $\mathbf{E}(\xi|\mathfrak{F}_t)$ for every $t \in \mathbb{T}$, then we have $\lim_{t \to \infty} X_t = \mathbf{E}(\xi|\mathfrak{F}_\infty)$ a.e. on $(\Omega, \mathfrak{F}_\infty, P)$. (See Remark 8.3). For our particular example

§5. MARTINGALE, SUBMARTINGALE AND SUPERMARTINGALE

here let us show that if f is continuous at $\omega_0 \in (0, 1]$ then $\lim_{n\to\infty} M_n(\omega_0) = f(\omega_0)$. Now the continuity of f at ω_0 implies that for every $\varepsilon < 0$ there exists $\delta > 0$ such that

(4) $\quad\quad |f(\omega) - f(\omega_0)| < \varepsilon \quad \text{for } \omega \in (\omega_0 - \delta, \omega_0 + \delta) \cap (0, 1].$

For every $n \in \mathbb{Z}_+$, from the disjointness of the members of $\mathfrak{I}_n$ there exists a unique $k(n) \in \mathbb{N}$ such that $\omega_0 \in I_{n,k(n)}$. For the sequence of intervals $\{I_{n,k(n)} : n \in \mathbb{Z}_+\}$ we have $I_{n,k(n)} \downarrow$ as $n \to \infty$ and $\cap_{n\in\mathbb{Z}_+} I_{n,k(n)} = \{\omega_0\}$. Since $m_L(I_{n,k(n)}) = 1/2^n \downarrow 0$ as $n \to \infty$ and since $\omega_0 \in I_{n,k(n)}$ for every $n \in \mathbb{Z}_+$, there exists $N \in \mathbb{Z}_+$ such that

(5) $\quad\quad I_{n,k(n)} \subset (\omega_0 - \delta, \omega_0 + \delta) \cap (0, 1] \quad \text{for } n \geq N.$

By (1)

(6) $\quad\quad M_n(\omega_0) = \dfrac{1}{m_L(I_{n,k(n)})} \int_{I_{n,k(n)}} f \, dm_L \quad \text{for } n \in \mathbb{Z}_+.$

Using (5) and (4) in (6) we have

$$f(\omega_0) - \varepsilon \leq M_n(\omega_0) \leq f(\omega_0) + \varepsilon \quad \text{for } n \geq N.$$

This shows that $\lim_{n\to\infty} M_n(\omega_0) = f(\omega_0)$.

Example 5.23. (Sums of Independent Random Variables with Mean 0)
Consider a sequence of independent random variables $X = \{X_n : n \in \mathbb{Z}_+\}$ on a probability space $(\Omega, \mathfrak{F}, P)$ with $\mathbf{E}(X_n) = 0$ for $n \in \mathbb{Z}_+$. Let

$$S_n = X_0 + \cdots + X_n \quad \text{and} \quad \mathfrak{F}_n = \sigma\{X_0, \ldots, X_n\} \quad \text{for } n \in \mathbb{Z}_+.$$

If we let $S = \{S_n : n \in \mathbb{Z}_+\}$ then S is an adapted L_1-process on the filtered space $(\Omega, \mathfrak{F}, \{\mathfrak{F}_n\}, P)$. To show that S is a martingale we verify that $\mathbf{E}(S_n|\mathfrak{F}_{n-1}) = S_{n-1}$ a.e. on $(\Omega, \mathfrak{F}_{n-1}, P)$ for every $n \in \mathbb{N}$. Now the independence of the system of random variables $\{X_0, \ldots, X_n\}$ implies the independence of the system $\{(X_0, \ldots, X_{n-1}), X_n\}$ of two random vectors, or equivalently, the independence of $\sigma\{(X_0, \ldots, X_{n-1})\}$ and X_n. Since $\sigma\{(X_0, \ldots, X_{n-1})\} = \sigma\{X_0, \ldots, X_{n-1}\} = \mathfrak{F}_{n-1}$ we have the independence of $\mathfrak{F}_{n-1}$ and X_n and this implies $\mathbf{E}(X_n|\mathfrak{F}_{n-1}) = \mathbf{E}(X_n) = 0$ a.e. on $(\Omega, \mathfrak{F}_{n-1}, P)$. Then

$$\mathbf{E}(S_n|\mathfrak{F}_{n-1}) = \mathbf{E}(S_{n-1} + X_n|\mathfrak{F}_{n-1}) = \mathbf{E}(S_{n-1}|\mathfrak{F}_{n-1}) + \mathbf{E}(X_n|\mathfrak{F}_{n-1}) = S_{n-1}$$

a.e. on $(\Omega, \mathfrak{F}_{n-1}, P)$. Therefore S is a martingale.

Example 5.24. (Products of Nonnegative Independent Random Variables with Mean 1)
Let $X = \{X_n : n \in \mathbb{Z}_+\}$ be a sequence of independent nonnegative random variables on a probability space $(\Omega, \mathfrak{F}, P)$ with $\mathbf{E}(X_n) = 1$ for $n \in \mathbb{Z}_+$. Let

$$M_n = X_0 \cdots X_n \quad \text{and} \quad \mathfrak{F}_n = \sigma\{X_0, \ldots, X_n\} \quad \text{for } n \in \mathbb{Z}_+.$$

Since $X_0, \ldots, X_n$ are nonnegative we have $\mathbf{E}(M_n) = \mathbf{E}(X_0) \cdots \mathbf{E}(X_n) = 1$ by the independence and by the Tonelli Theorem. Thus $M = \{M_n : n \in \mathbb{Z}_+\}$ is an adapted L_1-process on the filtered space $(\Omega, \mathfrak{F}, \{\mathfrak{F}_n\}, P)$. As we noted in Example 5.23, $\mathfrak{F}_{n-1}$ and X_n are independent and this implies $\mathbf{E}(X_n | \mathfrak{F}_{n-1}) = \mathbf{E}(X_n) = 1$ a.e. on $(\Omega, \mathfrak{F}_{n-1}, P)$ for $n \in \mathbb{N}$. Thus we have

$$\mathbf{E}(M_n | \mathfrak{F}_{n-1}) = \mathbf{E}(M_{n-1} X_n | \mathfrak{F}_{n-1}) = M_{n-1} \mathbf{E}(X_n | \mathfrak{F}_{n-1}) = M_{n-1}$$

a.e. on $(\Omega, \mathfrak{F}_{n-1}, P)$. This shows that M is a martingale.

Example 5.25. (Processes with Independent Increments with Mean 0)
Let $X = \{X_t : t \in \mathbb{R}_+\}$ be an L_1-process with independent increments on a probability space $(\Omega, \mathfrak{F}, P)$ with $\mathbf{E}(X_t - X_s) = 0$ for $s, t \in \mathbb{R}_+$ such that $s < t$. By independence of increments we mean that for every finite strictly increasing sequence $t_1 < \cdots < t_n$ in $\mathbb{R}_+$ the system of random variables $\{X_{t_1}, X_{t_2} - X_{t_1}, X_{t_3} - X_{t_2}, \ldots, X_{t_n} - X_{t_{n-1}}\}$ is an independent one. If we let $\mathfrak{F}_t^X = \sigma\{X_s : s \in [0, t]\}$ for $t \in \mathbb{R}_+$ then X is an adapted L_1-process on the filtered space $(\Omega, \mathfrak{F}, \{\mathfrak{F}_t^X\}, P)$. The independence of increments implies that for $s, t \in \mathbb{R}_+$, $s < t$, we have the independence of $\mathfrak{F}_s^X$ and $X_t - X_s$. (See Theorem 13.10.) Thus $\mathbf{E}(X_t - X_s | \mathfrak{F}_s^X) = \mathbf{E}(X_t - X_s) = 0$. Then

$$\mathbf{E}(X_t) = \mathbf{E}(X_t - X_s | \mathfrak{F}_s^X) + \mathbf{E}(X_s | \mathfrak{F}_s^X) = X_s \quad \text{a.e. on } (\Omega, \mathfrak{F}_s^X, P).$$

Therefore X is a martingale.

§6 Fundamental Submartingale Inequalities

A submartingale increases on average. From this monotonicity condition we derive some basic inequalities for estimating the behavior of sample functions of a submartingale. These inequalities are derived first for discrete time submartingales by means of truncation by stopping times and then extended to cover continuous time submartingales.

§6. FUNDAMENTAL SUBMARTINGALE INEQUALITIES

[I] Optional Stopping and Optional Sampling

In §3 we showed that if X is an adapted process and T is a stopping time on a filtered space then X_T is always a random variable when the time parameter is discrete and when the time parameter is continuous then X_T is a random variable if we assume that X is a measurable process. Let us consider the integrability of X_T.

Theorem 6.1. (Doob's Optional Stopping Theorem) *Let $X = \{X_n : n \in \mathbb{Z}_+\}$ be a submartingale and T be a stopping time on a filtered space $(\Omega, \mathfrak{F}, \{\mathfrak{F}_n\}, P)$. Assume that T is finite a.e. on $(\Omega, \mathfrak{F}_\infty, P)$. Then $\mathbf{E}(X_0) \leq \mathbf{E}(X_T) < \infty$ under each one of the following conditions.*
(a) *T is bounded, that is, there exists $m \in \mathbb{Z}_+$ such that $T(\omega) \leq m$ for $\omega \in \Omega$.*
(b) *X is bounded, that is, there exists $K \geq 0$ such that $|X_n(\omega)| \leq K$ for $(n,\omega) \in \mathbb{Z}_+ \times \Omega$.*
(c) *T is integrable and X has bounded increments, that is, there exists $L \geq 0$ such that $|X_n(\omega) - X_{n-1}(\omega)| \leq L$ for $(n,\omega) \in \mathbb{N} \times \Omega$.*
(d) *$X \leq 0$ on $\mathbb{R}_+ \times \Omega$.*
If X is a martingale, then under anyone of the conditions (a), (b), and (c), we have $\mathbf{E}(X_T) = \mathbf{E}(X_0)$.

Proof. If X is a submartingale then the stopped process $X^T = \{X_{T \wedge n} : n \in \mathbb{Z}_+\}$ is also a submartingale according to Theorem 5.21. Then $\{\mathbf{E}(X_{T \wedge n}) : n \in \mathbb{Z}_+\}$ is an increasing sequence in $\mathbb{R}$ which is bounded below by $\mathbf{E}(X_{T \wedge 0}) = \mathbf{E}(X_0)$.

Let us note that if $T(\omega) < \infty$ for some $\omega \in \Omega$, then there exists $N \in \mathbb{Z}_+$ such that $T(\omega) \wedge n = T(\omega)$ for $n \geq N$ so that

$$\lim_{n \to \infty} X_{T \wedge n}(\omega) = \lim_{n \to \infty} X(T(\omega) \wedge n, \omega) = X(T(\omega), \omega) = X_T(\omega).$$

Therefore if T is finite a.e. on $(\Omega, \mathfrak{F}_\infty, P)$ then

(1) $$\lim_{n \to \infty} X_{T \wedge n} = X_T \quad \text{a.e. on } (\Omega, \mathfrak{F}_\infty, P).$$

1) If we assume (a) then $T \wedge m = T$ so that $\mathbf{E}(X_0) \leq \mathbf{E}(X_{T \wedge m}) = \mathbf{E}(X_T)$ and then $\mathbf{E}(X_T) < \infty$. If X is a martingale then so is X^T by Theorem 5.21. This implies that $\mathbf{E}(X_{T \wedge m}) = \mathbf{E}(X_0)$ and therefore $\mathbf{E}(X_T) = \mathbf{E}(X_0)$.

2) Let us assume (b). Now the condition $|X_n(\omega)| \leq K$ for $(n,\omega) \in \mathbb{Z}_+ \times \Omega$ implies $|X_{T(\omega) \wedge n}(\omega)| \leq K$ for $(n,\omega) \in \mathbb{Z}_+ \times \Omega$. Then by (1), the Bounded Convergence Theorem is applicable and we have $\mathbf{E}(X_T) = \lim_{n \to \infty} \mathbf{E}(X_{T \wedge n}) \geq \mathbf{E}(X_0)$. Since $|\mathbf{E}(X_{T \wedge n})| \leq K$ for $n \in \mathbb{Z}_+$ we have $\mathbf{E}(X_0) \leq \mathbf{E}(X_T) \leq K$. If X is a martingale then X_T is a martingale so that $\mathbf{E}(X_{T \wedge n}) = \mathbf{E}(X_0)$ for $n \in \mathbb{Z}_+$. This implies $\mathbf{E}(X_T) = \mathbf{E}(X_0)$.

3) Assume (c). For every $(n,\omega) \in \mathbb{Z}_+ \times \Omega$ we have

$$X_{T\wedge n}(\omega) - X_0(\omega) = \sum_{k=1}^{T(\omega)\wedge n} \{X_k(\omega) - X_{k-1}(\omega)\}.$$

Then for every $n \in \mathbb{Z}_+$ we have

(2) $$|X_{T\wedge n} - X_0| \leq \sum_{k=1}^{T\wedge n} |X_k - X_{k-1}| \leq L \cdot (T \wedge n) \leq LT.$$

Now the integrability of T implies that T is finite a.e. on $(\Omega, \mathfrak{F}_\infty, P)$ so that (1) is applicable. Then we have $\lim_{n\to\infty}(X_{T\wedge n} - X_0) = X_T - X_0$ a.e. on $(\Omega, \mathfrak{F}_\infty, P)$. Thus by the Dominated Convergence Theorem, $X_T - X_0$ is integrable and $\mathbf{E}(X_T - X_0) = \lim_{n\to\infty} \mathbf{E}(X_{T\wedge n} - X_0) \geq 0$. This shows that X_T is integrable and $\mathbf{E}(X_0) \leq \mathbf{E}(X_T) < \infty$. If X is a martingale then $\mathbf{E}(X_T) = \mathbf{E}(X_0)$ by the same reason as in 2).

4) Assume (d). Then since $X(n,\omega) \leq 0$ for $(n,\omega) \in \mathbb{Z}_+ \times \Omega$ we have $X^T(n,\omega) = X(T(\omega) \wedge n, \omega) \leq 0$ for $(n,\omega) \in \mathbb{Z}_+ \times \Omega$. Also, since $T < \infty$ a.e. on $(\Omega, \mathfrak{F}_\infty, P)$, we have $X_T(\omega) = X(T(\omega), \omega) \leq 0$ for a.e. ω in $(\Omega, \mathfrak{F}_\infty, P)$. Then by (1) and by Fatou's Lemma for the limit superior of a sequence of nonpositive functions, we have the inequalities $0 \geq \mathbf{E}(X_T) \geq \limsup_{n\to\infty} \mathbf{E}(X_{T\wedge n}) \geq \mathbf{E}(X_0)$. ∎

Corollary 6.2. *Let $X = \{X_n : n \in \mathbb{Z}_+\}$ be a martingale with bounded increments, $C = \{C_n : n \in \mathbb{Z}_+\}$ be a bounded predictable process and T be an integrable stopping time on a filtered space $(\Omega, \mathfrak{F}, \{\mathfrak{F}_n\}, P)$. Then for the martingale transform $C \bullet X$ of X by C we have $\mathbf{E}[(C \bullet M)_T] = 0$.*

Proof. Let $K, L \geq 0$ be such that $|C_n| \leq K$ for $n \in \mathbb{Z}_+$ and $|X_n - X_{n-1}| \leq L$ for $n \in \mathbb{N}$. Since X is a martingale and C is a bounded predictable process, $C \bullet X$ is a null at 0 martingale by 2) Theorem 5.18. From Definition 5.16

$$|(C \bullet X)_n - (C \bullet X)_{n-1}| = |C_n(X_n - X_{n-1})| \leq KL \quad \text{for } n \in \mathbb{N}.$$

Thus (c) of Theorem 6.1 is satisfied and therefore $\mathbf{E}[(C \bullet X)_T] = \mathbf{E}[(C \bullet X)_0] = 0$. ∎

In the next theorem we extend part of Theorem 6.1 to bounded submartingales with continuous time. Extensions to unbounded, but uniformly integrable, submartingales will be given in §8.

§6. FUNDAMENTAL SUBMARTINGALE INEQUALITIES

Theorem 6.3. (Optional Stopping Theorem with Continuous Time) *Let $X = \{X_t : t \in \mathbb{R}_+\}$ be a right-continuous submartingale and T be a stopping time on a right-continuous filtered space $(\Omega, \mathfrak{F}, \{\mathfrak{F}_t\}, P)$. Assume that T is finite a.e. on $(\Omega, \mathfrak{F}_\infty, P)$. If X is bounded, that is, there exists $K \geq 0$ such that $|X(t, \omega)| \leq K$ for $(t, \omega) \in \mathbb{R}_+ \times \Omega$, then $\mathbf{E}(X_0) \leq \mathbf{E}(X_T) < \infty$. If X is a bounded right-continuous martingale then $\mathbf{E}(X_T) = \mathbf{E}(X_0)$.*

Proof. For each $n \in \mathbb{N}$, let

$$\vartheta_n(t) = \begin{cases} k2^{-n} & \text{for } t \in [(k-1)2^{-n}, k2^{-n}) \text{ for } k \in \mathbb{N} \\ \infty & \text{for } t = \infty \end{cases}$$

and let $T_n = \vartheta_n \circ T$. According to Theorem 3.20, $\{T_n : n \in \mathbb{N}\}$ is a decreasing sequence of stopping times on $(\Omega, \mathfrak{F}, \{\mathfrak{F}_t\}, P)$ with T_n assuming values in $\{k2^{-n} : k \in \mathbb{Z}_+\} \cup \{\infty\}$ and $T_n \downarrow T$ uniformly on Ω as $n \to \infty$. Let $\Lambda = \{T = \infty\}$ and $\Lambda_n = \{T_n = \infty\}$ for $n \in \mathbb{N}$. By the definition of ϑ_n we have

$$\Lambda_n = T_n^{-1}(\{\infty\}) = T^{-1} \circ \vartheta_n^{-1}(\{\infty\}) = T^{-1}(\{\infty\}) = \Lambda.$$

Since $P(\Lambda) = 0$ we have $P(\Lambda_n) = 0$ for every $n \in \mathbb{N}$. Thus T_n is finite a.e. on $(\Omega, \mathfrak{F}_\infty, P)$ for every $n \in \mathbb{N}$. Now for each fixed $n \in \mathbb{N}$, consider the filtered space $(\Omega, \mathfrak{F}, \{\mathfrak{F}_{k2^{-n}} : k \in \mathbb{Z}_+\}, P)$. Then T_n is a stopping time on this filtered space as we noted in 2) of Remark 3.23. Since X is a submartingale on $(\Omega, \mathfrak{F}, \{\mathfrak{F}_t : t \in \mathbb{R}_+\}, P)$, $\{X_{k2^{-n}} : k \in \mathbb{Z}_+\}$ is a submartingale on our discrete time filtered space above. Thus by Theorem 6.1, we have $\mathbf{E}(X_0) \leq \mathbf{E}(X_{T_n}) < \infty$. Since $T_n \downarrow T$ on Ω as $n \to \infty$ and since X is right-continuous, we have

$$\lim_{n \to \infty} X_{T_n}(\omega) = \lim_{n \to \infty} X(T_n(\omega), \omega) = X_T(\omega) \quad \text{for } \omega \in \Lambda_n^c.$$

Since X is bounded by K, we have $|X_{T_n}| \leq K$ on Λ_n^c. Thus by the Bounded Convergence Theorem, we have $\lim_{n \to \infty} \mathbf{E}(X_{T_n}) = \mathbf{E}(X_T)$. Since $|\mathbf{E}(X_{T_n})| \leq K$, we have $|\mathbf{E}(X_T)| \leq K$. Thus $\mathbf{E}(X_0) \leq \mathbf{E}(X_T) < \infty$. When X is a martingale, we have $\mathbf{E}(X_{T_n}) = \mathbf{E}(X_0)$ for every $n \in \mathbb{N}$ by Theorem 6.1 and thus $\mathbf{E}(X_T) = \mathbf{E}(X_0)$. ∎

Theorem 6.4. (Doob's Optional Sampling Theorem with Bounded Stopping Times, Discrete Case) *Let $X = \{X_n : n \in \mathbb{Z}_+\}$ be a submartingale and let S and T be stopping times satisfying $S \leq T \leq m$ for some $m \in \mathbb{Z}_+$ on a filtered space $(\Omega, \mathfrak{F}, \{\mathfrak{F}_n\}, P)$. Then*

(1) $$\mathbf{E}(X_T | \mathfrak{F}_S) \geq X_S \quad \text{a.e. on } (\Omega, \mathfrak{F}_S, P),$$

and in particular we have

(2) $$\mathbf{E}(X_T) \geq \mathbf{E}(X_S)$$

and

(3) $$\mathbf{E}(X_m) \geq \mathbf{E}(X_T) \geq \mathbf{E}(X_0).$$

If X is a martingale, then equalities in (1), (2), and (3) hold.

Proof. Since S and T are stopping times, X_S and X_T are $\mathfrak{F}_S$- and $\mathfrak{F}_T$-measurable respectively as we noted following Definition 3.21. The boundedness of S and T implies the integrability of X_S and X_T by Theorem 6.1.

Let us define a stochastic process $D^{(S,T)} = \{D_n^{(S,T)} : n \in \mathbb{Z}_+\}$ on the filtered space by setting
$$D_n^{(S,T)} = \mathbf{1}_{\{S < n \leq T\}} = \mathbf{1}_{\{n \leq T\}} - \mathbf{1}_{\{n \leq S\}} \quad \text{for } n \in \mathbb{Z}_+.$$
Note that since $\{S < 0 \leq T\} = \emptyset$, we have $D_0^{(S,T)} = 0$. Thus defined, $D^{(S,T)}$ is a bounded nonnegative process on the filtered space. Note that for every $n \in \mathbb{N}$ we have $\{n \leq T\} = \{T < n\}^c = (\cup_{k=0}^{n-1}\{T = k\})^c \in \mathfrak{F}_{n-1}$ so that $\mathbf{1}_{\{n \leq T\}}$ is $\mathfrak{F}_{n-1}$-measurable and similarly $\mathbf{1}_{\{n \leq S\}}$ is $\mathfrak{F}_{n-1}$-measurable. Thus $D_n^{(S,T)}$ is $\mathfrak{F}_{n-1}$-measurable for every $n \in \mathbb{N}$. This shows that $D^{(S,T)}$ is a predictable process. Since $D^{(S,T)}$ is a bounded nonnegative predictable process the martingale transform $D^{(S,T)} \bullet X$ of X by $D^{(S,T)}$ is a null at zero submartingale by Theorem 5.18. Now by Definition 5.16 for $n \in \mathbb{N}$ we have

$$\begin{aligned}(D^{(S,T)} \bullet X)_n &= \sum_{k=1}^{n} \mathbf{1}_{\{S < k \leq T\}}(X_k - X_{k-1}) \\ &= \sum_{k=1}^{n} \mathbf{1}_{\{k \leq T\}}(X_k - X_{k-1}) - \sum_{k=1}^{n} \mathbf{1}_{\{k \leq S\}}(X_k - X_{k-1}) \\ &= (X_{T \wedge n} - X_0) - (X_{S \wedge n} - X_0) \\ &= X_{T \wedge n} - X_{S \wedge n}\end{aligned}$$

by the computation made in the proof of Theorem 5.20. Since $S \leq T \leq m$ we have

$$(D^{(S,T)} \bullet X)_m = X_T - X_S.$$

Since $D^{(S,T)} \bullet X$ is a submartingale, we have

$$\mathbf{E}[(D^{(S,T)} \bullet X)_n] \geq \mathbf{E}[(D^{(S,T)} \bullet X)_0] = \mathbf{E}(0) = 0$$

and thus

$$\mathbf{E}(X_T - X_S) \geq 0,$$

§6. FUNDAMENTAL SUBMARTINGALE INEQUALITIES

proving (2).

To prove (1), note that since $\mathbf{E}(X_T|\mathfrak{F}_S)$ and X_S are both $\mathfrak{F}_S$-measurable it suffices to show

(4) $$\int_A X_T \, dP \geq \int_A X_S \, dP \quad \text{for } A \in \mathfrak{F}_S.$$

To prove (4), for each $A \in \mathfrak{F}_S$ define two random variables S_A and T_A on $(\Omega, \mathfrak{F}, P)$ by setting $S_A = S$ on A and $S_A = m$ on A^c and similarly $T_A = T$ on A and $T_A = m$ on A^c. Then S_A and T_A are stopping times. Indeed since $A \in \mathfrak{F}_S$ we have

$$\{S_A \leq n\} = \begin{cases} \Omega \in \mathfrak{F}_n & \text{for } n \geq m \\ \{S \leq n\} \cap A \in \mathfrak{F}_n & \text{for } n < m \end{cases}$$

and similarly for T_A since $A \in \mathfrak{F}_S \subset \mathfrak{F}_T$. (One can also argue that since S_A is $\mathfrak{F}_S$-measurable and $S_A \geq S$, S_A is a stopping time by a discrete time version of Theorem 3.6. Similarly for T_A.) Now that S_A and T_A are stopping times and $S_A \leq T_A \leq m$, (2) is applicable and therefore

(5) $$\mathbf{E}(X_{T_A}) \geq \mathbf{E}(X_{S_A}).$$

But $\mathbf{E}(X_{T_A}) = \int_A X_T \, dP + \int_{A^c} X_m \, dP$ and similarly $\mathbf{E}(X_{S_A}) = \int_A X_S \, dP + \int_{A^c} X_m \, dP$. By these two equations and (5) we have $\int_A X_T \, dP \geq \int_A X_S \, dP$. This proves (4).

To derive (3), note that since 0, T and m are bounded stopping times and $0 \leq T \leq m$, (3) is implied by (2).

If X is a supermartingale, then $-X$ is a submartingale so that the inequalities (1), (2), and (3) are reversed for a supermartingale. Then since a martingale is both a submartingale and a supermartingale, the equalities in (1), (2), and (3) hold. ∎

Optional sampling theorems with unbounded stopping times for uniformly integrable submartingales in both discrete and continuous time will be proved in §8.

Corollary 6.5. *Let T be a stopping time on a filtered space $(\Omega, \mathfrak{F}, \{\mathfrak{F}_n : n \in \mathbb{Z}_+\}, P)$ satisfying a boundedness condition $T \leq m$ for some $m \in \mathbb{Z}_+$.*
1) *If $X = \{X_n : n \in \mathbb{Z}_+\}$ is a submartingale on the filtered space, then*

(1) $$\mathbf{E}(|X_T|) \leq -\mathbf{E}(X_0) + 2\mathbf{E}(X_m^+) \leq 3 \sup_{n=0,\cdots,m} \mathbf{E}(|X_n|).$$

2) *If $X = \{X_n : n \in \mathbb{Z}_+\}$ is a supermartingale on the filtered space, then*

(2) $$\mathbf{E}(|X_T|) \leq \mathbf{E}(X_0) + 2\mathbf{E}(X_m^-) \leq 3 \sup_{n=0,\cdots,m} \mathbf{E}(|X_n|).$$

Proof. 1) Suppose X is a submartingale. Since $X_T = X_T^+ - X_T^-$ and $|X_T| = X_T^+ + X_T^-$, we have $|X_T| + X_T = 2X_T^+$. Thus

(3) $$\mathbf{E}(|X_T|) = -\mathbf{E}(X_T) + 2\mathbf{E}(X_T^+).$$

Since X is a submartingale, X^+ is a submartingale by Corollary 5.10. Applying (3) of Theorem 6.4 to X and X^+ we have $-\mathbf{E}(X_T) \leq -\mathbf{E}(X_0)$ and $\mathbf{E}(X_T^+) \leq \mathbf{E}(X_m^+)$. Using these in (3) we have (1).

2) When X is a supermartingale, $-X$ is a submartingale. Applying (1) to $-X$ we have $\mathbf{E}(|-X_T|) \leq -\mathbf{E}(-X_0) + 2\mathbf{E}((-X_m)^+)$, that is, $\mathbf{E}(|X_T|) \leq \mathbf{E}(X_0) + 2\mathbf{E}(X_m^-)$. ∎

Corollary 6.6. *Let $X = \{X_n : n \in \mathbb{Z}_+\}$ be an adapted L_1-process on a filtered space $(\Omega, \mathfrak{F}, \{\mathfrak{F}_n\}, P)$. Then X is a submartingale if and only if for any two bounded stopping times S and T such that $S \leq T$, we have $\mathbf{E}(X_T) \geq \mathbf{E}(X_S)$. In particular, X is a martingale if and only if $\mathbf{E}(X_T) = \mathbf{E}(X_S)$.*

Proof. The necessity of the condition is by (2) of Theorem 6.4. Let us prove the sufficiency of the condition. Suppose that $\mathbf{E}(X_T) \geq \mathbf{E}(X_S)$ for any two bounded stopping times S and T such that $S \leq T$. Since X is an adapted L_1-process, to show that it is a submartingale it suffices according to Observation 5.5 to show that for any $n, m \in \mathbb{Z}_+, n < m$, we have

(1) $$\int_E X_m \, dP \geq \int_E X_n \, dP \quad \text{for every } E \in \mathfrak{F}_n.$$

Now with $E \in \mathfrak{F}_n$ given, let us define two random variables S and T on $(\Omega, \mathfrak{F}, P)$ by setting $S = n$ on E, $S = m$ on E^c and $T = m$ on Ω. Then S is a stopping time since for every $k \in \mathbb{Z}_+$ we have

$$\{S \leq k\} = \begin{cases} \emptyset \in \mathfrak{F}_k & \text{for } k < n \\ E \in \mathfrak{F}_n \subset \mathfrak{F}_k & \text{for } n \leq k < m \\ \Omega \in \mathfrak{F}_k & \text{for } m \leq k. \end{cases}$$

T is trivially a stopping time. Thus we have two bounded stopping times S and T with $S \leq T$. Therefore by our assumption we have $\mathbf{E}(X_T) \geq \mathbf{E}(X_S)$. But

$$\mathbf{E}(X_T) = \int_E X_m \, dP + \int_{E^c} X_m \, dP$$

and

$$\mathbf{E}(X_S) = \int_E X_n \, dP + \int_{E^c} X_m \, dP.$$

Therefore (1) holds and X is a submartingale.

§6. FUNDAMENTAL SUBMARTINGALE INEQUALITIES

If X is a supermartingale, then $-X$ is a submartingale. Therefore by our results above, X is a supermartingale if and only if $\mathbf{E}(X_T) \leq \mathbf{E}(X_S)$ for any two bounded stopping times S and T such that $S \leq T$. Since a martingale is both a submartingale and a supermartingale, X is a martingale if and only if $\mathbf{E}(X_T) = \mathbf{E}(X_S)$ for any two bounded stopping times S and T such that $S \leq T$. ∎

[II] Maximal and Minimal Inequalities

For the sample path of a submartingale $X = \{X_n : n \in \mathbb{Z}_+\}$, Doob's maximal inequality gives an estimate of the probability that the maximum of $\{X_0(\omega), \cdots, X_m(\omega)\}$ exceeds a positive number λ in terms of the expectation of X_m.

Theorem 6.7. (Doob's Maximal and Minimal Inequalities, Finite Case) *Let* $X = \{X_n : n \in \mathbb{Z}_+\}$ *be a submartingale on a filtered space* $(\Omega, \mathfrak{F}, \{\mathfrak{F}_n\}, P)$. *Then for every* $m \in \mathbb{Z}_+$ *and* $\lambda > 0$ *we have*

(1) $$\lambda P\{\max_{n=0,\cdots,m} X_n \geq \lambda\} \leq \int_{\{\max_{n=0,\cdots,m} X_n \geq \lambda\}} X_m \, dP \leq \mathbf{E}(X_m^+)$$

and

(2) $$\lambda P\{\min_{n=0,\cdots,m} X_n \leq -\lambda\} \leq \int_{\{\min_{n=0,\cdots,m} X_n > -\lambda\}} X_m \, dP - \mathbf{E}(X_0) \leq \mathbf{E}(X_m^+) - \mathbf{E}(X_0).$$

Proof. 1) To prove (1), define a function T on Ω by setting

(3) $$T(\omega) = \begin{cases} \min\{n = 0, \cdots, m : X_n(\omega) \geq \lambda\}, \\ m \quad \text{if the set above is } \emptyset. \end{cases}$$

As we noted in Remark 3.23, to show that T is a stopping time it suffices to show that $\{T = n\} \in \mathfrak{F}_n$ for $n \in \mathbb{Z}_+$. For this observe that

$$\{T = n\} = \{X_0 < \lambda, \cdots, X_{n-1} < \lambda, X_n \geq \lambda\} \in \mathfrak{F}_n \quad \text{for } n = 0, \cdots, m-1,$$

$$\{T = m\} = \{X_0 < \lambda, \cdots, X_{m-1} < \lambda, X_m \geq \lambda\} \cup \{X_0 < \lambda, \cdots, X_m < \lambda\} \in \mathfrak{F}_m,$$

and

$$\{T = n\} = \emptyset \in \mathfrak{F}_n \quad \text{for } n > m.$$

For brevity let $A = \{\max_{n=0,\cdots,m} X_n \geq \lambda\}$. Since X is a submartingale and T is a stopping time bounded by m we have by (3) of Theorem 6.4

(4) $$\mathbf{E}(X_m) \geq \mathbf{E}(X_T) = \int_A X_T\, dP + \int_{A^c} X_T\, dP.$$

If $A = \emptyset$, then (1) holds trivially. If $A \neq \emptyset$, then on this set we have $X_T \geq \lambda$ by (3). By (3), we also have $T = m$ on A^c. Thus by (4) we have

$$\mathbf{E}(X_m) \geq \lambda P(A) + \int_{A^c} X_m\, dP$$

and therefore

$$\lambda P(A) \leq \mathbf{E}(X_m) - \int_{A^c} X_m\, dP = \int_A X_m\, dP \leq \int_\Omega X_m^+\, dP.$$

This proves (1).

2) To prove (2), let

(5) $$S(\omega) = \begin{cases} \min\{n = 0, \cdots, m : X_n(\omega) \leq -\lambda\} \\ m \quad \text{if the set above is } \emptyset. \end{cases}$$

The fact that S is a stopping time can be verified as we did for T above. For brevity, let $B = \{\min_{n=0,\cdots,m} X_n \leq -\lambda\}$. By (3) of Theorem 6.4 we have

(6) $$\mathbf{E}(X_0) \leq \mathbf{E}(X_S) = \int_B X_S\, dP + \int_{B^c} X_S\, dP.$$

If $B = \emptyset$, then (2) holds trivially. If $B \neq \emptyset$, then on this set we have $X_S \leq -\lambda$ by (5). Also $S = m$ on B^c. Thus by (6) we have

$$\mathbf{E}(X_0) \leq -\lambda P(B) + \int_{B^c} X_m\, dP.$$

Therefore
$$\lambda P(B) \leq -\mathbf{E}(X_0) + \int_{B^c} X_m\, dP \leq -\mathbf{E}(X_0) + \int_\Omega X_m^+\, dP.$$

This proves (2). ∎

Corollary 6.8 *If $X = \{X_n : n \in \mathbb{Z}_+\}$ is a supermartingale on a filtered space $(\Omega, \mathfrak{F}, \{\mathfrak{F}_n\}, P)$ and if $m \in \mathbb{Z}_+$ and $\lambda > 0$ then*

(1) $$\lambda P\{\max_{n=0,\cdots,m} X_n \geq \lambda\} \leq \mathbf{E}(X_0) - \int_{\{\max_{n=0,\cdots,m} X_n < \lambda\}} X_m\, dP \leq \mathbf{E}(X_0) + \mathbf{E}(X_m^-)$$

§6. FUNDAMENTAL SUBMARTINGALE INEQUALITIES

and

(2) $\quad \lambda P\{\min_{n=0,\cdots,m} X_n \leq -\lambda\} \leq -\int_{\{\min_{n=0,\cdots,m} X_n \leq -\lambda\}} X_m \leq \mathbf{E}(X_m^-).$

Proof. (1) and (2) follow respectively from (2) and (1) of Theorem 6.7 applied to the submartingale $-X$. ∎

Corollary 6.9. *Let* $X = \{X_n : n \in \mathbb{Z}_+\}$ *be an* L_2-*martingale on a filtered space* $(\Omega, \mathfrak{F}, \{\mathfrak{F}_n\}, P)$. *Then for every* $m \in \mathbb{Z}_+$ *and* $\lambda > 0$

(1) $\quad\quad\quad\quad\quad\quad \lambda^2 P\{\max_{n=0,\cdots,m} |X_n| \geq \lambda\} \leq \mathbf{E}(X_m^2).$

Proof. Since X is an L_2-martingale, $X^2 = \{X_n^2 : n \in \mathbb{Z}_+\}$ is a submartingale by Corollary 5.12. Then by (1) of Theorem 6.7,

$$\lambda^2 P\{\max_{n=0,\cdots,m} X_n^2 \geq \lambda^2\} \leq \mathbf{E}(X_m^2)$$

From this (1) follows. ∎

Let us consider a nonnegative submartingale $X = \{X_n : n \in \mathbb{Z}_+\}$ on a filtered space $(\Omega, \mathfrak{F}, \{\mathfrak{F}_n\}, P)$. Then for every $m \in \mathbb{Z}_+$ and $\lambda > 0$ the two random variables $\xi \equiv \max_{n=0,\cdots,m} X_n$ and $\eta \equiv X_m$ are nonnegative and satisfy according to (1) of Theorem 6.7 the following inequality

$$\lambda P\{\xi \geq \lambda\} \leq \int_{\{\xi \geq \lambda\}} \eta \, dP \quad \text{for } \lambda > 0.$$

In the next lemma we show that if ξ and η are two arbitrary nonnegative random variables satisfying the inequality above and if $\xi \in L_p(\Omega, \mathfrak{F}, P)$ for some $p \in (1, \infty)$ then we have $\|\xi\|_p \leq q\|\eta\|_p$ where q is the conjugate exponent of p. This result will be used in proving the Doob-Kolmogorov Inequality for nonnegative L_p-submartingales.

Lemma 6.10. *Let* X *and* Y *be nonnegative random variables on a probability space* $(\Omega, \mathfrak{F}, P)$ *satisfying the condition*

(1) $\quad\quad\quad\quad \lambda P\{X \geq \lambda\} \leq \int_{\{X \geq \lambda\}} Y \, dP \quad \text{for every } \lambda > 0.$

If $X \in L_p(\Omega, \mathfrak{F}, P)$ for some $p \in (1, \infty)$ and $q \in (1, \infty)$ is its conjugate exponent, then

(2) $$\|X\|_p \leq q\|Y\|_p.$$

Proof. Multiplying both sides of (1) by $p\lambda^{p-2}$ we have

(3) $$p\lambda^{p-1} P\{X \geq \lambda\} \leq p\lambda^{p-2} \int_{\{X \geq \lambda\}} Y\, dP.$$

Since $P\{X \geq \lambda\}$ and $\int_{\{X \geq \lambda\}} Y\, dP$ are decreasing functions of $\lambda \in (0, \infty)$, they are Borel measurable functions on $(0, \infty)$. Let m_L be the Lebesgue measure on $((0,\infty), \mathfrak{B}_{(0,\infty)})$. Integrating both sides of (3) we have

(4) $$\int_{(0,\infty)} p\lambda^{p-1} P\{X \geq \lambda\}\, m_L(d\lambda) \leq \int_{(0,\infty)} p\lambda^{p-2} \left\{ \int_{\{X \geq \lambda\}} Y\, dP \right\} m_L(d\lambda).$$

To change the order of integration in the iterated integral in (4) we need to verify the measurability of the integrands as functions on the product measure space $((0, \infty) \times \Omega, \sigma(\mathfrak{B}_{(0,\infty)} \times \mathfrak{F}), m_L \times P)$. Let ι be the identity mapping of $(0, \infty)$ into $\overline{\mathbb{R}}$. Since X and Y are $\mathfrak{F}/\mathfrak{B}_{\overline{\mathbb{R}}}$-measurable mappings of Ω into $\overline{\mathbb{R}}$, all three mappings ι, X, and Y may be regarded as $\sigma(\mathfrak{B}_{(0,\infty)} \times \mathfrak{F})/\mathfrak{B}_{\overline{\mathbb{R}}}$-measurable mappings of $(0,\infty) \times \Omega$ into $\overline{\mathbb{R}}$. This implies that $\{X \geq \lambda\} \in \sigma(\mathfrak{B}_{(0,\infty)} \times \mathfrak{F})$ and thus $\mathbf{1}_{\{X \geq \lambda\}}$ is a $\sigma(\mathfrak{B}_{(0,\infty)} \times \mathfrak{F})/\mathfrak{B}_{\overline{\mathbb{R}}}$-measurable mapping of $(0, \infty) \times \Omega$ into $\overline{\mathbb{R}}$. Therefore $p\lambda^{p-1}\mathbf{1}_{\{X \geq \lambda\}}$ and $p\lambda^{p-2}\mathbf{1}_{\{X \geq \lambda\}}Y$ are $\sigma(\mathfrak{B}_{(0,\infty)} \times \mathfrak{F})/\mathfrak{B}_{\overline{\mathbb{R}}}$-measurable mappings of $(0, \infty) \times \Omega$ into $\overline{\mathbb{R}}$.

Now for the left side of (4), we have by the Tonelli Theorem

$$\begin{aligned}
&\int_{(0,\infty)} p\lambda^{p-1} P\{X \geq \lambda\}\, m_L(d\lambda) \\
&= \int_{(0,\infty)} p\lambda^{p-1} \left\{ \int_\Omega \mathbf{1}_{\{X \geq \lambda\}}(\omega)\, P(d\omega) \right\} m_L(d\lambda) \\
&= \int_\Omega \left\{ \int_{(0,\infty)} p\lambda^{p-1} \mathbf{1}_{\{X \geq \lambda\}}(\omega)\, m_L(d\lambda) \right\} P(d\omega) \\
&= \int_\Omega \left\{ \int_{(0, X(\omega)]} p\lambda^{p-1}\, m_L(d\lambda) \right\} P(d\omega) \\
&= \int_\Omega X(\omega)^p\, P(d\omega) \\
&= \|X\|_p^p.
\end{aligned}$$

§6. FUNDAMENTAL SUBMARTINGALE INEQUALITIES

Similarly for the right side of (4), we have

$$\int_{(0,\infty)} p\lambda^{p-2} \left\{ \int_{\{X \geq \lambda\}} Y \, dP \right\} m_L(d\lambda)$$

$$= \int_{(0,\infty)} p\lambda^{p-2} \left\{ \int_\Omega \mathbf{1}_{\{X \geq \lambda\}}(\omega) Y(\omega) \, P(d\omega) \right\} m_L(d\lambda)$$

$$= \int_\Omega \left\{ \int_{(0,X(\omega)]} p\lambda^{p-2} m_L(d\lambda) \right\} Y(\omega) P(d\omega)$$

$$= \int_\Omega q \, X(\omega)^{p-1} Y(\omega) P(d\omega) \quad \text{since } p(p-1)^{-1} = q$$

$$\leq q \|Y\|_p \|X^{p-1}\|_q \quad \text{by Hölder's Inequality}$$

$$= q \|Y\|_p \{ \int_\Omega (X^{p-1})^q dP \}^{\frac{1}{q}}$$

$$= q \|Y\|_p \{ \int_\Omega X^p dP \}^{\frac{1}{q}} \quad \text{since } (p-1)q = p$$

$$= q \|Y\|_p \|X\|_p^{p/q}.$$

Thus (4) becomes

(5) $$\|X\|_p^p \leq q \|Y\|_p \|X\|_p^{p/q}.$$

Since $\|X\|_p < \infty$ we have $\|X\|_p^{p/q} < \infty$. If $\|X\|_p = 0$ then (2) holds trivially. If $\|X\|_p \neq 0$ then by dividing both sides of (5) by the finite positive number $\|X\|_p^{p/q}$ and recalling $p - p/q = 1$ we have (2). ∎

Theorem 6.11 (Doob-Kolmogorov Inequality) *Let $X = \{X_n : n \in \mathbb{Z}_+\}$ be a nonnegative L_p-submartingale on a filtered space $(\Omega, \mathfrak{F}, \{\mathfrak{F}_n\}, P)$ for some $p \in (1, \infty)$ with conjugate exponent $q \in (1, \infty)$. Then for every $m \in \mathbb{Z}_+$ and $\lambda > 0$*

(1) $$\lambda^p P\{ \max_{n=0,\cdots,m} X_n \geq \lambda \} \leq \int_{\{\max_{n=0,\cdots,m} X_n \geq \lambda\}} X_m^p \, dP \leq \mathbf{E}(X_m^p)$$

and

(2) $$\mathbf{E}(\max_{n=0,\cdots,m} X_n^p) \leq q^p \mathbf{E}(X_m^p).$$

Proof. Since X is a nonnegative L_p-submartingale, X^p is a nonnegative submartingale by Corollary 5.12. By applying (1) of Theorem 6.7 to X^p we have

$$\lambda^p P\{ \max_{n=0,\cdots,m} X_n^p \geq \lambda^p \} \leq \int_{\{\max_{n=0,\cdots,m} X_n^p \geq \lambda^p\}} X_m^p \, dP \leq \mathbf{E}(X_m^p),$$

which is equivalent to (1).

Applying (1) of Theorem 6.7 to the submartingale X we have

$$\lambda P\{\max_{n=0,\cdots,m} X_n \geq \lambda\} \leq \int_{\{\max_{n=0,\cdots,n} X_n \geq \lambda\}} X_m \, dP.$$

Thus the two nonnegative random variables $\max_{n=0,\cdots,m} X_n$ and X_m satisfy the condition (1) of Lemma 6.10. Also the fact that $X_0, \cdots, X_n$ are in $L_p(\Omega, \mathfrak{F}, P)$ implies that $\max_{n=0,\cdots,n} X_n$ is in $L_p(\Omega, \mathfrak{F}, P)$. Therefore by Lemma 6.10

$$\mathbf{E}(\max_{n=0,\cdots,m} X_n^p) = \mathbf{E}[(\max_{n=0,\cdots,m} X_n)^p] \leq q\mathbf{E}(X_m^p)$$

proving (2). ∎

Corollary 6.12. *Let $X = \{X_n : n \in \mathbb{Z}_+\}$ be a L_p-martingale on a filtered space $(\Omega, \mathfrak{F}, \{\mathfrak{F}_n\}, P)$ for some $p \in (1, \infty)$. With the conjugate exponent q of p we have for any $m \in \mathbb{Z}_+$ and $\lambda > 0$*

(1) $$\lambda^p P\{\max_{n=0,\cdots,m} |X_n| \geq \lambda\} \leq \int_{\{\max_{n=0,\cdots,m} |X_n| \geq \lambda\}} |X_m|^p \, dP \leq \mathbf{E}(|X_m|^p)$$

and

(2) $$\mathbf{E}(\max_{n=0,\cdots,m} |X_n|^p) \leq q^p \mathbf{E}(|X_m|^p).$$

Proof. Since X is an L_p-martingale, $|X|$ is a nonnegative L_p-submartingale by Corollary 5.12. Applying (1) and (2) of Theorem 6.11 to $|X|$ we have (1) and (2). ∎

Theorem 6.7 (Maximal and Minimal Inequalities, Finite Case) gave estimates for the probabilities of the maximum and minimum of finitely many elements in a submartingale. Let us extend these results to estimate probabilities of the supremum and infimum of the entire submartingale for both the discrete and the continuous case.

Theorem 6.13. (Maximal and Minimal Inequalities, Discrete Case) *Let $X = \{X_n : n \in \mathbb{Z}_+\}$ be a submartingale on a filtered space $(\Omega, \mathfrak{F}, \{\mathfrak{F}_n\}, P)$. Then for every $\lambda > 0$*

(1) $$\lambda P\{\sup_{n \in \mathbb{Z}_+} X_n > \lambda\} \leq \sup_{n \in \mathbb{Z}_+} \mathbf{E}(X_n^+)$$

§6. FUNDAMENTAL SUBMARTINGALE INEQUALITIES

and

(2) $$\lambda P\{\inf_{n\in\mathbb{Z}_+} X_n < -\lambda\} \leq \sup_{n\in\mathbb{Z}_+} \mathbf{E}(X_n^+) - \mathbf{E}(X_0).$$

Proof. To prove (1), note that $\max_{k=0,\cdots,n} X_k \uparrow \sup_{n\in\mathbb{Z}_+} X_n$ on Ω as $n \to \infty$. Thus

$$\{\sup_{n\in\mathbb{Z}_+} X_n > \lambda\} \subset \bigcup_{n\in\mathbb{Z}_+} \{\max_{k=0,\cdots,n} X_k \geq \lambda\} = \uparrow \lim_{n\to\infty} \{\max_{k=0,\cdots,n} X_k \geq \lambda\}$$

and therefore

(3) $$P\{\sup_{n\in\mathbb{Z}_+} X_n > \lambda\} \leq \uparrow \lim_{n\to\infty} P\{\max_{k=0,\cdots,n} X_k \geq \lambda\}.$$

By (1) of Theorem 6.7

(4) $$\lambda P\{\max_{k=0,\cdots,n} X_k \geq \lambda\} \leq \mathbf{E}(X_n^+) \leq \sup_{n\in\mathbb{Z}_+} \mathbf{E}(X_n^+).$$

From (3) and (4) we have (1). Similarly (2) is derived from (2) of Theorem 6.7. ∎

Theorem 6.14. (Maximal and Minimal Inequalities, Continuous Case) *Let $X = \{X_t : t \in \mathbb{R}_+\}$ be a submartingale on a filtered space $(\Omega, \mathfrak{F}, \{\mathfrak{F}_t\}, P)$. Let S be a countable dense subset of $\mathbb{R}_+$ and $I = [\alpha, \beta) \subset \mathbb{R}_+$. Then for every $\lambda > 0$*

(1) $$\lambda P\{\sup_{t\in I\cap S} X_t > \lambda\} \leq \mathbf{E}(X_\beta^+)$$

and

(2) $$\lambda P\{\inf_{t\in I\cap S} X_t < -\lambda\} \leq \mathbf{E}(X_\beta^+) - \mathbf{E}(X_\alpha).$$

If X is right-continuous then $I \cap S$ in (1) and (2) can be replaced by I.

Proof. Let $\{s_n : n \in \mathbb{Z}_+\}$ be an arbitrary renumbering of the elements of $I \cap S$. For each $N \in \mathbb{Z}_+$, let $t_0, \cdots, t_N$ be the rearrangement of $s_0, \cdots, s_N$ in increasing order. Then $\{X_{t_0}, \cdots, X_{t_N}\}$ is a submartingale with respect to $\{\mathfrak{F}_{t_0}, \cdots, \mathfrak{F}_{t_N}\}$. Thus by (1) of Theorem 6.7 and by the fact that X^+ is a submartingale so that $\mathbf{E}(X_t^+) \uparrow$ as $t \uparrow$, we have

(3) $$\lambda P\{\max_{t\in\{s_0,\cdots,s_N\}} X_t \geq \lambda\} = \lambda P\{\max_{t\in\{t_0,\cdots,t_N\}} X_t \geq \lambda\} \leq \mathbf{E}(X_\beta^+).$$

Since $\max_{t\in\{s_0,\cdots,s_N\}} X_t \uparrow \sup_{t\in S\cap I} X_t$ as $N \to \infty$ we have

$$\{\sup_{t\in I\cap S} X_t > \lambda\} \subset \bigcup_{N\in\mathbb{Z}_+} \{\max_{t\in\{s_0,\cdots,s_N\}} X_t \geq \lambda\}.$$

From this follows

(4) $$P\{\sup_{t\in I\cap S} X_t > \lambda\} \leq \uparrow \lim_{N\to\infty} P\{\max_{t\in\{s_0,\cdots,s_N\}} X_t \geq \lambda\}$$

Combining (3) and (4) we have (1).

When X is right-continuous, the right-continuity of the function $X_t(\omega), t \in \mathbb{R}_+$, implies $\sup_{t\in I} X_t(\omega) = \sup_{t\in I\cap S} X_t(\omega)$ for every $\omega \in \Omega$. Thus $\{\sup_{t\in I} X_t > \lambda\} = \{\sup_{t\in I\cap S} X_t > \lambda\}$ and therefore $I \cap S$ in (1) can be replaced by I. Similarly (2) is proved by using (2) of Theorem 6.7. ∎

Next we extend Theorem 6.11 (Doob-Kolmogorov Inequality) to estimates of the probabilities of the supremum and infimum of nonnegative L_p-submartingales for both the discrete and the continuous case.

Theorem 6.15. *Let $X = \{X_n : n \in \mathbb{Z}_+\}$ be a nonnegative L_p-submartingale on a filtered space $(\Omega, \mathfrak{F}, \{\mathfrak{F}_n\}, P)$ for some $p \in (1,\infty)$ with conjugate exponent $q \in (1,\infty)$. Then for every $\lambda > 0$*

(1) $$\lambda^p P\{\sup_{n\in\mathbb{Z}_+} X_n > \lambda\} \leq \sup_{n\in\mathbb{Z}_+} \mathbf{E}(X_n^p)$$

and

(2) $$\mathbf{E}(\sup_{n\in\mathbb{Z}_+} X_n^p) \leq q^p \sup_{n\in\mathbb{Z}_+} \mathbf{E}(X_n^p).$$

Proof. The inequality (1) is derived from (1) of Theorem 6.11 in the same way as (1) of Theorem 6.13 was derived from (1) of Theorem 6.7.

To prove (2) recall that by (2) of Theorem 6.11, for every $n \in \mathbb{Z}_+$ we have

(3) $$\mathbf{E}(\max_{k=0,\cdots,n} X_k^p) \leq q^p \mathbf{E}(X_n^p) \leq q^p \sup_{n\in\mathbb{Z}_+} \mathbf{E}(X_n^p).$$

Then since $\max_{k=0,\cdots,n} X_k^p \uparrow \sup_{n\in\mathbb{Z}_+} X_n^p$ as $n \to \infty$, if we let $n \to \infty$ in (3) then by the Monotone Convergence Theorem we have (2). ∎

§6. FUNDAMENTAL SUBMARTINGALE INEQUALITIES

Theorem 6.16. *Let $\{X_t : t \in \mathbb{R}_+\}$ be a nonnegative L_p-submartingale on a filtered space $(\Omega, \mathfrak{F}, \{\mathfrak{F}_t\}, P)$ for some $p \in (1, \infty)$ with conjugate exponent $q \in (1, \infty)$. Let S be a countable dense subset of $\mathbb{R}_+$ and $I = [\alpha, \beta] \subset \mathbb{R}_+$. Then for every $\lambda > 0$*

(1) $$\lambda^p P\{\sup_{t \in I \cap S} X_t > \lambda\} \leq \mathbf{E}(X_\beta^p)$$

and

(2) $$\mathbf{E}(\sup_{t \in I \cap S} X_t^p) \leq q^p \mathbf{E}(X_\beta^p).$$

If X is right-continuous then $I \cap S$ in (1) and (2) can be replaced by I.

Proof. This theorem is derived from Theorem 6.11 in exactly the same way Theorem 6.15 is derived from Theorem 6.7. ∎

[III] Upcrossing and Downcrossing Inequalities

Let $[a, b] \subset \mathbb{R}$. The number of times a sample path of a stochastic process traverses the interval $[a, b]$ is a measurement of the oscillation of the sample path. The upcrossing and downcrossing numbers of a sample path are defined to count that number.

Definition 6.17. *Let $X = \{X_n : n \in \mathbb{Z}_+\}$ be a stochastic processes on a probability space $(\Omega, \mathfrak{F}, P)$. Let $a, b \in \mathbb{R}$, $a < b$.*
1) Let us define a sequence of $\overline{\mathbb{Z}}_+$-valued functions $\{T_j : j \in \mathbb{N}\}$ on Ω by

$$T_1(\omega) = \inf\{n \in \mathbb{Z}_+ : X_n(\omega) \leq a\},$$
$$T_2(\omega) = \inf\{n > T_1(\omega) : X_n(\omega) \geq b\},$$
$$\vdots$$
$$T_{2k+1}(\omega) = \inf\{n > T_{2k}(\omega) : X_n(\omega) \leq a\},$$
$$T_{2k+2}(\omega) = \inf\{n > T_{2k+1}(\omega) : X_n(\omega) \geq b\},$$
$$\vdots$$

with the understanding that infimum on an empty set is ∞. For $N \in \mathbb{Z}_+$, the number of upcrossings by the sample path $\{X_n(\omega) : n \in \mathbb{Z}_+\}$ of the interval $[a, b]$ by time N is defined by

(1) $$(U_{[a,b]}^N X)(\omega) = \begin{cases} \max\{k \in \mathbb{N} : T_{2k}(\omega) \leq N\}, \\ 0 \quad \text{if the set above is } \emptyset. \end{cases}$$

The number of upcrossings of the interval $[a,b]$ by the sample path $\{X_n(\omega) : n \in \mathbb{Z}_+\}$ is defined by

(2) $$(U_{[a,b]}^\infty X)(\omega) = \lim_{N \to \infty} (U_{[a,b]}^N X)(\omega).$$

2) Define a sequence of $\overline{\mathbb{Z}}_+$-valued functions $\{S_j : j \in \mathbb{N}\}$ on Ω by

$$\begin{aligned} S_1(\omega) &= \inf\{n \in \mathbb{Z}_+ : X_n(\omega) \geq b\}, \\ S_2(\omega) &= \inf\{n > S_1(\omega) : X_n(\omega) \leq a\}, \\ &\vdots \\ S_{2k+1}(\omega) &= \inf\{n > S_{2k}(\omega) : X_n(\omega) \geq b\}, \\ S_{2k+2}(\omega) &= \inf\{n > S_{2k+1}(\omega) : X_n(\omega) \leq a\}, \\ &\vdots \end{aligned}$$

with the understanding that infimum on an empty set is ∞. For $N \in \mathbb{Z}_+$, the number of downcrossings by the sample path $\{X_n(\omega) : n \in \mathbb{Z}_+\}$ of the interval $[a,b]$ by time N is defined by

(3) $$(D_{[a,b]}^N X)(\omega) = \begin{cases} \max\{k \in \mathbb{N} : S_{2k}(\omega) \leq N\}, \\ 0 \quad \text{if the set above is } \emptyset. \end{cases}$$

The number of downcrossings of the interval $[a,b]$ by the sample path $\{X_n(\omega) : n \in \mathbb{Z}_+\}$ is defined by

(4) $$(D_{[a,b]}^\infty X)(\omega) = \lim_{N \to \infty} (D_{[a,b]}^N X)(\omega).$$

3) Let $Y^{(N)} = \{Y_n : n = 0, \ldots, N\}$ be a finite sequence of real valued random variables on $(\Omega, \mathfrak{F}, P)$. Define $\{T_j : j \in \mathbb{N}\}$ and $\{S_j : j \in \mathbb{N}\}$ as in 1) and 2) for the infinite sequence $Y = \{Y_n : n \in \mathbb{Z}_+\}$ with $Y_n = Y_N$ for $n > N$. We define the numbers of upcrossings and downcrossings of $[a,b]$ by $Y^{(N)}$ by equating

(5) $$(U_{[a,b]}^N Y^{(N)})(\omega) = (U_{[a,b]}^N Y)(\omega) \quad \text{and} \quad (D_{[a,b]}^N Y^{(N)})(\omega) = (D_{[a,b]}^N Y)(\omega).$$

Lemma 6.18. *Let $X = \{X_n : n \in \mathbb{Z}_+\}$ be an adapted process on a filtered space $(\Omega, \mathfrak{F}, \{\mathfrak{F}_n\}, P)$ and let $\{T_j : j \in \mathbb{N}\}$ and $\{S_j : j \in \mathbb{N}\}$ be as in Definition 6.17. Then*

§6. FUNDAMENTAL SUBMARTINGALE INEQUALITIES

1) $\{T_j : j \in \mathbb{N}\}$ is an increasing sequence of stopping times such that for every $\omega \in \Omega$ the sequence $\{T_j(\omega) : j \in \mathbb{N}\}$ is strictly increasing until the value ∞ is reached. Similarly for $\{S_j : j \in \mathbb{N}\}$.
2) For every $N \in \mathbb{Z}_+$, $U_{[a,b]}^N X$ and $D_{[a,b]}^N X$ are nonnegative $\mathfrak{F}_N$-measurable random variables on $(\Omega, \mathfrak{F}, P)$ bounded by $2^{-1}(N+1)$.
3) $U_{[a,b]}^\infty X$ and $D_{[a,b]}^\infty X$ are $\mathfrak{F}_\infty$-measurable random variables on $(\Omega, \mathfrak{F}, P)$ with values in $[0, \infty]$.

Proof. 1) Let us show that T_j is a stopping time for $j \in \mathbb{N}$. Since the time variable is discrete, to show that T_j is a stopping time it suffices to show that $\{T_j = k\} \in \mathfrak{F}_k$ for every $k \in \mathbb{Z}_+$ as we noted in Remark 3.23. Now for $k \in \mathbb{Z}_+$ we have by Definition 6.17

$$\{T_1 = k\} = \{X_0 > a\} \cap \cdots \cap \{X_{k-1} > a\} \cap \{X_k \leq a\} \in \mathfrak{F}_k.$$

Thus T_1 is a stopping time. Next suppose X_j is a stopping time for some $j \in \mathbb{N}$. To show that T_{j+1} is a stopping time let $k \in \mathbb{Z}_+$ and consider

$$\{T_{j+1} = k\} = \left[\bigcap_{i=0}^{k-1} \{T_j = i\}\right] \cap \{T_{j+1} = k\}.$$

Now for $i = 0, \ldots, k-1$, when j is odd we have

$$\{T_j = i\} \cap \{T_{j+1} = k\}$$
$$= \{T_j = i\} \cap \{X_{i+1} < b\} \cap \cdots \cap \{X_{k-1} < b\} \cap \{X_k \geq b\} \in \mathfrak{F}_k$$

and when j is odd we have

$$\{T_j = i\} \cap \{T_{j+1} = k\}$$
$$= \{T_j = i\} \cap \{X_{i+1} > a\} \cap \cdots \cap \{X_{k-1} > a\} \cap \{X_k \leq a\} \in \mathfrak{F}_k.$$

Thus $\{T_{j+1} = k\} \in \mathfrak{F}_k$ and this shows that T_{j+1} is a stopping time. Therefore by induction T_j is a stopping time for $j \in \mathbb{N}$. Similarly S_j is a stopping time for $j \in \mathbb{N}$.

2) For each $N \in \mathbb{Z}_+$, clearly $U_{[a,b]}^N X \leq 2^{-1}(N+1)$. To show that $U_{[a,b]}^N X$ is $\mathfrak{F}_N$-measurable, define for each $k \in \mathbb{N}$

$$C_{2k}(\omega) = \begin{cases} 1 & \text{if } T_{2k}(\omega) \leq N \\ 0 & \text{if } T_{2k}(\omega) > N. \end{cases}$$

Since T_{2k} is a stopping time, $\{T_{2k} \leq N\} \in \mathfrak{F}_N$ and thus C_{2k} is an $\mathfrak{F}_N$-measurable random variable. Now

$$\left(U_{[a,b]}^N X\right)(\omega) = \sum_{k \in \mathbb{N}} C_{2k}(\omega).$$

Then since C_{2k} is $\mathfrak{F}_N$-measurable for every $k \in \mathbb{N}$, $U_{[a,b]}^n X$ is $\mathfrak{F}_N$-measurable. Similarly for $D_{[a,b]}^N X$.

3) By definition $U_{[a,b]}^N X \uparrow U_{[a,b]}^\infty X$ as $N \to \infty$. Since $U_{[a,n]}^N X$ is $\mathfrak{F}_N$-measurable it is $\mathfrak{F}_\infty$-measurable. Since this holds for every $N \in \mathbb{Z}_+$, $U_{[a,b]}^\infty X$ is $\mathfrak{F}_\infty$-measurable. Similarly for $D_{[a,b]}^\infty X$. ∎

Theorem 6.19. (Doob's Upcrossing and Downcrossing Inequalities for Submartingales) Let $X = \{X_n : n \in \mathbb{Z}_+\}$ be a submartingale on a filtered space $(\Omega, \mathfrak{F}, \{\mathfrak{F}_n\}, P)$. Then for any $a, b \in \mathbb{R}$, $a < b$, and $N \in \mathbb{Z}_+$ we have

(1) $$\mathbf{E}(U_{[a,b]}^N X) \leq \frac{1}{b-a} \mathbf{E}[(X_N - a)^+ - (X_0 - a)^+],$$

and

(2) $$\mathbf{E}(D_{[a,b]}^N X) \leq \frac{1}{b-a} \mathbf{E}[(X_N - a)^+] + 1.$$

Proof. Let $Y = \{Y_n : n \in \mathbb{Z}_+\}$ be a stochastic process defined by $Y = (X - a)^+$, that is,

(3) $$Y_n = (X_n - a)^+ \quad \text{for } n \in \mathbb{Z}_+.$$

Since X is a submartingale, $X - a$ is a submartingale and then $Y = (X - a)^+$ is a nonnegative submartingale by Corollary 5.10. Note that

(4) $$U_{[a,b]}^N X = U_{[0,b-a]}^N Y \quad \text{and} \quad D_{[a,b]}^N X = D_{[0,b-a]}^N Y.$$

1) To prove (1), let $\{T_j : j \in \mathbb{N}\}$ be as in Definition 6.17 with a, b and X replaced by $0, b-a$ and Y respectively. Let $N \in \mathbb{Z}_+$ be given and let $k \in \mathbb{N}$ be such that $2k > N$. Then we have $T_{2k} \geq 2k - 1 > N - 1$ so that $T_{2k} \geq N$. Let $T_0^* = 0$ and $T_j^* = T_j \wedge N$ for $j \in \mathbb{N}$. Note that $\{T_j^* : j \in \mathbb{Z}_+\}$ is an increasing sequence of stopping times. Now since $T_0^* = 0$ and $T_{2k}^* \geq N$ we can write

(5) $$Y_N - Y_0 = \sum_{j=1}^{2k} \{Y_{T_j^*} - Y_{T_{j-1}^*}\} = \sum_{j=1}^{k} \{Y_{T_{2j}^*} - Y_{T_{2j-1}^*}\} + \sum_{j=0}^{k-1} \{Y_{T_{2j+1}^*} - Y_{T_{2j}^*}\}.$$

Regarding the first sum on the right side of (5), note that there is contribution to the sum only when $T_{2j-1}^* < N$, and note that when $T_{2j-1}^* < N$ then we have $T_{2j-1}^* = T_{2j-1}$ so that

§6. FUNDAMENTAL SUBMARTINGALE INEQUALITIES

$Y_{T^*_{2j-1}} = 0$. Thus we have

$$\sum_{j=1}^{k}\{Y_{T^*_{2j}} - Y_{T^*_{2j-1}}\} \geq (b-a)U^N_{[0,b-a]}Y$$

so that

(6) $$\sum_{j=1}^{k}\mathbf{E}\{Y_{T^*_{2j}} - Y_{T^*_{2j-1}}\} \geq (b-a)\mathbf{E}[U^N_{[0,b-a]}Y].$$

On the other hand since Y is a submartingale and $T^*_{2j} < T^*_{2j+1} \leq N$ for $j = 0, \cdots, k-1$, we have $\mathbf{E}(Y_{T^*_{2j+1}}) \geq \mathbf{E}(Y_{T^*_{2j}})$ by (2) of Theorem 6.4 so that

(7) $$\sum_{j=0}^{k-1}\mathbf{E}\{Y_{T^*_{2j+1}} - Y_{T^*_{2j}}\} \geq 0.$$

Using (6) and (7) in (5) we have

$$(b-a)\mathbf{E}(U^N_{[0,b-a]}Y) \leq \mathbf{E}(Y_N - Y_0).$$

By the fact that $Y_N - Y_0 = (X_N - a)^+ - (X_0 - a)^+$ and by (4) we have

$$\mathbf{E}(U^N_{[a,b]}X) \leq \frac{1}{b-a}\mathbf{E}[(X_N - a)^+ - (X_0 - a)^+],$$

which proves (1).

2) To prove (2), let $\{S_j : j \in \mathbb{N}\}$ be as in Definition 6.17 with a, b and X replaced by $0, b-a$ and Y respectively. Let $N \in \mathbb{Z}_+$ be given and let $k \in \mathbb{N}$ be such that $2k > N$. This implies that $S_{2k} \geq N$. Let $S^*_j = S_j \wedge N$ for $j \in \mathbb{N}$. Then we have

(8) $$\sum_{j=1}^{k}\{Y_{S^*_{2j}} - Y_{S^*_{2j-1}}\} \leq \{0 - (b-a)\}D^N_{[0,b-a]}Y + \{Y_N + (b-a)\}.$$

By the fact that $D^N_{[0,b-a]}Y = D^N_{[a,b]}X$ and by (8), we have

$$\sum_{j=1}^{k}\mathbf{E}\{Y_{S^*_{2j}} - Y_{S^*_{2j-1}}\} \leq (a-b)\mathbf{E}(D^N_{[a,b]}X) + \mathbf{E}[(X_N - a)^+] + (b-a).$$

Since Y is a submartingale and $S_{2j}^* \geq S_{2j-1}^*$, we have $\mathbf{E}\{Y_{S_{2j}^*} - Y_{S_{2j-1}^*}\} \geq 0$ by (2) of Theorem 6.4. Therefore we have

$$0 \leq (a-b)\mathbf{E}(D_{[a,b]}^N X) + \mathbf{E}[(X_N - a)^+] + (b-a),$$

and then

$$\mathbf{E}(D_{[a,b]}^N X) \leq \frac{1}{b-a}\mathbf{E}[(X_N - a)^+] + 1,$$

proving (2). ∎

Upcrossing and downcrossing inequalities for supermartingales can be derived from those for submartingales by the fact that if X is a supermartingale then $-X$ is a submartingale.

Theorem 6.20.. (Upcrossing and Downcrossing Inequalities for Supermartingales) *Let $\{X_n : n \in \mathbb{Z}_+\}$ be a supermartingale on a filtered space $(\Omega, \mathfrak{F}, \{\mathfrak{F}_n\}, P)$. Then for any $a, b \in \mathbb{R}$, $a < b$, and $N \in \mathbb{Z}_+$ we have*

(1) $$\mathbf{E}(U_{[a,b]}^N X) \leq \frac{1}{b-a}\mathbf{E}[(X_N - b)^-] + 1$$

and

(2) $$\mathbf{E}(D_{[a,b]}^N X) \leq \frac{1}{b-a}\mathbf{E}[(X_N - b)^- - (X_0 - b)^-].$$

Proof. Let us note that for any stochastic process X we have

$$U_{[a,b]}^N X = D_{[-b,-a]}^N(-X) \quad \text{and} \quad D_{[a,b]}^N X = U_{[-b,-a]}^N(-X).$$

Now if X is a supermartingale on the filtered space then $-X$ is a submartingale and therefore by (2) of Theorem 6.19 we have

$$\begin{aligned}\mathbf{E}(U_{[a,b]}^N X) &= \mathbf{E}[D_{[-b,-a]}^N(-X)] \leq \frac{1}{-a-(-b)}\mathbf{E}[(-X_N - (-b))^+] + 1 \\ &= \frac{1}{b-a}\mathbf{E}[(-X_N + b)^+] + 1 = \frac{1}{b-a}\mathbf{E}[(X_N - b)^-] + 1.\end{aligned}$$

§6. FUNDAMENTAL SUBMARTINGALE INEQUALITIES

This proves (1). Similarly by (1) of Theorem 6.19 we have

$$\begin{aligned}
\mathbf{E}(D^N_{[a,b]}X) &= \mathbf{E}[U^N_{[-b,-a]}(-X)] \\
&\leq \frac{1}{-a-(-b)}\mathbf{E}[(-X_N-(-b))^+ - (-X_0-(-b))^+] \\
&= \frac{1}{b-a}\mathbf{E}[(b-X_N)^+ - (b-X_0)^+] \\
&= \frac{1}{b-a}\mathbf{E}[(X_N-b)^- - (X_0-b)^-].
\end{aligned}$$

This proves (2). ∎

The following simplification of Theorem 6.19 and Theorem 6.20 is often useful.

Corollary 6.21. *If $X = \{X_n : n \in \mathbb{Z}_+\}$ is a submartingale on a filtered space $(\Omega, \mathfrak{F}, \{\mathfrak{F}_n\}, P)$ then for $a, b \in \mathbb{R}$, $a < b$, and $N \in \mathbb{Z}_+$, we have*

(1) $$\mathbf{E}(U^N_{[a,b]}X) \leq \frac{1}{b-a}\{\mathbf{E}(|X_N|) + \mathbf{E}(|X_0|) + 2|a|\}$$

and

(2) $$\mathbf{E}(D^N_{[a,b]}X) \leq \frac{1}{b-a}\{\mathbf{E}(|X_N|) + |a|\} + 1.$$

If X is a supermartingale, then

(3) $$\mathbf{E}(U^N_{[a,b]}X) \leq \frac{1}{b-a}\{\mathbf{E}(|X_N|) + |b|\} + 1$$

and

(4) $$\mathbf{E}(D^N_{[a,b]}X) \leq \frac{1}{b-a}\{\mathbf{E}(|X_N|) + \mathbf{E}(|X_0|) + 2|b|\}.$$

Proof. To prove (1) recall that by (1) of Theorem 6.19

$$\begin{aligned}
\mathbf{E}(U^N_{[a,b]}X) &\leq \frac{1}{b-a}\mathbf{E}[(X_N-a)^+ - (X_0-a)^+] \\
&\leq \frac{1}{b-a}\mathbf{E}[|X_N-a| + |X_0-a|] \\
&\leq \frac{1}{b-a}\mathbf{E}[|X_N| + |a| + |X_0| + |a|] \\
&= \frac{1}{b-a}\{\mathbf{E}(|X_N|) + \mathbf{E}(|X_0|) + 2|a|\}.
\end{aligned}$$

Similarly (2) is derived from (2) of Theorem 6.19 and (3) and (4) are derived from (1) and (2) of Theorem 6.20. ∎

Let us define upcrossing and downcrossing numbers for a stochastic process with continuous time.

Definition 6.22. *Let* $X = \{X_t : t \in \mathbb{R}_+\}$ *be a stochastic process on a probability space* $(\Omega, \mathfrak{F}, P)$, S *be a countable dense subset of* $\mathbb{R}_+$ *and* J *be an interval in* $\mathbb{R}_+$. *Let* $\tau = \{t_0, \cdots, t_N\}$ *be a strictly increasing finite sequence in* $J \cap S$. *Consider the finite sequence of real valued random variables* $X^{(\tau)} = \{X_{t_n} : n = 0, \ldots, N\}$ *and the numbers of upcrossings and downcrossings* $U^N_{[a,b]} X^{(\tau)}$ *and* $D^N_{[a,b]} X^{(\tau)}$ *by* $X^{(\tau)}$ *as defined by (5) of Definition 6.17. We define the numbers of upcrossings and downcrossings of* $[a, b]$ *by* X *on* τ *by setting*

(1) $$U^\tau_{[a,b]} X = U^N_{[a,b]} X^{(\tau)} \quad \text{and} \quad D^\tau_{[a,b]} X = D^N_{[a,b]} X^{(\tau)}$$

The numbers of upcrossings and downcrossings of $[a, b]$ *by* X *on* $J \cap S$ *are defined by*

(2) $$U^{J \cap S}_{[a,b]} X = \sup_{\{\tau\}} U^\tau_{[a,b]} X \quad \text{and} \quad D^{J \cap S}_{[a,b]} X = \sup_{\{\tau\}} D^\tau_{[a,b]} X,$$

where the suprema are over the collection of all strictly increasing finite sequences τ *in* $J \cap S$.

Note that if X is an adapted process on a filtered space $(\Omega, \mathfrak{F}, \{\mathfrak{F}_t\}, P)$ then by Lemma 6.18, $U^\tau_{[a,b]} X$ and $D^\tau_{[a,b]} X$ as defined above are $\mathfrak{F}_{t_N}$-measurable random variables on $(\Omega, \mathfrak{F}, P)$. Since $J \cap S$ is a countable set the collection of all strictly increasing finite sequences τ in $J \cap S$ is a countable collection and therefore the suprema over this collection are suprema of countably many $\mathfrak{F}_\infty$-measurable random variables on $(\Omega, \mathfrak{F}, P)$. Thus $U^{J \cap S}_{[a,b]} X$ and $D^{J \cap S}_{[a,b]} X$ are $\mathfrak{F}_\infty$-measurable.

Theorem 6.23. *Let* $X = \{X_t : t \in \mathbb{R}_+\}$ *be a submartingale on a filtered space* $(\Omega, \mathfrak{F}, \{\mathfrak{F}_t\}, P)$, S *be a countable dense subset of* $\mathbb{R}_+$ *and* J *be an interval in* $\mathbb{R}_+$. *Let* $a, b \in \mathbb{R}$, $a < b$. *Then for every strictly increasing finite sequence* $\tau = \{t_0, \cdots, t_N\}$ *in* $J \cap S$ *we have*

(1) $$\mathbf{E}(U^\tau_{[a,b]} X) \leq \frac{1}{b-a} \mathbf{E}[(X_{t_N} - a)^+ - (X_{t_0} - a)^+]$$

§6. FUNDAMENTAL SUBMARTINGALE INEQUALITIES

and

(2) $$\mathbf{E}(D^\tau_{[a,b]}X) \leq \frac{1}{b-a}\mathbf{E}[(X_{t_N}-a)^+] + 1.$$

If J has α and β as its end-points where $\alpha, \beta \in \mathbb{R}_+, \alpha < \beta$, then

(3) $$\mathbf{E}(U^{J\cap S}_{[a,b]}X) \leq \frac{1}{b-a}\mathbf{E}[(X_\beta - a)^+ - (X_\alpha - a)^+]$$

and

(4) $$\mathbf{E}(D^{J\cap S}_{[a,b]}X) \leq \frac{1}{b-a}\mathbf{E}[(X_\beta - a)^+] + 1.$$

Proof. Note that (1) and (2) are immediate from (1) of Definition 6.22 and Theorem 6.19. To prove (3), note that by (2) of Definition 6.22 there exists a sequence $\{\tau_n : n \in \mathbb{N}\}$ of strictly increasing finite sequences τ_n in $J \cap S$ such that $U^{J\cap S}_{[a,b]}X = \lim_{n\to\infty} U^{\tau_n}_{[a,b]}X$. Then by Fatou's Lemma

(5) $$\mathbf{E}(U^{J\cap S}_{[a,b]}X) \leq \liminf_{n\to\infty} \mathbf{E}(U^{\tau_n}_{[a,b]}X) \leq \sup_{\{\tau\}} \mathbf{E}(U^\tau_{[a,b]}X)$$

where the supremum is over the collection of all strictly increasing finite sequences τ in $J \cap S$. Now since X is a submartingale, $(X-a)^+$ is a submartingale so that $\mathbf{E}[(X_t-a)^+] \uparrow$ as $t \uparrow$. Thus from (1)

$$\mathbf{E}(U^\tau_{[a,b]}X) \leq \frac{1}{b-a}\mathbf{E}[(X_\beta-a)^+ - (X_\alpha-a)^+].$$

Using this in (5) we have (3). Similarly for (4). ∎

Theorem 6.24. *If X in Theorem 6.23 is a supermartingale, then*

(1) $$\mathbf{E}(U^\tau_{[a,b]}X) \leq \frac{1}{b-a}\mathbf{E}[(X_{t_N}-b)^-] + 1,$$

(2) $$\mathbf{E}(D^\tau_{[a,b]}X) \leq \frac{1}{b-a}\mathbf{E}[(X_{t_N}-b)^- - (X_{t_0}-b)^-],$$

(3) $$\mathbf{E}(U^{J\cap S}_{[a,b]}X) \leq \frac{1}{b-a}\mathbf{E}[(X_\beta-b)^-] + 1,$$

and

(4) $$\mathbf{E}(D_{[a,b]}^{J\cap S}X) \leq \frac{1}{b-a}\mathbf{E}[(X_\beta - b)^-(X_\alpha - b)^-].$$

Proof. (1) and (2) follow from (1) of Definition 6.22 and Theorem 6.20. Then (3) and (4) follow from (1) and (2) by the same argument as in Theorem 6.23. ∎

§7 Convergence of Submartingales

Let $X = \{X_t : t \in \mathbb{T}\}$ be a submartingale on a filtered space $(\Omega, \mathfrak{F}, \{\mathfrak{F}_t : t \in \mathbb{T}\}, P)$. Consider the process $X^+ = \{X_t^+ : t \in \mathbb{T}\}$. We shall show that if X^+ is L_1-bounded, that is, $\sup_{t \in \mathbb{T}} \mathbf{E}(X_t^+) < \infty$, then there exists an extended real valued integrable $\mathfrak{F}_\infty$-measurable random variable X_∞ on $(\Omega, \mathfrak{F}, P)$ such that $\lim_{t \to \infty} X_t = X_\infty$ a.e. on $(\Omega, \mathfrak{F}_\infty, P)$. Next we show that if we assume the uniform integrability of X^+ (which implies the L_1-boundedness of X^+), then X_∞ is a final element for the submartingale in the sense that $\mathbf{E}(X_\infty | \mathfrak{F}_t) \geq X_t$ a.e. on $(\Omega, \mathfrak{F}_t, P)$ for every $t \in \mathbb{T}$. In §8 we show that if X is uniformly integrable then we have $\lim_{t \to \infty} \|X_t - X_\infty\|_1 = 0$.

[I] Convergence of Submartingales with Discrete Time

Observation 7.1. If $X = \{X_t : t \in \mathbb{T}\}$ is a submartingale then X is L_1-bounded if and only if X^+ is, that is,

(1) $$\sup_{t \in \mathbb{T}} \mathbf{E}(|X_t|) < \infty \Leftrightarrow \sup_{t \in \mathbb{T}} \mathbf{E}(X_t^+) < \infty.$$

If X is a supermartingale then X is L_1-bounded if and only if X^- is, that is,

(2) $$\sup_{t \in \mathbb{T}} \mathbf{E}(|X_t|) < \infty \Leftrightarrow \sup_{t \in \mathbb{T}} \mathbf{E}(X_t^-) < \infty.$$

Proof. Note that since $|X_t| \geq X_t^+, X_t^-$, the condition $\sup_{t \in \mathbb{T}} \mathbf{E}(|X_t|) < \infty$ always implies both $\sup_{t \in \mathbb{T}} \mathbf{E}(X_t^+) < \infty$ and $\sup_{t \in \mathbb{T}} \mathbf{E}(X_t^-) < \infty$. Note also that we have $|X_t| = 2X_t^+ - X_t$ as well as $|X_t| = 2X_t^- + X_t$. Thus, if X is a submartingale then $\mathbf{E}(X_t) \uparrow$ as $t \uparrow$ and we have

$$\mathbf{E}(|X_t|) = 2\mathbf{E}(X_t^+) - \mathbf{E}(X_t) \leq 2\mathbf{E}(X_t^+) - \mathbf{E}(X_0)$$

§7. CONVERGENCE OF SUBMARTINGALES

so that

(3) $$\sup_{t \in \mathbf{T}} \mathbf{E}(|X_t|) \leq 2 \sup_{t \in \mathbf{T}} \mathbf{E}(X_t^+) - \mathbf{E}(X_0).$$

If X is a supermartingale then $\mathbf{E}(X_t) \downarrow$ as $t \uparrow$ and we have

$$\mathbf{E}(|X_t|) = 2\mathbf{E}(X_t^-) + \mathbf{E}(X_t) \leq 2\mathbf{E}(X_t^-) + \mathbf{E}(X_0)$$

so that

(4) $$\sup_{t \in \mathbf{T}} \mathbf{E}(|X_t|) \leq 2 \sup_{t \in \mathbf{T}} \mathbf{E}(X_t^-) + \mathbf{E}(X_0).$$

By (3) and (4) we have the implication $\Leftarrow$ in (1) and (2). ∎

Lemma 7.2. *Let $X = \{X_n : n \in \mathbb{Z}_+\}$ be an L_1-bounded submartingale or supermartingale on a filtered space $(\Omega, \mathfrak{F}, \{\mathfrak{F}_n\}, P)$. Then for any $a, b \in \mathbb{R}$, $a < b$, we have*

$$\left(U_{[a,b]}^\infty X\right)(\omega) < \infty \quad \text{and} \quad \left(D_{[a,b]}^\infty X\right)(\omega) < \infty \quad \text{for a.e. } \omega \text{ in } (\Omega, \mathfrak{F}_\infty, P).$$

Proof. Let X be an L_1-bounded submartingale. By (1) of Corollary 6.21 for every $N \in \mathbb{Z}_+$ we have

$$\begin{aligned} \mathbf{E}(U_{[a,b]}^N X) &\leq \frac{1}{b-a}\{\mathbf{E}(|X_N|) + \mathbf{E}(|X_0|) + 2|a|\} \\ &\leq \frac{2}{b-a}\{\sup_{n \in \mathbb{Z}_+} \mathbf{E}(|X_n|) + |a|\}. \end{aligned}$$

Since $U_{[a,b]}^N X \uparrow U_{[a,b]}^\infty X$ as $N \to \infty$, we have by the Monotone Convergence Theorem

$$\mathbf{E}(U_{[a,b]}^\infty X) \leq \frac{2}{b-a}\{\sup_{n \in \mathbb{Z}_+} \mathbf{E}(|X_n|) + |a|\} < \infty.$$

Thus $(U_{[a,b]}^\infty X)(\omega) < \infty$ for a.e. ω in $(\Omega, \mathfrak{F}_\infty, P)$. Similarly by using (2) of Corollary 6.21 we have $(D_{[a,b]}^\infty X)(\omega) < \infty$ for a.e. ω in $(\Omega, \mathfrak{F}_\infty, P)$. For an L_1-bounded supermartingale the same conclusion holds by means of (3) and (4) of Corollary 6.21. ∎

Lemma 7.3. *Let $X = \{X_n : n \in \mathbb{Z}_+\}$ be a stochastic process on a probability space $(\Omega, \mathfrak{F}, P)$. Let $a, b \in \mathbb{R}$, $a < b$. Then for every $\omega \in \Omega$*

$$\liminf_{n \to \infty} X_n(\omega) < a < b < \limsup_{n \to \infty} X_n(\omega) \Rightarrow (U_{[a,b]}^\infty X)(\omega) = \infty.$$

The same holds for $D^\infty_{[a,b]}X$.

Proof. Suppose $\liminf_{n\to\infty} X_n(\omega) < a < b < \limsup_{n\to\infty} X_n(\omega)$ for some $\omega \in \Omega$. Then since the limit inferior and the limit superior of a sequence are respectively the least and the greatest of the limit points of the sequence, there exists a subsequence $\{n_k\}$ of $\{n\}$ such that $\lim_{k\to\infty} X_{n_k}(\omega) = \liminf_{n\to\infty} X_n(\omega)$ and there exists a subsequence $\{n_\ell\}$ of $\{n\}$ such that $\lim_{\ell\to\infty} X_{n_\ell}(\omega) = \limsup_{n\to\infty} X_n(\omega)$. Thus there exist $n_1 < n_2 < n_3 < \cdots$ such that $X_{n_{2j-1}}(\omega) < a$ and $X_{n_{2j}}(\omega) > b$ for $j \in \mathbb{N}$. Therefore $(U^\infty_{[a,b]}X)(\omega) = \infty$. Similarly for $D^\infty_{[a,b]}X$. ∎

Theorem 7.4. (Doob's Martingale Convergence Theorem). *Let $X = \{X_n : n \in \mathbb{Z}_+\}$ be a submartingale or a supermartingale on a filtered space $(\Omega, \mathfrak{F}, \{\mathfrak{F}_n\}, P)$. Let us define an extended real valued $\mathfrak{F}_\infty$-measurable random variable X_∞ on $(\Omega, \mathfrak{F}, P)$ by setting $X_\infty(\omega) = \liminf_{n\to\infty} X_n(\omega)$ for $\omega \in \Omega$. If X is L_1-bounded then $\lim_{n\to\infty} X_n(\omega) = X_\infty(\omega)$ for a.e. ω in $(\Omega, \mathfrak{F}_\infty, P)$ and furthermore X_∞ is integrable so that X_∞ is real valued a.e. on $(\Omega, \mathfrak{F}_\infty, P)$.*

Proof. Let

$$\begin{aligned}\Lambda &= \{\omega \in \Omega : \lim_{n\to\infty} X_n(\omega) \text{ does not exist in } \overline{\mathbb{R}}\}\\ &= \{\omega \in \Omega : \liminf_{n\to\infty} X_n(\omega) < \limsup_{n\to\infty} X_n(\omega)\}.\end{aligned}$$

Let Q be the collection of all rational numbers. For $a, b, \in Q, a < b$, let

$$\Lambda_{a,b} = \{\omega \in \Omega : \liminf_{n\to\infty} X_n(\omega) < a < b < \limsup_{n\to\infty} X_n(\omega)\}.$$

Then

$$\Lambda = \bigcup_{a,b \in Q, a<b} \Lambda_{a,b}.$$

Note that since $\liminf_{n\to\infty} X_n$ and $\limsup_{n\to\infty} X_n$ are $\mathfrak{F}_\infty$-measurable, we have $\Lambda, \Lambda_{a,b} \in \mathfrak{F}_\infty$.

Now by Lemma 7.3, we have

$$\Lambda_{a,b} \subset \{\omega \in \Omega : (U^\infty_{[a,b]}X)(\omega) = \infty\},$$

and then by Lemma 7.2

$$P(\Lambda_{a,b}) \leq P\{\omega \in \Omega : (U^\infty_{[a,b]}X)(\omega) = \infty\} = 0.$$

§7. CONVERGENCE OF SUBMARTINGALES

Thus $\Lambda_{a,b}$ is a null set in $(\Omega, \mathfrak{F}_\infty, P)$ and as a countable union of such null sets Λ is a null set in $(\Omega, \mathfrak{F}_\infty, P)$. Therefore $\lim_{n\to\infty} X_n(\omega)$ exists in $\overline{\mathbb{R}}$ for a.e. ω in $(\Omega, \mathfrak{F}_\infty, P)$. Then

(1) $$\lim_{n\to\infty} X_n(\omega) = \liminf_{n\to\infty} X_n(\omega) = X_\infty(\omega) \quad \text{for a.e. } \omega \text{ in } (\Omega, \mathfrak{F}_\infty, P).$$

Thus by Fatou's Lemma and the L_1-boundedness of X,

$$\mathbf{E}(|X_\infty|) = \mathbf{E}(|\lim_{n\to\infty} X_n|) = \mathbf{E}(\lim_{n\to\infty} |X_n|) \leq \liminf_{n\to\infty} \mathbf{E}(|X_n|) \leq \sup_{n\in\mathbb{Z}_+} \mathbf{E}(|X_n|) < \infty.$$

Therefore X_∞ is integrable on Ω and consequently $X_\infty(\omega) \in \mathbb{R}$ for a.e. ω in $(\Omega, \mathfrak{F}_\infty, P)$. ∎

Corollary 7.5. *Let $X = \{X_n : n \in \mathbb{Z}_+\}$ be a nonpositive submartingale or a nonnegative supermartingale on a filtered space $(\Omega, \mathfrak{F}, \{\mathfrak{F}_n\}, P)$. Then X converges a.e. on $(\Omega, \mathfrak{F}_\infty, P)$.*

Proof. If X is a nonpositive submartingale, then $\mathbf{E}(X_0) \leq \mathbf{E}(X_n) \leq 0$ and $\mathbf{E}(|X_n|) = -\mathbf{E}(X_n) \leq -\mathbf{E}(X_0)$ for $n \in \mathbb{Z}_+$ so that $\sup_{n\in\mathbb{Z}_+} \mathbf{E}(|X_n|) \leq -\mathbf{E}(X_0) < \infty$, that is, X is L_1-bounded. Thus X converges a.e. on $(\Omega, \mathfrak{F}_\infty, P)$ by Theorem 7.4. If X is a nonnegative supermartingale then $-X$ is a nonpositive submartingale so that by the result above X converges a.e. on $(\Omega, \mathfrak{F}_\infty, P)$. ∎

Corollary 7.6. *Let $X = \{X_n : n \in \mathbb{Z}_+\}$ be an L_p-bounded nonnegative submartingale on a filtered space $(\Omega, \mathfrak{F}, \{\mathfrak{F}_n\}, P)$ for some $p \in (1, \infty)$. Let X_∞ be an extended real valued $\mathfrak{F}_\infty$-measurable random variable on $(\Omega, \mathfrak{F}, P)$ defined by $X_\infty = \liminf_{n\to\infty} X_n$. Then*

(1) $$\lim_{n\to\infty} X_n = X_\infty \quad \text{a.e. on } (\Omega, \mathfrak{F}_\infty, P),$$

(2) $$X_\infty \in L_p(\Omega, \mathfrak{F}_\infty, P) \quad \text{and} \quad \lim_{n\to\infty} \|X_n - X_\infty\|_p = 0,$$

and

(3) $$\|X_\infty\|_p = \uparrow \lim_{n\to\infty} \|X_n\|_p = \sup_{n\in\mathbb{Z}_+} \|X_n\|_p.$$

Proof. Since $\|X_n\|_1 \leq \|X_n\|_p$ for all $n \in \mathbb{Z}_+$ for $p \in (1, \infty)$, the L_p-boundedness of X implies its L_1-boundedness. Then by Theorem 7.4, X_∞ is integrable and (1) holds.

Since X is a nonnegative L_p-submartingale, if we write q for the conjugate exponent of p then we have $\mathbf{E}(\sup_{n\in\mathbb{Z}_+} X_n^p) \leq q^p \sup_{n\in\mathbb{Z}_+} \mathbf{E}(X_n^p)$ by (2) of Theorem 6.15. Since X is

L_p-bounded, that is, $\sup_{n \in \mathbb{Z}_+} \mathbf{E}(X_n^p) < \infty$, we have the integrability of $\sup_{n \in \mathbb{Z}_+} X_n^p$. This together with the fact that $0 \leq X_n^p \leq \sup_{n \in \mathbb{Z}_+} X_n^p$, implies the uniform integrability of $\{X_n^p : n \in \mathbb{Z}_+\}$ by Proposition 4.11. Since (1) implies $P \cdot \lim_{n \to \infty} X_n = X_\infty$, (2) holds by Theorem 4.16.

Finally the fact that X is a nonnegative L_p-submartingale implies that X^p is a nonnegative L_1-submartingale by Corollary 5.12. Therefore $\mathbf{E}(X_n^p) \uparrow$ as $n \to \infty$ and hence $\|X_n\|_p \uparrow \sup_{n \in \mathbb{Z}_+} \|X_n\|_p$. But (2) implies $\lim_{n \to \infty} \|X_n\|_p = \|X_\infty\|_p$. Thus $\|X_\infty\|_p = \uparrow \lim_{n \to \infty} \|X_n\|_p$. ∎

[II] Convergence of Submartingales with Continuous Time

For a stochastic process with continuous time $X = \{X_t : t \in \mathbb{R}_+\}$, the upcrossing number $U_{[a,b]}^{J \cap S} X$ where S is a countable dense subset of $\mathbb{R}_+$ and J is an interval in $\mathbb{R}_+$ was defined in Definition 6.22. In particular we have $U_{[a,b]}^S X = U_{[a,b]}^{\mathbb{R}_+ \cap S} X$.

Lemma 7.7. *Let $X = \{X_t : t \in \mathbb{R}_+\}$ be an L_1-bounded submartingale or supermartingale on a filtered space $(\Omega, \mathfrak{F}, \{\mathfrak{F}_t\}, P)$ and let S be a countable dense subset of $\mathbb{R}_+$. Then for every $a, b \in \mathbb{R}$ such that $a < b$, we have*

$$(U_{[a,b]}^S X)(\omega) < \infty \quad \text{and} \quad (D_{[a,b]}^S X)(\omega) < \infty \quad \text{for a.e. } \omega \text{ in } (\Omega, \mathfrak{F}_\infty, P).$$

Proof. Let $X = \{X_t : t \in \mathbb{R}_+\}$ be an L_1-bounded submartingale. For every $n \in \mathbb{N}$, let $J_n = [0, n]$. Then by (3) of Theorem 6.23 we have

$$\mathbf{E}\left(U_{[a,b]}^{J_n \cap S} X\right) \leq \frac{1}{b-a}\mathbf{E}[(X_n - a)^+ - (X_0 - a)^+]$$
$$\leq \frac{1}{b-a}\{\mathbf{E}(|X_n|) + \mathbf{E}(|X_0|) + 2|a|\}$$
$$\leq \frac{2}{b-a}\sup_{t \in \mathbb{R}_+}\{\mathbf{E}(|X_t|) + |a|\}.$$

Now since $U_{[a,b]}^{J_n \cap S} X \uparrow U_{[a,b]}^S X$ as $n \uparrow$, we have by Fatou's Lemma and the L_1-boundedness of X,

$$\mathbf{E}(U_{[a,b]}^S X) \leq \liminf_{n \to \infty} \mathbf{E}(U_{[a,b]}^{J_n \cap S} X) \leq \frac{2}{b-a}\sup_{t \in \mathbb{R}_+}\{\mathbf{E}(|X_t|) + |a|\} < \infty.$$

The integrability of $U_{[a,b]}^S X$ implies that it is finite a.e. on $(\Omega, \mathfrak{F}_\infty, P)$. Similarly for $D_{[a,b]}^S X$. Similarly for an L_1-bounded supermartingale by means of Theorem 6.24. ∎

§7. CONVERGENCE OF SUBMARTINGALES

Lemma 7.8. *Let $X = \{X_t : t \in \mathbb{R}_+\}$ be a stochastic process on a probability space $(\Omega, \mathfrak{F}, P)$ and let S be a countable dense subset of $\mathbb{R}_+$. Let $a, b \in \mathbb{R}$ such that $a < b$. Then for any $\omega \in \Omega$ we have*

$$\liminf_{t \to \infty} X_t(\omega) < a < b < \limsup_{t \to \infty} X_t(\omega) \Rightarrow \left(U_{[a,b]}^S X\right)(\omega) = \infty.$$

The same holds for $D_{[a,b]}^S X$.

Proof. Suppose $\liminf_{t \to \infty} X_t(\omega) < a < b < \limsup_{t \to \infty} X_t(\omega)$. We can select a strictly increasing sequence $\{t_n : n \in \mathbb{N}\}$ in S such that $\lim_{n \to \infty} t_n = \infty$ and such that $X_{t_{2k-1}}(\omega) < a$ and $X_{t_{2k}}(\omega) > b$ for $k \in \mathbb{N}$. Then for every $k \in \mathbb{N}$ if we let $\tau = \{t_1, \cdots, t_{2k}\}$, we have

$$k \leq (U_{[a,b]}^\tau X)(\omega) \leq (U_{[a,b]}^S X)(\omega).$$

Since this holds for every $k \in \mathbb{N}$, we have $(U_{[a,b]}^S X)(\omega) = \infty$. ∎

Theorem 7.9. *Let $X = \{X_t : t \in \mathbb{R}_+\}$ be a submartingale or supermartingale on a filtered space $(\Omega, \mathfrak{F}, \{\mathfrak{F}_t\}, P)$ and let X_∞ be an extended real valued $\mathfrak{F}_\infty$-measurable random variable on $(\Omega, \mathfrak{F}, P)$ defined by $X_\infty(\omega) = \liminf_{t \to \infty} X_t(\omega)$ for $\omega \in \Omega$. If X is L_1-bounded then X_∞ is integrable and $\lim_{t \to \infty} X_t(\omega) = X_\infty(\omega)$ for a.e. ω in $(\Omega, \mathfrak{F}_\infty, P)$.*

Proof. Let

$$\begin{aligned}\Lambda &= \{\omega \in \Omega : \lim_{t \to \infty} X_t(\omega) \text{ does not exist in } \overline{\mathbb{R}}\} \\ &= \{\omega \in \Omega : \liminf_{t \to \infty} X_t(\omega) < \limsup_{t \to \infty} X_t(\omega)\}\end{aligned}$$

and for any two rational numbers a and b such that $a < b$, let

$$\Lambda_{a,b} = \{\omega \in \Omega : \liminf_{t \to \infty} X_t(\omega) < a < b < \limsup_{t \to \infty} X_t(\omega)\}.$$

The proof then follows from Lemma 7.8 and Lemma 7.7 in the same way as Theorem 7.4 followed from Lemma 7.3 and Lemma 7.2. ∎

Corollary 7.10. *Let $\{X_t : t \in \mathbb{R}_+\}$ be an L_p-bounded nonnegative submartingale on a filtered space $(\Omega, \mathfrak{F}, \{\mathfrak{F}_t\}, P)$ for some $p \in (1, \infty)$. Let X_∞ be an extended real valued $\mathfrak{F}_\infty$-measurable random variable on $(\Omega, \mathfrak{F}, P)$ defined by $X_\infty = \liminf_{t \to \infty} X_t$. Then*

(1) $$\lim_{t \to \infty} X_t = X_\infty \quad \text{a.e. on } (\Omega, \mathfrak{F}_\infty, P),$$

(2) $\qquad X_\infty \in L_p(\Omega, \mathfrak{F}_\infty, P) \quad \text{and} \quad \lim_{t \to \infty} \|X_t - X_\infty\|_p = 0,$

and

(3) $\qquad \|X_\infty\|_p = \uparrow \lim_{t \to \infty} \|X_t\|_p = \sup_{t \in \mathbb{R}_+} \|X_t\|_p.$

Proof. Since X is L_p-bounded and $p \in (1, \infty)$, X is L_1-bounded. Therefore by Theorem 7.9 we have (1). To prove (2), let $\{t_n : n \in \mathbb{Z}_+\}$ be a strictly increasing sequence in $\mathbb{R}_+$ such that $t_n \uparrow \infty$ as $n \to \infty$. Then $X_\infty = \liminf_{n \to \infty} X_{t_n}$. Applying Corollary 7.6 to the L_p-bounded nonnegative submartingale $\{Y_n : n \in \mathbb{Z}_+\}$ with respect to $\{\mathfrak{G}_n : n \in \mathbb{Z}_+\}$ where $Y_n = X_{t_n}$ and $\mathfrak{G}_n = \mathfrak{F}_{t_n}$ for $n \in \mathbb{Z}_+$ we have $X_\infty \in L_p(\Omega, \mathfrak{F}_\infty, P)$ and $\lim_{n \to \infty} \|X_{t_n} - X_\infty\|_p = 0$. Since this holds for every strictly increasing sequence $\{t_n : n \in \mathbb{Z}_+\}$ such that $t_n \uparrow \infty$ we have $\lim_{t \to \infty} \|X_t - X_\infty\|_p = 0$. Thus (2) holds. Then (3) follows from (2) and the fact that X^p is a submartingale by 2) of Corollary 5.12 so that $\|X_t\|^p \uparrow$ as $t \to \infty$. ∎

[III] Closing a Submartingale with a Final Element

Definition 7.11. *Let $X = \{X_t : t \in \mathbb{T}\}$ be a submartingale, martingale or supermartingale on a filtered space $(\Omega, \mathfrak{F}, \{\mathfrak{F}_t : t \in \mathbb{T}\}, P)$. If there exists an extended real valued integrable $\mathfrak{F}_\infty$-measurable random variable ξ on $(\Omega, \mathfrak{F}, P)$ such that*

$$\mathbf{E}(\xi | \mathfrak{F}_t) \geq, =, \text{ or } \leq X_t \quad \text{a.e. on } (\Omega, \mathfrak{F}_t, P) \text{ for every } t \in \mathbb{T}$$

then we call ξ a final element for X.

If η is a random variable which satisfies all but the $\mathfrak{F}_\infty$-measurability condition on a final element for X, then an arbitrary version ξ of $\mathbf{E}(\eta | \mathfrak{F}_\infty)$ is a final element for X since ξ is $\mathfrak{F}_\infty$-measurable and since

$$\mathbf{E}(\xi | \mathfrak{F}_t) = \mathbf{E}[\mathbf{E}(\eta | \mathfrak{F}_\infty) | \mathfrak{F}_t] = \mathbf{E}(\eta | \mathfrak{F}_t) \geq, =, \text{ or } \leq X_t \quad \text{a.e. on } (\Omega, \mathfrak{F}_t, P).$$

If a final element ξ exists for a submartingale or supermartingale $X = \{X_t : t \in \mathbb{T}\}$, it may not be unique. Indeed if X is a submartingale and $\mathbf{E}(\xi | \mathfrak{F}_t) \geq X_t$ a.e. on $(\Omega, \mathfrak{F}_t, P)$ for every $t \in \mathbb{T}$, then for every $c \geq 0$ we have $\mathbf{E}(\xi + c | \mathfrak{F}_t) \geq X_t$ a.e. on $(\Omega, \mathfrak{F}_t, P)$ for every $t \in \mathbb{T}$ also. If however X is a martingale and a final element exits then it is unique according to Theorem 5.3.

Theorem 7.12. *Let $X = \{X_t : t \in \mathbb{T}\}$ be a submartingale, martingale or supermartingale on a filtered space $(\Omega, \mathfrak{F}, \{\mathfrak{F}_t : t \in \mathbb{T}\}, P)$.*

§7. CONVERGENCE OF SUBMARTINGALES

1) If X is a submartingale then a final element exists if and only if $X^+ = \{X_t^+ : t \in \mathbb{T}\}$ is uniformly integrable.
2) If X is a supermartingale then a final element exists if and only if $X^- = \{X_t^- : t \in \mathbb{T}\}$ is uniformly integrable.
3) If X is a martingale then a final element exists if and only if X is uniformly integrable.
In each of the three cases, if the uniform integrability condition is satisfied then there exists an extended real valued integrable $\mathfrak{F}_\infty$-measurable random variable X_∞ on $(\Omega, \mathfrak{F}, P)$ such that
1°. $\lim_{t \to \infty} X_t = X_\infty$ *a.e. on* $(\Omega, \mathfrak{F}_\infty, P)$.
2°. X_∞ *is a final element for X.*

Proof. If X is a supermartingale then $-X$ is a submartingale and $X^- = (-X)^+$. Thus if we prove the theorem for a submartingale then we have also a proof for a supermartingale. Furthermore since a martingale is both a submartingale and a supermartingale and since X is uniformly integrable if and only if both X^+ and X^- are, we also have a proof for a martingale. Therefore to prove the theorem it suffices to consider a submartingale. Let us assume that X is a submartingale.

Suppose X^+ is uniformly integrable. Then in particular X^+ is L_1-bounded. Since X is a submartingale the L_1-boundedness of X^+ implies that of X by Observation 7.1. Therefore by Theorem 7.4 in case $\mathbb{T} = \mathbb{Z}_+$ and by Theorem 7.9 in case $\mathbb{T} = \mathbb{R}_+$, there exists an extended real valued $\mathfrak{F}_\infty$-measurable integrable random variable X_∞ on $(\Omega, \mathfrak{F}, P)$ such that $\lim_{t \to \infty} X_t = X_\infty$ a.e. on $(\Omega, \mathfrak{F}_\infty, P)$. Then for every $a > 0$ we have

(1) $$\lim_{t \to \infty} X_t \vee (-a) = X_\infty \vee (-a) \quad \text{a.e. on } (\Omega, \mathfrak{F}_\infty, P).$$

Since $|X_t \vee (-a)| \leq X_t^+ + a$ and since X^+ is uniformly integrable, $X \vee (-a)$ is uniformly integrable by Proposition 4.11. The uniform integrability of $X \vee (-a)$ and the convergence in (1) imply according to Theorem 4.16 that

$$\lim_{t \to \infty} \|X_t \vee (-a) - X_\infty \vee (-a)\|_1 = 0.$$

Since convergence in L_1 of random variables implies convergence in L_1 of their conditional expectations (see Theorem B.26), for a fixed $t_0 \in \mathbb{T}$ we have

$$\lim_{t \to \infty} \|\mathbf{E}(X_t \vee (-a)|\mathfrak{F}_{t_0}) - \mathbf{E}(X_\infty \vee (-a)|\mathfrak{F}_{t_0})\|_1 = 0.$$

Since convergence of a sequence in L_1 implies the existence of a subsequence which converges a.e., there exists a sequence $\{t_n : n \in \mathbb{N}\}$, $t_n \uparrow \infty$ as $n \to \infty$, such that

(2) $$\lim_{n \to \infty} \mathbf{E}(X_{t_n} \vee (-a)|\mathfrak{F}_{t_0}) = \mathbf{E}(X_\infty \vee (-a)|\mathfrak{F}_{t_0}) \quad \text{a.e. on } (\Omega, \mathfrak{F}_{t_0}, P).$$

Since X is a submartingale, $X \vee (-a)$ is a submartingale by Theorem 5.9. Then for $n \in \mathbb{N}$ large enough so that $t_n > t_0$ we have by the submartingale property of $X \vee (-a)$

(3) $\quad\quad\quad\quad \mathbf{E}(X_{t_n} \vee (-a) | \mathfrak{F}_{t_0}) \geq X_{t_0} \vee (-a) \quad$ a.e. on $(\Omega, \mathfrak{F}_{t_0}, P)$.

By (2) and (3) we have

$$\mathbf{E}(X_\infty \vee (-a) | \mathfrak{F}_{t_0}) \geq X_{t_0} \vee (-a) \quad \text{a.e. on } (\Omega, \mathfrak{F}_{t_0}, P).$$

Letting $a \to \infty$, by the Conditional Monotone Convergence Theorem (see Theorem B.22) we have

$$\mathbf{E}(X_\infty | \mathfrak{F}_{t_0}) \geq X_{t_0} \quad \text{a.e. on } (\Omega, \mathfrak{F}_{t_0}, P).$$

Since this holds for every $t_0 \in \mathbb{T}$, the random variable X_∞ is a final element for X.

Conversely suppose that there exists a final element for X, that is, there exists an extended real valued integrable $\mathfrak{F}_\infty$-measurable random variable ξ on $(\Omega, \mathfrak{F}, P)$ such that for every $t \in \mathbb{T}$ we have

(4) $\quad\quad\quad\quad \mathbf{E}(\xi | \mathfrak{F}_t) \geq X_t \quad$ a.e. on $(\Omega, \mathfrak{F}_t, P)$.

Let us show that X^+ is uniformly integrable. Now (4) implies

$$(\mathbf{E}(\xi | \mathfrak{F}_t))^+ \geq X_t^+ \quad \text{a.e. on } (\Omega, \mathfrak{F}_t, P).$$

According to Jensen's Inequality for conditional expectations (see Theorem B.22), if η is an integrable random variable on $(\Omega, \mathfrak{F}, P)$, φ is a convex function on $\mathbb{R}$ such that $\varphi(\eta)$ is an integrable random variable and $\mathfrak{G}$ is an arbitrary sub-σ-algebra of $\mathfrak{F}$, then $\mathbf{E}(\varphi(\eta) | \mathfrak{G}) \geq \varphi(\mathbf{E}(\eta | \mathfrak{G}))$ a.e. on $(\Omega, \mathfrak{G}, P)$. Since $\varphi(x) = x \vee 0$ for $x \in \mathbb{R}$ is a convex function we have

$$\mathbf{E}(\xi^+ | \mathfrak{F}_t) \geq (\mathbf{E}(\xi | \mathfrak{F}_t))^+ \quad \text{a.e. on } (\Omega, \mathfrak{F}_t, P).$$

Combining the last two inequalities above we have

(5) $\quad\quad\quad\quad \mathbf{E}(\xi^+ | \mathfrak{F}_t) \geq X_t^+ \quad$ a.e. on $(\Omega, \mathfrak{F}_t, P)$.

By (5) we have $\mathbf{E}(X_t^+) \leq \mathbf{E}(\xi^+)$ for $t \in \mathbb{T}$ so that for every $\lambda > 0$ we have

$$P\{X_t^+ > \lambda\} \leq \frac{\mathbf{E}(X_t^+)}{\lambda} \leq \frac{\mathbf{E}(\xi^+)}{\lambda}$$

and therefore $\lim_{\lambda \to \infty} P\{X_t^+ > \lambda\} = 0$ uniformly in $t \in \mathbb{T}$, that is, for every $\eta > 0$ there exists some $\lambda_0 > 0$ such that $P\{X_t^+ > \lambda\} < \eta$ for all $t \in \mathbb{T}$ whenever $\lambda > \lambda_0$. Now the

§7. CONVERGENCE OF SUBMARTINGALES

integrability of ξ^+ implies that for every $\varepsilon > 0$ there exists $\eta > 0$ such that $\int_E \xi^+ dP < \varepsilon$ whenever $E \in \mathfrak{F}_\infty$ and $P(E) < \eta$. Thus for every $\varepsilon > 0$ there exists $\lambda_0 > 0$ such that

$$\sup_{t \in \mathbf{T}} \int_{\{X_t^+ > \lambda\}} \xi^+ \, dP \leq \varepsilon \quad \text{whenever } \lambda > \lambda_0.$$

The left side of the last inequality is a nonnegative decreasing function of λ and thus its limit as $\lambda \to \infty$ exists and is bounded between 0 and ε. By the arbitrariness of $\varepsilon > 0$ we have

(6) $$\lim_{\lambda \to \infty} \sup_{t \in \mathbf{T}} \int_{\{X_t^+ > \lambda\}} \xi^+ \, dP = 0.$$

Since $\{X_t^+ > \lambda\} \in \mathfrak{F}_t$, we have from (5)

(7) $$\int_{\{X_t^+ > \lambda\}} X_t^+ \, dP \leq \int_{\{X_t^+ > \lambda\}} \xi^+ \, dP.$$

By (6) and (7) we have

$$\lim_{\lambda \to \infty} \sup_{t \in \mathbf{T}} \int_{\{X_t^+ > \lambda\}} X_t^+ \, dP \leq \lim_{\lambda \to \infty} \sup_{t \in \mathbf{T}} \int_{\{X_t^+ > \lambda\}} \xi^+ \, dP = 0.$$

This proves the uniform integrability of X^+. ∎

[IV] Discrete Time L_2-Martingales

Let $(\Omega, \mathfrak{F}, P)$ be a probability space and consider $L_2(\Omega, \mathfrak{F}, P)$. If we define a binary function $\langle \cdot, \cdot \rangle$ on $L_2(\Omega, \mathfrak{F}, P)$ by $\langle \xi, \eta \rangle = \int_\Omega \xi \eta \, dP$ for $\xi, \eta \in L_2(\Omega, \mathfrak{F}, P)$, then $\langle \cdot, \cdot \rangle$ is an inner product and $L_2(\Omega, \mathfrak{F}, P)$ is a Hilbert space with respect to this inner product. For $\xi \in L_2(\Omega, \mathfrak{F}, P)$ we have $\|\xi\|_2 = \sqrt{\int_\Omega \xi^2 \, dP} = \sqrt{\langle \xi, \xi \rangle}$. If $\xi, \eta \in L_2(\Omega, \mathfrak{F}, P)$ and $\langle \xi, \eta \rangle = 0$, then we say that ξ and η are orthogonal.

Lemma 7.13. *An L_2-martingale $X = \{X_n : n \in \mathbb{Z}_+\}$ on a filtered space $(\Omega, \mathfrak{F}, \{\mathfrak{F}_n\}, P)$ is a stochastic process with orthogonal increments, that is, for any n, m, ℓ and k in $\mathbb{Z}_+$ such that $n > m \geq \ell > k$ we have*

(1) $$\langle X_n - X_m, X_\ell - X_k \rangle = 0.$$

In particular for every $n \in \mathbb{N}$ we have

(2) $$\mathbf{E}(X_n^2) = \mathbf{E}(X_0^2) + \sum_{k=1}^n \mathbf{E}[(X_k - X_{k-1})^2].$$

Proof. Let $n > m \geq \ell > k$. Then

$$\langle X_n - X_m, X_\ell \rangle = \mathbf{E}[(X_n - X_m)X_\ell] = \mathbf{E}[\mathbf{E}[(X_n - X_m)X_\ell | \mathfrak{F}_\ell]]$$
$$= \mathbf{E}[X_\ell \mathbf{E}[X_n - X_m | \mathfrak{F}_\ell]] = \mathbf{E}[X_\ell(X_\ell - X_\ell)] = 0.$$

Similarly we have $\langle X_n - X_m, X_k \rangle = 0$. Therefore (1) holds.

To prove (2), let us write $X_n = X_0 + \sum_{k=1}^{n}(X_k - X_{k-1})$. Then

$$\mathbf{E}(X_n^2) = \left\langle X_n = X_0 + \sum_{k=1}^{n}(X_k - X_{k-1}), X_n = X_0 + \sum_{k=1}^{n}(X_k - X_{k-1}) \right\rangle$$
$$= \langle X_0, X_0 \rangle + \sum_{k=1}^{n} \langle X_k - X_{k-1}, X_k - X_{k-1} \rangle = \mathbf{E}(X_0^2) + \sum_{k=1}^{n} \mathbf{E}[(X_k - X_{k-1})^2]. \blacksquare$$

Theorem 7.14. *Let $X = \{X_n : n \in \mathbb{Z}_+\}$ be an L_2-martingale on a filtered space $(\Omega, \mathfrak{F}, \{\mathfrak{F}_n\}, P)$. Then X is L_2-bounded if and only if*

(1) $$\sum_{k \in \mathbb{N}} \mathbf{E}[|X_k - X_{k-1}|^2] < \infty.$$

If X is L_2-bounded then there exists $X_\infty \in L_2(\Omega, \mathfrak{F}_\infty, P)$ such that

(2) $$\lim_{n \to \infty} X_n = X_\infty \quad a.e. \text{ on } (\Omega, \mathfrak{F}_\infty, P),$$

(3) $$\lim_{n \to \infty} \mathbf{E}[|X_n - X_\infty|^2] = 0,$$

and

(4) $$\mathbf{E}[|X_\infty - X_n|^2] = \sum_{k \geq n+1} \mathbf{E}[|X_k - X_{k-1}|^2].$$

Proof. Observe that $\sup_{n \in \mathbb{Z}_+} \mathbf{E}(X_n^2) < \infty$ if and only if $\sup_{n \in \mathbb{N}} \sum_{k=1}^{n} \mathbf{E}[|X_k - X_{k-1}|^2] < \infty$, that is, $\sum_{k \in \mathbb{N}} \mathbf{E}[|X_k - X_{k-1}|^2] < \infty$ by (2) of Lemma 7.13. Thus X is L_2-bounded if and only (1) holds.

Suppose X is L_2-bounded. Then X is L_1-bounded so that by Theorem 7.4 there exists $X_\infty \in L_1(\Omega, \mathfrak{F}_\infty, P)$ such that (2) holds. To show that $X_\infty \in L_2(\Omega, \mathfrak{F}_\infty, P)$ and (3) holds,

§7. CONVERGENCE OF SUBMARTINGALES

note that by (1) of Lemma 7.13 for any $n \in \mathbb{Z}_+$ and $p \in \mathbb{N}$ we have

$$(5) \quad \mathbf{E}[|X_{n+p} - X_n|^2] = \left\langle \sum_{k=n+1}^{n+p} (X_k - X_{k-1}), \sum_{k=n+1}^{n+p} (X_k - X_{k-1}) \right\rangle$$

$$= \sum_{k=n+1}^{n+p} \langle X_k - X_{k-1}, X_k - X_{k-1} \rangle = \sum_{k=n+1}^{n+p} \mathbf{E}[|X_k - X_{k-1}|^2].$$

By (2), Fatou's Lemma, and (5) we have

$$\mathbf{E}[|X_\infty - X_n|^2] = \mathbf{E}[|\lim_{p \to \infty} X_{n+p} - X_n|^2]$$

$$= \mathbf{E}[\lim_{p \to \infty} |X_{n+p} - X_n|^2] \leq \liminf_{p \to \infty} \mathbf{E}[|X_{n+p} - X_n|^2]$$

$$= \liminf_{p \to \infty} \sum_{k=n+1}^{n+p} \mathbf{E}[|X_k - X_{k-1}|^2] \leq \sum_{k \geq n+1} \mathbf{E}[|X_k - X_{k-1}|^2].$$

The last sum is finite by (1). This implies that $X_\infty - X_n \in L_2(\Omega, \mathfrak{F}_\infty, P)$ and then $X_\infty \in L_2(\Omega, \mathfrak{F}_\infty, P)$. Letting $n \to \infty$ in the last member of the inequalities above and recalling (1) we have (3).

Since $\lim_{n \to \infty} X_n = X_\infty$ in L_2 implies $\lim_{p \to \infty} (X_{n+p} - X_n) = X_\infty - X_n$ in L_2 and consequently $\lim_{p \to \infty} \mathbf{E}[|X_{n+p} - X_n|^2] = \mathbf{E}[|X_\infty - X_n|^2]$, by letting $p \to \infty$ on both sides of (5) we have (4). ∎

Let X be an L_2-martingale on a filtered space $(\Omega, \mathfrak{F}, \{\mathfrak{F}_n\}, P)$. Then X^2 is a submartingale so that according to Theorem 5.15 we have the Doob decomposition $X^2 = X_0^2 + M + A$ where M is a null at 0 martingale and A is a predictable almost surely increasing process and M and A are unique up to a null set in $(\Omega, \mathfrak{F}_\infty, P)$.

Definition 7.15. *Let X be an L_2-martingale on a filtered space $(\Omega, \mathfrak{F}, \{\mathfrak{F}_n\}, P)$. The predictable almost surely increasing process A in the Doob decomposition $X^2 = X_0^2 + M + A$ of the submartingale X^2 is called the quadratic variation process of X. It is uniquely determined up to a null set in $(\Omega, \mathfrak{F}_\infty, P)$.*

Theorem 7.16. *Let X be an L_2-martingale on a filtered space $(\Omega, \mathfrak{F}, \{\mathfrak{F}_n\}, P)$ and let A be the quadratic variational process of X. Let $A_\infty = \lim_{n \to \infty} A_n$. Then for every $n \in \mathbb{N}$ we have*

$$(1) \quad \mathbf{E}(X_n^2 - X_{n-1}^2 | \mathfrak{F}_{n-1}) = \mathbf{E}(|X_n - X_{n-1}|^2 | \mathfrak{F}_{n-1}) = A_n - A_{n-1} \quad \text{a.e. on } (\Omega, \mathfrak{F}_{n-1}, P)$$

and in particular

(2) $\quad E(X_n^2 - X_{n-1}^2) = E(|X_n - X_{n-1}|^2) = E(A_n - A_{n-1}).$

Also X is L_2-bounded if and only if A is integrable, that is, $E(A_\infty) < \infty$.

Proof. Let the Doob decomposition of the submartingale X^2 be given by

(3) $\quad X^2 = X_0^2 + M + A,$

where M is a null at 0 martingale. Then

$$E(X_n^2 - X_{n-1}^2 | \mathfrak{F}_{n-1}) = E[(M_n + A_n) - (M_{n-1} - A_{n-1}) | \mathfrak{F}_{n-1}]$$
$$= E(A_n - A_{n-1} | \mathfrak{F}_{n-1}) = A_n - A_{n-1}$$

by the martingale property of M and then by the predictability of A. On the other hand

$$E(|X_n - X_{n-1}|^2 | \mathfrak{F}_{n-1}) = E(X_n^2 - 2X_n X_{n-1} + X_{n-1}^2 | \mathfrak{F}_{n-1})$$
$$= E(X_n^2 | \mathfrak{F}_{n-1}) - 2X_{n-1} E(X_n | \mathfrak{F}_{n-1}) + X_{n-1}^2$$
$$= E(X_n^2 | \mathfrak{F}_{n-1}) - 2X_{n-1}^2 + X_{n-1}^2 = E(X_n^2 - X_{n-1}^2 | \mathfrak{F}_{n-1}).$$

This completes the proof of (1). Integrating both sides of (1) we have (2).

Since M is a null at 0 martingale we have $E(M_n) = E(M_0) = 0$ for $n \in \mathbb{N}$. Thus from (3) we have $E(X_n^2) = E(X_0^2) + E(A_n)$. Then

$$\sup_{n \in \mathbb{Z}_+} E(X_n^2) = E(X_0^2) + \sup_{n \in \mathbb{Z}_+} E(A_n)$$
$$= E(X_0^2) + \lim_{n \to \infty} E(A_n) = E(X_0^2) + E(A_\infty)$$

by the Monotone Convergence Theorem. Therefore $\sup_{n \in \mathbb{Z}_+} E(X_n^2) < \infty$ if and only if $E(A_\infty) < \infty$. ∎

§8 Uniformly Integrable Submartingales

[I] Convergence of Uniformly Integrable Submartingales

Let $X = \{X_t : t \in \mathbb{T}\}$ be a submartingale, martingale or supermartingale on a filtered space $(\Omega, \mathfrak{F}, \{\mathfrak{F}_t : t \in \mathbb{T}\}, P)$. In Theorem 7.9 we showed that if X is L_1-bounded then there

§8. UNIFORMLY INTEGRABLE SUBMARTINGALES

exists an extended real valued integrable $\mathfrak{F}_\infty$-measurable random variable X_∞ on $(\Omega, \mathfrak{F}, P)$ such that $\lim_{t \to \infty} X_t = X_\infty$ a.e. on $(\Omega, \mathfrak{F}_\infty, P)$. We show next that if X is uniformly integrable then $\lim_{t \to \infty} \|X_t - X_\infty\|_1 = 0$.

Theorem 8.1. *Let $X = \{X_t : t \in \mathbb{T}\}$ be a uniformly integrable submartingale, martingale or supermartingale on a filtered space $(\Omega, \mathfrak{F}, \{\mathfrak{F}_t : t \in \mathbb{T}\}, P)$. Then there exists an extended real valued integrable $\mathfrak{F}_\infty$-measurable random variable X_∞ on $(\Omega, \mathfrak{F}, P)$ such that*
1°. $\lim_{t \to \infty} X_t = X_\infty$ *a.e. on* $(\Omega, \mathfrak{F}_\infty, P)$.
2°. X_∞ *is a final element for X.*
3°. $\lim_{t \to \infty} \|X_t - X_\infty\|_1 = 0$.

Proof. If X is uniformly integrable then so are X^+ and X^-. Thus by Theorem 7.12 there exists an extended real valued integrable $\mathfrak{F}_\infty$-measurable random variable X_∞ on $(\Omega, \mathfrak{F}, P)$ satisfying 1° and 2°. Now 1° implies $P \cdot \lim_{t \to \infty} X_t = X_\infty$. This and the uniform integrability of X imply 3° according to Theorem 4.16. ∎

Let us note that under the assumption of uniform integrability of X, not just the uniform integrability of X^+ or X^- as in Theorem 7.12, 2° in Theorem 8.1 can be derived more directly than in the proof of Theorem 7.12. Let us show this for a submartingale X. Now for $s, t \in \mathbb{T}$, $t < s$, we have $\mathbf{E}(X_s | \mathfrak{F}_t) \geq X_t$ a.e. on $(\Omega, \mathfrak{F}_t, P)$ and thus

$$(1) \qquad \int_A X_s \, dP \geq \int_A X_t \, dP \quad \text{for } A \in \mathfrak{F}_t.$$

Now for $A \in \mathfrak{F}_t$ we have by 3° in Theorem 8.1

$$\lim_{s \to \infty} \left| \int_A X_s \, dP - \int_A X_\infty \, dP \right| \leq \lim_{s \to \infty} \int_A |X_s - X_\infty| \, dP \leq \lim_{s \to \infty} \|X_s - X_\infty\|_1 = 0$$

so that $\lim_{s \to \infty} \int_A X_s \, dP = \int_A X_\infty \, dP$. Thus, letting $s \to \infty$ in (1) we obtain

$$\int_A X_\infty \, dP \geq \int_A X_t \, dP \quad \text{for } A \in \mathfrak{F}_t.$$

Since this holds for every $A \in \mathfrak{F}_t$ we have $\mathbf{E}(X_\infty | \mathfrak{F}_t) \geq X_t$. Thus X_∞ is a final element for X.

The next theorem characterizes a uniformly integrable martingale as a martingale for which a final element exists.

Theorem 8.2. *Let $X = \{X_t : t \in \mathbb{T}\}$ be a martingale on a filtered space $(\Omega, \mathfrak{F}, \{\mathfrak{F}_t : t \in \mathbb{T}\}, P)$. Then X is uniformly integrable if and only if there exists a final element for X, that is, there exists an extended real valued integrable $\mathfrak{F}_\infty$-measurable random variable ξ on $(\Omega, \mathfrak{F}, P)$ such that $\mathbf{E}(\xi | \mathfrak{F}_t) = X_t$ a.e. on $(\Omega, \mathfrak{F}_t, P)$ for every $t \in \mathbb{T}$.*

Proof. If X is uniformly integrable then by Theorem 8.1 there exists a final element for X. Conversely if a final element ξ exists for X then $\mathbf{E}(\xi | \mathfrak{F}_t) = X_t$ a.e. on $(\Omega, \mathfrak{F}_t, P)$ for every $t \in \mathbb{T}$. Thus X is uniformly integrable by Theorem 4.24. ∎

Remark 8.3. As we noted following Definition 5.2, if $(\Omega, \mathfrak{F}, \{\mathfrak{F}_t : t \in \mathbb{T}\}, P)$ is a filtered space and ξ is an integrable random variable on a probability space $(\Omega, \mathfrak{F}, P)$ then by letting X_t be an arbitrary real valued version of $\mathbf{E}(\xi | \mathfrak{F}_t)$ for $t \in \mathbb{T}$ we obtain a uniformly integrable martingale $X = \{X_t : t \in \mathbb{T}\}$. Now the uniform integrability of X implies according to Theorem 8.1 the existence of an extended real valued integrable $\mathfrak{F}_\infty$-measurable random variable X_∞ which is a final element of the martingale X and to which X converges both almost surely and in L_1. But any version Y_∞ of $\mathbf{E}(\xi | \mathfrak{F}_\infty)$ is a final element of the martingale X. Thus by the uniqueness of the final element of a martingale according to Theorem 5.3 we have $Y_\infty = X_\infty$ a.e. on $(\Omega, \mathfrak{F}_\infty, P)$. In particular, we have $\lim_{t \to \infty} X_t = Y_\infty$ a.e. on $(\Omega, \mathfrak{F}_\infty, P)$ as well as $\lim_{t \to \infty} \|X_t - Y_\infty\|_1 = 0$, that is, we have

1°. $\lim_{t \to \infty} \mathbf{E}(\xi | \mathfrak{F}_t) = \mathbf{E}(\xi | \mathfrak{F}_\infty)$ a.e. on $(\Omega, \mathfrak{F}_\infty, P)$.

2°. $\lim_{t \to \infty} \|\mathbf{E}(\xi | \mathfrak{F}_t) - \mathbf{E}(\xi | \mathfrak{F}_\infty)\|_1 = 0$.

[II] Submartingales with Reversed Time

A submartingale $X = \{X_t : t \in \mathbb{T}\}$ has a first element but has no last element. We define a submartingale with reversed time as a process submartingale which has a last element but no first element. Our main question regarding a submartingale with reversed time is its convergence as the time parameter decreases. Here we confine ourselves to a discrete case only. We shall show that if a submartingale with reversed time $X = \{X_{-n} : n \in \mathbb{Z}_+\}$ is uniformly integrable then it has a first element to which X_{-n} converges both almost surely and in L_1 as $n \to \infty$.

Definition 8.4. *Let $(\Omega, \mathfrak{F}, P)$ be a probability space. A system of sub-σ-algebras of $\mathfrak{F}$, $\{\mathfrak{F}_{-n} : n \in \mathbb{Z}_+\}$ is called a filtration if it is an increasing system that is, $\mathfrak{F}_{-m} \subset \mathfrak{F}_{-n}$ for $m, n \in \mathbb{Z}_+$ such that $m > n$. We define $\mathfrak{F}_{-\infty} = \cap_{n \in \mathbb{Z}_+} \mathfrak{F}_{-n}$. A stochastic process $X = \{X_{-n} : n \in \mathbb{Z}_+\}$ on a filtered space $(\Omega, \mathfrak{F}, \{\mathfrak{F}_{-n} : n \in \mathbb{Z}_+\}, P)$ is said to be $\{\mathfrak{F}_{-n}\}$-adapted*

§8. UNIFORMLY INTEGRABLE SUBMARTINGALES

(or simply adapted) if X_{-n} is $\mathfrak{F}_{-n}$-measurable for every $n \in \mathbb{Z}_+$. An $\{\mathfrak{F}_{-n}\}$-adapted L_1-process $X = \{X_{-n} : n \in \mathbb{Z}_+\}$ is called a submartingale, martingale or supermartingale, with reversed time if $\mathbb{E}(X_{-n}|\mathfrak{F}_{-m}) \geq, =,$ or $\leq X_{-m}$ a.e. on $(\Omega, \mathfrak{F}_{-m}, P)$ for $n, m \in \mathbb{Z}_+$, $m > n$.

Lemma 8.5. *1) If $X = \{X_{-n} : n \in \mathbb{Z}_+\}$ is an adapted process on a filtered space $(\Omega, \mathfrak{F}, \{\mathfrak{F}_{-n}\}, P)$, then $\liminf_{n \to \infty} X_{-n}$ is an extended real valued $\mathfrak{F}_{-\infty}$-measurable random variable on $(\Omega, \mathfrak{F}, P)$.*
2) Every martingale, nonnegative submartingale and nonpositive supermartingale with reversed time is an L_1-bounded process.

Proof. 1) By definition, $\liminf_{n \to \infty} X_{-n} = \lim_{n \to \infty} \inf_{-m \leq -n} X_{-m}$. If $X = \{X_{-n} : n \in \mathbb{Z}_+\}$ is adapted with respect to $\{\mathfrak{F}_{-n} : n \in \mathbb{Z}_+\}$ then $\inf_{-m \leq -n} X_{-m}$ is $\mathfrak{F}_{-n}$-measurable for every $n \in \mathbb{Z}_+$ so that $\lim_{n \to \infty} \inf_{-m \leq -n} X_{-m}$ is $\mathfrak{F}_{-n}$-measurable for every $n \in \mathbb{Z}_+$ and thus it is $\cap_{n \in \mathbb{Z}_+} \mathfrak{F}_{-n}$-measurable. Therefore $\liminf_{n \to \infty} X_{-n}$ is $\mathfrak{F}_{-\infty}$-measurable.

2) If $X = \{X_{-n} : n \in \mathbb{Z}_+\}$ is a nonnegative submartingale with reversed time, then $\mathbb{E}(X_{-n}) \downarrow$ as $n \to \infty$ and thus $\mathbb{E}(X_{-n}) \leq \mathbb{E}(X_0)$ for all $n \in \mathbb{Z}_+$. Then by the nonnegativity of X we have $\sup_{n \in \mathbb{Z}_+} \mathbb{E}(|X_{-n}|) \leq \mathbb{E}(|X_0|) < \infty$. Thus X is L_1-bounded. If X is a martingale with reversed time, then $|X|$ is a nonnegative submartingale with reversed time so that $|X|$ is L_1-bounded, or equivalently, X is L_1-bounded. If X is a nonpositive supermartingale with reversed time, then $-X$ is a nonnegative submartingale with reversed time so that $-X$ is L_1-bounded, or equivalently, X is L_1-bounded. ∎

The next theorem shows that if $X = \{X_{-n} : n \in \mathbb{Z}_+\}$ is a submartingale with reversed time then $X_{-\infty} = \lim_{n \to \infty} X_{-n}$ always exists a.e. on $(\Omega, \mathfrak{F}_{-\infty}, P)$. If X is uniformly integrable, then $X_{-\infty}$ is integrable and X_{-n} converges to $X_{-\infty}$ as $n \to \infty$ both almost surely and in L_1 and furthermore $X_{-\infty}$ is a first element of the submartingale with reversed time.

Theorem 8.6. *Let $X = \{X_{-n} : n \in \mathbb{Z}_+\}$ be a submartingale with reversed time on a filtered space $(\Omega, \mathfrak{F}, \{\mathfrak{F}_{-n}\}, P)$. Let $X_{-\infty}$ be an extended real valued $\mathfrak{F}_{-\infty}$-measurable random variable on $(\Omega, \mathfrak{F}, P)$ defined by $X_{-\infty}(\omega) = \liminf_{n \to \infty} X_{-n}(\omega)$ for $\omega \in \Omega$. Then*

(1) $$\lim_{n \to \infty} X_{-n} = X_{-\infty} \quad \text{a.e. on } (\Omega, \mathfrak{F}_{-\infty}, P).$$

If X is uniformly integrable then

(2) $\quad X_{-\infty} \in L_1(\Omega, \mathfrak{F}_{-\infty}, P) \quad \text{and} \quad \lim_{n \to \infty} \|X_{-n} - X_{-\infty}\|_1 = 0,$

and for every $n \in \mathbb{Z}_+$ we have

(3) $\quad \mathbf{E}(X_{-n} | \mathfrak{F}_{-\infty}) \geq X_{-\infty} \quad \text{a.e. on } (\Omega, \mathfrak{F}_{-\infty}, P).$

If X is a supermartingale, then the inequality in (3) is reversed. If X is a martingale, then the inequality is replaced by an equality.

Proof. For $a, b \in \mathbb{R}$, $a < b$, consider the upcrossing number $U_{[a,b]}^{-N} X$ of $[a, b]$ by the finite sequence $\{X_{-N}, \ldots, X_0\}$. If X is a submartingale with reversed time then by Theorem 6.19

$$\mathbf{E}(U_{[a,b]}^{-N} X) \leq \frac{1}{b-a} \mathbf{E}[(X_0 - a)^+ - (X_{-N} - a)^+] \leq \frac{1}{b-a} \mathbf{E}[(X_0 - a)^+]$$

and if X is a supermartingale with reversed time then by Theorem 6.20 we have

$$\mathbf{E}(U_{[a,b]}^{-N} X) \leq \frac{1}{b-a} \mathbf{E}[(X_0 - a)^-].$$

Therefore for both cases we have

$$\mathbf{E}\left(U_{[a,b]}^{-N} X\right) \leq \frac{1}{b-a} \{\mathbf{E}(|X_0|) + |a|\}.$$

Let us define the upcrossing number of $[a, b]$ by $X = \{X_{-n} : n \in \mathbb{Z}_+\}$ by

$$U_{[a,b]}^{-\infty} X = \uparrow \lim_{N \to \infty} U_{[a,b]}^{-N} X.$$

Then by the Monotone Convergence Theorem

$$\mathbf{E}(U_{[a,b]}^{-\infty} X) = \lim_{N \to \infty} \mathbf{E}(U_{[a,b]}^{-N} X) \leq \frac{1}{b-a} \{\mathbf{E}(|X_0|) + |a|\} < \infty.$$

Thus $U_{[a,b]}^{-\infty} X < \infty$ a.e. on $(\Omega, \mathfrak{F}_{-\infty}, P)$. By the same argument as in Lemma 7.3 we have for any $a, b \in \mathbb{R}$, $a < b$

$$\liminf_{n \to \infty} X_{-n}(\omega) < a < b < \limsup_{n \to \infty} X_{-n}(\omega) \Rightarrow (U_{[a,b]}^{-\infty} X)(\omega) = \infty.$$

Then the fact that $U_{[a,b]}^{-\infty} X < \infty$ a.e. on $(\Omega, \mathfrak{F}_{-\infty}, P)$ implies that we have $\liminf_{n \to \infty} X_{-n}(\omega) = \limsup_{n \to \infty} X_{-n}(\omega)$ which in turn implies that $\lim_{n \to \infty} X_{-n}(\omega)$ exists in $\overline{\mathbb{R}}$ for

§8. UNIFORMLY INTEGRABLE SUBMARTINGALES

a.e. ω in $(\Omega, \mathfrak{F}_{-\infty}, P)$. Thus for our $X_{-\infty} = \liminf_{n \to \infty} X_{-n}$ we have $X_{-\infty} = \lim_{n \to \infty} X_{-n}$ a.e. in $(\Omega, \mathfrak{F}_{-\infty}, P)$. Now by Fatou's Lemma we have

$$\mathbf{E}(|X_{-\infty}|) = \mathbf{E}(\lim_{n \to \infty} |X_{-n}|) \leq \liminf_{n \to \infty} \mathbf{E}(|X_{-n}|) \leq \sup_{n \in \mathbb{Z}_+} \mathbf{E}(|X_{-n}|).$$

If we assume that X is uniformly integrable, then X is L_1-bounded so that the last member in the inequalities above is finite and therefore $X_{-\infty}$ is integrable. Then $X_{-\infty}$ is finite a.e. on $(\Omega, \mathfrak{F}_{-\infty}, P)$ and therefore X_{-n} converges to $X_{-\infty}$ a.e. on $(\Omega, \mathfrak{F}_{-\infty}, P)$ as $n \to \infty$. The uniform integrability of X then implies (2) by Theorem 4.16.

To prove (3) it suffices to show that for every $n \in \mathbb{Z}_+$

(4) $$\int_A X_{-n} \, dP \geq \int_A X_{-\infty} \, dP \quad \text{for } A \in \mathfrak{F}_{-\infty}.$$

Note that for any $A \in \mathfrak{F}$ and $k \in \mathbb{Z}_+$ we have

$$\left| \int_A X_{-k} \, dP - \int_A X_{-\infty} \, dP \right| \leq \int_A |X_{-k} - X_{-\infty}| \, dP \leq \|X_{-k} - X_{-\infty}\|_1$$

so that by (2) we have

(5) $$\lim_{k \to \infty} \int_A X_{-k} \, dP = \int_A X_{-\infty} \, dP \quad \text{for } A \in \mathfrak{F}.$$

Now for $m, n \in \mathbb{Z}_+$, $m > n$, since $\mathbf{E}(X_{-n} | \mathfrak{F}_{-m}) \geq X_{-m}$ a.e. on $(\Omega, \mathfrak{F}_{-m}, P)$ we have $\int_A X_{-n} \, dP \geq \int_A X_{-m} \, dP$ for $A \in \mathfrak{F}_{-m}$. Then since $\mathfrak{F}_{-\infty} \subset \mathfrak{F}_{-m}$ we have

(6) $$\int_A X_{-n} \, dP \geq \int_A X_{-m} \, dP \quad \text{for } A \in \mathfrak{F}_{-\infty}.$$

Then applying (5) to the right side of (6) we have (4). ∎

Let us remark that if a first element exists for a martingale with reversed time then it is unique. To show this, let $X = \{X_{-n} : n \in \mathbb{Z}_+\}$ be a martingale with reversed time on a filtered space $(\Omega, \mathfrak{F}, \{\mathfrak{F}_{-n} : n \in \mathbb{Z}_+\}, P)$. If $X_{-\infty}$ and $Y_{-\infty}$ are first elements of X then $\mathbf{E}(X_0 | \mathfrak{F}_{-\infty}) = X_{-\infty}$ and $\mathbf{E}(X_0 | \mathfrak{F}_{-\infty}) = Y_{-\infty}$ a.e. on $(\Omega, \mathfrak{F}_{-\infty}, P)$ so that $X_{-\infty} = Y_{-\infty}$ a.e. on $(\Omega, \mathfrak{F}_{-\infty}, P)$.

The next theorem concerns a martingale with reversed time generated by an integrable random variable.

Theorem 8.7. Let $(\Omega, \mathfrak{F}, \{\mathfrak{F}_{-n} : n \in \mathbb{Z}_+\}, P)$ be a filtered space and ξ be an extended real valued integrable random variable on $(\Omega, \mathfrak{F}, P)$. Let X_{-n} be an arbitrary version of $\mathbf{E}(\xi | \mathfrak{F}_{-n})$ for $n \in \mathbb{Z}_+$ and let $Y_{-\infty}$ be an arbitrary version of $\mathbf{E}(\xi | \mathfrak{F}_{-\infty})$. Then for the uniformly integrable martingale with reversed time $X = \{X_{-n} : n \in \mathbb{Z}_+\}$ we have

(1) $$\lim_{n \to \infty} X_{-n} = Y_{-\infty} \quad \text{a.e. on } (\Omega, \mathfrak{F}_{-\infty}, P)$$

and

(2) $$\lim_{n \to \infty} \|X_{-n} - Y_{-\infty}\|_1 = 0.$$

Proof. By Theorem 4.24, X is a uniformly integrable martingale with reversed time. Then there exists an extended real valued integrable $\mathfrak{F}_{-\infty}$-measurable random variable $X_{-\infty}$ such that $\lim_{n \to \infty} X_{-n} = X_{-\infty}$ a.e. on $(\Omega, \mathfrak{F}_{-\infty}, P)$ and $\lim_{n \to \infty} \|X_{-n} - X_{-\infty}\|_1 = 0$ by Theorem 8.6. Therefore, to prove (1) and (2) it suffices to show that $Y_{-\infty} = X_{-\infty}$ a.e. on $(\Omega, \mathfrak{F}_{-\infty}, P)$. Since $Y_{-\infty}$ and $X_{-\infty}$ are both $\mathfrak{F}_{-\infty}$-measurable it suffices to verify

(3) $$\int_A Y_{-\infty} \, dP = \int_A X_{-\infty} \, dP \quad \text{for } A \in \mathfrak{F}_{-\infty}.$$

Let $A \in \mathfrak{F}_{-\infty}$. Since $\mathfrak{F}_{-\infty} = \cap_{n \in \mathbb{Z}_+} \mathfrak{F}_{-n}$ we have $A \in \mathfrak{F}_{-n}$ for every $n \in \mathbb{Z}_+$. Since $\mathbf{E}(\xi | \mathfrak{F}_{-n}) = X_{-n}$ a.e. on $(\Omega, \mathfrak{F}_{-n}, P)$ and $A \in \mathfrak{F}_{-n}$ we have

(4) $$\int_A \xi \, dP = \int_A X_{-n} \, dP \quad \text{for all } n \in \mathbb{Z}_+.$$

Since $Y_{-\infty}$ is a version of $\mathbf{E}(\xi | \mathfrak{F}_{-\infty})$, we have

(5) $$\int_A Y_{-\infty} \, dP = \int_A \xi \, dP \quad \text{for } A \in \mathfrak{F}_{-\infty}.$$

But $\lim_{n \to \infty} \|X_{-n} - X_{-\infty}\|_1 = 0$ implies

$$\lim_{n \to \infty} \int_A X_{-n} \, dP = \int_A X_{-\infty} \, dP \quad \text{for } A \in \mathfrak{F}.$$

Thus, letting $n \to \infty$ in (4) we have

(6) $$\int_A \xi \, dP = \int_A X_{-\infty} \, dP \quad \text{for } A \in \mathfrak{F}_{-\infty}.$$

§8. UNIFORMLY INTEGRABLE SUBMARTINGALES

With (5) and (6) we have (3). ∎

In Theorem 8.6 we showed that if a submartingale with reversed time $X = \{X_{-n} : n \in \mathbb{Z}_+\}$ is uniformly integrable then it has a first element to which it converges both almost surely and in L_1 as $n \to \infty$. In the next theorem we give a sufficient condition for a submartingale with reversed time to be uniformly integrable. This condition is derived from the fact that a submartingale with reversed time has a last element, namely X_0.

Theorem 8.8. *If $X = \{X_{-n} : n \in \mathbb{Z}_+\}$ is a submartingale with reversed time on a filtered space $(\Omega, \mathfrak{F}, \{\mathfrak{F}_{-n}\}, P)$ and satisfies the condition*

$$(1) \qquad \inf_{n \in \mathbb{Z}_+} \mathbf{E}(X_{-n}) > -\infty,$$

then X is uniformly integrable. Similarly if X is a supermartingale with reversed time and satisfies the condition

$$(2) \qquad \sup_{n \in \mathbb{Z}_+} \mathbf{E}(X_{-n}) < \infty,$$

then X is uniformly integrable.

Proof. Let X be a submartingale with reversed time satisfying (1). To show that X is uniformly integrable, we show according to (3) of Definition 4.6 that for every $\varepsilon > 0$ there exists $\lambda > 0$ such that

$$(3) \qquad \int_{\{|X_{-n}| > \lambda\}} |X_{-n}| \, dP < \varepsilon \quad \text{for all } n \in \mathbb{Z}_+.$$

Now since X is a submartingale with reversed time, $\mathbf{E}(X_{-n}) \downarrow$ as $-n \downarrow$. By (1), $\mathbf{E}(X_{-n}) \downarrow c \in \mathbb{R}$ as $-n \downarrow -\infty$. Then for every $\varepsilon > 0$ there exists $N \in \mathbb{Z}_+$ such that $\mathbf{E}(X_{-N}) - c < \varepsilon/2$ and therefore

$$(4) \qquad \mathbf{E}(X_{-N}) - \mathbf{E}(X_{-n}) < \varepsilon/2 \quad \text{for } n > N.$$

Now for an arbitrary $\lambda > 0$ and $n \in \mathbb{Z}_+$, we have

$$(5) \qquad \int_{\{|X_{-n}| > \lambda\}} |X_{-n}| \, dP = \int_{\{X_{-n} > \lambda\}} X_{-n} \, dP - \int_{\{X_{-n} < -\lambda\}} X_{-n} \, dP$$
$$= \int_{\{X_{-n} > \lambda\}} X_{-n} \, dP + \int_{\{X_{-n} \geq -\lambda\}} X_{-n} \, dP - \int_{\Omega} X_{-n} \, dP.$$

By the submartingale property of X we have $\mathbf{E}(X_{-N}|\mathfrak{F}_{-n}) \geq X_{-n}$ a.e. on $(\Omega, \mathfrak{F}_{-n}, P)$ for $n > N$. Since $\{X_{-n} > \lambda\}, \{X_{-n} \geq -\lambda\} \in \mathfrak{F}_{-n}$ we have from (5) and (4)

$$
\begin{aligned}
(6) \quad & \int_{\{|X_{-n}|>\lambda\}} |X_{-n}| \, dP \\
& \leq \int_{\{X_{-n}>\lambda\}} X_{-N} \, dP + \int_{\{X_{-n}\geq -\lambda\}} X_{-N} \, dP - \int_{\Omega} X_{-N} \, dP + \frac{\varepsilon}{2} \\
& = \int_{\{X_{-n}>\lambda\}} X_{-N} \, dP - \int_{\{X_{-n}<-\lambda\}} X_{-N} \, dP + \frac{\varepsilon}{2} \\
& \leq \int_{\{X_{-n}>\lambda\}} |X_{-N}| \, dP + \int_{\{X_{-n}<-\lambda\}} |X_{-N}| \, dP + \frac{\varepsilon}{2} \\
& = \int_{\{|X_{-n}|>\lambda\}} |X_{-N}| \, dP + \frac{\varepsilon}{2} \quad \text{for } n > N.
\end{aligned}
$$

Since $X_0, \ldots, X_{-N}$ are all integrable there exists $\delta > 0$ such that

$$
(7) \quad \int_A |X_0| \, dP, \ldots, \int_A |X_{-N}| \, dP < \frac{\varepsilon}{2} \quad \text{for } A \in \mathfrak{F} \text{ with } P(A) < \delta.
$$

Since $|X_{-n}| = 2X_{-n}^+ - X_{-n}$ for every $n \in \mathbb{Z}_+$ we have

$$
P\{|X_{-n}| > \lambda\} \leq \frac{1}{\lambda} \mathbf{E}(|X_{-n}|) = \frac{1}{\lambda} \{2\mathbf{E}(X_{-n}^+) - \mathbf{E}(X_{-n})\}.
$$

Since X is a submartingale with reversed time so is X^+ and thus $\mathbf{E}(X_{-n}^+) \leq \mathbf{E}(X_0^+)$. Recalling that $\mathbf{E}(X_{-n}) \downarrow c$ as $-n \downarrow -\infty$, we have

$$
P\{|X_{-n}| > \lambda\} \leq \frac{1}{\lambda} \{2\mathbf{E}(X_0^+) - c\} < \infty \quad \text{for all } n \in \mathbb{Z}_+.
$$

Therefore there exists $\lambda > 0$ so large that $P\{|X_{-n}| > \lambda\} < \delta$ for all $n \in \mathbb{Z}_+$. For such λ, by (7) and (6) we have $\int_{\{|X_{-n}|>\lambda\}} |X_{-n}| \, dP < \varepsilon$ for all $n \in \mathbb{Z}_+$, which proves (3) and thus establishes the uniform integrability of the submartingale with reversed time.

If X is a supermartingale with reversed time and satisfies (2) then $-X$ is a submartingale with reversed time and satisfies $\inf_{n\in\mathbb{Z}_+} \mathbf{E}(-X_n) > -\infty$. Thus $-X$ is uniformly integrable and so is X. ∎

Corollary 8.9. *Let $X = \{X_t : t \in \mathbb{R}_+\}$ be a submartingale or a supermartingale on a filtered space $(\Omega, \mathfrak{F}, \{\mathfrak{F}_t\}, P)$. Let $\{t_{-n} : n \in \mathbb{Z}_+\}$ be a strictly decreasing sequence in $\mathbb{R}_+$ with $t_{-n} \downarrow t_{-\infty} \in \mathbb{R}_+$. Let $\mathfrak{F}_{t_{-\infty}} = \cap_{n\in\mathbb{Z}_+} \mathfrak{F}_{t_{-n}}$. Then $\{X_{t_{-n}} : n \in \mathbb{Z}_+\}$ is uniformly*

§8. UNIFORMLY INTEGRABLE SUBMARTINGALES

integrable and there exists $Y_{t_{-\infty}} \in L_1(\Omega, \mathfrak{F}_{t_{-\infty}}, P)$ such that $\lim_{n \to \infty} X_{t_{-n}} = Y_{t_{-\infty}}$ a.e. on $(\Omega, \mathfrak{F}_{t_{-\infty}}, P)$ and $\lim_{n \to \infty} \|X_{t_{-n}} - Y_{t_{-\infty}}\|_1 = 0$.

Proof. Since $t_{-n} \downarrow$ as $n \uparrow$, $\{X_{t_{-n}} : n \in \mathbb{Z}_+\}$ is a submartingale or a supermartingale with reversed time with respect to $\{\mathfrak{F}_{t_{-n}} : n \in \mathbb{Z}_+\}$ according as X is a submartingale or a supermartingale. Now if X is a submartingale then since $t_{-n} \geq 0$ for $n \in \mathbb{Z}_+$ we have

$$\inf_{n \in \mathbb{Z}_+} \mathbf{E}(X_{t_{-n}}) \geq \mathbf{E}(X_0) > -\infty,$$

and if X is a supermartingale then

$$\sup_{n \in \mathbb{Z}_+} \mathbf{E}(X_{t_{-n}}) \leq \mathbf{E}(X_0) < \infty.$$

Therefore by Theorem 8.8, $\{X_{t_{-n}} : n \in \mathbb{Z}_+\}$ is uniformly integrable. Thus by Theorem 8.6 (with the σ-algebra $\mathfrak{F}_{t_{-\infty}} = \cap_{n \in \mathbb{Z}_+} \mathfrak{F}_{t_{-n}}$ corresponding to $\mathfrak{F}_{-\infty} = \cap_{n \in \mathbb{Z}_+} \mathfrak{F}_{-n}$ in Theorem 8.6), there exists $Y_{t_{-\infty}} \in L_1(\Omega, \mathfrak{F}_{t_{-\infty}}, P)$ to which $\{X_{t_{-n}} : n \in \mathbb{Z}_+\}$ converges both almost surely and in L_1. ∎

[III] Optional Sampling by Unbounded Stopping Times

According to Theorem 6.4, if $X = \{X_n : n \in \mathbb{Z}_+\}$ is a submartingale and S and T are two bounded stopping times on a filtered space $(\Omega, \mathfrak{F}, \{\mathfrak{F}_n\}, P)$ and if $S \leq T$ on Ω then $\mathbf{E}(X_T | \mathfrak{F}_S) \geq X_S$ a.e. on $(\Omega, \mathfrak{F}_S, P)$ where the inequality is replaced by an equality when X is a martingale. Let us extend this result to the case $\mathbb{T} = \mathbb{R}_+$.

Theorem 8.10. (Optional Sampling with Bounded Stopping Times, Continuous Case) *Let $X = \{X_t : t \in \mathbb{R}_+\}$ be a right-continuous submartingale and S and T be two bounded stopping times on a right-continuous filtered space $(\Omega, \mathfrak{F}, \{\mathfrak{F}_t\}, P)$ such that $S \leq T$ on Ω. Then X_S and X_T are integrable and $\mathbf{E}(X_T | \mathfrak{F}_S) \geq X_S$ a.e. on $(\Omega, \mathfrak{F}_S, P)$. If X is a supermartingale, then the inequality is reversed. If X is a martingale, then the inequality is replaced by an equality.*

Proof. Let S and T be two bounded stopping times such that $S \leq T \leq m$ on Ω for some $m \in \mathbb{N}$. According to Theorem 3.20 there exist two sequences of stopping times $\{S_n : n \in \mathbb{N}\}$ and $\{T_n : n \in \mathbb{N}\}$ with respect to the filtration $\{\mathfrak{F}_t : t \in \mathbb{R}_+\}$ such that S_n and T_n assume values in $\{k2^{-n} : k = 1, \ldots, m2^n\}$, $S_n \leq T_n$ on Ω for each $n \in \mathbb{N}$, $S_n \downarrow S$ and $T_n \downarrow T$ on Ω as $n \to \infty$. Now since the ranges of S_n and T_n are contained in

$\{k2^{-n} : k = 1, \ldots, m2^n\}$, S_n and T_n are also stopping times with respect to the filtration $\{\mathfrak{F}_{k2^{-n}} : k \in \mathbb{Z}_+\}$. Therefore for each $n \in \mathbb{N}$ we have by Theorem 6.4, $\mathbf{E}(X_{T_n}|\mathfrak{F}_{S_n}) \geq X_{S_n}$ a.e. on $(\Omega, \mathfrak{F}_{S_n}, P)$ so that

(1) $$\int_A X_{T_n} dP \geq \int_A X_{S_n} dP \quad \text{for } A \in \mathfrak{F}_{S_n}.$$

To prove the theorem we show that

(2) $$\int_A X_T dP \geq \int_A X_S dP \quad \text{for } A \in \mathfrak{F}_S.$$

Now since $T_n \downarrow T$ and $S_n \downarrow S$ on Ω and since X is right-continuous we have

$$\lim_{n \to \infty} X_{T_n} = X_T \quad \text{and} \quad \lim_{n \to \infty} X_{S_n} = X_S \quad \text{on } \Omega.$$

If we show that $\{X_{T_n} : n \in \mathbb{N}\}$ and $\{X_{S_n} : n \in \mathbb{N}\}$ are uniformly integrable, then by Theorem 4.16 we have $X_T \in L_1(\Omega, \mathfrak{F}_T, P)$, $X_S \in L_1(\Omega, \mathfrak{F}_S, P)$, $\lim_{n \to \infty} \|X_{T_n} - X_T\|_1 = 0$ and $\lim_{n \to \infty} \|X_{S_n} - X_S\|_1 = 0$. From this, we have

$$\lim_{n \to \infty} \int_A X_{T_n} dP = \int_A X_T dP \quad \text{and} \quad \lim_{n \to \infty} \int_A X_{S_n} dP = \int_A X_S dP \quad \text{for } A \in \mathfrak{F}.$$

This and (1) then yield (2). Thus it remains to verify the uniform integrability of $\{X_{T_n} : n \in \mathbb{N}\}$ and $\{X_{S_n} : n \in \mathbb{N}\}$.

Now for each $n \in \mathbb{N}$, T_n is a stopping time with respect to the filtration $\{\mathfrak{F}_{k2^{-n}} : k \in \mathbb{Z}_+\}$. Since, by the construction in Theorem 3.20, the range of T_{n-1} is contained in that of T_n, T_{n-1} is also a stopping time with respect to the filtration $\{\mathfrak{F}_{k2^{-n}} : k \in \mathbb{Z}_+\}$. Then since $T_n \leq T_{n-1}$, Theorem 6.4 implies that $\mathbf{E}(X_{T_{n-1}}|\mathfrak{F}_{T_n}) \geq X_{T_n}$. Recall also that $T_n \leq T_{n-1}$ implies $\mathfrak{F}_{T_n} \subset \mathfrak{F}_{T_{n-1}}$. Thus $\{X_{T_n} : n \in \mathbb{N}\}$ is a submartingale with reversed time with respect to the filtration $\{\mathfrak{F}_{T_n} : n \in \mathbb{N}\}$. Note also that from $\mathbf{E}(X_{T_n}|\mathfrak{F}_0) \geq X_0$ a.e. on $(\Omega, \mathfrak{F}_0, P)$ we have $\mathbf{E}(X_{T_n}) \geq \mathbf{E}(X_0)$. Then $\inf_{n \in \mathbb{N}} \mathbf{E}(X_{T_n}) \geq \mathbf{E}(X_0) > -\infty$. Therefore by applying Theorem 8.8 to our submartingale with reversed time $\{X_{T_n} : n \in \mathbb{N}\}$ with respect to the filtration $\{\mathfrak{F}_{T_n} : n \in \mathbb{N}\}$ we have the uniform integrability of $\{X_{T_n} : n \in \mathbb{N}\}$. Similarly for $\{X_{S_n} : n \in \mathbb{N}\}$. ∎

Theorem 8.10 can be restated as follows.

Theorem 8.11. *Let $X = \{X_t : t \in \mathbb{R}_+\}$ be a right-continuous submartingale, martingale or supermartingale on a right-continuous filtered space $(\Omega, \mathfrak{F}, \{\mathfrak{F}_t\}, P)$. Let $\{S_t : t \in \mathbb{R}_+\}$ be*

§8. UNIFORMLY INTEGRABLE SUBMARTINGALES

an increasing system of stopping times on the filtered space each of which is bounded. Let $\mathfrak{G}_t = \mathfrak{F}_{S_t}$ and $Y_t = X_{S_t}$ for $t \in \mathbb{R}_+$. Then $Y = \{Y_t : t \in \mathbb{R}_+\}$ is a submartingale, martingale or supermartingale on the filtered space $(\Omega, \mathfrak{F}, \{\mathfrak{G}_t\}, P)$.

Let $X = \{X_t : t \in \mathbb{T}\}$ be an adapted process and T be a stopping time on a filtered space $(\Omega, \mathfrak{F}, \{\mathfrak{F}_t\}, P)$ and consider the stopped process $X^{T\wedge} = \{X_t^{T\wedge} : t \in \mathbb{T}\}$ where $X_t^{T\wedge}(\omega) = X(T(\omega) \wedge t, \omega)$ for $\omega \in \Omega$. In Theorem 3.31, we showed that if $\mathbb{T} = \mathbb{Z}_+$ then $X^{T\wedge}$ is adapted to the filtration $\{\mathfrak{F}_{T\wedge n} : n \in \mathbb{Z}_+\}$ and if $\mathbb{T} = \mathbb{R}_+$ then, under the assumption that both X and $\{\mathfrak{F}_t : t \in \mathbb{R}_+\}$ are right-continuous, $X^{T\wedge}$ is a right-continuous process adapted to the right-continuous filtration $\{\mathfrak{F}_{T\wedge t} : t \in \mathbb{R}_+\}$.

Theorem 8.12. *Let $X = \{X_t : t \in \mathbb{T}\}$ be a submartingale, martingale, or supermartingale and T be a stopping time on a filtered space $(\Omega, \mathfrak{F}, \{\mathfrak{F}_t\}, P)$.*
1) If $\mathbb{T} = \mathbb{Z}_+$, then $X^{T\wedge}$ is a submartingale, martingale, or supermartingale with respect to the filtration $\{\mathfrak{F}_{T\wedge n} : n \in \mathbb{Z}_+\}$ as well as the filtration $\{\mathfrak{F}_n : n \in \mathbb{Z}_+\}$.
2) If $\mathbb{T} = \mathbb{R}_+$, then under the assumption that both X and $\{\mathfrak{F}_t : t \in \mathbb{R}_+\}$ are right-continuous, $X^{T\wedge}$ is a right-continuous submartingale, martingale, or supermartingale with respect to the filtration $\{\mathfrak{F}_{T\wedge t} : t \in \mathbb{R}_+\}$ as well as the filtration $\{\mathfrak{F}_t : t \in \mathbb{R}_+\}$.

Proof. It suffices to prove the theorem for the case where X is a submartingale. Since $X^{T\wedge}$ is adapted to $\{\mathfrak{F}_{T\wedge t} : t \in \mathbb{T}\}$ according to Theorem 3.31 it is also adapted to $\{\mathfrak{F}_t : t \in \mathbb{T}\}$. Thus to show that $X^{T\wedge}$ is a submartingale with respect to $\{\mathfrak{F}_{T\wedge t} : t \in \mathbb{T}\}$ as well as $\{\mathfrak{F}_t : t \in \mathbb{T}\}$, we need only verify that for any pair $s, t \in \mathbb{T}$, $s < t$, we have

$$(1) \qquad \mathbf{E}(X_{T\wedge t}|\mathfrak{F}_{T\wedge s}) \geq X_{T\wedge s} \quad \text{a.e. on } (\Omega, \mathfrak{F}_{T\wedge s}, P),$$

and

$$(2) \qquad \mathbf{E}(X_{T\wedge t}|\mathfrak{F}_s) \geq X_{T\wedge s} \quad \text{a.e. on } (\Omega, \mathfrak{F}_s, P),$$

But (1) holds by the Optional Sampling Theorems, that is, Theorem 6.4 for $\mathbb{T} = \mathbb{Z}_+$ and Theorem 8.10 for $\mathbb{T} = \mathbb{R}_+$. Let us prove (2), or equivalently,

$$(3) \qquad \int_A X_{T\wedge t}\, dP \geq \int_A X_{T\wedge s}\, dP \quad \text{for } A \in \mathfrak{F}_s.$$

Note first that by (1) we have

$$(4) \qquad \int_{A_0} X_{T\wedge t}\, dP \geq \int_{A_0} X_{T\wedge s}\, dP \quad \text{for } A_0 \in \mathfrak{F}_{T\wedge s}.$$

Let $A \in \mathfrak{F}_s$. Decompose A into two subsets $A_1 = A \cap \{T > s\}$ and $A_2 = A \cap \{T \leq s\}$ both of which are in $\mathfrak{F}_s$ since $\{T \leq s\} \in \mathfrak{F}_s$. For $u \in \mathbb{T}$, consider the set

$$A_1 \cap \{T \wedge s \leq u\} = A \cap \{T > s\} \cap \{T \wedge s \leq u\}.$$

If $u < s$, then the set is equal to $\emptyset$ and hence in $\mathfrak{F}_u$. If on the other hand $u \geq s$, then the set is equal to $A \cap \{T > s\} = A_1 \in \mathfrak{F}_s \subset \mathfrak{F}_u$. In any case the set is in $\mathfrak{F}_u$ and therefore $A_1 \in \mathfrak{F}_{T \wedge s}$ by Definition 3.2. Then by (4)

$$(5) \qquad \int_{A_1} X_{T \wedge t}\, dP \geq \int_{A_1} X_{T \wedge s}\, dP.$$

On the other hand since $s < t$, on the set $\{T \leq s\}$ we have $X_{T \wedge t} = X_T = X_{T \wedge s}$ and thus by the fact that A_2 is a subset of $\{T \leq s\}$ we have

$$(6) \qquad \int_{A_2} X_{T \wedge t}\, dP = \int_{A_2} X_{T \wedge s}\, dP.$$

Adding (5) and (6) we have (3). ∎

By similar argument we have the following theorem.

Theorem 8.13. *Let $X = \{X_t : t \in \mathbb{T}\}$ be a submartingale, martingale, or supermartingale on a filtered space $(\Omega, \mathfrak{F}, \{\mathfrak{F}_t : t \in \mathbb{T}\}, P)$ and let S and T be two stopping times such that $S \leq T$ on the filtered space.*
1) If $\mathbb{T} = \mathbb{Z}_+$ then for every $n \in \mathbb{Z}_+$, $\{X_{S \wedge n}, X_{T \wedge n}\}$ is a two-term submartingale, martingale, or supermartingale with respect to $\{\mathfrak{F}_S, \mathfrak{F}_T\}$.
2) If $\mathbb{T} = \mathbb{R}_+$, then under the assumption that both X and $\{\mathfrak{F}_t : t \in \mathbb{R}_+\}$ are right-continuous, for every $t \in \mathbb{R}_+$, $\{X_{S \wedge t}, X_{T \wedge t}\}$ is a two-term submartingale, martingale, or supermartingale with respect to $\{\mathfrak{F}_S, \mathfrak{F}_T\}$.

Proof. It suffices to prove the theorem for the case where X is a submartingale. By Theorem 6.4 for the case $\mathbb{T} = \mathbb{Z}_+$ and by Theorem 8.10 for the case $\mathbb{T} = \mathbb{R}_+$, $\{X_{S \wedge t}, X_{T \wedge t}\}$ is a submartingale with respect to $\{\mathfrak{F}_{S \wedge t}, \mathfrak{F}_{T \wedge t}\}$, that is,

$$(1) \qquad \mathbf{E}(X_{T \wedge t} | \mathfrak{F}_{S \wedge t}) \geq X_{S \wedge t} \quad \text{a.e. on } (\Omega, \mathfrak{F}_{S \wedge t}, P).$$

Since $X_{S \wedge t}$ is $\mathfrak{F}_{S \wedge t}$-measurable it is $\mathfrak{F}_S$-measurable. Similarly $X_{T \wedge t}$ is $\mathfrak{F}_T$-measurable. Then to show that $\{X_{S \wedge t}, X_{T \wedge t}\}$ is a submartingale with respect to $\{\mathfrak{F}_S, \mathfrak{F}_T\}$ it remains to verify that $\mathbf{E}(X_{T \wedge t} | \mathfrak{F}_S) \geq X_{S \wedge t}$ a.e. on $(\Omega, \mathfrak{F}_S, P)$, or equivalently,

$$(2) \qquad \int_A X_{T \wedge t}\, dP \geq \int_A X_{S \wedge t} \quad \text{for } A \in \mathfrak{F}_S.$$

§8. UNIFORMLY INTEGRABLE SUBMARTINGALES

Note that by (1) we have

(3) $$\int_{A_0} X_{T \wedge t}\, dP \geq \int_{A_0} X_{S \wedge t} \quad \text{for } A_0 \in \mathfrak{F}_{S \wedge t}.$$

Let $A \in \mathfrak{F}_S$ and decompose A into two sets in $\mathfrak{F}_S$ by setting $A_1 = A \cap \{S > t\}$ and $A_2 = A \cap \{S \leq t\}$. For $u \in \mathbb{T}$, consider the set

$$A_1 \cap \{S \wedge t \leq u\} = A \cap \{S > t\} \cap \{S \wedge t \leq u\}.$$

If $u < t$ then the last intersection is an empty set so that $A_1 \cap \{S \wedge t \leq u\} = \emptyset \in \mathfrak{F}_u$. If $u \geq t$ then $\{S \wedge t \leq u\} = \Omega$ so that $A_1 \cap \{S \wedge t \leq u\} = A \cap \{S > t\} \in \mathfrak{F}_t$ since $A \in \mathfrak{F}_S$ and thus $A_1 \cap \{S \wedge t \leq u\} \in \mathfrak{F}_u$ since $t \leq u$. Therefore $A_1 \cap \{S \wedge t \leq u\} \in \mathfrak{F}_u$ for $u \in \mathbb{T}$. This shows that $A_1 \in \mathfrak{F}_{S \wedge t}$ and thus by (3) we have

(4) $$\int_{A_1} X_{T \wedge t}\, dP \geq \int_{A_1} X_{S \wedge t}\, dP.$$

On the set $\{S \leq t\}$, we have $X_{S \wedge t} = X_t = X_{T \wedge t}$ since $S \leq T$. Then since A_2 is a subset of $\{S \leq t\}$, we have

(5) $$\int_{A_2} X_{T \wedge t}\, dP = \int_{A_2} X_{S \wedge t}\, dP.$$

Adding (4) and (5) we have (2). ∎

Let us consider optional sampling by unbounded stopping times. Let $X = \{X_t : t \in \mathbb{T}\}$ be a submartingale on a filtered space $(\Omega, \mathfrak{F}, \{\mathfrak{F}_t : t \in \mathbb{T}\}, P)$ and let S and T be two stopping times on the filtered space such that $S \leq T$ on Ω. If S and T are bounded, then $\mathbf{E}(X_T | \mathfrak{F}_S) \geq X_S$ a.e. on $(\Omega, \mathfrak{F}_S, P)$ where the inequality is replaced by an equality when we have a martingale. This was proved in Theorem 6.4 for the case $\mathbb{T} = \mathbb{Z}_+$ and in Theorem 8.10 for the case $\mathbb{T} = \mathbb{R}_+$ under the assumption that both the filtered space and the submartingale were right-continuous. Let us extend these results to cases where the stopping times S and T are no longer assumed to be bounded. Since S and T may assume the value ∞, we need a random variable at infinity so that $X_S(\omega)$ can be defined for $\omega \in \Omega$ for which $S(\omega) = \infty$ and similarly for X_T. Suppose a final element exists for X, that is, there exists an extended real valued integrable $\mathfrak{F}_\infty$-measurable random variable ξ on $(\Omega, \mathfrak{F}, P)$ such that $\mathbf{E}(\xi | \mathfrak{F}_t) \geq X_t$ a.e. on $(\Omega, \mathfrak{F}_t, P)$ for every $t \in \mathbb{T}$. We showed in Theorem 7.12 that uniform integrability of X^+ is a necessary and sufficient condition for existence of a final element for X and if X^+ is uniformly integrable then there exists an extended real valued integrable

$\mathfrak{F}_\infty$-measurable random variable X_∞ such that X_∞ is a final element of X furthermore $\lim_{t\to\infty} X_t = X_\infty$ a.e. on $(\Omega, \mathfrak{F}_\infty, P)$. However X_∞ is not the only final element for X. In fact $X_\infty + c$ is a final element for X for any $c \geq 0$. With an arbitrary final element ξ for X, we define $X_S(\omega) = \xi(\omega)$ for $\omega \in \Omega$ such that $S(\omega) = \infty$ and similarly we define $X_T(\omega) = \xi(\omega)$ for $\omega \in \Omega$ such that $T(\omega) = \infty$. In Theorem 8.16 we show that $\mathbf{E}(X_T|\mathfrak{F}_S) \geq X_S$ a.e. on $(\Omega, \mathfrak{F}_S, P)$ independently of the choice of the final element ξ of X used in defining X_S and X_T.

Let us consider first a martingale with discrete time generated by an integrable random variable.

Theorem 8.14. *Let $(\Omega, \mathfrak{F}, \{\mathfrak{F}_n : n \in \mathbb{Z}_+\}, P)$ be a filtered space, ξ be an extended real valued integrable random variable on $(\Omega, \mathfrak{F}, P)$ and let $X = \{X_n : n \in \mathbb{Z}_+\}$ be a uniformly integrable martingale on the filtered space defined by letting X_n be an arbitrary real valued version of $\mathbf{E}(\xi|\mathfrak{F}_n)$ for $n \in \mathbb{Z}_+$. Let S and T be two stopping times on the filtered space such that $S \leq T$ on Ω. With an arbitrary version Y_∞ of $\mathbf{E}(\xi|\mathfrak{F}_\infty)$, define X_S and X_T with $X_S(\omega) = Y_\infty(\omega)$ for $\omega \in \{S = \infty\}$ and $X_T(\omega) = Y_\infty(\omega)$ for $\omega \in \{T = \infty\}$. Then*

(1) $$\mathbf{E}(\xi|\mathfrak{F}_T) = X_T \quad a.e. \text{ on } (\Omega, \mathfrak{F}_T, P),$$

so that in particular X_T is integrable and similarly for X_S. Also

(2) $$\mathbf{E}(X_T|\mathfrak{F}_S) = X_S \quad a.e. \text{ on } (\Omega, \mathfrak{F}_S, P).$$

Proof. Let $T_k = T \wedge k$ for $k \in \mathbb{Z}_+$. Then $\lim_{k\to\infty} X_{T_k}(\omega) = \lim_{k\to\infty} X_{T\wedge k}(\omega) = X_T(\omega)$ for $\omega \in \{T < \infty\}$. But we have $\lim_{k\to\infty} X_k(\omega) = Y_\infty(\omega)$ for a.e. ω in $(\Omega, \mathfrak{F}_\infty, P)$ by Remark 8.3 so that $\lim_{k\to\infty} X_{T_k}(\omega) = \lim_{k\to\infty} X_k(\omega) = Y_\infty(\omega) = X_T(\omega)$ for a.e. ω in $\{T = \infty\}$. Thus we have $\lim_{k\to\infty} X_{T_k} = X_T$ a.e. on $(\Omega, \mathfrak{F}_\infty, P)$.

To prove (1) we show that

(3) $$\int_A \xi\, dP = \int_A X_T\, dP \quad \text{for } A \in \mathfrak{F}_T.$$

Now since T_k and k are two stopping times and $T_k \leq k$ on Ω we have $\mathbf{E}(X_k|\mathfrak{F}_{T_k}) = X_{T_k}$ a.e. on $(\Omega, \mathfrak{F}_{T_k}, P)$ by Theorem 6.4 and thus

$$\int_A X_k\, dP = \int_A X_{T_k}\, dP \quad \text{for } A \in \mathfrak{F}_{T_k}.$$

§8. UNIFORMLY INTEGRABLE SUBMARTINGALES

Now if $A \in \mathfrak{F}_T$ then by Theorem 3.9, $A \cap \{T \leq k\} = A \cap \{T \leq T_k\} \in \mathfrak{F}_{T_k}$ so that

$$\int_{A \cap \{T \leq k\}} X_k \, dP = \int_{A \cap \{T \leq k\}} X_{T_k} \, dP \quad \text{for } A \in \mathfrak{F}_T.$$

Since $\mathbf{E}(\xi | \mathfrak{F}_k) = X_k$ a.e. on $(\Omega, \mathfrak{F}_k, P)$ and since $A \in \mathfrak{F}_T$ implies $A \cap \{T \leq k\} \in \mathfrak{F}_k$, we have

$$\int_{A \cap \{T \leq k\}} \xi \, dP = \int_{A \cap \{T \leq k\}} X_k \, dP = \int_{A \cap \{T \leq k\}} X_{T_k} \, dP = \int_{A \cap \{T \leq k\}} X_T \, dP \quad \text{for } A \in \mathfrak{F}_T.$$

Since this holds for every $k \in \mathbb{Z}_+$ we have

$$\int_{A \cap \{T=0\}} \xi \, dP = \int_{A \cap \{T=0\}} X_T \, dP$$

and

$$\int_{A \cap \{k-1 < T \leq k\}} \xi \, dP = \int_{A \cap \{k-1 < T \leq k\}} X_T \, dP.$$

From this and from the existence of $\int_{A \cap \{T < \infty\}} \xi \, dP$ we have by summing over $k \in \mathbb{N}$

(4) $$\int_{A \cap \{T < \infty\}} \xi \, dP = \int_{A \cap \{T < \infty\}} X_T \, dP \quad \text{for } A \in \mathfrak{F}_T.$$

Since $\mathbf{E}(\xi | \mathfrak{F}_\infty) = Y_\infty$ a.e. on $(\Omega, \mathfrak{F}_\infty, P)$ and since $\{T = \infty\} \in \mathfrak{F}_\infty$ and $A \in \mathfrak{F}_T \subset \mathfrak{F}_\infty$ we have $A \cap \{T = \infty\} \in \mathfrak{F}_\infty$ and therefore

(5) $$\int_{A \cap \{T=\infty\}} \xi \, dP = \int_{A \cap \{T=\infty\}} X_T \, dP \quad \text{for } A \in \mathfrak{F}_T.$$

From (4) and (5) we have (3). This proves (1). Finally, (2) follows from (1) since

$$\mathbf{E}(X_T | \mathfrak{F}_S) = \mathbf{E}[\mathbf{E}(\xi | \mathfrak{F}_T) | \mathfrak{F}_S] = \mathbf{E}(\xi | \mathfrak{F}_S) = X_S \quad \text{a.e. on } (\Omega, \mathfrak{F}_S, P).$$

This completes the proof. ∎

If $X = \{X_n : n \in \mathbb{Z}_+\}$ is a nonpositive submartingale on a filtered space $(\Omega, \mathfrak{F}, \{\mathfrak{F}_n\}, P)$ then $\mathbf{E}(0 | \mathfrak{F}_n) \geq X_n$ a.e. on $(\Omega, \mathfrak{F}_n, P)$ for $n \in \mathbb{Z}_+$. Thus in defining X_T for a stopping time T on the filtered space, we can use 0 as an integrable random variable ξ on the probability space $(\Omega, \mathfrak{F}, P)$ satisfying the condition $\mathbf{E}(\xi | \mathfrak{F}_n) \geq X_n$ a.e. on $(\Omega, \mathfrak{F}_n, P)$ for $n \in \mathbb{Z}_+$. This is precisely what we do in the next lemma.

Lemma 8.15. *Let $X = \{X_n : n \in \mathbb{Z}_+\}$ be a nonpositive submartingale and S and T be two stopping times on a filtered space $(\Omega, \mathfrak{F}, \{\mathfrak{F}_n\}, P)$ such that $S \leq T$ on Ω. Let us define X_S and X_T by setting $X_S(\omega) = 0$ for $\omega \in \{S = \infty\}$ and $X_T(\omega) = 0$ for $\omega \in \{T = \infty\}$. Then X_S and X_T are integrable and $\mathbf{E}(X_T | \mathfrak{F}_S) \geq X_S$ a.e. on $(\Omega, \mathfrak{F}_S, P)$.*

Proof. Let $T_k = T \wedge k$ for $k \in \mathbb{Z}_+$. Then T_k is a bounded stopping time so that by Theorem 6.1, X_{T_k} is integrable and $\mathbf{E}(X_0) \leq \mathbf{E}(X_{T_k}) < \infty$. Now

$$\lim_{k \to \infty} X_{T_k}(\omega) = \lim_{k \to \infty} X_{T \wedge k}(\omega) = X_T(\omega) \quad \text{for } \omega \in \{T < \infty\}$$

and

$$\limsup_{k \to \infty} X_{T_k}(\omega) = \limsup_{k \to \infty} X_{T \wedge k}(\omega) \leq 0 = X_T(\omega) \quad \text{for } \omega \in \{T = \infty\}.$$

Therefore by Fatou's Lemma for the limit superior of nonpositive functions we have

$$\mathbf{E}(X_0) \leq \limsup_{k \to \infty} \mathbf{E}(X_{T_k}) \leq \mathbf{E}(\limsup_{k \to \infty} X_{T_k}) \leq \mathbf{E}(X_T) \leq 0$$

so that $\mathbf{E}(X_T) \in \mathbb{R}$, that is, X_T is integrable. Similarly X_S is integrable.

To show that $\mathbf{E}(X_T | \mathfrak{F}_S) \geq X_S$ a.e. on $(\Omega, \mathfrak{F}_S, P)$ we show that

$$(1) \qquad \int_A X_T \, dP \geq \int_A X_S \, dP \quad \text{for } A \in \mathfrak{F}_S.$$

Let $S_k = S \wedge k$ for $k \in \mathbb{Z}_+$. Then $S_k \leq T_k$ on Ω so that by Theorem 6.4 we have $\mathbf{E}(X_{T_k} | \mathfrak{F}_{S_k}) \geq X_{S_k}$ a.e. on $(\Omega, \mathfrak{F}_{S_k}, P)$ and then

$$(2) \qquad \int_A X_{T_k} \, dP \geq \int_A X_{S_k} \, dP \quad \text{for } A \in \mathfrak{F}_{S_k}.$$

Let $A \in \mathfrak{F}_S$. Then by Theorem 3.9 we have $A \cap \{S \leq k\} = A \cap \{S \leq S_k\} \in \mathfrak{F}_{S_k}$ so that by (2)

$$\int_{A \cap \{S \leq k\}} X_{T_k} \, dP \geq \int_{A \cap \{S \leq k\}} X_{S_k} \, dP \quad \text{for } A \in \mathfrak{F}_S.$$

Since $\{S \leq k\} \supset \{T \leq k\}$ and since X_{T_k} is nonpositive, the integral on the left side is increased if the domain of integration $A \cap \{S \leq k\}$ is replaced by $A \cap \{T \leq k\}$. Thus

$$\int_{A \cap \{T \leq k\}} X_T \, dP \geq \int_{A \cap \{S \leq k\}} X_S \, dP \quad \text{for } A \in \mathfrak{F}_S.$$

Letting $k \to \infty$ we have by the Monotone Convergence Theorem

$$\int_{A \cap \{T < \infty\}} X_T \, dP \geq \int_{A \cap \{S < \infty\}} X_S \, dP \quad \text{for } A \in \mathfrak{F}_S.$$

§8. UNIFORMLY INTEGRABLE SUBMARTINGALES

Since $X_T(\omega) = 0$ for $\omega \in \{T = \infty\}$ and $X_S(\omega) = 0$ for $\omega \in \{S = \infty\}$, we have (1). ∎

Theorem 8.16. (Optional Sampling with Unbounded Stopping Times, Discrete Case) *Let $X = \{X_n : n \in \mathbb{Z}_+\}$ be a submartingale, a martingale, or a supermartingale on a filtered space $(\Omega, \mathfrak{F}, \{\mathfrak{F}_n\}, P)$. Suppose there exists an extended real valued integrable random variable ξ on $(\Omega, \mathfrak{F}, P)$ such that for every $n \in \mathbb{Z}_+$ we have*

(1) $$\mathbf{E}(\xi|\mathfrak{F}_n) \geq, =, \text{ or } \leq X_n \quad \text{a.e. on } (\Omega, \mathfrak{F}_n, P)$$

according as X is a submartingale, a martingale, or a supermartingale. Let S and T be two stopping times on the filtered space such that $S \leq T$ on Ω. Let Y_∞ be an arbitrary version of $\mathbf{E}(\xi|\mathfrak{F}_\infty)$ and define X_S and X_T with $X_S(\omega) = Y_\infty(\omega)$ for $\omega \in \{S = \infty\}$ and $X_T(\omega) = Y_\infty(\omega)$ for $\omega \in \{T = \infty\}$. Then X_S and X_T are integrable and satisfy

(2) $$\mathbf{E}(\xi|\mathfrak{F}_T) \geq, =, \text{ or } \leq X_T \quad \text{a.e. on } (\Omega, \mathfrak{F}_T, P)$$

and

(3) $$\mathbf{E}(X_T|\mathfrak{F}_S) \geq, =, \text{ or } \leq X_S \quad \text{a.e. on } (\Omega, \mathfrak{F}_S, P),$$

according as X is a submartingale, a martingale, or a supermartingale.

Proof. It suffices to prove the theorem for a submartingale.

If we let Y_n be an arbitrary real valued version of $\mathbf{E}(\xi|\mathfrak{F}_n)$ for $n \in \mathbb{Z}_+$ then by Remark 8.3, $Y = \{Y_n : n \in \mathbb{Z}_+\}$ is a uniformly integrable martingale with $\lim_{n\to\infty} Y_n = Y_\infty$ a.e. on $(\Omega, \mathfrak{F}_\infty, P)$ and $\lim_{n\to\infty} \|Y_n - Y_\infty\|_1 = 0$. If we define Y_S and Y_T with $Y_S(\omega) = Y_\infty(\omega)$ for $\omega \in \{S = \infty\}$ and $Y_T(\omega) = Y_\infty(\omega)$ for $\omega \in \{T = \infty\}$, then by Theorem 8.14 both Y_S and Y_T are integrable and satisfy

(4) $$\mathbf{E}(\xi|\mathfrak{F}_T) = Y_T \quad \text{a.e. on } (\Omega, \mathfrak{F}_T, P)$$

and

(5) $$\mathbf{E}(Y_T|\mathfrak{F}_S) = Y_S \quad \text{a.e. on } (\Omega, \mathfrak{F}_S, P).$$

Let us define a stochastic process Z by setting $Z = X - Y$. Then since X is a submartingale and Y is a martingale, Z is a submartingale. Also Z is nonpositive since by (1) we have
$$Z_n = X_n - Y_n = X_n - \mathbf{E}(\xi|\mathfrak{F}_n) \leq 0 \quad \text{a.e. on } (\Omega, \mathfrak{F}_n, P).$$

If we define Z_S and Z_T with $Z_S(\omega) = 0$ for $\omega \in \{S = \infty\}$ and $Z_T(\omega) = 0$ for $\omega \in \{T = \infty\}$, then by Lemma 8.15 the two random variables Z_S and Z_T are integrable and

(6) $$\mathbf{E}(0|\mathfrak{F}_T) \geq Z_T \quad \text{a.e. on } (\Omega, \mathfrak{F}_T, P)$$

and

(7) $$\mathbf{E}(Z_T|\mathfrak{F}_S) \geq Z_S \quad \text{a.e. on } (\Omega, \mathfrak{F}_S, P).$$

Then $X_T = (Y + Z)_T = Y_T + Z_T$ is integrable and similarly for X_S. Adding (4) and (6) side by side we have (2) and adding (5) and (7) side by side we have (3) for the submartingale X. ∎

We observe that Theorem 8.14 can be extended to a right-continuous martingale $X = \{X_t : t \in \mathbb{R}_+\}$ generated by an integrable random variable ξ on a right-continuous filtered space $(\Omega, \mathfrak{F}, \{\mathfrak{F}_t\}, P)$ by approximating a stopping time T with a sequence of discrete valued stopping times $\{T_n : n \in \mathbb{N}\}$ according to Theorem 3.20 and, in passing to the limit, using the uniform integrability of the resulting martingale with reversed time $\{X_{T_n} : n \in \mathbb{N}\}$ according to Theorem 8.8. The proof parallels that of Theorem 8.10 and the argument need not be repeated here. Let us state the result for later reference.

Theorem 8.17. *Let $X = \{X_t : t \in \mathbb{R}_+\}$ be a right-continuous uniformly integrable martingale on a right-continuous filtered space $(\Omega, \mathfrak{F}, \{\mathfrak{F}_t\}, P)$ where X_t is a version of $\mathbf{E}(\xi|\mathfrak{F}_t)$ for $t \in \mathbb{R}_+$ for an extended real valued integrable random variable ξ on $(\Omega, \mathfrak{F}, P)$. Let S and T be two stopping times on the filtered space such that $S \leq T$ on Ω. With an arbitrary real valued version Y_∞ of $\mathbf{E}(\xi|\mathfrak{F}_\infty)$, define X_S and X_T with $X_S(\omega) = Y_\infty(\omega)$ for $\omega \in \{S = \infty\}$ and $X_T(\omega) = Y_\infty(\omega)$ for $\omega \in \{T = \infty\}$. Then $\mathbf{E}(\xi|\mathfrak{F}_T) = X_T$ a.e. on $(\Omega, \mathfrak{F}_T, P)$ and thus $\mathbf{E}(X_T|\mathfrak{F}_S) = X_S$ a.e. on $(\Omega, \mathfrak{F}_S, P)$.*

Similarly Theorem 8.16 can be extended to a right-continuous submartingale or supermartingale on a right-continuous filtered space by using Theorem 3.20 and Theorem 8.8 exactly in the same way as in the proof of Theorem 8.10. Thus we have the following theorem.

Theorem 8.18. (Optional Sampling with Unbounded Stopping Times, Continuous Case). *Let $X = \{X_t : t \in \mathbb{R}_+\}$ be a right-continuous submartingale, martingale, or supermartingale on a right-continuous filtered space $(\Omega, \mathfrak{F}, \{\mathfrak{F}_t\}, P)$. Suppose there exists an extended*

§8. UNIFORMLY INTEGRABLE SUBMARTINGALES

real valued integrable random variable ξ on $(\Omega, \mathfrak{F}, P)$ such that for every $t \in \mathbb{R}_+$

(1) $\qquad \mathbf{E}(\xi|\mathfrak{F}_t) \geq, =, \text{ or } \leq X_t \quad a.e. \text{ on } (\Omega, \mathfrak{F}_t, P),$

according as X is a submartingale, martingale, or supermartingale. Let S and T be two stopping times on the filtered space such that $S \leq T$ on Ω. Let Y_∞ be an arbitrary real valued version of $\mathbf{E}(\xi|\mathfrak{F}_\infty)$ and define X_S and X_T with $X_S(\omega) = Y_\infty(\omega)$ for $\omega \in \{S = \infty\}$ and $X_T(\omega) = Y_\infty(\omega)$ for $\omega \in \{T = \infty\}$. Then X_S and X_T are integrable and satisfy

(2) $\qquad \mathbf{E}(\xi|\mathfrak{F}_T) \geq, =, \text{ or } \leq X_T \quad a.e. \text{ on } (\Omega, \mathfrak{F}_T, P)$

and

(3) $\qquad \mathbf{E}(X_T|\mathfrak{F}_S) \geq, =, \text{ or } \leq X_S \quad a.e. \text{ on } (\Omega, \mathfrak{F}_S, P),$

according as X is a submartingale, martingale, or supermartingale.

As a restatement of Theorem 8.18 we have:

Theorem 8.19. *Let $X = \{X_t : t \in \mathbb{R}_+\}$ be a right-continuous submartingale, martingale, or supermartingale on a right-continuous filtered space $(\Omega, \mathfrak{F}, \{\mathfrak{F}_t\}, P)$ satisfying condition (1) in Theorem 8.18 and let Y_∞ be a real valued version of $\mathbf{E}(\xi|\mathfrak{F}_\infty)$. Let $\{S_t : t \in \mathbb{R}_+\}$ be an increasing system of stopping times on the filtered space. Let $\mathfrak{G}_t = \mathfrak{F}_{S_t}$ and $Y_t = X_{S_t}$ for $t \in \mathbb{R}_+$ where $X_{S_t}(\omega) = Y_\infty(\omega)$ for $\omega \in \{S_t = \infty\}$. Then $Y = \{Y_t : t \in \mathbb{R}_+\}$ is a submartingale or a supermartingale on the filtered space $(\Omega, \mathfrak{F}, \{\mathfrak{G}_t\}, P)$ according as X is a submartingale, martingale, or supermartingale.*

[IV] Uniform Integrability of Random Variables at Stopping Times

If a submartingale $X = \{X_t : t \in \mathbb{T}\}$ on a filtered space $(\Omega, \mathfrak{F}, \{\mathfrak{F}_t\}, P)$ is uniformly integrable and **S** is the collection of all finite stopping times on the filtered space, is the system of random variables $\{X_T : T \in \mathbf{S}\}$ also uniformly integrable? We shall show that this is the case when $\mathbb{T} = \mathbb{Z}_+$.

Definition 8.20. *Let **S** be the collection of all finite stopping times on a filtered space $(\Omega, \mathfrak{F}, \{\mathfrak{F}_t : t \in \mathbb{T}\}, P)$. For every $a \in \mathbb{T}$ let $\mathbf{S}_a$ be the subcollection of **S** consisting of those stopping times which are bounded by a. A submartingale $X = \{X_t : t \in \mathbb{T}\}$ on the filtered space is said to belong to the class (D) if $\{X_T : T \in \mathbf{S}\}$ is uniformly integrable.*

X is said to belong to the class (DL), if $\{X_T : T \in \mathbf{S}_a\}$ is uniformly integrable for every $a \in \mathbf{T}$.

Theorem 8.21. *A submartingale $X = \{X_n : n \in \mathbb{Z}_+\}$ on a filtered space $(\Omega, \mathfrak{F}, \{\mathfrak{F}_n\}, P)$ belongs to the class (D) if and only if it is uniformly integrable.*

Proof. Let $\mathbf{S}$ be the collection of all finite stopping times on the filtered space. Since every deterministic time is a member of $\mathbf{S}$, if X belongs to the class (D) then X is uniformly integrable.

Conversely, suppose X is uniformly integrable. Then by Theorem 8.1 there exists an extended real valued $\mathfrak{F}_\infty$-measurable integrable random variable X_∞ on $(\Omega, \mathfrak{F}, P)$ such that $\lim_{n \to \infty} X_n = X_\infty$ a.e. on $(\Omega, \mathfrak{F}_\infty, P)$, $\lim_{n \to \infty} \|X_n - X_\infty\|_1 = 0$, and $\mathbf{E}(X_\infty | \mathfrak{F}_n) \geq X_n$ a.e. on $(\Omega, \mathfrak{F}_n, P)$ for $n \in \mathbb{Z}_+$.

If we let Y_n be an arbitrary real valued version of $\mathbf{E}(X_\infty | \mathfrak{F}_n)$ for $n \in \mathbb{Z}_+$ then by Remark 8.3, $Y = \{Y_n : n \in \mathbb{Z}_+\}$ is a uniformly integrable martingale with $\lim_{n \to \infty} Y_n = X_\infty$ a.e. on $(\Omega, \mathfrak{F}_\infty, P)$ and $\lim_{n \to \infty} \|X_n - X_\infty\|_1 = 0$.

If we define a process $Z = \{Z_n : n \in \mathbb{Z}_+\}$ by letting $Z = X - Y$ then since X is a submartingale and Y is a martingale, Z is a submartingale. Also Z is nonpositive since $Z_n = X_n - Y_n = X_n - \mathbf{E}(X_\infty | \mathfrak{F}_n) \leq 0$ a.e. on $(\Omega, \mathfrak{F}_n, P)$ for $n \in \mathbb{Z}_+$. Note also that since $\lim_{n \to \infty} X_n = X_\infty = \lim_{n \to \infty} Y_n$ a.e. on $(\Omega, \mathfrak{F}_\infty, P)$ we have $\lim_{n \to \infty} Z_n = 0$ a.e. on $(\Omega, \mathfrak{F}_\infty, P)$.

From $X = Y + Z$ we have $X_T = Y_T + Z_T$ for $T \in \mathbf{S}$. Thus, to show the uniform integrability of $\{X_T : T \in \mathbf{S}\}$ we show that of $\{Y_T : T \in \mathbf{S}\}$ and $\{Z_T : T \in \mathbf{S}\}$. By (1) of Theorem 8.14, we have $\mathbf{E}(X_\infty | \mathfrak{F}_T) = Y_T$ a.e. on $(\Omega, \mathfrak{F}_T, P)$ for every $T \in \mathbf{S}$. Thus by Theorem 4.24, $\{Y_T : T \in \mathbf{S}\}$ is uniformly integrable.

To prove the uniform integrability of $\{Z_T : T \in \mathbf{S}\}$, let us note that since $\lim_{n \to \infty} Z_n = 0$ a.e. on $(\Omega, \mathfrak{F}_\infty, P)$ and since $Z = X - Y$ is uniformly integrable, we have $\lim_{n \to \infty} \|Z_n\|_1 = 0$ by Theorem 4.16. Therefore for every $\varepsilon > 0$ there exists $k \in \mathbb{Z}_+$ such that $\mathbf{E}(|Z_k|) < \varepsilon$. For any $T \in \mathbf{S}$ and $\lambda > 0$ we have

$$(1) \quad \int_{\{|Z_T| > \lambda\}} |Z_T| \, dP = \sum_{i=1}^{k} \int_{\{|Z_T| > \lambda\} \cap \{T=i\}} |Z_T| \, dP + \int_{\{|Z_T| > \lambda\} \cap \{T>k\}} |Z_T| \, dP$$

$$\leq \sum_{i=1}^{k} \int_{\{|Z_i| > \lambda\}} |Z_i| \, dP + \int_{\{T>k\}} |Z_T| \, dP.$$

§8. UNIFORMLY INTEGRABLE SUBMARTINGALES

For the finitely many integrable random variables $Z_1, \ldots, Z_k$ we have

$$(2) \qquad \lim_{\lambda \to \infty} \sum_{i=1}^{k} \int_{\{|Z_i| > \lambda\}} |Z_i| \, dP = 0.$$

For the two stopping times k and $T \vee k$ satisfying $k \leq T \vee k$ on Ω we have $\mathbf{E}(Z_{T \vee k} | \mathfrak{F}_k) \geq Z_k$ a.e. on $(\Omega, \mathfrak{F}_k, P)$ by Theorem 8.16. Since $\{T > k\} = \{T \leq k\}^c \in \mathfrak{F}_k$, we have

$$\int_{\{T>k\}} Z_T \, dP = \int_{\{T>k\}} Z_{T \vee k} \, dP \geq \int_{\{T>k\}} Z_k \, dP.$$

Since Z is nonpositive, Z_T is nonpositive and thus by the fact that $\mathbf{E}(|Z_k|) < \varepsilon$ we have

$$(3) \qquad \int_{\{T>k\}} |Z_T| \, dP \leq \int_{\{T>k\}} |Z_k| \, dP < \varepsilon.$$

By (1), (2) and (3) we have

$$\limsup_{\lambda \to \infty} \left\{ \sup_{T \in \mathbf{S}} \int_{\{|Z_T| > \lambda\}} |Z_T| \, dP \right\} \leq \varepsilon.$$

From the arbitrariness of ε the limit superior above is equal to 0. Therefore

$$\lim_{\lambda \to \infty} \sup_{T \in \mathbf{S}} \int_{\{|Z_T| > \lambda\}} |Z_T| \, dP = 0,$$

proving the uniform integrability of $\{Z_T : T \in \mathbf{S}\}$. ∎

For $\mathbb{T} = \mathbb{R}_+$, a uniformly integrable right-continuous submartingale X may not belong to the class (D). For a counter example see [16] Johnson and Helms. If however X is a uniformly integrable right-continuous martingale then it is in the class (D) as we show next.

Theorem 8.22. *On a right-continuous filtered space* $(\Omega, \mathfrak{F}, \{\mathfrak{F}_t : t \in \mathbb{R}_+\}, P)$,
1) every right-continuous martingale belongs to the class (DL),
2) every right-continuous nonnegative submartingale belongs to the class (DL),
3) every uniformly integrable right-continuous martingale belongs to the class (D).

Proof. 1) Let X be a right-continuous martingale. Then for every $a \in \mathbb{R}_+$ we have by Theorem 8.10, $\mathbf{E}(X_a | \mathfrak{F}_T) = X_T$ a.e. on $(\Omega, \mathfrak{F}_T, P)$ for $T \in \mathbf{S}_a$. Thus by Theorem 4.24 $\{X_T : T \in \mathbf{S}_a\}$ is uniformly integrable.

2) Let X be a nonnegative right-continuous submartingale. Let $a \in \mathbb{R}_+$. For every $T \in \mathbf{S}_a$ we have $\mathbf{E}(X_a | \mathfrak{F}_T) \geq X_T$ a.e. on $(\Omega, \mathfrak{F}_T, P)$ by Theorem 8.10. For every $\lambda > 0$ we have $\{X_T > \lambda\} \in \mathfrak{F}_T$ so that

(1) $$\int_{\{X_T > \lambda\}} X_a \, dP \geq \int_{\{X_T > \lambda\}} X_T \, dP.$$

Also $\lambda P\{X_T > \lambda\} \leq \mathbf{E}(X_T) \leq \mathbf{E}(X_a)$ so that

(2) $$\lim_{\lambda \to \infty} P\{X_T > \lambda\} = 0 \quad \text{uniformly in } T \in \mathbf{S}_a.$$

By (1) and (2)

$$\lim_{\lambda \to \infty} \sup_{T \in \mathbf{S}_a} \int_{\{X_T > \lambda\}} X_T \, dP \leq \lim_{\lambda \to \infty} \sup_{T \in \mathbf{S}_a} \int_{\{X_T > \lambda\}} X_a \, dP = 0.$$

Therefore $\{X_T : T \in \mathbf{S}_a\}$ is uniformly integrable and X belongs to the class (DL).

3) If X is a uniformly integrable right-continuous martingale then by Theorem 8.1 there exists an extended real valued integrable random variable ξ such that $\mathbf{E}(\xi | \mathfrak{F}_t) = X_t$ a.e. on $(\Omega, \mathfrak{F}_t, P)$ for every $t \in \mathbb{R}_+$. Thus according to Theorem 8.17 we have $\mathbf{E}(\xi | \mathfrak{F}_T) = X_T$ a.e. on $(\Omega, \mathfrak{F}_T, P)$ for every $T \in \mathbf{S}$. Therefore by Theorem 4.24, $\{X_T : T \in \mathbf{S}\}$ is uniformly integrable, that is, X belongs to the class (D). ∎

Remark 8.23. If $X = \{X_t : t \in \mathbb{R}_+\}$ is a right-continuous submartingale on a right-continuous filtered space $(\Omega, \mathfrak{F}, \{\mathfrak{F}_t\}, P)$ and if X is in the class (D), then not only is $\{X_T : T \in \mathbf{S}\}$ uniformly integrable but also $\{X_T : T \in \mathbf{S}_\infty\}$ where $\mathbf{S}_\infty$ is the collection of all stopping times, finite or not, on the filtered space and X_T is defined with $X_\infty \equiv \lim_{t \to \infty} X_t$.

Proof. The uniform integrability of $\{X_T : T \in \mathbf{S}\}$ implies the uniform integrability of its closure $\overline{\{X_T : T \in \mathbf{S}\}}$ in $L_1(\Omega, \mathfrak{F}_\infty, P)$ according to Theorem 4.19. Thus, to show the uniform integrability of $\{X_T : T \in \mathbf{S}_\infty\}$ we show that it is contained in $\overline{\{X_T : T \in \mathbf{S}\}}$. To show this we note that for any $T \in \mathbf{S}_\infty$, $\{T \wedge n : n \in \mathbb{N}\}$ is a sequence in $\mathbf{S}$ and show that the sequence $\{X_{T \wedge n} : n \in \mathbb{N}\}$ in $\{X_T : T \in \mathbf{S}\}$ satisfies $\lim_{n \to \infty} \|X_{T \wedge n} - X_T\|_1 = 0$.

Now the uniform integrability of $\{X_T : T \in \mathbf{S}\}$ implies that of $\{X_t : t \in \mathbb{R}_+\}$ since deterministic times are particular cases of finite stopping times. The uniform integrability of X then implies according to Theorem 8.1 that $X_\infty = \lim_{t \to \infty} X_t$ exists a.e. on $(\Omega, \mathfrak{F}_\infty, P)$, and is an integrable random variable satisfying $\lim_{t \to \infty} \|X_t - X_\infty\|_1 = 0$. Then for any $T \in \mathbf{S}_\infty$, finite or not, X_T is defined with X_∞ and $\lim_{n \to \infty} X_{T \wedge n} = X_T$ a.e. on $(\Omega, \mathfrak{F}_\infty, P)$. Now since $T \wedge n \in \mathbf{S}$ for $n \in \mathbb{N}$, $\{X_{T \wedge n} : n \in \mathbb{N}\}$ is uniformly integrable by our assumption that X is in the class (D). Then $\lim_{n \to \infty} \|X_{T \wedge n} - X_T\|_1 = 0$ by Theorem 4.16. ∎

§9 Regularity of Sample Functions of Submartingales

[I] Sample Functions of Right-Continuous Submartingales

The Maximal and Minimal Inequalities and the Upcrossing Inequality for submartingales imply certain regularity properties for their sample functions. We show in Theorem 9.2 that if $\{X_t : t \in \mathbb{R}_+\}$ is a right-continuous submartingale on a filtered space $(\Omega, \mathfrak{F}, \{\mathfrak{F}_t\}, P)$ then almost every sample function is bounded on every finite interval in $\mathbb{R}_+$, has a finite left limit everywhere on $\mathbb{R}_+$, and has at most countably many points of discontinuity.

Proposition 9.1. *Let $X = \{X_t : t \in \mathbb{R}_+\}$ be a submartingale on a filtered space $(\Omega, \mathfrak{F}, \{\mathfrak{F}_t\}, P)$ and let Q_+ be the collection of all nonnegative rational numbers.*
1) *There exists a null set Λ_∞ in $(\Omega, \mathfrak{F}_\infty, P)$ such that if $\omega \in \Lambda_\infty^c$ then $X(\cdot, \omega)$ is a bounded function on $[0, \beta) \cap Q_+$ for every $\beta \in \mathbb{R}_+$.*
2) *There exists a null set Λ in $(\Omega, \mathfrak{F}_\infty, P)$, $\Lambda \supset \Lambda_\infty$, such that if $\omega \in \Lambda^c$ then $\lim_{s \uparrow t, s \in Q_+} X_s(\omega)$ and $\lim_{s \downarrow t, s \in Q_+} X_s(\omega)$ exist in $\mathbb{R}$ for every $t \in \mathbb{R}_+$.*

Proof. 1) For every $n \in \mathbb{N}$, let $Q_n = [0, n) \cap Q_+$. Then for every $n \in \mathbb{N}$ and $\lambda > 0$ we have by Theorem 6.14

$$\lambda P\{\sup_{t \in Q_n} X_t > \lambda\} \leq \mathbf{E}(|X_n|)$$

and

$$\lambda P\{\inf_{t \in Q_n} X_t < -\lambda\} \leq \mathbf{E}(|X_n|) + \mathbf{E}(|X_0|).$$

Since

$$\{\sup_{t \in Q_n} |X_t| > \lambda\} = \{\sup_{t \in Q_n} X_t > \lambda\} \cup \{\inf_{t \in Q_n} X_t < -\lambda\},$$

we have

$$P\{\sup_{t \in Q_n} |X_t| > \lambda\} \leq \frac{1}{\lambda}\{\mathbf{E}(|X_0|) + 2\mathbf{E}(|X_n|)\}$$

and therefore

(1) $$\lim_{\lambda \to \infty} P\{\sup_{t \in Q_n} |X_t| > \lambda\} = 0.$$

For every $n \in \mathbb{N}$, let

$$\Lambda_n = \{\omega \in \Omega : X(\cdot, \omega) \text{ not bounded on } Q_n\} = \bigcap_{k \in \mathbb{N}} \left\{\sup_{t \in Q_n} |X_t| > k\right\} \in \mathfrak{F}_\infty.$$

Then
$$P(\Lambda_n) \leq P\left\{\sup_{t \in Q_n} |X_t| > k\right\} \quad \text{for every } k \in \mathbb{N},$$

and thus by (1) we have $P(\Lambda_n) = 0$. Let $\Lambda_\infty = \cup_{n \in \mathbb{N}} \Lambda_n$, a null set in $(\Omega, \mathfrak{F}_\infty, P)$. If $\omega \in \Lambda_\infty^c$, then $\omega \in \Lambda_n^c$ for every $n \in \mathbb{N}$ so that $X(\cdot, \omega)$ is bounded on Q_n for every $n \in \mathbb{N}$. Then $X(\cdot, \omega)$ is bounded on $[0, \beta) \cap Q_+$ for every $\beta \in \mathbb{R}_+$.

2) Let Q be the collection of all rational numbers. For $n \in \mathbb{N}$ and $a, b \in Q$ such that $a < b$, let
$$A_{n,a,b} = \{\omega \in \Omega : (U_{[a,b]}^{Q_n} X)(\omega) = \infty\} \in \mathfrak{F}_\infty.$$

According to (3) of Theorem 6.23 we have
$$\mathbf{E}(U_{[a,b]}^{Q_n} X) \leq \frac{1}{b-a} \mathbf{E}[(X_n - a)^+ - (X_0 - a)^+] < \infty$$

so that $U_{[a,b]}^{Q_n} X < \infty$ a.e. on $(\Omega, \mathfrak{F}_\infty, P)$ and thus we have $P(A_{n,a,b}) = 0$. Let
$$A = \bigcup_{n \in \mathbb{N}} \bigcup_{a,b \in Q, a<b} A_{n,a,b}.$$

As a countable union of null sets, A is a null set in $(\Omega, \mathfrak{F}_\infty, P)$. Let us show that if $\omega \in A^c$ then $\lim_{s \uparrow t, s \in Q_+} X_s(\omega)$ exists in $\overline{\mathbb{R}}$ for all $t \in \mathbb{R}_+$. Assume the contrary. Then there exists some $t \in \mathbb{R}_+$ such that $\lim_{s \uparrow t, s \in Q_+} X_s(\omega)$ does not exist in $\overline{\mathbb{R}}$. Let $n \in \mathbb{N}$ be such that $t \leq n$. Then $\lim_{s \uparrow t, s \in Q_n} X_s(\omega)$ does not exist in $\overline{\mathbb{R}}$, that is, we have
$$\liminf_{s \uparrow t, s \in Q_n} X_s(\omega) < \limsup_{s \uparrow t, s \in Q_n} X_s(\omega),$$

and therefore there exist $a, b \in Q$, $a < b$, such that
$$\liminf_{s \uparrow t, s \in Q_n} X_s(\omega) < a < b < \limsup_{s \uparrow t, s \in Q_n} X_s(\omega).$$

Then there exists a strictly increasing sequence $\{s_m : m \in \mathbb{N}\}$ in Q_n such that $s_m \uparrow t$, $X_{s_{2k-1}}(\omega) < a$ and $X_{s_{2k}}(\omega) > b$ for $k \in \mathbb{N}$ so that $\left(U_{[a,b]}^{Q_n} X\right)(\omega) = \infty$. Then $\omega \in A_{n,a,b} \subset A$, contradicting the assumption that $\omega \in A^c$. This shows that if $\omega \in A^c$ then $\lim_{s \uparrow t, s \in Q_+} X_s(\omega)$ exists in $\overline{\mathbb{R}}$ for every $t \in \mathbb{R}_+$. Consider the null set $\Lambda = A \cup \Lambda_\infty$ in $(\Omega, \mathfrak{F}_\infty, P)$. For $\omega \in \Lambda^c$ we have $\omega \in \Lambda_\infty^c$ so that $X(\cdot, \omega)$ is bounded on $[0, \beta) \cap Q_+$ for every $\beta \in \mathbb{R}_+$ and this

§9. REGULARITY OF SAMPLE FUNCTIONS OF SUBMARTINGALES

implies that $\lim_{s\uparrow t, s\in Q_+} X_s(\omega)$ exists in $\mathbb{R}$ for every $t \in \mathbb{R}_+$. We show similarly that if $\omega \in \Lambda^c$ then $\lim_{s\downarrow t, s\in Q_+} X_s(\omega)$ exists in $\mathbb{R}$ for every $t \in \mathbb{R}_+$. ∎

Theorem 9.2. *Let $X = \{X_t : t \in \mathbb{R}_+\}$ be a right-continuous submartingale on a filtered space $(\Omega, \mathfrak{F}, \{\mathfrak{F}_t\}, P)$. Then there exists a null set Λ in $(\Omega, \mathfrak{F}_\infty, P)$ such that for every $\omega \in \Lambda^c$ the sample function $X(\cdot, \omega)$ is bounded on every finite interval in $\mathbb{R}_+$, has finite left limit everywhere on $\mathbb{R}_+$, and has at most countably many points of discontinuity.*

Proof. Let Λ be the null set in $(\Omega, \mathfrak{F}_\infty, P)$ in Proposition 9.1 and let $\omega \in \Lambda^c$.

1) To show that $X(\cdot, \omega)$ is bounded on an arbitrary finite interval, let $\beta \in \mathbb{R}_+$. By Proposition 9.1, $X(\cdot, \omega)$ is bounded on $[0, \beta) \cap Q_+$ so that there exists $K > 0$ such that $|X(r, \omega)| \leq K$ for $r \in [0, \beta) \cap Q_+$. Let $t \in [0, \beta)$. Then there exists a sequence $\{r_n : n \in \mathbb{N}\}$ in $[0, \beta) \cap Q_+$ such that $r_n \downarrow t$. By the right-continuity of $X(\cdot, \omega)$, we have $X(t, \omega) = \lim_{n \to \infty} X(r_n, \omega)$ and thus $|X(t, \omega)| \leq K$, that is, $X(\cdot, \omega)$ is bounded on $[0, \beta)$.

2) To show that $\lim_{t \uparrow t_0} X(t, \omega)$ exists in $\mathbb{R}$ for every $t_0 \in (0, \infty)$, assume the contrary, that is, $\lim_{t \uparrow t_0} X(t, \omega)$ does not exist in $\mathbb{R}$ for some $t_0 \in (0, \infty)$. Then since $X(\cdot, \omega)$ is bounded on every finite interval, $\lim_{t \uparrow t_0} X(t, \omega)$ can not exist in $\overline{\mathbb{R}}$ either. Thus there exist $a, b \in \mathbb{R}$, $a < b$, such that

$$\liminf_{t \uparrow t_0} X(t, \omega) < a < b < \limsup_{t \uparrow t_0} X(t, \omega).$$

Then we can select a strictly increasing sequence $\{t_n : n \in \mathbb{N}\}$ such that $t_n \uparrow t_0$ and $X(t_n, \omega) < a$ for odd n and $X(t_n, \omega) > b$ for even n. By the right-continuity of $X(\cdot, \omega)$ there exists a rational number $s_n \in (t_n, t_{n+1})$ such that $X(s_n, \omega) < a$ for odd n and $X(s_n, \omega) > b$ for each $n \in \mathbb{N}$. Then $\{s_n : n \in \mathbb{N}\}$ is a strictly increasing sequence of rational numbers such that $s_n \uparrow t_0$ and $\lim_{n \to \infty} X(s_n, \omega)$ does not exist. But according to Proposition 9.1, $\lim_{s\uparrow t_0, s\in Q_+} X_s(\omega)$ exists in $\mathbb{R}$. This is a contradiction. Thus $\lim_{t \uparrow t_0} X(t, \omega)$ exists in $\mathbb{R}$ for every $t_0 \in (0, \infty)$.

3) The fact that $X(\cdot, \omega)$ is real valued, right-continuous and has finite left limit everywhere on $\mathbb{R}_+$ implies that $X(\cdot, \omega)$ has at most countably many points of discontinuity. This implication, which is unrelated to the submartingale, is proved in Proposition 9.3 below. ∎

According to Theorem 9.2, almost every sample function of a right-continuous submartingale is bounded on every finite interval. An arbitrary real valued right-continuous function does not have this property. A real valued continuous function f on $\mathbb{R}$ is bounded on every finite interval of $\mathbb{R}$, but if f is only right-continuous, then f may not be bounded on every finite interval. To construct such a function, let $t_0 = 0$ and $t_k = \sum_{j=1}^{k} 2^{-j}$ for $k \in \mathbb{N}$. Decompose $[0,1)$ into subintervals $I_k = [t_{k-1}, t_k)$ for $k \in \mathbb{N}$. Define f on $[0,1)$ by setting $f(x) = k$ for $x \in I_k$ for $k \in \mathbb{N}$. Thus defined, f is right-continuous but unbounded on $[0, 1)$. Extend the definition of f periodically with period 1 to the entire $\mathbb{R}$. Then f is right-continuous on $\mathbb{R}$ but unbounded on any finite interval containing an integer in its interior.

Proposition 9.3. *Let f be a real valued function on $\mathbb{R}$ such that $f(t-) \equiv \lim_{s \uparrow t} f(s)$ and $f(t+) \equiv \lim_{s \downarrow t} f(s)$ exist in $\mathbb{R}$ for every $t \in \mathbb{R}$. Let*

$$E = \{t \in \mathbb{R} : f(t-) \neq f(t+)\},$$

and for every $k \in \mathbb{N}$ let

$$E_k = \{t \in \mathbb{R} : |f(t-) - f(t+)| \geq \frac{1}{k}\}.$$

Then E is a countable set and for every finite interval $[a, b]$ in $\mathbb{R}$, $E_k \cap [a, b]$ is a finite set. In particular, if f is a real valued function which is right-continuous and has finite left limit everywhere on $\mathbb{R}$, then f has at most countably many points of discontinuity.

Proof. Note that $E = \cup_{k \in \mathbb{N}} E_k$. Suppose E is an uncountable set. Then there exists $k_0 \in \mathbb{N}$ such that E_{k_0} is an uncountable set. Let $E_{k_0, m} = E_{k_0} \cap [m, m+1]$ for $m \in \mathbb{Z}$. Then there exists $m_0 \in \mathbb{Z}$ such that E_{k_0, m_0} is an uncountable set. By partitioning $[m_0, m_0 + 1]$ into two closed intervals of equal length (with one end-point in common) and repeating the process to the two resulting closed intervals indefinitely we obtain a decreasing sequence of closed intervals $\{I_n : n \in \mathbb{Z}_+\}$ such that I_n has length $\ell(I_n) = 2^{-n}$ and $E_{k_0} \cap I_n$ is an uncountable set for every $n \in \mathbb{Z}_+$. By the Nested Interval Theorem there exists $t^* \in \mathbb{R}$ such that $\cap_{n \in \mathbb{Z}_+} I_n = \{t^*\}$. Let $\varepsilon \in (0, \frac{1}{2k_0})$. Since $f(t^*-)$ and $f(t^*+)$ exist in $\mathbb{R}$ there exists $\delta > 0$ such that $|f(t') - f(t'')| < \varepsilon$ for $t', t'' \in (t^* - \delta, t^*)$ and $|f(t') - f(t'')| < \varepsilon$ for $t', t'' \in (t^*, t^* + \delta)$. Since $t^* \in I_n$ for every $n \in \mathbb{Z}_+$ and since $\lim_{n \to \infty} \ell(I_n) = 0$, there exists $n_0 \in \mathbb{Z}_+$ such that $I_{n_0} \subset (t^* - \delta, t^* + \delta)$. Then since $E_{k_0} \cap I_{n_0}$ is an uncountable set so is $E_{k_0} \cap (t^* - \delta, t^* + \delta)$ and thus at least one of the two sets $E_{k_0} \cap (t^* - \delta, t^*)$ and $E_{k_0} \cap (t^*, t^* + \delta)$ is not empty. Suppose for instance $E_{k_0} \cap (t^* - \delta, t^*) \neq \emptyset$. Let $t_0 \in E_{k_0} \cap (t^* - \delta, t^*)$. Then $|f(t_0-) - f(t_0+)| \geq \frac{1}{k_0}$ and thus there exist $t', t'' \in (t^* - \delta, t^*)$ such that $|f(t') - f(t'')| \geq \frac{1}{2k_0} > \varepsilon$, contradicting the selection of our $\delta > 0$. This completes the proof that E is a countable set.

Let $k \in \mathbb{N}$ and let $[a, b] \subset \mathbb{R}$. To show that $E_k \cap [a, b]$ is a finite set by contradiction, assume that it is an infinite set. Then there exists a sequence $\{t_n : n \in \mathbb{N}\}$ of distinct points

§9. REGULARITY OF SAMPLE FUNCTIONS OF SUBMARTINGALES

in $E_k \cap [a,b]$ which converges to some $t_0 \in [a,b]$. Now we have either a subsequence $\{t_{n_\ell} : \ell \in \mathbb{N}\}$ such that $t_{n_\ell} \uparrow t_0$ or a subsequence $\{t_{n_m} : m \in \mathbb{N}\}$ such that $t_{n_m} \downarrow t_0$. Consider the latter case for instance. Since $t_{n_m} \in E_k$ and $t_{n_m} > t_0$ (a consequence of distinctness of t_{n_m} for $m \in \mathbb{N}$ and the fact that $t_{n_m} \downarrow t_0$), there exist t'_{n_m} and t''_{n_m} in the interval $(t_{n_m} - \frac{1}{2}(t_{n_m} - t_0), t_{n_m} + \frac{1}{2}(t_{n_m} - t_0))$ such that $|f(t'_{n_m}) - f(t''_{n_m})| \geq \frac{1}{2k}$ for $m \in \mathbb{N}$. Now since $t'_{n_m}, t''_{n_m} > t_0$ and $\lim_{m \to \infty} t'_{n_m} = t_0$ and $\lim_{m \to \infty} t''_{n_m} = t_0$, this contradicts the existence of $f(t_0+)$. Similarly $t_{n_\ell} \uparrow t_0$ contradicts the existence of $f(t_0+)$. This shows that $E_k \cap [a,b]$ is a finite set. ∎

[II] Right-Continuous Modification of a Submartingale

Let $X = \{X_t : t \in \mathbb{R}_+\}$ be a submartingale on a filtered space $(\Omega, \mathfrak{F}, \{\mathfrak{F}_t\}, P)$. We are interested in the existence of a submartingale Y which is a modification of X in the sense that $Y(t, \cdot) = X(t, \cdot)$ a.e. on $(\Omega, \mathfrak{F}_t, P)$ for every $t \in \mathbb{R}_+$ but has the property that every sample function is right-continuous everywhere on $\mathbb{R}_+$. We shall show that if the filtration $\{\mathfrak{F}_t : t \in \mathbb{R}_+\}$ is right-continuous and if $\mathfrak{F}_0$ is augmented, that is, if $\mathfrak{F}_0$ contains all the null sets of the probability space $(\Omega, \mathfrak{F}, P)$, then the right-continuity of the function $\mathbf{E}(X_t), t \in \mathbb{R}_+$, implies the existence of such a modification Y.

Let us remark that if $X = \{X_t : t \in \mathbb{R}_+\}$ is a right-continuous submartingale on a right-continuous filtered space $(\Omega, \mathfrak{F}, \{\mathfrak{F}_t\}, P)$, then the function $\mathbf{E}(X_t), t \in \mathbb{R}_+$, is indeed right-continuous. To show this, let $t \in \mathbb{R}_+$ be arbitrarily fixed. Let $\{t_{-n} : n \in \mathbb{Z}_+\}$ be a strictly decreasing sequence in $\mathbb{R}_+$ with $t_{-n} \downarrow t$ as $n \to \infty$. The process $\{X_{t_{-n}} : n \in \mathbb{Z}_+\}$ is then a submartingale with reversed time with respect to the filtration $\{\mathfrak{F}_{t_{-n}} : n \in \mathbb{Z}_+\}$. According to Corollary 8.9, there exists $Y_{t_{-\infty}} \in L_1(\Omega, \mathfrak{F}_{t_{-\infty}}, P)$, where $\mathfrak{F}_{t_{-\infty}} = \cap_{n \in \mathbb{Z}_+} \mathfrak{F}_{t_{-n}}$, such that $\lim_{n \to \infty} X_{t_{-n}} = Y_{t_{-\infty}}$ a.e. on $(\Omega, \mathfrak{F}_{t_{-\infty}}, P)$ and $\lim_{n \to \infty} \|X_{t_{-n}} - Y_{t_{-\infty}}\|_1 = 0$. Now the right-continuity of the filtration implies that $\mathfrak{F}_{t_{-\infty}} = \mathfrak{F}_t$ and the right-continuity of X implies that we have $X_t = \lim_{n \to \infty} X_{t_{-n}} = Y_{t_{-\infty}}$ a.e. on $(\Omega, \mathfrak{F}_t, P)$. Thus we have $\lim_{n \to \infty} \|X_{t_{-n}} - X_t\|_1 = 0$. This implies $\lim_{n \to \infty} \mathbf{E}(X_{t_{-n}}) = \mathbf{E}(X_t)$. From the arbitrariness of the sequence $\{t_{-n} : n \in \mathbb{Z}_+\}$ we have the right-continuity of the function $\mathbf{E}(X_t), t \in \mathbb{R}_+$, at t.

Definition 9.4. *If $X = \{X_t : t \in \mathbb{R}_+\}$ is a submartingale on a filtered space $(\Omega, \mathfrak{F}, \{\mathfrak{F}_t\}, P)$ and if there exists a right-continuous submartingale $X^{(r)} = \{X_t^{(r)} : t \in \mathbb{R}_+\}$ on the filtered space such that $X_t^{(r)} = X_t$ a.e. on $(\Omega, \mathfrak{F}_t, P)$ for every $t \in \mathbb{R}_+$, then $X^{(r)}$ is called a right-continuous modification of X.*

Definition 9.5. *A filtration $\{\mathfrak{F}_t : t \in \mathbb{R}_+\}$ of a probability space $(\Omega, \mathfrak{F}, P)$ is called aug-*

mented if $\mathfrak{F}_t$ *is an augmented sub-σ-algebra, that is,* $\mathfrak{F}_t$ *contains all the null sets in* $(\Omega, \mathfrak{F}, P)$, *for every* $t \in \mathbb{R}_+$. *A filtered space* $(\Omega, \mathfrak{F}, \{\mathfrak{F}_t\}, P)$ *is called an augmented filtered space if the filtration* $\{\mathfrak{F}_t : t \in \mathbb{R}_+\}$ *is augmented.*

Note that since a filtration $\{\mathfrak{F}_t : t \in \mathbb{R}_+\}$ is an increasing system of sub-σ-algebras, it is augmented if and only if $\mathfrak{F}_0$ is augmented.

Theorem 9.6. *If* $X = \{X_t : t \in \mathbb{R}_+\}$ *is a submartingale on an augmented right-continuous filtered space* $(\Omega, \mathfrak{F}, \{\mathfrak{F}_t\}, P)$ *and if* $\mathbf{E}(X_t), t \in \mathbb{R}_+$ *is a right-continuous function, then* X *has a right-continuous modification* $X^{(r)} = \{X_t^{(r)} : t \in \mathbb{R}_+\}$.

The proof of this theorem is based on the following three Propositions.

Proposition 9.7. *Let* $X = \{X_t : t \in \mathbb{R}_+\}$ *be a submartingale on a filtered space* $(\Omega, \mathfrak{F}, \{\mathfrak{F}_t\}, P)$, *let* Q_+ *be the collection of all nonnegative rational numbers and let* Λ *be the null set in* $(\Omega, \mathfrak{F}_\infty, P)$ *in Proposition 9.1. Let* $X^{(r)} = \{X_t^{(r)} : t \in \mathbb{R}_+\}$ *and* $X^{(\ell)} = \{X_t^{(\ell)} : t \in \mathbb{R}_+\}$ *be two processes defined by setting for every* $t \in \mathbb{R}_+$

$$\begin{cases} X_t^{(r)}(\omega) = \lim_{s \downarrow t, s \in Q_+} X_s(\omega) & \text{for } \omega \in \Lambda^c, \\ X_t^{(\ell)}(\omega) = \lim_{s \uparrow t, s \in Q_+} X_s(\omega) & \text{for } \omega \in \Lambda^c, \\ X_t^{(r)}(\omega) = X_t^{(\ell)}(\omega) = 0 & \text{for } \omega \in \Lambda. \end{cases}$$

Then $X^{(r)}$ *is right-continuous with finite left limit, and* $X^{(\ell)}$ *is left-continuous with finite right limit on* $\mathbb{R}_+$. *For both* $X^{(r)}$ *and* $X^{(\ell)}$ *every sample function is bounded on every finite interval in* $\mathbb{R}_+$. *The process* $X^{(r)}$ *is an L_1-processes.*

Proof. Let us prove the existence of finite left limits for $X^{(r)}$. It suffices to prove the existence of finite left limit of $X^{(r)}(\cdot, \omega)$ on $\mathbb{R}_+$ for $\omega \in \Lambda^c$. Let $t_0 \in \mathbb{R}_+$. To show that $\lim_{t \uparrow t_0} X^{(r)}(t, \omega)$ exists in $\mathbb{R}$ we show that for every $\varepsilon > 0$ there exists $\delta > 0$ such that

$$|X^{(r)}(t', \omega) - X^{(r)}(t'', \omega)| < \varepsilon \quad \text{for } t', t'' \in (t_0 - \delta, t_0) \cap \mathbb{R}_+.$$

Now since $X^{(\ell)}(t_0, \omega) = \lim_{s \uparrow t_0, s \in Q_+} X(s, \omega) \in \mathbb{R}$, for every $\varepsilon > 0$ there exists $\delta > 0$ such that

$$|X(s, \omega) - X^{(\ell)}(t_0, \omega)| < \frac{\varepsilon}{2} \quad \text{for } s \in (t_0 - \delta, t_0) \cap Q_+.$$

Let $t', t'' \in (t_0 - \delta, t_0) \cap \mathbb{R}_+$ and let $s_n', s_n'' \in (t_0 - \delta, t_0) \cap Q_+$ be such that $s_n' \downarrow t'$ and $s_n'' \downarrow t''$ as $n \to \infty$. Then by the definition of $X^{(r)}$

$$|X^{(r)}(t', \omega) - X^{(r)}(t'', \omega)|$$

§9. REGULARITY OF SAMPLE FUNCTIONS OF SUBMARTINGALES

$$= |\lim_{n\to\infty} X(s'_n,\omega) - \lim_{n\to\infty} X(s''_n,\omega)|$$

$$= \lim_{n\to\infty} |X(s'_n,\omega) - X(s''_n,\omega)|$$

$$\leq \limsup_{n\to\infty}\{|X(s'_n,\omega) - X^{(\ell)}(t_0,\omega)| + |X(s''_n,\omega) - X^{(\ell)}(t_0,\omega)|\}$$

$$\leq \frac{\varepsilon}{2} + \frac{\varepsilon}{2} = \varepsilon.$$

This proves the existence of finite left limits for $X^{(r)}$. The existence of finite right limit of $X^{(\ell)}$, the right-continuity of $X^{(r)}$ and the left-continuity of $X^{(\ell)}$ can be proved likewise.

By the definition of $X^{(r)}$ and by Proposition 9.1, every sample function of $X^{(r)}$ is bounded on every finite interval in $\mathbb{R}_+$. Similarly for $X^{(\ell)}$.

To show that $X_t^{(r)}$ is integrable for every $t \in \mathbb{R}_+$, let $s_n \in Q_+$ be such that $s_n \downarrow t$ as $n \to \infty$. Then $\lim_{n\to\infty} X_{s_n} = X_t^{(r)}$ a.e. on Ω. Thus by Corollary 8.9, $X_t^{(r)}$ is integrable. ∎

Proposition 9.8. *In the same setting as in Proposition 9.7, assume further that the filtered space $(\Omega, \mathfrak{F}, \{\mathfrak{F}_t\}, P)$ is augmented. Let us define two filtrations of $(\Omega, \mathfrak{F}, P)$, $\{\mathfrak{F}_t^{(r)} : t \in \mathbb{R}_+\}$ and $\{\mathfrak{F}_t^{(\ell)} : t \in \mathbb{R}_+\}$ by setting for every $t \in \mathbb{R}_+$*

$$\mathfrak{F}_t^{(r)} = \bigcap_{s>t, s\in Q_+} \mathfrak{F}_s \quad \text{and} \quad \mathfrak{F}_t^{(\ell)} = \sigma\left(\bigcup_{s<t, s\in Q_+} \mathfrak{F}_s\right)$$

(Note that $\mathfrak{F}_t^{(r)} = \bigcap_{u>t} \mathfrak{F}_u$ and $\mathfrak{F}_t^{(\ell)} = \sigma(\bigcup_{u<t} \mathfrak{F}_u)$.)
1) $X^{(r)}$ is adapted to $\{\mathfrak{F}_t^{(r)} : t \in \mathbb{R}_+\}$ and $X^{(\ell)}$ is adapted to $\{\mathfrak{F}_t^{(\ell)} : t \in \mathbb{R}_+\}$.
2) Furthermore for every $t \in \mathbb{R}_+$ we have

(1) $$E(X_t^{(r)}|\mathfrak{F}_t) \geq X_t \quad \text{a.e. on } (\Omega, \mathfrak{F}_t, P)$$

and

(2) $$E(X_t|\mathfrak{F}_t^{(\ell)}) \geq X_t^{(\ell)} \quad \text{a.e. on } (\Omega, \mathfrak{F}_t^{(\ell)}, P).$$

3) $X^{(r)}$ is a submartingale with respect to $\{\mathfrak{F}_t^{(r)} : t \in \mathbb{R}_+\}$. If X is a martingale to start with then $X^{(r)}$ is a martingale.

Proof. For each $s \in Q_+$ consider a random variable Y_s on $(\Omega, \mathfrak{F}, P)$ defined by $Y_s(\omega) = X_s(\omega)$ for $\omega \in \Lambda^c$ and $Y_s(\omega) = 0$ for $\omega \in \Lambda$. Since $\mathfrak{F}_0$ contains all null sets in $(\Omega, \mathfrak{F}, P)$ we have in particular $\Lambda \in \mathfrak{F}_0 \subset \mathfrak{F}_s$. Then from the $\mathfrak{F}_s$-measurability of X_s follows the

$\mathfrak{F}_s$-measurability of Y_s. Since Y_s is $\mathfrak{F}_s$-measurable for every $s \in Q_+$, $\lim_{s \downarrow t, s \in Q_+} Y_s$ is $\mathfrak{F}_s$-measurable for every $s \in Q_+$, $s > t$, and thus it is $\mathfrak{F}_t^{(r)}$-measurable. But by the definitions of $X_t^{(r)}$ and Y_s we have $X_t^{(r)} = \lim_{s \downarrow t, s \in Q_+} Y_s$. Thus $X_t^{(r)}$ is $\mathfrak{F}_t^{(r)}$-measurable. This shows that $X^{(r)}$ is adapted to $\{\mathfrak{F}_t^{(r)} : t \in \mathbb{R}_+\}$. Similarly $X^{(\ell)}$ is adapted to $\{\mathfrak{F}_t^{(\ell)} : t \in \mathbb{R}_+\}$.

To prove (1), let $t \in \mathbb{R}_+$ and let $\{s_n : n \in \mathbb{Z}_+\}$ be a strictly decreasing sequence in Q_+ such that $s_n \downarrow t$ as $n \to \infty$. By Corollary 8.9, $\{X_{s_n} : n \in \mathbb{Z}_+\}$ is uniformly integrable and there exists $Y \in L_1(\Omega, \mathfrak{F}_t^{(r)}, P)$ such that $\lim_{n \to \infty} X_{s_n} = Y$ a.e. on $(\Omega, \mathfrak{F}_t^{(r)}, P)$ and $\lim_{n \to \infty} \|X_{s_n} - Y\|_1 = 0$. Since $\lim_{n \to \infty} X_{s_n} = X_t^{(r)}$ a.e. on $(\Omega, \mathfrak{F}_t^{(r)}, P)$ by the definition of $X_t^{(r)}$ we have $X_t^{(r)} = Y$ a.e. on $(\Omega, \mathfrak{F}_t^{(r)}, P)$ and therefore $\lim_{n \to \infty} \|X_{s_n} - X_t^{(r)}\|_1 = 0$. This convergence in L_1 implies $\lim_{n \to \infty} \|E(X_{s_n}|\mathfrak{F}_t) - E(X_t^{(r)}|\mathfrak{F}_t)\|_1 = 0$. Then there exists a subsequence $\{n_k\}$ of $\{n\}$ such that $\lim_{k \to \infty} E(X_{s_{n_k}}|\mathfrak{F}_t) = E(X_t^{(r)}|\mathfrak{F}_t)$ a.e. on $(\Omega, \mathfrak{F}_t, P)$. By the submartingale property of X we have $E(X_{s_{n_k}}|\mathfrak{F}_t) \geq X_t$ a.e. on $(\Omega, \mathfrak{F}_t, P)$. Therefore $E(X_t^{(r)}|\mathfrak{F}_t) \geq X_t$ a.e. on $(\Omega, \mathfrak{F}_t, P)$. This proves (1).

To prove (2), let $t \in \mathbb{R}_+$ and $\{s_n : n \in \mathbb{Z}_+\}$ be a strictly increasing sequence in Q_+ such that $s_n \uparrow t$ as $n \to \infty$. Let Y_∞ be an arbitrary version of $E(X_t|\mathfrak{F}_t^{(\ell)})$ and for each $n \in \mathbb{Z}_+$ let Y_n be an arbitrary version of $E(X_t|\mathfrak{F}_{s_n})$. Then by Remark 8.3, we have $\lim_{n \to \infty} Y_n = Y_\infty$ a.e. on $(\Omega, \mathfrak{F}_t^{(\ell)}, P)$. By the submartingale property of X we have $Y_n = E(X_t|\mathfrak{F}_{s_n}) \geq X_{s_n}$ a.e. on $(\Omega, \mathfrak{F}_{s_n}, P)$. Letting $n \to \infty$ and recalling the definition of $X_t^{(\ell)}$ we have $Y_\infty \geq X_t^{(\ell)}$ a.e. on $(\Omega, \mathfrak{F}_t^{(\ell)}, P)$. This proves (2).

Since $X^{(r)}$ is an L_1-process adapted to $\{\mathfrak{F}_t^{(r)} : t \in \mathbb{R}_+\}$, to show that it is a submartingale with respect to $\{\mathfrak{F}_t^{(r)} : t \in \mathbb{R}_+\}$ it remains only to verify that for $s, t \in \mathbb{R}_+$, $s < t$, we have $E(X_t^{(r)}|\mathfrak{F}_s^{(r)}) \geq X_s^{(r)}$ a.e. on $(\Omega, \mathfrak{F}_s^{(r)}, P)$, or equivalently,

$$(3) \qquad \int_E X_t^{(r)} \, dP \geq \int_E X_s^{(r)} \, dP \quad \text{for every } E \in \mathfrak{F}_s^{(r)}.$$

Let $\{\varepsilon_n : n \in \mathbb{Z}_+\}$ be a strictly decreasing sequence such that $\varepsilon_n \downarrow 0$ and $t + \varepsilon_n \in Q_+$ for every $n \in \mathbb{Z}_+$. Then by the definition of $X_t^{(r)}$, $\lim_{n \to \infty} X_{t+\varepsilon_n} = X_t^{(r)}$ a.e. on $(\Omega, \mathfrak{F}_t^{(r)}, P)$ and by Corollary 8.9, $\{X_{t+\varepsilon_n} : n \in \mathbb{Z}_+\}$ is uniformly integrable. Therefore by Theorem 4.16, $\lim_{n \to \infty} \|X_{t+\varepsilon_n} - X_t^{(r)}\|_1 = 0$. This convergence in L_1 implies

$$(4) \qquad \lim_{n \to \infty} \int_E X_{t+\varepsilon_n} \, dP = \int_E X_t^{(r)} \, dP \quad \text{for every } E \in \mathfrak{F}.$$

Similarly, with a strictly decreasing sequence $\{\eta_n : n \in \mathbb{Z}_+\}$ such that $\eta_n \downarrow 0$ and $s + \eta_n \in$

§9. REGULARITY OF SAMPLE FUNCTIONS OF SUBMARTINGALES

Q_+ for every $n \in \mathbb{Z}_+$, we obtain

(5) $$\lim_{n \to \infty} \int_E X_{s+\eta_n} \, dP = \int_E X_s^{(r)} \, dP \quad \text{for every } E \in \mathfrak{F}.$$

By choosing η_n so that $\eta_n < \varepsilon_n$, we have $\mathbf{E}(X_{t+\varepsilon_n} | \mathfrak{F}_{s+\eta_n}) \geq X_{s+\eta_n}$ a.e. on $(\Omega, \mathfrak{F}_{s+\eta_n}, P)$ by the submartingale property of X, that is,

(6) $$\int_E X_{t+\varepsilon_n} \, dP \geq \int_E X_{s+\eta_n} \, dP \quad \text{for } E \in \mathfrak{F}_{s+\eta_n}.$$

Then for $E \in \mathfrak{F}_s^{(r)} \subset \mathfrak{F}_{s+\eta_n}$, we have by (4), (6) and (5)

$$\int_E X_t^{(r)} \, dP = \lim_{n \to \infty} \int_E X_{t+\varepsilon_n} \, dP \geq \lim_{n \to \infty} \int_E X_{s+\eta_n} \, dP = \int_E X_s^{(r)} \, dP,$$

proving (3). If X is a martingale, then the inequality in (6) is replaced by an equality and therefore the inequality in (3) is replaced by an equality so that $X^{(r)}$ is a martingale. ∎

Proposition 9.9. *In the same setting as in Proposition 9.7, assume that the filtered space $(\Omega, \mathfrak{F}, \{\mathfrak{F}_t\}, P)$ is augmented and right-continuous. Then for every $t_0 \in \mathbb{R}_+$ we have $X_{t_0}^{(r)} \geq X_{t_0}$ a.e. on $(\Omega, \mathfrak{F}_{t_0}, P)$ and $X_{t_0}^{(r)} = X_{t_0}$ a.e. on $(\Omega, \mathfrak{F}_{t_0}, P)$ if and only the function $\mathbf{E}(X_t), t \in \mathbb{R}_+$, is right-continuous at t_0.*

Proof. Let us assume the right-continuity of the filtration $\{\mathfrak{F}_t : t \in \mathbb{R}_+\}$. Then $\mathfrak{F}_t^{(r)} = \mathfrak{F}_t$ for every $t \in \mathbb{R}_+$.

Let $t_0 \in \mathbb{R}_+$. To show that $X_{t_0}^{(r)} \geq X_{t_0}$ a.e. on $(\Omega, \mathfrak{F}_{t_0}, P)$, let $\{\varepsilon_n : n \in \mathbb{Z}_+\}$ be a strictly decreasing sequence such that $\varepsilon_n \downarrow 0$ and $t_0 + \varepsilon_n \in Q_+$ for every $n \in \mathbb{Z}_+$. By the submartingale property of X we have

$$\int_E X_{t_0+\varepsilon_n} \, dP \geq \int_E X_{t_0} \, dP \quad \text{for } E \in \mathfrak{F}_{t_0}.$$

Recalling (4) in the proof of Proposition 9.8, we have

$$\int_E X_{t_0}^{(r)} \, dP \geq \int_E X_{t_0} \, dP \quad \text{for } E \in \mathfrak{F}_{t_0}.$$

Since $\mathfrak{F}_{t_0}^{(r)} = \mathfrak{F}_{t_0}$, $X_{t_0}^{(r)}$ and X_{t_0} are both $\mathfrak{F}_{t_0}$-measurable. Then the last inequality implies that $X_{t_0}^{(r)} \geq X_{t_0}$ a.e. on $(\Omega, \mathfrak{F}_{t_0}, P)$.

Let $\{s_n : n \in \mathbb{Z}_+\}$ be a strictly decreasing sequence in Q_+ such that $s_n \downarrow t_0$ as $n \to \infty$. Then as we saw in the proof of Proposition 9.8, $\lim_{n \to \infty} \|X_{s_n} - X_{t_0}^{(r)}\|_1 = 0$ so that

$\lim_{n\to\infty} \mathbf{E}(X_{s_n}) = \mathbf{E}(X_{t_0}^{(r)})$. Since X is a submartingale, $\mathbf{E}(X_t)$ decreases as t decreases. Thus the sequential convergence above implies that $\mathbf{E}(X_t) \downarrow \mathbf{E}(X_{t_0}^{(r)})$ as $t \downarrow t_0$. Now if $X_{t_0}^{(r)} = X_{t_0}$ a.e. on $(\Omega, \mathfrak{F}_{t_0}, P)$ then $\mathbf{E}(X_{t_0}^{(r)}) = \mathbf{E}(X_{t_0})$ so that $\mathbf{E}(X_t) \downarrow \mathbf{E}(X_{t_0})$ as $t \downarrow t_0$, that is, $\mathbf{E}(X_t)$ is right-continuous at t_0. Conversely, if $\mathbf{E}(X_t)$ is right-continuous at t_0, then $\mathbf{E}(X_t) \downarrow \mathbf{E}(X_{t_0})$ as $t \downarrow t_0$ so that $\mathbf{E}(X_{t_0}^{(r)}) = \mathbf{E}(X_{t_0})$. This, together with the fact that $X_{t_0}^{(r)} \geq X_{t_0}$ a.e. on $(\Omega, \mathfrak{F}_{t_0}, P)$, implies that $X_{t_0}^{(r)} = X_{t_0}$ a.e. on $(\Omega, \mathfrak{F}_{t_0}, P)$. ∎

Proof of Theorem 9.6. Let $X = \{X_t : t \in \mathbb{R}_+\}$ be a submartingale on an augmented right-continuous filtered space $(\Omega, \mathfrak{F}, \{\mathfrak{F}_t\}, P)$ such that $\mathbf{E}(X_t)$, $t \in \mathbb{R}_+$, is a right-continuous function. Let Q_+ be the collection of all nonnegative rational numbers. According to Proposition 9.1, there exists a null set Λ in $(\Omega, \mathfrak{F}_\infty, P)$ such that $\lim_{s\uparrow t, s\in Q_+} X_s(\omega)$ and $\lim_{s\downarrow t, s\in Q_+} X_s(\omega)$ exist in $\mathbb{R}$ for every $t \in \mathbb{R}_+$ and for every $\omega \in \Lambda^c$. If we define $X^{(r)} = \{X_t^{(r)} : t \in \mathbb{R}_+\}$ by setting for every $t \in \mathbb{R}_+$

$$\begin{cases} X_t^{(r)}(\omega) = \lim_{s\downarrow t, s\in Q_+} X_s(\omega) & \text{for } \omega \in \Lambda^c \\ X_t^{(r)}(\omega) = 0 & \text{for } \omega \in \Lambda, \end{cases}$$

then according to Proposition 9.7, $X^{(r)}$ is an L_1-process and every sample function of $X^{(r)}$ is bounded on every finite interval in $\mathbb{R}_+$ and is right-continuous with finite left limit everywhere on $\mathbb{R}_+$. The right-continuity of the filtration $\{\mathfrak{F}_t : t \in \mathbb{R}_+\}$ implies that $X^{(r)}$ is $\{\mathfrak{F}_t\}$-adapted according to Proposition 9.8. The right-continuity of $\mathbf{E}(X_t)$, $t \in \mathbb{R}_+$, implies that $X_t^{(r)} = X_t$ a.e. on $(\Omega, \mathfrak{F}_t, P)$ for every $t \in \mathbb{R}_+$ according to Proposition 9.9. Thus $X^{(r)}$ is a right-continuous modification of X. ∎

Corollary 9.10. *Let $(\Omega, \mathfrak{F}, \{\mathfrak{F}_t\}, P)$ be an augmented right-continuous filtered space and let ξ be an integrable random variable on the probability space $(\Omega, \mathfrak{F}, P)$. Then for every $t \in \mathbb{R}_+$ there exists a version X_t of $\mathbf{E}(\xi|\mathfrak{F}_t)$ such that $X = \{X_t : t \in \mathbb{R}_+\}$ is a right-continuous uniformly integrable martingale on the filtered space. Furthermore if $|\xi| \leq K$ a.e. on $(\Omega, \mathfrak{F}, P)$ for some $K > 0$, then X_t can be so chosen that besides being right-continuous X satisfies the condition $|X(t, \omega)| \leq K$ for $(t, \omega) \in \mathbb{R}_+ \times \Omega$.*

Proof. Let Y_t be an arbitrary version of $\mathbf{E}(\xi|\mathfrak{F}_t)$ for $t \in \mathbb{R}_+$. Then $Y = \{Y_t : t \in \mathbb{R}_+\}$ is a martingale on the filtered space and according to Theorem 4.24 it is uniformly integrable. Now $\mathbf{E}(Y_t)$ is constant function of $t \in \mathbb{R}_+$ since $\mathbf{E}(Y_t) = \mathbf{E}[\mathbf{E}(\xi|\mathfrak{F}_t)] = \mathbf{E}(\xi)$. Thus by Theorem 9.6 there exists a right-continuous martingale $X = \{X_t : t \in \mathbb{R}_+\}$ on the filtered space such that $X_t = Y_t$ a.e. on $(\Omega, \mathfrak{F}_t, P)$ for every $t \in \mathbb{R}_+$. This implies that X_t is a version

§10. INCREASING PROCESSES

of $\mathbf{E}(\xi|\mathfrak{F}_t)$ for every $t \in \mathbb{R}_+$. Thus for every $t \in \mathbb{R}_+$ there exists a version X_t of $\mathbf{E}(\xi|\mathfrak{F}_t)$ such that $X = \{X_t : t \in \mathbb{R}_+\}$ is a right-continuous martingale on the filtered space. The uniform integrability of $\{Y_t : t \in \mathbb{R}_+\}$ implies the uniform integrability of $\{X_t : t \in \mathbb{R}_+\}$.

If $|\xi| \leq K$ a.e. on $(\Omega, \mathfrak{F}, P)$ then for every $t \in \mathbb{R}_+$ we have $|\mathbf{E}(\xi|\mathfrak{F}_t)| \leq K$ a.e. on $(\Omega, \mathfrak{F}_t, P)$. Let Y_t be an arbitrary version of $\mathbf{E}(\xi|\mathfrak{F}_t)$. Then there exists a null set Λ_{Y_t} in $(\Omega, \mathfrak{F}_t, P)$ such that $|Y_t| \leq K$ for $\omega \in \Lambda_{Y_t}^c$. If we let $Z_t = Y_t$ on $\Lambda_{Y_t}^c$ and $Z_t = 0$ on Λ_{Y_t}, then Z_t is a version of $\mathbf{E}(\xi|\mathfrak{F}_t)$ and $|Z_t| \leq K$ on Ω. According to Proposition 9.1, there exists a null set Λ in $(\Omega, \mathfrak{F}_\infty, P)$ such that if $\omega \in \Lambda^c$ then $\lim_{s\uparrow t, s\in Q_+} Z_s(\omega)$ and $\lim_{s\downarrow t, s\in Q_+} Z_s(\omega)$ exist in $\mathbb{R}$ for every $t \in \mathbb{R}_+$. By defining X_t for every $t \in \mathbb{R}_+$ by

$$\begin{cases} X_t(\omega) = \lim_{s\downarrow t, s\in Q_+} Z_s(\omega) & \text{for } \omega \in \Lambda^c \\ X_t(\omega) = 0 & \text{for } \omega \in \Lambda, \end{cases}$$

we obtain a version X_t of $\mathbf{E}(\xi|\mathfrak{F}_t)$ such that $X = \{X_t : t \in \mathbb{R}_+\}$ is a right-continuous martingale bounded by K. ∎

§10 Increasing Processes

[I] The Lebesgue Stieltjes Integral

If we set aside technical refinements in its definition, then an increasing process is a stochastic process $A = \{A_t : t \in \mathbb{R}_+\}$ on a probability space $(\Omega, \mathfrak{F}, P)$ whose sample functions $A(\cdot, \omega)$, $\omega \in \Omega$, are real valued monotone increasing functions on $\mathbb{R}_+$. For each $\omega \in \Omega$, let $\mu(\cdot, \omega)$ be the Lebesgue-Stieltjes measure on $(\mathbb{R}_+, \mathfrak{B}_{\mathbb{R}_+})$ determined by the monotone increasing function $A(\cdot, \omega)$. If $X = \{X_t : t \in \mathbb{R}_+\}$ is a stochastic process on $(\Omega, \mathfrak{F}, P)$ whose sample functions $X(\cdot, \omega)$, $\omega \in \Omega$, are Borel measurable real valued functions on $\mathbb{R}_+$, then for any $t \in \mathbb{R}_+$, the Lebesgue-Stieltjes integral $\int_{[0,t]} X(s,\omega)\,\mu(ds,\omega)$ exists for every $\omega \in \Omega$ such that at least one of $\int_{[0,t]} X^+(s,\omega)\,\mu(ds,\omega)$ and $\int_{[0,t]} X^-(s,\omega)\,\mu(ds,\omega)$ is finite.

Let us review the definition of the Lebesgue-Stieltjes measure determined by a real valued monotone increasing function on $\mathbb{R}$.

A collection $\mathfrak{I}$ of subsets of a set S is called a *semialgebra* of subsets of S if it satisfies the following conditions:

1°. $\emptyset, S \in \mathfrak{I}$,

2°. $I \cap J \in \mathfrak{I}$ for $I, J \in \mathfrak{I}$,

3°. for every $I \in \mathfrak{I}$ there exists a finite disjoint collection $\{I_k : k = 0, \ldots, n\}$ in $\mathfrak{I}$ such that $I_0 = I$ and $\cup_{j=0}^{k} I_j \in \mathfrak{I}$ for $k = 0, \ldots, n$ with $\cup_{j=0}^{n} I_j = S$.

In $\mathbb{R}$, for instance, the collection of all intervals which are open on the left and closed on the right and $\emptyset$ constitute a semialgebra.

If $\mathfrak{I}$ is a semialgebra, then I^c is a finite disjoint union of members of $\mathfrak{I}$ for every $I \in \mathfrak{I}$. Also every finite union of members of $\mathfrak{I}$ is equal to a finite disjoint union of members of $\mathfrak{I}$. Let us write $\alpha(\mathfrak{I})$ for the algebra generated by $\mathfrak{I}$, that is, the smallest algebra of subsets of S containing $\mathfrak{I}$. It follows immediately that $\alpha(\mathfrak{I})$ is the collection of all finite unions of members of $\mathfrak{I}$.

Let us call a set function μ on a semialgebra $\mathfrak{I}$ of subsets of a set S a measure on a semialgebra if it is $\overline{\mathbb{R}}_+$-valued with $\mu(\emptyset) = 0$ and countably additive on $\mathfrak{I}$, that is, if $\{I_n : n \in \mathbb{N}\}$ is a disjoint collection in $\mathfrak{I}$ with $\cup_{n \in \mathbb{N}} I_n \in \mathfrak{I}$ then $\mu(\cup_{n \in \mathbb{N}} I_n) = \sum_{n \in \mathbb{N}} \mu(I_n)$. Similarly we call a set function on an algebra of subsets of a set a measure on an algebra if it is $\overline{\mathbb{R}}_+$-valued with $\mu(\emptyset) = 0$ and countably additive on the algebra.

Let μ be a measure on a semialgebra $\mathfrak{I}$ of subsets of a set S. Since every $A \in \alpha(\mathfrak{I})$ is the union of finitely many disjoint members $I_1, \ldots, I_n$ of $\mathfrak{I}$, if we define $\mu(A) = \sum_{j=1}^{n} \mu(I_j)$ then μ is well-defined, that is, it does not depend on the way A is decomposed, and furthermore it is countably additive on $\alpha(\mathfrak{I})$ and is thus a measure on $\alpha(\mathfrak{I})$. The extension of a measure from a semialgebra $\mathfrak{I}$ to the algebra $\alpha(\mathfrak{I})$ generated by it is always unique, that is, if μ and ν are two measures on $\alpha(\mathfrak{I})$ such that $\mu = \nu$ on $\mathfrak{I}$ then $\mu = \nu$ on $\alpha(\mathfrak{I})$.

A measure μ on an algebra $\mathfrak{A}$ of subsets of a set S can always be extended to a measure on $\sigma(\mathfrak{A})$, the σ-algebra generated by $\mathfrak{A}$ by means of an outer measure derived from μ as follows. Let us define a set function μ^* on the σ-algebra of all subsets of S by setting for an arbitrary $E \subset S$

$$\mu^*(E) = \inf\{\sum_{n \in \mathbb{N}} \mu(A_n) : A_n \in \mathfrak{A}, n \in \mathbb{N}, E \subset \bigcup_{n \in \mathbb{N}} A_n\},$$

where the infimum is over the collection of all coverings of E by countably many members of $\mathfrak{A}$. Thus defined, μ^* is an outer measure satisfying the conditions that it is an $\overline{\mathbb{R}}_+$-valued set function defined on the σ-algebra of all subsets of S, vanishes for $\emptyset$, monotone increasing in the sense that $\mu^*(E_1) \leq \mu^*(E_2)$ for $E_1 \subset E_2 \subset S$, and countably subadditive, that is, for an arbitrary collection $\{E_n : n \in \mathbb{N}\}$ of subsets of S we have $\mu^*(\cup_{n \in \mathbb{N}} E_n) \leq \sum_{n \in \mathbb{N}} \mu^*(E_n)$. The collection $\mathfrak{A}^*_\mu$ of all subsets E of S satisfying the Carathéodory criterion

$$\mu^*(T) = \mu^*(T \cap E) + \mu^*(T \cap E^c) \quad \text{for every } T \subset S$$

§10. INCREASING PROCESSES

is a σ-algebra of subsets of S containing the algebra $\mathfrak{A}$ and the restriction of μ^* to this σ-algebra which we denote by μ is countably additive on this σ-algebra. Thus we have extended the measure μ on the algebra $\mathfrak{A}$ to a measure on the σ-algebra $\mathfrak{A}_\mu^*$. This is the Hopf Extension Theorem. Note also that the measure space $(S, \mathfrak{A}_\mu^*, \mu)$ is always a complete measure space. The σ-algebra $\mathfrak{A}_\mu^*$ depends on μ. However, since $\mathfrak{A} \subset \mathfrak{A}_\mu^*$ we always have $\sigma(\mathfrak{A}) \subset \mathfrak{A}_\mu^*$ for all μ. The restriction of μ on $\mathfrak{A}_\mu^*$ to $\sigma(\mathfrak{A})$ is then an extension of the original measure μ on $\mathfrak{A}$ to a measure on $\sigma(\mathfrak{A})$. The extension of a measure μ on an algebra $\mathfrak{A}$ to a measure on the σ-algebra $\sigma(\mathfrak{A})$ is unique provided μ is σ-finite on $\mathfrak{A}$.

Let g be a real valued monotone increasing function on $\mathbb{R}$. Then for every $a \in \mathbb{R}$, we have $g(a-) \equiv \lim_{x \uparrow a} g(x) \in \mathbb{R}$ and $g(a+) \equiv \lim_{x \downarrow a} g(x) \in \mathbb{R}$; g is continuous at $a \in \mathbb{R}$ if and only if $g(a-) = g(a+)$; and g has at most countably many points of discontinuity. Let $g(-\infty) = \lim_{x \to -\infty} g(x)$ and $g(\infty) = \lim_{x \to +\infty} g(x)$. Let μ_g be a set function defined on the collection of all open intervals and singletons in $\mathbb{R}$ by setting

$$\mu_g((a,b)) = g(b-) - g(a+) \quad and \quad \mu_g(\{a\}) = g(a+) - g(a-).$$

The definition of μ_g is then extended to intervals of other types by setting

$$\mu_g((a,b]) = \mu_g((a,b)) + \mu_g(\{b\}) = g(b+) - g(a+),$$
$$\mu_g([a,b)) = \mu_g(\{a\}) + \mu_g((a,b)) = g(b-) - g(a-),$$
$$\mu_g([a,b]) = \mu_g(\{a\}) + \mu_g((a,b)) + \mu_g(\{b\}) = g(b+) - g(a-).$$

Note in particular that if g is right-continuous, then $\mu_g((a,b]) = g(b) - g(a)$.

Let $\mathfrak{I}$ be the semialgebra of subsets of $\mathbb{R}$ consisting of $\emptyset$, $(-\infty, \infty)$, and all left-open and right-closed intervals, that is, all intervals of the types $(a, b]$, $(-\infty, b]$, and (a, ∞). Let g be a real valued monotone increasing function on $\mathbb{R}$ and let μ_g be a set function on $\mathfrak{I}$ defined by $\mu_g(a, b] = g(b+) - g(a+)$. Then μ_g is a σ-finite measure on the semialgebra $\mathfrak{I}$. Let us extend μ_g to a σ-finite measure on the algebra $\mathfrak{A} = \alpha(\mathfrak{I})$ by the procedure described above. Let μ_g^* be the outer-measure derived from the measure μ_g on the algebra $\mathfrak{A}$. Let $\mathfrak{A}_g^*$ be the σ-algebra containing $\mathfrak{A}$ and consisting of all subsets of $\mathbb{R}$ satisfying the Carathéodory criterion. The complete measure space $(\mathbb{R}, \mathfrak{A}_g^*, \mu_g)$ is the *Lebesgue-Stieltjes measure space* on $\mathbb{R}$ determined by the real valued monotone increasing function g on $\mathbb{R}$. In particular, the Lebesgue-Stieltjes measure space on $\mathbb{R}$ determined by the function $g(x) = x$ for $x \in \mathbb{R}$ is the Lebesgue measure space on $\mathbb{R}$. Note that the σ-algebra $\mathfrak{A}_g^*$ depends on the function g. For instance, when $g(x) = x$ for $x \in \mathbb{R}$ then $\mathfrak{A}_g^*$ is the σ-algebra of the Lebesgue-measurable sets in $\mathbb{R}$. But when g is the identically vanishing function on $\mathbb{R}$, then $\mu_g(\mathbb{R}) = 0$ so that $\mathbb{R}$ is a null set in the complete measure space $(\mathbb{R}, \mathfrak{A}_g^*, \mu_g)$, and this implies that every subset of the null set $\mathbb{R}$ is in the σ-algebra $\mathfrak{A}_g^*$ and therefore $\mathfrak{A}_g^*$ in this case is the collection of all subsets of $\mathbb{R}$. When considering a family of Lebesgue-Stieltjes measures μ_g on $\mathbb{R}$ corresponding to a family of real valued monotone increasing functions g it is convenient to have a common σ-algebra of subsets of $\mathbb{R}$ on which all the Lebesgue-Stieltjes measures in the family are defined. Now since $\mathfrak{A} \subset \mathfrak{A}_g^*$ and since $\sigma(\mathfrak{A}) = \sigma(\alpha(\mathfrak{I})) = \mathfrak{B}_\mathbb{R}$ we have $\mathfrak{B}_\mathbb{R} \subset \mathfrak{A}_g^*$. The Borel σ-algebra $\mathfrak{B}_\mathbb{R}$ then can serve as the common σ-algebra. The measure spaces $(\mathbb{R}, \mathfrak{B}_\mathbb{R}, \mu_g)$, corresponding to real valued monotone increasing functions g on $\mathbb{R}$ have a common σ-algebra $\mathfrak{B}_\mathbb{R}$ on which all the measures μ_g are defined.

The Lebesgue-Stieltjes measure μ_g on $(\mathbb{R}, \mathfrak{A}_g^*)$ determined by a real valued monotone increasing function g on $\mathbb{R}$ is always a σ-finite measure. It is a finite measure if and only if g is bounded on $\mathbb{R}$. If $E \in \mathfrak{A}_g^*$ and E is a bounded set, that is, contained in a finite interval, then $\mu_g(E) < \infty$. For any $a \in \mathbb{R}$, we have $\mu_g(\{a\}) = g(a+) - g(a-)$ so that $\mu_g(\{a\}) = 0$ if and only g is continuous at $a \in \mathbb{R}$. The value of $\mu_g(\{a\})$ does not depend on the value of g at $a \in \mathbb{R}$. Furthermore, from the definition of μ_g on the semialgebra $\mathfrak{I}$ of left-open and right-closed intervals in $\mathbb{R}$ by setting $\mu_g((a, b]) = g(b+) - g(a+)$ which is independent of the values $g(a)$ and $g(b)$, it is clear that if we redefine g at its points of discontinuity in such a way that its monotone increasing property is not destroyed then the redefined function determines the same Lebesgue-Stieltjes measure space as the original function g. In particular the right-continuous modification of g defined by $h(x) = g(x+)$ for $x \in \mathbb{R}$ determines the same Lebesgue-Stieltjes measure space as g, that is, $\mu_h = \mu_g$ on $\mathfrak{A}_h^* = \mathfrak{A}_g^*$. For the same reason as above, for any $c \in \mathbb{R}$ we have $\mu_{g+c} = \mu_g$ on $\mathfrak{A}_{g+c}^* = \mathfrak{A}_g^*$.

[II] Integration with Respect to Increasing Processes

Definition 10.1. *A stochastic process $A = \{A_t : t \in \mathbb{R}_+\}$ on a filtered space $(\Omega, \mathfrak{F}, \{\mathfrak{F}_t\}, P)$ is called an increasing process if it satisfies the following conditions.*

1°. *A is $\{\mathfrak{F}_t\}$-adapted.*

2°. *A is an L_1-process.*

3°. *$A(\cdot, \omega)$ is a real valued right-continuous monotone increasing function on $\mathbb{R}_+$ with $A(0, \omega) = 0$ for every $\omega \in \Omega$.*

A is called an almost surely increasing process if it satisfies conditions 1°, 2° and the following condition.

§10. INCREASING PROCESSES

4°. *There exists a null set Λ_A in $(\Omega, \mathfrak{F}_\infty, P)$ such that 3° holds for every $\omega \in \Lambda_A^c$.*

We call Λ_A an exceptional null set for the almost surely increasing process A.

Note that an exceptional null set Λ_A for an almost surely increasing process A is not unique. In fact any null set in $(\Omega, \mathfrak{F}_\infty, P)$ containing Λ_A is an exceptional null set for A.

For an almost surely increasing process $A = \{A_t : t \in \mathbb{R}_+\}$, $A_\infty(\omega) = \lim_{t \to \infty} A_t(\omega)$ exists for $\omega \in \Lambda_A^c$, that is, A_∞ exists a.e. on $(\Omega, \mathfrak{F}_\infty, P)$. Let $A_\infty(\omega) = 0$ for $\omega \in \Lambda_A$ so that A_∞ is an extended real valued $\mathfrak{F}_\infty$-measurable random variable defined on the entire space Ω. Regarding the integrability of A_∞ we have the following.

Lemma 10.2. *For an almost surely increasing process $A = \{A_t : t \in \mathbb{R}_+\}$ on a filtered space $(\Omega, \mathfrak{F}, \{\mathfrak{F}_t\}, P)$ the following conditions are equivalent.*
1°. A_∞ *is integrable.*
2°. A *is uniformly integrable.*
3°. A *is L_1-bounded.*

Proof. If A is an almost surely increasing process, then $0 \leq A_t \leq A_\infty$ a.e. on $(\Omega, \mathfrak{F}_\infty, P)$. Thus, if A_∞ is integrable, then by 4) of Proposition 4.11, A is uniformly integrable and in particular L_1-bounded. Conversely, if A is L_1-bounded, that is, $\sup_{t \in \mathbb{R}_+} \mathbf{E}(A_t) < \infty$, then by the Monotone Convergence Theorem we have $\mathbf{E}(A_\infty) = \lim_{t \to \infty} \mathbf{E}(A_t) \leq \sup_{t \in \mathbb{R}_+} \mathbf{E}(A_t) < \infty$ so that A_∞ is integrable. ∎

To consider the Lebesgue-Stieltjes measures on $(\mathbb{R}_+, \mathfrak{B}_{\mathbb{R}_+})$ determined by the sample functions of an increasing process $A = \{A_t : t \in \mathbb{R}_+\}$ on a filtered space $(\Omega, \mathfrak{F}, \{\mathfrak{F}_t\}, P)$ let us assume that the definition of a sample function $A(\cdot, \omega)$ is always extended to all of $\mathbb{R}$ by setting $A(t, \omega) = 0$ for $t \in (-\infty, 0)$. Let $\mu_A(\cdot, \omega)$ be the Lebesgue-Stieltjes measure on $(\mathbb{R}, \mathfrak{B}_\mathbb{R})$ determined by the real valued right-continuous increasing function $A(\cdot, \omega)$ on $\mathbb{R}$. By our extension of the definition of $A(\cdot, \omega)$, we have $\mu_A(\{0\}, \omega) = A(0, \omega) - A(0-, \omega) = 0$. We then restrict $\mu_A(\cdot, \omega)$ to $(\mathbb{R}_+, \mathfrak{B}_{\mathbb{R}_+})$. For an almost surely increasing process A, we let $\mu_A(\cdot, \omega) = 0$ for ω in an exceptional null set Λ_A for A. By this convention, corresponding to an almost surely increasing process A there exists a family $\{\mu_A(\cdot, \omega) : \omega \in \Omega\}$ of Lebesgue-Stieltjes measures on $(\mathbb{R}_+, \mathfrak{B}_{\mathbb{R}_+})$.

Definition 10.3. *For an almost surely increasing process $A = \{A_t : t \in \mathbb{R}_+\}$ on a filtered space $(\Omega, \mathfrak{F}, \{\mathfrak{F}_t\}, P)$ let $\mu_A(\cdot, \omega)$ be the Lebesgue-Stieltjes measure on $(\mathbb{R}_+, \mathfrak{B}_{\mathbb{R}_+})$ deter-*

mined by $A(\cdot,\omega)$ for $\omega \in \Omega$. Let X be a stochastic process on the filtered space such that $X(\cdot,\omega)$ is a Borel measurable function on $\mathbb{R}_+$ for every $\omega \in \Omega$. We define the integral of X with respect to A on $[0,t]$ for $t \in \mathbb{R}_+$ by

$$\int_{[0,t]} X(s,\omega)\, dA(s,\omega) = \int_{[0,t]} X(s,\omega)\, \mu_A(ds,\omega) \quad \text{for } \omega \in \Omega,$$

provided the Lebesgue-Stieltjes integral on the right hand side exists.

The Lebesgue-Stieltjes integral $\int_{[0,t]} X(s,\omega)\, \mu_A(ds,\omega)$ exists if and only at least one of the two nonnegative Lebesgue-Stieltjes integrals $\int_{[0,t]} X^+(s,\omega)\, \mu_A(ds,\omega)$ and $\int_{[0,t]} X^-(s,\omega)\, \mu_A(ds,\omega)$ is finite. In particular, since $\mu_A([0,t],\omega) < \infty$ for every $t \in \mathbb{R}_+$ and $\omega \in \Omega$, if $X(\cdot,\omega)$ is bounded on $[0,t]$ for some $\omega \in \Omega$ then the two nonnegative Lebesgue-Stieltjes integrals are finite and $\int_{[0,t]} X(s,\omega)\, \mu_A(ds,\omega)$ exists as a real number. Also, if $X(\cdot,\omega)$ is nonnegative on $[0,t]$, then $\int_{[0,t]} X(s,\omega)\, \mu_A(ds,\omega)$ exists.

Let us consider the measurability of the integral of a stochastic process with respect to an almost surely increasing process. The integral $\int_{[0,t]} X(s,\omega)\, \mu_A(ds,\omega)$ of a bounded stochastic process X with Borel measurable sample functions with respect to an almost surely increasing process A on a filtered space $(\Omega, \mathfrak{F}, \{\mathfrak{F}_t\}, P)$ in Definition 10.3 is defined for each $\omega \in \Omega$. Recall that according to our convention, for ω in an exceptional null set Λ_A of the almost surely increasing process A we set $\mu_A(\cdot,\omega) = 0$. The integral is thus a function defined on Ω. Then the question as to the measurability of this function arises. We shall show in Lemma 10.10 and Theorem 10.11 that if X is an adapted measurable process and A is an almost surely increasing process on an augmented filtered space $(\Omega, \mathfrak{F}, \{\mathfrak{F}_t\}, P)$, then the fact that A is an adapted process implies that the integral is an $\mathfrak{F}_t$-measurable random variable on $(\Omega, \mathfrak{F}, P)$.

Let us consider integration with respect to a family of measures in general.

Definition 10.4. *Let $(S, \mathfrak{A})$ and $(T, \mathfrak{B})$ be two measurable spaces. A family of measures $\{\mu(\cdot, y) : y \in T\}$ on $(S, \mathfrak{A})$ is said to be $\mathfrak{B}$-measurable if $\mu(A, \cdot)$ is an $\overline{\mathbb{R}}_+$-valued $\mathfrak{B}$-measurable function on T for every $A \in \mathfrak{A}$.*

Theorem 10.5. *1) Let $(S, \mathfrak{A})$ and $(T, \mathfrak{B})$ be two measurable spaces, let $\{\mu(\cdot, y) : y \in T\}$ be a $\mathfrak{B}$-measurable family of measures on $(S, \mathfrak{A})$, and let ν be a measure on $(T, \mathfrak{B})$. Then*

§10. INCREASING PROCESSES

the set function λ on the semialgebra $\mathfrak{A} \times \mathfrak{B}$ defined by

$$\lambda(A \times B) = \int_B \mu(A, y) \nu(dy) \quad \text{for } A \times B \in \mathfrak{A} \times \mathfrak{B}$$

is countably additive on $\mathfrak{A} \times \mathfrak{B}$ with $\lambda(\emptyset) = 0$ and thus can be extended to a measure on $\sigma(\mathfrak{A} \times \mathfrak{B})$.

2) If there exists a disjoint collection $\{A_n : n \in \mathbb{N}\}$ in $\mathfrak{A}$ such that $\cup_{n \in \mathbb{N}} A_n = S$ and $\int_T \mu(A_n, y) \nu(dy) < \infty$ for $n \in \mathbb{N}$, then λ is σ-finite on $\mathfrak{A} \times \mathfrak{B}$ and its extension as a measure to $\sigma(\mathfrak{A} \times \mathfrak{B})$ is unique. If $\int_T \mu(S, y) \nu(dy) < \infty$, then λ is finite on $\mathfrak{A} \times \mathfrak{B}$ and so is its extension to $\sigma(\mathfrak{A} \times \mathfrak{B})$.

Proof. Since $\mathfrak{A}$ and $\mathfrak{B}$ are σ-algebras, $\mathfrak{A} \times \mathfrak{B}$ is a semialgebra. Indeed $\emptyset, S \times T \in \mathfrak{A} \times \mathfrak{B}$; $\mathfrak{A} \times \mathfrak{B}$ is closed under intersections; and if $E \in \mathfrak{A} \times \mathfrak{B}$ given by $E = A \times B$ with $A \in \mathfrak{A}$ and $B \in \mathfrak{B}$, then by letting $E_0 = A \times B$, $E_1 = A \times B^c$ and $E_2 = A^c \times T$ we have $E_0, E_1, E_2 \in \mathfrak{A} \times \mathfrak{B}$, $E_0 \cup E_1 = A \times T \in \mathfrak{A} \times \mathfrak{B}$ and $E_0 \cup E_1 \cup E_2 = S \times T$. This shows that $\mathfrak{A} \times \mathfrak{B}$ is a semialgebra.

To verify the countable additivity of λ on the semialgebra $\mathfrak{A} \times \mathfrak{B}$, let $\{E_n : n \in \mathbb{N}\}$ be a disjoint collection in $\mathfrak{A} \times \mathfrak{B}$ such that $E = \cup_{n \in \mathbb{N}} E_n \in \mathfrak{A} \times \mathfrak{B}$. Let $E_n = A_n \times B_n$ where $A_n \in \mathfrak{A}$, $B_n \in \mathfrak{B}$ for $n \in \mathbb{N}$ and let $E = A \times B$ with $A \in \mathfrak{A}$ and $B \in \mathfrak{B}$. Now

$$\sum_{n \in \mathbb{N}} \lambda(E_n) = \sum_{n \in \mathbb{N}} \lambda(A_n \times B_n) = \sum_{n \in \mathbb{N}} \int_{B_n} \mu(A_n, y) \nu(dy)$$
$$= \sum_{n \in \mathbb{N}} \int_T \mathbf{1}_{B_n}(y) \mu(A_n, y) \nu(dy) = \int_T \sum_{n \in \mathbb{N}} \mathbf{1}_{B_n}(y) \mu(A_n, y) \nu(dy)$$

by the Monotone Convergence Theorem. For $D \subset S \times T$ and $y \in T$, let $D_{\cdot,y}$ be the S-section of D at $y \in T$ defined by $D_{\cdot,y} = \{x \in S : (x, y) \in D\}$. Since $E_n = A_n \times B_n$, we have

$$(E_n)_{\cdot,y} = \begin{cases} A_n & \text{if } y \in B_n \\ \emptyset & \text{if } y \in B_n^c, \end{cases}$$

and thus

$$\mathbf{1}_{B_n}(y) \mu(A_n, y) = \mu((E_n)_{\cdot,y}, y).$$

Therefore

$$\sum_{n \in \mathbb{N}} \lambda(E_n) = \int_T \sum_{n \in \mathbb{N}} \mu((E_n)_{\cdot,y}, y) \nu(dy).$$

Since $\{E_n : n \in \mathbb{N}\}$ is a disjoint collection, so is $\{(E_n)_{\cdot,y} : n \in \mathbb{N}\}$. Also from $\cup_{n \in \mathbb{N}} E_n = E$ we have

$$\bigcup_{n \in \mathbb{N}} (E_n)_{\cdot,y} = E_{\cdot,y} = \begin{cases} A & \text{if } y \in B \\ \emptyset & \text{if } x \in B^c. \end{cases}$$

Thus

$$\sum_{n\in\mathbb{N}} \lambda(E_n) = \int_T \mu((\cup_{n\in\mathbb{N}} E_n)_{\cdot,y}, y)\,\nu(dy) = \int_T \mu(E_{\cdot,y}, y)\,\nu(dy)$$
$$= \int_T \mathbf{1}_B(y)\mu(A,y)\,\nu(dy) = \int_B \mu(A,y)\,\nu(dy) = \lambda(E).$$

This proves the countable additivity of λ on the semialgebra $\mathfrak{A} \times \mathfrak{B}$. Then λ can be extended to a measure on the σ-algebra $\sigma(\mathfrak{A} \times \mathfrak{B})$.

If there exists a disjoint collection $\{A_n : n \in \mathbb{N}\}$ in $\mathfrak{A}$ such that $\cup_{n\in\mathbb{N}} A_n = S$ and $\int_T \mu(A_n, y)\,\nu(dy) < \infty$ for $n \in \mathbb{N}$, then $\{A_n \times T : n \in \mathbb{N}\}$ is a disjoint collection in $\mathfrak{A} \times \mathfrak{B}$ with $\cup_{n\in\mathbb{N}}(A_n \times T) = S \times T$ and $\lambda(A_n \times T) < \infty$ for $n \in \mathbb{N}$. Thus λ is σ-finite on $\mathfrak{A} \times \mathfrak{B}$. This implies that the extension of λ as measure to $\sigma(\mathfrak{A} \times \mathfrak{B})$ is unique. ∎

Let us prepare a variant of Theorem 1.10, a monotone class theorem for functions, adapted to the type of functions that enter as integrands in an integral with respect to an almost surely increasing process as defined in Definition 10.3.

Proposition 10.6. *Let $(S, \mathfrak{A})$ and $(T, \mathfrak{B})$ be two measurable spaces. Let $\mathbf{V}$ be a collection of real valued functions f on $S \times T$ such that $f(\cdot, y)$ is bounded on S for every $y \in T$. Suppose $\mathbf{V}$ satisfies the following conditions.*

1°. $\mathbf{1}_E \in \mathbf{V}$ *for every* $E \in \mathfrak{A} \times \mathfrak{B}$,

2°. $\mathbf{V}$ *is a linear space,*

3°. *if $\{f_n : n \in \mathbb{N}\}$ is an increasing sequence of nonnegative functions in $\mathbf{V}$ such that $f \equiv \lim_{n\to\infty} f_n$ is real valued and $f(\cdot, y)$ is bounded on S for every $y \in T$, then $f \in \mathbf{V}$.*

Then $\mathbf{V}$ contains all real valued $\sigma(\mathfrak{A} \times \mathfrak{B})$-measurable functions f on $S \times T$ such that $f(\cdot, y)$ is bounded on S for every $y \in T$.

Proof. Clearly $\mathfrak{A} \times \mathfrak{B}$ is a π-class of subsets of $S \times T$ with the property that $S \times T \in \mathfrak{A} \times \mathfrak{B}$. Using this fact and conditions 1°, 2°, and 3° we show by the same argument as in the proof of Theorem 1.10 that $\mathbf{1}_E \in \mathbf{V}$ for every $E \in \sigma(\mathfrak{A} \times \mathfrak{B})$.

Let f be a simple function on the measurable space $(S \times T, \sigma(\mathfrak{A} \times \mathfrak{B}))$, that is, $f = \sum_{i=1}^k c_i \mathbf{1}_{E_i}$ where $E_i \in \sigma(\mathfrak{A} \times \mathfrak{B})$ and $c_i \in \mathbb{R}$ for $i = 1, \ldots, k$. Note that $f(\cdot, y)$ is bounded on S for every $y \in T$ and $f \in \mathbf{V}$ by 1° and 2°.

§10. INCREASING PROCESSES

Let f be an arbitrary real valued $\sigma(\mathfrak{A} \times \mathfrak{B})$-measurable function on $S \times T$ such that $f(\cdot, y)$ is bounded on S for every $y \in T$. Let $f = f^+ - f^-$. Then f^+ is $\sigma(\mathfrak{A} \times \mathfrak{B})$-measurable and $f^+(\cdot, y)$ is bounded on S for every $y \in T$. Then there exists an increasing sequence of nonnegative simple functions $\{f_n : n \in \mathbb{N}\}$ on $(S \times T, \sigma(\mathfrak{A} \times \mathfrak{B}))$ such that $f_n \uparrow f^+$. As we noted above, $f_n \in \mathbf{V}$ for $n \in \mathbb{N}$. Therefore by 3°, $f^+ \in \mathbf{V}$. Similarly for f^-. Then $f \in \mathbf{V}$ by 2°. Therefore $\mathbf{V}$ contains all real valued $\sigma(\mathfrak{A} \times \mathfrak{B})$-measurable functions f on $S \times T$ such that $f(\cdot, y)$ is bounded on S for every $y \in T$. ∎

For functions of the type considered in Proposition 10.6 which are nonnegative we have the following corollary which can be proved in the same way as Proposition 10.6.

Corollary 10.7. *Let $(S, \mathfrak{A})$ and $(T, \mathfrak{B})$ be two measurable spaces. Let $\mathbf{V}$ be a collection of real valued nonnegative functions f on $S \times T$ such that $f(\cdot, y)$ is bounded on S for every $y \in T$. Suppose $\mathbf{V}$ satisfies the following conditions.*

1°. $1_E \in \mathbf{V}$ *for every* $E \in \mathfrak{A} \times \mathfrak{B}$.

2°. *if $f, g \in \mathbf{V}$ and $\alpha, \beta \geq 0$ then $\alpha f + \beta g \in \mathbf{V}$; if $f, g \in \mathbf{V}$ and $f \leq g$ then $g - f \in \mathbf{V}$.*

3°. *if $\{f_n; n \in \mathbb{N}\}$ is an increasing sequence of nonnegative functions in $\mathbf{V}$ such that $f \equiv \lim_{n \to \infty} f_n$ is a real valued nonnegative function and $f(\cdot, y)$ is bounded on S for every $y \in T$, then $f \in \mathbf{V}$.*

Then $\mathbf{V}$ contains all real valued nonnegative $\sigma(\mathfrak{A} \times \mathfrak{B})$-measurable functions f on $S \times T$ such that $f(\cdot, y)$ is bounded on S for every $y \in T$.

Theorem 10.8. *Let $(S, \mathfrak{A})$ and $(T, \mathfrak{B})$ be two measurable spaces and let $\{\mu(\cdot, y) : y \in T\}$ be a $\mathfrak{B}$-measurable family of finite measures on $(S, \mathfrak{A})$. For an extended real valued $\sigma(\mathfrak{A} \times \mathfrak{B})$-measurable function f on $S \times T$, let f^* be defined by*

$$f^*(y) = \int_S f(x, y) \, \mu(dx, y)$$

for $y \in T$ for which the integral exists.
1) If f is such that $f(\cdot, y)$ is bounded on S for every $y \in T$, then f^* is a real valued $\mathfrak{B}$-measurable function on T.
2) If f is a nonnegative extended real valued function, then f^* is a nonnegative extended real valued $\mathfrak{B}$-measurable function on T.

Proof. The $\sigma(\mathfrak{A} \times \mathfrak{B})$-measurability of f on $S \times T$ implies the $\mathfrak{A}$-measurability of $f(\cdot, y)$ on S for every $y \in T$.

1) Suppose f is such that $f(\cdot, y)$ is bounded on S for every $y \in T$. The boundedness of $f(\cdot, y)$ on S and the finiteness of the measure $\mu(\cdot, y)$ implies the finiteness of $f^*(y)$ for every $y \in T$. Thus f^* is a real valued function on T. We apply Proposition 10.6 to prove the $\mathfrak{B}$-measurability of f^*. To do this, let **V** be the collection of all real valued $\sigma(\mathfrak{A} \times \mathfrak{B})$-measurable functions f on $S \times T$ such that $f(\cdot, y)$ is bounded on S for every $y \in T$ and such that f^* is real valued and $\mathfrak{B}$-measurable on T. Let us verify that **V** satisfies conditions 1°, 2° and 3° in Proposition 10.6. Let $f = \mathbf{1}_E$ with $E = A \times B$ where $A \in \mathfrak{A}$ and $B \in \mathfrak{B}$. Clearly f is a real valued $\sigma(\mathfrak{A} \times \mathfrak{B})$-measurable function on $S \times T$ and $f(\cdot, y)$ is bounded on S for every $y \in T$. Also

$$f^*(y) = \int_S \mathbf{1}_E(x,y)\,\mu(dx,y) = \int_A \mathbf{1}_B(y)\,\mu(dx,y) = \mathbf{1}_B(y)\mu(A,y).$$

Since both $\mathbf{1}_B$ and $\mu(A, \cdot)$ are $\mathfrak{B}$-measurable functions on T, so is their product f^*. Therefore $f \in \mathbf{V}$. Clearly **V** is a linear space. If $\{f_n : n \in \mathbb{N}\}$ is an increasing sequence of nonnegative functions in **V** such that $f \equiv \lim_{n\to\infty} f_n$ is real valued and $f(\cdot, y)$ is bounded on S for every $y \in T$, then

$$f^*(y) = \int_S \lim_{n\to\infty} f_n(x,y)\,\mu(dx,y) = \lim_{n\to\infty} \int_S f_n(x,y)\,\mu(dx,y) = \lim_{n\to\infty} f_n^*(y) \quad \text{for } y \in T,$$

by the Monotone Convergence Theorem. Then the $\mathfrak{B}$-measurability of f_n^* for every $n \in \mathbb{N}$ implies that of f^*. Thus $f \in \mathbf{V}$. This verifies the conditions in Proposition 10.6 for our **V**. Therefore **V** contains all real valued $\sigma(\mathfrak{A} \times \mathfrak{B})$-measurable functions f on $S \times T$ such that $f(\cdot, y)$ is bounded on S for every $y \in T$. Thus for such a function f, f^* is real valued and $\mathfrak{B}$-measurable on T.

2) If f is a nonnegative extended real valued function, then f^* is a nonnegative extended real valued function on T. It remains to show the $\mathfrak{B}$-measurability of f^*. Let $f_n = f \wedge n$ for $n \in \mathbb{N}$. Then $f_n \uparrow f$ on $S \times T$. By 1), f_n^* is a real valued $\mathfrak{B}$-measurable function on T for $n \in \mathbb{N}$. By the Monotone Convergence Theorem

$$f^*(y) = \int_S \lim_{n\to\infty} f_n(x,y)\,\mu(dx,y) = \lim_{n\to\infty} f_n^*(y) \quad \text{for } y \in T.$$

Since f_n^* is $\mathfrak{B}$-measurable on T for every $n \in \mathbb{N}$, so is f^*. ∎

Theorem 10.9. (*An Extension of Fubini's Theorem*) *Let $(S, \mathfrak{A})$ and $(T, \mathfrak{B})$ be two measurable spaces, let $\{\mu(\cdot, y) : y \in T\}$ be a $\mathfrak{B}$-measurable family of finite measures on $(S, \mathfrak{A})$*

§10. INCREASING PROCESSES

and let ν be a measure on $(T, \mathfrak{B})$. Let λ be a measure on $\sigma(\mathfrak{A} \times \mathfrak{B})$ such that

(1) $$\lambda(A \times B) = \int_B \mu(A, y) \nu(dy) \quad \text{for } A \times B \in \mathfrak{A} \times \mathfrak{B}.$$

1) Let f be a real valued $\sigma(\mathfrak{A} \times \mathfrak{B})$-measurable function on $S \times T$ such that $f(\cdot, y)$ is bounded on S for every $y \in T$. If $\int_{S \times T} f(x, y) \lambda(d(x, y))$ exists in $\overline{\mathbb{R}}$ then

(2) $$\int_{S \times T} f(x, y) \lambda(d(x, y)) = \int_T \left\{ \int_S f(x, y) \mu(dx, y) \right\} \nu(dy),$$

in the sense that the right side of (2) also exists and the equality holds. On the other hand, if $\int_T \{\int_S |f(x, y)| \mu(dx, y)\} \nu(dy) < \infty$, then both sides of (2) exist and the equality holds.
2) If f is a nonnegative extended real valued $\sigma(\mathfrak{A} \times \mathfrak{B})$-measurable function on $S \times T$ then (2) holds.

Proof. 0) Let us show first that if f is a nonnegative valued $\sigma(\mathfrak{A} \times \mathfrak{B})$-measurable function on $S \times T$ such that $f(\cdot, y)$ is bounded on S for every $y \in T$, then (2) holds. Let $\mathbf{V}$ be the collection of all nonnegative valued $\sigma(\mathfrak{A} \times \mathfrak{B})$-measurable functions f on $S \times T$ such that $f(\cdot, y)$ is bounded on S for every $y \in T$ and satisfies (2). Let us verify that our $\mathbf{V}$ satisfies conditions 1°, 2° and 3° of Corollary 10.7. Let $E \in \mathfrak{A} \times \mathfrak{B}$ be given by $E = A \times B$ where $A \in \mathfrak{A}$ and $B \in \mathfrak{B}$. To show that $\mathbf{1}_E \in \mathbf{V}$ we show that $\mathbf{1}_E$ satisfies (2). Now

$$\int_{S \times T} \mathbf{1}_E(x, y) \lambda(d(x, y)) = \lambda(E) = \lambda(A \times B).$$

On the other hand,

$$\int_T \left\{ \int_S \mathbf{1}_E(x, y) \mu(dx, y) \right\} \nu(dy) = \int_T \mathbf{1}_B(y) \mu(A, y) \nu(dy)$$
$$= \int_B \mu(A, y) \nu(dy) = \lambda(A \times B).$$

This shows that $\mathbf{1}_E \in \mathbf{V}$. Clearly $\mathbf{V}$ satisfies 2° of Corollary 10.7. Let $\{f_n : n \in \mathbb{N}\}$ be an increasing sequence of nonnegative functions in $\mathbf{V}$ such that $f \equiv \lim_{n \to \infty} f_n$ is a real valued nonnegative function and $f(\cdot, y)$ is bounded on S for every $y \in T$. To show that $f \in \mathbf{V}$ we show that f satisfies (2). By the Monotone Convergence Theorem we have

$$\int_{S \times T} f(x, y) \lambda(d(x, y)) = \lim_{n \to \infty} \int_{S \times T} f_n(x, y) \lambda(d(x, y))$$
$$= \lim_{n \to \infty} \int_T \left\{ \int_S f_n(x, y) \mu(dx, y) \right\} \nu(dy) = \int_T \left\{ \int_S f(x, y) \mu(dx, y) \right\} \nu(dy).$$

This shows that $f \in \mathbf{V}$. Therefore by Corollary 10.7, $\mathbf{V}$ contains all nonnegative valued $\sigma(\mathfrak{A} \times \mathfrak{B})$-measurable function f on $S \times T$ such that $f(\cdot, y)$ is bounded on S for every $y \in T$. Thus (2) holds for such functions f.

1) Let f be a real valued $\sigma(\mathfrak{A} \times \mathfrak{B})$-measurable function on $S \times T$ such that $f(\cdot, y)$ is bounded on S for every $y \in T$. Suppose $\int_{S \times T} f(x,y) \lambda(d(x,y))$ exists. Then at least one of the two integrals $\int_{S \times T} f^+(x,y) \lambda(d(x,y))$ and $\int_{S \times T} f^-(x,y) \lambda(d(x,y))$ is finite. According to the results in 0) above, we have

$$(3) \qquad \int_{S \times T} f^+(x,y) \lambda(d(x,y)) = \int_T \left\{ \int_S f^+(x,y) \mu(dx,y) \right\} \nu(dy)$$

and

$$(4) \qquad \int_{S \times T} f^-(x,y) \lambda(d(x,y)) = \int_T \left\{ \int_S f^-(x,y) \mu(dx,y) \right\} \nu(dy).$$

Since at least one of (3) and (4) is finite, the difference of the two exists in $\overline{\mathbb{R}}$. Subtracting (4) from (3), we obtain (2).

If on the other hand we have $\int_T \left\{ \int_S |f(x,y)| \mu(dx,y) \right\} \nu(dy) < \infty$, then applying our result in 0) to $|f|$ we have

$$\int_{S \times T} |f(x,y)| \lambda(d(x,y)) = \int_T \left\{ \int_S |f(x,y)| \mu(dx,y) \right\} \nu(dy) < \infty.$$

Thus $\int_{S \times T} |f(x,y)| \lambda(d(x,y))$ exists and this implies according to what we showed above the existence of the right side of (2) and the equality.

2) For a nonnegative extended real valued function f on $S \times T$, let $f_n = f \wedge n$ for $n \in \mathbb{N}$. Then $f_n \uparrow f$ on $S \times T$. By 0) we have

$$(5) \qquad \int_{S \times T} f_n(x,y) \lambda(d(x,y)) = \int_T \left\{ \int_S f_n(x,y) \mu(dx,y) \right\} \nu(dy).$$

As we saw in the proof of Theorem 10.8, $\int_S f_n(x,y) \mu(dx,y) \uparrow \int_S f(x,y) \mu(dx,y)$ for $y \in T$. Letting $n \to \infty$ on both sides of (5), we have (2) by the Monotone Convergence Theorem. ∎

Let us remark that the existence or even the finiteness of the right side of (2) in Theorem 10.9 does not imply the existence of the left side. To construct an example, let $(S, \mathfrak{A})$ and $(T, \mathfrak{B})$ be copies of $([-1, 1], \mathfrak{B}_{[-1,1]})$, let $\mu(\cdot, y)$ be the Lebesgue measure m_L on $([-1, 1], \mathfrak{B}_{[-1,1]})$ for every $y \in T$, let $\nu = m_L$ also and let $\lambda = m_L \times m_L$. Let s be the

§10. INCREASING PROCESSES

sign function, that is, $s(x) = 1$ for $x \geq 0$ and $s(x) = -1$ for $x < 0$. Define f on $S \times T$ by setting

$$f(x,y) = \begin{cases} s(x)|y|^{-\frac{3}{2}} & \text{for } x \in S \text{ and } y \in T - \{0\} \\ 0 & \text{for } x \in S \text{ and } y = 0. \end{cases}$$

Our function f is $\sigma(\mathfrak{A} \times \mathfrak{B})$-measurable and $f(\cdot, y)$ is bounded on S for every $y \in T$. Simple calculations show that $\int_T \{\int_S f(x,y) \mu(dx,y)\} \nu(dy) = 0$. On the other hand the integral $\int_{S \times T} f(x,y) \lambda(d(x,y))$ does not exist since we have both $\int_{S \times T} f^+(x,y) \lambda(d(x,y)) = \infty$ and $\int_{S \times T} f^-(x,y) \lambda(d(x,y)) = \infty$.

Let us consider the measurability of an integral with respect to an increasing process.

Lemma 10.10. *Let $A = \{A_t : t \in \mathbb{R}_+\}$ be an almost surely increasing process on a filtered space $(\Omega, \mathfrak{F}, \{\mathfrak{F}_t\}, P)$. The family of Lebesgue-Stieltjes measures $\{\mu_A(\cdot, \omega) : \omega \in \Omega\}$ on the measurable space $(\mathbb{R}_+, \mathfrak{B}_{\mathbb{R}_+})$ determined by the sample functions of A is $\mathfrak{F}_\infty$-measurable. Also, for every $t \in \mathbb{R}_+$, the restrictions of the Lebesgue-Stieltjes measures to the measurable space $([0, t], \mathfrak{B}_{[0,t]})$ constitute an $\mathfrak{F}_\infty$-measurable family of finite measures. Furthermore, if A is an increasing process or if A is only an almost surely increasing process but the filtered space is augmented, then the family of measures on $([0, t], \mathfrak{B}_{[0,t]})$ is $\mathfrak{F}_t$-measurable for every $t \in \mathbb{R}_+$.*

Proof. Consider the two measurable spaces $([0, t], \mathfrak{B}_{[0,t]})$ and $(\Omega, \mathfrak{F}_t)$. By restricting the measure $\mu_A(\cdot, \omega)$ to $([0, t], \mathfrak{B}_{[0,t]})$, we have a family of finite measures $\{\mu_A(\cdot, \omega) : \omega \in \Omega\}$ on $([0, t], \mathfrak{B}_{[0,t]})$. Let us show that this family is $\mathfrak{F}_t$-measurable in the sense of Definition 10.4, that is, for every $E \in \mathfrak{B}_{[0,t]}$, $\mu_A(E, \cdot)$ is an $\mathfrak{F}_t$-measurable function on Ω. We show this by applying Theorem 1.7, Dynkin's monotone class theorem for sets. Let $\mathfrak{J}$ be the semialgebra of subsets of $\mathbb{R}$ consisting of all the left-open and right-closed intervals in $\mathbb{R}$ and $\emptyset$. Then $\mathfrak{E} = \mathfrak{J} \cap [0, t]$ is a semialgebra of subsets of $[0, t]$. Note that $\mathfrak{E}$ consists of sets of the types: $(a, b]$ where $0 \leq a < b \leq t$; $[0, b]$ where $0 \leq b \leq t$; and $\emptyset$. Let $\mathfrak{O}$ be the collection of the open sets in $\mathbb{R}$ and let $\mathfrak{O}_{[0,t]}$ be the collection of all the open sets in $[0, t]$ in its relative topology, that is, $\mathfrak{O}_{[0,t]} = \mathfrak{O} \cap [0, t]$. By Theorem 1.5 we have

$$\sigma_{[0,t]}(\mathfrak{E}) = \sigma_{[0,t]}(\mathfrak{J} \cap [0,t]) = \sigma_{\mathbb{R}}(\mathfrak{J}) \cap [0,t]$$
$$= \sigma_{\mathbb{R}}(\mathfrak{O}) \cap [0,t] = \sigma_{[0,t]}(\mathfrak{O} \cap [0,t]) = \sigma_{[0,t]}(\mathfrak{O}_{[0,t]}) = \mathfrak{B}_{[0,t]}.$$

Now let $E \in \mathfrak{E}$ be given by $E = (a, b]$ where $0 \leq a < b \leq t$. If A is an increasing process, then $\mu_A(E, \omega) = \mu_A((a, b], \omega) = A(b, \omega) - A(a, \omega)$ for every $\omega \in \Omega$. Since A is an adapted process and since $a, b \leq t$, $A(a, \cdot)$ and $A(b, \cdot)$ are $\mathfrak{F}_t$-measurable random variables and thus $\mu_A(E,\cdot)$ is $\mathfrak{F}_t$-measurable. On the other hand, if A is only an almost surely increasing process, then there exists a null set Λ_A in $(\Omega, \mathfrak{F}_\infty, P)$ such that $A(\cdot,\omega)$ is a real valued right-continuous monotone increasing function on $\mathbb{R}_+$ with $A(0,\omega) = 0$ for every $\omega \in \Lambda_A^c$. According to our convention in the definition of μ_A, we set $\mu_A(\cdot,\omega) = 0$ for $\omega \in \Lambda_A$. If the filtered space is augmented, then $\Lambda_A \in \mathfrak{F}_t$ and this implies the $\mathfrak{F}_t$-measurability of $\mu_A((a,b],\cdot)$. Similarly when $E = [0,b]$ with $0 \leq b \leq t$ and when $E = \emptyset$. Next, let $\mathfrak{D}$ be the collection of all members E of $\mathfrak{B}_{[0,t]}$ such that $\mu_A(E,\cdot)$ is an $\mathfrak{F}_t$-measurable random variable on Ω. We have shown above that $\mathfrak{D}$ contains the semialgebra $\mathfrak{E}$. If $E_1, E_2 \in \mathfrak{D}$ and $E_1 \subset E_2$ then $E_2 - E_1 \in \mathfrak{D}$ since $\mu_A(E_2 - E_1, \omega) = \mu_A(E_2, \omega) - \mu_A(E_1, \omega)$ which is an $\mathfrak{F}_t$-measurable function of $\omega \in \Omega$. If $E_n \in \mathfrak{D}$ for $n \in \mathbb{N}$ and $E_n \uparrow E$, then $E \in \mathfrak{B}_{[0,t]}$ and $\mu_A(E,\omega) = \lim_{n\to\infty} \mu_A(E_n,\omega)$, which is an $\mathfrak{F}_t$-measurable function of $\omega \in \Omega$. Thus $E \in \mathfrak{D}$. Thus $\mathfrak{D}$ is a d-class. Since $\mathfrak{D}$ contains the semialgebra $\mathfrak{E}$, it contains the σ-algebra generated by $\mathfrak{E}$, namely $\mathfrak{B}_{[0,t]}$. Therefore $\mu_A(E,\cdot)$ is $\mathfrak{F}_t$-measurable for every $E \in \mathfrak{B}_{[0,t]}$.

The $\mathfrak{F}_\infty$-measurability of the family $\{\mu_A(\cdot,\omega) : \omega \in \Omega\}$ is shown by the same argument as above. ∎

Theorem 10.11. *Let $A = \{A_t : t \in \mathbb{R}_+\}$ be an almost surely increasing process on an augmented filtered space $(\Omega, \mathfrak{F}, \{\mathfrak{F}_t\}, P)$ and let $\{\mu_A(\cdot,\omega) : \omega \in \Omega\}$ be the family of Lebesgue-Stieltjes measures on $(\mathbb{R}_+, \mathfrak{B}_{\mathbb{R}_+})$ determined by A. Let X be an adapted measurable process on the filtered space so that its sample functions are Borel measurable functions on $\mathbb{R}_+$. For $t \in \mathbb{R}_+$, let*

$$\int_{[0,t]} X(s,\omega)\, dA(s,\omega) \equiv \int_{[0,t]} X(s,\omega)\, \mu_A(ds,\omega)$$

for every $\omega \in \Omega$ for which the Lebesgue-Stieltjes integral exists.
1) If every sample function of X is bounded on every finite interval in $\mathbb{R}_+$, then $\int_{[0,t]} X_s\, dA_s$ is a real valued $\mathfrak{F}_t$-measurable random variable on $(\Omega, \mathfrak{F}, P)$.
2) If the sample functions of X are nonnegative functions on $\mathbb{R}_+$, then $\int_{[0,t]} X_s\, dA_s$ is a nonnegative extended real valued $\mathfrak{F}_t$-measurable random variable on $(\Omega, \mathfrak{F}, P)$.
If we do not assume that the filtered space is augmented then under the condition on X in 1) or 2), $\int_{[0,t]} X_s\, dA_s$ is $\mathfrak{F}_\infty$-measurable.

Proof. By Lemma 10.10, Theorem 10.11 is a particular case of Theorem 10.8. ∎

If $A = \{A_t : t \in \mathbb{R}_+\}$ is an almost surely increasing process on a filtered space $(\Omega, \mathfrak{F}, \{\mathfrak{F}_t\}, P)$ such that $\mathbf{E}(A_\infty) < \infty$, then there exists a null set Λ in $(\Omega, \mathfrak{F}_\infty, P)$ such that $A_\infty(\omega) < \infty$, and thus the Lebesgue-Stieltjes measure $\mu_A(\cdot,\omega)$ on $(\mathbb{R}_+, \mathfrak{B}_{\mathbb{R}_+})$ is a finite measure for $\omega \in \Lambda^c$. Let us adopt a convention that $\mu_A(\cdot,\omega) = 0$ for $\omega \in \Lambda$. Then

§10. INCREASING PROCESSES

by Lemma 10.10, the family of Lebesgue-Stieltjes measures $\{\mu_A(\cdot,\omega) : \omega \in \Omega\}$ on the measurable space $(\mathbb{R}_+, \mathfrak{B}_{\mathbb{R}_+})$ determined by the sample functions of A is an $\mathfrak{F}_\infty$-measurable family of finite measures. Thus the following theorem is a particular case of Theorem 10.8.

Theorem 10.12. *Let $A = \{A_t : t \in \mathbb{R}_+\}$ be an almost surely increasing process with $\mathbf{E}(A_\infty) < \infty$ on a filtered space $(\Omega, \mathfrak{F}, \{\mathfrak{F}_t\}, P)$. Let X be an adapted measurable process on the filtered space so that its sample functions are real valued Borel measurable functions on $\mathbb{R}_+$. Let*

$$\int_{\mathbb{R}_+} X(s,\omega)\, dA(s,\omega) \equiv \int_{\mathbb{R}_+} X(s,\omega)\, \mu_A(ds,\omega)$$

for every $\omega \in \Omega$ for which the Lebesgue-Stieltjes integral exists.
1) If X is a bounded process, then $\int_{\mathbb{R}_+} X_s\, dA_s$ is a real valued $\mathfrak{F}_\infty$-measurable random variable on $(\Omega, \mathfrak{F}, P)$.
2) If X is a nonnegative process, then $\int_{\mathbb{R}_+} X_s\, dA_s$ is a nonnegative extended real valued $\mathfrak{F}_\infty$-measurable random variable on $(\Omega, \mathfrak{F}, P)$.

Theorem 10.13. *Let $A = \{A_t : t \in \mathbb{R}_+\}$ be an almost surely increasing process and $M = \{M_t : t \in \mathbb{R}_+\}$ be a right-continuous bounded martingale on a filtered space $(\Omega, \mathfrak{F}, \{\mathfrak{F}_t\}, P)$. Then for every $t \in \mathbb{R}_+$, we have*

(1) $$\mathbf{E}(M_t A_t) = \mathbf{E}\left[\int_{[0,t]} M(s,\cdot)\, dA(s,\cdot)\right].$$

If $\mathbf{E}(A_\infty) < \infty$, then

(2) $$\mathbf{E}(M_\infty A_\infty) = \mathbf{E}\left[\int_{\mathbb{R}_+} M(s,\cdot)\, dA(s,\cdot)\right].$$

Proof. Let $|M(t,\omega)| \leq K$ for $(t,\omega) \in \mathbb{R}_+ \times \Omega$ for some $K \geq 0$. Let $t \in \mathbb{R}_+$ be fixed. With $n \in \mathbb{N}$ fixed, let $s_k = k2^{-n}t$ for $k = 0, \ldots, 2^n$ and let $I_k = (s_{k-1}, s_k]$ for $k = 1, \ldots, 2^n$. Let $M^{(n)}$ be defined by

(3) $$M_0^{(n)} = M_0 \quad \text{and} \quad M_s^{(n)} = M_{s_k} \quad \text{for } s \in I_k,\ k = 1, \ldots, 2^n.$$

Since $M(\cdot,\omega)$ is a right-continuous function on $\mathbb{R}_+$ for every $\omega \in \Omega$, we have

(4) $$\lim_{n \to \infty} M^{(n)}(s,\omega) = M(s,\omega) \quad \text{for } (s,\omega) \in [0,t] \times \Omega.$$

Now

$$E(M_t A_t) = E\left[M(t) \sum_{k=1}^{2^n} \{A(s_k) - A(s_{k-1})\}\right]$$

$$= \sum_{k=1}^{2^n} E\left[M(t)\{A(s_k) - A(s_{k-1})\}\right]$$

$$= \sum_{k=1}^{2^n} E\left[E\left[M(t)\{A(s_k) - A(s_{k-1})\} | \mathfrak{F}_{s_k}\right]\right]$$

$$= \sum_{k=1}^{2^n} E\left[M(s_k)\{A(s_k) - A(s_{k-1})\}\right]$$

where the last equality is by the fact that $A(s_k) - A(s_{k-1})$ is $\mathfrak{F}_{s_k}$-measurable and then by the fact that M is a martingale. Then

$$E(M_t A_t) = E\left[\sum_{k=1}^{2^n} M(s_k)\{A(s_k) - A(s_{k-1})\}\right] = E\left[\int_{[0,t]} M^{(n)}(s)\, dA(s)\right],$$

since for each $\omega \in \Omega$, $M^{(n)}(\cdot, \omega)$ is a step function assuming the value $M(s_k, \omega)$ on the interval I_k and since $\mu_A(I_k, \omega) = A(s_k, \omega) - A(s_{k-1}, \omega)$ by the right-continuity of $A(\cdot, \omega)$ for $\omega \in \Lambda_A^c$ where Λ_A is an exceptional null set of the almost surely increasing process A. Thus

(5) $$E(M_t A_t) = E\left[\int_{[0,t]} M^{(n)}(s)\, dA(s)\right].$$

Now since $|M(t, \omega)| \leq K$ for $(t, \omega) \in \mathbb{R}_+ \times \Omega$ and $\mu_A([0,t], \omega) < \infty$, and since (4) holds, we have by the Bounded Convergence Theorem,

(6) $$\lim_{n \to \infty} \int_{[0,t]} M^{(n)}(s, \omega)\, dA(s, \omega) = \int_{[0,t]} M(s, \omega)\, dA(s, \omega) \quad \text{for } \omega \in \Omega.$$

Since

$$\left|\int_{[0,t]} M^{(n)}(s, \omega)\, dA(s, \omega)\right| \leq K A(t, \omega) \quad \text{for } \omega \in \Lambda_A^c$$

with $A(t) \in L_1(\Omega, \mathfrak{F}, P)$ and since (6) holds, by applying the Dominated Convergence Theorem to the right side of (5) we obtain (1).

Since M is a bounded martingale it is L_1-bounded and thus by Theorem 7.9 the extended real valued $\mathfrak{F}_\infty$-measurable random variable and $\lim_{t \to \infty} M_t = M_\infty$ a.e. on $(\Omega, \mathfrak{F}_\infty, P)$.

§10. INCREASING PROCESSES

Suppose $\mathbf{E}(A_\infty) < \infty$. Since $\lim_{t\to\infty} M_t A_t = M_\infty A_\infty$ a.e. on $(\Omega, \mathfrak{F}_\infty, P)$ and since $|M_t A_t| \leq K A_\infty$ a.e. on $(\Omega, \mathfrak{F}_\infty, P)$, we have by the Dominated Convergence Theorem

$$\lim_{t\to\infty} \mathbf{E}[M_t A_t] = \mathbf{E}[M_\infty A_\infty]. \tag{7}$$

On the other hand

$$\lim_{t\to\infty} \int_{[0,t]} M(s,\cdot)\, dA(s,\cdot) = \lim_{t\to\infty} \int_{\mathbb{R}_+} \mathbf{1}_{[0,t]} M(s,\cdot)\, dA(s,\cdot) = \int_{\mathbb{R}_+} M(s,\cdot)\, dA(s,\cdot).$$

We also have $|\int_{[0,t]} M(s,\cdot)\, dA(s,\cdot)| \leq K A_\infty$ a.e. on $(\Omega, \mathfrak{F}_\infty, P)$. Since $K A_\infty$ is integrable, by the Dominated Convergence Theorem we have

$$\lim_{t\to\infty} \mathbf{E}\left[\int_{[0,t]} M(s,\cdot)\, dA(s,\cdot)\right] = \mathbf{E}\left[\int_{\mathbb{R}_+} M(s,\cdot)\, dA(s,\cdot)\right]. \tag{8}$$

Letting $t \to \infty$ on both sides of (1), we obtain (2) by (7) and (8). ∎

[III] Doob-Meyer Decomposition

Let $M = \{M_t : t \in \mathbb{R}_+\}$ be a right-continuous bounded martingale on a filtered space $(\Omega, \mathfrak{F}, \{\mathfrak{F}_t\}, P)$. According to Theorem 9.2 the right-continuity of M implies that there exists a null set Λ in $(\Omega, \mathfrak{F}_\infty, P)$ such that for every $\omega \in \Lambda^c$ the sample function $M(\cdot, \omega)$ has finite left limit everywhere on $\mathbb{R}_+$ and has at most countably many points of discontinuity.

Definition 10.14. *For a right-continuous bounded martingale $M = \{M_t : t \in \mathbb{R}_+\}$ on a filtered space $(\Omega, \mathfrak{F}, \{\mathfrak{F}_t\}, P)$ with the exceptional null set Λ in $(\Omega, \mathfrak{F}_\infty, P)$ as described above, let $M_- = \{M_-(t) : t \in \mathbb{R}_+\}$ be defined by*

$$M_-(t,\omega) = \begin{cases} \lim_{s \uparrow t} M(s,\omega) & \text{for } t \in (0,\infty) \text{ and } \omega \in \Lambda^c, \\ M(0,\omega) & \text{for } t = 0 \text{ and } \omega \in \Lambda^c, \\ 0 & \text{for } t \in \mathbb{R}_+ \text{ and } \omega \in \Lambda. \end{cases}$$

Thus defined, $M_-(t)$ is $\mathfrak{F}_\infty$-measurable for every $t \in \mathbb{R}_+$. Let $K > 0$ be a bound for the right-continuous bounded martingale M. Then $|M(t,\omega)| \leq K$ for all $(t,\omega) \in \mathbb{R}_+ \times \Omega$ so that $|M_-(t)| \leq K$ and therefore $M_-(t)$ is integrable for every $t \in \mathbb{R}_+$.

Lemma 10.15. *Let $M = \{M_t : t \in \mathbb{R}_+\}$ be a right-continuous bounded martingale on an augmented filtered space $(\Omega, \mathfrak{F}, \{\mathfrak{F}_t\}, P)$. Then $M_- = \{M_-(t) : t \in \mathbb{R}_+\}$ is a left-continuous bounded martingale on the filtered space.*

Proof. As we have noted, $M_-(t)$ is an $\mathfrak{F}_\infty$-measurable integrable random variable on the probability space $(\Omega, \mathfrak{F}, P)$ for every $t \in \mathbb{R}_+$.

Let Λ be the null set in $(\Omega, \mathfrak{F}_\infty, P)$ in Definition 10.14. Since the filtered space is augmented, $\Lambda \in \mathfrak{F}_t$ for every $t \in \mathbb{R}_+$. According to Definition 10.14, $M_-(t)$ is the limit of $\mathfrak{F}_t$-measurable functions $M(s)$ for $s < t$ on the $\mathfrak{F}_t$-measurable set Λ^c and is thus $\mathfrak{F}_t$-measurable on Λ^c. Also $M_-(t)$ is trivially $\mathfrak{F}_t$-measurable on the $\mathfrak{F}_t$-measurable set Λ. Thus $M_-(t)$ is $\mathfrak{F}_t$-measurable on Ω.

To show that M_- is a martingale, let $0 \leq t' < s < t''$. Since M is a martingale, we have

(1) $$\mathbf{E}(M_s | \mathfrak{F}_{t'}) = M_{t'} \quad \text{a.e. on} (\Omega, \mathfrak{F}_{t'}, P).$$

Since $M_-(t'') = \lim_{s \uparrow t''} M(s)$ a.e. on $(\Omega, \mathfrak{F}_{t''}, P)$ and since $|M(s)| \leq K$ for $s \in \mathbb{R}_+$ where $K > 0$ is a bound for M, we have by the Conditional Bounded Convergence Theorem

(2) $$\lim_{s \uparrow t''} \mathbf{E}[M(s) | \mathfrak{F}_{t'}] = \mathbf{E}[\lim_{s \uparrow t''} M(s) | \mathfrak{F}_{t'}] = \mathbf{E}[M_-(t'') | \mathfrak{F}_{t'}] \quad \text{a.e. on} (\Omega, \mathfrak{F}_{t'}, P).$$

From (1) and (2) we have

$$\mathbf{E}[M_-(t'') | \mathfrak{F}_{t'}] = M_{t'} \quad \text{a.e. on} (\Omega, \mathfrak{F}_{t'}, P).$$

This shows that M_- is a martingale. By Definition 10.14, M_- is bounded by K and is left-continuous. ∎

Lemma 10.16. *Let $M = \{M_t : t \in \mathbb{R}_+\}$ be a right-continuous bounded martingale and $A = \{A_t : t \in \mathbb{R}_+\}$ be an almost surely increasing process on a filtered space $(\Omega, \mathfrak{F}, \{\mathfrak{F}_t\}, P)$. For $t > 0$, let $t_{n,k} = k 2^{-n} t$ for $k = 0, \ldots, 2^n$ and $n \in \mathbb{N}$. Then there exists a null set Λ in $(\Omega, \mathfrak{F}_\infty, P)$, independent of t, such that for every $\omega \in \Lambda^c$ we have*

(1) $$\lim_{n \to \infty} \sum_{k=1}^{2^n} M(t_{n,k-1}, \omega) \{A(t_{n,k}, \omega) - A(t_{n,k-1}, \omega)\} = \int_{[0,t]} M_-(s, \omega) \, dA(s, \omega).$$

Also

(2) $$\lim_{n \to \infty} \mathbf{E}\left[\sum_{k=1}^{2^n} M(t_{n,k-1})\{A(t_{n,k}) - A(t_{n,k-1})\}\right] = \mathbf{E}\left[\int_{[0,t]} M_-(s) \, dA(s)\right].$$

§10. INCREASING PROCESSES

Proof. For every $n \in \mathbb{N}$, let us define $M^{(n)} = \{M_s^{(n)} : s \in [0,t]\}$ by setting

(3) $\quad \begin{cases} M^{(n)}(0) = M(0) \\ M^{(n)}(s) = M(t_{n,k-1}) \quad \text{for } s \in (t_{n,k-1}, t_{n,k}] \text{ for } k = 1, \ldots, 2^n. \end{cases}$

Every sample function of $M^{(n)}$ is a left-continuous step function on $[0,t]$. Now the right-continuity of the martingale M implies according to Theorem 9.2 the existence of a null set Λ_M in $(\Omega, \mathfrak{F}_\infty, P)$ such that $M(\cdot, \omega)$ has finite left limit everywhere on $\mathbb{R}_+$ for $\omega \in \Lambda_M^c$. Since A is an almost surely increasing process, there exists a null set Λ_A in $(\Omega, \mathfrak{F}_\infty, P)$ such that $A(\cdot, \omega)$ is a real valued right-continuous monotone increasing function on $\mathbb{R}_+$ with $A(0, \omega) = 0$ for $\omega \in \Lambda_A^c$. Let $\Lambda = \Lambda_M \cup \Lambda_A$. By Definition 10.14, for $\omega \in \Lambda_M^c$ we have

(4) $\quad \lim_{n \to \infty} M^{(n)}(s, \omega) = M_-(s, \omega).$

For $\omega \in \Lambda^c$, we have

(5) $\quad \sum_{k=1}^{2^n} M(t_{n,k-1}, \omega)\{A(t_{n,k}, \omega) - A(t_{n,k-1}, \omega)\} = \int_{[0,t]} M^{(n)}(s, \omega) \, dA(s, \omega).$

By (4) and the Bounded Convergence Theorem, we have

(6) $\quad \lim_{n \to \infty} \int_{[0,t]} M^{(n)}(s, \omega) \, dA(s, \omega) = \int_{[0,t]} M_-(s, \omega) \, dA(s, \omega).$

Combining (5) and (6) we have (1).

To prove (2), note that if we let $K > 0$ be a bound for the bounded martingale M then

$$\left| \sum_{k=1}^{2^n} M(t_{n,k-1}, \omega)\{A(t_{n,k}, \omega) - A(t_{n,k-1}, \omega)\} \right| \leq K A(t, \omega).$$

Since $A(t)$ is integrable and since (5) and (6) hold, we can apply the Dominated Convergence Theorem to have

$$\lim_{n \to \infty} \mathbf{E}\left[\sum_{k=1}^{2^n} M(t_{n,k-1})\{A(t_{n,k}) - A(t_{n,k-1})\} \right]$$
$$= \mathbf{E}\left[\lim_{n \to \infty} \int_{[0,t]} M^{(n)}(s) \, dA(s) \right] = \mathbf{E}\left[\int_{[0,t]} M_-(s) \, dA(s) \right].$$

This proves (2). ∎

Definition 10.17. *An almost surely increasing process $A = \{A_t : t \in \mathbb{R}_+\}$ on a filtered space $(\Omega, \mathfrak{F}, \{\mathfrak{F}_t\}, P)$ is called natural if for every right-continuous bounded martingale $M = \{M_t : t \in \mathbb{R}_+\}$ on the filtered space we have*

$$\mathbf{E}\left[\int_{[0,t]} M(s)\,dA(s)\right] = \mathbf{E}\left[\int_{[0,t]} M_-(s)\,dA(s)\right] \quad \text{for } t \in \mathbb{R}_+,$$

or equivalently

$$\mathbf{E}[M(t)A(t)] = \mathbf{E}\left[\int_{[0,t]} M_-(s)\,dA(s)\right] \quad \text{for } t \in \mathbb{R}_+.$$

Let us observe that the sample functions of M and M_- are bounded Borel measurable functions on $\mathbb{R}_+$ so that the two Lebesgue-Stieltjes integrals $\int_{[0,t]} M(s,\omega)\,dA(s,\omega)$ and $\int_{[0,t]} M_-(s,\omega)\,dA(s,\omega)$ exist as real numbers for every $\omega \in \Omega$. Also $\int_{[0,t]} M(s)\,dA(s)$ and $\int_{[0,t]} M_-(s)\,dA(s)$ are $\mathfrak{F}_\infty$-measurable (and $\mathfrak{F}_t$-measurable in case the filtered space is augmented) random variables according to Theorem 10.11 and furthermore if $K > 0$ is a bound for M then

$$\mathbf{E}\left[\int_{[0,t]} M(s)\,dA(s)\right], \mathbf{E}\left[\int_{[0,t]} M_-(s)\,dA(s)\right] \leq K\mathbf{E}(A_t).$$

Theorem 10.18. *Let $A = \{A_t : t \in \mathbb{R}_+\}$ be an almost surely increasing process on a filtered space $(\Omega, \mathfrak{F}, \{\mathfrak{F}_t\}, P)$. If A is almost surely continuous, then it is natural.*

Proof. Let $M = \{M_t : t \in \mathbb{R}_+\}$ be a right-continuous bounded martingale on the filtered space. By Theorem 9.2 there exists a null set Λ_M in $(\Omega, \mathfrak{F}_\infty, P)$ such that $M(\cdot, \omega)$ has finite left limit everywhere and has at most countably many points of discontinuity on $\mathbb{R}_+$ for every $\omega \in \Lambda_M^c$. Since A is an almost surely continuous almost surely increasing process, there exists a null set Λ_A in $(\Omega, \mathfrak{F}_\infty, P)$ such that $A(\cdot, \omega)$ is a continuous monotone increasing function on $\mathbb{R}_+$ with $A(0, \omega) = 0$ for $\omega \in \Lambda_A^c$. Let $\Lambda = \Lambda_A \cup \Lambda_M$. Let $\omega \in \Lambda^c$ and let $\{t_n : n \in \mathbb{N}\}$ be the points of discontinuity of $M(\cdot, \omega)$ on $\mathbb{R}_+$. Then $M(\cdot, \omega) - M_-(\cdot, \omega)$ vanishes on $\mathbb{R}_+$ except at t_n where it assumes the value $M(t_n, \omega) - \lim_{s \uparrow t_n} M(s, \omega)$. Thus for any $t \in (0, \infty)$ we have

$$\int_{[0,t]} \{M(s,\omega) - M_-(s,\omega)\}\,dA(s,\omega)$$
$$= \sum_{n \in \mathbb{N}: t_n \leq t} \{M(t_n, \omega) - \lim_{s \uparrow t_n} M(s, \omega)\}\mu_A(\{t_n\}, \omega) = 0,$$

§10. INCREASING PROCESSES

since the continuity of $A(\cdot, \omega)$ implies $\mu_A(\{s\}, \omega) = 0$ for every $s \in (0, \infty)$. Since the last equality holds for every $\omega \in \Lambda^c$ and since $P(\Lambda) = 0$, we have

$$\mathbf{E}\left[\int_{[0,t]} \{M(s) - M_-(s)\} \, dA(s)\right] = 0.$$

This shows that A is natural. ∎

Lemma 10.19. *1) An almost surely increasing process $A = \{A_t : t \in \mathbb{R}_+\}$ on a filtered space $(\Omega, \mathfrak{F}, \{\mathfrak{F}_t : t \in \mathbb{R}_+\}, P)$ is a submartingale of class (DL). If $\mathbf{E}(A_\infty) < \infty$, then A of class (D).*
2) If M is a right-continuous martingale and A is an almost surely increasing process on a right-continuous filtered space $(\Omega, \mathfrak{F}, \{\mathfrak{F}_t\}, P)$, then $M + A$ is a submartingale of class (DL).
3) If M is a uniformly integrable right-continuous martingale and A is an almost surely increasing process with $\mathbf{E}(A_\infty) < \infty$ on a right-continuous filtered space $(\Omega, \mathfrak{F}, \{\mathfrak{F}_t\}, P)$, then $M + A$ is a submartingale of class (D).

Proof. 1) Since A is an almost surely increasing process, there exists a null set Λ_A in $(\Omega, \mathfrak{F}_\infty, P)$ such that $A(\cdot, \omega)$ is a real valued right-continuous monotone increasing function on $\mathbb{R}_+$ with $A(0, \omega) = 0$ for every $\omega \in \Lambda_A^c$. Thus for every pair $s, t \in \mathbb{R}_+$ such that $s < t$ we have $A_t \geq A_s$ a.e. on $(\Omega, \mathfrak{F}_\infty, P)$ and therefore $\mathbf{E}(A_t | \mathfrak{F}_s) \geq \mathbf{E}(A_s | \mathfrak{F}_s) = A_s$ a.e. on $(\Omega, \mathfrak{F}_s, P)$. This shows that A is a submartingale.

To show that A is of class (DL), let $a \in \mathbb{R}_+$ and consider the collection $\mathbf{S}_a$ of all stopping times on the filtered space which are bounded by a. Then for every $T \in \mathbf{S}_a$ we have $0 \leq A_T(\omega) \leq A_a(\omega)$ for $\omega \in \Lambda_A^c$, that is, $0 \leq A_T \leq A_a$ a.e. on $(\Omega, \mathfrak{F}_\infty, P)$. Then since A_a is integrable, $\{A_T : T \in \mathbf{S}_a\}$ is uniformly integrable by 4) of Proposition 4.11. Since this holds for every $a \in \mathbb{R}_+$, A is of class (DL). Suppose $\mathbf{E}(A_\infty) < \infty$ where $A_\infty = \lim_{t \to \infty} A_t$ on Λ_A^c. Let $\mathbf{S}$ be the collection of all finite stopping times on the filtered space. Since $0 \leq A_T(\omega) \leq A_\infty(\omega)$ for $\omega \in \Lambda_A^c$, that is, $0 \leq A_T \leq A_\infty$ a.e. on $(\Omega, \mathfrak{F}_\infty, P)$, for every $T \in \mathbf{S}$, the integrability of A_∞ implies the uniform integrability of $\{A_T : T \in \mathbf{S}\}$. Therefore A is of class (D).

2) A right-continuous martingale M is always of class (DL) according to Theorem 8.22. An almost surely increasing process A is a submartingale of class (DL) according to 1). Then $M + A$ is a submartingale. The fact that $M + A$ is of class (DL) follows from the fact that $(M + A)_T = M_T + A_T$ for every $T \in \mathbf{S}$ and from Proposition 4.9.

3) A uniformly integrable right-continuous martingale M is always of class (D) by Theorem 8.22. An almost surely increasing process A with $\mathbf{E}(A_\infty) < \infty$ is of class (D) by

1). Then $M + A$ is of class (D) by the same argument as in 2). ∎

Lemma 10.20. *Let $A = \{A_t : t \in \mathbb{R}_+\}$ and $A' = \{A'_t : t \in \mathbb{R}_+\}$ be two almost surely increasing processes on a filtered space $(\Omega, \mathfrak{F}, \{\mathfrak{F}_t\}, P)$ such that*

(1) $$\mathbf{E}[A_t - A_s | \mathfrak{F}_s] = \mathbf{E}[A'_t - A'_s | \mathfrak{F}_s] \quad \text{a.e. on } (\Omega, \mathfrak{F}_s, P),$$

for every pair $s, t \in \mathbb{R}_+$ such that $s < t$. Let Y be a left-continuous bounded adapted process on the filtered space. Then for every $t \in \mathbb{R}_+$ we have

(2) $$\mathbf{E}\left[\int_{[0,t]} Y(s)\, dA(s)\right] = \mathbf{E}\left[\int_{[0,t]} Y(s)\, dA'(s)\right].$$

Proof. Let $t > 0$ be fixed. For every $n \in \mathbb{N}$, let $t_{n,k} = k2^{-n}t$ for $k = 0, \ldots, 2^n$ and let $I_{n,k} = (t_{n,k-1}, t_{n,k}]$ for $k = 1, \ldots, 2^n$. Define a left-continuous bounded adapted process $Y^{(n)}$ on $[0, t] \times \Omega$ by setting

(3) $$\begin{cases} Y^{(n)}(s, \omega) = Y(t_{n,k-1}, \omega) & \text{for } s \in I_{n,k},\ k = 1, \ldots, 2^n,\ \text{and } \omega \in \Omega \\ Y^{(n)}(0, \omega) = Y(0, \omega) & \text{for } \omega \in \Omega. \end{cases}$$

By the same argument as in Lemma 2.8 we have

(4) $$\lim_{n \to \infty} Y^{(n)}(s, \omega) = Y(s, \omega) \quad \text{for } (s, \omega) \in [0, t] \times \Omega.$$

Since A is an almost surely increasing process there exists a null set Λ_A such that $A(\cdot, \omega)$ is a real valued right-continuous monotone increasing function on $\mathbb{R}_+$ with $A(0, \omega) = 0$ for $\omega \in \Lambda_A^c$. Then for every $\omega \in \Lambda_A^c$ we have by (3)

(5) $$\int_{[0,t]} Y^{(n)}(s, \omega)\, dA(s, \omega) = \sum_{k=1}^{2^n} Y(t_{n,k-1}, \omega)\{A(t_{n,k}, \omega) - A(t_{n,k-1}, \omega)\}.$$

For $\omega \in \Lambda_A^c$, (4) and the Bounded Convergence Theorem imply

(6) $$\lim_{n \to \infty} \int_{[0,t]} Y^{(n)}(s, \omega)\, dA(s, \omega) = \int_{[0,t]} Y(s, \omega)\, dA(s, \omega).$$

If $K > 0$ is a bound for Y, then for $\omega \in \Lambda_A^c$ we have

(7) $$\left| \int_{[0,t]} Y(s, \omega)\, dA(s, \omega) \right| \leq K A(t, \omega).$$

§10. INCREASING PROCESSES

Now from (5) we have

$$\mathbf{E}\left[\int_{[0,t]} Y^{(n)}(s)\,dA(s)\right] = \sum_{k=1}^{2^n} \mathbf{E}\left[Y(t_{n,k-1})\{A(t_{n,k}) - A(t_{n,k-1})\}\right]$$

$$= \sum_{k=1}^{2^n} \mathbf{E}\left[Y(t_{n,k-1})\mathbf{E}[A(t_{n,k}) - A(t_{n,k-1})|\mathfrak{F}_{t_{n,k-1}}]\right]$$

We have a similar expression for $\mathbf{E}\left[\int_{[0,t]} Y^{(n)}(s)\,dA'(s)\right]$. But (1) implies that

$\mathbf{E}[A(t_{n,k}) - A(t_{n,k-1})|\mathfrak{F}_{t_{n,k-1}}] = \mathbf{E}[A'(t_{n,k}) - A'(t_{n,k-1})|\mathfrak{F}_{t_{n,k-1}}]$ a.e. on $(\Omega, \mathfrak{F}_{t_{n,k-1}}, P)$.

Therefore we have

(8) $$\mathbf{E}\left[\int_{[0,t]} Y^{(n)}(s)\,dA(s)\right] = \mathbf{E}\left[\int_{[0,t]} Y^{(n)}(s)\,dA'(s)\right].$$

Since A_t is integrable, (6), (7) and the Dominated Convergence Theorem imply

$$\lim_{n\to\infty} \mathbf{E}\left[\int_{[0,t]} Y^{(n)}(s)\,dA(s)\right] = \mathbf{E}\left[\int_{[0,t]} Y(s)\,dA(s)\right].$$

Similarly for $\mathbf{E}\left[\int_{[0,t]} Y^{(n)}(s)\,dA'(s)\right]$. Thus letting $n \to \infty$ on both sides of (8) we have (2). ∎

Theorem 10.21. *Let $X = \{X_t : t \in \mathbb{R}_+\}$ be a right-continuous submartingale on an augmented right-continuous filtered space $(\Omega, \mathfrak{F}, \{\mathfrak{F}_t\}, P)$. If $A = \{A_t : t \in \mathbb{R}_+\}$ and $A' = \{A'_t : t \in \mathbb{R}_+\}$ are two natural almost surely increasing processes on the filtered space such that both $M = X - A$ and $M' = X - A'$ are martingales then A and A' are equivalent, that is, there exists a null set Λ in $(\Omega, \mathfrak{F}, P)$ such that $A(\cdot, \omega) = A'(\cdot, \omega)$ for $\omega \in \Lambda^c$.*

Proof. Since $A - A' = (X - M) - (X - M') = M' - M$, $A - A'$ is a martingale. Thus for every pair $s, t \in \mathbb{R}_+$ such that $s < t$, we have $\mathbf{E}[(A - A')_t|\mathfrak{F}_s] = (A - A')_s$ a.e. on $(\Omega, \mathfrak{F}_s, P)$, that is, we have $\mathbf{E}(A_t - A_s|\mathfrak{F}_s) = \mathbf{E}(A'_t - A'_s|\mathfrak{F}_s)$ a.e. on $(\Omega, \mathfrak{F}_s, P)$. Thus A and A' satisfy condition (1) in Lemma 10.20.

Let ξ be a bounded random variable on $(\Omega, \mathfrak{F}, P)$. According to Corollary 9.10, for each $t \in \mathbb{R}_+$ we can select a version Z_t of $\mathbf{E}(\xi|\mathfrak{F}_t)$ so that $Z = \{Z_t : t \in \mathbb{R}_+\}$ is a right-continuous martingale. Then Z_- defined as in Definition 10.14 is a left-continuous bounded martingale. Thus by Lemma 10.20 for every $t \in \mathbb{R}_+$ we have

$$\mathbf{E}\left[\int_{[0,t]} Z_-(s)\,dA(s)\right] = \mathbf{E}\left[\int_{[0,t]} Z_-(s)\,dA'(s)\right].$$

Now since A is natural, we have $\mathbf{E}\left[\int_{[0,t]} Z_-(s)\, dA(s)\right] = \mathbf{E}(Z_t A_t)$ and similarly $\mathbf{E}\left[\int_{[0,t]} Z_-(s)\, dA'(s)\right] = \mathbf{E}(Z_t A'_t)$. Therefore we have $\mathbf{E}(Z_t A_t) = \mathbf{E}(Z_t A'_t)$. Recalling that $Z_t = \mathbf{E}(\xi | \mathfrak{F}_t)$ we have $\mathbf{E}[\mathbf{E}(\xi | \mathfrak{F}_t) A_t] = \mathbf{E}[\mathbf{E}(\xi | \mathfrak{F}_t) A'_t]$. Then by the $\mathfrak{F}_t$-measurability of A_t and A'_t we have $\mathbf{E}[\mathbf{E}(\xi A_t | \mathfrak{F}_t)] = \mathbf{E}[\mathbf{E}(\xi A'_t | \mathfrak{F}_t)]$ that is, $\mathbf{E}(\xi A_t) = \mathbf{E}(\xi A'_t)$. In particular by letting $\xi = \mathbf{1}_E$ for $E \in \mathfrak{F}_t$, we have $\int_E A_t\, dP = \int_E A'_t\, dP$ for every $E \in \mathfrak{F}_t$. This shows that $A_t = A'_t$ a.e. on $(\Omega, \mathfrak{F}_t, P)$. Then since A and A' are almost surely right-continuous processes, there exists a null set Λ in $(\Omega, \mathfrak{F}, P)$ such that $A(\cdot, \omega) = A'(\cdot, \omega)$ for $\omega \in \Lambda^c$ according to Theorem 2.3. ∎

Lemma 10.22. *Let $X = \{X_t : t \in \mathbb{R}_+\}$ be a right-continuous submartingale of class (DL) on an augmented right-continuous filtered space $(\Omega, \mathfrak{F}, \{\mathfrak{F}_t\}, P)$. For $a > 0$ define a stochastic process $Y = \{Y_t : t \in [0, a]\}$ by setting*

(1) $$Y_t = X_t - \mathbf{E}(X_a | \mathfrak{F}_t) \quad \text{for } t \in [0, a].$$

For each $n \in \mathbb{N}$, let $t_{n,k} = k2^{-n}a$ for $k = 0, \ldots, 2^n$, let $\mathbb{T}_n = \{t_{n,k} : k = 0, \ldots, 2^n\}$ and define a discrete time stochastic process $A^{(n)} = \{A^{(n)}_t : t \in \mathbb{T}_n\}$ by setting

(2) $$\begin{cases} A^{(n)}(t_{n,k}) = \sum_{j=1}^{k} \mathbf{E}[Y(t_{n,j}) - Y(t_{n,j-1}) | \mathfrak{F}_{t_{n,j-1}}] & \text{for } k = 1, \ldots, 2^n \\ A^{(n)}(t_{n,0}) = 0. \end{cases}$$

Then $\{A^{(n)}_a : n \in \mathbb{N}\}$ is uniformly integrable.

Proof. The stochastic process $\{\mathbf{E}(X_a | \mathfrak{F}_t) : t \in [0, a]\}$, where $\mathbf{E}(X_a | \mathfrak{F}_t)$ is a real valued version, is a uniformly integrable martingale with respect to the filtration $\{\mathfrak{F}_t : t \in [0, a]\}$. Since the filtered space $(\Omega, \mathfrak{F}, \{\mathfrak{F}_t\}, P)$ is augmented and right-continuous, by Corollary 9.10 we can select a version of $\mathbf{E}(X_a | \mathfrak{F}_t)$ for each $t \in [0, a]$ so that the martingale $\{\mathbf{E}(X_a | \mathfrak{F}_t) : t \in [0, a]\}$ is right-continuous. The right-continuity implies that this martingale is of class (DL) by Theorem 8.22.

The process Y defined by (1) is then a right-continuous submartingale. Since X is a submartingale, we have $\mathbf{E}(X_a | \mathfrak{F}_t) \geq X_t$ a.e. on $(\Omega, \mathfrak{F}_t, P)$ for every $t \in [0, a]$ and thus Y is nonpositive. In particular, $Y_a = X_a - \mathbf{E}(X_a | \mathfrak{F}_a) = 0$ a.e. on $(\Omega, \mathfrak{F}_a, P)$. Since X is of class (DL) and the martingale $\{\mathbf{E}(X_a | \mathfrak{F}_t) : t \in [0, a]\}$ is also of class (DL), Y is of class (DL).

Since Y is a submartingale, each summand in (2) is a.e. nonnegative and thus $A^{(n)}$ is a discrete time almost surely increasing process with respect to the filtration $\{\mathfrak{F}_{t_{n,k}} : k = 0, \ldots, 2^n\}$. Also $A^{(n)}$ is predictable since by (2), $A^{(n)}(t_{n,k})$ is $\mathfrak{F}_{t_{n,k-1}}$-measurable for $k = 1, \ldots, 2^n$.

§10. INCREASING PROCESSES

To prove the uniform integrability of $\{A_a^{(n)} : n \in \mathbb{N}\}$, let us show first that there is a version of $\mathbf{E}[A_a^{(n)}|\mathfrak{F}_{t_{n,k}}]$ such that

(3) $\quad\quad\quad \mathbf{E}[A_a^{(n)}|\mathfrak{F}_{t_{n,k}}] = A^{(n)}(t_{n,k}) - Y(t_{n,k}) \quad \text{for } k = 0, \ldots, 2^n.$

Now by (2) we have

$$\mathbf{E}[A_a^{(n)}|\mathfrak{F}_{t_{n,k}}] = \mathbf{E}[\sum_{j=1}^{2^n} \mathbf{E}[Y(t_{n,j}) - Y(t_{n,j-1})|\mathfrak{F}_{t_{n,j-1}}]|\mathfrak{F}_{t_{n,k}}]$$

$$= \sum_{j=1}^{k} \mathbf{E}[Y(t_{n,j}) - Y(t_{n,j-1})|\mathfrak{F}_{t_{n,j-1}}]\mathbf{E}[1|\mathfrak{F}_{t_{n,k}}]$$

$$+ \sum_{j=k+1}^{2^n} \mathbf{E}[Y(t_{n,j}) - Y(t_{n,j-1})|\mathfrak{F}_{t_{n,k}}]$$

$$= A^{(n)}(t_{n,k})\mathbf{E}[1|\mathfrak{F}_{t_{n,k}}] + \mathbf{E}[Y(a) - Y(t_{n,k})|\mathfrak{F}_{t_{n,k}}]$$

$$= A^{(n)}(t_{n,k})\mathbf{E}[1|\mathfrak{F}_{t_{n,k}}] - Y(t_{n,k}) \quad \text{a.e. on } (\Omega, \mathfrak{F}_{t_{n,k}}, P).$$

By selecting the constant 1 from among all the versions of $\mathbf{E}[1|\mathfrak{F}_{t_{n,k}}]$, we have (3).

With $c > 0$ and the convention that $t_{n,-1} = 0$, let us define a $\mathbb{T}_n$-valued function on Ω by setting

(4) $\quad\quad\quad S_c^{(n)} = \begin{cases} \inf\{t_{n,k-1} \in \mathbb{T}_n : A^{(n)}(t_{n,k}) > c\} \\ a \quad \text{if the set above is } \emptyset. \end{cases}$

To show that $S_c^{(n)}$ is a stopping time with respect to the filtration $\{\mathfrak{F}_{t_{n,k}} : k = 0, \ldots, 2^n\}$, we verify that $\{S_c^{(n)} = t_{n,k}\} \in \mathfrak{F}_{t_{n,k}}$ for $k = 0, \ldots, 2^n$. Now recalling that $t_{n,-1} = 0$, we have

$$\{S_c^{(n)} = t_{n,0}\} = \{A(t_{n,0}) > c\} \cup \{A(t_{n,0}) \leq c, A(t_{n,1}) > c\} \in \mathfrak{F}_{t_{n,0}},$$

by the fact that $A^{(n)}$ is predictable, and similarly for $k = 1, \ldots, 2^n - 1$ we have

$$\{S_c^{(n)} = t_{n,k}\} = \{A(t_{n,0}) \leq c, \ldots, A(t_{n,k}) \leq c, A(t_{n,k+1}) > c\} \in \mathfrak{F}_{t_{n,k}},$$

and finally,

$$\{S_c^{(n)} = t_{n,2^n}\} = \{A(t_{n,0}) \leq c, \ldots, A(t_{n,2^n}) \leq c\} \in \mathfrak{F}_{t_{n,2^n}}.$$

Now if we apply the Optional Sampling Theorem in the form of (2) of Theorem 8.16 to the martingale $\{\mathbf{E}[A_a^{(n)}|\mathfrak{F}_{t_{n,k}}] : k = 0, \ldots, 2^n\}$ given by (3) with respect to the filtration $\{\mathfrak{F}_{t_{n,k}} : k = 0, \ldots, 2^n\}$, then

(5) $\quad\quad\quad \mathbf{E}[A_a^{(n)}|\mathfrak{F}_{S_c^{(n)}}] = A^{(n)}(S_c^{(n)}) - Y(S_c^{(n)}).$

180   CHAPTER 2. MARTINGALES

Now for any $\omega \in \Omega$, if $A_a^{(n)}(\omega) > c$ then by (4) we have $S_c^{(n)}(\omega) \leq t_{n,2^n-1} < a$. Conversely, if $S_c^{(n)}(\omega) < a$ then there exists $k \in \{0, \ldots, 2^n\}$ such that $A_{t_{n,k}}^{(n)}(\omega) > c$ and then since $A^{(n)}$ is an almost surely increasing process we have $A_a^{(n)}(\omega) > c$ for a.e. ω in $(\Omega, \mathfrak{F}_a, P)$. Thus we have

(6) $$A_a^{(n)}(\omega) > c \Leftrightarrow S_c^{(n)}(\omega) < a \quad \text{for a.e. } \omega \text{ in } (\Omega, \mathfrak{F}_a, P).$$

Therefore

(7) $$\int_{\{A_a^{(n)}>c\}} A_a^{(n)} dP = \int_{\{S_c^{(n)}<a\}} A_a^{(n)} dP$$
$$= \int_{\{S_c^{(n)}<a\}} E[A_a^{(n)}|\mathfrak{F}_{S_c^{(n)}}] dP \quad \text{since } \{S_c^{(n)} < a\} \in \mathfrak{F}_{S_c^{(n)}}$$
$$= \int_{\{S_c^{(n)}<a\}} A^{(n)}(S_c^{(n)}) dP - \int_{\{S_c^{(n)}<a\}} Y(S_c^{(n)}) dP \quad \text{by (5)}$$
$$\leq cP\{S_c^{(n)} < a\} - \int_{\{S_c^{(n)}<a\}} Y(S_c^{(n)}) dP,$$

since $A^{(n)}(S_c^{(n)}) \leq c$. On the other hand

(8) $$-\int_{\{S_{c/2}^{(n)}<a\}} Y(S_{c/2}^{(n)}) dP$$
$$= \int_{\{S_{c/2}^{(n)}<a\}} E[A_a^{(n)}|\mathfrak{F}_{S_{c/2}^{(n)}}] dP - \int_{\{S_{c/2}^{(n)}<a\}} A^{(n)}(S_{c/2}^{(n)}) dP \quad \text{by (5)}$$
$$= \int_{\{S_{c/2}^{(n)}<a\}} A_a^{(n)} dP - \int_{\{S_{c/2}^{(n)}<a\}} A^{(n)}(S_{c/2}^{(n)}) dP$$
$$\geq \int_{\{S_c^{(n)}<a\}} \{A_a^{(n)} - A^{(n)}(S_{c/2}^{(n)})\} dP,$$

since the last integrand is nonnegative and $\{S_c^{(n)} < a\} \subset \{S_{c/2}^{(n)} < a\}$. Now $A^{(n)}(S_{c/2}^{(n)}) \leq c/2$. Also $S_c^{(n)}(\omega) < a$ implies that there exists $k \in \{0, \ldots, 2^n\}$ such that $A_{t_{n,k}}^{(n)}(\omega) > c$ and then since $A_{t_{n,k}}^{(n)}(\omega)$ increases as k increases for a.e. ω in $(\Omega, \mathfrak{F}_a, P)$ we have $A_a^{(n)}(\omega) > c$ for a.e. ω in $\{S_c^{(n)} < a\}$. Using these facts in the last member of (8) we have

(9) $$-\int_{\{S_{c/2}^{(n)}<a\}} Y(S_{c/2}^{(n)}) dP \geq \frac{c}{2} P\{S_c^{(n)} < a\}.$$

Using (9) in (7) we have

(10) $$\int_{\{A_a^{(n)}>c\}} A_a^{(n)} dP \leq -2 \int_{\{S_{c/2}^{(n)}<a\}} Y(S_{c/2}^{(n)}) dP - \int_{\{S_c^{(n)}<a\}} Y(S_c^{(n)}) dP.$$

§10. INCREASING PROCESSES

To show the uniform integrability of $\{A_a^{(n)} : n \in \mathbb{N}\}$ we show

(11) $$\lim_{c \to \infty} \sup_{n \in \mathbb{N}} \int_{\{A_a^{(n)} > c\}} A_a^{(n)} \, dP = 0.$$

If we show that

(12) $$\lim_{c \to \infty} \sup_{n \in \mathbb{N}} \left\{ -\int_{\{S_c^{(n)} < a\}} Y(S_c^{(n)}) \, dP \right\} = 0,$$

then we also have

(13) $$\lim_{c \to \infty} \sup_{n \in \mathbb{N}} \left\{ -2 \int_{\{S_{c/2}^{(n)} < a\}} Y(S_{c/2}^{(n)}) \, dP \right\} = 0.$$

Applying (12) and (13) to (10) we obtain (11).

Let us prove (12). Recall that Y is a submartingale of class (DL) with respect to the filtration $\{\mathfrak{F}_t : t \in [0, a]\}$. Since $S_c^{(n)}, n \in \mathbb{N}$ are stopping times with respect to the filtration $\{\mathfrak{F}_t : t \in \mathbb{T}_n\}$ and since $\{\mathfrak{F}_t : t \in \mathbb{T}_n\} \subset \{\mathfrak{F}_t : t \in [0, a]\}$, they are also stopping times with respect to the filtration $\{\mathfrak{F}_t : t \in [0, a]\}$. Since these stopping times are all bounded by a, the fact that Y is of class (DL) implies that $\{Y_{S_c^{(n)}} : n \in \mathbb{N}\}$ is uniformly integrable. Thus by Theorem 4.7 for every $\varepsilon > 0$ there exists some $\delta > 0$ such that

(14) $$-\int_E Y(S_c^{(n)}) \, dP < \varepsilon \quad \text{for all } n \in \mathbb{N} \text{ whenever } E \in \mathfrak{F} \text{ and } P(E) < \delta.$$

Now

$$\mathbf{E}(A_a^{(n)}) \geq \int_{\{A_a^{(n)} > c\}} A_a^{(n)} \, dP \geq c P\{A_a^{(n)} > c\} = c P\{S_c^{(n)} < a\}$$

by (6). On the other hand from (3) and (2) we have $\mathbf{E}[A_a^{(n)} | \mathfrak{F}_0] = A_0^{(n)} - Y_0 = -Y_0$ so that $\mathbf{E}[A_a^{(n)}] = \mathbf{E}(-Y_0)$. Thus we have

$$P\{S_c^{(n)} < a\} \leq \frac{1}{c} \mathbf{E}(-Y_0) \quad \text{for all } c > 0 \text{ and } n \in \mathbb{N}.$$

Thus there exists $c_0 > 0$ such that $P\{S_c^{(n)} < a\} < \delta$ for all $n \in \mathbb{N}$ when $c > c_0$. Then by (14)

$$-\int_{\{S_c^{(n)} < a\}} Y(S_c^{(n)}) \, dP < \varepsilon \quad \text{for all } n \in \mathbb{N} \text{ when } c > c_0.$$

Therefore

$$\sup_{n \in \mathbb{N}} \left\{ -\int_{\{S_c^{(n)} < a\}} Y(S_c^{(n)}) \, dP \right\} \leq \varepsilon \quad \text{for } c > c_0.$$

Thus
$$\limsup_{c\to\infty} \sup_{n\in\mathbb{N}} \left\{ -\int_{\{S_c^{(n)}<a\}} Y(S_c^{(n)}) \, dP \right\} \le \varepsilon.$$

Since this holds for an arbitrary $\varepsilon > 0$, we have (12). ∎

Theorem 10.23. (Doob-Meyer Decomposition) *Let $X = \{X_t : t \in \mathbb{R}_+\}$ be a right-continuous submartingale of class (DL) on an augmented right-continuous filtered space. Then there exists a right-continuous martingale $M = \{M_t : t \in \mathbb{R}_+\}$ and a natural almost surely increasing process $A = \{A_t : t \in \mathbb{R}_+\}$ on the filtered space such that $X(\cdot,\omega) = M(\cdot,\omega) + A(\cdot,\omega)$ for $\omega \in \Lambda^c$ where Λ is a null set in $(\Omega, \mathfrak{F}, P)$. Moreover such a decomposition is unique in the sense that if M and M' are right-continuous martingales and A and A' are natural almost surely increasing processes such that $X = M + A = M' + A'$ then M and M' are equivalent and A and A' are equivalent, that is, there exists a null set Λ in $(\Omega, \mathfrak{F}, P)$ such that $M(\cdot,\omega) = M'(\cdot,\omega)$ and $A(\cdot,\omega) = A'(\cdot,\omega)$ for $\omega \in \Lambda^c$.*

Proof. 1) Uniqueness of decomposition. Suppose M and M' are two right-continuous martingales and A and A' are two natural almost surely increasing processes such that $X = M + A$ and $X = A' + M'$ also. Then $X - A$ and $X - A'$ are both martingales so that by Theorem 10.21 there exists a null set Λ in $(\Omega, \mathfrak{F}, P)$ such that $A(\cdot,\omega) = A'(\cdot,\omega)$ for $\omega \in \Lambda^c$. Then since $M - M' = A' - A$ we have $(M - M')(\cdot,\omega) = (A - A')(\cdot,\omega) = 0$, that is, $M(\cdot,\omega) = M'(\cdot,\omega)$ for $\omega \in \Lambda^c$.

2) Existence of decomposition on a finite interval. Let us show that for an arbitrary $a > 0$, there exists a right-continuous martingale $M = \{M_t : t \in [0, a]\}$, a natural almost surely increasing process $A = \{A_t : t \in [0, a]\}$ and a null set Λ in $(\Omega, \mathfrak{F}_a, P)$ such that $X(t,\omega) = M(t,\omega) + A(t,\omega)$ for $t \in [0, a]$ and $\omega \in \Lambda^c$.

With $a > 0$ fixed, let $Y = \{Y_t : t \in [0, a]\}$, $\mathbb{T}_n$ and $A^{(n)} = \{A_t^{(n)} : t \in \mathbb{T}_n\}$ for $n \in \mathbb{N}$ be as defined by (1) and (2) of Lemma 10.22. Then $\{A_a^{(n)} : n \in \mathbb{N}\}$ is uniformly integrable so that by Theorem 4.21 there exist $A_a^* \in L_1(\Omega, \mathfrak{F}_a, P)$ and a subsequence $\{n_\ell\}$ of $\{n\}$ such that

(1) $$\lim_{\ell\to\infty} \int_\Omega A_a^{(n_\ell)} \xi \, dP = \int_\Omega A_a^* \xi \, dP \quad \text{for every } \xi \in L_\infty(\Omega, \mathfrak{F}_a, P).$$

According to Corollary 9.10, for each $t \in \mathbb{R}_+$ we can select a version of $\mathbf{E}(A_a^* | \mathfrak{F}_t)$ so that the martingale $\{\mathbf{E}(A_a^* | \mathfrak{F}_t) : t \in [0, a]\}$ with respect to the filtration $\{\mathfrak{F}_t : t \in [0, a]\}$ is right-continuous. Define a stochastic process $A = \{A_t : t \in [0, a]\}$ by setting

(2) $$A_t = Y_t + \mathbf{E}(A_a^* | \mathfrak{F}_t) \quad \text{for } t \in [0, a].$$

§10. INCREASING PROCESSES

Since Y too is right-continuous, A is right-continuous. If we define a stochastic process M on $[0, a]$ by setting $M = X - A$, then by (1) of Lemma 10.22 and (2) above we have

$$(3) \qquad M_t = \mathbf{E}(X_a - A_a^* | \mathfrak{F}_t) \quad \text{for } t \in [0, a]$$

so that M is a martingale with respect to the filtration $\{\mathfrak{F}_t : t \in [0, a]\}$. The right-continuity of X and A implies that of M. It remains to show that A is a natural almost surely increasing process with respect to the filtration $\{\mathfrak{F}_t : t \in [0, a]\}$.

Let us show that A defined by (2) is an almost surely increasing process. For $n \in \mathbb{N}$ recall the discrete time almost surely increasing process $A^{(n)} = \{A_t^{(n)} : t \in \mathbb{T}_n\}$ with respect to the filtration $\{\mathfrak{F}_t : t \in \mathbb{T}_n\}$ defined by (2) of Lemma 10.22. According to (3) of Lemma 10.22, at any $t_{n,k} \in \mathbb{T}_n$ we have

$$(4) \qquad A^{(n)}(t_{n,k}) = Y(t_{n,k}) + \mathbf{E}[A_a^{(n)} | \mathfrak{F}_{t_{n,k}}],$$

and thus

$$(5) \qquad \begin{aligned} & Y(t_{n,k}) - Y(t_{n,k-1}) + \mathbf{E}[A_a^{(n)} | \mathfrak{F}_{t_{n,k}}] - \mathbf{E}[A_a^{(n)} | \mathfrak{F}_{t_{n,k-1}}] \\ & = A^{(n)}(t_{n,k}) - A^{(n)}(t_{n,k-1}) \geq 0 \quad \text{a.e. on } (\Omega, \mathfrak{F}_a, P). \end{aligned}$$

According to (2) and (3) of Theorem 4.23, there exists a subsequence $\{n_\ell\}$ of $\{n\}$ such that for every $\xi \in L_\infty(\Omega, \mathfrak{F}_a, P)$ we have

$$(6) \qquad \lim_{\ell \to \infty} \int_\Omega \mathbf{E}[A_a^{(n_\ell)} | \mathfrak{F}_{t_{n,k}}] \xi \, dP = \int_\Omega \mathbf{E}[A_a^* | \mathfrak{F}_{t_{n,k}}] \xi \, dP$$

and

$$(7) \qquad \lim_{\ell \to \infty} \int_\Omega \mathbf{E}[A_a^{(n_\ell)} | \mathfrak{F}_{t_{n,k-1}}] \xi \, dP = \int_\Omega \mathbf{E}[A_a^* | \mathfrak{F}_{t_{n,k-1}}] \xi \, dP.$$

Multiplying both sides of (5) by $\xi \in L_\infty(\Omega, \mathfrak{F}_a, P)$, $\xi \geq 0$, and integrating, we have

$$(8) \qquad \int_\Omega \left\{ Y(t_{n,k}) - Y(t_{n,k-1}) + \mathbf{E}[A_a^{(n)} | \mathfrak{F}_{t_{n,k}}] - \mathbf{E}[A_a^{(n)} | \mathfrak{F}_{t_{n,k-1}}] \right\} \xi \, dP \geq 0.$$

For $n_\ell \geq n$ we have $\mathbb{T}_n \subset \mathbb{T}_{n_\ell}$ so that $t_{n,k-1}, t_{n,k} \in \mathbb{T}_{n_\ell}$ and (8) holds with $A_a^{(n)}$ replaced by $A_a^{(n_\ell)}$. Letting $\ell \to \infty$ in (8) and using (6) and (7) we have

$$\int_\Omega \left\{ Y(t_{n,k}) - Y(t_{n,k-1}) + \mathbf{E}[A_a^* | \mathfrak{F}_{t_{n,k}}] - \mathbf{E}[A_a^* | \mathfrak{F}_{t_{n,k-1}}] \right\} \xi \, dP \geq 0.$$

By (2), the last inequality is reduced to

$$\int_\Omega \{A(t_{n,k}) - A(t_{n,k-1})\} \xi \, dP \geq 0.$$

In particular, with $\xi = \mathbf{1}_E$ where $E \in \mathfrak{F}_a$, we have

$$\int_E \{A(t_{n,k}) - A(t_{n,k-1})\} \, dP \geq 0 \quad \text{for every } E \in \mathfrak{F}_a.$$

Since both $A(t_{n,k})$ and $A(t_{n,k-1})$ are $\mathfrak{F}_a$-measurable, the last inequality implies that $A(t_{n,k}) \geq A(t_{n,k-1})$ a.e. on $(\Omega, \mathfrak{F}_a, P)$. Then since $\cup_{n \in \mathbb{N}} \mathbb{T}_n$ is a countable set, there exists a null set Λ_∞ in $(\Omega, \mathfrak{F}_a, P)$ such that for every $\omega \in \Lambda_\infty^c$ we have $A(s, \omega) \leq A(t, \omega)$ for all $s, t \in \cup_{n \in \mathbb{N}} \mathbb{T}_n$ such that $s < t$. Then by the right-continuity of A we have for every $\omega \in \Lambda_\infty^c$, $A(s, \omega) \leq A(t, \omega)$ for all $s, t \in [0, a]$ such that $s < t$. Next, let us show that $A_0 = 0$ a.e. on $(\Omega, \mathfrak{F}_0, P)$. By (2), $A_0 = \{Y_0 + \mathbf{E}(A_a^* | \mathfrak{F}_0)\}$. Now for $E \in \mathfrak{F}_0$ we have

$$\int_\Omega \mathbf{1}_E \mathbf{E}(A_a^* | \mathfrak{F}_0) \, dP = \int_\Omega \mathbf{E}(\mathbf{1}_E A_a^* | \mathfrak{F}_0) \, dP$$

$$= \int_\Omega \mathbf{1}_E A_a^* \, dP = \lim_{\ell \to \infty} \int_\Omega \mathbf{1}_E A_a^{(n_\ell)} \, dP \quad \text{by (1)}$$

$$= \lim_{\ell \to \infty} \int_\Omega \mathbf{1}_E \sum_{j=1}^{2^{n_\ell}} \mathbf{E}[Y(t_{n_\ell,j}) - Y(t_{n_\ell,j-1}) | \mathfrak{F}_{t_{n_\ell,j-1}}] \, dP \quad \text{by (2) of Lemma 10.22}$$

$$= \lim_{\ell \to \infty} \sum_{j=1}^{2^{n_\ell}} \int_\Omega \mathbf{E}[\mathbf{1}_E \{Y(t_{n_\ell,j}) - Y(t_{n_\ell,j-1})\} | \mathfrak{F}_{t_{n_\ell,j-1}}] \, dP$$

$$= \lim_{\ell \to \infty} \sum_{j=1}^{2^{n_\ell}} \int_\Omega \mathbf{1}_E \{Y(t_{n_\ell,j}) - Y(t_{n_\ell,j-1})\} \, dP$$

$$= \lim_{\ell \to \infty} \int_\Omega \mathbf{1}_E \{Y(a) - Y(0)\} \, dP$$

$$= -\int_\Omega \mathbf{1}_E Y(0) \, dP$$

since $Y(a) = 0$ a.e. on $(\Omega, \mathfrak{F}_a, P)$. Thus we have $\int_E \{Y(0) + \mathbf{E}(A_a^* | \mathfrak{F}_0)\} \, dP = 0$ for every $E \in \mathfrak{F}_0$. This implies that $Y(0) + \mathbf{E}(A_a^* | \mathfrak{F}_0) = 0$ a.e. on $(\Omega, \mathfrak{F}_0, P)$. Thus there exists a null set Λ_0 in $(\Omega, \mathfrak{F}_0, P)$ such that $A(0, \omega) = 0$ for $\omega \in \Lambda_0^c$. Let $\Lambda = \Lambda_0 \cup \Lambda_\infty$. Then for every $\omega \in \Lambda^c$, $A(\cdot, \omega)$ is real valued right-continuous monotone increasing with $A(0, \omega) = 0$. This shows that A is an almost surely increasing process on $[0, a]$. ∎

§10. INCREASING PROCESSES

To show that the almost surely increasing process $A = \{A_t : t \in [0, a]\}$ is natural, we show that for every right-continuous bounded martingale $M = \{M_t : t \in [0, a]\}$ we have

$$(9) \qquad \mathbf{E}[M_t A_t] = \mathbf{E}\left[\int_{[0,t]} M_-(s)\, dA(s)\right] \quad \text{for } t \in [0, a].$$

Since a is arbitrary it suffices to prove (9) for the case $t = a$. Note first that

$$\begin{aligned}
\mathbf{E}[M_a A_a^{(n)}] &= \sum_{j=1}^{2^n} \mathbf{E}[M(a)\{A^{(n)}(t_{n,j}) - A^{(n)}(t_{n,j-1})\}] \\
&= \sum_{j=1}^{2^n} \mathbf{E}[\mathbf{E}[M(a)\{A^{(n)}(t_{n,j}) - A^{(n)}(t_{n,j-1})\} | \mathfrak{F}_{t_{n,j-1}}]] \\
&= \sum_{j=1}^{2^n} \mathbf{E}[\{A^{(n)}(t_{n,j}) - A^{(n)}(t_{n,j-1})\} \mathbf{E}[M(a) | \mathfrak{F}_{t_{n,j-1}}]] \\
&= \sum_{j=1}^{2^n} \mathbf{E}[M(t_{n,j-1})\{A^{(n)}(t_{n,j}) - A^{(n)}(t_{n,j-1})\}] \\
&= \sum_{j=1}^{2^n} \mathbf{E}[M(t_{n,j-1}) \mathbf{E}[Y(t_{n,j}) - Y(t_{n,j-1}) | \mathfrak{F}_{t_{n,j-1}}]] \\
&= \sum_{j=1}^{2^n} \mathbf{E}[M(t_{n,j-1})\{Y(t_{n,j}) - Y(t_{n,j-1})\}] \\
&= \sum_{j=1}^{2^n} \mathbf{E}[M(t_{n,j-1})\{A(t_{n,j}) - A(t_{n,j-1})\}] \\
&\quad - \sum_{j=1}^{2^n} \mathbf{E}[M(t_{n,j-1})\{\mathbf{E}(A_a^* | \mathfrak{F}_{t_{n,j}}) - \mathbf{E}(A_a^* | \mathfrak{F}_{t_{n,j-1}})\}],
\end{aligned}$$

where the third equality is by the predictability of the process $A^{(n)}$, the fifth equality is by (2) of Lemma 10.22 and the last equality is by (2). Note then that for the summands in the last sum we have

$$\begin{aligned}
&\mathbf{E}[M(t_{n,j-1})\{\mathbf{E}(A_a^* | \mathfrak{F}_{t_{n,j}}) - \mathbf{E}(A_a^* | \mathfrak{F}_{t_{n,j-1}})\}] \\
&= \mathbf{E}[\mathbf{E}[M(t_{n,j-1}) A_a^* | \mathfrak{F}_{t_{n,j}}]] - \mathbf{E}[\mathbf{E}[M(t_{n,j-1}) A_a^* | \mathfrak{F}_{t_{n,j-1}}]] = 0.
\end{aligned}$$

Thus we have

$$(10) \qquad \mathbf{E}[M_a A_a^{(n)}] = \mathbf{E}[\sum_{j=1}^{2^n} M(t_{n,j-1})\{A(t_{n,j}) - A(t_{n,j-1})\}].$$

Since $M_a \in L_\infty(\Omega, \mathfrak{F}_a, P)$ we have by (1) and then by (2) for the definition of A_a and the fact that $Y_a = 0$ a.e. on $(\Omega, \mathfrak{F}_a, P)$ by (1) of Lemma 10.22 we have

(11) $$\lim_{\ell \to \infty} \mathbf{E}[M_a A_a^{(n_\ell)}] = \mathbf{E}[M_a A_a^*] = \mathbf{E}[M_a \mathbf{E}(A_a^* | \mathfrak{F}_a)] = \mathbf{E}[M_a A_a].$$

On the other hand according to Lemma 10.16 we have

(12) $$\lim_{\ell \to \infty} \mathbf{E}[\sum_{j=1}^{2^{n_\ell}} M(t_{n_\ell, j-1})\{A(t_{n_\ell, j}) - A(t_{n_\ell, j-1})\}] = \mathbf{E}\left[\int_{[0,a]} M_-(s)\, dA(s)\right].$$

Using (11) and (12) in (10) we have (9). This completes the proof that the decomposition exists on an arbitrary finite interval $[0, a]$.

3) Existence of decomposition on $\mathbb{R}_+$. According to our results in 2), for every $n \in \mathbb{N}$ there exists a right-continuous martingale $M^{(n)} = \{M_t^{(n)} : t \in [0, n]\}$, a natural almost surely increasing process $A^{(n)} = \{A_t^{(n)} : t \in [0, n]\}$ and a null set $\lambda^{(n)}$ such that $X(t, \omega) = M^{(n)}(t, \omega) + A^{(n)}(t, \omega)$ for $t \in [0, n]$ and $\omega \in (\lambda^{(n)})^c$. Then by the same argument as in Theorem 10.21, there exists a null set Λ_n in $(\Omega, \mathfrak{F}, P)$ such that $M^{(n)} = M^{(n+1)}$ and consequently $A^{(n)} = A^{(n+1)}$ also on $[0, n] \times \Lambda_n^c$. Let $\Lambda = \cup_{n \in \mathbb{N}} \Lambda_n$. Define two processes M and A on $\mathbb{R}_+ \times \Omega$ by setting

$$\begin{cases} M(t, \omega) = M^{(n)}(t, \omega) & \text{for } t \in [0, n] \text{ and } \omega \in \Lambda^c \\ M(t, \omega) = 0 & \text{for } t \in [0, n] \text{ and } \omega \in \Lambda, \end{cases}$$

and similarly

$$\begin{cases} A(t, \omega) = A^{(n)}(t, \omega) & \text{for } t \in [0, n] \text{ and } \omega \in \Lambda^c \\ A(t, \omega) = 0 & \text{for } t \in [0, n] \text{ and } \omega \in \Lambda. \end{cases}$$

M and A are well-defined and satisfy the requirements of the theorem. ∎

Corollary 10.24. *Let $X = \{X_t : t \in \mathbb{R}_+\}$ be a right-continuous submartingale of class (D) on an augmented right-continuous filtered space $(\Omega, \mathfrak{F}, \{\mathfrak{F}_t\}, P)$. Then in the Doob-Meyer decomposition $X = M + A$, the right-continuous martingale M is uniformly integrable and the natural almost surely increasing process A is uniformly integrable.*

Proof. If X is a submartingale of class (D), then X is uniformly integrable since every deterministic time is a finite stopping time. Then by Theorem 8.1, there exists an integrable $\mathfrak{F}_\infty$-measurable random variable X_∞ to which X converges both a.e. on $(\Omega, \mathfrak{F}_\infty, P)$ and in L_1. In particular, we have $\lim_{t \to \infty} \mathbf{E}(X_t) = \mathbf{E}(X_\infty)$.

§10. INCREASING PROCESSES

If $X = M + A$ is a Doob-Meyer decomposition of X, then since M is a martingale we have $\mathbf{E}(M_t) = \mathbf{E}(M_0) \in \mathbb{R}$ for every $t \in \mathbb{R}_+$. By the Monotone Convergence Theorem we have

$$\mathbf{E}(A_\infty) = \lim_{t \to \infty} \mathbf{E}(A_t) = \lim_{t \to \infty} \{\mathbf{E}(X_t) - \mathbf{E}(M_t)\} = \mathbf{E}(X_\infty) - \mathbf{E}(M_0) \in \mathbb{R}.$$

Thus $\mathbf{E}(A_\infty) < \infty$. Then by Lemma 10.2, A is uniformly integrable. The uniform integrability of both X and A implies that of $M = X - A$. ∎

[IV] Regular Submartingales

Definition 10.25. *A submartingale $X = \{X_t : t \in \mathbb{R}_+\}$ on a filtered space $(\Omega, \mathfrak{F}, \{\mathfrak{F}_t\}, P)$ is said to be regular if for every $a > 0$ and every increasing sequence $\{T_n : n \in \mathbb{N}\}$ in the collection $\mathbf{S}_a$ of all stopping times bounded by a converging to some $T \in \mathbf{S}_a$, we have $\lim_{n \to \infty} \mathbf{E}(X_{T_n}) = \mathbf{E}(X_T)$.*

Remark 10.26. From the definition above it follows immediately that a submartingale $\{X_t : t \in \mathbb{R}_+\}$ on a filtered space $(\Omega, \mathfrak{F}, \{\mathfrak{F}_t\}, P)$ is regular if and only if for any stopping times $T_n, n \in \mathbb{N}$, and T such that $T_n \uparrow T$ on Ω we have $\lim_{n \to \infty} \mathbf{E}[X_{T_n \wedge t}] = \mathbf{E}[X_{T \wedge t}]$ for every $t \in \mathbb{R}_+$.

Observation 10.27. Every right-continuous martingale $X = \{X_t : t \in \mathbb{R}_+\}$ on a right-continuous filtered space $(\Omega, \mathfrak{F}, \{\mathfrak{F}_t\}, P)$ is regular.

Proof. Let $a > 0$. For any $S, T \in \mathbf{S}_a$ such that $S \leq T$ on Ω we have by Theorem 8.10 (Optional Sampling with Bounded Stopping Times), $\mathbf{E}(X_T | \mathfrak{F}_S) = X_S$ a.e. on $(\Omega, \mathfrak{F}_S, P)$ and thus $\mathbf{E}(X_T) = \mathbf{E}(X_S)$. Then for an increasing sequence $\{T_n : n \in \mathbb{N}\}$ and $T \in \mathbf{S}_a$ such that $T_n \leq T$ for $n \in \mathbb{N}$ we have $\mathbf{E}(X_{T_n}) = \mathbf{E}(X_T)$ for every $n \in \mathbb{N}$. ∎

Observation 10.28. An almost surely continuous and nonnegative right-continuous submartingale $X = \{X_t : t \in \mathbb{R}_+\}$ on a right-continuous filtered space $(\Omega, \mathfrak{F}, \{\mathfrak{F}_t\}, P)$ is always regular.

Proof. Let $a > 0$. For any stopping times $T_n, n \in \mathbb{N}$, and T in $\mathbf{S}_a$ such that $T_n \uparrow T$ on Ω, we have $\lim_{n \to \infty} X_{T_n} = X_T$ a.e. on $(\Omega, \mathfrak{F}, P)$ since almost every sample function is continuous. Thus by Fatou's Lemma we have

(1) $$\mathbf{E}[X_T] = \mathbf{E}[\lim_{n \to \infty} X_{T_n}] \leq \liminf_{n \to \infty} \mathbf{E}[X_{T_n}].$$

By Theorem 8.10, $\{X_{T_n}, n \in \mathbb{N}, X_T\}$ is a submartingale with respect to the filtration $\{\mathfrak{F}_{T_n}, n \in \mathbb{N}, \mathfrak{F}_T\}$ so that $\mathbf{E}[X_{T_n}] \leq \mathbf{E}[X_T]$ for $n \in \mathbb{N}$ and $\mathbf{E}[X_{T_n}] \uparrow$ as $n \to \infty$. Thus $\lim_{n \to \infty} \mathbf{E}[X_{T_n}] \leq \mathbf{E}[X_T]$. On the other hand (1) implies that $\mathbf{E}[X_T] \leq \lim_{n \to \infty} \mathbf{E}[X_{T_n}]$. Therefore we have $\lim_{n \to \infty} \mathbf{E}[X_{T_n}] = \mathbf{E}[X_T]$, proving the regularity of X. ∎

Observation 10.29. An almost surely continuous almost surely increasing process $A = \{A_t : t \in \mathbb{R}_+\}$ on a filtered space $(\Omega, \mathfrak{F}, \{\mathfrak{F}_t\}, P)$ is always regular.

Proof. If A is an almost surely continuous almost surely increasing process, then there exists a null set Λ_A in $(\Omega, \mathfrak{F}_\infty, P)$ such that $A(\cdot, \omega)$ is a real valued continuous monotone increasing function on $\mathbb{R}_+$ with $A(0, \omega) = 0$ for $\omega \in \Lambda_A^c$. If $\{T_n : n \in \mathbb{N}\}$ is an increasing sequence in $\mathbf{S}_a$ converging to $T \in \mathbf{S}_a$ then we have $A_{T_n} \uparrow A_T$ as $n \to \infty$ on Λ_A^c by the continuity of $A(\cdot, \omega)$. Thus by the Monotone Convergence Theorem we have $\lim_{n \to \infty} \mathbf{E}(A_{T_n}) = \mathbf{E}(A_T)$. ∎

An almost surely continuous almost surely increasing process is natural by Theorem 10.18 and regular as we have just shown. We shall show that if an almost surely increasing process is both natural and regular then it is almost surely continuous provided that the filtered space is an augmented right-continuous filtered space on a complete probability space.

Remark 10.30. If an almost surely increasing process $A = \{A_t : t \in \mathbb{R}_+\}$ on a filtered space $(\Omega, \mathfrak{F}, \{\mathfrak{F}_t\}, P)$ is regular then for every increasing sequence $\{T_n : n \in \mathbb{N}\}$ in $\mathbf{S}_a$ converging to $T \in \mathbf{S}_a$ we have $\lim_{n \to \infty} A_{T_n} = A_T$ a.e. on $(\Omega, \mathfrak{F}_\infty, P)$.

Proof. Since A is an almost surely increasing process there exists a null set Λ_A in $(\Omega, \mathfrak{F}_\infty, P)$ such that $A(\cdot, \omega)$ is a real valued right-continuous monotone increasing function on $\mathbb{R}_+$ with $A(0, \omega) = 0$ for $\omega \in \Lambda_A^c$. Thus $A_{T_n} \uparrow$ as $n \to \infty$ and $A_{T_n} \leq A_T$ on Λ_A^c. By the Monotone Convergence Theorem we have $\lim_{n \to \infty} \mathbf{E}(A_{T_n}) = \mathbf{E}(\lim_{n \to \infty} A_{T_n})$ and by the regularity of A we have $\lim_{n \to \infty} \mathbf{E}(A_{T_n}) = \mathbf{E}(A_T)$. Therefore we have
$\mathbf{E}[A_T - \lim_{n \to \infty} A_{T_n}] = 0$. Since $A_T - \lim_{n \to \infty} A_{T_n} \geq 0$ on Λ_A^c, the last equality implies that the integrand is equal to 0, that is, $\lim_{n \to \infty} A_{T_n} = A_T$, a.e. on $(\Omega, \mathfrak{F}_\infty, P)$. ∎

Lemma 10.31. *Let $A = \{A_t : t \in \mathbb{R}_+\}$ be an almost surely increasing process on a filtered space $(\Omega, \mathfrak{F}, \{\mathfrak{F}_t\}, P)$.*
1) If for every $a > 0$ and every increasing sequence $\{T_n : n \in \mathbb{N}\}$ in $\mathbf{S}_a$ converging to

§10. INCREASING PROCESSES

$T \in \mathbf{S}_a$ we have $\lim_{n\to\infty} \|A_{T_n} - A_T\|_1 = 0$, then A is regular.

2) Conversely, if A is regular then for T_n and T as in 1), A_{T_n} converges to A_T both a.e. on $(\Omega, \mathfrak{F}_\infty, P)$ and in L_1 as $n \to \infty$.

3) If A is regular then for every $c > 0$ the process $A \wedge c = \{(A \wedge c)(t) : t \in \mathbb{R}_+\}$, where $(A \wedge c)(t) = A(t) \wedge c$, is also regular.

Proof. 1) is immediate since $\lim_{n\to\infty} \|A_{T_n} - A_T\|_1 = 0$ implies $\lim_{n\to\infty} \mathbf{E}(A_{T_n}) = \mathbf{E}(A_T)$.

2) The regularity of A implies that A_{T_n} converges to A_T a.e. on $(\Omega, \mathfrak{F}_\infty, P)$ as $n \to \infty$ by Remark 10.30. Also, since $0 \leq A_{T_n} \leq A_T$ a.e. on $(\Omega, \mathfrak{F}_\infty, P)$, the convergence $\lim_{n\to\infty} \mathbf{E}(A_{T_n}) = \mathbf{E}(A_T)$ is equivalent to $\lim_{n\to\infty} \mathbf{E}(|A_T - A_{T_n}|) = 0$, that is, A_{T_n} converges to A_T in L_1 as $n \to \infty$.

3) If A is regular then with T_n and T as in 1) we have $\lim_{n\to\infty} A_{T_n} = A_T$ a.e. on $(\Omega, \mathfrak{F}_\infty, P)$ as we saw in 2). Then $\lim_{n\to\infty} A_{T_n} \wedge c = A_T \wedge c$, that is, $\lim_{n\to\infty}(A \wedge c)_{T_n} = (A \wedge c)_T$, a.e. on $(\Omega, \mathfrak{F}_\infty, P)$. According to 2), A_{T_n} converges to A_T in L_1 and thus $\{A_{T_n} : n \in \mathbb{N}\}$ is uniformly integrable by Theorem 4.16. This implies that $\{A_{T_n} \wedge c : n \in \mathbb{N}\}$ is uniformly integrable by 4) of Proposition 4.11. Therefore by Theorem 4.16 $A_{T_n} \wedge c$ converges to $A_T \wedge c$ in L_1. Then by 1), $A \wedge c$ is regular. ∎

Lemma 10.32. *Let $A = \{A_t : t \in \mathbb{R}_+\}$ be a natural and regular almost surely increasing process on an augmented right-continuous filtered space on a complete probability space $(\Omega, \mathfrak{F}, \{\mathfrak{F}_t\}, P)$. For $a, c > 0$, define a stochastic process $A \wedge c = \{(A \wedge c)(t) : t \in [0, a]\}$ where $(A \wedge c)(t) = A(t) \wedge c$. Let $t_{n,k} = k2^{-n}a$ for $k = 0, \ldots, 2^n$, $I_{n,k} = [t_{n,k-1}, t_{n,k})$ for $k = 1, \ldots, 2^n - 1$, $I_{n,2^n} = [t_{n,2^n-1}, a]$ and $n \in \mathbb{N}$. For $n \in \mathbb{N}$, define a right-continuous step function ϑ_n on $[0, a]$ by setting*

(1) $$\vartheta_n(t) = t_{n,k} \quad \text{for } t \in I_{n,k}, k = 1, \ldots, 2^n.$$

For $n \in \mathbb{N}$, define a stochastic process $A^{(n)} = \{A^{(n)}(t) : t \in [0, a]\}$ by

(2) $$A^{(n)}(t) = \mathbf{E}[(A \wedge c)(\vartheta_n(t))|\mathfrak{F}_t] \quad \text{for } t \in [0, a],$$

where we choose a version of the conditional expectation at each $t \in [0, a]$ so that $A^{(n)}$ is a right-continuous martingale on each $I_{n,k}$. Then there exist a subsequence $\{n_\ell\}$ of $\{n\}$ and a null set L_a in $(\Omega, \mathfrak{F}_a, P)$ such that

(3) $$\lim_{\ell\to\infty} \sup_{t\in[0,a]} |A^{(n_\ell)}(t, \omega) - (A \wedge c)(t, \omega)| = 0 \quad \text{for } \omega \in L_a^c.$$

Proof. Note that by (1) and (2), we have

(4) $\qquad A^{(n)}(t) = \mathbf{E}[(A \wedge c)(t_{n,k})|\mathfrak{F}_t] \quad \text{for } t \in I_{n,k}, k = 1, \ldots, 2^n.$

Thus $A^{(n)}$ is a martingale on each $I_{n,k}$ and we can select a version of the conditional expectation at each $t \in I_{n,k}$ so that the martingale is right-continuous on each $I_{n,k}$ by applying Corollary 9.10. Note also that

Let us show first that there exists a null set L in $(\Omega, \mathfrak{F}_a, P)$ such that $A^{(n)}$ is a right-continuous process on $[0, a]$.

(5) $\qquad A^{(n)}(t, \omega) \downarrow \quad \text{as } n \to \infty \text{ for all } t \in [0, a] \text{ when } \omega \in L^c,$

and

(6) $\qquad A^{(n)}(t, \omega) \geq (A \wedge c)(t, \omega) \quad \text{for all } n \in \mathbb{N} \text{ and all } t \in [0, a] \text{ when } \omega \in L^c.$

Since A is an almost surely increasing process, there exists a null set Λ_A in $(\Omega, \mathfrak{F}_\infty, P)$ such that $A(\cdot, \omega)$ and then $(A \wedge c)(\cdot, \omega)$ also are real valued right-continuous monotone increasing functions on $[0, a]$ and vanishing at $t = 0$ when $\omega \in \Lambda_A^c$. Since $\vartheta_n(t) \downarrow t$ as $n \to \infty$ for every $t \in [0, a]$, we have $(A \wedge c)(\vartheta_{n+1}(t)) \leq (A \wedge c)(\vartheta_n(t))$ on Λ_A^c. Then there exists a null set $M_{t,n}$ in $(\Omega, \mathfrak{F}_t, P)$ such that

$$\mathbf{E}[(A \wedge c)(\vartheta_{n+1}(t))|\mathfrak{F}_t] \leq \mathbf{E}[(A \wedge c)(\vartheta_n(t))|\mathfrak{F}_t] \quad \text{on } M_{t,n}^c.$$

Let $M_t = \cup_{n \in \mathbb{N}} M_{t,n}$. Then $\mathbf{E}[(A \wedge c)(\vartheta_n(t))|\mathfrak{F}_t] \downarrow$ on M_t^c as $n \to \infty$. Let $\{r_m : m \in \mathbb{N}\}$ be the collection of all the rationals in $[0, a]$. For the null set $M = \cup_{m \in \mathbb{N}} M_{r_m}$ we have

$$A^{(n)}(r_m, \omega) \downarrow \quad \text{as } n \to \infty \text{ for all } m \in \mathbb{N} \text{ when } \omega \in M^c.$$

Then by the right-continuity of $A^{(n)}$ on $[0, a]$ we have (5) for $\omega \in M^c$. Since $\vartheta_n(t) \geq t$ and since $A \wedge c$ is increasing on Λ_A^c we have

$$(A \wedge c)(\vartheta_n(t)) \geq (A \wedge c)(t) \quad \text{for } t \in [0, a] \text{ on } \Lambda_A^c.$$

Thus

$$A^{(n)}(t) = \mathbf{E}[(A \wedge c)(\vartheta_n(t))|\mathfrak{F}_t] \geq (A \wedge c)(t) \quad \text{a.e. on } (\Omega, \mathfrak{F}_t, P).$$

Therefore there exists a null set N_{n,r_m} in $(\Omega, \mathfrak{F}_a, P)$ such that $A^{(n)}(r_m) \geq (A \wedge c)(r_m)$ on N_{n,r_m}^c. Then for the null set $N_n = \cup_{m \in \mathbb{N}} N_{n,r_m}$ in $(\Omega, \mathfrak{F}_a, P)$ we have

$$A^{(n)}(r_m) \geq (A \wedge c)(r_m) \quad \text{for all } m \in \mathbb{N} \text{ on } N_n^c.$$

§10. INCREASING PROCESSES

By the right-continuity of $A^{(n)}$ on $[0, a]$ and that of $A \wedge c$ on Λ_A^c we have

$$A^{(n)}(t) \geq (A \wedge c)(t) \quad \text{for all } t \in [0, a] \text{ on } (N_n \cup \Lambda_A)^c.$$

Let $N = \cup_{n \in \mathbb{N}}(N_n \cup \Lambda_A)$. Then (6) holds on N^c. Finally, the null set $L = M \cup N$ in $(\Omega, \mathfrak{F}_a, P)$ satisfies both (5) and (6).

For $\varepsilon > 0$, let

(7) $$T_{n,\varepsilon}(\omega) = \begin{cases} \inf\{t \in [0, a] : A^{(n)}(t, \omega) - (A \wedge c)(t, \omega) > \varepsilon\} \\ a \quad \text{if the set above is } \emptyset. \end{cases}$$

Since $T_{n,\varepsilon}$ is defined as the first passage time of the open set (ε, ∞) by the right-continuous adapted process $A^{(n)} - (A \wedge c)$ with respect to the right-continuous filtration $\{\mathfrak{F}_t : t \in [0, a]\}$, $T_{n,\varepsilon}$ is a stopping time by Proposition 3.5. Thus $T_{n,\varepsilon}$ is in the collection $\mathbf{S}_a$ of all stopping times that are bounded by a. From (7), we have for any $\omega \in \Omega$

(8) $$T_{n,\varepsilon}(\omega) = a \Leftrightarrow A^{(n)}(t, \omega) - (A \wedge c)(t, \omega) \leq \varepsilon \quad \text{for all } t \in [0, a].$$

By (5) we have $T_{n,\varepsilon}(\omega) \uparrow$ as $n \to \infty$ for $\omega \in L^c$. Let T_ε be defined by

(9) $$T_\varepsilon(\omega) = \begin{cases} \lim_{n \to \infty} T_{n,\varepsilon}(\omega) & \text{for } \omega \in L^c \\ a & \text{for } \omega \in L. \end{cases}$$

Since our filtered space is an augmented right continuous filtered space on a complete probability space, T_ε defined by (9) is a stopping time by Theorem 3.19. Clearly $T_\varepsilon \in \mathbf{S}_a$. Now since $\vartheta_n(t) \geq t$, we have

(10) $$\vartheta_n(T_{n,\varepsilon}) \geq T_{n,\varepsilon} \quad \text{on } \Omega.$$

Also, since $T_{n,\varepsilon}$ is $\mathfrak{F}_{T_{n,\varepsilon}}/\mathfrak{B}_{\mathbb{R}_+}$-measurable and ϑ_n is $\mathfrak{B}_{\mathbb{R}_+}/\mathfrak{B}_{\mathbb{R}_+}$-measurable, the composite mapping $\vartheta_n(T_{n,\varepsilon})$ is $\mathfrak{F}_{T_{n,\varepsilon}}/\mathfrak{B}_{\mathbb{R}_+}$-measurable. This fact and (10) imply that $\vartheta_n(T_{n,\varepsilon})$ is a stopping time by Theorem 3.6. Therefore $\vartheta_n(T_{n,\varepsilon}) \in \mathbf{S}_a$. By (10) and by the fact that ϑ_n is an increasing function on $[0, a]$ and $T_{n,\varepsilon} \leq T_\varepsilon$ we have

$$T_{n,\varepsilon} \leq \vartheta_n(T_{n,\varepsilon}) \leq \vartheta_n(T_\varepsilon).$$

Since $T_{n,\varepsilon} \uparrow T_\varepsilon$ as $n \to \infty$ on L^c and since $\vartheta_n(t) \downarrow t$ implies $\vartheta_n(T_\varepsilon) \downarrow T_\varepsilon$ as $n \to \infty$, we have

(11) $$\vartheta_n(T_{n,\varepsilon}) \uparrow T_\varepsilon \quad \text{as } n \to \infty \text{ on } L^c.$$

Now since $A^{(n)}$ is a martingale on $(t_{n,k-1}, t_{n,k}]$, for the stopping time $T_{n,\varepsilon} \wedge t_{n,k}$ we have by Theorem 8.10 (Optional Sampling for Bounded Stopping Times)

(12) $\quad\quad \mathbf{E}[A^{(n)}(t_{n,k})|\mathfrak{F}_{T_{n,\varepsilon}\wedge t_{n,k}}] = A^{(n)}(T_{n,\varepsilon}\wedge t_{n,k}) \quad$ a.e. on $\mathfrak{F}_{T_{n,\varepsilon}\wedge t_{n,k}}$,

and then recalling (4) we have

(13) $\quad\quad \mathbf{E}[\mathbf{E}[(A\wedge c)(t_{n,k})|\mathfrak{F}_{t_{n,k}}]|\mathfrak{F}_{T_{n,\varepsilon}\wedge t_{n,k}}] = A^{(n)}(T_{n,\varepsilon}\wedge t_{n,k}) \quad$ a.e. on $\mathfrak{F}_{T_{n,\varepsilon}\wedge t_{n,k}}$.

For $k = 1, \ldots, 2^n$ and $n \in \mathbb{N}$, let

(14) $\quad\quad G_{n,k} = \{T_{n,\varepsilon} \in (t_{n,k-1}, t_{n,k}]\} = \{\vartheta_n(T_{n,\varepsilon}) = t_{n,k}\}.$

Since $\{G_{n,k} : k = 1, \ldots, 2^n\}$ is a disjoint collection in $\mathfrak{F}_a$ whose union is equal to Ω, we have

$$\begin{aligned}
\mathbf{E}[A^{(n)}(T_{n,\varepsilon})] &= \sum_{k=1}^{2^n} \int_{G_{n,k}} A^{(n)}(T_{n,\varepsilon})\,dP \\
&= \sum_{k=1}^{2^n} \int_{G_{n,k}} A^{(n)}(T_{n,\varepsilon}\wedge t_{n,k})\,dP \quad \text{since } T_{n,\varepsilon} \leq t_{n,k} \text{ on } G_{n,k} \\
&= \sum_{k=1}^{2^n} \int_{G_{n,k}} \mathbf{E}[(A\wedge c)(t_{n,k})|\mathfrak{F}_{T_{n,\varepsilon}\wedge t_{n,k}}] \quad \text{by (13)}.
\end{aligned}$$

By (14) we have $G_{n,k} \in \mathfrak{F}_{T_{n,\varepsilon}}$ as well as $G_{n,k} \in \mathfrak{F}_{t_{n,k}}$. Therefore according to Theorem 3.12 we have $G_{n,k} \in \mathfrak{F}_{T_{n,\varepsilon}} \cap \mathfrak{F}_{t_{n,k}} = \mathfrak{F}_{T_{n,\varepsilon}\wedge t_{n,k}}$. Thus

(15) $\quad\quad \begin{aligned}
\mathbf{E}[A^{(n)}(T_{n,\varepsilon})] &= \sum_{k=1}^{2^n} \int_{G_{n,k}} (A\wedge c)(t_{n,k})\,dP \\
&= \sum_{k=1}^{2^n} \int_{G_{n,k}} (A\wedge c)(\vartheta_n(T_{n,\varepsilon}))\,dP \quad \text{by (14)} \\
&= \mathbf{E}[(A\wedge c)(\vartheta_n(T_{n,\varepsilon}))].
\end{aligned}$

Since $A \wedge c$ is monotone increasing on Λ_A^c, (10) implies

(16) $\quad\quad (A\wedge c)(\vartheta_n(T_{n,\varepsilon})) - (A\wedge c)(T_{n,\varepsilon}) \geq 0 \quad \text{on } \Lambda_A^c.$

§10. INCREASING PROCESSES

Now

(17)
$$\mathbf{E}[(A \wedge c)(\vartheta_n(T_{n,\varepsilon})) - (A \wedge c)(T_{n,\varepsilon})]$$
$$= \mathbf{E}[A^{(n)}(T_{n,\varepsilon}) - (A \wedge c)(T_{n,\varepsilon})] \quad \text{by (15)}$$
$$\geq \int_{\{T_{n,\varepsilon} < a\}} \{A^{(n)}(T_{n,\varepsilon}) - (A \wedge c)(T_{n,\varepsilon})\} \, dP \quad \text{by (16)}$$
$$\geq \varepsilon P\{T_{n,\varepsilon} < a\}$$

by the fact that the integrand is bounded below by ε on $\{T_{n,\varepsilon} < a\}$ according to (7). The regularity of A implies that of $A \wedge c$ according to Lemma 10.31. Now $T_{n,\varepsilon}$, $\vartheta_n(T_{n,\varepsilon})$ and T_ε are all in $\mathbf{S}_a$ and since $T_{n,\varepsilon} \uparrow T_\varepsilon$ and $\vartheta_n(T_{n,\varepsilon}) \uparrow T_\varepsilon$ on Ω as $n \to \infty$, we have $\lim_{n \to \infty} \mathbf{E}[(A \wedge c)(T_{n,\varepsilon})] = \mathbf{E}[(A \wedge c)(T_\varepsilon)]$ and $\lim_{n \to \infty} \mathbf{E}[(A \wedge c)(\vartheta_n(T_{n,\varepsilon}))] = \mathbf{E}[(A \wedge c)(T_\varepsilon)]$. Thus letting $n \to \infty$ in (17), we obtain

(18)
$$\lim_{n \to \infty} P\{T_{n,\varepsilon} < a\} = 0.$$

By (7) we have

$$\{T_{n,\varepsilon} < a\} = \{A^{(n)}(t) - (A \wedge c)(t) > \varepsilon \text{ for some } t \in [0,a]\}$$
$$= \left\{\sup_{t \in [0,a]} \{A^{(n)}(t) - (A \wedge c)(t)\} > \varepsilon\right\}.$$

By (6) we have $A^{(n)} - (A \wedge c) \geq 0$ on $[0,a] \times L^c$. Thus

(19)
$$\{T_{n,\varepsilon} < a\} \cap L^c = \left\{\sup_{t \in [0,a]} |A^{(n)}(t) - (A \wedge c)(t)| > \varepsilon\right\} \cap L^c.$$

Let us define a function ξ_n on Ω by

(20)
$$\xi_n(\omega) = \sup_{t \in [0,a]} |A^{(n)}(t,\omega) - (A \wedge c)(t,\omega)|.$$

Since $A^{(n)} - (A \wedge c)$ is right-continuous on Ω we have,

$$\xi_n(\omega) = \sup_{m \in \mathbb{N}} |A^{(n)}(r_m,\omega) - (A \wedge c)(r_m,\omega)|.$$

As the supremum of countably many $\mathfrak{F}_a$-measurable random variables, ξ_n as defined above is an $\mathfrak{F}_a$-measurable random variable. Thus by (19) and (20) we have

$$P\{|\xi_n| > \varepsilon\} = P(\{\xi_n > \varepsilon\} \cap L^c) = P(\{T_{n,\varepsilon} < a\} \cap L^c) = P\{T_{n,\varepsilon} < a\}$$

and therefore by (18) we have $\lim_{n\to\infty} P\{|\xi_n| > \varepsilon\} = 0$. Since this holds for every $\varepsilon > 0$, ξ_n converges to 0 in probability as $n \to \infty$. This implies that there exists a subsequence $\{n_\ell\}$ of $\{n\}$ such that ξ_{n_ℓ} converges to 0 a.e. on $(\Omega, \mathfrak{F}_a, P)$ as $\ell \to \infty$. Therefore there exists a null set L_a in $(\Omega, \mathfrak{F}_a, P)$ such that $\lim_{\ell\to\infty} \xi_{n_\ell}(\omega) = 0$ for $\omega \in L_a^c$. This proves (3). ∎

Theorem 10.33. *Let $A = \{A_t : t \in \mathbb{R}_+\}$ be an almost surely increasing process on an augmented right-continuous filtered space on a complete probability space $(\Omega, \mathfrak{F}, \{\mathfrak{F}_t\}, P)$. If A is both natural and regular then A is almost surely continuous.*

Proof. Suppose we show that for every $a > 0$ there exists a null set Λ_a in $(\Omega, \mathfrak{F}_a, P)$ such that $A(\cdot, \omega)$ is continuous on $[0, a]$ for $\omega \in \Lambda_a^c$. Then for every $n \in \mathbb{N}$ there exists a null set Λ_n in $(\Omega, \mathfrak{F}_n, P)$ such that $A(\cdot, \omega)$ is continuous on $[0, n]$ for $\omega \in \Lambda_n^c$. Thus for the null set $\Lambda_\infty = \cup_{n\in\mathbb{N}} \Lambda_n$, $A(\cdot, \omega)$ is continuous on $\mathbb{R}_+$ for $\omega \in \Lambda_\infty^c$ and we are done. It remains to show the existence of Λ_a.

Since A is an almost surely increasing process, there exists a null set Λ_A in $(\Omega, \mathfrak{F}_\infty, P)$ such that $A(\cdot, \omega)$ is a real valued right-continuous monotone increasing function on $\mathbb{R}_+$ with $A(0, \omega) = 0$ for $\omega \in \Lambda_A^c$. Let us define $A_- = \{A_-(t) : t \in \mathbb{R}_+\}$ by setting

(1) $$A_-(t, \omega) = \begin{cases} \lim_{s\uparrow t} A(s, \omega) & \text{for } t \in (0, \infty) \text{ and } \omega \in \Lambda_A^c \\ A(0, \omega) & \text{for } t = 0 \text{ and } \omega \in \Lambda_A^c \\ 0 & \text{for } t \in \mathbb{R}_+ \text{ and } \omega \in \Lambda_A. \end{cases}$$

Since our filtered space is an augmented filtered space, the null set Λ_A is in $\mathfrak{F}_t$ for every $t \in \mathbb{R}_+$. Thus $A_-(t)$ is $\mathfrak{F}_t$-measurable for every $t \in \mathbb{R}_+$ and therefore A_- is an adapted process.

Let $a > 0$ be fixed. For every $c > 0$, let $A \wedge c = \{(A \wedge c)(t) : t \in [0, a]\}$ be defined by setting $(A \wedge c)(t) = A(t) \wedge c$ and similarly define $A_- \wedge c = \{(A_- \wedge c)(t) : t \in [0, a]\}$ by setting $(A_- \wedge c)(t) = A_-(t) \wedge c$. For $\omega \in \Lambda_A^c$, $(A \wedge c)(\cdot, \omega)$ is real valued right-continuous monotone increasing and bounded by c on $[0, a]$ with $(A \wedge c)(0, \omega) = 0$. In particular it has at most countably many points of discontinuity and a discontinuity occurs at $t \in [0, a]$ if and only if $(A_- \wedge c)(t, \omega) < (A \wedge c)(t, \omega)$. Thus for every $\omega \in \Lambda_A^c$, the sum

(2) $$\sum_{t\in[0,a]} \{(A \wedge c)(t, \omega) - (A_- \wedge c)(t, \omega)\}$$

is a sum of at most countably many positive terms. Let us show that

(3) $$\mathbf{E}\left[\sum_{t\in[0,a]} \{(A \wedge c)(t, \omega) - (A_- \wedge c)(t, \omega)\}^2\right] = 0,$$

§10. INCREASING PROCESSES

where the integrand is undefined on the null set Λ_A. Now for $\omega \in \Lambda_A^c$ we have

$$\sum_{t \in [0,a]} \{(A \wedge c)(t, \omega) - (A_- \wedge c)(t, \omega)\}^2$$

$$= \int_{[0,a]} \{(A \wedge c)(t, \omega) - (A_- \wedge c)(t, \omega)\} \, d(A \wedge c)(t)$$

$$\leq \int_{[0,a]} \{(A \wedge c)(t, \omega) - (A_- \wedge c)(t, \omega)\} \, dA(t).$$

Thus, to show (3), it suffices to show

$$\mathbf{E}\left[\int_{[0,a]} \{(A \wedge c)(t, \omega) - (A_- \wedge c)(t, \omega)\} \, dA(t)\right] = 0,$$

or equivalently

(4) $$\mathbf{E}\left[\int_{[0,a]} (A \wedge c)(t, \omega) \, dA(t)\right] = \mathbf{E}\left[\int_{[0,a]} (A_- \wedge c)(t, \omega) \, dA(t)\right].$$

Consider the process $A^{(n)}$ defined in Lemma 10.32. Since $A^{(n)}$ is a bounded right-continuous martingale on $I_{n,k}$ for $k = 1, \ldots, 2^n$ into which $[0, a]$ is decomposed, the fact that A is natural implies

(5) $$\mathbf{E}\left[\int_{[0,a]} A^{(n)}(t) \, dA(t)\right] = \sum_{k=1}^{2^n} \mathbf{E}\left[\int_{I_{n,k}} A^{(n)}(t) \, dA(t)\right]$$

$$= \sum_{k=1}^{2^n} \mathbf{E}\left[\int_{I_{n,k}} A_-^{(n)}(t) \, dA(t)\right] = \mathbf{E}\left[\int_{[0,a]} A_-^{(n)}(t) \, dA(t)\right].$$

According to Lemma 10.32, there exist a subsequence $\{n_\ell\}$ of $\{n\}$ and a null set L_a in $(\Omega, \mathfrak{F}_a, P)$ such that for $\omega \in L_a^c$ we have

(6) $$\lim_{\ell \to \infty} A^{(n_\ell)}(t, \omega) = (A \wedge c)(t, \omega) \quad \text{uniformly for } t \in [0, a].$$

Then since $A^{(n_\ell)}(t) \downarrow$ as $\ell \to \infty$ for all $t \in [0, a]$ except on a null set as we showed in Lemma 10.32, we have $\int_{[0,a]} A^{(n_\ell)}(t) \, dA(t) \downarrow$ except on a null set in $(\Omega, \mathfrak{F}_a, P)$, the Monotone Convergence Theorem implies

(7) $$\lim_{\ell \to \infty} \mathbf{E}\left[\int_{[0,a]} A^{(n_\ell)}(t) \, dA(t)\right] = \mathbf{E}\left[\lim_{\ell \to \infty} \int_{[0,a]} A^{(n_\ell)}(t) \, dA(t)\right]$$

$$= \mathbf{E}\left[\int_{[0,a]} (A \wedge c)(t) \, dA(t)\right],$$

where the second equality is by the uniform convergence (6). Since (6) implies that for $\omega \in L_a^c$ we have also $\lim_{\ell \to \infty} A_-^{(n_\ell)}(t, \omega) = (A_- \wedge c)(t, \omega)$ uniformly for $t \in [0, a]$, we have by the same argument as in (7),

$$\text{(8)} \quad \lim_{\ell \to \infty} \mathbf{E}\left[\int_{[0,a]} A_-^{(n_\ell)}(t)\, dA(t)\right] = \mathbf{E}\left[\int_{[0,a]} (A_- \wedge c)(t)\, dA(t)\right].$$

Using (7) and (8) in (5) we have (4) and thus (3) holds.

Now (3) implies that there exists a null set $\Lambda_{a,c}$ in $(\Omega, \mathfrak{F}_a, P)$ such that for $\omega \in \Lambda_{a,c}^c$ we have

$$\sum_{t \in [0,a]} \{(A \wedge c)(t, \omega) - (A_- \wedge c)(t, \omega)\}^2 = 0.$$

Then since $(A \wedge c)(t, \omega) \geq (A_- \wedge c)(t, \omega) \geq 0$ for $\omega \in \Lambda_A^c$ we have

$$\sum_{t \in [0,a]} \{(A \wedge c)(t, \omega) - (A_- \wedge c)(t, \omega)\} = 0,$$

for $\omega \in (\Lambda_{a,c} \cup \Lambda_A)^c$. Thus $(A \wedge c)(\cdot, \omega)$ is continuous on $[0, a]$ for $\omega \in (\Lambda_{a,c} \cup \Lambda_A)^c$. Let $\Lambda_a = (\cup_{m \in \mathbb{N}} \Lambda_{a,m}) \cup \Lambda_A$. Then $(A \wedge m)(\cdot, \omega)$ is continuous on $[0, a]$ for all $m \in \mathbb{N}$ when $\omega \in \Lambda_a^c$. Thus by the fact that $A(t, \omega) \leq A(a, \omega) < \infty$ when $\omega \in \Lambda_A^c$ we have the continuity of $A(\cdot, \omega)$ on $[0, a]$ when $\omega \in \Lambda_a^c$. ∎

Theorem 10.34. *Let $A = \{A_t : t \in \mathbb{R}_+\}$ be an almost surely increasing process on an augmented right-continuous filtered space on a complete probability space $(\Omega, \mathfrak{F}, \{\mathfrak{F}_t\}, P)$. Then A is almost surely continuous if and only if A is both natural and regular.*

Proof. If A is almost surely continuous, then A is natural by Theorem 10.18 and regular by Observation 10.29. Conversely if A is both natural and regular then A is almost surely continuous by Theorem 10.33. ∎

Theorem 10.35. *Let $X = \{X_t : t \in \mathbb{R}_+\}$ be a right-continuous submartingale of class (DL) on an augmented right-continuous filtered space on a complete probability space $(\Omega, \mathfrak{F}, \{\mathfrak{F}_t\}, P)$ and let $X = M + A$ be a Doob-Meyer decomposition. Then A is almost surely continuous if and only if X is regular.*

Proof. Since M is a right-continuous martingale on a right-continuous filtered space, it is regular by Observation 10.27. Thus X is regular if and only if A is. Since A is natural, it is regular if and only if it is almost surely continuous by Theorem 10.34. Therefore X is regular if and only if A is almost surely continuous. ∎

Chapter 3

Stochastic Integrals

§11 L_2-Martingales and Quadratic Variation Processes

[I] The Space of Right-Continuous L_2-Martingales

Definition 11.1. *A filtered space* $(\Omega, \mathfrak{F}, \{\mathfrak{F}_t : t \in \mathbb{R}_+\}, P)$ *is called a standard filtered space if it satisfies the following conditions.*

1°. *The probability space* $(\Omega, \mathfrak{F}, P)$ *is a complete measure space.*

2°. *The filtration* $\{\mathfrak{F}_t : t \in \mathbb{R}_+\}$ *is right-continuous.*

3°. *The filtration* $\{\mathfrak{F}_t : t \in \mathbb{R}_+\}$ *is augmented.*

Let us observe that if $(\Omega, \mathfrak{F}, \{\mathfrak{F}_t\}, P)$ is a standard filtered space, then $(\Omega, \mathfrak{F}_t, P)$ is a complete measure space for every $t \in \mathbb{R}_+$. In fact if Λ is a null set in $(\Omega, \mathfrak{F}_t, P)$ then it is a null set in $(\Omega, \mathfrak{F}, P)$, and then by the completeness of $(\Omega, \mathfrak{F}, P)$ every subset Λ_0 of Λ is in $\mathfrak{F}$ and therefore in $\mathfrak{F}_t$ since $\mathfrak{F}_t$ is augmented.

Recall that two stochastic processes $X = \{X_t : t \in \mathbb{R}_+\}$ and $Y = \{Y_t : t \in \mathbb{R}_+\}$ on a probability space $(\Omega, \mathfrak{F}, P)$ are said to be equivalent if there exists a null set Λ in $(\Omega, \mathfrak{F}, P)$ such that $X(\cdot, \omega) = Y(\cdot, \omega)$ for $\omega \in \Lambda^c$.

Observation 11.2. Let $X = \{X_t : t \in \mathbb{R}_+\}$ be a submartingale, martingale, or supermartingale on an augmented filtered space $(\Omega, \mathfrak{F}, \{\mathfrak{F}_t\}, P)$ in which $(\Omega, \mathfrak{F}, P)$ is a complete measure space. If Y is an equivalent process of X, then Y is a submartingale, martingale, or supermartingale, according as X is.

Proof. It suffices to prove this for a submartingale. Let Λ be a null set in $(\Omega, \mathfrak{F}, P)$ such that $X(\cdot, \omega) = Y(\cdot, \omega)$ for $\omega \in \Lambda^c$. Suppose X is a submartingale. Then X is an adapted process so that X_t is $\mathfrak{F}_t$-measurable for every $t \in \mathbb{R}_+$. Since the filtered space is augmented, Λ is in $\mathfrak{F}_t$. Since $Y_t = X_t$ on Λ^c, Y_t is $\mathfrak{F}_t$-measurable on Λ^c. Since $(\Omega, \mathfrak{F}, P)$ is a complete measure space, Y_t is $\mathfrak{F}_t$-measurable on Λ. Thus Y_t is $\mathfrak{F}_t$-measurable on Ω. This shows that Y is an adapted process. Since $Y_t = X_t$ on Λ^c and X_t is integrable, Y_t is integrable. This shows that Y is an L_1-process. Now since X is a submartingale, for any pair $s, t \in \mathbb{R}_+$ such that $s < t$ we have $\mathbf{E}(X_t | \mathfrak{F}_s) \geq X_s$ a.e. on $(\Omega, \mathfrak{F}_s, P)$. Since $X_t = Y_t$ a.e. on $(\Omega, \mathfrak{F}_t, P)$, we have $\mathbf{E}(X_t | \mathfrak{F}_s) = \mathbf{E}(Y_t | \mathfrak{F}_s)$ a.e. on $(\Omega, \mathfrak{F}_s, P)$. But $X_s = Y_s$ a.e. on $(\Omega, \mathfrak{F}_s, P)$. Thus $\mathbf{E}(Y_t | \mathfrak{F}_s) \geq Y_s$ a.e. on $(\Omega, \mathfrak{F}_s, P)$. This shows that Y is a submartingale. ∎

Definition 11.3. *Let* $\mathbf{M}_2(\Omega, \mathfrak{F}, \{\mathfrak{F}_t\}, P)$, *or briefly* $\mathbf{M}_2$, *be the linear space of equivalence classes of all right-continuous L_2-martingales* $X = \{X_t : t \in \mathbb{R}_+\}$ *with* $X_0 = 0$ *a.s. on a standard filtered space* $(\Omega, \mathfrak{F}, \{\mathfrak{F}_t\}, P)$. *Let* $\mathbf{M}_2^c(\Omega, \mathfrak{F}, \{\mathfrak{F}_t\}, P)$, *or briefly* $\mathbf{M}_2^c$, *be the linear subspace consisting of almost surely continuous members of* $\mathbf{M}_2$.

Unless otherwise stated, the filtered space $(\Omega, \mathfrak{F}, \{\mathfrak{F}_t\}, P)$ in $\mathbf{M}_2(\Omega, \mathfrak{F}, \{\mathfrak{F}_t\}, P)$ is always a standard filtered space.

Note that $0 \in \mathbf{M}_2$ is the equivalence class consisting of right-continuous L_2-martingales $X = \{X_t : t \in \mathbb{R}_+\}$ such that there exists a null set Λ in $(\Omega, \mathfrak{F}, P)$ such that $X(\cdot, \omega) = 0$ on $\mathbb{R}_+$ for $\omega \in \Lambda^c$.

Recall that according to Theorem 9.2, for every right-continuous martingale $X = \{X_t : t \in \mathbb{R}_+\}$ on a filtered space $(\Omega, \mathfrak{F}, \{\mathfrak{F}_t\}, P)$, there exists a null set Λ in $(\Omega, \mathfrak{F}_\infty, P)$ such that the sample function $X(\cdot, \omega)$ is bounded on every finite interval in $\mathbb{R}_+$, has finite left limit everywhere on $\mathbb{R}_+$ and has at most countably many points of discontinuity for $\omega \in \Lambda^c$.

Definition 11.4. *On* $\mathbf{M}_2(\Omega, \mathfrak{F}, \{\mathfrak{F}_t\}, P)$ *where* $(\Omega, \mathfrak{F}, \{\mathfrak{F}_t\}, P)$ *is an arbitrary filtered space, let*

(1) $\qquad\qquad\qquad \|X\|_t = \mathbf{E}(X_t^2)^{1/2} \qquad \text{for } X \in \mathbf{M}_2, t \in \mathbb{R}_+,$

(2) $\qquad\qquad\qquad \|X\|_\infty = \sum_{m \in \mathbb{N}} 2^{-m} \{\|X\|_m \wedge 1\} \quad \text{for } X \in \mathbf{M}_2.$

Note that since $X \in \mathbf{M}_2$ is a martingale, X^2 is a submartingale and therefore $\|X\|_t \uparrow$

§11. L_2-MARTINGALES AND QUADRATIC VARIATION PROCESSES

as $t \to \infty$.

Recall that a on a linear space **V** over scalars $\mathbb{K}$, where $\mathbb{K}$ is either $\mathbb{R}$ or $\mathbb{C}$, is a function p on **V** such that $p(x) \in [0, \infty)$ for $x \in \mathbf{V}$ and $p(0) = 0$ for $0 \in \mathbf{V}$, $p(\alpha x) = |\alpha| p(x)$ for $x \in \mathbf{V}$ and $\alpha \in \mathbb{K}$, and $p(x+y) \leq p(x) + p(y)$ for $x, y \in \mathbf{V}$.

Remark 11.5. $\|\cdot\|_t$ and $\|\cdot\|_\infty$ have the following properties.
1) $\|\cdot\|_t$ is a *seminorm* on $\mathbf{M}_2$ for every $t \in \mathbb{R}_+$.
2) For $X, X^{(n)} \in \mathbf{M}_2$, $n \in \mathbb{N}$, we have $\lim_{n \to \infty} \|X^{(n)} - X\|_\infty = 0$ if and only if we have $\lim_{n \to \infty} \|X^{(n)} - X\|_m = 0$ for every $m \in \mathbb{N}$.
3) $\|\cdot\|_\infty$ is a quasinorm on $\mathbf{M}_2$.

Proof. 1) is obvious. Note that $\|\cdot\|_t$ is not a norm since $\|M\|_t = 0$ does not imply $M = 0 \in \mathbf{M}_2$.

2) Since $2^m \|X\|_\infty \geq \|X\|_m \wedge 1$ for every $m \in \mathbb{N}$, $\lim_{n \to \infty} \|X^{(n)} - X\|_\infty = 0$ implies $\lim_{n \to \infty} \{\|X^{(n)} - X\|_m \wedge 1\} = 0$, which in turn implies $\lim_{n \to \infty} \|X^{(n)} - X\|_m = 0$. Conversely if $\lim_{n \to \infty} \|X^{(n)} - X\|_m = 0$ for every $m \in \mathbb{N}$, then

$$\lim_{n \to \infty} \|X^{(n)} - X\|_\infty = \lim_{n \to \infty} \sum_{m \in \mathbb{N}} 2^{-m} \{\|X^{(n)} - X\|_m \wedge 1\} = 0$$

by interpreting the sum as the Lebesgue integral of a step function on $[0, 1)$ with steps of lengths 2^{-m} and with values $\|X^{(n)} - X\|_m \wedge 1$ on these steps, and then by applying the Bounded Convergence Theorem to pass to the limit under summation.

3) Clearly $\|X\|_\infty \in [0, 1]$ for $X \in \mathbf{M}_2$ and $\|0\|_\infty = 0$. Conversely, if $X \in \mathbf{M}_2$ and $\|X\|_\infty = 0$ then $\|X\|_m = 0$ for every $m \in \mathbb{N}$ and consequently $\|X\|_t = 0$ for every $t \in \mathbb{R}_+$. Then $X_t = 0$ a.e. on $(\Omega, \mathfrak{F}_t, P)$ for every $t \in \mathbb{R}_+$. The right-continuity of X then implies that there exists a null set Λ in $(\Omega, \mathfrak{F}_\infty, P)$ such that $X(\cdot, \omega) = 0$ on $\mathbb{R}_+$ for $\omega \in \Lambda^c$ by Theorem 2.3. Thus $X = 0 \in \mathbf{M}_2$. Clearly $\|-X\|_\infty = \|X\|_\infty$ for $X \in \mathbf{M}_2$. The triangle inequality for $\|\cdot\|_\infty$ follows from the triangle inequalities for $\|\cdot\|_m$ for $m \in \mathbb{N}$, and the fact that for every $a, b \geq 0$ we have $(a+b) \wedge 1 \leq (a \wedge 1) + (b \wedge 1)$. Thus $\|\cdot\|_\infty$ is a quasinorm on $X \in \mathbf{M}_2$. ∎

Proposition 11.6. *Let $X, X^{(n)} \in \mathbf{M}_2(\Omega, \mathfrak{F}, \{\mathfrak{F}_t\}, P)$ for $n \in \mathbb{N}$ where $(\Omega, \mathfrak{F}, \{\mathfrak{F}_t\}, P)$ is an augmented right-continuous filtered space. If $\lim_{n \to \infty} \|X^{(n)} - X\|_\infty = 0$, then $X^{(n)}$ converges* to X in probability uniformly on $[0, m)$ for every $m \in \mathbb{N}$, that is,

$$(1) \qquad P \cdot \lim_{n \to \infty} \left\{ \sup_{t \in [0,m)} |X_t^{(n)} - X_t| \right\} = 0.$$

Furthermore there exists a subsequence $\{n_k\}$ of $\{n\}$ and a null set Λ in $(\Omega, \mathfrak{F}_\infty, P)$ such that for every $\omega \in \Lambda^c$ the sample functions $X^{(n_k)}(\cdot, \omega)$ converge to $X(\cdot, \omega)$ uniformly on every finite interval in $\mathbb{R}_+$.

Proof. Note that since $X^{(n)} - X$ is right-continuous, the supremum over $[0, m)$ in (1) is equal to the supremum over a countable dense subset of $[0, m)$ and is therefore a random variable, in fact an $\mathfrak{F}_m$-measurable random variable, on $(\Omega, \mathfrak{F}, P)$. Now $\lim_{n \to \infty} \| X^{(n)} - X \|_\infty = 0$ implies that $\lim_{n \to \infty} \| X^{(n)} - X \|_m = 0$, that is, $\lim_{n \to \infty} \mathbf{E}[|X_m^{(n)} - X_m|^2]^{1/2} = 0$, for every $m \in \mathbb{N}$. Since $|X^{(n)} - X|$ is a right-continuous nonnegative L_2-submartingale, for every $\eta > 0$ we have by (1) of Theorem 6.16

$$P\left\{ \sup_{t \in [0,m)} |X_t^{(n)} - X_t| > \eta \right\} \leq \eta^{-2} \mathbf{E}\left[|X_m^{(n)} - X_m|^2 \right]$$

and therefore

$$\lim_{n \to \infty} P \left\{ \sup_{t \in [0,m)} |X_t^{(n)} - X_t| > \eta \right\} = 0.$$

This proves (1).

Since convergence in probability of a sequence of random variables implies the existence of a subsequence which converges almost surely, (1) implies that for each $m \in \mathbb{N}$ there exist a subsequence $\{n_{m,k} : k \in \mathbb{N}\}$ of $\{n\}$ and a null set Λ_m in $(\Omega, \mathfrak{F}_m, P)$ such that

$$\lim_{n \to \infty} \sup_{t \in [0,m)} |X^{(n_{m,k})}(t, \omega) - X(t, \omega)| = 0 \quad \text{for } \omega \in \Lambda_m^c.$$

For $m \geq 2$, let $\{n_{m,k} : k \in \mathbb{N}\}$ be a subsequence of $\{n_{m-1,k} : k \in \mathbb{N}\}$ and let $\Lambda = \cup_{m \in \mathbb{N}} \Lambda_m$, a null set in $(\Omega, \mathfrak{F}_\infty, P)$. Then for the subsequence $\{n_{k,k} : k \in \mathbb{N}\}$ of $\{n\}$ we have

$$\lim_{k \to \infty} \sup_{s \in [0,t)} |X^{(n_{k,k})}(s, \omega) - X(s, \omega)| = 0 \quad \text{for every } t \in \mathbb{R}_+ \text{ when } \omega \in \Lambda^c.$$

This completes the proof. ∎

To show that $\mathbf{M}_2$ is a complete metric space with respect to the metric associated with the quasinorm $\| \cdot \|_\infty$ we need the following proposition.

§11. L_2-MARTINGALES AND QUADRATIC VARIATION PROCESSES

Proposition 11.7. *Let $\{X^{(n)} : n \in \mathbb{N}\}$, where $X^{(n)} = \{X_t^{(n)} : t \in \mathbb{R}_+\}$, be a sequence of left- or right-continuous processes on a probability space $(\Omega, \mathfrak{F}, P)$ and let $D \subset \mathbb{R}_+$. If the sequence is a Cauchy sequence with respect to uniform convergence in probability on D, that is, if for every $\eta > 0$ and every $\varepsilon > 0$ there exists $N \in \mathbb{N}$ such that*

(1) $$P\left\{\sup_{t \in D} |X_t^{(n)} - X_t^{(\ell)}| > \eta \right\} < \varepsilon \quad \text{for } n, \ell \geq N,$$

then there exists a left- or right-continuous process X on $D \times \Omega$ such that

(2) $$P \cdot \lim_{n \to \infty} \left\{ \sup_{t \in D} |X_t^{(n)} - X_t| \right\} = 0,$$

and thus there exist a subsequence $\{n_k\}$ of $\{n\}$ and a null set Λ in $(\Omega, \mathfrak{F}, P)$ such that

(3) $$\lim_{k \to \infty} \sup_{t \in D} |X_t^{(n_k)} - X_t| = 0 \quad \text{for } \omega \in \Lambda^c.$$

Proof. Observe that the left- or right-continuity of $X^{(n)}$ for $n \in \mathbb{N}$ implies the measurability of the supremum over D in (1) as noted in the proof of Proposition 11.6. Now by (1) we can select a subsequence $\{n_k\}$ of $\{n\}$ such that

(4) $$P\left\{\sup_{t \in D} |X_t^{(n_{k+1})} - X_t^{(n_k)}| > 2^{-k} \right\} < 2^{-k},$$

and then

$$\sum_{k \in \mathbb{N}} P\left\{\sup_{t \in D} |X_t^{(n_{k+1})} - X_t^{(n_k)}| > 2^{-k} \right\} < \infty.$$

Let $A_k = \{\sup_{t \in D} |X_t^{(n_{k+1})} - X_t^{(n_k)}| \geq 2^{-k}\}$ and $A = \limsup_{k \to \infty} A_k$. Since $\sum_{k \in \mathbb{N}} P(A_k) < \infty$ we have $P(A) = 0$, that is, $P\{\omega \in \Omega : \omega \in A_k \text{ for infinitely many } k \in \mathbb{N}\} = 0$, by the Borel-Cantelli Theorem. Therefore there exists a null set Λ in $(\Omega, \mathfrak{F}, P)$ such that

(5) $$\sum_{k \in \mathbb{N}} \sup_{t \in D} |X^{(n_{k+1})}(t, \omega) - X^{(n_k)}(t, \omega)| < \infty \quad \text{for } \omega \in \Lambda^c.$$

If we define a real valued function Y on $D \times \Lambda^c$ by setting

(6) $$Y(t, \omega) = \sum_{k \in \mathbb{N}} \{X^{(n_{k+1})}(t, \omega) - X^{(n_k)}(t, \omega)\} \quad \text{for } (t, \omega) \in D \times \Lambda^c,$$

then we have

(7) $$Y(t,\omega) = \lim_{k \to \infty} X^{(n_{k+1})}(t,\omega) - X^{(n_1)}(t,\omega) \quad \text{for } (t,\omega) \in D \times \Lambda^c.$$

Define a real valued function X on $D \times \Omega$ by

(8) $$X(t,\omega) = \begin{cases} Y(t,\omega) + X^{(n_1)}(t,\omega) & \text{for } (t,\omega) \in D \times \Lambda^c \\ 0 & \text{for } (t,\omega) \in D \times \Lambda. \end{cases}$$

By (6), $Y(t,\cdot)$ is $\mathfrak{F}$-measurable on Λ^c and thus $X(t,\cdot)$ is $\mathfrak{F}$-measurable on Ω for $t \in D$. By the uniform convergence on D in (7) which is implied by (5), $Y(\cdot,\omega)$ is left- or right-continuous on D for $\omega \in \Lambda^c$ and then $X(\cdot,\omega)$ is left- or right-continuous on D for $\omega \in \Omega$.

To prove (2), let us recall that for an arbitrary $\varepsilon > 0$ there exists $N_1 \in \mathbb{N}$ such that

(9) $$P\left\{\sup_{t \in D} |X_t^{(n)} - X_t^{(\ell)}| > \frac{\varepsilon}{2}\right\} < \frac{\varepsilon}{2} \quad \text{for } n, \ell \geq N_1.$$

By the uniform convergence of $X^{(n_k)}(\cdot,\omega)$ to $X(\cdot,\omega)$ on D for $\omega \in \Lambda^c$, we have $\lim_{k \to \infty} \sup_{t \in D} |X_t^{(n_k)} - X_t| = 0$ on Λ^c. Since almost sure convergence implies convergence in probability, there exists $N_2 \in \mathbb{N}$ such that

(10) $$P\left\{\sup_{t \in D} |X_t^{(n_k)} - X_t| > \frac{\varepsilon}{2}\right\} < \frac{\varepsilon}{2} \quad \text{for } n_k \geq N_2.$$

Note that since $\sup_{t \in D} |X_t^{(n)} - X_t| \leq \sup_{t \in D} |X_t^{(n)} - X_t^{(n_k)}| + \sup_{t \in D} |X_t^{n_k} - X_t|$ we have

$$\left\{\sup_{t \in D} |X_t^{(n)} - X_t| \leq \varepsilon\right\} \supset \left\{\sup_{t \in D} |X_t^{(n)} - X_t^{(n_k)}| \leq \frac{\varepsilon}{2}\right\} \cup \left\{\sup_{t \in D} |X_t^{(n_k)} - X_t| \leq \frac{\varepsilon}{2}\right\}$$

and therefore

(11) $$P\left\{\sup_{t \in D} |X_t^{(n)} - X_t| > \varepsilon\right\}$$
$$\leq P\left\{\sup_{t \in D} |X_t^{(n)} - X_t^{(n_k)}| > \frac{\varepsilon}{2}\right\} + P\left\{\sup_{t \in D} |X_t^{(n_k)} - X_t| > \frac{\varepsilon}{2}\right\}.$$

Let $N = \max\{N_1, N_2\}$. If we take $n_k \geq N$, then by (11), (9) and (10) we have for $n \geq N$

$$P\left\{\sup_{t \in D} |X_t^n - X_t| > \varepsilon\right\} < \varepsilon.$$

§11. L_2-MARTINGALES AND QUADRATIC VARIATION PROCESSES

This proves (2). Finally, (3) is a direct consequence of (2). ∎

Theorem 11.8. *Let $(\Omega, \mathfrak{F}, \{\mathfrak{F}_t\}, P)$ be an augmented right-continuous filtered space. Then $\mathbf{M}_2(\Omega, \mathfrak{F}, \{\mathfrak{F}_t\}, P)$ is complete with respect to the metric associated with the quasinorm $\|\cdot\|_\infty$ in Definition 11.4. Furthermore, $\mathbf{M}_2^c(\Omega, \mathfrak{F}, \{\mathfrak{F}_t\}, P)$ is a closed linear subspace of $\mathbf{M}_2(\Omega, \mathfrak{F}, \{\mathfrak{F}_t\}, P)$.*

Proof. Let $\{X^{(n)} : n \in \mathbb{N}\}$ be a Cauchy sequence with respect to the metric associated with the quasinorm $\|\cdot\|_\infty$ on $\mathbf{M}_2$. Then for every $\delta > 0$ there exists $N_\delta \in \mathbb{N}$ such that

$$\text{(1)} \qquad \|X^{(n)} - X^{(\ell)}\|_\infty < \delta \quad \text{for } n, \ell \geq N_\delta.$$

Let us show that this implies that for fixed $m \in \mathbb{N}$, $\{X^{(n)} : n \in \mathbb{N}\}$ is a Cauchy sequence with respect to uniform convergence in probability on $[0, m)$, that is, for every $\eta > 0$ and $\varepsilon > 0$, there exists $N_{\eta,\varepsilon} \in \mathbb{N}$ such that

$$\text{(2)} \qquad P\left\{\sup_{t \in [0,m)} |X_t^{(n)} - X_t^{(\ell)}| > \eta\right\} \leq \varepsilon \quad \text{for } n, \ell \geq N_{\eta,\varepsilon}.$$

Now for any $n, \ell \in \mathbb{N}$, $X^{(n)} - X^{(\ell)}$ is a right-continuous L_2-martingale so that $|X^{(n)} - X^{(\ell)}|$ is a right-continuous nonnegative L_2-submartingale. Then by (1) of Theorem 6.16 we have

$$\eta^2 P\left\{\sup_{t \in [0,m)} |X_t^{(n)} - X_t^{(\ell)}| > \eta\right\} \leq \mathbf{E}[|X_m^{(n)} - X_m^{(\ell)}|^2],$$

in other words,

$$\eta P\left\{\sup_{t \in [0,m)} |X_t^{(n)} - X_t^{(\ell)}| > \eta\right\}^{1/2} \leq \|X^{(n)} - X^{(\ell)}\|_m.$$

Then if $\eta \in (0, 1]$, we have

$$\eta P\left\{\sup_{t \in [0,m)} |X_t^{(n)} - X_t^{(\ell)}| > \eta\right\}^{1/2}$$
$$= \eta P\left\{\sup_{t \in [0,m)} |X_t^{(n)} - X_t^{(\ell)}| > \eta\right\}^{1/2} \wedge 1 \leq \|X^{(n)} - X^{(\ell)}\|_m \wedge 1.$$

Thus by (2) of Definition 11.4 and (1) we have

$$2^{-m} \eta P\left\{\sup_{t \in [0,m)} |X_t^{(n)} - X_t^{(\ell)}| > \eta\right\}^{1/2} \leq \|X^{(n)} - X^{(\ell)}\|_\infty < \delta \quad \text{for } n, \ell \geq N_\delta,$$

that is,
$$P\left\{\sup_{t\in[0,m)}|X_t^{(n)}-X_t^{(\ell)}|>\eta\right\}<(2^m\eta^{-1}\delta)^2 \quad \text{for } n,\ell\geq N_\delta.$$

For an arbitrary $\varepsilon>0$, let $\delta(\varepsilon)>0$ be so small that $(2^m\eta^{-1}\delta(\varepsilon))^2<\varepsilon$. Then
$$P\left\{\sup_{t\in[0,m)}|X_t^{(n)}-X_t^{(\ell)}|>\eta\right\}<\varepsilon \quad \text{for } n,\ell\geq N_{\delta(\varepsilon)}$$

for our $\eta\in(0,1]$. On the other hand for $\eta>1$ we have
$$P\left\{\sup_{t\in[0,m)}|X_t^{(n)}-X_t^{(\ell)}|>\eta\right\}\leq P\left\{\sup_{t\in[0,m)}|X_t^{(n)}-X_t^{(\ell)}|>1\right\}<\varepsilon \quad \text{for } n,\ell\geq N_{\delta(\varepsilon)}.$$

This proves (2) with $N_{\eta,\varepsilon}=N_{\delta(\varepsilon)}$.

Now the fact that $\{X^{(n)}:n\in\mathbb{N}\}$ is a Cauchy sequence with respect to uniform convergence in probability on $[0,m)$ implies according to Proposition 11.7 that there exists a subsequence $\{n_{m,k}:k\in\mathbb{N}\}$ of $\{n\}$ and a null set Λ_m in $(\Omega,\mathfrak{F},P)$ such that
$$\lim_{k\to\infty}\sup_{t\in[0,m)}|X^{(n_{m,k})}(t,\omega)-X(t,\omega)|=0 \quad \text{for } \omega\in\Lambda_m^c,$$

where X is a right-continuous process on $(\Omega,\mathfrak{F},P)$. We can select the subsequences inductively so that $\{n_{m,k}:k\in\mathbb{N}\}$ is a subsequence of $\{n_{m-1,k}:k\in\mathbb{N}\}$ for $m\geq 2$. Then for the subsequence $\{n_{k,k}:k\in\mathbb{N}\}$ of $\{n\}$ and the null set $\Lambda=\cup_{m\in\mathbb{N}}\Lambda_m$ in $(\Omega,\mathfrak{F},P)$ we have

(3) $$\lim_{k\to\infty}\sup_{t\in[0,m)}|X^{(n_{k,k})}(t,\omega)-X(t,\omega)|=0 \quad \text{for } m\in\mathbb{N} \text{ and } \omega\in\Lambda^c.$$

Let us redefine X on $\mathbb{R}_+\times\Lambda$ by setting

(4) $$X(t,\omega)=0 \quad \text{for } (t,\omega)\in\mathbb{R}_+\times\Lambda.$$

Note that the definition of X on $\mathbb{R}_+\times\Lambda^c$ is independent of $m\in\mathbb{N}$. To show that X is an adapted process, that is, X_t is $\mathfrak{F}_t$-measurable for every $t\in\mathbb{R}_+$, note first that since the filtered space is augmented, the null set Λ in $(\Omega,\mathfrak{F},P)$ is in $\mathfrak{F}_t$ for every $t\in\mathbb{R}_+$. Now (3) implies that the sequence of $\mathfrak{F}_t$-measurable random variables $X_t^{(n_{k,k})}$ for $k\in\mathbb{N}$ converge to X_t on $\Lambda^c\in\mathfrak{F}_t$ so that X_t is $\mathfrak{F}_t$-measurable on $\Lambda^c\in\mathfrak{F}_t$. Then since $X_t=0$ on Λ by (4), X_t is $\mathfrak{F}_t$-measurable on Ω. This shows that X is an adapted process. Since $X_0^{(n_{k,k})}=0$ a.e. for $k\in\mathbb{N}$, we have $X_0=0$ a.e.

To show that X is an L_2-process, that is, $\mathbf{E}(X_t^2) < \infty$ for every $t \in \mathbb{R}_+$, let $t \in \mathbb{R}_+$ be arbitrarily selected and let $m \in \mathbb{N}$ be so large that $t \in [0, m)$. Since $X^{(n_i,i)} - X^{(n_j,j)}$ is a martingale, $|X^{(n_i,i)} - X^{(n_j,j)}|^2$ is a submartingale and therefore $\mathbf{E}[|X_t^{(n_i,i)} - X_t^{(n_j,j)}|^2]$ increases with t. Thus by (1), for every $\varepsilon > 0$ there exists $N \in \mathbb{N}$ such that

$$\mathbf{E}[|X_t^{(n_i,i)} - X_t^{(n_j,j)}|^2]^{1/2} \leq \| X^{(n_i,i)} - X^{(n_j,j)} \|_m < \varepsilon \quad \text{for } i, j \geq N.$$

Thus by the completeness of $L_2(\Omega, \mathfrak{F}_t, P)$ with respect to the metric of the L_2 norm, there exists $Y_t \in L_2(\Omega, \mathfrak{F}_t, P)$ such that $\lim_{k \to \infty} X_t^{(n_k,k)} = Y_t$ in L_2. Then there exists a subsequence of $\{X_t^{(n_k,k)} : k \in \mathbb{N}\}$ which converges to Y_t a.e. on $(\Omega, \mathfrak{F}_t, P)$. But the subsequence converges to X_t a.e. on $(\Omega, \mathfrak{F}_t, P)$. Thus $X_t = Y_t$ a.e. on $(\Omega, \mathfrak{F}_t, P)$. Therefore $X_t \in L_2(\Omega, \mathfrak{F}_t, P)$. This shows that X is an L_2-process. Note also that $\lim_{k \to \infty} X_t^{(n_k,k)} = X_t$ in L_2.

Let us show that X is a martingale. Let $s, t \in \mathbb{R}_+$ and $s < t$. Since $X_t^{(n_k,k)}$ converges to X_t in L_2, $\mathbf{E}(X_t^{(n_k,k)}|\mathfrak{F}_s)$ converges to $\mathbf{E}(X_t|\mathfrak{F}_s)$ in L_2. (See appendix) Since $X^{(n_k,k)}$ is a martingale we have $\mathbf{E}(X_t^{(n_k,k)}|\mathfrak{F}_s) = X_s^{(n_k,k)}$ a.e. on $(\Omega, \mathfrak{F}_s, P)$. Thus $X_s^{(n_k,k)}$ converges to $\mathbf{E}(X_t|\mathfrak{F}_s)$ in L_2. Then there exists a subsequence $\{n_\ell\}$ of $\{n_{k,k}\}$ such that $X_s^{(n_\ell)}$ converges to $\mathbf{E}(X_t|\mathfrak{F}_s)$ a.e. on $(\Omega, \mathfrak{F}_s, P)$. But $X_s^{(n_\ell)}$ converges to X_s a.e. on $(\Omega, \mathfrak{F}_s, P)$ also. Thus $\mathbf{E}(X_t|\mathfrak{F}_s) = X_s$ a.e. on $(\Omega, \mathfrak{F}_s, P)$. This shows that X is a martingale.

To show that $\lim_{n \to \infty} \| X^{(n)} - X \|_\infty = 0$, note that

$$\| X^{(n)} - X \|_\infty \leq \| X^{(n)} - X^{(n_k,k)} \|_\infty + \| X^{(n_k,k)} - X \|_\infty.$$

Since $\lim_{k \to \infty} \| X^{(n_k,k)} - X \|_m = 0$ for every $m \in \mathbb{N}$, we have $\lim_{k \to \infty} \| X^{(n_k,k)} - X \|_\infty = 0$ by Remark 11.5. From this fact and the fact that $\{X^{(n)} : n \in \mathbb{N}\}$ is a Cauchy sequence with respect to the metric associated with $\| \cdot \|_\infty$, we have $\lim_{n \to \infty} \| X^{(n)} - X \|_\infty = 0$.

Finally to show that $\mathbf{M}_2^c$ is a closed subset of $\mathbf{M}_2$, we show that if $\{X^{(n)} : n \in \mathbb{N}\}$ is a sequence in $\mathbf{M}_2^c$ and if $\lim_{n \to \infty} \| X^{(n)} - X \|_\infty = 0$ for some $X \in \mathbf{M}_2$, then $X \in \mathbf{M}_2^c$. But according to Proposition 11.6, $\lim_{n \to \infty} \| X^{(n)} - X \|_\infty = 0$ implies that there exists a subsequence $\{n_k\}$ of $\{n\}$ and a null set Λ in $(\Omega, \mathfrak{F}, P)$ such that $X^{(n_k)}(\cdot, \omega)$ converges to $X(\cdot, \omega)$ uniformly on every finite interval in $\mathbb{R}_+$ for $\omega \in \Lambda^c$. Thus $X(\cdot, \omega)$ is continuous on $\mathbb{R}_+$ for $\omega \in \Lambda^c$, that is, $X \in \mathbf{M}_2^c$. ∎

[II] Signed Lebesgue-Stieltjes Measures

Observation 11.9. Let us give a brief review of signed Lebesgue-Stieltjes measures. Let a' and a'' be two real valued right-continuous monotone increasing functions on $\mathbb{R}_+$ and let

$v = a' - a''$, a real valued right-continuous function on $\mathbb{R}_+$ with bounded variation on $[0, t]$ for every $t \in \mathbb{R}_+$. Consider the total variation function $\|v\|$ of v defined by

(1) $$\|v\|(t) = \sup\left\{\sum_{k=1}^{n} |v(t_k) - v(t_{k-1})| : 0 = t_0 < \cdots < t_n = t\right\},$$

for every $t \in \mathbb{R}_+$. $\|v\|$ is a real valued right-continuous monotone increasing function on $\mathbb{R}_+$ with $\|v\|(0) = 0$. Since v is right-continuous, $\|v\|(t)$ can be computed as

(2) $$\|v\|(t) = \sup_{n\in\mathbb{N}} \sum_{k=1}^{2^n} |v(t_{n,k}) - v(t_{n,k-1})| = \lim_{n\to\infty} \sum_{k=1}^{2^n} |v(t_{n,k}) - v(t_{n,k-1})|,$$

where $t_{n,k} = k 2^{-n} t$, $k = 0, \ldots, 2^n$, and $n \in \mathbb{N}$.

Let $\mu_{a'}$, $\mu_{a''}$ and $\mu_{\|v\|}$ be the Lebesgue-Stieltjes measures on $(\mathbb{R}_+, \mathfrak{B}_{\mathbb{R}_+})$ determined by a', a'', and $\|v\|$ respectively. Let us call $\mu_{\|v\|}$ the total variation measure of v. For every $t \in \mathbb{R}_+$, $\mu_v = \mu_{a'} - \mu_{a''}$ is a signed Lebesgue-Stieltjes measure on $([0, t], \mathfrak{B}_{[0,t]})$. μ_v is defined on $(\mathbb{R}_+, \mathfrak{B}_{\mathbb{R}_+})$ if and only if at least one of $\mu_{a'}(\mathbb{R}_+)$ and $\mu_{a''}(\mathbb{R}_+)$ is finite. Consider the signed Lebesgue-Stieltjes measure μ_v on $([0, t], \mathfrak{B}_{[0,t]})$. According to the Hahn Decomposition Theorem, there exist two sets $A, B \in \mathfrak{B}_{[0,t]}$ such that $A \cap B = \emptyset$, $A \cup B = [0, t]$, $\mu_v(E \cap A) \geq 0$ and $\mu_v(E \cap B) \leq 0$ for every $E \in \mathfrak{B}_{[0,t]}$ and moreover such a decomposition is unique in the sense that if A' and B' are another pair of such sets then $\mu_v(E \cap A) = \mu_v(E \cap A')$ and $\mu_v(E \cap B) = \mu_v(E \cap B')$ for every $E \in \mathfrak{B}_{[0,t]}$. The two measures μ_v^+ and μ_v^- on $([0, t], \mathfrak{B}_{[0,t]})$ defined by

(3) $$\mu_v^+(E) = \mu_v(E \cap A) \text{ and } \mu_v^-(E) = -\mu_v(E \cap B) \quad \text{for } E \in \mathfrak{B}_{[0,t]}$$

are called the positive and negative parts of μ_v. Note that

(4) $$\mu_v = \mu_v^+ - \mu_v^-.$$

and

(5) $$\mu_{\|v\|} = \mu_v^+ + \mu_v^-.$$

and in particular,

(6) $$\|v\|(t) = \mu_{\|v\|}([0, t]) = (\mu_v^+ + \mu_v^-)([0, t]).$$

For a real valued Borel-measurable function f on $[0, t]$, we define

$$\int_{[0,t]} f(s)\, d\mu_v(s) = \int_{[0,t]} f(s) \mu_v(ds),$$

and
$$\int_{[0,t]} f(s)\, d\mu_{|v|}(s) = \int_{[0,t]} f(s)\mu_{|v|}(ds),$$
provided the Lebesgue-Stieltjes integrals on the right hand sides exist. Note that with $A, B \in \mathfrak{B}_{[0,t]}$ as above in the definition of μ_v^+ and μ_v^-, we have

$$\begin{aligned}
&\int_{[0,t]} f(s)\mu_{|v|}(ds) \\
&= \int_{[0,t]} f(s)\mu_v^+(ds) + \int_{[0,t]} f(s)\mu_v^-(ds) \\
&= \int_{[0,t]} f(s)\{\mathbf{1}_A(s) - \mathbf{1}_B(s)\}\mu_v^+(ds) - \int_{[0,t]} f(s)\{\mathbf{1}_A(s) - \mathbf{1}_B(s)\}\mu_v^-(ds),
\end{aligned}$$

since we have $\int_{[0,t]} f\mathbf{1}_B \, d\mu_v^+ = \int_{[0,t]} f\mathbf{1}_A \, d\mu_v^- = 0$, $\int_{[0,t]} f\mathbf{1}_A \, d\mu_v^+ = \int_{[0,t]} f \, d\mu_v^+$ and $\int_{[0,t]} f\mathbf{1}_B \, d\mu_v^- = \int_{[0,t]} f \, d\mu_v^-$. Thus

(7) $$\int_{[0,t]} f(s)\mu_{|v|}(ds) = \int_{[0,t]} f(s)\{\mathbf{1}_A(s) - \mathbf{1}_B(s)\}\mu_v(ds).$$

[III] Locally Bounded Variation Processes

Definition 11.10. *A stochastic process* $V = \{V_t : t \in \mathbb{R}_+\}$ *on a filtered space* $(\Omega, \mathfrak{F}, \{\mathfrak{F}_t\}, P)$ *is called a locally bounded variation process if it satisfies the following conditions.*

1°. V *is* $\{\mathfrak{F}_t\}$*-adapted.*

2°. $V(\cdot, \omega)$ *is a right-continuous function on* $\mathbb{R}_+$ *and is of bounded variation on* $[0, t]$ *for every* $t \in \mathbb{R}_+$ *with* $V(0, \omega) = 0$ *for every* $\omega \in \Omega$.

3°. *The process* $|V| = \{|V|_t : t \in \mathbb{R}_+\}$ *defined by*

(1) $$|V|(t, \omega) = \sup_{n \in \mathbb{N}} \sum_{k=1}^{2^n} |V(t_{n,k}, \omega) - V(t_{n,k-1}, \omega)| \quad \text{for } (t, \omega) \in \mathbb{R}_+ \times \Omega,$$

where $t_{n,k} = k 2^{-n} t$ *for* $k = 0, \cdots, 2^n$ *and* $n \in \mathbb{N}$, *is an* L_1*-process.*

A stochastic process V *on an augmented filtered space* $(\Omega, \mathfrak{F}, \{\mathfrak{F}_t\}, P)$ *is called an almost surely locally bounded variation process if it satisfies* 1° *and the following conditions.*

4°. *There exists a null set* Λ_V *in* $(\Omega, \mathfrak{F}, P)$ *such that* $V(\cdot, \omega)$ *satisfies condition* 2° *when* $\omega \in \Lambda_V^c$.

5°. The process $\|V\|$ which is defined by (1) for $\omega \in \Lambda_V^c$ and by setting $\|V\|(\cdot,\omega) = 0$ for $\omega \in \Lambda_V$ is an L_1-process.

The set Λ_V is called an exceptional null set for the almost surely locally bounded variation process V. The process $\|V\|$ is called the total variation process of V.

Lemma 11.11. *If $V = \{V_t : t \in \mathbb{R}_+\}$ is an almost surely locally bounded variation process on an augmented filtered space $(\Omega, \mathfrak{F}, \{\mathfrak{F}_t\}, P)$, then its total variation process $\|V\| = \{\|V\|_t : t \in \mathbb{R}_+\}$ is an increasing process.*

Proof. Since the filtered space is augmented, an exceptional null set Λ_V for V is in $\mathfrak{F}_t$ for every $t \in \mathbb{R}_+$. For $t \in \mathbb{R}_+$, the sum in (1) of Definition 11.10 is $\mathfrak{F}_t$-measurable and so is the supremum over $n \in \mathbb{N}$. Thus $\|V\|(t, \cdot)$ is $\mathfrak{F}_t$-measurable on $\Lambda_V^c \in \mathfrak{F}_t$. Since $\|V\|(t, \cdot) = 0$ on Λ_V, $\|V\|(t, \cdot)$ is $\mathfrak{F}_t$-measurable on Ω. This shows that $\|V\|$ is an L_1-process. Also every sample function of $\|V\|$ is a right-continuous monotone increasing function on $\mathbb{R}_+$ vanishing at $t = 0$. This shows that $\|V\|$ is an increasing process. ∎

Theorem 11.12. *A stochastic process $V = \{V_t : t \in \mathbb{R}_+\}$ on an augmented filtered space $(\Omega, \mathfrak{F}, \{\mathfrak{F}_t\}, P)$ is a locally bounded variation process if and only if $V = A' - A''$ where A' and A'' are two increasing processes on the filtered space. Similarly a stochastic process on an augmented filtered space is an almost surely locally bounded variation process if and only it is the difference of two almost surely increasing processes on the filtered space.*

Proof. 1) Suppose $V = A' - A''$ where A' and A'' are two increasing processes on a filtered space. The fact that V satisfies the conditions in Definition 11.10 can be verified easily. In particular for the sum in (1) in Definition 11.10, we have $\sum_{k=1}^{2^n} |V(t_{n,k}, \omega) - V(t_{n,k-1}, \omega)| \leq A'(t, \omega) + A''(t, \omega)$ so that $\|V\|(t, \omega) \leq A'(t, \omega) + A''(t, \omega)$ for $(t, \omega) \in \mathbb{R}_+ \times \Omega$. Since A'_t and A''_t are integrable so is $\|V\|_t$.

2) Suppose V is a locally bounded variation process on a filtered space. As we saw in Lemma 11.11, $\|V\|$ is an increasing process. Let $A = \|V\| - V$. Then A is a right-continuous adapted L_1-process vanishing at $t = 0$. Also for $s, t \in \mathbb{R}_+$, $s < t$, we have

$$A(t, \omega) - A(s, \omega) = \{\|V\|(t, \omega) - \|V\|(s, \omega)\} - \{V(t, \omega) - V(s, \omega)\} \geq 0.$$

Thus A is an increasing process. With $A' = \|V\|$ and $A'' = A$, we have $V = A' - A''$ where A' and A'' are increasing processes.

3) The statement regarding an almost surely bounded variation process on an augmented filtered space is proved likewise. ∎

§11. L_2-MARTINGALES AND QUADRATIC VARIATION PROCESSES

Definition 11.13. *An almost surely locally bounded variation process V on an augmented filtered space is called natural if $V = A' - A''$ where A' and A'' are two natural almost surely increasing processes.*

Let $A = \{A_t : t \in \mathbb{R}_+\}$ be an almost surely increasing process on a filtered space $(\Omega, \mathfrak{F}, \{\mathfrak{F}_t\}, P)$ and let μ_A be the family of Lebesgue-Stieltjes measures on $(\mathbb{R}_+, \mathfrak{B}_{\mathbb{R}_+})$ determined by the sample functions of A. For a real valued function X on $\mathbb{R}_+ \times \Omega$ such that $X(\cdot, \omega)$ is a Borel measurable function on $\mathbb{R}_+$ for every $\omega \in \Omega$, we defined in Definition 10.3

$$\int_{[0,t]} X(s,\omega) \, dA(s,\omega) = \int_{[0,t]} X(s,\omega) \mu_A(ds,\omega) \quad \text{for } \omega \in \Omega,$$

provided the Lebesgue-Stieltjes integral on the right hand side exists.

Definition 11.14. *Let $V = \{V_t : t \in \mathbb{R}_+\}$ be an almost surely locally bounded variation process on a filtered space $(\Omega, \mathfrak{F}, \{\mathfrak{F}_t\}, P)$ given by $V = A' - A''$ where A' and A'' are two almost surely increasing processes. For every $t \in \mathbb{R}_+$, the family of signed Lebesgue-Stieltjes measures on $([0,t], \mathfrak{B}_{[0,t]})$ determined by V is defined by $\mu_V = \mu_{A'} - \mu_{A''}$.*

Note that if $V = A' - A''$ and $V = B' - B''$ where A', A'', B' and B'' are almost surely increasing processes on the filtered space $(\Omega, \mathfrak{F}, \{\mathfrak{F}_t\}, P)$, there exists a null set Λ in $(\Omega, \mathfrak{F}_\infty, P)$ such that $\mu_{A'}(\cdot, \omega) - \mu_{A''}(\cdot, \omega) = \mu_{B'}(\cdot, \omega) - \mu_{B''}(\cdot, \omega)$ for $\omega \in \Lambda^c$, so that the family of signed Lebesgue-Stieltjes measures μ_V on $([0,t], \mathfrak{B}_{[0,t]})$ is independent of the decomposition of V into two almost surely increasing processes up to a null set in $(\Omega, \mathfrak{F}_\infty, P)$.

If at least one of A' and A'', say A' satisfies the integrability condition $\mathbf{E}(A'_\infty) < \infty$, then there exists a null set Λ in $(\Omega, \mathfrak{F}_\infty, P)$ such that for every $\omega \in \Lambda^c$ we have $A'_\infty(\omega) < \infty$ so that $\mu_{A'}(\mathbb{R}_+, \omega) < \infty$ and thus $\mu_V(\mathbb{R}_+, \omega) = \mu_{A'}(\mathbb{R}_+, \omega) - \mu_{A''}(\mathbb{R}_+, \omega)$ is defined. Recall that the condition $\mathbf{E}(A'_\infty) < \infty$ is equivalent to the L_1-boundedness and the uniform integrability of A' according to Lemma 10.2.

For an almost surely locally bounded variation process $V = \{V_t : t \in \mathbb{R}_+\}$ on a filtered space $(\Omega, \mathfrak{F}, \{\mathfrak{F}_t\}, P)$ and a real valued function X on $\mathbb{R}_+ \times \Omega$ such that $X(\cdot, \omega)$ is a Borel measurable function on $\mathbb{R}_+$, we define for every $t \in \mathbb{R}_+$ and $\omega \in \Omega$,

$$\int_{[0,t]} X(s,\omega) \, dV(s,\omega) = \int_{[0,t]} X(s,\omega) \mu_V(ds,\omega)$$

provided the integral with respect to the signed Lebesgue-Stieltjes measure on the right side exists. Note that since $|\mu_V(E,\omega)| \leq \mu_{|V|}(E,\omega)$ for $E \in \mathfrak{B}_{\mathbb{R}_+}$, we have

$$\left| \int_{[0,t]} X(s,\omega)\, dV(s,\omega) \right| \leq \int_{[0,t]} |X(s,\omega)|\, d|V|(s,\omega).$$

Definition 11.15. *Let* $\mathbf{A}(\Omega, \mathfrak{F}, \{\mathfrak{F}_t\}, P)$ *be the collection of equivalence classes of all almost surely increasing processes and* $\mathbf{V}(\Omega, \mathfrak{F}, \{\mathfrak{F}_t\}, P)$ *be the linear space of equivalence classes of all almost surely locally bounded variation processes on a standard filtered space* $(\Omega, \mathfrak{F}, \{\mathfrak{F}_t\}, P)$. *Let* $\mathbf{A}^c(\Omega, \mathfrak{F}, \{\mathfrak{F}_t\}, P)$ *be the subcollection of* $\mathbf{A}(\Omega, \mathfrak{F}, \{\mathfrak{F}_t\}, P)$ *consisting of almost surely continuous members and let* $\mathbf{V}^c(\Omega, \mathfrak{F}, \{\mathfrak{F}_t\}, P)$ *be the linear subspace of* $\mathbf{V}(\Omega, \mathfrak{F}, \{\mathfrak{F}_t\}, P)$ *consisting of almost surely continuous members. If an almost surely increasing process on* $(\Omega, \mathfrak{F}, \{\mathfrak{F}_t\}, P)$ *is natural, then so are all its equivalent processes. In this case we call the equivalence class natural. Similarly for an equivalence class of almost surely locally bounded variation processes on the filtered space.*

In what follows we write $A \in \mathbf{A}(\Omega, \mathfrak{F}, \{\mathfrak{F}_t\}, P)$ to mean both an almost surely increasing process and an equivalence class of such processes. Whether a process or an equivalence class of processes is meant should be clear from the context. Similarly for $V \in \mathbf{V}(\Omega, \mathfrak{F}, \{\mathfrak{F}_t\}, P)$.

Observation 11.16. Let $A \in \mathbf{A}(\Omega, \mathfrak{F}, \{\mathfrak{F}_t\}, P)$ and $p \in [1, \infty)$ be fixed. Consider the collection of all measurable processes $X = \{X_t : t \in \mathbb{R}_+\}$ on $(\Omega, \mathfrak{F}, P)$ satisfying the condition

(1) $$\mathbf{E}\left[\int_{[0,t]} |X(s)|^p\, dA(s) \right] < \infty \quad \text{for every } t \in \mathbb{R}_+.$$

The family of Lebesgue-Stieltjes measures $\{\mu_A(\cdot, \omega) : \omega \in \Omega\}$ on $(\mathbb{R}_+, \mathfrak{B}_{\mathbb{R}_+})$ is an $\mathfrak{F}_\infty$-measurable family and the restrictions of these measures to $([0,t], \mathfrak{B}_{[0,t]})$ constitute an $\mathfrak{F}_\infty$-measurable family of finite measures for every $t \in \mathbb{R}_+$ according to Lemma 10.10. By Theorem 10.11, $\int_{[0,t]} |X(s)|^p\, dA(s)$ is an $\mathfrak{F}_\infty$-measurable random variable on $(\Omega, \mathfrak{F}, P)$ and, if X is an adapted process, then $\int_{[0,t]} |X(s)|^p\, dA(s)$ is an $\mathfrak{F}_t$-measurable random variable.

For two measurable processes X and Y on $(\Omega, \mathfrak{F}, P)$ each satisfying condition (1), the condition

(2) $$\mathbf{E}\left[\int_{[0,t]} |X(s) - Y(s)|^p\, dA(s) \right] = 0 \quad \text{for every } t \in \mathbb{R}_+,$$

is an equivalence relation. Let $\mathbf{L}_{p,\infty}(\mathbb{R}_+ \times \Omega, \mu_A, P)$ be the collection of the equivalence classes of all measurable processes on $(\Omega, \mathfrak{F}, P)$ satisfying condition (1) with respect to this

§11. L_2-MARTINGALES AND QUADRATIC VARIATION PROCESSES

equivalence relation. From the fact that $|a+b|^p = 2^p\{|a|^p + |b|^p\}$ for $a, b \in \mathbb{R}$, it follows immediately that if $X, Y \in \mathbf{L}_{p,\infty}(\mathbb{R}_+ \times \Omega, \mu_A, P)$ then $X + Y \in \mathbf{L}_{p,\infty}(\mathbb{R}_+ \times \Omega, \mu_A, P)$. Clearly if $X \in \mathbf{L}_{p,\infty}(\mathbb{R}_+ \times \Omega, \mu_A, P)$ and $c \in \mathbb{R}$ then $cX \in \mathbf{L}_{p,\infty}(\mathbb{R}_+ \times \Omega, \mu_A, P)$. Thus $\mathbf{L}_{p,\infty}(\mathbb{R}_+ \times \Omega, \mu_A, P)$ is a linear space.

Let us observe that condition (1) is equivalent to

$$(3) \qquad \mathbf{E}\left[\int_{[0,m]} |X(s)|^p \, dA(s)\right] < \infty \quad \text{for every } m \in \mathbb{N}.$$

Condition (3) implies that for every $m \in \mathbb{N}$, there exists a null set Λ_m in $(\Omega, \mathfrak{F}, P)$ such that $\int_{[0,m]} |X(s,\omega)|^p \, dA(s,\omega) < \infty$ for $\omega \in \Lambda_m^c$. Then $\Lambda = \cup_{m \in \mathbb{N}} \Lambda_m$ is a null set in $(\Omega, \mathfrak{F}, P)$ such that for every $\omega \in \Lambda^c$ we have

$$(4) \qquad \int_{[0,t]} |X(s,\omega)|^p \, dA(s,\omega) < \infty \quad \text{for every } t \in \mathbb{R}_+.$$

The element $0 \in \mathbf{L}_{p,\infty}(\mathbb{R}_+ \times \Omega, \mu_A, P)$ is the equivalence class of measurable processes X on $(\Omega, \mathfrak{F}, P)$ such that

$$(5) \qquad \mathbf{E}\left[\int_{[0,t]} |X(s)|^p \, dA(s)\right] = 0 \quad \text{for every } t \in \mathbb{R}_+.$$

Note that when a measurable process X satisfies condition (5) then by the same argument as in (4) there exists a null set Λ in $(\Omega, \mathfrak{F}, P)$ such that for every $\omega \in \Lambda^c$ we have

$$(6) \qquad \int_{[0,t]} |X(s,\omega)|^p \, dA(s,\omega) = 0 \quad \text{for every } t \in \mathbb{R}_+.$$

Conversely if there exists a null set Λ in $(\Omega, \mathfrak{F}, P)$ such that for every $\omega \in \Lambda^c$ condition (6) holds, then (5) holds. Thus (5) and (6) are equivalent. Note that (6) does not imply that $X(\cdot, \omega) = 0$ for $\omega \in \Lambda^c$ since we may have $A(\cdot, \omega) = 0$. Thus the equivalence condition (2) is less stringent than the equivalence condition that $X(\cdot, \omega) = Y(\cdot, \omega)$ for $\omega \in \Lambda^c$ where Λ is a null set in $(\Omega, \mathfrak{F}, P)$.

Definition 11.17. *For $A \in \mathbf{A}(\Omega, \mathfrak{F}, \{\mathfrak{F}_t\}, P)$ and $p \in [0, \infty)$, let $\mathbf{L}_{p,\infty}(\mathbb{R}_+ \times \Omega, \mu_A, P)$ be the linear space of the equivalence classes of all measurable processes $X = \{X_t : t \in \mathbb{R}_+\}$ on $(\Omega, \mathfrak{F}, P)$ satisfying the condition*

$$(1) \qquad \mathbf{E}\left[\int_{[0,t]} |X(s)|^p \, dA(s)\right] < \infty \quad \textit{for every } t \in \mathbb{R}_+$$

For $X \in \mathbf{L}_{p,\infty}(\mathbb{R}_+ \times \Omega, \mu_A, P)$, define for every $t \in \mathbb{R}_+$

(2) $$\|X\|_{p,t}^{A,P} = \mathbf{E}\left[\int_{[0,t]} |X(s)|^p \, dA(s)\right]^{1/p}$$

and

(3) $$\|X\|_{p,\infty}^{A,P} = \sum_{m \in \mathbb{N}} 2^{-m} \{\|X\|_{p,m}^{A,P} \wedge 1\}.$$

Remark 11.18. For $A \in \mathbf{A}(\Omega, \mathfrak{F}, \{\mathfrak{F}_t\}, P)$, the functions $\|\cdot\|_{p,t}^{A,P}$ for $t \in \mathbb{R}_+$ and $\|\cdot\|_{p,\infty}^{A,P}$ on $\mathbf{L}_{p,\infty}(\mathbb{R}_+ \times \Omega, \mu_A, P)$ defined above have the following properties.
1) $\|\cdot\|_{p,t}^{A,P}$ is a seminorm on $\mathbf{L}_{p,\infty}(\mathbb{R}_+ \times \Omega, \mu_A, P)$ for every $t \in \mathbb{R}_+$.
2) For $X, X^{(n)} \in \mathbf{L}_{p,\infty}(\mathbb{R}_+ \times \Omega, \mu_A, P)$, $n \in \mathbb{N}$, we have $\lim_{n \to \infty} \|X^{(n)} - X\|_{p,\infty}^{A,P} = 0$ if and only if $\lim_{n \to \infty} \|X^{(n)} - X\|_{p,m}^{A,P} = 0$ for every $m \in \mathbb{N}$.
3) $\|\cdot\|_{p,\infty}^{A,P}$ is a quasinorm on $\mathbf{L}_{2,\infty}(\mathbb{R}_+ \times \Omega, \mu_A, P)$.

Proof. These statements are proved in the same way as Remark 11.5 for the seminorm $|\cdot|_t$ and the quasinorm $|\cdot|_\infty$ on the space $\mathbf{M}_2(\Omega, \mathfrak{F}, \{\mathfrak{F}_t\}, P)$. ∎

Observation 11.19. Let $A \in \mathbf{A}(\Omega, \mathfrak{F}, \{\mathfrak{F}_t\}, P)$ and $p \in [0, \infty)$. Then every bounded measurable process $X = \{X_t : t \in \mathbb{R}_+\}$ on $(\Omega, \mathfrak{F}, P)$ is in $\mathbf{L}_{p,\infty}(\mathbb{R}_+ \times \Omega, \mu_A, P)$.

Proof. Let X be a measurable process on $(\Omega, \mathfrak{F}, P)$ such that $|X(t,\omega)| \leq K$ for $(t,\omega) \in \mathbb{R}_+ \times \Omega$ for some $K \geq 0$. Then for any $(t,\omega) \in \mathbb{R}_+ \times \Omega$, we have

$$\int_{[0,t]} |X(s,\omega)|^p \mu_A(ds,\omega) \leq K^p \mu_A([0,t],\omega) = K^p A_t(\omega),$$

for a.e. ω in $(\Omega, \mathfrak{F}_\infty, P)$ so that

$$\mathbf{E}\left[\int_{[0,t]} |X(s)|^p \, dA(s)\right] \leq K^p \mathbf{E}(A_t) < \infty.$$

Thus $X \in \mathbf{L}_{p,\infty}(\mathbb{R}_+ \times \Omega, \mu_A, P)$. ∎

[IV] Quadratic Variation Processes

Proposition 11.20. Let $M \in \mathbf{M}_2(\Omega, \mathfrak{F}, \{\mathfrak{F}_t\}, P)$. There exists an equivalence class $A \in \mathbf{A}(\Omega, \mathfrak{F}, \{\mathfrak{F}_t\}, P)$ such that $M^2 - A$ is a right-continuous null at 0 martingale. Moreover such A can be chosen to be natural and under this condition A is unique.

§11. L_2-MARTINGALES AND QUADRATIC VARIATION PROCESSES

Proof. Since M is a right-continuous L_2-martingale, M^2 is a right-continuous nonnegative submartingale. Since our filtered space is right-continuous, M^2 belongs to the class (DL) by Theorem 8.22. Therefore by Theorem 10.23 (Doob-Meyer Decomposition), there exists a unique natural almost surely increasing process A such that $M^2 - A$ is a right-continuous martingale. Since both M and A are null at 0, so is $M^2 - A$. ∎

Lemma 11.21. *Let M be a right-continuous martingale and A be a natural almost surely increasing process on an augmented right-continuous filtered space $(\Omega, \mathfrak{F}, \{\mathfrak{F}_t\}, P)$. If $M + A$ is a natural almost surely increasing process, then $M = 0$.*

Proof. Since M is a right-continuous martingale on a right-continuous filtered space, it belongs to the class (DL) by Theorem 8.22. Since A is an almost surely increasing process, it is a submartingale of class (DL) by Lemma 10.19. Thus $M + A$ is a right-continuous submartingale of class (DL) on an augmented right-continuous filtered space and therefore by Theorem 10.23 (Doob-Meyer Decomposition) $M + A = M' + A'$ where M' is a right-continuous martingale and A' is a natural almost surely increasing process and furthermore the decomposition is unique. Now by assumption $M + A$ is itself a natural almost surely increasing process. Thus by the uniqueness of the decomposition, we have $M + A = A'$ and $M' = 0$. On the other hand, $M + A$ is a Doob-Meyer decomposition of $M' + A'$. Thus $M = M' = 0$. ∎

Proposition 11.22. *Let $M, N \in \mathbf{M}_2(\Omega, \mathfrak{F}, \{\mathfrak{F}_t\}, P)$. There exists an equivalence class $V \in \mathbf{V}(\Omega, \mathfrak{F}, \{\mathfrak{F}_t\}, P)$ such that $MN - V$ is a right-continuous null at 0 martingale. Moreover such V can be chosen to be natural and under this condition V is unique.*

Proof. Let $M' = (M + N)/2$ and $M'' = (M - N)/2$. Then both M' and M'' are in $\mathbf{M}_2(\Omega, \mathfrak{F}, \{\mathfrak{F}_t\}, P)$ so that by Proposition 11.20 there exist natural almost surely increasing processes A' and A'' such that $(M')^2 - A'$ and $(M'')^2 - A''$ are right-continuous null at 0 martingales. Now $MN = (M')^2 - (M'')^2$ so that

$$(1) \qquad MN - (A' - A'') = \{(M')^2 - A'\} - \{(M'')^2 - A''\}.$$

Since $A' - A''$ is a natural almost surely locally bounded variation process and $\{(M')^2 - A'\} - \{(M'')^2 - A''\}$ is a right-continuous null at 0 martingale, the equality (1) proves the existence of a natural almost surely locally bounded variation process V such that $MN - V$ is a right continuous null at 0 martingale.

To prove the uniqueness, suppose V' and V'' are two natural almost surely locally bounded variation process such that $MN - V'$ and $MN - V''$ are right-continuous null at

0 martingales. Then $V' - V'' = (MN - V'') - (MN - V')$ is a right-continuous null at 0 martingale. On the other hand, since V' and V'' are natural almost surely locally bounded variation process, so is $V' - V''$. Thus $V' - V'' = A' - A''$ where A' and A'' are two natural almost surely increasing processes. Now $(V' - V'') + A'' = A'$ so that the sum of the right-continuous martingale $V' - V''$ and the natural almost surely increasing process A'' is equal to a natural almost surely increasing process A'. Then by Lemma 11.21, $V' - V'' = 0$, that is, $V' = V''$. ∎

Definition 11.23. *Let $M \in \mathbf{M}_2(\Omega, \mathfrak{F}, \{\mathfrak{F}_t\}, P)$. We write $[M]$ for $A \in \mathbf{A}(\Omega, \mathfrak{F}, \{\mathfrak{F}_t\}, P)$ such that $M^2 - A$ is a right-continuous null at 0 martingale and call $[M]$ the quadratic variation process of M. For $M, N \in \mathbf{M}_2(\Omega, \mathfrak{F}, \{\mathfrak{F}_t\}, P)$, we write $[M, N]$ for $V \in \mathbf{V}(\Omega, \mathfrak{F}, \{\mathfrak{F}_t\}, P)$ such that $MN - V$ is a right-continuous null at 0 martingale and call $[M, N]$ the quadratic variation process of M and N.*

In Definition 11.23, the existence of $[M]$ and $[M, N]$ and their uniqueness under the condition that they be natural have been proved in Proposition 11.20 and Proposition 11.22. In what follows we write $[M]$ for both an equivalence class of processes and an arbitrary representatives of an equivalence class. Similarly for $[M, N]$.

Remark 11.24. Let M and N be right-continuous L_2-martingales on a standard filtered space $(\Omega, \mathfrak{F}, \{\mathfrak{F}_t\}, P)$ which may not be null at 0. Then $M - M_0$ and $N - N_0$ are in $\mathbf{M}_2(\Omega, \mathfrak{F}, \{\mathfrak{F}_t\}, P)$ so that there exists $X \equiv [M - M_0, N - N_0]$ in $\mathbf{V}(\Omega, \mathfrak{F}, \{\mathfrak{F}_t\}, P)$ such that $(M - M_0)(N - N_0) - X$ is a null at 0 martingale. Let $Y = M_0 N_0 + X$. Then $MN - Y$ is a null at 0 martingale.

Proof. Since X is a null at 0 process, so is $MN - Y = MN - M_0 N_0 - X$. Since $(M - M_0)(N - N_0) - X$ is a martingale, to show that $MN - Y$ is a martingale it suffices to show that $\{MN - Y\} - \{(M - M_0)(N - N_0) - X\}$ is a martingale. Now

$$\{MN - Y\} - \{(M - M_0)(N - N_0) - X\}$$
$$= (MN - M_0 N_0) - (M - M_0)(N - N_0)$$
$$= N_0 M + M_0 N.$$

Since M and N are L_2-processes, $N_0 M$ and $M_0 N$ are L_1 processes and in fact martingales and so is the sum. ∎

The family $\mu_{[M]}$ of Lebesgue-Stieltjes measures on $(\mathbb{R}_+, \mathfrak{B}_{\mathbb{R}_+})$ determined by

§11. L_2-MARTINGALES AND QUADRATIC VARIATION PROCESSES

$[M] \in \mathbf{A}(\Omega, \mathfrak{F}, \{\mathfrak{F}_t\}, P)$ and the family of signed Lebesgue-Stieltjes measures $\mu_{[M,N]}$ on $([0, t], \mathfrak{B}_{[0,t]})$ for $t \in \mathbb{R}_+$ determined by $[M, N] \in \mathbf{V}(\Omega, \mathfrak{F}, \{\mathfrak{F}_t\}, P)$ are defined. The space $\mathbf{L}_{2,\infty}(\mathbb{R}_+ \times \Omega, \mu_{[M]}, P)$ is defined by Definition 11.17. As an immediate consequence of the Definition 11.23, we have the following lemma.

Lemma 11.25. *Let $M, N \in \mathbf{M}_2(\Omega, \mathfrak{F}, \{\mathfrak{F}_t\}, P)$. Then for $s, t \in \mathbb{R}_+$, $s < t$, we have*

(1) $$\mathbf{E}[(M_t - M_s)(N_t - N_s)|\mathfrak{F}_s] = \mathbf{E}[M_t N_t - M_s N_s | \mathfrak{F}_s]$$
$$= \mathbf{E}[[M, N]_t - [M, N]_s | \mathfrak{F}_s],$$

and in particular

(2) $$\mathbf{E}[(M_t - M_s)^2|\mathfrak{F}_s] = \mathbf{E}[M_t^2 - M_s^2|\mathfrak{F}_s] = \mathbf{E}[[M]_t - [M]_s|\mathfrak{F}_s],$$

a.e. on $(\Omega, \mathfrak{F}_s, P)$.

Proof. To prove the first equality in (1), note that by the martingale property of M and N we have

$$\mathbf{E}[(M_t - M_s)(N_t - N_s)|\mathfrak{F}_s] = \mathbf{E}[M_t N_t - M_s N_t - M_t N_s + M_s N_s | \mathfrak{F}_s]$$
$$= \mathbf{E}[M_t N_t - M_s N_s | \mathfrak{F}_s] \quad \text{a.e. on } (\Omega, \mathfrak{F}_s, P).$$

To prove the second equality in (1), note that since $MN - [M, N]$ is a martingale, we have

$$\mathbf{E}[\{M_t N_t - [M, N]_t\} - \{M_s N_s - [M, N]_s\}|\mathfrak{F}_s] = 0 \quad \text{a.e. on } (\Omega, \mathfrak{F}_s, P).$$

This proves the second equality in (1). ∎

Quadratic variation processes have the following algebraic properties.

Proposition 11.26. *Let $M, M', M'', N \in \mathbf{M}_2(\Omega, \mathfrak{F}, \{\mathfrak{F}_t\}, P)$. Then*
(1) $[M] = 0 \Leftrightarrow M = 0$,
(2) $[M, M] = [M]$,
(3) $[M, N] = [N, M]$,
(4) $[c'M' + c''M'', N] = c'[M', N] + c''[M'', N]$ for $c', c'' \in \mathbb{R}$,
(5) $[cM] = c^2[M]$ for $c \in \mathbb{R}$,
(6) $[M, N] = 4^{-1}[M + N] - 4^{-1}[M - N]$.

Proof. 1) If $M = 0$, then clearly 0 is a quadratic variation process of M. Conversely suppose $[M] = 0$. By the definition of $[M]$, $X \equiv M^2 - [M]$ is a right-continuous null at 0 martingale. Thus when $[M] = 0$ then for any $t \in \mathbb{R}_+$ we have

$$\mathbf{E}(M_t^2) = \mathbf{E}(X_t) + \mathbf{E}([M]_t) = \mathbf{E}(X_0) + \mathbf{E}(0) = 0,$$

and therefore there exists a null set Λ_t in $(\Omega, \mathfrak{F}, P)$ such that $M_t = 0$ on Λ_t^c. Let $\{r_m : m \in \mathbb{N}\}$ be the collection of all rational numbers in $\mathbb{R}_+$ and let $\Lambda = \cup_{m \in \mathbb{N}} \Lambda_m$. Then for the null set Λ we have $M(r_m, \omega) = 0$ for all $m \in \mathbb{N}$ when $\omega \in \Lambda^c$. Then by the right-continuity of $M(\cdot, \omega)$ we have $M(t, \omega) = 0$ for all $t \in \mathbb{R}_+$ when $\omega \in \Lambda^c$.

2) Since an almost surely increasing process is a particular case of an almost surely locally bounded variation process, we have (2).

3) (3) follows from the fact that $MN = NM$.

4) (4) follows from the definition of $[\cdot, \cdot]$ and the fact that we have $c'M' + c''M'' \in \mathbf{M}_2(\Omega, \mathfrak{F}, \{\mathfrak{F}_t\}, P)$. (5) is a particular case of (4).

5) With A', A'', M', M'' as defined in the Proof of Proposition 11.22, we have

$$[M, N] = A' - A'' = [M'] - [M'']$$
$$= [2^{-1}(M + N)] - [2^{-1}(M - N)] = 4^{-1}[M + N] - 4^{-1}[M - N],$$

proving (6). ∎

Let us consider the continuity of a quadratic variation process. Recall that if $M \in \mathbf{M}_2(\Omega, \mathfrak{F}, \{\mathfrak{F}_t\}, P)$ then M^2 is a right-continuous submartingale of class (DL) by Theorem 8.22. Then according to Theorem 10.35, under the assumption that the filtered space is a standard filtered space, $[M]$ is almost surely continuous if and only if M^2 is a regular submartingale. In Proposition 11.27 and Proposition 11.30, we give sufficient conditions for M^2 to be regular.

Proposition 11.27. *Let $M, N \in \mathbf{M}_2^c(\Omega, \mathfrak{F}, \{\mathfrak{F}_t\}, P)$. Then M^2 and N^2 are regular submartingales and $[M]$, $[N]$ and $[M, N]$ are almost surely continuous.*

Proof. M^2 is a right-continuous submartingale of class (DL) by Theorem 8.22. It is also a continuous nonnegative submartingale on a right-continuous filtered space so that it is a regular submartingale by Observation 10.28. Therefore $[M]$ is almost surely continuous by Theorem 10.35. Similarly for $[N]$. Now since both $2^{-1}(M + N)$ and $2^{-1}(M - N)$ are in $\mathbf{M}_2^c(\Omega, \mathfrak{F}, \{\mathfrak{F}_t\}, P)$, both $[2^{-1}(M + N)]$ and $[2^{-1}(M - N)]$ are almost surely continuous by our result above. Then by (6) of Proposition 11.26, $[M, N]$ is almost surely continuous. ∎

§11. L_2-MARTINGALES AND QUADRATIC VARIATION PROCESSES

Definition 11.28. *A filtration $\{\mathfrak{F}_t : t \in \mathbb{R}_+\}$ on a probability space $(\Omega, \mathfrak{F}, \{\mathfrak{F}_t\}, P)$ is said to have no time of discontinuity if for any stopping times T_n, $n \in \mathbb{N}$, and T with respect to the filtration such that $T_n \uparrow T$ on Ω as $n \to \infty$ we have $\sigma(\cup_{n \in \mathbb{N}} \mathfrak{F}_{T_n}) = \mathfrak{F}_T$.*

Lemma 11.29. *Let $M = \{M_t : t \in \mathbb{R}_+\}$ be an L_2-martingale on a filtered space $(\Omega, \mathfrak{F}, \{\mathfrak{F}_t\}, P)$ in which the filtration has no time of discontinuity. Then the submartingale M^2 is regular.*

Proof. Let $a > 0$ and let T_n, $n \in \mathbb{N}$, and T be stopping times bounded by a such that $T_n \uparrow T$ on Ω as $n \to \infty$. To show the regularity of M^2 we verify

(1) $$\lim_{n \to \infty} \mathbf{E}(M_{T_n}^2) = \mathbf{E}(M_T^2).$$

Let $X_n = \mathbf{E}[M_a | \mathfrak{F}_{T_n}]$ for $n \in \mathbb{N}$ and let $X_\infty = \mathbf{E}[M_a | \mathfrak{G}_\infty]$ where $\mathfrak{G}_\infty = \sigma(\cup_{n \in \mathbb{N}} \mathfrak{F}_{T_n})$. According to Remark 8.3, $\{X_n : n \in \mathbb{N}\}$ is a uniformly integrable martingale with respect to the filtration $\{\mathfrak{F}_{T_n} : n \in \mathbb{N}\}$ having X_∞ as a final element and in particular X_n converges to X_∞ both almost surely and in L_1. Since M is a martingale we have $X_n = M_{T_n}$ by Theorem 6.4 (Doob's Optional Sampling Theorem with Bounded Stopping Times). Since the filtration has no time of discontinuity we have $\mathfrak{G}_\infty = \mathfrak{F}_T$ and thus $X_\infty = M_T$. Thus M_{T_n} converges to M_T almost surely. Thus by Fatou's Lemma

(2) $$\mathbf{E}(M_T^2) = \mathbf{E}(\lim_{n \to \infty} M_{T_n}^2) \leq \liminf_{n \to \infty} \mathbf{E}(M_{T_n}^2).$$

on the other hand since $\{X_n; n \in \mathbb{N}\}$ is a martingale with X_∞ as a final element, $\{X_n^2; n \in \mathbb{N}\}$ is a submartingale with X_∞^2 as a final element. Thus $\mathbf{E}(M_{T_n}^2) \uparrow$ as $n \to \infty$ and $\mathbf{E}(M_{T_n}^2) \leq \mathbf{E}(M_T^2)$. Thus (1) follows from (2). ∎

Proposition 11.30 *Let $M, N \in \mathbf{M}_2(\Omega, \mathfrak{F}, \{\mathfrak{F}_t\}, P)$ where the filtered space is a standard filtered space with no time of discontinuity. Then M^2 and N^2 are regular submartingales and $[M]$, $[N]$ and $[M, N]$ are almost surely continuous.*

Proof. By Lemma 11.29, M^2 and N^2 are regular right-continuous submartingales of class (DL). Thus by Theorem 10.35, $[M]$ and $[N]$ are almost surely continuous. Since $2^{-1}(M+N)$ and $2^{-1}(M-N)$ are in $\mathbf{M}_2(\Omega, \mathfrak{F}, \{\mathfrak{F}_t\}, P)$, $[2^{-1}(M+N)]$ and $[2^{-1}(M-N)]$ are almost surely continuous by than same reason as above. Then by (5) of Proposition 11.26, $[M, N]$ is almost surely continuous. ∎

For $M, N \in \mathbf{M}_2(\Omega, \mathfrak{F}, \{\mathfrak{F}_t\}, P)$, the quadratic variation processes $[M]$ and $[M, N]$ are in $\mathbf{A}(\Omega, \mathfrak{F}, \{\mathfrak{F}_t\}, P)$ and $\mathbf{V}(\Omega, \mathfrak{F}, \{\mathfrak{F}_t\}, P)$ respectively and thus $|[M, N]|$ is in

$\mathbf{A}(\Omega, \mathfrak{F}, \{\mathfrak{F}_t\}, P)$ by Lemma 11.11 and the families of Lebesgue-Stieltjes measures $\mu_{[M]}$ and $\mu_{[M,N]}$ are defined on $(\mathbb{R}_+, \mathfrak{B}_{\mathbb{R}_+})$ and the family of signed Lebesgue-Stieltjes measures $\mu_{[M,N]}$ is defined on $([0,t], \mathfrak{B}_{[0,t]})$ for every $t \in \mathbb{R}_+$ according to Definition 11.14. For a real valued function X on $\mathbb{R}_+ \times \Omega$ such that $X(\cdot, \omega)$ is a Borel measurable function on $\mathbb{R}_+$, we define for every $t \in \mathbb{R}_+$ and $\omega \in \Omega$,

$$\int_{[0,t]} X(s,\omega)\,d[M](s,\omega) = \int_{[0,t]} X(s,\omega)\mu_{[M]}(ds,\omega)$$

and

$$\int_{[0,t]} X(s,\omega)\,d[M,N](s,\omega) = \int_{[0,t]} X(s,\omega)\mu_{[M,N]}(ds,\omega)$$

provided the integral with respect to the signed Lebesgue-Stieltjes measure on the right sides exist. Note that by (5) of Proposition 11.26, we have

$$\mu_{[M]} = 4^{-1}\mu_{[M+N]} + 4^{-1}\mu_{[M-N]}$$

on $([0,t], \mathfrak{B}_{[0,t]})$.

Theorem 11.31. *Let* $A \in \mathbf{A}^c(\Omega, \mathfrak{F}, \{\mathfrak{F}_t\}, P)$. *For* $t \in \mathbb{R}_+$ *and* $n \in \mathbb{N}$, *let* Δ_n *be the partition of* $[0,t]$ *by* $0 = t_{n,0} < \cdots < t_{n,p_n} = t$. *Let* $|\Delta_n| = \max_{k=1,\ldots,p_n} (t_{n,k} - t_{n,k-1})$ *and* $\lim_{n \to \infty} |\Delta_n| = 0$. *Then*

(1) $$\lim_{n \to \infty} \Big\| \sum_{k=1}^{p_n} \mathbf{E}[A_{t_{n,k}} - A_{t_{n,k-1}} | \mathfrak{F}_{t_{n,k-1}}] - A_t \Big\|_1 = 0.$$

If $A_t \in L_2(\Omega, \mathfrak{F}, P)$, *then*

(2) $$\lim_{n \to \infty} \Big\| \sum_{k=1}^{p_n} \mathbf{E}[A_{t_{n,k}} - A_{t_{n,k-1}} | \mathfrak{F}_{t_{n,k-1}}] - A_t \Big\|_2 = 0.$$

Proof. Let us prove (2) first. For brevity let us write

(3) $$\begin{cases} \alpha_{n,k} = A_{t_{n,k}} - A_{t_{n,k-1}} & \text{for } k = 1,\ldots,p_n \\ \beta_{n,k} = \mathbf{E}[\alpha_{n,k} | \mathfrak{F}_{t_{n,k-1}}] & \text{for } k = 1,\ldots,p_n \\ S_n = \sum_{k=1}^{p_n} \beta_{n,k} & \text{for } n \in \mathbb{N}. \end{cases}$$

Then

(4) $$\begin{aligned} \mathbf{E}[\{S_n - A_t\}^2] &= \mathbf{E}[\{\sum_{k=1}^{p_n}(\alpha_{n,k} - \beta_{n,k})\}^2] \\ &= \sum_{j,k=1,\ldots,p_n} \mathbf{E}[(\alpha_{n,j} - \beta_{n,j})(\alpha_{n,k} - \beta_{n,k})]. \end{aligned}$$

§11. L_2-MARTINGALES AND QUADRATIC VARIATION PROCESSES

Now for $j < k$ we have

$$\mathbf{E}[(\alpha_{n,j} - \beta_{n,j})(\alpha_{n,k} - \beta_{n,k})] = \mathbf{E}[\mathbf{E}[(\alpha_{n,j} - \beta_{n,j})(\alpha_{n,k} - \beta_{n,k})|\mathfrak{F}_{t_{n,k-1}}]]$$

where the conditional expectation is equal to

$$\mathbf{E}[\alpha_{n,j}\alpha_{n,k} - \alpha_{n,j}\beta_{n,k} - \beta_{n,j}\alpha_{n,k} + \beta_{n,j}\beta_{n,k}|\mathfrak{F}_{t_{n,k-1}}]$$
$$= \alpha_{n,j}\mathbf{E}[\alpha_{n,k}|\mathfrak{F}_{t_{n,k-1}}] - \alpha_{n,j}\mathbf{E}[\beta_{n,k}|\mathfrak{F}_{t_{n,k-1}}]$$
$$- \beta_{n,j}\mathbf{E}[\alpha_{n,k}|\mathfrak{F}_{t_{n,k-1}}] + \beta_{n,j}\mathbf{E}[\beta_{n,k}|\mathfrak{F}_{t_{n,k-1}}]$$
$$= 0 \quad \text{a.e. on } (\Omega, \mathfrak{F}_{t_{n,k-1}}, P),$$

since $\mathbf{E}[\beta_{n,k}|\mathfrak{F}_{t_{n,k-1}}] = \mathbf{E}[\alpha_{n,k}|\mathfrak{F}_{t_{n,k-1}}]$. Thus the expectations of the cross products in (4) are all equal to 0 and consequently

$$\mathbf{E}[\{S_n - A_t\}^2] = \mathbf{E}[\sum_{k=1}^{p_n}(\alpha_{n,k} - \beta_{n,k})^2]$$
$$= \sum_{k=1}^{p_n}\{\mathbf{E}[\alpha_{n,k}^2] - 2\mathbf{E}[\alpha_{n,k}\beta_{n,k}] + \mathbf{E}[\beta_{n,k}^2]\}.$$

Now

$$\mathbf{E}[\alpha_{n,k}\beta_{n,k}] = \mathbf{E}[\mathbf{E}[\alpha_{n,k}\beta_{n,k}|\mathfrak{F}_{t_{n,k-1}}]] = \mathbf{E}[\beta_{n,k}\mathbf{E}[\alpha_{n,k}|\mathfrak{F}_{t_{n,k-1}}]] = \mathbf{E}[\beta_{n,k}^2].$$

Thus we have

(5) $$\mathbf{E}[\{S_n - A_t\}^2] = \sum_{k=1}^{p_n}\{\mathbf{E}[\alpha_{n,k}^2] - \mathbf{E}[\beta_{n,k}^2]\} \leq \sum_{k=1}^{p_n}\mathbf{E}[\alpha_{n,k}^2]$$
$$\leq \mathbf{E}[\sum_{k=1}^{p_n}\{A_{t_{n,k}} - A_{t_{n,k-1}}\}^2] \leq \mathbf{E}[\sup_{k=1,\ldots,p_n}\{A_{t_{n,k}} - A_{t_{n,k-1}}\}A_t].$$

Now $\lim_{n\to\infty}\sup_{k=1,\ldots,p_n}\{A_{t_{n,k}} - A_{t_{n,k-1}}\} = 0$ a. e. in $(\Omega, \mathfrak{F}, P)$ since $A(\cdot, \omega)$ is uniformly continuous on $[0, t]$ for almost every $\omega \in \Omega$. Also since almost every sample function of A is an increasing function, we have $\sup_{k=1,\ldots,p_n}\{A_{t_{n,k}} - A_{t_{n,k-1}}\}A_t \leq A_t^2$ a.e. in $(\Omega, \mathfrak{F}, P)$. Then since $A_t \in L_2(\Omega, \mathfrak{F}, P)$, the limit as $n \to \infty$ of the last member in (5) is equal to 0 by the Dominated Convergence Theorem. Thus $\lim_{n\to\infty}\mathbf{E}[\{S_n - A_t\}^2] = 0$. This proves (2).

Consider now the general case where $A_t \in L_1(\Omega, \mathfrak{F}, P)$. For $m \in \mathbb{N}$, let $A^{(m)} = A \wedge m$. Then both $A^{(m)}$ and $A - A^{(m)}$ are in $\mathbf{A}^c(\Omega, \mathfrak{F}, \{\mathfrak{F}_t\}, P)$ and $A^{(m)}$ is bounded by m. Let us write $A = A^{(m)} + \{A - A^{(m)}\}$ and let

(6) $$\begin{cases} S'_n = \sum_{k=1}^{p_n}\mathbf{E}[A^{(m)}_{t_{n,k}} - A^{(m)}_{t_{n,k-1}}|\mathfrak{F}_{t_{n,k-1}}] & \text{for } n \in \mathbb{N} \\ S''_n = \sum_{k=1}^{p_n}\mathbf{E}[(A - A^{(m)})_{t_{n,k}} - (A - A^{(m)})_{t_{n,k-1}}|\mathfrak{F}_{t_{n,k-1}}] & \text{for } n \in \mathbb{N}. \end{cases}$$

Then $A_t = A_t^{(m)} + (A - A^{(m)})_t$ and $S_n = S_n' + S_n''$. Thus

(7) $$\mathbf{E}(|S_n - A_t|) \leq \mathbf{E}(|S_n' - A_t^{(m)}|) + \mathbf{E}(S_n'') + \mathbf{E}[(A - A^{(m)})_t]$$

since $(A - A^{(m)})_t \geq 0$ and $S_n'' \geq 0$ a.e. Now since $A^{(m)} \in \mathbf{A}^c(\Omega, \mathfrak{F}, \{\mathfrak{F}_t\}, P)$ and $\mathbf{E}[(A_t^{(m)})^2] \leq m^2 < \infty$ we have $\lim_{n \to \infty} \mathbf{E}[\{S_n' - A_t^{(m)}\}^2] = 0$ by (2) and therefore we have $\lim_{n \to \infty} \mathbf{E}(|S_n' - A_t^{(m)}|) = 0$. Note that the second and third expectations on the right side of (7) are equal since by (6) we have

$$\mathbf{E}(S_n'') = \sum_{k=1}^{p_n} \mathbf{E}[(A - A^{(m)})_{t_{n,k}} - (A - A^{(m)})_{t_{n,k-1}}] = \mathbf{E}[(A - A^{(m)})_t].$$

Now
$$\mathbf{E}[(A - A^{(m)})_t] = \int_{\{A_t > m\}} \{A_t - A_t^{(m)}\} \, dP \leq \int_{\{A_t > m\}} A_t \, dP.$$

Let $\varepsilon > 0$ be arbitrarily given. Since A_t is integrable, we have $\int_{\{A_t > m\}} A_t \, dP < \varepsilon$ for sufficiently large $m \in \mathbb{N}$. For such m we have $\mathbf{E}(|S_n - A_t|) \leq \mathbf{E}(|S_n' - A_t^{(m)}|) + 2\varepsilon$ by (7). Then we have $\limsup_{n \to \infty} \mathbf{E}(|S_n - A_t|) \leq 2\varepsilon$. From the arbitrariness of $\varepsilon > 0$ we have $\lim_{n \to \infty} \mathbf{E}(|S_n - A_t|) = 0$. This proves (1). ∎

Theorem 11.32. *Let $M, N \in \mathbf{M}_2^c(\Omega, \mathfrak{F}, \{\mathfrak{F}_t\}, P)$. For $t \in \mathbb{R}_+$ and $n \in \mathbb{N}$, let Δ_n be the partition of $[0, t]$ by $0 = t_{n,0} < \cdots < t_{n,p_n} = t$. Let $|\Delta_n| = \max_{k=1,\ldots,p_n} (t_{n,k} - t_{n,k-1})$ and $\lim_{n \to \infty} |\Delta_n| = 0$. Then*

(1) $$\lim_{n \to \infty} \| \sum_{k=1}^{p_n} \mathbf{E}[\{M_{t_{n,k}} - M_{t_{n,k-1}}\}\{N_{t_{n,k}} - N_{t_{n,k-1}}\} | \mathfrak{F}_{t_{n,k-1}}] - [M, N]_t \|_1 = 0.$$

and in particular

(2) $$\lim_{n \to \infty} \| \sum_{k=1}^{p_n} \mathbf{E}[\{M_{t_{n,k}} - M_{t_{n,k-1}}\}^2 | \mathfrak{F}_{t_{n,k-1}}] - [M]_t \|_1 = 0.$$

Proof. Since $M \in \mathbf{M}_2^c(\Omega, \mathfrak{F}, \{\mathfrak{F}_t\}, P)$ we have $[M] \in \mathbf{A}^c(\Omega, \mathfrak{F}, \{\mathfrak{F}_t\}, P)$ by Proposition 11.27. Applying (2) of Lemma 11.25 and identifying $[M]$ with A in Theorem 11.30 we have (2). To derive (1) from (2), note that for any $s, u \in \mathbb{R}_+$ such that $s < u$ we have

$$(M_u - M_s)(N_u - N_s)$$

$$= \frac{1}{4}[\{(M_u - M_s) + (N_u - N_s)\}^2 - \{(M_u - M_s) - (N_u - N_s)\}^2]$$
$$= \frac{1}{4}[\{(M+N)_u - (M+N)_s\}^2 - \{(M-N)_u - (M-N)_s\}^2].$$

Also by Proposition 11.26 we have

$$[M, N]_t = \frac{1}{4}\{[M+N]_t - [M-N]_t\}.$$

By these two equalities we have

$$\| \sum_{k=1}^{p_n} \mathbf{E}[\{M_{t_{n,k}} - M_{t_{n,k-1}}\}\{N_{t_{n,k}} - N_{t_{n,k-1}}\} | \mathfrak{F}_{t_{n,k-1}}] - [M,N]_t \|_1$$
$$\leq \frac{1}{4} \| \sum_{k=1}^{p_n} \mathbf{E}[\{(M+N)(t_{n,k}) - (M+N)(t_{n,k-1})\}^2 | \mathfrak{F}_{t_{n,k-1}}] - [M+N]_t \|_1$$
$$+ \frac{1}{4} \| \sum_{k=1}^{p_n} \mathbf{E}[\{(M-N)(t_{n,k}) - (M-N)(t_{n,k-1})\}^2 | \mathfrak{F}_{t_{n,k-1}}] - [M-N]_t \|_1.$$

By (1) the two terms on the right side of the last inequality tend to 0 as $n \to \infty$. This proves (1). ∎

In connection with Theorem 11.31, let us remark that we have also

$$\lim_{n \to \infty} \| \sum_{k=1}^{p_n} \{M_{t_{n,k}} - M_{t_{n,k-1}}\}\{N_{t_{n,k}} - N_{t_{n,k-1}}\} - [M,N]_t \|_1 = 0.$$

This equality will be proved in §12.

§12 Stochastic Integrals with Respect to Martingales

Let $X = \{X_t : t \in \mathbb{R}_+\}$ be a stochastic process and $M = \{M_t : t \in \mathbb{R}_+\}$ be a martingale on a filtered space $(\Omega, \mathfrak{F}, \{\mathfrak{F}_t\}, P)$. An integral of $X(\cdot, \omega)$ with respect to $M(\cdot, \omega)$ on an interval $[0, t]$, $\int_{[0,t]} X(s, \omega) \, dM(s, \omega)$ for $t \in \mathbb{R}_+$ and $\omega \in \Omega$, can not be defined as a Lebesgue-Stieltjes integral of $X(\cdot, \omega)$ with respect to a signed Lebesgue-Stieltjes measure on $([0,t], \mathfrak{B}_{[0,t]})$ since $M(\cdot, \omega)$ may not be a function of locally bounded variation on $\mathbb{R}_+$ and a corresponding signed Lebesgue-Stieltjes measure may not exist. If however the sample functions of X are step functions, then $\int_{[0,t]} X(s,\omega)\,dM(s,\omega)$ can be defined as a Riemann-Stieltjes sum of $X(\cdot,\omega)$ with respect to $M(\cdot,\omega)$ for every $\omega \in \Omega$.

We shall show that if X is a bounded adapted left-continuous simple process and M is a right-continuous L_2-martingale on a standard filtered space, then the Riemann-Stieltjes sum of X with respect to M is a right-continuous L_2-martingale on the filtered space. We then extend the definition to cover processes X which are predictable processes and satisfy the integrability conditions of the space $\mathbf{L}_{2,\infty}(\mathbb{R}_+ \times \Omega, \mu_{[M]}, P)$.

[I] Stochastic Integral of Bounded Left-Continuous Adapted Simple Processes with Respect to L_2-Martingales

Definition 12.1. *Let $\mathbf{L}_0(\Omega, \mathfrak{F}, \{\mathfrak{F}_t\}, P)$ be the collection of all bounded adapted left-continuous simple processes on a standard filtered space $(\Omega, \mathfrak{F}, \{\mathfrak{F}_t\}, P)$, that is, for every $X \in \mathbf{L}_0(\Omega, \mathfrak{F}, \{\mathfrak{F}_t\}, P)$ there exist a strictly increasing sequence $\{t_k : k \in \mathbb{Z}_+\}$ in $\mathbb{R}_+$ with $t_0 = 0$ and $\lim_{k \to \infty} t_k = \infty$ and a bounded sequence of real valued random variables $\{\xi_k : k \in \mathbb{Z}_+\}$, that is $|\xi_k(\omega)| \leq K$ for all $\omega \in \Omega$ and $k \in \mathbb{Z}_+$ for some $K \geq 0$, such that ξ_0 is $\mathfrak{F}_{t_0}$-measurable, ξ_k is $\mathfrak{F}_{t_{k-1}}$-measurable for $k \in \mathbb{N}$ and X is given as*

(1) $$\begin{cases} X(t,\omega) = \xi_k(\omega) & \text{for } t \in (t_{k-1}, t_k],\, k \in \mathbb{N},\, \omega \in \Omega \\ X(0,\omega) = \xi_0(\omega) & \text{for } \omega \in \Omega, \end{cases}$$

that is,

(2) $$X(t,\omega) = \xi_0(\omega)\mathbf{1}_{\{0\}}(t) + \sum_{k \in \mathbb{N}} \xi_k(\omega)\mathbf{1}_{(t_{k-1},t_k]}(t) \quad \text{for } (t,\omega) \in \mathbb{R}_+ \times \Omega.$$

Note that in the definition of $\mathbf{L}_0(\Omega, \mathfrak{F}, \{\mathfrak{F}_t\}, P)$, the probability measure P of the underlying standard filtered space $(\Omega, \mathfrak{F}, \{\mathfrak{F}_t\}, P)$ plays no role.

Observation 12.2. 1) Every $X \in \mathbf{L}_0(\Omega, \mathfrak{F}, \{\mathfrak{F}_t\}, P)$ is a predictable process.
2) $\mathbf{L}_0(\Omega, \mathfrak{F}, \{\mathfrak{F}_t\}, P) \subset \mathbf{L}_{2,\infty}(\mathbb{R}_+ \times \Omega, \mu_{[M]}, P)$ for every M in $\mathbf{M}_2(\Omega, \mathfrak{F}, \{\mathfrak{F}_t\}, P)$.

Proof. 1) Since X is a left-continuous adapted process on the filtered space, it is a predictable process on the filtered space.

2) If $X \in \mathbf{L}_0(\Omega, \mathfrak{F}, \{\mathfrak{F}_t\}, P)$, then X is a left-continuous process so that it is a measurable process on $(\Omega, \mathfrak{F}, P)$ by Theorem 2.10. Also for every $t \in \mathbb{R}_+$, say $t \in [t_{k-1}, t_k]$ for

§12. STOCHASTIC INTEGRALS WITH RESPECT TO MARTINGALES

some $k \in \mathbb{N}$, we have

$$\mathbf{E}\left[\int_{[0,t]} X^2(s)\,d[M](s)\right] \leq \mathbf{E}\left[\int_{[0,t_k]} X^2(s)\,d[M](s)\right]$$
$$= \mathbf{E}\left[\sum_{i=1}^{k} \xi_i^2\{[M]_{t_i} - [M]_{t_{i-1}}\}\right] \leq K^2 \mathbf{E}[[M]_{t_k}] < \infty,$$

where K bound of the sequence of random variables $\{\xi_k : k \in \mathbb{Z}_+\}$. This shows that $X \in \mathbf{L}_{2,\infty}(\mathbb{R}_+ \times \Omega, \mu_{[M]}, P)$. ∎

Definition 12.3. *Let* $M \in \mathbf{M}_2(\Omega, \mathfrak{F}, \{\mathfrak{F}_t\}, P)$. *For* $X \in \mathbf{L}_0(\Omega, \mathfrak{F}, \{\mathfrak{F}_t\}, P)$ *given by*

(1) $$X(t) = \xi_0 \mathbf{1}_{\{0\}}(t) + \sum_{k \in \mathbb{N}} \xi_k \mathbf{1}_{(t_{k-1}, t_k]}(t) \quad \text{for } t \in \mathbb{R}_+$$

as in Definition 12.1, we define a function $X \bullet M$ *on* $\mathbb{R}_+ \times \Omega$ *by setting*

(2) $$(X \bullet M)(t) = \sum_{i=1}^{k-1} \xi_i\{M(t_i) - M(t_{i-1})\} + \xi_k\{M(t) - M(t_{k-1})\}$$

for $t \in [t_{k-1}, t_k]$ *and* $k \in \mathbb{N}$ *with the understanding that* $\sum_{i=1}^{0} = 0$, *that is,*

(3) $$(X \bullet M)(t) = \sum_{i \in \mathbb{N}} \xi_i\{M(t_i \wedge t) - M(t_{i-1} \wedge t)\} \quad \text{for } t \in \mathbb{R}_+.$$

We call $X \bullet M$ *the stochastic integral of* X *with respect to* M.

Note that $X(0)$ and ξ_0 play no roles in the definition of $X \bullet M$ and

$$\begin{cases} (X \bullet M)(0) = \xi_1\{M(0) - M(0)\} = 0, \\ (X \bullet M)(t_k) = \sum_{i=1}^{k} \xi_i\{M(t_i) - M(t_{i-1})\} & \text{for } k \in \mathbb{N}. \end{cases}$$

Note also that $X \bullet M$ is well-defined, that is, it does not depend on the different ways an element X of $\mathbf{L}_0(\Omega, \mathfrak{F}, \{\mathfrak{F}_t\}, P)$ can be expressed in the form (1).

Proposition 12.4. *Let* $X \in \mathbf{L}_0(\Omega, \mathfrak{F}, \{\mathfrak{F}_t\}, P)$ *and* $M \in \mathbf{M}_2(\Omega, \mathfrak{F}, \{\mathfrak{F}_t\}, P)$ *Then for* $X \bullet M$ *we have*
1) $X \bullet M \in \mathbf{M}_2(\Omega, \mathfrak{F}, \{\mathfrak{F}_t\}, P)$.

2) $X \bullet M \in \mathbf{M}_2^c(\Omega, \mathfrak{F}, \{\mathfrak{F}_t\}, P)$ if $M \in \mathbf{M}_2^c(\Omega, \mathfrak{F}, \{\mathfrak{F}_t\}, P)$.
3) If $Y \in \mathbf{L}_0(\Omega, \mathfrak{F}, \{\mathfrak{F}_t\}, P)$ and $N \in \mathbf{M}_2(\Omega, \mathfrak{F}, \{\mathfrak{F}_t\}, P)$, then for every $t \in \mathbb{R}_+$, we have

(1) $$\mathbf{E}[(X \bullet M)_t (Y \bullet N)_t] = \mathbf{E}\left[\int_{[0,t]} X(s) Y(s)\, d[M,N](s)\right]$$

and in particular

(2) $$\|X \bullet M\|_t = \mathbf{E}\left[\int_{[0,t]} X^2(s)\, d[M](s)\right]^{1/2} = \|X\|_{2,t}^{[M],P}$$

and

(3) $$\|X \bullet M\|_\infty = \|X\|_{2,\infty}^{[M],P}.$$

Proof. By (2) of Definition 12.3, every sample function of $X \bullet M$ is right-continuous since M has this property. Also $(X \bullet M)(0) = 0$ on Ω. Since ξ_i is $\mathfrak{F}_{t_{i-1}}$-measurable for $i \in \mathbb{N}$, $(X \bullet M)(t)$ is $\mathfrak{F}_t$-measurable for every $t \in \mathbb{R}_+$, that is, $X \bullet M$ is an adapted process. Since M is an L_2-process and ξ_i is a bounded random variable for $i \in \mathbb{N}$, $X \bullet M$ is an L_2-process.

To show that $X \bullet M$ is a martingale, we show that for every pair $s, t \in \mathbb{R}_+$ such that $s < t$ we have

(4) $$\mathbf{E}[(X \bullet M)(t) - (X \bullet M)(s) | \mathfrak{F}_s] = 0 \quad \text{a.e. on } (\Omega, \mathfrak{F}_s, P).$$

We may add two more points t_k in the expression of X by (2) of Definition 12.1 if necessary so that $s = t_k$ and $t = t_{k+p}$ for some $k \in \mathbb{Z}_+$ and $p \in \mathbb{N}$. Then

$$(X \bullet M)(t) - (X \bullet M)(s) = \sum_{i=k+1}^{k+p} \xi_i \{M(t_i) - M(t_{i-1})\}.$$

Now for $i = k+1, \ldots, k+p$, we have

$$\mathbf{E}[\xi_i \{M(t_i) - M(t_{i-1})\} | \mathfrak{F}_s] = \mathbf{E}[\mathbf{E}[\xi_i \{M(t_i) - M(t_{i-1})\} | \mathfrak{F}_{t_{i-1}}] | \mathfrak{F}_s]$$
$$= \mathbf{E}[\xi_i \mathbf{E}[M(t_i) - M(t_{i-1}) | \mathfrak{F}_{t_{i-1}}] | \mathfrak{F}_s] = 0 \quad \text{a.e. on } (\Omega, \mathfrak{F}_s, P)$$

by the $\mathfrak{F}_{t_{i-1}}$-measurability of ξ_i and the martingale property of M. Thus (4) holds.

To prove (1), let $X, Y \in \mathbf{L}_0(\Omega, \mathfrak{F}, \{\mathfrak{F}_t\}, P)$ be expressed with a common strictly increasing sequence $\{t_k : k \in \mathbb{Z}_+\}$ with $t_0 = 0$ and $\lim_{k \to \infty} t_k = \infty$ as

(5) $$X(t) = \xi_0 \mathbf{1}_{\{0\}}(t) + \sum_{k \in \mathbb{N}} \xi_k \mathbf{1}_{(t_{k-1}, t_k]}(t) \quad \text{for } t \in \mathbb{R}_+$$

§12. STOCHASTIC INTEGRALS WITH RESPECT TO MARTINGALES

and

(6) $$Y(t) = \eta_0 \mathbf{1}_{\{0\}}(t) + \sum_{k \in \mathbb{N}} \eta_k \mathbf{1}_{(t_{k-1}, t_k]}(t) \quad \text{for } t \in \mathbb{R}_+,$$

where $\{\xi_k : k \in \mathbb{Z}_+\}$ and $\{\eta_k : k \in \mathbb{Z}_+\}$ are two bounded sequences of random variables on $(\Omega, \mathfrak{F}, P)$ such that ξ_0, η_0 are $\mathfrak{F}_{t_0}$-measurable and ξ_k, η_k are $\mathfrak{F}_{t_{k-1}}$-measurable for $k \in \mathbb{N}$. Let $t \in (0, \infty)$ be given. We add another point to the sequence $\{t_k : k \in \mathbb{Z}_+\}$ if necessary so that our $t = t_k$ for some $k \in \mathbb{N}$. For brevity, let $\alpha_i = M(t_i) - M(t_{i-1})$ and $\beta_i = N(t_i) - N(t_{i-1})$ for $i \in \mathbb{N}$. Then $(X \bullet M)_t = \sum_{i=1}^k \xi_i \alpha_i$ and $(Y \bullet N)_t = \sum_{i=1}^k \eta_i \beta_i$ and therefore

(7) $$\mathbf{E}[(X \bullet M)_t (Y \bullet N)_t] = \sum_{i=1}^k \sum_{j=1}^k \mathbf{E}[\xi_i \eta_j \alpha_i \beta_j].$$

Now for $i = j$, we have

$$\mathbf{E}[\xi_i \eta_i \alpha_i \beta_i] = \mathbf{E}[\mathbf{E}[\xi_i \eta_i \alpha_i \beta_i | \mathfrak{F}_{t_{i-1}}]]$$
$$= \mathbf{E}[\xi_i \eta_i \mathbf{E}[\alpha_i \beta_i | \mathfrak{F}_{t_{i-1}}]] \quad \text{by the } \mathfrak{F}_{t_{i-1}}\text{-measurability of } \xi_i, \eta_i$$
$$= \mathbf{E}[\xi_i \eta_i \mathbf{E}[[M,N]_{t_i} - [M,N]_{t_{i-1}} | \mathfrak{F}_{t_{i-1}}]] \quad \text{by Lemma 11.25}$$
$$= \mathbf{E}[\mathbf{E}[\xi_i \eta_i \{[M,N]_{t_i} - [M,N]_{t_{i-1}}\} | \mathfrak{F}_{t_{i-1}}]]$$
$$= \mathbf{E}[\xi_i \eta_i \{[M,N]_{t_i} - [M,N]_{t_{i-1}}\}].$$

On the other hand for $i \neq j$, say $i < j$, we have

$$\mathbf{E}[\xi_i \eta_j \alpha_i \beta_j] = \mathbf{E}[\mathbf{E}[\xi_i \eta_j \alpha_i \beta_j | \mathfrak{F}_{t_{j-1}}]]$$
$$= \mathbf{E}[\xi_i \eta_j \alpha_i \mathbf{E}[\beta_j | \mathfrak{F}_{t_{j-1}}]] \quad \text{by the } \mathfrak{F}_{t_{j-1}}\text{-measurability of } \xi_i, \eta_j, \alpha_i$$
$$= 0 \quad \text{by the martingale property of } N.$$

Using these computations in (7), we have

$$\mathbf{E}[(X \bullet M)_t (Y \bullet N)_t] = \mathbf{E}\left[\sum_{i=1}^k \xi_i \eta_i \{[M,N]_{t_i} - [M,N]_{t_{i-1}}\}\right]$$
$$= \mathbf{E}\left[\int_{[0,t]} X(s) Y(s) \, d[M,N](s)\right]$$

since the sample functions of $X \cdot Y$ are step functions. This proves (1). In particular when $X = Y$ and $M = N$, (1) reduces to

(8) $$\mathbf{E}[(X \bullet M)_t^2] = \mathbf{E}\left[\int_{[0,t]} X^2(s) \, d[M](s)\right].$$

Recalling the definition of $\lVert\cdot\rVert_t$ in Definition 11.4 and the definition of $\lVert\cdot\rVert_{2,t}^{[M],P}$ in Definition 11.17, we have (2). By Definition 11.3 and Definition 11.17 again we have

$$\lVert X \bullet M \rVert_\infty = \sum_{m\in\mathbb{N}} 2^{-m}\{\lVert X \bullet M \rVert_m \wedge 1\} = \sum_{m\in\mathbb{N}} 2^{-m}\{\lVert X \rVert_{2,m}^{[M],P} \wedge 1\} = \lVert X \rVert_{2,\infty}^{[M],P},$$

proving (3). ∎

As we noted in Observation 12.2, $\mathbf{L}_0(\Omega, \mathfrak{F}, \{\mathfrak{F}_t\}, P) \subset \mathbf{L}_{2,\infty}(\mathbb{R}_+ \times \Omega, \mu_{[M]}, P)$ for every $M \in \mathbf{M}_2(\Omega, \mathfrak{F}, \{\mathfrak{F}_t\}, P)$. According to (3) of Proposition 12.4, the mapping of $X \in \mathbf{L}_0(\Omega, \mathfrak{F}, \{\mathfrak{F}_t\}, P)$ into $X \bullet M \in \mathbf{M}_2(\Omega, \mathfrak{F}, \{\mathfrak{F}_t\}, P)$ is an isometry with respect to the metric associated with the quasinorm $\lVert\cdot\rVert_{2,\infty}^{[M],P}$ on $\mathbf{L}_{2,\infty}(\mathbb{R}_+ \times \Omega, \mu_{[M]}, P)$ and the metric associated with the quasinorm $\lVert\cdot\rVert_\infty$ on $\mathbf{M}_2(\Omega, \mathfrak{F}, \{\mathfrak{F}_t\}, P)$. We shall show in Theorem 12.8 that if X is a predictable process on the filtered space $(\Omega, \mathfrak{F}, \{\mathfrak{F}_t\}, P)$ and if X is also in $\mathbf{L}_{2,\infty}(\mathbb{R}_+ \times \Omega, \mu_{[M]}, P)$, then there exists a sequence $\{X^{(n)} : n \in \mathbb{N}\}$ in $\mathbf{L}_0(\Omega, \mathfrak{F}, \{\mathfrak{F}_t\}, P)$ such that $\lim_{n\to\infty} \lVert X^{(n)} - X \rVert_{2,\infty}^{[M],P} = 0$. Now $\{X^{(n)} : n \in \mathbb{N}\}$ is a Cauchy sequence in $\mathbf{L}_{2,\infty}(\mathbb{R}_+ \times \Omega, \mu_{[M]}, P)$ and the isometry implies that $\{X^{(n)} \bullet M : n \in \mathbb{N}\}$ is a Cauchy sequence in $\mathbf{M}_2(\Omega, \mathfrak{F}, \{\mathfrak{F}_t\}, P)$. Then the completeness of $\mathbf{M}_2(\Omega, \mathfrak{F}, \{\mathfrak{F}_t\}, P)$ implies that there exists some $Y \in \mathbf{M}_2(\Omega, \mathfrak{F}, \{\mathfrak{F}_t\}, P)$ such that $\lim_{n\to\infty} \lVert X^{(n)} \bullet M - Y \rVert_\infty = 0$. We then define $X \bullet M = Y$.

According to Proposition 12.4, $X \bullet M$ and $Y \bullet N$ are in $\mathbf{M}_2(\Omega, \mathfrak{F}, \{\mathfrak{F}_t\}, P)$ for X and Y in $\mathbf{L}_0(\Omega, \mathfrak{F}, \{\mathfrak{F}_t\}, P)$ and thus the quadratic variation processes $[X \bullet M]$, $[Y \bullet N]$, and $[X \bullet M, Y \bullet N]$ exist. The next theorem shows that these processes are obtained by integrating the sample functions of X^2, Y^2, and XY with respect to the families of Lebesgue-Stieltjes measures $\mu_{[M]}$, $\mu_{[N]}$, and $\mu_{[M,N]}$ determined by the quadratic variation processes $[M]$, $[N]$, and $[M, N]$.

Proposition 12.5. *Let $M, N \in \mathbf{M}_2(\Omega, \mathfrak{F}, \{\mathfrak{F}_t\}, P)$ and $X, Y \in \mathbf{L}_0(\Omega, \mathfrak{F}, \{\mathfrak{F}_t\}, P)$. Then there exists a null set Λ in $(\Omega, \mathfrak{F}, P)$ such that on Λ^c, for every $t \in \mathbb{R}_+$, we have*

(1) $$[X \bullet M, Y \bullet N]_t = \int_{[0,t]} X(s)Y(s)\, d[M,N](s),$$

and thus for any $s, t \in \mathbb{R}_+$, $s < t$, we have

(2) $\quad\quad\quad\quad \mathbf{E}[\{(X \bullet M)_t - (X \bullet M)_s\}\{(Y \bullet N)_t - (Y \bullet N)_s\} | \mathfrak{F}_s]$

$\quad = \mathbf{E}[(X \bullet M)_t(Y \bullet N)_t - (X \bullet M)_s(Y \bullet N)_s | \mathfrak{F}_s]$

$\quad = \mathbf{E}\left[\int_{(s,t]} X(u)Y(u)\, d[M,N](u) \Big| \mathfrak{F}_s\right] \quad \text{a.e. on } (\Omega, \mathfrak{F}_s, P).$

§12. STOCHASTIC INTEGRALS WITH RESPECT TO MARTINGALES

In particular

(3) $$[X \bullet M]_t = \int_{[0,t]} X^2(s)\, d[M](s),$$

and

(4) $$E[\{(X \bullet M)_t - (X \bullet M)_s\}^2 | \mathfrak{F}_s] = E[(X \bullet M)_t^2 - (X \bullet M)_s^2 | \mathfrak{F}_s]$$
$$= E\left[\int_{(s,t]} X^2(u)\, d[M](u) \Big| \mathfrak{F}_s\right] \quad \text{a.e. on } (\Omega, \mathfrak{F}_s, P).$$

Proof. Let $V = \{V_t : t \in \mathbb{R}_+\}$ where

(5) $$V(t) = \int_{[0,t]} X(s) Y(s)\, d[M, N](s).$$

To prove (1), we show that $V \in \mathbf{V}(\Omega, \mathfrak{F}, \{\mathfrak{F}_t\}, P)$ and that $(X \bullet M)(Y \bullet N) - V$ is a null at 0 right-continuous martingale. Now since $[M, N] \in \mathbf{V}(\Omega, \mathfrak{F}, \{\mathfrak{F}_t\}, P)$, we have $[M, N] = A' - A''$ where $A', A'' \in \mathbf{A}(\Omega, \mathfrak{F}, \{\mathfrak{F}_t\}, P)$ according to Theorem 11.12. Then

(6) $$V(t) = \int_{[0,t]} (X \cdot Y)(s)\, dA'(s) - \int_{[0,t]} (X \cdot Y)(s)\, dA''(s).$$

Let us define two processes $A^{(1)} = \{A^{(1)}_t : t \in \mathbb{R}_+\}$ and $A^{(2)} = \{A^{(2)}_t : t \in \mathbb{R}_+\}$ on $(\Omega, \mathfrak{F}, P)$ by setting

(7) $$A^{(1)}(t) = \int_{[0,t]} (X \cdot Y)^+(s)\, dA'(s) + \int_{[0,t]} (X \cdot Y)^-(s)\, dA''(s),$$

and

(8) $$A^{(2)}(t) = \int_{[0,t]} (X \cdot Y)^-(s)\, dA'(s) + \int_{[0,t]} (X \cdot Y)^+(s)\, dA''(s).$$

Then from (6) we have

(9) $$V(t) = A^{(1)}(t) - A^{(2)}(t).$$

Note that since X and Y are bounded processes, $A^{(1)}(t)$ and $A^{(2)}(t)$ defined by (7) and (8) are finite on Ω and thus $V(t)$ is defined and finite on Ω. By Theorem 10.11, $A^{(1)}$ and $A^{(2)}$ are adapted processes and so is V. Since A' and A'' are L_1-processes and X and Y are bounded processes, $A^{(1)}$ and $A^{(2)}$ are L_1-processes. The right-continuity of $A^{(1)}$ and $A^{(2)}$ is implied by that of A' and A''. Also the sample functions of $A^{(1)}$ and $A^{(2)}$ are real valued monotone increasing and vanish at 0. Thus $A^{(1)}, A^{(2)} \in \mathbf{A}(\Omega, \mathfrak{F}, \{\mathfrak{F}_t\}, P)$ and therefore $V \in \mathbf{V}(\Omega, \mathfrak{F}, \{\mathfrak{F}_t\}, P)$ by Theorem 11.12.

To prove that $(X \bullet M)(Y \bullet N) - V$ is a null at 0 right-continuous martingale is equivalent to proving the second equality in (2). Note that the first equality in (2) is an immediate consequence of the martingale property of $X \bullet M$ and $Y \bullet N$. To prove the second equality in (2), let X and Y be expressed as in (5) and (6) in the proof of Proposition 12.4. We may assume without loss of generality that $s = t_k$ and $t = t_{k+p}$ for some $k \in \mathbb{Z}_+$ and $p \in \mathbb{N}$. Then with $\alpha_i = M(t_i) - M(t_{i-1})$ and $\beta_i = N(t_i) - N(t_{i-1})$ for $i \in \mathbb{N}$ as in the proof of Proposition 12.4, we have $(X \bullet M)_t - (X \bullet M)_s = \sum_{i=k+1}^{k+p} \xi_i \alpha_i$ and similarly $(Y \bullet N)_t - (Y \bullet N)_s = \sum_{i=k+1}^{k+p} \eta_i \beta_i$. Thus

$$\{(X \bullet M)_t - (X \bullet M)_s\}\{(Y \bullet N)_t - (Y \bullet N)_s\} = \sum_{i=k+1}^{k+p} \sum_{j=k+1}^{k+p} \xi_i \eta_j \alpha_i \beta_j.$$

Now for $k+1 \leq i < j \leq k+p$, we have

$$\mathbf{E}[\xi_i \eta_j \alpha_i \beta_j | \mathfrak{F}_s] = \mathbf{E}[\mathbf{E}[\xi_i \eta_j \alpha_i \beta_j | \mathfrak{F}_{t_{j-1}}] | \mathfrak{F}_s] = \mathbf{E}[\xi_i \eta_j \alpha_i \mathbf{E}(\beta_j | \mathfrak{F}_{t_{j-1}}) | \mathfrak{F}_s] = 0$$

since $\mathbf{E}(\beta_j | \mathfrak{F}_{t_{j-1}}) = 0$ by the martingale property of N. On the other hand

$$\mathbf{E}[\xi_i \eta_i \alpha_i \beta_i | \mathfrak{F}_s] = \mathbf{E}[\mathbf{E}[\xi_i \eta_i \alpha_i \beta_i | \mathfrak{F}_{t_{i-1}}] | \mathfrak{F}_s] = \mathbf{E}[\xi_i \eta_i \mathbf{E}(\alpha_i \beta_i | \mathfrak{F}_{t_{i-1}}) | \mathfrak{F}_s]$$
$$= \mathbf{E}[\xi_i \eta_i \mathbf{E}[[M,N]_{t_i} - [M,N]_{t_{i-1}} | \mathfrak{F}_{t_{i-1}}] | \mathfrak{F}_s] = \mathbf{E}[\xi_i \eta_i \{[M,N]_{t_i} - [M,N]_{t_{i-1}}\} | \mathfrak{F}_s],$$

by Lemma 11.25. Thus we have

$$\mathbf{E}[\{(X \bullet M)_t - (X \bullet M)_s\}\{(Y \bullet N)_t - (Y \bullet N)_s\} | \mathfrak{F}_s]$$
$$= \sum_{i=k+1}^{k+p} \mathbf{E}[\xi_i \eta_i \{[M,N]_{t_i} - [M,N]_{t_{i-1}}\} | \mathfrak{F}_s]$$
$$= \mathbf{E}\left[\sum_{i=k+1}^{k+p} \xi_i \eta_i \{[M,N]_{t_i} - [M,N]_{t_{i-1}}\} \Big| \mathfrak{F}_s\right]$$
$$= \mathbf{E}\left[\int_{(s,t]} X(u) Y(u) \, d[M,N](u) \Big| \mathfrak{F}_s\right],$$

proving the second equality in (2). ∎

Regarding integrals with respect to the families of Lebesgue-Stieltjes measures $\mu_{[M]}$, $\mu_{[N]}$, $\mu_{[M,N]}$, and $\mu_{|[M,N]|}$ we have the following inequalities.

§12. STOCHASTIC INTEGRALS WITH RESPECT TO MARTINGALES

Theorem 12.6. *Let* $M, N \in \mathbf{M}_2^c(\Omega, \mathfrak{F}, \{\mathfrak{F}_t\}, P)$.

1) There exists a null set Λ_∞ *in* $(\Omega, \mathfrak{F}, P)$ *such that for any two bounded real valued functions* X *and* Y *on* $\mathbb{R}_+ \times \Omega$ *with Borel measurable sample functions* $X(\cdot, \omega)$ *and* $Y(\cdot, \omega)$ *for every* $\omega \in \Omega$ *we have for every* $t \in \mathbb{R}_+$

(1) $$\left| \int_{[0,t]} X(s) Y(s) \, d[M, N](s) \right| \leq \int_{[0,t]} |X(s) Y(s)| \, d\|[M, N]\|(s)$$
$$\leq \left\{ \int_{[0,t]} X^2(s) \, d[M](s) \right\}^{1/2} \left\{ \int_{[0,t]} Y^2(s) \, d[N](s) \right\}^{1/2} < \infty \quad \text{on } \Lambda_\infty^c.$$

2) Let X *and* Y *be two adapted measurable processes on the filtered space whose sample functions are almost surely locally square integrable on* $\mathbb{R}_+$ *with respect to* $\mu_{[M]}$ *and* $\mu_{[N]}$ *respectively, that is,* $\int_{[0,t]} X^2(s, \omega) \, d[M](s, \omega) < \infty$ *for every* $t \in \mathbb{R}_+$ *when* $\omega \in \Lambda_X^c$ *where* Λ_X *is a null set in* $(\Omega, \mathfrak{F}, P)$ *and similarly* $\int_{[0,t]} Y^2(s, \omega) \, d[N](s, \omega) < \infty$ *for every* $t \in \mathbb{R}_+$ *when* $\omega \in \Lambda_Y^c$ *where* Λ_Y *is a null set in* $(\Omega, \mathfrak{F}, P)$. *Then there exists a null set* $\Lambda_{\infty, X, Y}$ *in* $(\Omega, \mathfrak{F}, P)$ *such that (1) holds for every* $t \in \mathbb{R}_+$ *and* $\omega \in \Lambda_{\infty, X, Y}^c$ *and furthermore*

(2) $$\mathbf{E}\left[\left| \int_{[0,t]} X(s) Y(s) \, d[M, N](s) \right| \right] \leq \mathbf{E}\left[\int_{[0,t]} |X(s) Y(s)| \, d\|[M, N]\|(s) \right]$$
$$\leq \mathbf{E}\left[\int_{[0,t]} X^2(s) \, d[M](s) \right]^{1/2} \mathbf{E}\left[\int_{[0,t]} Y^2(s) \, d[N](s) \right]^{1/2}.$$

Proof. 1) Consider first the case where $X, Y \in \mathbf{L}_0(\Omega, \mathfrak{F}, \{\mathfrak{F}_t\}, P)$. Then $X \bullet M, Y \bullet N \in \mathbf{M}_2^c(\Omega, \mathfrak{F}, \{\mathfrak{F}_t\}, P)$ so that we have $a(X \bullet M) + b(Y \bullet N) \in \mathbf{M}_2^c(\Omega, \mathfrak{F}, \{\mathfrak{F}_t\}, P)$ for any $a, b \in \mathbb{R}_+$ and thus the quadratic variation process $[a(X \bullet M) + b(Y \bullet N)]$ exists. Let Q be the collection of all rational numbers. Then for every $a, b \in Q$, there exists a null set $\Lambda_{X,Y,a,b}$ in $(\Omega, \mathfrak{F}, P)$ such that $[a(X \bullet M) + b(Y \bullet N)](\cdot, \omega)$ is a continuous monotone increasing function on $\mathbb{R}_+$ vanishing at $t = 0$ for $\omega \in \Lambda_{X,Y,a,b}^c$. Thus by Proposition 11.26 we have for $t \in \mathbb{R}_+$ and $\omega \in \Lambda_{X,Y,a,b}^c$,

(3) $$0 \leq [a(X \bullet M) + b(Y \bullet N)](t, \omega)$$
$$= a^2 [X \bullet M](t, \omega) + 2ab[X \bullet M, Y \bullet N](t, \omega) + b^2 [Y \bullet N](t, \omega).$$

Then for the null set $\Lambda_{X,Y} = \cup_{a,b \in Q} \Lambda_{X,Y,a,b}$, the inequality (3) holds for all $a, b \in Q$ and $t \in \mathbb{R}_+$ when $\omega \in \Lambda^c$. Since the right side of (3) is a continuous function in a and b and since Q is dense in $\mathbb{R}$, (3) holds for all $a, b \in \mathbb{R}$ and $t \in \mathbb{R}_+$ when $\omega \in \Lambda_{X,Y}^c$. Thus we have a nonnegative definite quadratic form in a and b for each fixed $t \in \mathbb{R}_+$ and $\omega \in \Lambda_{X,Y}^c$ and this implies that the determinant of the matrix of the coefficients of the quadratic form is nonnegative, that is, for $(t,\omega) \in \mathbb{R}_+ \times \Lambda_{X,Y}^c$ we have

$$\{[X \bullet M, Y \bullet N](t,\omega)\}^2 \leq [X \bullet M](t,\omega)[Y \bullet N](t,\omega).$$

Now since $M, N \in \mathbf{M}_2^c(\Omega, \mathfrak{F}, \{\mathfrak{F}_t\}, P)$, the quadratic variation processes $[M]$, $[N]$, and $[M, N]$ are almost surely continuous according to Proposition 11.27. This fact and Proposition 12.5 imply that there exists a null set $\Lambda'_{X,Y}$ in $(\Omega, \mathfrak{F}, P)$ containing $\Lambda_{X,Y}$ such that for $\omega \in (\Lambda'_{X,Y})^c$ the functions $[M](\cdot,\omega)$, $[N](\cdot,\omega)$, and $[M, N](\cdot,\omega)$ are continuous on $\mathbb{R}_+$ and furthermore for $(t,\omega) \in \mathbb{R}_+ \times (\Lambda'_{X,Y})^c$ we have

$$(4) \quad \left\{\int_{[0,t]} X(s,\omega)Y(s,\omega)\,d[M,N](s,\omega)\right\}^2$$
$$\leq \left\{\int_{[0,t]} X^2(s,\omega)\,d[M](s,\omega)\right\}\left\{\int_{[0,t]} Y^2(s,\omega)\,d[N](s,\omega)\right\}.$$

Let $\mathbb{Q}$ be the subcollection of $\mathbf{L}_0(\Omega, \mathfrak{F}, \{\mathfrak{F}_t\}, P)$ consisting of bounded adapted left-continuous simple processes X of the type

$$X(t,\omega) = a_0 \mathbf{1}_{\{t_0\}}(t) + \sum_{k \in \mathbb{N}} a_k \mathbf{1}_{(t_{k-1}, t_k]}(t) \quad \text{for } (t,\omega) \in \mathbb{R}_+ \times \Omega,$$

where $\{t_k : k \in \mathbb{Z}_+\}$ is a strictly increasing sequence of rational numbers in $\mathbb{R}_+$ with $t_0 = 0$ and $\{a_k : k \in \mathbb{Z}_+\}$ is a bounded sequence of rational numbers. Note that for every $X \in \mathbb{Q}$ the sample functions are all identical and that $\mathbb{Q}$ is a countable collection. Let Λ_∞ be the null set in $(\Omega, \mathfrak{F}, P)$ which is the union of the null sets $\Lambda'_{X,Y}$ for $X, Y \in \mathbb{Q}$. Then (4) holds for $(t,\omega) \in \mathbb{R}_+ \times \Lambda_\infty^c$.

Let $\omega \in \Lambda_\infty^c$ be fixed. Since $[M](\cdot,\omega)$, $[N](\cdot,\omega)$, and $[M,N](\cdot,\omega)$ are continuous on $\mathbb{R}_+$, for every $t \in \mathbb{R}_+$ we have $\mu_{[M]}(\{t\},\omega) = \mu_{[N]}(\{t\},\omega) = \mu_{[M,N]}(\{t\},\omega) = 0$. Then for two arbitrary bounded Borel measurable functions f and g and for an arbitrary $\varepsilon > 0$, there exist two bounded left-continuous step functions f_ε and g_ε assuming only rational values and having only rational points of discontinuities such that $|I_1 - I_{1,\varepsilon}| < \varepsilon$, $|I_2 - I_{2,\varepsilon}| < \varepsilon$, and $|I_3 - I_{3,\varepsilon}| < \varepsilon$, where

$$I_1 = \int_{[0,t]} f^2(s)\,d[M](s,\omega), \qquad I_{1,\varepsilon} = \int_{[0,t]} f_\varepsilon^2(s)\,d[M](s,\omega),$$
$$I_2 = \int_{[0,t]} g^2(s)\,d[N](s,\omega), \qquad I_{2,\varepsilon} = \int_{[0,t]} g_\varepsilon^2(s)\,d[N](s,\omega),$$
$$I_3 = \left|\int_{[0,t]} f(s)g(s)\,d[M,N](s,\omega)\right|^2, \quad I_{3,\varepsilon} = \left|\int_{[0,t]} f_\varepsilon(s)g_\varepsilon(s)\,d[M,N](s,\omega)\right|^2.$$

§12. STOCHASTIC INTEGRALS WITH RESPECT TO MARTINGALES   231

Now since f_ε and g_ε are bounded left-continuous step functions assuming only rational values and having only rational points of discontinuities, there exist $X, Y \in \mathbb{Q}$ such that $X = f_\varepsilon$ and $Y = g_\varepsilon$. Thus by (4) we have $I_{3,\varepsilon} \leq I_{1,\varepsilon} I_{2,\varepsilon}$. Then

$$I_3 \leq I_{3,\varepsilon} + \varepsilon \leq I_{1,\varepsilon} I_{2,\varepsilon} + \varepsilon \leq \{I_1 + \varepsilon\}\{I_2 + \varepsilon\} + \varepsilon.$$

By the arbitrariness of $\varepsilon > 0$, we have $I_3 \leq I_1 I_2$, that is, for arbitrary bounded Borel measurable functions f and g on $\mathbb{R}_+$, we have for every $t \in \mathbb{R}_+$

$$(5) \left\{\int_{[0,t]} f(s)g(s)[M,N](s,\omega)\right\}^2 \leq \left\{\int_{[0,t]} f^2(s)\,d[M](s,\omega)\right\}\left\{\int_{[0,t]} g^2(s)\,d[N](s,\omega)\right\}.$$

Clearly we have

$$(6) \qquad \left|\int_{[0,t]} f(s)g(s)[M,N](s,\omega)\right| \leq \int_{[0,t]} |f(s)g(s)|\,d\mathbf{|}[M,N]\mathbf{|}(s,\omega).$$

Let $t \in \mathbb{R}_+$ be fixed and let $\{A_\omega, B_\omega\} \subset \mathfrak{B}_{[0,t]}$ be the Hahn decomposition for the signed Lebesgue-Stieltjes measure $\mu_{[M,N]}(\cdot,\omega)$ on $([0,t], \mathfrak{B}_{[0,t]})$ for our $\omega \in \Lambda_\infty^c$. Then by (7) in Observation 11.9, we have

$$(7) \qquad \int_{[0,t]} |f(s)g(s)|\,d\mathbf{|}[M,N]\mathbf{|}(s,\omega)$$

$$= \int_{[0,t]} |f(s)g(s)|\{\mathbf{1}_{A_\omega}(s) - \mathbf{1}_{B_\omega}(s)\}\,d[M,N](s,\omega)$$

$$\leq \left\{\int_{[0,t]} f^2(s)\,d[M](s,\omega)\right\}^{1/2}\left\{\int_{[0,t]} g^2(s)\{\mathbf{1}_{A_\omega}(s) - \mathbf{1}_{B_\omega}(s)\}^2\,d[N](s,\omega)\right\}^{1/2}$$

$$= \left\{\int_{[0,t]} f^2(s)\,d[M](s,\omega)\right\}^{1/2}\left\{\int_{[0,t]} g^2(s)\,d[N](s,\omega)\right\}^{1/2}$$

where the inequality above is obtained by applying (5) to the two bounded Borel measurable functions $|f|$ and $|g|\{\mathbf{1}_{A_\omega} - \mathbf{1}_{B_\omega}\}$.

We have shown so far that if $\omega \in \Lambda_\infty^c$, then for any two bounded Borel measurable functions f and g on $\mathbb{R}_+$ and any $t \in \mathbb{R}_+$ we have

$$(8) \qquad \left|\int_{[0,t]} f(s)g(s)\,d[M,N](s,\omega)\right| \leq \int_{[0,t]} |f(s)g(s)|\,d\mathbf{|}[M,N]\mathbf{|}(s,\omega)$$

$$\leq \left\{\int_{[0,t]} f^2(s)\,d[M](s,\omega)\right\}^{1/2}\left\{\int_{[0,t]} g^2(s)\,d[N](s,\omega)\right\}^{1/2} < \infty.$$

Let X and Y be two bounded real valued functions on $\mathbb{R}_+ \times \Omega$ such that $X(\cdot, \omega)$ and $Y(\cdot, \omega)$ are Borel measurable functions on $\mathbb{R}_+$ for every $\omega \in \Omega$. Then by (8), (1) holds for every $t \in \mathbb{R}_+$ and $\omega \in \Lambda_\infty^c$.

2) Let X and Y be two adapted measurable processes on the filtered space whose sample functions are almost surely locally square integrable on $\mathbb{R}_+$ with respect to $\mu_{[M]}$ and $\mu_{[M]}$ respectively. Recall that every sample function of a measurable process is a Borel measurable function. For every $n \in \mathbb{N}$, let $X^{(n)} = \mathbf{1}_{[-n,n]}(X)$ and $Y^{(n)} = \mathbf{1}_{[-n,n]}(Y)$. We have

(9) $\quad \lim_{n\to\infty} X^{(n)}(t,\omega) = X(t,\omega)$ and $\lim_{n\to\infty} Y^{(n)}(t,\omega) = Y(t,\omega) \quad$ for $(t,\omega) \in \mathbb{R}_+ \times \Omega$.

Now $X^{(n)}$ and $Y^{(n)}$ are bounded measurable processes so that $X^{(n)}(\cdot,\omega)$ and $Y^{(n)}(\cdot,\omega)$ are Borel measurable functions for every $\omega \in \Omega$. Thus by 1), for every $t \in \mathbb{R}_+$ we have

(10) $\quad \left| \int_{[0,t]} X^{(n)}(s) Y^{(n)}(s) \, d[M,N](s) \right| \leq \int_{[0,t]} |X^{(n)}(s) Y^{(n)}(s)| \, d\|[M,N]\|(s)$

$\leq \left\{ \int_{[0,t]} (X^{(n)})^2(s) \, d[M](s) \right\}^{1/2} \left\{ \int_{[0,t]} (Y^{(n)})^2(s) \, d[N](s) \right\}^{1/2} < \infty \quad$ on Λ_∞^c.

Now there exists a null set $\Lambda_{\infty,X,Y}$, which we may assume to contain Λ_∞ without loss of generality, such that

(11) $\quad \int_{[0,t]} X^2(s,\omega) \, d[M](s,\omega), \int_{[0,t]} Y^2(s,\omega) \, d[N](s,\omega) < \infty,$

for $(t,\omega) \in \mathbb{R}_+ \times \Lambda_{\infty,X,Y}^c$. Since $(X^{(n)})^2 \leq X^2$ and $(Y^{(n)})^2 \leq Y^2$, we have by the Dominated Convergence Theorem

(12) $\quad \begin{cases} \int_{[0,t]} (X^{(n)})^2(s,\omega) \, d[M](s,\omega) \uparrow \int_{[0,t]} X^2(s,\omega) \, d[M](s,\omega) < \infty, \\ \int_{[0,t]} (Y^{(n)})^2(s,\omega) \, d[N](s,\omega) \uparrow \int_{[0,t]} Y^2(s,\omega) \, d[N](s,\omega) < \infty, \end{cases}$

for $(t,\omega) \in \mathbb{R}_+ \times \Lambda_{\infty,X,Y}^c$. Since $|X^{(n)} Y^{(n)}| \uparrow |XY|$ as $n \to \infty$, the Monotone Convergence Theorem implies that for every $t \in \mathbb{R}_+$ and $\omega \in \Lambda_{\infty,X,Y}^c$ we have

(13) $\quad \lim_{n\to\infty} \int_{[0,t]} |X^{(n)}(s,\omega) Y^{(n)}(s,\omega)| \, d\|[M,N]\|(s,\omega)$

$= \int_{[0,t]} |X(s,\omega) Y(s,\omega)| \, d\|[M,N]\|(s,\omega)$

$\leq \left\{ \int_{[0,t]} X^2(s,\omega) \, d[M](s,\omega) \right\}^{1/2} \left\{ \int_{[0,t]} Y^2(s,\omega) \, d[N](s,\omega) \right\}^{1/2}$

$< \infty,$

§12. STOCHASTIC INTEGRALS WITH RESPECT TO MARTINGALES

where the first inequality is by (10) and (12).

To prove the first equality in (1), let us decompose $X^{(n)}$ in its positive part and negative part as $X^{(n)} = X^{(n)+} - X^{(n)-}$ and similarly $Y^{(n)} = Y^{(n)+} - Y^{(n)-}$. Then

$$X^{(n)}Y^{(n)} = \{X^{(n)+}Y^{(n)+} + X^{(n)-}Y^{(n)-}\} - \{X^{(n)+}Y^{(n)-} + X^{(n)-}Y^{(n)+}\}.$$

Since $\mu_{[M,N]} = \mu^+_{[M,N]} - \mu^-_{[M,N]}$, we have

(14)
$$\int_{[0,t]} X^{(n)}Y^{(n)} \, d\mu_{[M,N]}$$
$$= \int_{[0,t]} \{X^{(n)+}Y^{(n)+} + X^{(n)-}Y^{(n)-}\} \, d\mu^+_{[M,N]}$$
$$+ \int_{[0,t]} \{X^{(n)+}Y^{(n)-} + X^{(n)-}Y^{(n)+}\} \, d\mu^-_{[M,N]}$$
$$- \int_{[0,t]} \{X^{(n)+}Y^{(n)+} + X^{(n)-}Y^{(n)-}\} \, d\mu^-_{[M,N]}$$
$$- \int_{[0,t]} \{X^{(n)+}Y^{(n)-} + X^{(n)-}Y^{(n)+}\} \, d\mu^+_{[M,N]}.$$

Now $X^{(n)-} \uparrow X^+$, $X^{(n)-} \uparrow X^-$, $Y^{(n)+} \uparrow Y^+$, and $Y^{(n)-} \uparrow Y^-$ as $n \to \infty$. Letting $n \to \infty$ and applying the Monotone Convergence Theorem to each one of the four integrals on the right side of (14) we obtain

(15)
$$\lim_{n \to \infty} \int_{[0,t]} X^{(n)}Y^{(n)} \, d\mu_{[M,N]}$$
$$= \int_{[0,t]} \{X^+Y^+ + X^-Y^-\} \, d\mu^+_{[M,N]} + \int_{[0,t]} \{X^+Y^- + X^-Y^+\} \, d\mu^-_{[M,N]}$$
$$- \int_{[0,t]} \{X^+Y^+ + X^-Y^-\} \, d\mu^-_{[M,N]} - \int_{[0,t]} \{X^+Y^- + X^-Y^+\} \, d\mu^+_{[M,N]}$$
$$= \int_{[0,t]} XY \, d\mu_{[M,N]}.$$

Letting $n \to \infty$ in the first inequality in (10), we obtain the first inequality in (1) for every $t \in \mathbb{R}_+$ and $\omega \in \Lambda^c_{\infty,X,Y}$ by (15) and the equality in (13).

Now the fact that X and Y are adapted measurable processes on the filtered space implies that $\int_{[0,t]} X(s)Y(s) \, d[M,N](s)$, $\int_{[0,t]} |X(s)Y(s)| \, d|[M,N]|(s)$, $\int_{[0,t]} X^2(s) \, d[M](s)$, and $\int_{[0,t]} Y^2(s) \, d[N](s)$ are all random variables, in fact $\mathfrak{F}_t$-measurable random variables, on $(\Omega, \mathfrak{F}, P)$ by Theorem 10.11. Thus integrating (1) with respect to P and applying Schwarz's Inequality we have (2). ∎

For the families of Lebesgue-Stieltjes measures $\mu_{[M]}$, $\mu_{[N]}$, and $\mu_{\|[M,N]\|}$ on the measurable space $(\mathbb{R}_+, \mathfrak{B}_{\mathbb{R}_+})$ determined by $[M]$, $[N]$, and $\|[M,N]\|$, we have the following inequality.

Theorem 12.7. *Let $M, N \in \mathbf{M}_2^c(\Omega, \mathfrak{F}, \{\mathfrak{F}_t\}, P)$. Then there exists a null set Λ in $(\Omega, \mathfrak{F}, P)$ such that for every $\omega \in \Lambda^c$, $t \in \mathbb{R}_+$ and $E \in \mathfrak{B}_{[0,t]}$ we have*

$$\{\mu_{\|[M,N]\|}(E, \omega)\}^2 \leq \mu_{[M]}(E, \omega) \mu_{[N]}(E, \omega).$$

Proof. Let $\mathfrak{J}$ be the semialgebra consisting of all intervals of the type $(s', s'']$ in $(0, t]$ and $\emptyset$ and let $\mathfrak{A}$ be the algebra of subsets of $(0, t]$ generated by $\mathfrak{J}$, that is, $\mathfrak{A}$ is the collection of all finite disjoint unions of members of $\mathfrak{J}$. Let $A \in \mathfrak{A}$. Then $A = (s_1', s_1''] \cup \cdots \cup (s_n', s_n'']$ where $0 \leq s_1' < s_1'' < \cdots < s_n' < s_n'' \leq t$. Let us define two processes X and Y in $\mathbf{L}_0(\Omega, \mathfrak{F}, \{\mathfrak{F}_t\}, P)$ by setting

$$X(s, \omega) = Y(s, \omega) = \begin{cases} 1 & \text{for } (s, \omega) \in A \times \Omega \\ 0 & \text{for } (s, \omega) \in A^c \times \Omega. \end{cases}$$

Let Λ_∞ be as defined in Proposition 12.6. The on Λ_∞^c we have

$$\left\{ \int_{[0,t]} |X(s,\omega) Y(s,\omega)| \, d\|[M,N]\|(s,\omega) \right\}^2$$
$$\leq \left\{ \int_{[0,t]} X^2(s,\omega) \, d[M](s,\omega) \right\} \left\{ \int_{[0,t]} Y^2(s,\omega) \, d[N](s,\omega) \right\},$$

that is,

$$\left\{ \int_{[0,t]} \mathbf{1}_A(s) \, d\|[M,N]\|(s,\omega) \right\}^2 \leq \left\{ \int_{[0,t]} \mathbf{1}_A(s) \, d[M](s,\omega) \right\} \left\{ \int_{[0,t]} \mathbf{1}_A(s) \, d[N](s,\omega) \right\}.$$

In other words, for $\omega \in \Lambda_\infty^c$, we have

(1) $\quad \{\mu_{\|[M,N]\|}(A, \omega)\}^2 \leq \mu_{[M]}(A, \omega) \mu_{[N]}(A, \omega) \quad$ for every $A \in \mathfrak{A}$.

For each $\omega \in \Lambda_\infty^c$, consider a measure on $\mathfrak{B}_{(0,t]}$ defined by

$$\nu(\cdot, \omega) = \mu_{\|[M,N]\|}(\cdot, \omega) + \mu_{[M]}(\cdot, \omega) + \mu_{[N]}(\cdot, \omega).$$

Since $\sigma(\mathfrak{A}) = \mathfrak{B}_{(0,t]}$, according to Theorem 1.11 for every $E \in \mathfrak{B}_{(0,t]}$ and $\varepsilon > 0$ there exists $A \in \mathfrak{A}$ such that $\nu(E \triangle A, \omega) < \varepsilon$. Then

(2) $\quad \mu_{\|[M,N]\|}(E \triangle A, \omega), \mu_{[M]}(E \triangle A, \omega), \mu_{[N]}(E \triangle A, \omega) < \varepsilon.$

§12. STOCHASTIC INTEGRALS WITH RESPECT TO MARTINGALES

By (1) and (2), we have

$$\{\mu_{[\![M,N]\!]}(E,\omega)\}^2 \leq \{\mu_{[\![M,N]\!]}(A,\omega)+\varepsilon\}^2$$
$$= \{\mu_{[\![M,N]\!]}(A,\omega)\}^2 + 2\varepsilon\mu_{[\![M,N]\!]}(A,\omega) + \varepsilon^2$$
$$\leq \mu_{[M]}(A,\omega)\mu_{[N]}(A,\omega) + 2\varepsilon\mu_{[\![M,N]\!]}((0,t],\omega) + \varepsilon^2$$
$$\leq \{\mu_{[M]}(E,\omega)+\varepsilon\}\{\mu_{[N]}(E,\omega)+\varepsilon\} + 2\varepsilon\mu_{[\![M,N]\!]}((0,t],\omega) + \varepsilon^2$$
$$\leq \mu_{[M]}(E,\omega)\mu_{[N]}(E,\omega) + \varepsilon\{\mu_{[M]}((0,t],\omega) + \mu_{[N]}((0,t],\omega) + 2\mu_{[\![M,N]\!]}((0,t],\omega)\}$$
$$+ 2\varepsilon^2.$$

Since this holds for every $\varepsilon > 0$, we have

$$(3) \qquad \{\mu_{[\![M,N]\!]}(E,\omega)\}^2 \leq \mu_{[M]}(E,\omega)\mu_{[N]}(E,\omega).$$

By our definition of the family of Lebesgue-Stieltjes measures μ_A on $(\mathbb{R}_+, \mathfrak{B}_{\mathbb{R}_+})$ determined by an almost surely increasing process A, there exists a null set Λ_0 in $(\Omega, \mathfrak{F}, P)$ such that $\mu_{[M]}(\{0\},\omega) = \mu_{[N]}(\{0\},\omega) = \mu_{[M,N]}(\{0\},\omega) = 0$ for $\omega \in \Lambda_0^c$. Let $\Lambda = \Lambda_\infty \cup \Lambda_0$. Then for every $\omega \in \Lambda^c$, (3) holds for every $E \in \mathfrak{B}_{[0,t]}$. ∎

[II] Stochastic Integral of Predictable Processes with Respect to L_2-Martingales

Let us show first that if X is a predictable process on $(\Omega, \mathfrak{F}, \{\mathfrak{F}_t\}, P)$ which is also in $\mathbf{L}_{p,\infty}(\mathbb{R}_+ \times \Omega, \mu_{[M]}, P)$ for some $M \in \mathbf{M}_2(\Omega, \mathfrak{F}, \{\mathfrak{F}_t\}, P)$ and $p \in [1, \infty)$, then there exists a sequence in $\mathbf{L}_0(\Omega, \mathfrak{F}, \{\mathfrak{F}_t\}, P)$ which converges to X in the metric of the quasinorm $\|\cdot\|_{p,\infty}^{[M],P}$ on $\mathbf{L}_{p,\infty}(\mathbb{R}_+ \times \Omega, \mu_{[M]}, P)$.

Theorem 12.8. *Let $M \in \mathbf{M}_2(\Omega, \mathfrak{F}, \{\mathfrak{F}_t\}, P)$. Let $X = \{X_t : t \in \mathbb{R}_+\}$ be a predictable process on the filtered space $(\Omega, \mathfrak{F}, \{\mathfrak{F}_t\}, P)$ which is in $\mathbf{L}_{p,\infty}(\mathbb{R}_+ \times \Omega, \mu_{[M]}, P)$ for some $p \in [0.\infty)$. Then there exists a sequence $\{X^{(n)} : n \in \mathbb{N}\}$ in $\mathbf{L}_0(\Omega, \mathfrak{F}, \{\mathfrak{F}_t\}, P)$ such that $\lim_{n\to\infty} \|X^{(n)} - X\|_{p,\infty}^{[M],P} = 0$.*

Proof. We prove this theorem by applying Theorem 2.22, a monotone class theorem for predictable processes. By Observation 11.19, all bounded measurable processes on $(\Omega, \mathfrak{F}, P)$ are in $\mathbf{L}_{p,\infty}(\mathbb{R}_+ \times \Omega, \mu_{[M]}, P)$. Let $\mathbf{V}$ be the collection of all bounded measurable processes Y on $(\Omega, \mathfrak{F}, P)$ for which there exist sequences $\{Y^{(n)} : n \in \mathbb{N}\}$ in $\mathbf{L}_0(\Omega, \mathfrak{F}, \{\mathfrak{F}_t\}, P)$ such that $\lim_{n\to\infty} \|Y^{(n)} - Y\|_{p,\infty}^{[M],P} = 0$. Let us verify that $\mathbf{V}$ satisfies conditions 1°, 2° and 3° in Theorem 2.22.

1°. Let us show that **V** contains all bounded left-continuous adapted processes. Let Y be such a process and suppose $|Y| \leq K$ on $\mathbb{R}_+ \times \Omega$ for some $K \geq 0$. For each $n \in \mathbb{N}$, define $Y^{(n)}$ by

(1) $\quad Y^{(n)}(t,\omega) = \begin{cases} Y((k-1)2^{-n},\omega) & \text{for } t \in ((k-1)2^{-n}, k2^{-n}] \text{ for } k \in \mathbb{N}, \omega \in \Omega \\ Y(0,\omega) & \text{for } \omega \in \Omega. \end{cases}$

Then $Y^{(n)} \in \mathbf{L}_0(\Omega, \mathfrak{F}, \{\mathfrak{F}_t\}, P)$ and for every $m \in \mathbb{N}$ we have

(2) $\quad \|Y^{(n)} - Y\|_{p,m}^{[M],P} = \left[\int_\Omega \left\{ \int_{[0,m]} |Y^{(n)}(s,\omega) - Y(s,\omega)|^p \, d[M](s,\omega) \right\} P(d\omega) \right]^{1/p}.$

According to Lemma 2.8, $\lim_{n \to \infty} Y^{(n)}(t,\omega) = Y(t,\omega)$ for $(t,\omega) \in \mathbb{R}_+ \times \Omega$. Thus by the Bounded Convergence Theorem we have

$$\lim_{n \to \infty} \int_{[0,m]} |Y^{(n)}(s,\omega) - Y(s,\omega)|^p \, d[M](s,\omega) = 0 \quad \text{for every } \omega \in \Omega.$$

Now

$$\int_{[0,m]} |Y^{(n)}(s,\omega) - Y(s,\omega)|^p \, d[M](s,\omega) \leq 2^p K^p [M]_m(\omega),$$

and $\mathbf{E}[[M]_m] < \infty$ since $[M]$ is an L_1-process. Thus in letting $n \to \infty$ in (2) we can apply the Dominated Convergence Theorem for the integrals with respect to P and then apply the Bounded Convergence Theorem for the integrals with respect to $\mu_{[M]}(\cdot, \omega)$ for each $\omega \in \Omega$. Then we have $\lim_{n \to \infty} \|Y^{(n)} - Y\|_{p,m}^{[M],P} = 0$. Since this holds for every $m \in \mathbb{N}$, we have $\lim_{n \to \infty} \|Y^{(n)} - Y\|_{p,\infty}^{[M],P} = 0$ by Remark 11.18. This shows that $Y \in \mathbf{V}$.

2°. Clearly **V** is a linear space.

3°. Suppose $\{Y^{(n)} : n \in \mathbb{N}\}$ is an increasing sequence of nonnegative members of **V** such that $Y \equiv \lim_{n \to \infty} Y^{(n)}$ is bounded. Then by the Dominated Convergence Theorem and then by the Bounded Convergence Theorem as in (2), we have $\lim_{n \to \infty} \|Y - Y^{(n)}\|_{p,m}^{[M],P} = 0$ for every $m \in \mathbb{N}$ and therefore $\lim_{n \to \infty} \|Y - Y^{(n)}\|_{p,\infty}^{[M],P} = 0$ by Remark 11.18. Since $Y^{(n)} \in \mathbf{V}$ there exists $Z^{(n)} \in \mathbf{L}_0(\Omega, \mathfrak{F}, \{\mathfrak{F}_t\}, P)$ such that $\|Y^{(n)} - Z^{(n)}\|_{p,\infty}^{[M],P} < 2^{-n}$. Then for the sequence $\{Z^{(n)} : n \in \mathbb{N}\}$ in $\mathbf{L}_0(\Omega, \mathfrak{F}, \{\mathfrak{F}_t\}, P)$ we have

$$\lim_{n \to \infty} \|Y - Z^{(n)}\|_{p,\infty}^{[M],P} \leq \lim_{n \to \infty} \left\{ \|Y - Y^{(n)}\|_{p,\infty}^{[M],P} + \|Y^{(n)} - Z^{(n)}\|_{p,\infty}^{[M],P} \right\} = 0.$$

Thus $Y \in \mathbf{V}$.

This shows that **V** satisfies the conditions in Theorem 2.22. Thus **V** contains all bounded predictable processes on the filtered space. Therefore for every bounded predictable process

§12. STOCHASTIC INTEGRALS WITH RESPECT TO MARTINGALES

X on the filtered space, there exists a sequence $\{X^{(n)} : n \in \mathbb{N}\}$ in $\mathbf{L}_0(\Omega, \mathfrak{F}, \{\mathfrak{F}_t\}, P)$ such that $\lim_{n \to \infty} \|X^{(n)} - X\|_{p,\infty}^{[M],P} = 0$.

Finally, let X be a predictable process which is in $\mathbf{L}_{p,\infty}(\mathbb{R}_+ \times \Omega, \mu_{[M]}, P)$ also. For each $n \in \mathbb{N}$, let us define $X^{(n)}$ by setting

$$X^{(n)}(t, \omega) = \mathbf{1}_{[-n,n]}(X(t, \omega))X(t, \omega).$$

Since X is a predictable process, it is an $\mathfrak{S}/\mathfrak{B}_\mathbb{R}$-measurable mapping of $\mathbb{R}_+ \times \Omega$ into $\mathbb{R}$. Since $\mathbf{1}_{[-n,n]}$ is a $\mathfrak{B}_\mathbb{R}/\mathfrak{B}_\mathbb{R}$-measurable mapping of $\mathbb{R}$ into $\mathbb{R}$, $X^{(n)} = \mathbf{1}_{[-n,n]}(X)$ is an $\mathfrak{S}/\mathfrak{B}_\mathbb{R}$-measurable mapping of $\mathbb{R}_+ \times \Omega$ into $\mathbb{R}$, that is, it is a predictable process. Thus $X^{(n)}$ is a bounded predictable process. Now for every $m \in \mathbb{N}$ we have

$$(3) \quad \|X^{(n)} - X\|_{p,m}^{[M],P} = \left[\int_\Omega \left\{ \int_{[0,m]} |X^{(n)}(s, \omega) - X(s, \omega)|^p \, d[M](s, \omega) \right\} P(d\omega) \right]^{1/p}.$$

Since $X \in \mathbf{L}_{p,\infty}(\mathbb{R}_+ \times \Omega, \mu_{[M]}, P)$, according to (3) of Observation 11.16 there exists a null set Λ in $(\Omega, \mathfrak{F}, P)$ such that for every $\omega \in \Lambda^c$ we have $\int_{[0,m]} |X(s, \omega)|^p \, d[M](s, \omega) < \infty$ for every $m \in \mathbb{N}$. Since $|X^{(n)}(s, \omega) - X(s, \omega)|^p \leq |X(s, \omega)|^p$, we have by the Dominated Convergence Theorem

$$\lim_{n \to \infty} \int_{[0,m]} |X^{(n)}(s, \omega) - X(s, \omega)|^p \, d[M](s, \omega) = 0 \quad \text{for every } \omega \in \Lambda^c.$$

On the other hand,

$$\int_{[0,m]} |X^{(n)}(s, \omega) - X(s, \omega)|^p \, d[M](s, \omega) \leq \int_{[0,m]} |X(s, \omega)|^p \, d[M](s, \omega),$$

and $\mathbf{E}[\int_{[0,m]} |X(s)|^p \, d[M](s)] < \infty$. Thus by applying the Dominated Convergence Theorem twice in letting $n \to \infty$ in (3), we have $\lim_{n \to \infty} \|X - X^{(n)}\|_{p,m}^{[M],P} = 0$ for every $m \in \mathbb{N}$ and therefore by Remark 11.18 we have $\lim_{n \to \infty} \|X - X^{(n)}\|_{p,\infty}^{[M],P} = 0$. Now since $X^{(n)}$ is a bounded predictable process, according to what we showed above there exists $Z^{(n)}$ in $\mathbf{L}_0(\Omega, \mathfrak{F}, \{\mathfrak{F}_t\}, P)$ such that $\|X^{(n)} - Z^{(n)}\|_{p,\infty}^{[M],P} < 2^{-n}$. From this and the triangle inequality

$$\|X - Z^{(n)}\|_{p,\infty}^{[M],P} \leq \|X - X^{(n)}\|_{p,\infty}^{[M],P} + \|X^{(n)} - Z^{(n)}\|_{p,\infty}^{[M],P},$$

we have $\lim_{n \to \infty} \|X - Z^{(n)}\|_{p,\infty}^{[M],P} = 0$. ∎

Let us extend the definition of $X \bullet M$ to a predictable process X on the filtered space $(\Omega, \mathfrak{F}, \{\mathfrak{F}_t\}, P)$ which is also in $\mathbf{L}_{2,\infty}(\mathbb{R}_+ \times \Omega, \mu_{[M]}, P)$. According to Theorem 12.8, there exists a sequence $\{X^{(n)} : n \in \mathbb{N}\}$ in $\mathbf{L}_0(\Omega, \mathfrak{F}, \{\mathfrak{F}_t\}, P)$ such that $\lim_{n \to \infty} \|X^{(n)} - X\|_{2,\infty}^{[M],P} = 0$. For $m, n \in \mathbb{N}$, by Proposition 12.4, we have

$$\| X^{(m)} \bullet M - X^{(n)} \bullet M \|_\infty = \|X^{(m)} - X^{(n)}\|_{2,\infty}^{[M],P}.$$

Then since $\{X^{(n)} : n \in \mathbb{N}\}$ is a Cauchy sequence in the metric of the quasinorm $\|\cdot\|_{2,\infty}^{[M],P}$, the sequence $\{X^{(n)} \bullet M : n \in \mathbb{N}\}$ is a Cauchy sequence in the metric of the quasinorm $\|\cdot\|_\infty$. The completeness of $\mathbf{M}_2(\Omega, \mathfrak{F}, \{\mathfrak{F}_t\}, P)$ with respect to the metric of the quasinorm $\|\cdot\|_\infty$ implies that there exists $Y \in \mathbf{M}_2(\Omega, \mathfrak{F}, \{\mathfrak{F}_t\}, P)$ such that $\lim_{n \to \infty} \| X^{(n)} \bullet M - Y \|_\infty = 0$. We define $X \bullet M = Y$. It is clear that Y does not depend on the choice of the sequence $\{X^{(n)} : n \in \mathbb{N}\}$ in $\mathbf{L}_0(\Omega, \mathfrak{F}, \{\mathfrak{F}_t\}, P)$ satisfying the condition $\lim_{n \to \infty} \|X^{(n)} - X\|_{2,\infty}^{[M],P} = 0$. In fact if $\{X^{(1,n)} : n \in \mathbb{N}\}$ and $\{X^{(2,n)} : n \in \mathbb{N}\}$ are two such sequences and if we let $\{X^{(n)} : n \in \mathbb{N}\}$ be the sequence $X^{(1,1)}, X^{(2,1)}, X^{(1,2)}, X^{(2,2)}, X^{(1,3)}, X^{(2,3)}, \ldots$, then for $Y \in \mathbf{M}_2(\Omega, \mathfrak{F}, \{\mathfrak{F}_t\}, P)$ such that $\lim_{n \to \infty} \| X^{(n)} \bullet M - Y \|_\infty = 0$, we have $\lim_{n \to \infty} \| X^{(i,n)} \bullet M - Y \|_\infty = 0$ for $i = 1$ and 2.

Definition 12.9 Let $M \in \mathbf{M}_2(\Omega, \mathfrak{F}, \{\mathfrak{F}_t\}, P)$ and let $X = \{X_t : t \in \mathbb{R}_+\}$ be a predictable process on the filtered space $(\Omega, \mathfrak{F}, \{\mathfrak{F}_t\}, P)$ such that $X \in \mathbf{L}_{2,\infty}(\mathbb{R}_+ \times \Omega, \mu_{[M]}, P)$. The element $X \bullet M \in \mathbf{M}_2(\Omega, \mathfrak{F}, \{\mathfrak{F}_t\}, P)$ such that $\lim_{n \to \infty} \| X^{(n)} \bullet M - X \bullet M \|_\infty = 0$ for an arbitrary sequence $\{X^{(n)} : n \in \mathbb{N}\}$ in $\mathbf{L}_0(\Omega, \mathfrak{F}, \{\mathfrak{F}_t\}, P)$ satisfying the condition $\lim_{n \to \infty} \|X^{(n)} - X\|_{2,\infty}^{[M],P} = 0$ is called the stochastic integral of X with respect to M. For every $t \in \mathbb{R}_+$, the random variable $(X \bullet M)(t)$ has an alternate notation $\int_{[0,t]} X(s) \, dM(s)$. Its value at $\omega \in \Omega$ is denoted by $(\int_{[0,t]} X(s) \, dM(s))(\omega)$.

Note that the stochastic integral $X \bullet M$ is defined as a stochastic process. Note also that although we use the notation $\int_{[0,t]} X(s) \, dM(s)$ for the random variable $(X \bullet M)(t)$, it is not defined as the Lebesgue-Stieltjes integral of the individual sample functions of X. The random variable $\int_{[0,t]} X(s) \, dM(s)$ is sometimes called the stochastic integral of X with respect to M on $[0, t]$.

Observation 12.10. If $M \in \mathbf{M}_2^c(\Omega, \mathfrak{F}, \{\mathfrak{F}_t\}, P)$ then $X \bullet M \in \mathbf{M}_2^c(\Omega, \mathfrak{F}, \{\mathfrak{F}_t\}, P)$ also. This follows from the fact that for an arbitrary sequence $\{X^{(n)} : n \in \mathbb{N}\}$ in $\mathbf{L}_0(\Omega, \mathfrak{F}, \{\mathfrak{F}_t\}, P)$ such that $\lim_{n \to \infty} \|X^{(n)} - X\|_{2,\infty}^{[M],P} = 0$, $X^{(n)} \bullet M$ is in $\mathbf{M}_2^c(\Omega, \mathfrak{F}, \{\mathfrak{F}_t\}, P)$ according to Proposition 12.4 so that our $X \bullet M$ satisfying the condition $\lim_{n \to \infty} \| X^{(n)} \bullet M - X \bullet M \|_\infty = 0$ is in the closed linear subspace $\mathbf{M}_2^c(\Omega, \mathfrak{F}, \{\mathfrak{F}_t\}, P)$ of $\mathbf{M}_2(\Omega, \mathfrak{F}, \{\mathfrak{F}_t\}, P)$.

§12. STOCHASTIC INTEGRALS WITH RESPECT TO MARTINGALES 239

Observation 12.11. For $X \bullet M$ in Definition 12.9, we have $\lfloor X \bullet M \rfloor_\infty = \|X\|_{2,\infty}^{[M],P}$, that is, the mapping of a predictable process X in $\mathbf{L}_{2,\infty}(\mathbb{R}_+ \times \Omega, \mu_{[M]}, P)$ into $X \bullet M$ in $\mathbf{M}_2(\Omega, \mathfrak{F}, \{\mathfrak{F}_t\}, P)$ preserves the quasinorm. Also for every $t \in \mathbb{R}_+$, we have $\lfloor X \bullet M \rfloor_t = \|X\|_{2,t}^{[M],P}$.

Proof. Let $\{X^{(n)} : n \in \mathbb{N}\}$ be a sequence in $\mathbf{L}_0(\Omega, \mathfrak{F}, \{\mathfrak{F}_t\}, P)$ satisfying the condition $\lim_{n \to \infty} \|X^{(n)} - X\|_{2,\infty}^{[M],P} = 0$. Then $\lim_{n \to \infty} \lfloor X^{(n)} \bullet M - X \bullet M \rfloor_\infty = 0$ and this implies that $\lim_{n \to \infty} \lfloor X^{(n)} \bullet M \rfloor_\infty = \lfloor X \bullet M \rfloor_\infty$. But $\lfloor X^{(n)} \bullet M \rfloor_\infty = \|X^{(n)}\|_{2,\infty}^{[M],P}$ by Proposition 12.4. Also $\lim_{n \to \infty} \|X^{(n)} - X\|_{2,\infty}^{[M],P} = 0$ implies $\lim_{n \to \infty} \|X^{(n)}\|_{2,\infty}^{[M],P} = \|X\|_{2,\infty}^{[M],P}$. Therefore $\lfloor X \bullet M \rfloor_\infty = \|X\|_{2,\infty}^{[M],P}$.

For every $t \in \mathbb{R}_+$, $\lim_{n \to \infty} \|X^{(n)} - X\|_{2,\infty}^{[M],P} = 0$ implies $\lim_{n \to \infty} \|X^{(n)} - X\|_{2,t}^{[M],P} = 0$ and thus $\lim_{n \to \infty} \|X^{(n)}\|_{2,t}^{[M],P} = \|X\|_{2,t}^{[M],P}$. Similarly $\lim_{n \to \infty} \lfloor X^{(n)} \bullet M - X \bullet M \rfloor_\infty = 0$ implies $\lim_{n \to \infty} \lfloor X^{(n)} \bullet M - X \bullet M \rfloor_t = 0$ and then $\lim_{n \to \infty} \lfloor X^{(n)} \bullet M \rfloor_t = \lfloor X \bullet M \rfloor_t$. But according to Proposition 12.4, $\lfloor X^{(n)} \bullet M \rfloor_t = \|X^{(n)}\|_{2,t}^{[M],P}$. Thus $\lfloor X \bullet M \rfloor_t = \|X\|_{2,t}^{[M],P}$. ∎

Proposition 12.12. Let $M \in \mathbf{M}_2(\Omega, \mathfrak{F}, \{\mathfrak{F}_t\}, P)$ where $(\Omega, \mathfrak{F}, P)$ is not assumed to be complete. For predictable processes X and Y on the filtered space $(\Omega, \mathfrak{F}, \{\mathfrak{F}_t\}, P)$ which are also in $\mathbf{L}_{2,\infty}(\mathbb{R}_+ \times \Omega, \mu_{[M]}, P)$ and for $\alpha, \beta \in \mathbb{R}$ we have
(1) $X = Y \in \mathbf{L}_{2,\infty}(\mathbb{R}_+ \times \Omega, \mu_{[M]}, P) \Rightarrow X \bullet M = Y \bullet M \in \mathbf{M}_2(\Omega, \mathfrak{F}, \{\mathfrak{F}_t\}, P)$,
(2) $\{\alpha X + \beta Y\} \bullet M = \alpha(X \bullet M) + \beta(Y \bullet M)$,
(3) $(\{\alpha X + \beta Y\} \bullet M)_t = \alpha(X \bullet M)_t + \beta(Y \bullet M)_t$ a.e. on $(\Omega, \mathfrak{F}_t, P)$ for $t \in \mathbb{R}_+$,
(4) $1 \bullet M = M$.

Proof. To prove (1), suppose $X = Y$ as elements of $\mathbf{L}_{2,\infty}(\mathbb{R}_+ \times \Omega, \mu_{[M]}, P)$, that is, $\|X - Y\|_{2,\infty}^{[M],P} = 0$ for every $m \in \mathbb{N}$ according to (4) of Observation 11.16. Let $\{X^{(n)} : n \in \mathbb{N}\}$ be a sequence in $\mathbf{L}_0(\Omega, \mathfrak{F}, \{\mathfrak{F}_t\}, P)$ such that $\lim_{n \to \infty} \|X^{(n)} - X\|_{2,\infty}^{[M],P} = 0$. This implies $\lim_{n \to \infty} \|X^{(n)} - X\|_{2,m}^{[M],P} = 0$ for every $m \in \mathbb{N}$ according to Remark 11.18. Since $\|X^{(n)} - Y\|_{2,m}^{[M],P} \leq \|X^{(n)} - X\|_{2,m}^{[M],P} + \|X - Y\|_{2,m}^{[M],P}$, we have $\lim_{n \to \infty} \|X^{(n)} - Y\|_{2,m}^{[M],P} = 0$ for every $m \in \mathbb{N}$ and therefore $\lim_{n \to \infty} \|X^{(n)} - Y\|_{2,\infty}^{[M],P} = 0$ by Remark 11.18. Thus we have $\lim_{n \to \infty} \lfloor X^{(n)} \bullet M - X \bullet M \rfloor_\infty = 0$ as well as $\lim_{n \to \infty} \lfloor X^{(n)} \bullet M - Y \bullet M \rfloor_\infty = 0$. Therefore $X \bullet M = Y \bullet M$ as elements of $\mathbf{M}_2(\Omega, \mathfrak{F}, \{\mathfrak{F}_t\}, P)$.

To prove (2), let $\{X^{(n)} : n \in \mathbb{N}\}$ and $\{Y^{(n)} : n \in \mathbb{N}\}$ be two sequences of processes in $\mathbf{L}_0(\Omega, \mathfrak{F}, \{\mathfrak{F}_t\}, P)$ such that $\lim_{n \to \infty} \|X^{(n)} - X\|_{2,\infty}^{[M],P} = 0$ and $\lim_{n \to \infty} \|Y^{(n)} - Y\|_{2,\infty}^{[M],P} = 0$. It

follows from Definition 12.3 that
$$\{\alpha X^{(n)} + \beta Y^{(n)}\} \bullet M = \alpha(X^{(n)} \bullet M) + \beta(Y^{(n)} \bullet M).$$
Letting $n \to \infty$ we obtain (2).

Now (2) implies that there exists a null set Λ in $(\Omega, \mathfrak{F}, P)$ such that for $\omega \in \Lambda^c$ we have $(\{\alpha X + \beta Y\} \bullet M)(\cdot, \omega) = \alpha(X \bullet M)(\cdot, \omega) + \beta(Y \bullet M)(\cdot, \omega)$ and in particular $(\{\alpha X + \beta Y\} \bullet M)(t, \omega) = \alpha(X \bullet M)(t, \omega) + \beta(Y \bullet M)(t, \omega)$ for every $t \in \mathbb{R}_+$. Since $\Lambda \in \mathfrak{F}_0 \subset \mathfrak{F}_t$, we have (3).

By the fact that $1 \in \mathbf{L}_0(\Omega, \mathfrak{F}, \{\mathfrak{F}_t\}, P)$, (4) follows from Definition 12.3. ∎

Proposition 12.13. *Let $M \in \mathbf{M}_2(\Omega, \mathfrak{F}, \{\mathfrak{F}_t\}, P)$. Let $X^{(n)}$, $n \in \mathbb{N}$, and X be predictable processes which are in $\mathbf{L}_{2,\infty}(\mathbb{R}_+ \times \Omega, \mu_{[M]}, P)$ such that $\lim_{n \to \infty} \|X^{(n)} - X\|_{2,\infty}^{[M],P} = 0$. Then $\lim_{n \to \infty} \mathbf{I} X^{(n)} \bullet M - X \bullet M \mathbf{I}_\infty = 0$.*

Proof. There exists $Y^{(n)} \in \mathbf{L}_0(\Omega, \mathfrak{F}, \{\mathfrak{F}_t\}, P)$ such that $\|Y^{(n)} - X^{(n)}\|_{2,\infty}^{[M],P} < 2^{-n}$ for each $n \in \mathbb{N}$ by Theorem 12.8. Then by the triangle inequality of the metric of the quasinorm,
$$\|Y^{(n)} - X\|_{2,\infty}^{[M],P} \leq \|Y^{(n)} - X^{(n)}\|_{2,\infty}^{[M],P} + \|X^{(n)} - X\|_{2,\infty}^{[M],P},$$
we have $\lim_{n \to \infty} \|Y^{(n)} - X\|_{2,\infty}^{[M],P} = 0$. Now
$$\begin{aligned} &\mathbf{I} X^{(n)} \bullet M - X \bullet M \mathbf{I}_\infty \\ \leq\ & \mathbf{I} X^{(n)} \bullet M - Y^{(n)} \bullet M \mathbf{I}_\infty + \mathbf{I} Y^{(n)} \bullet M - X \bullet M \mathbf{I}_\infty \\ =\ & \|X^{(n)} - Y^{(n)}\|_{2,\infty}^{[M],P} + \|Y^{(n)} - X\|_{2,\infty}^{[M],P}. \end{aligned}$$
by Observation 12.11 and thus $\lim_{n \to \infty} \mathbf{I} X^{(n)} \bullet M - X \bullet M \mathbf{I}_\infty = 0$. ∎

Remark 12.14. Let $M \in \mathbf{M}_2(\Omega, \mathfrak{F}, \{\mathfrak{F}_t\}, P)$. For predictable processes $X^{(n)}$, $n \in \mathbb{N}$, and X, which are in $\mathbf{L}_{2,\infty}(\mathbb{R}_+ \times \Omega, \mu_{[M]}, P)$ and such that $\lim_{n \to \infty} \|X^{(n)} - X\|_{2,\infty}^{[M],P} = 0$, we have $\{X^{(n)} \bullet M, n \in \mathbb{N}, X \bullet M\} \subset \mathbf{M}_2(\Omega, \mathfrak{F}, \{\mathfrak{F}_t\}, P)$ and by Proposition 12.13 we have $\lim_{n \to \infty} \mathbf{I} X^{(n)} \bullet M - X \bullet M \mathbf{I}_\infty = 0$. This implies according to Proposition 11.6 that for every $m \in \mathbb{N}$ we have
$$P \cdot \lim_{n \to \infty} \sup_{t \in [0,m)} |(X^{(n)} \bullet M)_t - (X \bullet M)_t| = 0,$$
and moreover there exists a subsequence $\{n_k\}$ of $\{n\}$ and a null set Λ in $(\Omega, \mathfrak{F}, P)$ such that for every $\omega \in \Lambda^c$ the sample functions $(X^{(n_k)} \bullet M)(\cdot, \omega)$ converge to the sample function $(X \bullet M)(\cdot, \omega)$ uniformly on every finite interval in $\mathbb{R}_+$.

§12. STOCHASTIC INTEGRALS WITH RESPECT TO MARTINGALES

Remark 12.15. For $M \in \mathbf{M}_2(\Omega, \mathfrak{F}, \{\mathfrak{F}_t\}, P)$ and a predictable process X on the filtered space which is also in $\mathbf{L}_{2,\infty}(\mathbb{R}_+ \times \Omega, \mu_{[M]}, P)$, the stochastic integral $X \bullet M$ is in $M \in \mathbf{M}_2(\Omega, \mathfrak{F}, \{\mathfrak{F}_t\}, P)$. In particular it is a right-continuous martingale and thus according to Theorem 9.2 almost every sample function of $X \bullet M$ is not only right-continuous but also has finite left limit everywhere on $\mathbb{R}_+$ and is bounded on every finite interval in $\mathbb{R}_+$ and has at most countably many points of discontinuity.

$(X \bullet M)(t, \omega)$ is related to the limit of Riemann-Stieltjes sums of $X(\cdot, \omega)$ with respect to $M(\cdot, \omega)$ in the following way.

Proposition 12.16 *Let $M \in \mathbf{M}_2(\Omega, \mathfrak{F}, \{\mathfrak{F}_t\}, P)$ and X be a bounded left-continuous adapted process on the filtered space. For $n \in \mathbb{N}$, let Δ_n be a partition of $\mathbb{R}_+$ into subintervals by a strictly increasing sequence $\{t_{n,k} : k \in \mathbb{Z}_+\}$ in $\mathbb{R}_+$ such that $t_{n,0} = 0$ and $t_{n,k} \uparrow \infty$ as $k \to \infty$ and $\lim_{n \to \infty} |\Delta_n| = 0$ where $|\Delta_n| = \sup_{k \in \mathbb{N}}(t_{n,k} - t_{n,k-1})$. Then for every $t \in \mathbb{R}_+$ we have*

$$(1) \quad \lim_{n \to \infty} \|\sum_{k=1}^{p_n} X(t_{n,k-1})\{M(t_{n,k}) - M(t_{n,k-1})\} - (X \bullet M)(t)\|_2 = 0,$$

where $t_{n,p_n} = t$ and $t_{n,k} < t$ for $k = 0, \ldots, p_n - 1$ for $n \in \mathbb{N}$. In particular

$$(2) \quad P \cdot \lim_{n \to \infty} \sum_{k=1}^{p_n} X(t_{n,k-1})\{M(t_{n,k}) - M(t_{n,k-1})\} = (X \bullet M)(t).$$

Proof. Note that since X is a left-continuous and adapted process, it is a predictable process. The fact that X is bounded implies that X is in $\mathbf{L}_{2,\infty}(\mathbb{R}_+ \times \Omega, \mu_{[M]}, P)$ by Observation 11.19. Thus $X \bullet M$ exists in $\mathbf{M}_2(\Omega, \mathfrak{F}, \{\mathfrak{F}_t\}, P)$. For $n \in \mathbb{N}$, define a process $X^{(n)}$ in $\mathbf{L}_0(\Omega, \mathfrak{F}, \{\mathfrak{F}_t\}, P)$ by setting

$$X^{(n)}(s) = X(0)\mathbf{1}_{\{0\}}(s) + \sum_{k \in \mathbb{N}} X(t_{n,k-1})\mathbf{1}_{(t_{n,k-1}, t_{n,k}]}(s) \quad \text{for } s \in \mathbb{R}_+.$$

The left-continuity of every sample function of X implies that $\lim_{n \to \infty} X^{(n)}(s, \omega) = X(s, \omega)$ for $(s, \omega) \in \mathbb{R}_+ \times \Omega$. Now $[M] \in \mathbf{A}(\Omega, \mathfrak{F}, \{\mathfrak{F}_t\}, P)$ and in particular $[M]_t$ is integrable for every $t \in \mathbb{R}_+$. From this it follows by applying the Bounded Convergence Theorem and the Dominated Convergence Theorem that for every $m \in \mathbb{N}$ we have $\lim_{n \to \infty} \|X^{(n)} - X\|_{2,m}^{[M],P} = 0$ and hence $\lim_{n \to \infty} \|X^{(n)} - X\|_{2,\infty}^{[M],P} = 0$ by Remark 11.18. Thus by Definition 12.9, we have

$\lim_{n\to\infty} \| X^{(n)} \bullet M - X \bullet M \|_\infty = 0$. This implies that $\lim_{n\to\infty} \| X^{(n)} \bullet M - X \bullet M \|_t = 0$ for every $t \in \mathbb{R}_+$. Now since $X^{(n)} \in \mathbf{L}_0(\Omega, \mathfrak{F}, \{\mathfrak{F}_t\}, P)$ we have according to Definition 12.3

$$(X^{(n)} \bullet M)(t) = \sum_{k=1}^{p_n} X(t_{n,k-1})\{M(t_{n,k}) - M(t_{n,k-1})\}.$$

Thus (1) holds. ∎

For $M, N \in \mathbf{M}_2(\Omega, \mathfrak{F}, \{\mathfrak{F}_t\}, P)$ and predictable processes $X \in \mathbf{L}_{2,\infty}(\mathbb{R}_+ \times \Omega, \mu_{[M]}, P)$ and $Y \in \mathbf{L}_{2,\infty}(\mathbb{R}_+ \times \Omega, \mu_{[N]}, P)$, both of the stochastic integrals $X \bullet M$ and $Y \bullet N$ are in $\mathbf{M}_2(\Omega, \mathfrak{F}, \{\mathfrak{F}_t\}, P)$ so that the quadratic variation processes $[X \bullet M]$, $[Y \bullet N]$, and $[X \bullet M, Y \bullet N]$ exist. In next theorem under the assumption that X and Y are almost surely continuous, we show that these quadratic variation processes are obtained by integrating the sample functions of X^2, Y^2, and XY with respect to the families of Lebesgue-Stieltjes measures $\mu_{[M]}$, $\mu_{[N]}$, and $\mu_{[M,N]}$ determined by the quadratic variation processes $[M]$, $[N]$, and $[M,N]$.

Theorem 12.17. *Let $M, N \in \mathbf{M}_2^c(\Omega, \mathfrak{F}, \{\mathfrak{F}_t\}, P)$ and let X and Y be two predictable processes on the filtered space $(\Omega, \mathfrak{F}, \{\mathfrak{F}_t\}, P)$ such that $X \in \mathbf{L}_{2,\infty}(\mathbb{R}_+ \times \Omega, \mu_{[M]}, P)$ and $Y \in \mathbf{L}_{2,\infty}(\mathbb{R}_+ \times \Omega, \mu_{[N]}, P)$. Then there exists a null set Λ in $(\Omega, \mathfrak{F}, P)$ such that on Λ^c, for every $t \in \mathbb{R}_+$, we have*

(1) $$[X \bullet M, Y \bullet N]_t = \int_{[0,t]} X(s)Y(s)\, d[M,N](s),$$

and thus for any $s, t \in \mathbb{R}_+$, $s < t$, we have

(2) $$\mathbf{E}[\{(X \bullet M)_t - (X \bullet M)_s\}\{(Y \bullet N)_t - (Y \bullet N)_s\}|\mathfrak{F}_s]$$
$$= \mathbf{E}[(X \bullet M)_t(Y \bullet N)_t - (X \bullet M)_s(Y \bullet N)_s|\mathfrak{F}_s]$$
$$= \mathbf{E}\left[\int_{(s,t]} X(u)Y(u)\, d[M,N](u)\Big|\mathfrak{F}_s\right] \quad \text{a.e. on } (\Omega, \mathfrak{F}_s, P).$$

In particular there exists a null set Λ in $(\Omega, \mathfrak{F}, P)$ such that on Λ^c we have for every $t \in \mathbb{R}_+$

(3) $$[X \bullet M]_t = \int_{[0,t]} X^2(s)\, d[M](s),$$

and

(4) $$\mathbf{E}[\{(X \bullet M)_t - (X \bullet M)_s\}^2|\mathfrak{F}_s] = \mathbf{E}[(X \bullet M)_t^2 - (X \bullet M)_s^2|\mathfrak{F}_s]$$
$$= \mathbf{E}\left[\int_{(s,t]} X^2(u)\, d[M](u)\Big|\mathfrak{F}_s\right] \quad \text{a.e. on } (\Omega, \mathfrak{F}_s, P).$$

§12. STOCHASTIC INTEGRALS WITH RESPECT TO MARTINGALES

Proof. Let Λ be the union of the two null sets Λ_∞ and $\Lambda_{\infty,X,Y}$ in the statement of Theorem 12.6. Let $V = \{V_t : t \in \mathbb{R}_+\}$ be defined by

$$(5) \quad V(t,\omega) = \begin{cases} \int_{[0,t]} X(s,\omega) Y(s,\omega)\, d[M,N](s,\omega) & \text{for } (t,\omega) \in \mathbb{R}_+ \times \Lambda^c \\ 0 & \text{for } (t,\omega) \in \mathbb{R}_+ \times \Lambda. \end{cases}$$

By (1) of Theorem 12.6, V is real valued function on $\mathbb{R}_+ \times \Omega$.

To prove (1), we show that $V \in \mathbf{V}(\Omega, \mathfrak{F}, \{\mathfrak{F}_t\}, P)$ and that $(X \bullet M)(Y \bullet N) - V$ is a null at 0 right-continuous martingale. Note that the first equality in (2) is an immediate consequence of the martingale property of $X \bullet M$ and $Y \bullet N$. Note also that proving that $(X \bullet M)(Y \bullet N) - V$ is a martingale is equivalent to proving the second equality in (2).

To show that $V \in \mathbf{V}(\Omega, \mathfrak{F}, \{\mathfrak{F}_t\}, P)$, note first that by writing $V = A' - A''$ where $A', A'' \in \mathbf{A}(\Omega, \mathfrak{F}, \{\mathfrak{F}_t\}, P)$ by Theorem 11.12 and by applying Theorem 10.11 we have the $\mathfrak{F}_t$-measurability of V_t for every $t \in \mathbb{R}_+$. Thus V is an adapted process. The right-continuity of V follows from that of $[M,N]$. For $t \in \mathbb{R}_+$, let $t_{n,k} = k2^{-n}t$ for $k = 0, \ldots, 2^n$ and $n \in \mathbb{N}$. Then by (5) we have

$$\sum_{k=1}^{2^n} |V(t_{n,k}) - V(t_{n,k-1})| = \sum_{k=1}^{2^n} \left| \int_{(t_{n,k-1}, t_{n,k}]} X(s) Y(s)\, d[M,N](s) \right|$$
$$\leq \int_{[0,t]} |X(s)Y(s)|\, d\|[M,N]\|(s) < \infty \quad \text{on } \Lambda^c$$

by (1) of Theorem 12.6, and thus

$$\|V\|(t) = \sup_{n \in \mathbb{N}} \sum_{k=1}^{2^n} |V(t_{n,k}) - V(t_{n,k-1})| < \infty \quad \text{on } \Lambda^c,$$

that is, $V(\cdot, \omega)$ is a right-continuous function of bounded variation on $[0,t]$ for every $t \in \mathbb{R}_+$ and vanishes at $t = 0$ for $\omega \in \Lambda^c$. By (2) of Theorem 12.6, we have

$$\mathbf{E}(\|V\|_t) \leq \mathbf{E}\left[\int_{[0,t]} |X(s)Y(s)|\, d\|[M,N]\|(s)\right] < \infty,$$

and thus $\|V\|_t$ is integrable for every $t \in \mathbb{R}_+$. This shows that $\|V\|$ is an L_1-process and completes the proof that $V \in \mathbf{V}(\Omega, \mathfrak{F}, \{\mathfrak{F}_t\}, P)$.

Let us prove the second equality in (2). Now by Theorem 12.8, there exist two sequences $\{X^{(n)} : n \in \mathbb{N}\}$ and $\{Y^{(n)} : n \in \mathbb{N}\}$ in $\mathbf{L}_0(\Omega, \mathfrak{F}, \{\mathfrak{F}_t\}, P)$ with $\lim_{n \to \infty} \|X^{(n)} - X\|_{2,\infty}^{[M],P} = 0$

and $\lim_{n\to\infty} \|Y^{(n)} - Y\|_{2,\infty}^{[N],P} = 0$. Since $X^{(n)}$ and $Y^{(n)}$ are in $\mathbf{L}_0(\Omega, \mathfrak{F}, \{\mathfrak{F}_t\}, P)$, by Proposition 12.5 for any $s, t \in \mathbb{R}_+, s < t$, we have

(6) $\quad \mathbf{E}[(X^{(n)} \bullet M)_t(Y^{(n)} \bullet N)_t - (X^{(n)} \bullet M)_s(Y^{(n)} \bullet N)_s | \mathfrak{F}_s]$
$$= \mathbf{E}\left[\int_{(s,t]} X^{(n)}(u)Y^{(n)}(u)\, d[M,N](u) \Big| \mathfrak{F}_s\right] \quad \text{a.e. on } (\Omega, \mathfrak{F}_s, P).$$

Let us show that if we restrict (6) to a certain subsequence $\{n_\ell\}$ of $\{n\}$ and let $\ell \to \infty$, then we obtain the second equality in (2). Now the convergence $\lim_{n\to\infty} \|X^{(n)} - X\|_{2,\infty}^{[M],P} = 0$ implies $\lim_{n\to\infty} \mathbf{I} X^{(n)} \bullet M - X \bullet M \mathbf{I}_\infty = 0$ according to Definition 12.9 and this in turn implies $\lim_{n\to\infty} \mathbf{I} X^{(n)} \bullet M - X \bullet M \mathbf{I}_t = 0$ for every $t \in \mathbb{R}_+$ by Remark 11.4. Thus $(X^{(n)} \bullet M)_t$ converges to $(X \bullet M)_t$ in $L_2(\Omega, \mathfrak{F}, P)$. Similarly $(X^{(n)} \bullet M)_s$, $(Y^{(n)} \bullet M)_t$ and $(Y^{(n)} \bullet M)_s$ converge to $(X \bullet M)_s$, $(Y \bullet M)_t$ and $(Y \bullet M)_s$ respectively in $L_2(\Omega, \mathfrak{F}, P)$. Thus $(X^{(n)} \bullet M)_t(Y^{(n)} \bullet N)_t - (X^{(n)} \bullet M)_s(Y^{(n)} \bullet N)_s$ converges to $(X \bullet M)_t(Y \bullet N)_t - (X \bullet M)_s(Y \bullet N)_s$ in $L_1(\Omega, \mathfrak{F}, P)$. Now the convergence of a sequence of random variables to a random variable in $L_1(\Omega, \mathfrak{F}, P)$ implies the convergence of the sequence of the conditional expectations with respect to an arbitrary sub-σ-algebra of $\mathfrak{F}$ in $L_1(\Omega, \mathfrak{F}, P)$. (See Theorem B.26.) Since convergence in $L_1(\Omega, \mathfrak{F}, P)$ of a sequence of random variables implies the existence of a subsequence which converges a.e. on $(\Omega, \mathfrak{F}, P)$, there exists a subsequence $\{n_k\}$ of $\{n\}$ such that

(7) $\quad \lim_{k\to\infty} \mathbf{E}[(X^{(n_k)} \bullet M)_t(Y^{(n_k)} \bullet N)_t - (X^{(n_k)} \bullet M)_s(Y^{(n_k)} \bullet N)_s | \mathfrak{F}_s]$
$$= \mathbf{E}[(X \bullet M)_t(Y \bullet N)_t - (X \bullet M)_s(Y \bullet N)_s | \mathfrak{F}_s] \quad \text{a.e. on } (\Omega, \mathfrak{F}_s, P).$$

Let us show next that

(8) $\quad \lim_{n\to\infty} \mathbf{E}\left\{\left|\int_{(s,t]} X^{(n_k)}(u)Y^{(n_k)}(u)\, d[M,N](u) - \int_{(s,t]} X(u)Y(u)\, d[M,N](u)\right|\right\} = 0.$

By the algebraic equality

$$X^{(n_k)}Y^{(n_k)} - XY = \{X^{(n_k)} - X\}\{Y^{(n_k)} - Y\} + \{X^{(n_k)} - X\}Y + X\{Y^{(n_k)} - Y\}$$

and by (2) of Theorem 12.6, we have

(9) $\quad \mathbf{E}\left\{\left|\int_{(s,t]} X^{(n_k)}(u)Y^{(n_k)}(u)\, d[M,N](u) - \int_{(s,t]} X(u)Y(u)\, d[M,N](u)\right|\right\}$
$$\leq \mathbf{E}\left\{\int_{(s,t]} |X^{(n_k)}(u) - X(u)|^2\, d[M](u)\right\}^{1/2} \mathbf{E}\left[\int_{(s,t]} |Y^{(n_k)}(u) - Y(u)|^2\, d[N](u)\right]^{1/2}$$

§12. STOCHASTIC INTEGRALS WITH RESPECT TO MARTINGALES 245

$$+ \mathbf{E}\left\{\int_{(s,t]} |X^{(n_k)}(u) - X(u)|^2 \, d[M](u)\right\}^{1/2} \mathbf{E}\left[\int_{(s,t]} Y^2(u) \, d[N](u)\right]^{1/2}$$

$$+ \mathbf{E}\left\{\int_{(s,t]} X^2(u) \, d[M](u)\right\}^{1/2} \mathbf{E}\left[\int_{(s,t]} |Y^{(n_k)}(u) - Y(u)|^2 \, d[N](u)\right]^{1/2}$$

$$= \|X^{(n_k)} - X\|_{2,\infty}^{[M],P} \|Y^{(n_k)} - Y\|_{2,\infty}^{[N],P} + \|X^{(n_k)} - X\|_{2,\infty}^{[M],P} \|Y\|_{2,\infty}^{[N],P}$$

$$+ \|X\|_{2,\infty}^{[M],P} \|Y^{(n_k)} - Y\|_{2,\infty}^{[N],P}.$$

Now $\lim_{k \to \infty} \|X^{(n_k)} - X\|_{2,\infty}^{[M],P} = 0$ implies $\lim_{n \to \infty} \|X^{(n_k)} - X\|_{2,t}^{[M],P} = 0$. Similarly we have $\lim_{k \to \infty} \|Y^{(n_k)} - Y\|_{2,t}^{[N],P} = 0$. Thus if we let $k \to \infty$ in (9) we have (8). Now (8) implies the existence of a subsequence $\{n_\ell\}$ of $\{n_k\}$ such that

(10)
$$\lim_{n \to \infty} \mathbf{E}\left[\int_{(s,t]} X^{(n_\ell)}(u) Y^{(n_\ell)}(u) \, d[M,N](u) \big| \mathfrak{F}_s\right]$$
$$= \mathbf{E}\left[\int_{(s,t]} X(u) Y(u) \, d[M,N](u) \big| \mathfrak{F}_s\right] \quad \text{a.e. on } (\Omega, \mathfrak{F}_s, P).$$

Restricting (6) to the subsequence $\{n_\ell\}$ and letting $\ell \to \infty$ and applying by (7) and (10), we have the second equality in (2). This completes the proof of (1) and (2). ∎

As a consequence of Theorem 12.17, we have the following characterization of the stochastic integral of a process.

Proposition 12.18. *Let $M \in \mathbf{M}_2^c(\Omega, \mathfrak{F}, \{\mathfrak{F}_t\}, P)$ and let X be a predictable processes on the filtered space $(\Omega, \mathfrak{F}, \{\mathfrak{F}_t\}, P)$ such that $X \in \mathbf{L}_{2,\infty}(\mathbb{R}_+ \times \Omega, \mu_{[M]}, P)$. Then $X \bullet M$ is the unique element in $\mathbf{M}_2(\Omega, \mathfrak{F}, \{\mathfrak{F}_t\}, P)$ such that for every $N \in \mathbf{M}_2(\Omega, \mathfrak{F}, \{\mathfrak{F}_t\}, P)$ there exists a null set Λ in $(\Omega, \mathfrak{F}, P)$ such that on Λ^c for every $t \in \mathbb{R}_+$ we have*

(1)
$$[X \bullet M, N]_t = \int_{[0,t]} X(s) \, d[M,N](s).$$

Proof. By taking $Y = 1 \in \mathbf{L}_{2,\infty}(\mathbb{R}_+ \times \Omega, \mu_{[N]}, P)$ in Theorem 12.17, we see that $X \bullet M$ satisfies (1). To show the uniqueness, suppose $Z \in \mathbf{M}_2(\Omega, \mathfrak{F}, \{\mathfrak{F}_t\}, P)$ has the property that for every $N \in \mathbf{M}_2(\Omega, \mathfrak{F}, \{\mathfrak{F}_t\}, P)$ there exists a null set Λ in $(\Omega, \mathfrak{F}, P)$ such that on Λ^c for every $t \in \mathbb{R}_+$ we have

(2)
$$[Z, N]_t = \int_{[0,t]} X(s) \, d[M,N](s).$$

Now (1) and (2) imply according to Proposition 11.26 that $[X \bullet M - Z, N] = 0$ for every $N \in \mathbf{M}_2(\Omega, \mathfrak{F}, \{\mathfrak{F}_t\}, P)$. Thus with $N = X \bullet M - Z$ we have $[X \bullet M - Z] = 0$ and hence $Z = X \bullet M$. ∎

For $M \in \mathbf{M}_2(\Omega, \mathfrak{F}, \{\mathfrak{F}_t\}, P)$ and a predictable process X on the filtered space which is in $\mathbf{L}_{2,\infty}(\mathbb{R}_+ \times \Omega, \mu_{[M]}, P)$, the stochastic integral $X \bullet M$ exists in $\mathbf{M}_2(\Omega, \mathfrak{F}, \{\mathfrak{F}_t\}, P)$. Thus if Y is a predictable process on the filtered space and is also in $\mathbf{L}_{2,\infty}(\mathbb{R}_+ \times \Omega, \mu_{[X \bullet M]}, P)$, then $Y \bullet (X \bullet M)$ is defined and is in $\mathbf{M}_2(\Omega, \mathfrak{F}, \{\mathfrak{F}_t\}, P)$. Concerning $Y \bullet (X \bullet M)$ we have the following theorem.

Theorem 12.19. *Let $M \in \mathbf{M}_2^c(\Omega, \mathfrak{F}, \{\mathfrak{F}_t\}, P)$ and let X and Y be two predictable processes on the filtered space such that $X, YX \in \mathbf{L}_{2,\infty}(\mathbb{R}_+ \times \Omega, \mu_{[M]}, P)$. Then $Y \in \mathbf{L}_{2,\infty}(\mathbb{R}_+ \times \Omega, \mu_{[X \bullet M]}, P)$ and there exists a null set Λ in $(\Omega, \mathfrak{F}, P)$ such that for $\omega \in \Lambda^c$ we have*

$$(Y \bullet (X \bullet M))(\cdot, \omega) = (YX \bullet M)(\cdot, \omega).$$

Proof. Since Y is a predictable process, it is a measurable process. Thus to show that it is in $\mathbf{L}_{2,\infty}(\mathbb{R}_+ \times \Omega, \mu_{[X \bullet M]}, P)$ it remains to show that for every $t \in \mathbb{R}_+$ we have

(1) $$\mathbf{E}\left[\int_{[0,t]} Y^2(s) \, d[X \bullet M](s)\right] < \infty.$$

Now according to Theorem 12.17 there exists a null set Λ in $(\Omega, \mathfrak{F}, P)$ such that for $\omega \in \Lambda^c$ we have for every $t \in \mathbb{R}_+$

$$[X \bullet M](t, \omega) = \int_{[0,t]} X^2(s, \omega) \, d[M](s, \omega).$$

Thus the Lebesgue-Stieltjes measure $\mu_{[X \bullet M]}(\cdot, \omega)$ on $(\mathbb{R}_+, \mathfrak{B}_{\mathbb{R}_+})$ is absolutely continuous with respect to the Lebesgue-Stieltjes measure $\mu_{[M]}(\cdot, \omega)$ with the Radon-Nikodym derivative

$$\frac{d\mu_{[X \bullet M]}}{d\mu_{[M]}}(s, \omega) = X^2(s, \omega) \quad \text{for } s \in \mathbb{R}_+.$$

Thus by the Lebesgue-Radon-Nikodym Theorem, we have

$$\int_{[0,t]} Y^2(s, \omega) \, d[X \bullet M](s, \omega) = \int_{[0,t]} Y^2(s, \omega) X^2(s, \omega) \, d[M](s, \omega)$$

§12. STOCHASTIC INTEGRALS WITH RESPECT TO MARTINGALES

and therefore

$$\mathbf{E}\left[\int_{[0,t]} Y^2(s)\, d[X \bullet M](s)\right] = \mathbf{E}\left[\int_{[0,t]} Y^2(s) X^2(s)\, d[M](s)\right] < \infty$$

since $YX \in \mathbf{L}_{2,\infty}(\mathbb{R}_+ \times \Omega, \mu_{[M]}, P)$. This proves (1).

Now that Y is in $\mathbf{L}_{2,\infty}(\mathbb{R}_+ \times \Omega, \mu_{[X \bullet M]}, P)$, the stochastic integral $Y \bullet (X \bullet M)$ is defined and is in $\mathbf{M}_2^c(\Omega, \mathfrak{F}, \{\mathfrak{F}_t\}, P)$. On the other hand since X and Y are predictable processes, so is XY. Then since YX is in $\mathbf{L}_{2,\infty}(\mathbb{R}_+ \times \Omega, \mu_{[M]}, P)$, the stochastic integral $YX \bullet M$ is defined and is in $\mathbf{M}_2^c(\Omega, \mathfrak{F}, \{\mathfrak{F}_t\}, P)$. Now according to Proposition 12.18, $Y \bullet (X \bullet M)$ is the unique element in $\mathbf{M}_2(\Omega, \mathfrak{F}, \{\mathfrak{F}_t\}, P)$ such that for every N in $\mathbf{M}_2(\Omega, \mathfrak{F}, \{\mathfrak{F}_t\}, P)$ there exists a null set Λ_1 in $(\Omega, \mathfrak{F}, P)$ such that on Λ_1^c we have for every $t \in \mathbb{R}_+$

$$[Y \bullet (X \bullet M), N]_t = \int_{[0,t]} Y(s)\, d[X \bullet M, N](s), \tag{2}$$

and similarly $YX \bullet M$ is the unique element in $\mathbf{M}_2(\Omega, \mathfrak{F}, \{\mathfrak{F}_t\}, P)$ such that for every N in $\mathbf{M}_2(\Omega, \mathfrak{F}, \{\mathfrak{F}_t\}, P)$ there exists a null set Λ_2 in $(\Omega, \mathfrak{F}, P)$ such that on Λ_2^c we have for every $t \in \mathbb{R}_+$

$$[YX \bullet M, N]_t = \int_{[0,t]} Y(s) X(s)\, d[M, N](s). \tag{3}$$

Thus if we show that there exists a null set Λ_3 in $(\Omega, \mathfrak{F}, P)$ such that on Λ_3^c we have for every $t \in \mathbb{R}_+$

$$\int_{[0,t]} Y(s)\, d[X \bullet M, N](s) = \int_{[0,t]} Y(s) X(s)\, d[M, N](s), \tag{4}$$

then we have $\{Y \bullet (X \bullet M)\}(\cdot, \omega) = (YX \bullet M)(\cdot, \omega)$ for $\omega \in \Lambda^c$ where $\Lambda = \cup_{i=1}^3 \Lambda_i$ and we are done. To prove (4), note that according to Theorem 12.17 there exists a null set Λ_3 in $(\Omega, \mathfrak{F}, P)$ such that on Λ_3^c we have for every $t \in \mathbb{R}_+$

$$[X \bullet M, N]_t = [X \bullet M, 1 \bullet N]_t = \int_{[0,t]} X(s)\, d[M, N](s).$$

Thus for $\omega \in \Lambda_3^c$, $\mu_{[X \bullet M, N]}(\cdot, \omega)$ is absolutely continuous with respect to $\mu_{[M,N]}(\cdot, \omega)$ with Radon-Nikodym derivative $X(\cdot, \omega)$ so that by the Lebesgue-Radon-Nikodym Theorem we have (4). ∎

[III] Truncating the Integrand and Stopping the Integrator by Stopping Times

Theorem 12.20. Let $M, N \in \mathbf{M}_2(\Omega, \mathfrak{F}, \{\mathfrak{F}_t\}, P)$ and let X and Y be two predictable processes on the filtered space $(\Omega, \mathfrak{F}, \{\mathfrak{F}_t\}, P)$ such that $X \in \mathbf{L}_{2,\infty}(\mathbb{R}_+ \times \Omega, \mu_{[M]}, P)$ and $Y \in \mathbf{L}_{2,\infty}(\mathbb{R}_+ \times \Omega, \mu_{[N]}, P)$. Let S and T be two stopping times on the filtered space such that $S \leq T$. Then for every $t \in \mathbb{R}_+$, we have

(1) $\quad\quad\quad \mathbf{E}[(X \bullet M)_{T \wedge t} - (X \bullet M)_{S \wedge t} | \mathfrak{F}_S] = 0 \quad$ a.e. on $(\Omega, \mathfrak{F}_S, P)$,

and

(2) $\quad\quad\quad \mathbf{E}[\{(X \bullet M)_{T \wedge t} - (X \bullet M)_{S \wedge t}\}\{(Y \bullet N)_{T \wedge t} - (Y \bullet N)_{S \wedge t}\} | \mathfrak{F}_S]$
$= \mathbf{E}[(X \bullet M)_{T \wedge t}(Y \bullet N)_{T \wedge t} - (X \bullet M)_{S \wedge t}(Y \bullet N)_{S \wedge t} | \mathfrak{F}_S]$
$= \mathbf{E}[\int_{(S \wedge t, T \wedge t]} X(u) Y(u) \, d[M, N](u) | \mathfrak{F}_S] \quad$ a.e. on $(\Omega, \mathfrak{F}_S, P)$,

and in particular

(3) $\quad\quad\quad \mathbf{E}[\{(X \bullet M)_{T \wedge t} - (X \bullet M)_{S \wedge t}\}^2 | \mathfrak{F}_S]$
$= \mathbf{E}[(X \bullet M)^2_{T \wedge t} - (X \bullet M)^2_{S \wedge t} | \mathfrak{F}_S]$
$= \mathbf{E}[\int_{(S \wedge t, T \wedge t]} X^2(u) \, d[M](u) | \mathfrak{F}_S] \quad$ a.e. on $(\Omega, \mathfrak{F}_S, P)$.

Proof. Since $X \bullet M$ is a right-continuous martingale with respect to a right-continuous filtration, $\{(X \bullet M)_{S \wedge t}, (X \bullet M)_{T \wedge t}\}$ is a two-term martingale with respect to $\{\mathfrak{F}_S, \mathfrak{F}_T\}$ for every $t \in \mathbb{R}_+$ according to Theorem 8.13, that is,

(4) $\quad\quad\quad \mathbf{E}[(X \bullet M)_{T \wedge t} | \mathfrak{F}_S] = (X \bullet M)_{S \wedge t} \quad$ a.e. on $(\Omega, \mathfrak{F}_S, P)$.

From this follows (1). Similarly we have

(5) $\quad\quad\quad \mathbf{E}[(Y \bullet N)_{T \wedge t} | \mathfrak{F}_S] = (Y \bullet N)_{S \wedge t} \quad$ a.e. on $(\Omega, \mathfrak{F}_S, P)$.

The first equality in (2) then follows from (4) and (5).

Since $(X \bullet M)(Y \bullet N) - [X \bullet M, Y \bullet N]$ is a right-continuous martingale with respect to a right-continuous filtration, by Theorem 8.13 again we have

$\mathbf{E}[(X \bullet M)_{T \wedge t}(Y \bullet N)_{T \wedge t} - [X \bullet M, Y \bullet N]_{T \wedge t} | \mathfrak{F}_S]$
$= (X \bullet M)_{S \wedge t}(Y \bullet N)_{S \wedge t} - [X \bullet M, Y \bullet N]_{S \wedge t} \quad$ a.e. on $(\Omega, \mathfrak{F}_S, P)$,

that is,

(6) $\quad\quad\quad \mathbf{E}[(X \bullet M)_{T \wedge t}(Y \bullet N)_{T \wedge t} - (X \bullet M)_{S \wedge t}(Y \bullet N)_{S \wedge t} | \mathfrak{F}_S]$
$= \mathbf{E}[[X \bullet M, Y \bullet N]_{T \wedge t} - [X \bullet M, Y \bullet N]_{S \wedge t} | \mathfrak{F}_S] \quad$ a.e. on $(\Omega, \mathfrak{F}_S, P)$.

§12. STOCHASTIC INTEGRALS WITH RESPECT TO MARTINGALES 249

According to (1) of Theorem 12.17, $[X \bullet M, Y \bullet N](t, \omega)$ is equal to the integral of $X(\cdot, \omega)Y(\cdot, \omega)$ with respect to the signed Lebesgue-Stieltjes measure $\mu_{[M,N]}(\cdot, \omega)$ on $[0, t]$ for every $t \in \mathbb{R}_+$ when $\omega \in \Lambda^c$ where Λ is a null set in $(\Omega, \mathfrak{F}, P)$. Thus we have

$$[X \bullet M, Y \bullet N]_{T \wedge t}(\omega) - [X \bullet M, Y \bullet N]_{S \wedge t}(\omega) = \int_{(S \wedge t, T \wedge t]} X(u, \omega)Y(u, \omega) \, d[M, N](u, \omega)$$

for every $t \in \mathbb{R}_+$ when $\omega \in \Lambda^c$. Using this equality in (6), we have the second equality in (2). ∎

Let $M \in \mathbf{M}_2(\Omega, \mathfrak{F}, \{\mathfrak{F}_t\}, P)$ and let X be a predictable process on $(\Omega, \mathfrak{F}, \{\mathfrak{F}_t\}, P)$ such that $X \in \mathbf{L}_{2,\infty}(\mathbb{R}_+ \times \Omega, \mu_{[M]}, P)$. We shall show that for every stopping time T on the filtered space we have $X^{[T]} \bullet M = (X \bullet M)^{T \wedge}$, that is, the stochastic integral of the truncated process by the stopping time is equal to the stochastic integral of the process stopped by the stopping time. We shall prove this first for bounded adapted left-continuous simple processes, and then extend to the general case.

Lemma 12.21. *Let $M \in \mathbf{M}_2(\Omega, \mathfrak{F}, \{\mathfrak{F}_t\}, P)$ and $X \in \mathbf{L}_0(\Omega, \mathfrak{F}, \{\mathfrak{F}_t\}, P)$. Then for any stopping time T on the filtered space $(\Omega, \mathfrak{F}, \{\mathfrak{F}_t\}, P)$ there exists a null set Λ in $(\Omega, \mathfrak{F}, P)$ such that $(X^{[T]} \bullet M)(\cdot, \omega) = (X \bullet M)^{T \wedge}(\cdot, \omega)$ for $\omega \in \Lambda^c$.*

Proof. Let $X \in \mathbf{L}_0(\Omega, \mathfrak{F}, \{\mathfrak{F}_t\}, P)$ be given by

(1) $\quad X(t, \omega) = \xi_0(\omega)\mathbf{1}_{\{0\}}(t) + \sum_{k \in \mathbb{N}} \xi_k(\omega)\mathbf{1}_{(t_{k-1}, t_k]}(t) \quad \text{for } (t, \omega) \in \mathbb{R}_+ \times \Omega$

where $\{t_k : k \in \mathbb{R}_+\}$ is a strictly increasing sequence in $\mathbb{R}_+$ with $t_0 = 0$ and $\lim_{k \to \infty} t_k = \infty$; $\{\xi_k : k \in \mathbb{Z}_+\}$ is a bounded sequence of random variables on $(\Omega, \mathfrak{F}, P)$ such that ξ_0 is $\mathfrak{F}_{t_0}$-measurable and ξ_k is $\mathfrak{F}_{t_{k-1}}$-measurable for $k \in \mathbb{N}$. Let T be a stopping time on the filtered space. The sample functions of the truncated process $X^{[T]}$ are still left-continuous step functions but $X^{[T]}$ may not be a simple process anymore since $T(\omega) \in \mathbb{R}_+$ may be a point of discontinuity for $X^{[T]}(\cdot, \omega)$ and $T(\omega)$ varies with $\omega \in \Omega$. Let us construct a sequence of discrete valued stopping times $\{T_n : n \in \mathbb{N}\}$ which converges to T and for which $X^{[T_n]}$ is in $\mathbf{L}_0(\Omega, \mathfrak{F}, \{\mathfrak{F}_t\}, P)$ for every $n \in \mathbb{N}$. For $n \in \mathbb{N}$, let $s_{n,k} = k2^{-n}$, $k \in \mathbb{Z}_+$, and $\{u_{n,k} : k \in \mathbb{Z}_+\} = \{t_k : k \in \mathbb{Z}_+\} \cup \{s_{n,k} : k \in \mathbb{Z}_+\}$ where $u_{n,k}$ are distinct and numbered in increasing order. Let us define an $\overline{\mathbb{R}}_+$-valued function ϑ_n on $\overline{\mathbb{R}}_+$ by setting

(2) $\quad \vartheta_n(t) = \begin{cases} u_{n,1} & \text{for } t \in [u_{n,0}, u_{n,1}] \\ u_{n,k} & \text{for } t \in (u_{n,k-1}, u_{n,k}] \text{ and } k \geq 2 \\ \infty & \text{for } t = \infty, \end{cases}$

and let

(3) $$T_n = \vartheta_n \circ T.$$

Since T is an $\mathfrak{F}_T/\mathfrak{B}_{\overline{\mathbb{R}}_+}$-measurable mapping of Ω into $\overline{\mathbb{R}}_+$ and ϑ_n is a $\mathfrak{B}_{\overline{\mathbb{R}}_+}/\mathfrak{B}_{\overline{\mathbb{R}}_+}$-measurable mapping of $\overline{\mathbb{R}}_+$ into $\overline{\mathbb{R}}_+$, T_n is an $\mathfrak{F}_T/\mathfrak{B}_{\overline{\mathbb{R}}_+}$-measurable mapping of Ω into $\overline{\mathbb{R}}_+$. Also since $\vartheta_n(t) \geq t$ for $t \in \overline{\mathbb{R}}_+$, we have $T_n \geq T$ on Ω. Thus by Theorem 3.6, T_n is a stopping time. Let us show that $X^{[T_n]}$ is in $\mathbf{L}_0(\Omega, \mathfrak{F}, \{\mathfrak{F}_t\}, P)$ and

(4) $$(X^{[T_n]} \bullet M) = (X \bullet M)^{T_n \wedge} \quad \text{on } \mathbb{R}_+ \times \Omega.$$

The process X given by (1) can also be written as

(5) $$X(t,\omega) = \eta_{n,0}(\omega)\mathbf{1}_{\{0\}}(t) + \sum_{k \in \mathbb{N}} \eta_{n,k}(\omega)\mathbf{1}_{(u_{n,k-1},u_{n,k}]}(t) \quad \text{for } (t,\omega) \in \mathbb{R}_+ \times \Omega.$$

where for each $n \in \mathbb{N}$, $\{\eta_{n,k} : k \in \mathbb{Z}_+\}$ is a bounded sequence of random variables on $(\Omega, \mathfrak{F}, P)$ such that $\eta_{n,0}$ is $\mathfrak{F}_{u_{n,0}}$-measurable and $\eta_{n,k}$ is $\mathfrak{F}_{u_{n,k-1}}$-measurable for $k \in \mathbb{N}$. By (2) of Observation 3.34, we have

(6) $$X^{[T_n]}(t,\omega) = \begin{cases} X(t\omega) & \text{for } t \leq T_n(\omega) \\ 0 & \text{for } t > T_n(\omega). \end{cases}$$

Clearly $X^{[T_n]}$ is bounded process. By Observation 3.35, $X^{[T_n]}$ is an adapted process. It remains to show that $X^{[T_n]}$ is a left-continuous simple process. Now T_n assumes values in $\{u_{n,k} : k \in \mathbb{N}\} \cup \{\infty\}$ and in fact we have

(7) $$\begin{cases} \{T_n = u_{n,1}\} = \{T \in [u_{n,0}, u_{n,1}]\} \\ \{T_n = u_{n,k}\} = \{T \in (u_{n,k-1}, u_{n,k}]\} & \text{for } k \geq 2 \\ \{T_n = \infty\} = \{T = \infty\}. \end{cases}$$

Define a sequence of real valued random variables $\{\zeta_{n,k} : k \in \mathbb{Z}_+\}$ by setting for $k = 0$

(8) $$\zeta_{n,0} = \eta_{n,0},$$

and for $k \geq 1$

(9) $$\zeta_{n,k} = \begin{cases} \eta_{n,k} & \text{on } \{T_n \geq u_{n,k}\} = \{T > u_{n,k-1}\} \\ 0 & \text{on } \{T_n \leq u_{n,k-1}\} = \{T \leq u_{n,k-1}\}. \end{cases}$$

Since $\{T_n \leq u_{n,k-1}\} \in \mathfrak{F}_{u_{n,k-1}}$ by the fact that T_n is a stopping time, we have $\{T_n \geq u_{n,k}\} = \{T_n \leq u_{n,k-1}\}^c \in \mathfrak{F}_{u_{n,k-1}}$. Then by the $\mathfrak{F}_{u_{n,k-1}}$-measurability of $\eta_{n,k}$, our $\zeta_{n,k}$ is

§12. STOCHASTIC INTEGRALS WITH RESPECT TO MARTINGALES

$\mathfrak{F}_{u_{n,k-1}}$-measurable. By (5) and (6) and the definition of $\{\zeta_{n,k} : k \in \mathbb{Z}_+\}$ by (8) and (9), we have

(10) $\quad X^{[T_n]}(t,\omega) = \zeta_{n,0}(\omega)\mathbf{1}_{\{0\}}(t) + \sum_{k \in \mathbb{N}} \zeta_{n,k}(\omega)\mathbf{1}_{(u_{n,k-1}, u_{n,k}]}(t) \quad \text{for } (t,\omega) \in \mathbb{R}_+ \times \Omega.$

Thus $X^{[T_n]}$ is a bounded adapted left-continuous simple process, that is, X is in $\mathbf{L}_0(\Omega, \mathfrak{F}, \{\mathfrak{F}_t\}, P)$.

Now by (3) in Definition 12.3 and (10), we have

(11) $\quad (X^{[T_n]} \bullet M)(t) = \sum_{k \in \mathbb{N}} \zeta_{n,k}\{M(u_{n,k} \wedge t) - M(u_{n,k-1} \wedge t)\}$

and similarly by (5) we have

(12) $\quad (X \bullet M)(T_n \wedge t) = \sum_{k \in \mathbb{N}} \eta_{n,k}\{M(u_{n,k} \wedge T_n \wedge t) - M(u_{n,k-1} \wedge T_n \wedge t)\}.$

Take an arbitrary $\omega \in \Omega$. Recalling (7), let us assume that $T_n(\omega) = u_{n,k_0}$ for some $k_0 \in \mathbb{Z}_+$. Then $T_n(\omega) \geq u_{n,k}$ and thus $\zeta_{n,k}(\omega) = \eta_{n,k}(\omega)$ for $k = 1, \ldots, k_0$, and similarly $T_n(\omega) \leq u_{n,k-1}$ and thus $\zeta_{n,k}(\omega) = 0$ for $k \geq k_0 + 1$ by (9). Therefore from (11), we have

(13) $\quad (X^{[T_n]} \bullet M)(t,\omega) = \sum_{k=1}^{k_0} \eta_{n,k}(\omega)\{M(u_{n,k} \wedge t, \omega) - M(u_{n,k-1} \wedge t, \omega)\}.$

On the other hand, since $T_n(\omega) = u_{n,k_0}$ we have $u_{n,k-1} \wedge T_n(\omega) = T_n(\omega) = u_{n,k} \wedge T_n(\omega)$ for $k \geq k_0 + 1$. Thus from (12) we have

(14) $\quad (X \bullet M)(T_n(\omega) \wedge t, \omega) = \sum_{k=1}^{k_0} \eta_{n,k}(\omega)\{M(u_{n,k} \wedge t, \omega) - M(u_{n,k-1} \wedge t, \omega)\}.$

By (13) and (14), we have (4) for the case where $T_n(\omega) = u_{n,k_0}$ for some $k_0 \in \mathbb{Z}_+$. For the case where $T_n(\omega) = \infty$, (4) holds likewise by applying (9).

Now for any $t \in \mathbb{R}_+$, we have

$$\|X^{[T_n]} - X^{[T]}\|_{2,t}^{[M],P} = \left[\int_\Omega \left\{\int_{[0,t]} |\mathbf{1}_{\{(\cdot) \leq T_n\}} - \mathbf{1}_{\{(\cdot) \leq T\}}|X^2 \, d[M]\right\} dP\right]^{1/2}.$$

According to (1) in the proof of Theorem 3.38, the first factor in the integrand above converges to 0 on $\mathbb{R}_+ \times \Omega$. Since X is a bounded process the integrand is bounded. Thus by the Bounded Convergence Theorem and the Dominated Convergence Theorem, we have $\lim_{n\to\infty}\|X^{[T_n]} - X^{[T]}\|_{2,t}^{[M],P} = 0$ for every $t \in \mathbb{R}_+$. Thus $\lim_{n\to\infty}\|X^{[T_n]} - X^{[T]}\|_{2,\infty}^{[M],P} = 0$ according to Remark 11.16. Then $\lim_{n\to\infty} \|X^{[T_n]} \bullet M - X^{[T]} \bullet M\|_\infty = 0$ by Definition 12.9. Thus by Proposition 11.5, there exist a null set Λ in $(\Omega, \mathfrak{F}, P)$ and a subsequence $\{n_k\}$ of $\{n\}$ such that

(15) $\quad \lim_{k\to\infty}(X^{[T_{n_k}]} \bullet M)(t,\omega) = (X^{[T]} \bullet M)(t,\omega) \quad$ for $(t,\omega) \in \mathbb{R}_+ \times \Lambda^c$.

Since $\vartheta_n(t) \downarrow t$ for $t \in \overline{\mathbb{R}}_+$, we have $T_{n_k} \downarrow T$ and then $T_{n_k} \wedge t \downarrow T \wedge t$ on Ω for every $t \in \mathbb{R}_+$. Since $X \bullet M$ is right-continuous, we have

(16) $\quad \lim_{n\to\infty}(X \bullet M)(T_{n_k}(\omega) \wedge t, \omega) = (X \bullet M)(T(\omega) \wedge t, \omega) \quad$ for $(t,\omega) \in \mathbb{R}_+ \times \Omega$.

Taking the subsequence $\{n_k\}$ in (4), letting $k \to \infty$ and using (15) and (16), we have the proof of the lemma. ∎

Theorem 12.22. *Let $M \in \mathbf{M}_2(\Omega, \mathfrak{F}, \{\mathfrak{F}_t\}, P)$ and let X be a predictable processes on the filtered space $(\Omega, \mathfrak{F}, \{\mathfrak{F}_t\}, P)$ such that $X \in \mathbf{L}_{2,\infty}(\mathbb{R}_+ \times \Omega, \mu_{[M]}, P)$. Let T be a stopping time on the filtered space. Then there exists a null set Λ in $(\Omega, \mathfrak{F}, P)$ such that $(X^{[T]} \bullet M)(\cdot, \omega) = (X \bullet M)^{T\wedge}(\cdot, \omega)$ for $\omega \in \Lambda^c$.*

Proof. By Theorem 12.8, there exists a sequence $\{X^{(n)} : n \in \mathbb{N}\}$ in $\mathbf{L}_0(\Omega, \mathfrak{F}, \{\mathfrak{F}_t\}, P)$ such that $\lim_{n\to\infty}\|X^{(n)} - X\|_{2,\infty}^{[M],P} = 0$ and thus $\lim_{n\to\infty}\|X^{(n)} \bullet M - X \bullet M\|_\infty = 0$ by Definition 12.9. Then by Remark 12.14, there exist a null set Λ_1 in $(\Omega, \mathfrak{F}, P)$ and a subsequence $\{n_k\}$ of $\{n\}$ such that $\lim_{k\to\infty}(X^{(n_k)} \bullet M)(t,\omega) = (X \bullet M)(t,\omega)$ for $(t,\omega) \in \mathbb{R}_+ \times \Lambda_1^c$. This implies that

(1) $\quad \lim_{k\to\infty}(X^{(n_k)} \bullet M)(T(\omega) \wedge t, \omega) = (X \bullet M)(T(\omega) \wedge t, \omega) \quad$ for $(t,\omega) \in \mathbb{R}_+ \times \Lambda_1^c$.

Since

$$|(X^{(n)})^{[T]} - X^{[T]}| = |\mathbf{1}_{\{\cdot \leq T\}}\{X^{(n)} - X\}| \leq |X^{(n)} - X|,$$

the convergence $\lim_{n\to\infty}\|X^{(n)} - X\|_{2,\infty}^{[M],P} = 0$ implies $\lim_{n\to\infty}\|(X^{(n)})^{[T]} - X^{[T]}\|_{2,\infty}^{[M],P} = 0$. Now since $X^{(n)}$ is a predictable process, $(X^{(n)})^{[T]}$ is a predictable process according to Observation 3.35. Clearly $(X^{(n)})^{[T]} \in \mathbf{L}_{2,\infty}(\mathbb{R}_+ \times \Omega, \mu_{[M]}, P)$. Thus by Proposition 12.13 we have $\lim_{n\to\infty}\|(X^{(n)})^{[T]} \bullet M - X^{[T]} \bullet M\|_\infty = 0$. According to Lemma 12.21 we have $(X^{(n)})^{[T]} \bullet M = (X^{(n)} \bullet M)^{T\wedge}$. Thus we have $\lim_{n\to\infty}\|(X^{(n)} \bullet M)^{T\wedge} - X^{[T]} \bullet M\|_\infty = 0$.

§12. STOCHASTIC INTEGRALS WITH RESPECT TO MARTINGALES

This implies according to Remark 12.14 the existence of a null set Λ_2 and a subsequence $\{n_\ell\}$ of $\{n_k\}$ such that

(2) $\qquad \lim_{\ell \to \infty} (X^{(n_\ell)} \bullet M)(T(\omega) \wedge t, \omega) = (X^{[T]} \bullet M)(t, \omega) \quad \text{for } (t, \omega) \in \mathbb{R}_+ \times \Lambda_2^c.$

With $\Lambda = \Lambda_1 \cup \Lambda_2$, we have the proof of the theorem by (1) and (2). ∎

In Theorem 12.22, we showed that $X^{[T]} \bullet M = (X \bullet M)^{T \wedge}$. We show next that $X \bullet M^{T \wedge} = (X \bullet M)^{T \wedge}$. For this we need the following lemma.

Lemma 12.23. *Let $M \in \mathbf{M}_2(\Omega, \mathfrak{F}, \{\mathfrak{F}_t\}, P)$ and let T be a stopping time on the filtered space. Then for the two families of Lebesgue-Stieltjes measures $\mu_{[M]}$ and $\mu_{[M^{T \wedge}]}$ on $(\mathbb{R}_+, \mathfrak{B}_{\mathbb{R}_+})$ there exists a null set Λ in $(\Omega, \mathfrak{F}, P)$ such that*

(1) $\qquad \mu_{[M^{T \wedge}]}(E, \omega) \leq \mu_{[M]}(E, \omega) \quad \text{for every } E \in \mathfrak{B}_{\mathbb{R}_+} \text{ when } \omega \in \Lambda^c.$

Proof. Since M is in $\mathbf{M}_2(\Omega, \mathfrak{F}, \{\mathfrak{F}_t\}, P)$, $M^{T \wedge}$ is in $\mathbf{M}_2(\Omega, \mathfrak{F}, \{\mathfrak{F}_t\}, P)$ by Theorem 8.12. Thus the quadratic variation process $[M^{T \wedge}]$ and the family of Lebesgue-Stieltjes measures $\mu_{[M^{T \wedge}]}$ on $(\mathbb{R}_+, \mathfrak{B}_{\mathbb{R}_+})$ are defined.

If $X \in \mathbf{L}_0(\Omega, \mathfrak{F}, \{\mathfrak{F}_t\}, P)$, then $X \bullet M$ and $X \bullet M^{T \wedge}$ are defined. For $t \in \mathbb{R}_+$, integrating the equality (4) in Proposition 12.5, we have

(2) $\qquad \mathbf{E}\left[(X \bullet M)_t^2 - (X \bullet M)_0^2\right] = \mathbf{E}\left[\int_{[0,t]} X^2(u) \, d[M](u)\right],$

and similarly

(3) $\qquad \mathbf{E}\left[(X \bullet M^{T \wedge})_t^2 - (X \bullet M^{T \wedge})_0^2\right] = \mathbf{E}\left[\int_{[0,t]} X^2(u) \, d[M^{T \wedge}](u)\right].$

Now for $X \in \mathbf{L}_0(\Omega, \mathfrak{F}, \{\mathfrak{F}_t\}, P)$ given by (1) of Definition 12.3 as

$$X(t) = \xi_0 \mathbf{1}_{\{0\}}(t) + \sum_{k \in \mathbb{N}} \xi_k \mathbf{1}_{(t_{k-1}, t_k]}(t) \quad \text{for } t \in \mathbb{R}_+$$

we have by (3) of Definition 12.3

$$(X \bullet M)(t) = \sum_{i \in \mathbb{N}} \xi_i \{M(t_i \wedge t) - M(t_{i-1} \wedge t)\} \quad \text{for } t \in \mathbb{R}_+.$$

Replacing M with $M^{T\wedge}$, we have for $t \in \mathbb{R}_+$

$$(X \bullet M^{T\wedge})(t) = \sum_{i \in \mathbb{N}} \xi_i \{M^{T\wedge}(t_i \wedge t) - M^{T\wedge}(t_{i-1} \wedge t)\}$$
$$= \sum_{i \in \mathbb{N}} \xi_i \{M(t_i \wedge T \wedge t) - M(t_{i-1} \wedge T \wedge t)\} = (X \bullet M)(T \wedge t),$$

that is,

(4) $$X \bullet M^{T\wedge} = (X \bullet M)^{T\wedge}.$$

Thus for the left side of (3) we have

(5) $$\mathbf{E}\left[(X \bullet M^{T\wedge})_t^2 - (X \bullet M^{T\wedge})_0^2\right] = \mathbf{E}\left[\{(X \bullet M)^{T\wedge}\}_t^2 - \{(X \bullet M)^{T\wedge}\}_0^2\right]$$
$$= \mathbf{E}\left[(X \bullet M)_{T\wedge t}^2 - (X \bullet M)_0^2\right] \leq \mathbf{E}\left[(X \bullet M)_t^2 - (X \bullet M)_0^2\right],$$

where the last inequality is from the fact that since $X \bullet M$ is an L_2-martingale $(X \bullet M)^2$ is a submartingale so that $\mathbf{E}[(X \bullet M)_{T\wedge t}^2] \leq \mathbf{E}[(X \bullet M)_t^2]$. Using (3) and (2) in (5), we have

$$\mathbf{E}\left[\int_{[0,t]} X^2(u)\, d[M^{T\wedge}](u)\right] \leq \mathbf{E}\left[\int_{[0,t]} X^2(u)\, d[M](u)\right].$$

Letting $t \to \infty$ we have by the Monotone Convergence Theorem

(6) $$\mathbf{E}\left[\int_{\mathbb{R}_+} X^2(u)\, d[M^{T\wedge}](u)\right] \leq \mathbf{E}\left[\int_{\mathbb{R}_+} X^2(u)\, d[M](u)\right],$$

for every stopping time T and $X \in \mathbf{L}_0(\Omega, \mathfrak{F}, \{\mathfrak{F}_t\}, P)$.

Let Q be the collection of all rational numbers in $\mathbb{R}_+$. Let $a, b \in Q$, $a < b$, and $A \in \mathfrak{F}_a$. The mapping X of $\mathbb{R}_+ \times \Omega$ into $\mathbb{R}$ defined by setting $X(t, \omega) = \mathbf{1}_A(\omega)\mathbf{1}_{(a,b]}(t)$ for $(t, \omega) \in \mathbb{R}_+ \times \Omega$ is then a member of $\mathbf{L}_0(\Omega, \mathfrak{F}, \{\mathfrak{F}_t\}, P)$ so that by applying (6) to our X we have

$$\int_A \mu_{[M^{T\wedge}]}((a, b], \cdot)\, dP \leq \int_A \mu_{[M]}((a, b], \cdot)\, dP.$$

Since this holds for an arbitrary $A \in \mathfrak{F}_a$, there exists a null set $\Lambda_{a,b}$ in $(\Omega, \mathfrak{F}, P)$ such that for $\omega \in \Lambda_{a,b}^c$ we have

(7) $$\mu_{[M^{T\wedge}]}((a, b], \omega) \leq \mu_{[M]}((a, b], \omega).$$

Let $\Lambda = \cup_{a,b \in Q, a<b} \Lambda_{a,b}$. Then (7) holds for every pair $a, b \in Q$ such that $a < b$ when $\omega \in \Lambda^c$. Let $\omega_0 \in \Lambda^c$ be arbitrarily fixed. The collection $\mathfrak{I}$ consisting of $\emptyset$, $(0, \infty)$ and

§12. STOCHASTIC INTEGRALS WITH RESPECT TO MARTINGALES

intervals of the type $(a, b]$ where $a, b \in Q$, $a < b$, is a semialgebra of subsets of $(0, \infty)$. Let $\mathfrak{A}$ be the algebra generated by $\mathfrak{J}$, that is, the collection of all finite disjoint unions of members of $\mathfrak{J}$. Now since $\mu_{[M^{T\wedge}]}(\cdot, \omega_0)$ and $\mu_{[M]}(\cdot, \omega_0)$ are measures on $((0, \infty), \mathfrak{B}_{(0,\infty)})$ and since $\mu_{[M^{T\wedge}]}(E, \omega_0) \leq \mu_{[M]}(E, \omega_0)$ for every $E \in \mathfrak{J}$, the inequality holds for every $E \in \mathfrak{A}$ by the additivity of the measures. Now $\mathfrak{B}_{(0,\infty)} = \sigma(\mathfrak{A})$. To extend the inequality from $\mathfrak{A}$ to $\mathfrak{B}_{(0,\infty)}$ by applying Theorem 1.13, let $\mathfrak{O}$ be the collection of all open sets in $(0, \infty)$ and let $\mathfrak{O}_n$ be the collection of all open sets in $(0, n]$ for $n \in \mathbb{N}$. Then by Theorem 1.5 we have

$$\mathfrak{B}_{(0,n]} = \sigma(\mathfrak{O}_n) = \sigma(\mathfrak{O} \cap (0, n]) = \sigma(\mathfrak{O}) \cap (0, n]$$
$$= \mathfrak{B}_{(0,\infty)} \cap (0, n] = \sigma(\mathfrak{A}) \cap (0, n] = \sigma(\mathfrak{A} \cap (0, n]).$$

Note that $\mathfrak{A} \cap (0, n]$ is an algebra of subsets of $(0, n]$ and at the same time $\mathfrak{A} \cap (0, n] \subset \mathfrak{A}$. For $E \in \mathfrak{B}_{(0,\infty)}$, let $E_n = E \cap (0, n] \in \sigma(\mathfrak{A}) \cap (0, n] = \mathfrak{B}_{(0,n]}$ for $n \in \mathbb{N}$. Since $\mu_{[M^{T\wedge}]}(\cdot, \omega_0) + \mu_{[M]}(\cdot, \omega_0)$ is a finite measure on $((0, n], \mathfrak{B}_{(0,n]})$, according to Theorem 1.13 for every $\varepsilon > 0$ there exists some F in the algebra $\mathfrak{A} \cap (0, n]$ of subsets of $(0, n]$ such that

$$\mu_{[M^{T\wedge}]}(E_n \triangle F, \omega_0) + \mu_{[M]}(E_n \triangle F, \omega_0) < \varepsilon.$$

Then $\mu_{[M^{T\wedge}]}(E_n \triangle F, \omega_0), \mu_{[M]}(E_n \triangle F, \omega_0) < \varepsilon$ so that

$$\mu_{[M^{T\wedge}]}(E_n, \omega_0) \leq \mu_{[M^{T\wedge}]}(F, \omega_0) + \varepsilon$$
$$\leq \mu_{[M]}(F, \omega_0) + \varepsilon \leq \mu_{[M]}(E_n, \omega_0) + 2\varepsilon,$$

where the second inequality is from the fact that $F \in \mathfrak{A} \cap (0, n] \subset \mathfrak{A}$. From the arbitrariness of $\varepsilon > 0$, we have $\mu_{[M^{T\wedge}]}(E_n, \omega_0) \leq \mu_{[M]}(E_n, \omega_0)$. Letting $n \to \infty$, we have $\mu_{[M^{T\wedge}]}(E, \omega_0) \leq \mu_{[M]}(E, \omega_0)$. Thus $\mu_{[M^{T\wedge}]}(\cdot, \omega_0) \leq \mu_{[M]}(\cdot, \omega_0)$ on $\mathfrak{B}_{(0,\infty)}$. Since $\mu_{[M^{T\wedge}]}(\{0\}, \omega_0) = \mu_{[M]}(\{0\}, \omega_0) = 0$, we have $\mu_{[M^{T\wedge}]}(E, \omega_0) \leq \mu_{[M]}(E, \omega_0)$ for every $E \in \mathfrak{B}_{\mathbb{R}_+}$. Since ω_0 is an arbitrary element of Λ^c, this completes the proof of (1). ∎

Theorem 12.24. *Let $M \in \mathbf{M}_2(\Omega, \mathfrak{F}, \{\mathfrak{F}_t\}, P)$, let X be a predictable process on the filtered space which is in $\mathbf{L}_{2,\infty}(\mathbb{R}_+ \times \Omega, \mu_{[M]}, P)$ and let T be a stopping time on the filtered space. Then X is in $\mathbf{L}_{2,\infty}(\mathbb{R}_+ \times \Omega, \mu_{[M^{T\wedge}]}, P)$ so that $X \bullet M^{T\wedge}$ is defined. For $X \bullet M^{T\wedge}$ there exists a null set Λ in $(\Omega, \mathfrak{F}, P)$ such that $(X \bullet M^{T\wedge})(\cdot, \omega) = (X \bullet M)^{T\wedge}(\cdot, \omega)$ for $\omega \in \Lambda^c$.*

Proof. Since $M \in \mathbf{M}_2(\Omega, \mathfrak{F}, \{\mathfrak{F}_t\}, P)$, we have $M^{T\wedge} \in \mathbf{M}_2(\Omega, \mathfrak{F}, \{\mathfrak{F}_t\}, P)$ by Theorem 8.12. Let Λ be the null set in Lemma 12.23. Let $X \in \mathbf{L}_{2,\infty}(\mathbb{R}_+ \times \Omega, \mu_{[M]}, P)$. Then for $\omega \in \Lambda^c$ we have for every $t \in \mathbb{R}_+$

$$\int_{[0,t]} X^2(s, \omega) \, d[M^{T\wedge}](s, \omega) \leq \int_{[0,t]} X^2(s, \omega) \, d[M](s, \omega)$$

so that
$$\mathbf{E}\left[\int_{[0,t]} X^2(s)\,d[M^{T\wedge}](s)\right] \leq \mathbf{E}\left[\int_{[0,t]} X^2(s)\,d[M](s)\right].$$

Since X is in $\mathbf{L}_{2,\infty}(\mathbb{R}_+ \times \Omega, \mu_{[M]}, P)$, the right side of the last inequality is finite so that the left side is finite and thus X is in $\mathbf{L}_{2,\infty}(\mathbb{R}_+ \times \Omega, \mu_{[M^{T\wedge}]}, P)$. Then $X \bullet M^{T\wedge}$ is defined and is in $\mathbf{M}_2(\Omega, \mathfrak{F}, \{\mathfrak{F}_t\}, P)$.

Let X be a predictable process which is in $\mathbf{L}_{2,\infty}(\mathbb{R}_+ \times \Omega, \mu_{[M]}, P)$. By Theorem 12.8 and Definition 12.9, there exists a sequence $\{X^{(n)} : n \in \mathbb{N}\}$ in $\mathbf{L}_0(\Omega, \mathfrak{F}, \{\mathfrak{F}_t\}, P)$ such that $\lim_{n\to\infty} \|X^{(n)} - X\|_{2,\infty}^{[M],P} = 0$ and

(1) $$\lim_{n\to\infty} \mathbf{I} X^{(n)} \bullet M - X \bullet M \mathbf{I}_\infty = 0.$$

Now for every $t \in \mathbb{R}_+$, Lemma 12.23 implies that

$$\|X^{(n)} - X\|_{2,t}^{[M^{T\wedge}],P} = \mathbf{E}\left[\int_{[0,t]} |X^{(n)}(s) - X(s)|^2\,d[M^{T\wedge}](s)\right]^{1/2}$$
$$\leq \mathbf{E}\left[\int_{[0,t]} |X^{(n)}(s) - X(s)|^2\,d[M](s)\right]^{1/2} = \|X^{(n)} - X\|_{2,t}^{[M],P}.$$

Since $\lim_{n\to\infty} \|X^{(n)} - X\|_{2,\infty}^{[M],P} = 0$ implies $\lim_{n\to\infty} \|X^{(n)} - X\|_{2,t}^{[M],P} = 0$ the last inequality implies $\lim_{n\to\infty} \|X^{(n)} - X\|_{2,t}^{[M^{T\wedge}],P} = 0$. Then $\lim_{n\to\infty} \|X^{(n)} - X\|_{2,\infty}^{[M^{T\wedge}],P} = 0$ by applying Remark 11.18. Consequently by Definition 12.9 we have

(2) $$\lim_{n\to\infty} \mathbf{I} X^{(n)} \bullet M^{T\wedge} - X \bullet M^{T\wedge} \mathbf{I}_\infty = 0.$$

Now (4) implies $\lim_{n\to\infty} \mathbf{I} X^{(n)} \bullet M - X \bullet M \mathbf{I}_t = 0$ for every $t \in \mathbb{R}_+$ according to Remark 11.5, that is,

(3) $$\lim_{n\to\infty} \mathbf{E}\left[\{(X^{(n)} \bullet M) - (X \bullet M)\}^2(t)\right] = 0.$$

Since $X^{(n)} \bullet M - X \bullet M$ is an L_2-martingale, $\{X^{(n)} \bullet M - X \bullet M\}^2$ is a submartingale and this implies that

$$\mathbf{E}[\{X^{(n)} \bullet M - X \bullet M\}^2(T \wedge t)] \leq \mathbf{E}[\{X^{(n)} \bullet M - X \bullet M\}^2(t)].$$

Therefore by (3) we have $\lim_{n\to\infty} \mathbf{E}[(X^{(n)} \bullet M - X \bullet M)^2(T \wedge t)] = 0$, that is,

$$\lim_{n\to\infty} \mathbf{E}[\{(X^{(n)} \bullet M)^{T\wedge} - (X \bullet M)^{T\wedge}\}^2(t)] = 0$$

§12. STOCHASTIC INTEGRALS WITH RESPECT TO MARTINGALES

Now since $X^{(n)} \in \mathbf{L}_0(\Omega, \mathfrak{F}, \{\mathfrak{F}_t\}, P)$, $(X^{(n)} \bullet M)^{T\wedge} = X^{(n)} \bullet M^{T\wedge}$ as we saw in the proof of Lemma 12.23. Thus

$$\lim_{n\to\infty} \mathbf{E}[\{X^{(n)} \bullet M^{T\wedge} - (X \bullet M)^{T\wedge}\}^2(t)] = 0,$$

that is,

$$\lim_{n\to\infty} \| X^{(n)} \bullet M^{T\wedge} - (X \bullet M)^{T\wedge} \|_t = 0.$$

Since this holds for every $t \in \mathbb{R}_+$, we have according to Remark 11.5

(4) $$\lim_{n\to\infty} \| X^{(n)} \bullet M^{T\wedge} - (X \bullet M)^{T\wedge} \|_\infty = 0.$$

By (2) and (4), we have $X \bullet M^{T\wedge} = (X \bullet M)^{T\wedge}$. ∎

Proposition 12.25. *Let $M, N \in \mathbf{M}_2^c(\Omega, \mathfrak{F}, \{\mathfrak{F}_t\}, P)$ and let T be a stopping time on the filtered space. Then there exists a null set Λ in $(\Omega, \mathfrak{F}, P)$ such that for every $\omega \in \Lambda^c$ and $t \in \mathbb{R}_+$ we have*

(1) $$\mu_{[M^{T\wedge}, N^{T\wedge}]}(\cdot, \omega) = \mu_{[M,N]^{T\wedge}}(\cdot, \omega) \quad \text{on } ([0, t], \mathfrak{B}_{[0,t]}),$$

and thus

(2) $$[M^{T\wedge}, N^{T\wedge}](\cdot, \omega) = [M, N]^{T\wedge}(\cdot, \omega).$$

In particular

(3) $$\mu_{[M^{T\wedge}]}(\cdot, \omega) = \mu_{[M]^{T\wedge}}(\cdot, \omega) \quad \text{on } (\mathbb{R}_+, \mathfrak{B}_{\mathbb{R}_+}),$$

and

(4) $$[M^{T\wedge}](\cdot, \omega) = [M]^{T\wedge}(\cdot, \omega).$$

Proof. Since $M, N \in \mathbf{M}_2^c(\Omega, \mathfrak{F}, \{\mathfrak{F}_t\}, P)$, we have $M^{T\wedge}, N^{T\wedge} \in \mathbf{M}_2^c(\Omega, \mathfrak{F}, \{\mathfrak{F}_t\}, P)$. Then for $X \in \mathbf{L}_0(\Omega, \mathfrak{F}, \{\mathfrak{F}_t\}, P)$, $X \bullet M^{T\wedge}$ and $X \bullet N^{T\wedge}$ are defined. According to Proposition 12.5 there exists a null set Λ_1 in $(\Omega, \mathfrak{F}, P)$ such that on Λ_1^c we have for every $t \in \mathbb{R}_+$

(5) $$[X \bullet M^{T\wedge}, X \bullet N^{T\wedge}]_t = \int_{[0,t]} X^2(s) \, d[M^{T\wedge}, N^{T\wedge}](s).$$

By Theorem 12.22 and Theorem 12.24, $X \bullet M^{T\wedge} = X^{[T]} \bullet M$ and $X \bullet N^{T\wedge} = X^{[T]} \bullet N$. Thus there exists a null set Λ_2 in $(\Omega, \mathfrak{F}, P)$ such that on Λ_2^c we have for every $t \in \mathbb{R}_+$

(6) $\quad [X \bullet M^{T\wedge}, X \bullet N^{T\wedge}]_t = [X^{[T]} \bullet M, X^{[T]} \bullet N]_t = \int_{[0,t]} (X^{[T]})^2(s) \, d[M,N](s)$

$= \int_{[0,t]} \mathbf{1}_{\{(\cdot) \le T\}} X^2(s) \, d[M,N](s) = \int_{[0,t]} X^2(s) \, d[M,N]^{T\wedge}(s),$

where the second equality is by Theorem 12.17 and third equality is by Definition 3.33. Let $\Lambda = \Lambda_1 \cup \Lambda_2$. Then by (5) and (6), we have on Λ^c and for every $t \in \mathbb{R}_+$

$$\int_{[0,t]} X^2(s) \, d[M^{T\wedge}, N^{T\wedge}](s) = \int_{[0,t]} X^2(s) \, d[M,N]^{T\wedge}(s).$$

In particular with $a, b \in \mathbb{R}_+$, $a < b \le t$ and $X \in \mathbf{L}_0(\Omega, \mathfrak{F}, \{\mathfrak{F}_t\}, P)$ defined by $X(s, \omega) = \mathbf{1}_{(a,b]}(s)$ for $(s, \omega) \in \mathbb{R}_+ \times \Omega$, we have for $\omega \in \Lambda^c$

$$\int_{[0,t]} \mathbf{1}_{(a,b]}(s) \, d[M^{T\wedge}, N^{T\wedge}](s, \omega) = \int_{[0,t]} \mathbf{1}_{(a,b]}(s) \, d[M,N]^{T\wedge}(s, \omega).$$

From the arbitrariness of $t \in \mathbb{R}_+$ in the last equality, for any $a, b \in \mathbb{R}_+$, $a < b$, and $\omega \in \Lambda^c$, we have $\mu_{[M^{T\wedge}, N^{T\wedge}]}((a,b], \omega) = \mu_{[M,N]^{T\wedge}}((a,b], \omega)$. Now since $\mu_{[M]^{T\wedge}}(\cdot, \omega)$ is a finite signed measure on $([0,t], \mathfrak{B}_{[0,t]})$ and since the collection $\mathfrak{I}$ of intervals of the type $(a,b]$ in $(0,t]$ and $\emptyset$ is a semialgebra and $\sigma(\mathfrak{I}) = \mathfrak{B}_{(0,t]}$, the last equality implies according to Corollary 1.8 that (1) holds for every $\omega \in \Lambda^c$. Then for every $t \in \mathbb{R}_+$ we have

$$[M^{T\wedge}, N^{T\wedge}](t, \omega) = \mu_{[M^{T\wedge}, N^{T\wedge}]}([0,t], \omega) = \mu_{[M,N]^{T\wedge}}([0,t], \omega) = [M,N]^{T\wedge}(t, \omega),$$

proving (2). ∎

Lemma 12.26. *Let $M = \{M_t : t \in \mathbb{R}_+\}$ be a null at 0 martingale on a filtered space $(\Omega, \mathfrak{F}, \{\mathfrak{F}_t\}, P)$ such that $|M| \le K$ on $[0,t] \times \Omega$ for some $t \in \mathbb{R}_+$ and $K \ge 0$. For $0 = t_0 < \cdots < t_n = t$, let $S = \sum_{j=1}^n \{M(t_j) - M(t_{j-1})\}^2$. Then $\mathbf{E}(S^2) \le 12K^4$.*

Proof. For brevity in notation, let $\xi_j = M(t_j)$ and $\alpha_j = \{\xi_j - \xi_{j-1}\}^2$ for $j = 1, \ldots, n$. Then $S = \sum_{j=1}^n \alpha_j$. Now

(1) $\quad S^2 = \{\sum_{j=1}^n \alpha_j\}^2 = \sum_{j=1}^n \alpha_j^2 + 2 \sum_{i,j=1,\ldots,n; i<j} \alpha_i \alpha_j = \sum_{j=1}^n \alpha_j^2 + 2 \sum_{i=1}^n \alpha_i \{\sum_{j=i+1}^n \alpha_j\}.$

§12. STOCHASTIC INTEGRALS WITH RESPECT TO MARTINGALES

Note that by the martingale property of M we have for $j = 1, \ldots, n$

(2) $\quad \mathbf{E}[\alpha_j | \mathfrak{F}_{t_{j-1}}] = \mathbf{E}[(\xi_j - \xi_{j-1})^2 | \mathfrak{F}_{t_{j-1}}] = \mathbf{E}[(\xi_j^2 - \xi_{j-1}^2) | \mathfrak{F}_{t_{j-1}}]$ a.e. on $(\Omega, \mathfrak{F}_{t_{j-1}}, P)$.

Now

(3) $\quad \displaystyle\sum_{j=i+1}^{n} \mathbf{E}[\alpha_j | \mathfrak{F}_{t_i}] = \sum_{j=i+1}^{n} \mathbf{E}[\alpha_j | \mathfrak{F}_{t_{j-1}} | \mathfrak{F}_{t_i}]$

$\quad = \displaystyle\sum_{j=i+1}^{n} \mathbf{E}[(\xi_j^2 - \xi_{j-1}^2) | \mathfrak{F}_{t_{j-1}} | \mathfrak{F}_{t_i}] \quad \text{by (2)}$

$\quad = [\displaystyle\sum_{j=i+1}^{n} \mathbf{E}[(\xi_j^2 - \xi_{j-1}^2) | \mathfrak{F}_{t_i}] = \mathbf{E}[\xi_n^2 - \xi_i^2 | \mathfrak{F}_{t_i}]$

$\quad = \mathbf{E}[(\xi_n - \xi_i)^2 | \mathfrak{F}_{t_i}] \leq (2K)^2 \quad$ a.e. on $(\Omega, \mathfrak{F}_{t_i}, P)$.

Then

(4) $\quad \mathbf{E}[\displaystyle\sum_{i=1}^{n} \alpha_i \{\sum_{j=i+1}^{n} \alpha_j\}] = \sum_{i=1}^{n} \mathbf{E}[\mathbf{E}[\alpha_i \{\sum_{j=i+1}^{n} \alpha_j\} | \mathfrak{F}_{t_i}]]$

$\quad = \displaystyle\sum_{i=1}^{n} \mathbf{E}[\alpha_i \mathbf{E}[\sum_{j=i+1}^{n} \alpha_j | \mathfrak{F}_{t_i}]] \leq (2K)^2 \sum_{i=1}^{n} \mathbf{E}(\alpha_i) = (2K)^2 \mathbf{E}(S)$,

where the inequality is by (3). But

(5) $\quad \mathbf{E}(S) = \mathbf{E}[\displaystyle\sum_{i=1}^{n} \alpha_i] = \sum_{i=1}^{n} \mathbf{E}\left[\mathbf{E}[\alpha_i | \mathfrak{F}_{t_{i-1}}]\right] = \sum_{i=1}^{n} \mathbf{E}[\mathbf{E}[(\xi_i^2 - \xi_{i-1}^2) | \mathfrak{F}_{t_{i-1}}]]$

$\quad = \displaystyle\sum_{i=1}^{n} \mathbf{E}(\xi_i^2 - \xi_{i-1}^2) = \mathbf{E}(\xi_n^2) \leq K^2$,

where the third equality is by (2) and the last equality is from the fact that $\xi_0 = 0$. By (4) and (5) we have

(6) $\quad \mathbf{E}[2 \displaystyle\sum_{i=1}^{n} \alpha_i \{\sum_{j=i+1}^{n} \alpha_j\}] \leq 8K^4$.

Finally since $\alpha_j \leq (2K)^2$ for $j = 1, \ldots, n$, we have

(7) $\quad \mathbf{E}[\displaystyle\sum_{j=1}^{n} \alpha_j^2] \leq (2K)^2 \mathbf{E}[\sum_{j=1}^{n} \alpha_j] \leq 4K^4$,

260   CHAPTER 3. STOCHASTIC INTEGRALS

by (5). Using (6) and (7) in (1), we have $E(S^2) \leq 12K^4$. ∎

Theorem 12.27. Let $M, N \in \mathbf{M}_2^c(\Omega, \mathfrak{F}, \{\mathfrak{F}_t\}, P)$. For $t \in \mathbb{R}_+$ and $n \in \mathbb{N}$, let Δ_n be the partition of $[0, t]$ by $0 = t_{n,0} < \cdots < t_{n,p_n} = t$. Let $|\Delta_n| = \max_{k=1,\ldots,p_n}(t_{n,k} - t_{n,k-1})$ and $\lim_{n\to\infty} |\Delta_n| = 0$. Then

(1) $$\lim_{n\to\infty} \Big\| \sum_{k=1}^{p_n} \{M_{t_{n,k}} - M_{t_{n,k-1}}\}\{N_{t_{n,k}} - N_{t_{n,k-1}}\} - [M, N]_t \Big\|_1 = 0$$

and in particular

(2) $$\lim_{n\to\infty} \Big\| \sum_{k=1}^{p_n} \{M_{t_{n,k}} - M_{t_{n,k-1}}\}^2 - [M]_t \Big\|_1 = 0.$$

Proof. Let us prove (2) and then derive (1) from it. Consider first the case where both M and $[M]$ are bounded by a constant $K \geq 0$. For brevity let us write

(3) $$\begin{cases} A_{n,k} = \{M_{n,k} - M_{n,k-1}\}^2 \text{ and } B_{n,k} = [M]_{t_{n,k}} - [M]_{t_{n,k-1}} & \text{for } k = 1, \ldots, p_n \\ \alpha_n = \max_{k=1,\ldots,p_n} A_{n,k} \text{ and } \beta_n = \max_{k=1,\ldots,p_n} B_{n,k} & \text{for } n \in \mathbb{N} \\ S_n = \sum_{k=1}^{p_n} A_{n,k} & \text{for } n \in \mathbb{N} \end{cases}$$

Then

$$E[\{S_n - [M]_t\}^2] = E[\{\textstyle\sum_{k=1}^{p_n}(A_{n,k} - B_{n,k})\}^2] = \sum_{j,k=1,\ldots,p_n} E[(A_{n,j} - B_{n,j})(A_{n,k} - B_{n,k})].$$

Now for $j < k$ we have

$$E[(A_{n,j} - B_{n,j})(A_{n,k} - B_{n,k})] = E[E[(A_{n,j} - B_{n,j})(A_{n,k} - B_{n,k})|\mathfrak{F}_{t_{n,k-1}}]]$$
$$= E[(A_{n,j} - B_{n,j})E[(A_{n,k} - B_{n,k})|\mathfrak{F}_{t_{n,k-1}}]] = 0,$$

since $E[(A_{n,k} - B_{n,k})|\mathfrak{F}_{t_{n,k-1}}] = 0$ by (2) of Lemma 11.25. Thus

(4) $$E[\{S_n - [M]_t\}^2] = E[\textstyle\sum_{k=1}^{p_n}(A_{n,k} - B_{n,k})^2] \leq 2E[\textstyle\sum_{k=1}^{p_n}(A_{n,k}^2 + B_{n,k}^2)]$$
$$\leq 2E[\alpha_n \textstyle\sum_{k=1}^{p_n} A_{n,k}] + 2E[\beta_n B \textstyle\sum_{k=1}^{p_n} B_{n,k}]$$
$$\leq 2E(\alpha_n^2)^{1/2} E(S_n^2)^{1/2} + 2E(\beta_n [M]_t)$$
$$\leq 2E(\alpha_n^2)^{1/2} \sqrt{12} K^2 + 2E(\beta_n [M]_t)$$

§12. STOCHASTIC INTEGRALS WITH RESPECT TO MARTINGALES

by Lemma 12.26. Since almost every sample function of M and $[M]$ is continuous and hence uniformly continuous on $[0, t]$, $\lim_{n\to\infty} \alpha_n^2 = 0$ and $\lim_{n\to\infty} \beta_n = 0$ a.e. on Ω. Since M and $[M]$ are bounded, we have $\lim_{n\to\infty} \mathbf{E}(\alpha_n^2) = 0$ and $\lim_{n\to\infty} \mathbf{E}(\beta_n[M]_t) = 0$ by the Bounded Convergence Theorem. Thus we have $\lim_{n\to\infty} \mathbf{E}[\{S_n - [M]_t\}^2] = 0$ and therefore $\lim_{n\to\infty} \mathbf{E}(|S_n - [M]_t|) = 0$. This proves (2) for the case where M and $[M]$ are bounded.

To prove (2) for the general case we may assume without loss of generality that every sample function of M and $[M]$ is continuous. For $m \in \mathbb{N}$, let

$$\begin{cases} S_m = \inf\{t \in \mathbb{R}_+ : |M_t| > m\} \wedge m \\ T_m = \inf\{t \in \mathbb{R}_+ : [M]_t > m\} \wedge m \\ R_m = S_m \wedge T_m. \end{cases}$$

As we showed in Proposition 3.5, the first passage time of an open set $(-\infty, -m) \cup (m, \infty)$ in $\mathbb{R}$ by a right-continuous adapted process is a stopping time. Thus S_m, T_m, and then $S_m \wedge T_m$, are stopping times. From the fact that every sample function of M and $[M]$ is real valued it follows that $S_m \uparrow \infty$ and $S_m \uparrow \infty$ and thus $R_m \uparrow \infty$ as $m \to \infty$ on Ω. By the continuity of the sample functions of M and $[M]$, $M^{R_m \wedge}$ and $[M]^{R_m \wedge}$ are bounded by m. By Proposition 12.25, we have $[M]^{R_m \wedge} = [M^{R_m \wedge}]$ and thus $[M]^{R_m \wedge}$ is bounded. Let

$$\begin{cases} A_{n,k}^{(m)} = \{(M^{R_m \wedge})_{t_{n,k}} - (M^{R_m \wedge})_{t_{n,k-1}}\}^2 & \text{for } k = 1, \ldots, p-n \\ B_{n,k}^{(m)} = [M^{R_m \wedge}]_{t_{n,k}} - [M^{R_m \wedge}]_{t_{n,k-1}} & \text{for } k = 1, \ldots, p-n \end{cases}$$

Now

(5) $\quad \mathbf{E}(|\sum_{k=1}^{p_n}(A_{n,k} - B_{n,k})|) \leq \mathbf{E}(|\sum_{k=1}^{p_n}(A_{n,k} - A_{n,k}^{(m)})|)$
$\quad + \mathbf{E}(|\sum_{k=1}^{p_n}(A_{n,k}^{(m)} - B_{n,k}^{(m)})|) + \mathbf{E}(|\sum_{k=1}^{p_n}(B_{n,k}^{(m)} - B_{n,k})|).$

Since $M^{R_m \wedge}$ and $[M^{R_m \wedge}]$ are bounded, we have $\lim_{n\to\infty} \mathbf{E}(|\sum_{k=1}^{p_n}(A_{n,k}^{(m)} - B_{n,k}^{(m)})|) = 0$ by our result above for the bounded case. Regarding the first term on the right side of (5), note that $|(M^{R_m \wedge})_{t_{n,k}} - (M^{R_m \wedge})_{t_{n,k-1}}| \leq |M_{t_{n,k}} - M_{t_{n,k-1}}|$ so that $A_{n,k}^{(m)} \leq A_{n,k}$ and therefore

$$\mathbf{E}(|\sum_{k=1}^{p_n}(A_{n,k} - A_{n,k}^{(m)})|) = \mathbf{E}(\sum_{k=1}^{p_n}(A_{n,k} - A_{n,k}^{(m)}))$$
$$= \sum_{k=1}^{p_n} \mathbf{E}([M]_{t_{n,k}} - [M]_{t_{n,k-1}}) - \sum_{k=1}^{p_n} \mathbf{E}([M^{R_m \wedge}]_{t_{n,k}} - [M^{R_m \wedge}]_{t_{n,k-1}})$$
$$= \mathbf{E}([M]_t) - \mathbf{E}([M^{R_m \wedge}]_t),$$

where the second equality is by Lemma 11.25. But $[M^{R_m \wedge}]_t = ([M]^{R_m \wedge})_t \leq [M]_t$ and $\lim_{m\to\infty}([M]^{R_m \wedge})_t = [M]_t$ since $R_m \uparrow \infty$ on Ω. By the Dominated Convergence Theorem we have $\lim_{m\to\infty} \mathbf{E}(([M]^{R_m \wedge})_t) = \mathbf{E}([M]_t)$. Then for every $\varepsilon > 0$ there exists $m_\varepsilon \in \mathbb{N}$ such that $\mathbf{E}(|\sum_{k=1}^{p_n}(A_{n,k} - A_{n,k}^{(m)})|) < \varepsilon$ for all $n \in \mathbb{N}$ when $m \geq m_\varepsilon$. For the third term on the right side of (5), we have according to Lemma 12.23

$$B_{n,k}^{(m)} = [M^{R_m \wedge}]_{t_{n,k}} - [M^{R_m \wedge}]_{t_{n,k-1}} \leq [M]_{t_{n,k}} - [M]_{t_{n,k-1}} = B_{n,k}.$$

Thus

$$\mathbf{E}(|\textstyle\sum_{k=1}^{p_n}(B_{n,k}^{(m)} - B_{n,k})|) = \mathbf{E}(\textstyle\sum_{k=1}^{p_n}(B_{n,k}^{(m)} - B_{n,k})) \leq \mathbf{E}([M]_t) - \mathbf{E}([M^{R_m \wedge}]_t).$$

Then for every $\varepsilon > 0$ there exists $m_\varepsilon \in \mathbb{N}$ such that $\mathbf{E}(|\sum_{k=1}^{p_n}(B_{n,k}^{(m)} - B_{n,k})|) < \varepsilon$ for all $n \in \mathbb{N}$ when $m \geq m_\varepsilon$. Therefore for $m \geq m_\varepsilon$ we have

$$\mathbf{E}(|\textstyle\sum_{k=1}^{p_n}(A_{n,k} - B_{n,k})|) \leq \mathbf{E}(|\textstyle\sum_{k=1}^{p_n}(A_{n,k}^{(m)} - B_{n,k}^{(m)})|) + 2\varepsilon \quad \text{for all } n \in \mathbb{N},$$

and thus $\limsup_{n\to\infty} \mathbf{E}(|\sum_{k=1}^{p_n}(A_{n,k} - B_{n,k})|) \leq 2\varepsilon$. From the arbitrariness of $\varepsilon > 0$ we have $\lim_{n\to\infty} \mathbf{E}(|\sum_{k=1}^{p_n}(A_{n,k} - B_{n,k})|) = 0$, proving (2).

Finally (1) is derived from (2) by using the same argument as in (1) and (2) in Theorem 11.32. ∎

§13 Adapted Brownian Motions

[I] Processes with Independent Increments

Definition 13.1. *Let $(\Omega, \mathfrak{F}, P)$ be a probability space and let $d \in \mathbb{N}$. A mapping ξ of Ω into $\mathbb{R}^d$ which is $\mathfrak{F}/\mathfrak{B}_{\mathbb{R}^d}$-measurable is called a d-dimensional random vector. A d-dimensional stochastic process on $(\Omega, \mathfrak{F}, P)$ is a mapping X of $\mathbb{R}_+ \times \Omega$ into $\mathbb{R}^d$ such that $X(t, \cdot)$ is a d-dimensional random vector on the probability space for every $t \in \mathbb{R}_+$.*

Regarding the Borel σ-algebra $\mathfrak{B}_{\mathbb{R}^d}$ on $\mathbb{R}^d$ we have the following proposition.

Proposition 13.2. *for any $d \in \mathbb{N}$ we have*

(1) $$\mathfrak{B}_{\mathbb{R}^d} = \sigma(\mathfrak{B}_{\mathbb{R}_1} \times \cdots \times \mathfrak{B}_{\mathbb{R}_d}).$$

More generally, for $d = d_1 + \cdots + d_k$ where $d_1, \ldots, d_k \in \mathbb{N}$ we have

(2) $$\mathfrak{B}_{\mathbb{R}^d} = \sigma(\mathfrak{B}_{\mathbb{R}^{d_1}} \times \cdots \times \mathfrak{B}_{\mathbb{R}^{d_k}}).$$

§13. ADAPTED BROWNIAN MOTIONS

Proof. The open intervals with rational endpoints in $\mathbb{R}$ constitute a countable base for the open sets in $\mathbb{R}$. Thus (1) follows from Theorem 1.4.

To prove (2), let $\mathfrak{R}$ be the countable collection of subsets of $\mathbb{R}^d$ of the type $(a_1, b_1) \times \cdots \times (a_d, b_d)$ where the $a_1, b_1, \ldots, a_d, b_d$ are all rational numbers. Clearly $\mathfrak{R} \subset \mathfrak{B}_{\mathbb{R}^{d_1}} \times \cdots \times \mathfrak{B}_{\mathbb{R}^{d_k}}$. Since every open sets in $\mathbb{R}^d$ is a union of members of the countable collection $\mathfrak{R}$, the collection $\mathfrak{O}_{\mathbb{R}^d}$ of all open sets in $\mathbb{R}^d$ is contained in $\sigma(\mathfrak{B}_{\mathbb{R}^{d_1}} \times \cdots \times \mathfrak{B}_{\mathbb{R}^{d_k}})$. Therefore

$$(3) \qquad \mathfrak{B}_{\mathbb{R}^d} = \sigma(\mathfrak{O}_{\mathbb{R}^d}) \subset \sigma(\mathfrak{B}_{\mathbb{R}^{d_1}} \times \cdots \times \mathfrak{B}_{\mathbb{R}^{d_k}}).$$

On the other hand, writing $\mathfrak{O}_{\mathbb{R}^{d_i}}$ for the collection of open sets in $\mathbb{R}^{d_i}$ for $i = 1, \ldots, d$, we have

$$\mathfrak{B}_{\mathbb{R}^{d_1}} \times \cdots \times \mathfrak{B}_{\mathbb{R}^{d_k}} = \sigma(\mathfrak{O}_{\mathbb{R}^{d_1}}) \times \cdots \times \sigma(\mathfrak{O}_{\mathbb{R}^{d_k}})$$
$$\subset \sigma(\mathfrak{O}_{\mathbb{R}^{d_1}} \times \cdots \times \mathfrak{O}_{\mathbb{R}^{d_k}}) \subset \sigma(\mathfrak{O}_{\mathbb{R}^d}) = \mathfrak{B}_{\mathbb{R}^d}$$

where the first set inclusion is by Lemma 1.3. Thus

$$(4) \qquad \sigma(\mathfrak{B}_{\mathbb{R}^{d_1}} \times \cdots \times \mathfrak{B}_{\mathbb{R}^{d_k}}) \subset \mathfrak{B}_{\mathbb{R}^d}.$$

With (3) and (4), we have (2). ∎

Let ξ be a d-dimensional random vector on a probability space $(\Omega, \mathfrak{F}, P)$. Let π_i be the projection of $\mathbb{R}^d = \mathbb{R}_1 \times \cdots \times \mathbb{R}_d$ onto $\mathbb{R}_i$. Consider the ith component $\xi_i = \pi_i \circ \xi$ of ξ. Since ξ is $\mathfrak{F}/\mathfrak{B}_{\mathbb{R}^d}$-measurable and π_i is $\mathfrak{B}_{\mathbb{R}^d}/\mathfrak{B}_{\mathbb{R}_i}$-measurable, ξ_i is $\mathfrak{F}/\mathfrak{B}_{\mathbb{R}_i}$-measurable, that is, ξ_i is a random variable. Conversely, if $\xi_1, \ldots, \xi_d$ are d real valued random variables on a probability space $(\Omega, \mathfrak{F}, P)$, then for the mapping $\xi = (\xi_1, \ldots, \xi_d)$ of Ω into $\mathbb{R}^d$, we have

$$\begin{aligned}\xi^{-1}(\mathfrak{B}_{\mathbb{R}^d}) &= \xi^{-1}(\sigma(\mathfrak{B}_{\mathbb{R}_1} \times \cdots \times \mathfrak{B}_{\mathbb{R}_d})) \quad \text{by Proposition 13.1} \\ &= \sigma(\xi^{-1}(\mathfrak{B}_{\mathbb{R}_1} \times \cdots \times \mathfrak{B}_{\mathbb{R}_d})) \quad \text{by Theorem 1.1} \\ &= \sigma\{\xi_1^{-1}(E_1) \cap \cdots \cap \xi_d^{-1}(E_d) : E_1 \in \mathfrak{B}_{\mathbb{R}_1}, \ldots, E_d \in \mathfrak{B}_{\mathbb{R}_d}\}.\end{aligned}$$

Now since $\xi_i^{-1}(E_i) \in \mathfrak{F}$ for $i = 1, \ldots, d$, we have $\xi_1^{-1}(E_1) \cap \cdots \cap \xi_d^{-1}(E_d) \in \mathfrak{F}$. Thus $\xi^{-1}(\mathfrak{B}_{\mathbb{R}^d}) \subset \mathfrak{F}$ and therefore ξ is a random vector. We summarize these observations in the following proposition.

Proposition 13.3. *Given a probability space $(\Omega, \mathfrak{F}, P)$, let X_i be a mapping of Ω into $\mathbb{R}$ for $i = 1, \ldots, d$ and let $X = (X_1, \ldots, X_d)$, a mapping of Ω into $\mathbb{R}^d$. Then X is a random vector on $(\Omega, \mathfrak{F}, P)$ if and only if X_i is a real valued random variable on $(\Omega, \mathfrak{F}, P)$ for $i = 1, \ldots, d$.*

Proposition 13.4. *1) Let $X = \{X_t : t \in \mathbb{R}_+\}$ be a d-dimensional stochastic process on a filtered space $(\Omega, \mathfrak{F}, \{\mathfrak{F}_t\}, P)$. Let π_i be the projection of $\mathbb{R}^d = \mathbb{R}_1 \times \cdots \times \mathbb{R}_d$ onto $\mathbb{R}_i$ for $i = 1, \ldots, d$. The ith component of X defined by $X^{(i)} = \pi_i \circ X$ is then a 1-dimensional stochastic process on the filtered space and it is adapted if X is.*

2) Conversely let $X^{(i)}$ be a 1-dimensional stochastic process on a filtered space $(\Omega, \mathfrak{F}, \{\mathfrak{F}_t\}, P)$ for $i = 1, \ldots, d$. Then the mapping X of $\mathbb{R}_+ \times \Omega$ into $\mathbb{R}^d = \mathbb{R}_1 \times \cdots \times \mathbb{R}_d$ defined by $X = (X^{(1)}, \ldots, X^{(d)})$ is a d-dimensional stochastic process on the filtered space and X is adapted if $X^{(1)}, \ldots, X^{(d)}$ are.

Proof. 1) If X is a d-dimensional stochastic process and $X^{(i)} = \pi_i \circ X$, then $X^{(i)}$ is a mapping of $\mathbb{R}_+ \times \Omega$ into $\mathbb{R}$ and for every $t \in \mathbb{R}_+$ we have

$$X_t^{(i)}(\cdot) = X^{(i)}(t, \cdot) = (\pi_i \circ X)(t, \cdot) = (\pi_i \circ X_t)(\cdot).$$

The fact that $X^{(i)}$ is a stochastic process on the probability space follows from the fact that

$$\begin{aligned}(X_t^{(i)})^{-1}(\mathfrak{B}_\mathbb{R}) &= (X_t^{-1} \circ \pi_i^{-1})(\mathfrak{B}_\mathbb{R}) \\ &= X_t^{-1}(\mathbb{R}_1 \times \cdots \times \mathbb{R}_{i-1} \times \mathfrak{B}_\mathbb{R} \times \mathbb{R}_{i+1} \times \cdots \times \mathbb{R}_d) \\ &\subset X_t^{-1}(\mathfrak{B}_{\mathbb{R}^d}) \subset \mathfrak{F}.\end{aligned}$$

If X is $\{\mathfrak{F}_t\}$-adapted, then $\mathfrak{F}$ in the last expression is replaced by $\mathfrak{F}_t$ and $X^{(i)}$ is $\{\mathfrak{F}_t\}$-adapted.

2) Conversely suppose $X^{(1)}, \ldots, X^{(d)}$ are 1-dimensional stochastic processes on a filtered space $(\Omega, \mathfrak{F}, \{\mathfrak{F}_t\}, P)$. To show that $X = (X^{(1)}, \ldots, X^{(d)})$ is a d-dimensional stochastic process on the filtered space, let $E_1, \ldots, E_d \in \mathfrak{B}_\mathbb{R}$. Then

$$\begin{aligned}X_t^{(-1)}(E_1 \times \cdots \times E_d) &= (X_t^{(1)}, \ldots, X_t^{(d)})(E_1 \times \cdots \times E_d) \\ &= (X_t^{(1)})^{-1}(E_1) \cap \cdots \cap (X_t^{(d)})^{-1}(E_d) \in \mathfrak{F}.\end{aligned}$$

From this we have

$$\begin{aligned}X_t^{-1}(\mathfrak{B}_{\mathbb{R}^d}) &= X_t^{-1}(\sigma(\mathfrak{B}_{\mathbb{R}_1} \times \cdots \times \mathfrak{B}_{\mathbb{R}_d})) \quad \text{by Proposition 13.1} \\ &= \sigma(X_t^{-1}(\mathfrak{B}_{\mathbb{R}_1} \times \cdots \times \mathfrak{B}_{\mathbb{R}_d})) \quad \text{by Theorem 1.1} \\ &= \sigma\{X_t^{-1}(E_1 \times \cdots \times E_d) : E_i \in \mathfrak{B}_{\mathbb{R}_i}, i = 1, \ldots, d\} \\ &\subset \mathfrak{F}.\end{aligned}$$

This shows that X is a d-dimensional stochastic process on the filtered space. If $X^{(1)}, \ldots, X^{(d)}$ are $\{\mathfrak{F}_t\}$-adapted then $\mathfrak{F}$ in the last two expressions are replaced by $\mathfrak{F}_t$ and X is $\{\mathfrak{F}_t\}$-adapted. ∎

§13. ADAPTED BROWNIAN MOTIONS

Observation 13.5. Let $(\Omega, \mathfrak{F}, P)$ be a probability space and let $\{(S_\alpha, \mathfrak{S}_\alpha) : \alpha \in A\}$ be a collection of measurable spaces. Let X_α be an $\mathfrak{F}/\mathfrak{S}_\alpha$-measurable mapping of Ω into S_α for each $\alpha \in A$. Then $\sigma\{X_\alpha : \alpha \in A\} = \sigma(\cup_{\alpha \in A} \sigma(X_\alpha))$.

Proof. Since X_α is $\sigma(X_\alpha)/\mathfrak{S}_\alpha$-measurable, it is $\sigma(\cup_{\alpha \in A} \sigma(X_\alpha))/\mathfrak{S}_\alpha$-measurable for every $\alpha \in A$. Thus $\sigma\{X_\alpha : \alpha \in A\} \subset \sigma(\cup_{\alpha \in A} \sigma(X_\alpha))$. On the other hand, $\sigma(X_\alpha) \subset \sigma\{X_\alpha : \alpha \in A\}$ for every $\alpha \in A$ so that $\cup_{\alpha \in A} \sigma(X_\alpha) \subset \sigma\{X_\alpha : \alpha \in A\}$ and therefore $\sigma(\cup_{\alpha \in A} \sigma(X_\alpha)) \subset \sigma\{X_\alpha : \alpha \in A\}$. ∎

Lemma 13.6. Let $\{X_{1,1}, \ldots, X_{1,d_1}; \ldots; X_{k,1}, \ldots, X_{k,d_k}\}$, where $k \in \mathbb{N}$ and $d_1, \ldots, d_k \in \mathbb{N}$, be a system of real valued random variables on a probability space $(\Omega, \mathfrak{F}, P)$. Let $X_i = (X_{i,1}, \ldots, X_{i,d_i})$ for $i = 1, \ldots, k$ and $X = (X_1, \ldots, X_k)$. Then

$$\sigma(X) = \sigma\{X_1, \ldots, X_k\} = \sigma\{X_{1,1} \ldots, X_{k,d_k}\}.$$

Proof. Let $d = d_1 + \cdots + d_k$. Now

$$\begin{aligned}
\sigma(X) &= X^{-1}(\mathfrak{B}_{\mathbb{R}^d}) \\
&= X^{-1}(\sigma(\mathfrak{B}_{\mathbb{R}^{d_1}} \times \cdots \times \mathfrak{B}_{\mathbb{R}^{d_k}})) \quad \text{by Proposition 13.2} \\
&= \sigma(X^{-1}(\mathfrak{B}_{\mathbb{R}^{d_1}} \times \cdots \times \mathfrak{B}_{\mathbb{R}^{d_k}})) \quad \text{by Theorem 1.1} \\
&= \sigma\{X^{-1}(E_1 \times \cdots \times E_k) : E_i \in \mathfrak{B}_{\mathbb{R}^{d_i}}, i = 1, \ldots, k\} \\
&= \sigma\{X_1^{-1}(E_1) \cap \cdots \cap X_k^{-1}(E_k) : E_i \in \mathfrak{B}_{\mathbb{R}^{d_i}}, i = 1, \ldots, k\} \\
&\subset \sigma\{\sigma(X_1) \cup \cdots \cup \sigma(X_k)\} \\
&= \sigma\{X_1, \ldots, X_k\} \quad \text{by Observation 13.5.}
\end{aligned}$$

On the other hand

$$\begin{aligned}
\sigma(X_i) &= X_i^{-1}(\mathfrak{B}_{\mathbb{R}^{d_i}}) \\
&= X^{-1}(\mathbb{R}_1 \times \cdots \times \mathbb{R}_{i-1} \times \mathfrak{B}_{\mathbb{R}^{d_i}} \times \mathbb{R}_{i+1} \times \cdots \times \mathbb{R}_k) \\
&\subset X^{-1}(\mathfrak{B}_{\mathbb{R}^{d_1}} \times \cdots \times \mathfrak{B}_{\mathbb{R}^{d_k}}) \\
&\subset \sigma(X^{-1}(\mathfrak{B}_{\mathbb{R}^{d_1}} \times \cdots \times \mathfrak{B}_{\mathbb{R}^{d_k}})) \\
&= X^{-1}(\sigma(\mathfrak{B}_{\mathbb{R}^{d_1}} \times \cdots \times \mathfrak{B}_{\mathbb{R}^{d_k}})) \quad \text{by Theorem 1.1} \\
&= X^{-1}(\mathfrak{B}_{\mathbb{R}^d}) \quad \text{by Proposition 13.2} \\
&= \sigma(X).
\end{aligned}$$

Recalling Observation 13.5, we have

$$\sigma\{X_1,\ldots,X_k\} = \sigma(\cup_{i=1}^k \sigma(X_i)) \subset \sigma(X).$$

Therefore we have

(1) $$\sigma(X) = \sigma\{X_1,\ldots,X_k\}.$$

Now

$$\begin{aligned}
\sigma\{X_1,\ldots,X_k\} &= \sigma(\cup_{i=1}^k \sigma(X_i)) \quad \text{by Observation 13.5} \\
&= \sigma(\cup_{i=1}^k \sigma\{X_{i,1},\ldots,X_{i,d_i}\}) \quad \text{by applying (1)} \\
&= \sigma[\cup_{i=1}^k \{\cup_{j=1}^{d_i} \sigma(X_{i,j})\}] \quad \text{by Observation 13.5} \\
&= \sigma(\cup_{i=1}^k \cup_{j=1}^{d_i} \sigma(X_{i,j})) \\
&= \sigma\{X_{1,1},\ldots,X_{k,d_k}\} \quad \text{by Observation 13.5,}
\end{aligned}$$

where the fourth equality is from the fact that on the one hand $\sigma\{\cup_{j=1}^{d_i}\sigma(X_{i,j})\} \subset \sigma[\cup_{i=1}^k \cup_{j=1}^{d_i} \sigma(X_{i,j})]$ and on the other hand $\cup_{j=1}^{d_i}\sigma(X_{i,j}) \subset \sigma\{\cup_{j=1}^{d_i}\sigma(X_{i,j})\}$. ∎

Lemma 13.7. *Let $X = \{X_t : t \in \mathbb{R}_+\}$ be a d-dimensional stochastic process on a probability space $(\Omega,\mathfrak{F},P)$. For $t \in \mathbb{R}_+$, let T_t be the collection of all finite strictly increasing sequences in $[0,t]$. For $\tau \in T_t$ given by $\tau = \{t_1,\ldots,t_n\}$, let $X_\tau = (X_{t_1},\ldots,X_{t_n})$. Then*
1) $\cup_{\tau \in T_t}\sigma(X_\tau)$ is a π-class of subsets of Ω,
2) $\sigma\{X_s : s \in [0,t]\} = \sigma(\cup_{\tau \in T_t}\sigma(X_\tau))$.
Similarly if we let T be the collection of all finite strictly increasing sequences in $\mathbb{R}_+$ and for $\tau \in T$ define X_τ in the same way as above, then $\cup_{\tau \in T}\sigma(X_\tau)$ is a π-class of subsets of Ω, and $\sigma\{X_t : t \in \mathbb{R}_+\} = \sigma(\cup_{\tau \in T}\sigma(X_\tau))$.

Proof. 1) Let $A_1, A_2 \in \cup_{\tau \in T_t}\sigma(X_\tau)$. Then $A_1 \in \sigma(X_{\tau_1})$ and $A_2 \in \sigma(X_{\tau_2})$ for some $\tau_1, \tau_2 \in T_t$. Suppose τ_1 and τ_2 have n_1 and n_2 entries respectively. Let τ be the element in T_t obtained by combining τ_1 and τ_2 and let n be the number of its entries. If we let π_1 be the projection of $\mathbb{R}_1^d \times \cdots \times \mathbb{R}_n^d$ onto $\mathbb{R}_1^d \times \cdots \times \mathbb{R}_{n_1}^d$, then $X_{\tau_1} = \pi_1 \circ X_\tau$ so that

$$\sigma(X_{\tau_1}) = X_{\tau_1}^{-1}(\mathfrak{B}_{\mathbb{R}_1^d \times \cdots \times \mathbb{R}_{n_1}^d}) = X_\tau^{-1} \circ \pi_1^{-1}(\mathfrak{B}_{\mathbb{R}_1^d \times \cdots \times \mathbb{R}_{n_1}^d}) \subset X_\tau^{-1}(\mathfrak{B}_{\mathbb{R}_1^d \times \cdots \times \mathbb{R}_n^d}),$$

and similarly for $\sigma(X_{\tau_2})$. Thus both A_1 and A_2 are in $X_\tau^{-1}(\mathfrak{B}_{\mathbb{R}_1^d \times \cdots \times \mathbb{R}_n^d})$. Hence there exist B_1 and B_2 in $\mathfrak{B}_{\mathbb{R}_1^d \times \cdots \times \mathbb{R}_n^d}$ such that $A_1 = X_\tau^{-1}(B_1)$ and $A_2 = X_\tau^{-1}(B_2)$. Then

$$A_1 \cap A_2 = X_\tau^{-1}(B_1 \cap B_2) \in X_\tau^{-1}(\mathfrak{B}_{\mathbb{R}_1^d \times \cdots \times \mathbb{R}_n^d}) = \sigma(X_\tau) \subset \cup_{\tau \in T_t}\sigma(X_\tau).$$

§13. ADAPTED BROWNIAN MOTIONS

This shows that $\cup_{\tau \in \mathcal{T}_t} \sigma(X_\tau)$ is a π-class of subsets of Ω.

2) By Observation 13.5 and by the fact that $\{s\} \in \mathcal{T}_t$ for $s \leq t$, we have

$$\sigma\{X_s : s \in [0,t]\} = \sigma(\cup_{s \in [0,t]} \sigma(X_s)) \subset \sigma(\cup_{\tau \in \mathcal{T}_t} \sigma(X_\tau)).$$

To show the reverse inclusion, it suffices to show $\cup_{\tau \in \mathcal{T}_t} \sigma(X_\tau) \subset \sigma\{X_s : s \in [0,t]\}$. To show this last inclusion, it suffices to show $\sigma(X_\tau) \subset \sigma\{X_s : s \in [0,t]\}$ for every $\tau \in \mathcal{T}_t$. Let $\tau = \{t_1, \ldots, t_n\}$. Then we have

$$\sigma(X_\tau) = \sigma\{(X_{t_1}, \ldots, X_{t_n})\} = \sigma\{X_{t_1}, \ldots, X_{t_n}\} \subset \sigma\{X_s : s \in [0,t]\},$$

where the second equality is by Lemma 13.6. This completes the proof of 2). ∎

Lemma 13.8. *Let $X = \{X_t : t \in \mathbb{R}_+\}$ be a d-dimensional stochastic process on a probability space $(\Omega, \mathfrak{F}, P)$. For a finite strictly increasing sequence $\tau = \{t_1, \ldots, t_n\}$ in $\mathbb{R}_+$, let $X_\tau = (X_{t_1}, \ldots, X_{t_n})$ and let*

$$T(X_\tau) = (X_{t_1}, X_{t_2} - X_{t_1}, X_{t_3} - X_{t_2}, \ldots, X_{t_n} - X_{t_{n-1}}).$$

Then $\sigma(T(X_\tau)) = \sigma(X_\tau)$.

Proof. Let $X_{i,1}, \ldots, X_{i,d}$ be the components of X_{t_i} for $i = 1, \ldots, n$. Then

$$X_\tau = (X_{1,1}, \ldots, X_{1,d}; X_{2,1}, \ldots, X_{2,d}; \ldots; X_{n,1}, \ldots, X_{n,d}),$$

and $T(X_\tau)$ has components given by

$$(X_{1,1}, \ldots, X_{1,d}; X_{2,1} - X_{1,1}, \ldots, X_{2,d} - X_{1,d}; \ldots; X_{n,1} - X_{n-1,1}, \ldots, X_{n,d} - X_{n-1,d}).$$

The transformation T is represented by a nonsingular $nd \times nd$ matrix A with entries $a_{i,j} = 1$ when $i = j$, $a_{i,j} = -1$ when $i - j = d$, and $a_{i,j} = 0$ otherwise. Thus the linear transformation T of $\mathbb{R}^{nd}$ into $\mathbb{R}^{nd}$ is a homeomorphism. Now

(1) $$\sigma(T(X_\tau)) = (T(X_\tau))^{-1}(\mathfrak{B}_{\mathbb{R}^{nd}}) = X_\tau^{-1}(T^{-1}(\mathfrak{B}_{\mathbb{R}^{nd}})).$$

Let $\mathfrak{O}^{(nd)}$ be the collection of all open sets in $\mathbb{R}^{nd}$. Since T is a homeomorphism of $\mathbb{R}^{nd}$ we have $T^{-1}(\mathfrak{O}^{(nd)}) = \mathfrak{O}^{(nd)}$. Then

(2) $$T^{-1}(\mathfrak{B}_{\mathbb{R}^{nd}}) = T^{-1}(\sigma(\mathfrak{O}^{(nd)})) = \sigma(T^{-1}(\mathfrak{O}^{(nd)})) = \sigma(\mathfrak{O}^{(nd)}) = \mathfrak{B}_{\mathbb{R}^{nd}},$$

where the second equality is by Theorem 1.1. Using (2) in (1), we have $\sigma(T(X_\tau)) = X_\tau^{-1}(\mathfrak{B}_{\mathbb{R}^{nd}}) = \sigma(X_\tau)$. ∎

Definition 13.9. Let $X = \{X_t : t \in \mathbb{R}_+\}$ be a d-dimensional stochastic process on a probability space $(\Omega, \mathfrak{F}, P)$. We say that X is a process with independent increments if for every finite strictly increasing sequence $\{t_1, \ldots, t_n\}$ in $\mathbb{R}_+$ the system of random vectors $\{X_{t_1}, X_{t_2} - X_{t_1}, X_{t_3} - X_{t_2}, \ldots, X_{t_n} - X_{t_{n-1}}\}$ is an independent system.

Theorem 13.10. Let $X = \{X_t : t \in \mathbb{R}_+\}$ be a d-dimensional stochastic process with independent increments on a probability space $(\Omega, \mathfrak{F}, P)$. Let $\{\mathfrak{F}_t^X : t \in \mathbb{R}_+\}$ be the filtration on the probability space $(\Omega, \mathfrak{F}, P)$ generated by X, that is, $\mathfrak{F}_t^X = \sigma\{X_s : s \in [0, t]\}$ for $t \in \mathbb{R}_+$. Then for every pair $s, t \in \mathbb{R}_+$ such that $s < t$, the system $\{\mathfrak{F}_s^X, X_t - X_s\}$ is independent, that is, $\{\mathfrak{F}_s^X, \sigma(X_t - X_s)\}$ is an independent system.

Proof. Let $\mathcal{T}_s$ be the collection of all finite strictly increasing sequences τ in $[0, s]$. By Lemma 13.7, $\cup_{\tau \in \mathcal{T}_s} \sigma(X_\tau)$ is a π-class of subsets of Ω and $\mathfrak{F}_s^X = \sigma\{X_u : u \in [0, s]\} = \sigma(\cup_{\tau \in \mathcal{T}_s} \sigma(X_\tau))$. Thus to show the independence of the system $\{\mathfrak{F}_s^X, \sigma(X_t - X_s)\}$, it suffices to show the independence of the system $\{\cup_{\tau \in \mathcal{T}_s} \sigma(X_\tau), \sigma(X_t - X_s)\}$. (See Theorem A.4.) Thus it remains to show that for any $A \in \cup_{\tau \in \mathcal{T}_s} \sigma(X_\tau)$ and $B \in \sigma(X_t - X_s)$ we have $P(A \cap B) = P(A)P(B)$. Now if $A \in \cup_{\tau \in \mathcal{T}_s} \sigma(X_\tau)$, then there exists $\tau = \{t_1, \ldots, t_n\} \in \mathcal{T}_s$ such that $A \in \sigma(X_\tau)$. Let T be as in Lemma 13.8. Then $\sigma(X_\tau) = \sigma(T(X_\tau))$ and thus $A \in \sigma(T(X_\tau))$. Since X is a process with independent increments, the system of random vectors $\{X_{t_1}, X_{t_2} - X_{t_1}, \ldots, X_{t_n} - X_{t_{n-1}}, X_s - X_{t_n}, X_t - X_s\}$ is an independent system and so is its subsystem $\{X_{t_1}, X_{t_2} - X_{t_1}, \ldots, X_{t_n} - X_{t_{n-1}}, X_t - X_s\}$. Then $\{(X_{t_1}, X_{t_2} - X_{t_1}, \ldots, X_{t_n} - X_{t_{n-1}}), X_t - X_s\}$ is an independent system of two random. (See Theorem A.11.) Then $\{T(X_\tau), X_t - X_s\}$ is an independent system (see Theorem A.8), or equivalently, $\{\sigma(T(X_\tau)), \sigma(X_t - X_s)\}$ is an independent system. Then for $A \in \sigma(T(X_\tau))$ and $B \in \sigma(X_t - X_s)$, we have $P(A \cap B) = P(A)P(B)$. ∎

The following theorem contains the converse of Theorem 13.10.

Theorem 13.11. Let $X = \{X_t : t \in \mathbb{R}_+\}$ be a d-dimensional adapted process on a filtered space $(\Omega, \mathfrak{F}, \{\mathfrak{F}_t\}, P)$. If for every pair $s, t \in \mathbb{R}_+$ such that $s < t$, $\{\mathfrak{F}_s, X_t - X_s\}$ is an independent system, then X is a process with independent increments.

Proof. Let $t_1 < \cdots < t_n$. Let us prove the independence of the system $\{X_{t_1}, X_{t_2} - X_{t_1}, \ldots, X_{t_n} - X_{t_{n-1}}\}$. Let $\xi_1 = X_{t_1}, \xi_2 = X_{t_2} - X_{t_1}, \ldots, \xi_n = X_{t_n} - X_{t_{n-1}}$ for brevity. Since

§13. ADAPTED BROWNIAN MOTIONS

ξ_i is an $\mathfrak{F}_{t_{n-1}}/\mathfrak{B}_{\mathbb{R}^d}$-measurable mapping of Ω into $\mathbb{R}^d$ for $i = 1, \ldots, n-1$, $(\xi_1, \ldots, \xi_{n-1})$ is an $\mathfrak{F}_{t_{n-1}}/\mathfrak{B}_{\mathbb{R}^{(n-1)d}}$-measurable mapping of Ω into $\mathbb{R}^{(n-1)d}$. Thus $\sigma\{(\xi_1, \ldots, \xi_{n-1})\} \subset \mathfrak{F}_{t_{n-1}}$. Then since $\{\mathfrak{F}_{t_{n-1}}, \sigma(\xi_n)\}$ is an independent system by the assumption of the theorem, $\{\sigma\{(\xi_1, \ldots, \xi_{n-1})\}, \sigma(\xi_n)\}$ is an independent system. Then for the probability distribution $P_{(\xi_1,\ldots,\xi_{n-1},\xi_n)}$ of the random vector $(\xi_1, \ldots, \xi_{n-1}, \xi_n)$ on $(\mathbb{R}^{nd}, \mathfrak{B}_{\mathbb{R}^{nd}})$ we have $P_{(\xi_1,\ldots,\xi_{n-1},\xi_n)} = P_{(\xi_1,\ldots,\xi_{n-1})} \times P_{\xi_n}$. By the same argument we have $P_{(\xi_1,\ldots,\xi_{n-1})} = P_{(\xi_1,\ldots,\xi_{n-2})} \times P_{\xi_{n-1}}$ and so on. Thus we have $P_{(\xi_1,\ldots,\xi_{n-1},\xi_n)} = P_{\xi_1} \times \cdots \times P_{\xi_n}$. This proves the independence of the system $\{\xi_1, \ldots, \xi_n\}$. (See Theorem A.9.) ∎

Theorem 13.12. *Let $X = \{X_t : t \in \mathbb{R}_+\}$ be a d-dimensional adapted process on a filtered space $(\Omega, \mathfrak{F}, \{\mathfrak{F}_t\}, P)$. If for every pair $s, t \in \mathbb{R}_+$ such that $s < t$, $\{\mathfrak{F}_s, X_t - X_s\}$ is an independent system, then for any $s \leq t_0 < t_1 < \cdots < t_n$, the system $\{\mathfrak{F}_s, (X_{t_1} - X_{t_0}, \ldots, X_{t_n} - X_{t_{n-1}})\}$ is independent.*

Proof. To show the independence of $\{\mathfrak{F}_s, (X_{t_1} - X_{t_0}, \ldots, X_{t_n} - X_{t_{n-1}})\}$, it suffices to show the independence of $\{Z, (X_{t_1} - X_{t_0}, \ldots, X_{t_n} - X_{t_{n-1}})\}$ for every $\mathfrak{F}_s$-measurable random variable Z. (See Theorem A.13.) Let us show first the independence of $\{Z, X_{t_1} - X_{t_0}, \ldots, X_{t_n} - X_{t_{n-1}}\}$ by induction. Now since $\{\mathfrak{F}_{t_0}, X_{t_1} - X_{t_0}\}$ is an independent system and since Z is $\mathfrak{F}_s$-measurable and hence $\mathfrak{F}_{t_0}$-measurable, we have the independence of $\{Z, X_{t_1} - X_{t_0}\}$. (See Theorem A.13.) Suppose we have shown the independence of $\{Z, X_{t_1} - X_{t_0}, \ldots, X_{t_k} - X_{t_{k-1}}\}$ for some $k < n$. By the assumption in the theorem, $\{\mathfrak{F}_{t_k}, X_{t_{k+1}} - X_{t_k}\}$ is an independent system. Since $Z, X_{t_1} - X_{t_0}, \ldots, X_{t_k} - X_{t_{k-1}}$ are all $\mathfrak{F}_{t_k}$-measurable, so is $(Z, X_{t_1} - X_{t_0}, \ldots, X_{t_k} - X_{t_{k-1}})$. Thus $\{(Z, X_{t_1} - X_{t_0}, \ldots, X_{t_k} - X_{t_{k-1}}), X_{t_{k+1}} - X_{t_k}\}$ is an independent system. (See Theorem A.13.) Then the independence of the system $\{Z, X_{t_1} - X_{t_0}, \ldots, X_{t_k} - X_{t_{k-1}}\}$ implies the independence of the system $\{Z, X_{t_1} - X_{t_0}, \ldots, X_{t_k} - X_{t_{k-1}}, X_{t_{k+1}} - X_{t_k}\}$. (See Theorem A.11.) Therefore by induction we have the independence of the system $\{Z, X_{t_1} - X_{t_0}, \ldots, X_{t_n} - X_{t_{n-1}}\}$. Then $\{Z, (X_{t_1} - X_{t_0}, \ldots, X_{t_n} - X_{t_{n-1}})\}$ is an independent system. (See Theorem A.11.) ∎

Theorem 13.13. *Let $X = \{X_t : t \in \mathbb{R}_+\}$ be a d-dimensional adapted process on a filtered space $(\Omega, \mathfrak{F}, \{\mathfrak{F}_t\}, P)$ such that every $s, t \in \mathbb{R}_+$, $s < t$, the system $\{\mathfrak{F}_s, X_t - X_s\}$ is independent. Let*

$$\mathfrak{G}_s = \sigma\{X_{t''} - X_{t'} : s \leq t' < t'' < \infty\} \quad \text{for } s \in \mathbb{R}_+.$$

Then $\{\mathfrak{F}_s, \mathfrak{G}_s\}$ is an independent system.

Proof. Let **Y** be the collection of all random vectors Y of the form

(1) $$Y = (X_{t_1''} - X_{t_1'}, \ldots, X_{t_n''} - X_{t_n'}),$$

where $s \leq t_i' < t_i''$ for $i = 1, \ldots, n$ and $n \in \mathbb{N}$. By the same argument as in the Proof of Lemma 13.7, it is easily verified that $\cup_{Y \in \mathbf{Y}} \sigma(Y)$ is a π-class of subsets of Ω and

(2) $$\mathfrak{G}_s = \sigma(\cup_{Y \in \mathbf{Y}} \sigma(Y)).$$

Let **Z** be the collection of all random vectors of the form

(3) $$Z = (X_{t_1} - X_{t_0}, \ldots, X_{t_m} - X_{t_{m-1}})$$

where $s \leq t_0 < \cdots < t_m$ and $m \in \mathbb{N}$. Let us show that

(4) $$\cup_{Z \in \mathbf{Z}} \sigma(Z) = \cup_{Y \in \mathbf{Y}} \sigma(Y).$$

Let $Y \in \mathbf{Y}$ be as given by (1). Let $t_i', t_i'', i = 1, \ldots, n$ be arranged in the increasing order and let the resulting finite strictly increasing sequence be $\{t_0, \ldots, t_m\}$ and let Z be given by (3). Then each component of Y is the sum of some components of Z and hence $\sigma(Y) \subset \sigma(Z)$. On the other hand, every member of **Z** is also a member of **Y**. This proves (4). Thus $\cup_{Z \in \mathbf{Z}} \sigma(Z)$ is a π-class of subsets of Ω and by (4) and (2) we have

(5) $$\mathfrak{G}_s = \sigma(\cup_{Z \in \mathbf{Z}} \sigma(Z)).$$

To show the independence of $\{\mathfrak{F}_s, \mathfrak{G}_s\}$ it suffices to show the independence of the system $\{\mathfrak{F}_s, \cup_{Z \in \mathbf{Z}} \sigma(Z)\}$. (See Theorem A.4.) Let $A \in \mathfrak{F}_s$ and $B \in \cup_{Z \in \mathbf{Z}} \sigma(Z)$. Then $B \in \sigma(Z)$ for some $Z \in \mathbf{Z}$. But according to Theorem 13.12, $\{\mathfrak{F}_s, Z\}$ is an independent system. Thus $P(A \cap B) = P(A)P(B)$. ∎

[II] Brownian Motions in $\mathbb{R}^d$

Given a probability space $(\Omega, \mathfrak{F}, P)$ and a measurable space $(S, \mathfrak{G})$. Let X be an $\mathfrak{F}/\mathfrak{G}$-measurable mapping of Ω into S. The probability distribution of X on $(S, \mathfrak{G})$ is the probability measure P_X on $(S, \mathfrak{G})$ defined by

$$P_X(E) = (P \circ X^{-1})(E) = P(X^{-1}(E)) \quad \text{for } E \in \mathfrak{G}.$$

Let f be an extended real valued $\mathfrak{G}$-measurable function on S and let $E \in \mathfrak{G}$. According to the Image Probability Law we have

$$\int_{X^{-1}(E)} f[X(\omega)] P(d\omega) = \int_E f(x) P_X(dx),$$

§13. ADAPTED BROWNIAN MOTIONS

in the sense that the existence of one side implies that of the other and the equality of the two. In particular with $E = S$, we have

$$\int_\Omega f[X(\omega)] \, P(d\omega) = \int_S f(x) \, P_X(dx).$$

Definition 13.14. *A d-dimensional Brownian motion is a d-dimensional stochastic process $X = \{X_t : t \in \mathbb{R}_+\}$ on a probability space $(\Omega, \mathfrak{F}, P)$ such that*

1°. *X is a process with independent increments,*

2°. *for every $s, t \in \mathbb{R}_+$ such that $s < t$, the probability distribution of the random vector $X_t - X_s$ is the d-dimensional normal distribution $N_d(0, (t-s) \cdot I)$,*

3°. *every sample function of X is an $\mathbb{R}^d$-valued continuous function on $\mathbb{R}_+$.*

Clearly a Brownian motion does not exist on an arbitrary probability space. For the proof of its existence on the space of continuous functions by means of Kolmogorov's extension theorem we refer to [31] J. Yeh.

Notations 13.15. According to Definition 13.14, for a d-dimensional Brownian motion $X = \{X_t : t \in \mathbb{R}_+\}$ on a probability space $(\Omega, \mathfrak{F}, P)$, the probability distribution $P_{X_t - X_s}$ of the random vector $X_t - X_s$ for $s, t \in \mathbb{R}_+$, $s < t$, is the d-dimensional normal distribution $N_d(0, (t-s) \cdot I)$ with mean vector 0 and covariance matrix $(t-s) \cdot I$ where I is the $d \times d$ identity matrix. Thus $P_{X_t - X_s}$ is absolutely continuous with respect to the Lebesgue measure m_L^d on $(\mathbb{R}^d, \mathfrak{B}_{\mathbb{R}^d})$ with Radon-Nikodym derivative given by

(1) $$\frac{dP_{X_t - X_s}}{dm_L^d}(x) = \{2\pi(t-s)\}^{-d/2} \exp\left\{-\frac{1}{2}\frac{|x|^2}{t-s}\right\} \quad \text{for } x \in \mathbb{R}^d,$$

where we write $|x|$ for the Euclidean norm of $x \in \mathbb{R}^d$. If we let p be a function on $(0, \infty) \times \mathbb{R}^d$ defined by

(2) $$p(t, x) = (2\pi t)^{-d/2} \exp\left\{-\frac{1}{2}\frac{|x|^2}{t}\right\} \quad \text{for } (t, x) \in (0, \infty) \times \mathbb{R}^d,$$

then

(3) $$\frac{dP_{X_t - X_s}}{dm_L^d}(x) = p(t - s, x) \quad \text{for } x \in \mathbb{R}^d.$$

The initial distribution of X, that is, the probability distribution P_{X_0} of the initial random variable X_0 of the Brownian motion X, is an arbitrary probability measure on $(\mathbb{R}^d, \mathfrak{B}_{\mathbb{R}^d})$. As a particular case we have a unit mass at some $a \in \mathbb{R}^d$, that is, $P_{X_0}(\{a\}) = 1$ and consequently $P_{X_0}(\mathbb{R}^d - \{a\}) = 0$.

Proposition 13.16. *Let $X = \{X_t : t \in \mathbb{R}_+\}$ be a d-dimensional Brownian motion on a probability space $(\Omega, \mathfrak{F}, P)$. Then for $0 = t_0 < \cdots < t_n$, the probability distribution $P_{(X_{t_0},\ldots,X_{t_n})}$ on $(\mathbb{R}^{(n+1)d}, \mathfrak{B}_{\mathbb{R}^{(n+1)d}})$ of the random vector $(X_{t_0}, \ldots, X_{t_n})$ is given by*

(1) $\quad P\{(X_{t_0}, \ldots, X_{t_n}) \in E\} = P_{(X_{t_0},\ldots,X_{t_n})}(E)$

$\quad = \int_E p(t_1 - t_0, x_1 - x_0) \cdots p(t_n - t_{n-1}, x_n - x_{n-1})(P_{X_0} \times m_L^{nd})(d(x_0, \ldots, x_n))$

$\quad = \{(2\pi)^n \prod_{j=1}^n (t_j - t_{j-1})\}^{-d/2} \int_E \exp\left\{-\frac{1}{2} \sum_{j=1}^n \frac{|x_j - x_{j-1}|^2}{t_j - t_{j-1}}\right\}$

$\quad \cdot (P_{X_0} \times m_L^d \times \cdots \times m_L^d)(d(x_0, x_1, \ldots, x_n)),$

for $E \in \mathfrak{B}_{\mathbb{R}^{(n+1)d}}$. In particular when $E = E_0 \times \cdots \times E_n$ where $E_j \in \mathfrak{B}_{\mathbb{R}^d}$ for $j = 0, \ldots, n$, we have

(2) $\quad P\{X_{t_0} \in E_0, \ldots, X_{t_n} \in E_n\} = \{(2\pi)^n \prod_{j=1}^n (t_j - t_{j-1})\}^{-d/2}$

$\quad \cdot \int_{E_0} P_{X_0}(dx_0) \int_{E_1} m_L^d(dx_1) \cdots \int_{E_n} m_L^d(dx_n) \exp\left\{-\frac{1}{2} \sum_{j=1}^n \frac{|x_j - x_{j-1}|^2}{t_j - t_{j-1}}\right\}.$

For $t > 0$ and $E \in \mathfrak{B}_{\mathbb{R}^d}$, we have

(3) $\quad P\{X_t \in E\} = (2\pi t)^{-d/2} \int_{\mathbb{R}^d} P_{X_0}(dx_0) \int_E \exp\left\{-\frac{1}{2} \frac{|x - x_0|^2}{t - t_0}\right\} m_L^d(dx).$

In particular when P_{X_0} is a unit mass at $a \in \mathbb{R}^d$, we have

(4) $\quad P\{X_t \in E\} = (2\pi t)^{-d/2} \int_E \exp\left\{-\frac{1}{2} \frac{|x - a|^2}{t - t_0}\right\} m_L^d(dx).$

Proof. Consider the mapping T of $\mathbb{R}^{(n+1)d}$ onto $\mathbb{R}^{(n+1)d}$ defined by

$$(y_0, \ldots, y_n) = T(x_0, \ldots, x_n) = (x_0, x_1 - x_0, \ldots, x_n - x_{n-1}).$$

§13. ADAPTED BROWNIAN MOTIONS

As we noted in the proof of Lemma 13.8, T is a homeomorphism of $\mathbb{R}^{(n+1)d}$. Now

$$T(X_{t_0}, \ldots, X_{t_n}) = (X_{t_0}, X_{t_1} - X_{t_0}, \ldots, X_{t_n} - X_{t_{n-1}}),$$

and

$$(X_{t_0}, \ldots, X_{t_n}) = T^{-1}(X_{t_0}, X_{t_1} - X_{t_0}, \ldots, X_{t_n} - X_{t_{n-1}}).$$

For any $E \in \mathfrak{B}_{\mathbb{R}^{(n+1)d}}$ we have

$$\begin{aligned}
P_{(X_{t_0}, \ldots, X_{t_n})}(E) &= P \circ (X_{t_0}, \ldots, X_{t_n})^{-1}(E) \\
&= P \circ (X_{t_0}, X_{t_1} - X_{t_0}, \ldots, X_{t_n} - X_{t_{n-1}})^{-1}(T(E)) \\
&= P_{(X_{t_0}, X_{t_1} - X_{t_0}, \ldots, X_{t_n} - X_{t_{n-1}})}(T(E)) \\
&= (P_{X_{t_0}} \times P_{X_{t_1} - X_{t_0}} \times \cdots \times P_{X_{t_n} - X_{t_{n-1}}})(T(E)) \\
&= \int_{T(E)} p(t_1 - t_0, y_1) \cdots p(t_n - t_{n-1}, y_n)(P_{X_{t_0}} \times m_L^{nd})(d(y_0, \ldots, y_n)) \\
&= \int_E p(t_1 - t_0, x_1 - x_0) \cdots p(t_n - t_{n-1}, x_n - x_{n-1})(P_{X_{t_0}} \times m_L^{nd})(d(x_0, \ldots, x_n))
\end{aligned}$$

where the fourth equality is by the independence of increments, the fifth equality is by (3) of Notations 13.15, and the last equality is by the fact that the Jacobian of T is equal to 1. This proves the second equality in (1). The third equality in (1) then follows from the definition of p by (2) of Notations 13.15. ∎

Definition 13.17. *By an $\{\mathfrak{F}_t\}$-adapted d-dimensional Brownian motion we mean a d-dimensional stochastic process $X = \{X_t : t \in \mathbb{R}_+\}$ on a standard filtered space $(\Omega, \mathfrak{F}, \{\mathfrak{F}_t\}, P)$ such that*

1°. *X is an $\{\mathfrak{F}_t\}$-adapted process, that is, X_t is an $\mathfrak{F}_t/\mathfrak{B}_{\mathbb{R}^d}$-measurable mapping of Ω into $\mathbb{R}^d$ for every $t \in \mathbb{R}_+$,*

2°. *for every $s, t \in \mathbb{R}_+$, $s < t$, the system $\{\mathfrak{F}_s, X_t - X_s\}$ is independent,*

3°. *for every $s, t \in \mathbb{R}_+$, $s < t$, $P_{X_t - X_s} = N(0, (t-s) \cdot I)$,*

4°. *every sample function of X is an $\mathbb{R}^d$-valued continuous function on $\mathbb{R}_+$.*

Remark 13.18. An $\{\mathfrak{F}_t\}$-adapted d-dimensional Brownian motion on a standard filtered space $(\Omega, \mathfrak{F}, \{\mathfrak{F}_t\}, P)$ is always a d-dimensional Brownian motion on the probability space

$(\Omega, \mathfrak{F}, P)$ in the sense of Definition 13.14. This follows from the fact that conditions 1° and 2° of Definition 13.17 imply condition 1° of Definition 13.14 according to Theorem 13.11.

Conversely if $X = \{X_t : t \in \mathbb{R}_+\}$ is a d-dimensional Brownian motion on a complete probability space $(\Omega, \mathfrak{F}, P)$, then a filtration can be constructed on the probability space so that X is an adapted Brownian motion on a standard filtered space. This is done as follows. Let $\{\mathfrak{F}_t^X : t \in \mathbb{R}_+\}$ be the filtration generated by X, that is, $\mathfrak{F}_t^X = \sigma\{X_s : s \in [0,t]\}$ for $t \in \mathbb{R}_+$. Let $\mathfrak{N}$ be the collection of all the null sets in the complete probability space $(\Omega, \mathfrak{F}, P)$ and let $\overline{\mathfrak{F}}_t^X = \sigma(\mathfrak{F}_t^X \cup \mathfrak{N})$ for $t \in \mathbb{R}_+$. Then $\{\overline{\mathfrak{F}}_t^X : t \in \mathbb{R}_+\}$ is an augmented filtration on the complete probability space. The fact that this filtration is right-continuous will be proved in Proposition 13.22. Then $(\Omega, \mathfrak{F}, \{\overline{\mathfrak{F}}_t^X\}, P)$ is a standard filtered space and X is an $\overline{\mathfrak{F}}_t^X$-adapted process. It remains to verify the independence of $\{\overline{\mathfrak{F}}_s^X, X_t - X_s\}$ for every $s, t \in \mathbb{R}_+$ such that $s < t$. Now according to Theorem 13.10, the fact that X is a process with independent increments implies the independence of $\{\mathfrak{F}_s^X, X_t - X_s\}$, that is, for every $A \in \mathfrak{F}_s^X$ and $B \in \sigma(X_t - X_s)$ we have $P(A \cap B) = P(A)P(B)$. For any $N \in \mathfrak{N}$, we have $P(N \cap B) = 0 = P(N)P(B)$. Thus $\{\mathfrak{F}_s^X \cup \mathfrak{N}, X_t - X_s\}$ is an independent system. Now since $(\Omega, \mathfrak{F}, P)$ is a complete probability space, an arbitrary subset of a member of $\mathfrak{N}$ is again a member of $\mathfrak{N}$. Thus $\mathfrak{F}_s^X \cup \mathfrak{N}$ is closed under intersection, that is, it is a π-class of subsets of Ω. Then the independence of $\{\mathfrak{F}_s^X \cup \mathfrak{N}, \sigma(X_t - X_s)\}$ implies that of $\{\sigma(\mathfrak{F}_s^X \cup \mathfrak{N}), \sigma(X_t - X_s)\}$. (See Theorem A.4.) Thus $\{\overline{\mathfrak{F}}_s^X, X_t - X_s\}$ is an independent system. Therefore X is an $\{\overline{\mathfrak{F}}_t^X\}$-adapted d-dimensional Brownian motion on the standard filtered space $(\Omega, \mathfrak{F}, \{\overline{\mathfrak{F}}_t^X\}, P)$.

Lemma 13.19. *Let $X = \{X_t : t \in \mathbb{R}_+\}$ be a d-dimensional Brownian motion on a complete probability space $(\Omega, \mathfrak{F}, P)$. Let $\mathfrak{N}$ be the collection of all the null sets in $(\Omega, \mathfrak{F}, P)$ and let $\mathfrak{F}_t^X = \sigma\{X_s : s \in [0,t]\}$ and $\overline{\mathfrak{F}}_t^X = \sigma(\mathfrak{F}_t^X \cup \mathfrak{N})$ for $t \in \mathbb{R}_+$. Let $0 = t_0 < \cdots < t_{k-1} \leq s < t_k < \cdots < t_n$ and let f_j, $j = 0, \ldots, n$, be bounded real valued continuous functions on $\mathbb{R}^d$. Then*

(1) $\quad \mathbf{E}[f_0(X_{t_0}) \cdots f_n(X_{t_n}) | \overline{\mathfrak{F}}_s^X] = f_0(X_{t_0}) \cdots f_{k-1}(X_{t_{k-1}}) \varphi(X_s)$ *a.e. on* $(\Omega, \overline{\mathfrak{F}}_s^X, P)$,

where φ is a real valued function on $\mathbb{R}^d$ defined by

(2) $\quad \varphi(x) = \{(2\pi)^{n-k+1}(t_k - s) \prod_{j=k+1}^{n}(t_j - t_{j-1})\}^{-d/2}$

$\quad \cdot \int_{\mathbb{R}^{(n-k+1)d}} \exp\left\{-\frac{1}{2}\left(\frac{|x_k - x|^2}{t_k - s} + \sum_{j=k+1}^{n} \frac{|x_j - x_{j-1}|^2}{t_j - t_{j-1}}\right)\right\}$

§13. ADAPTED BROWNIAN MOTIONS

$$\cdot \; f_k(x_k) \cdots f_n(x_n) \, m_L^{(n-k+1)d}(d(x_k, \ldots, x_n)) \quad \text{for } x \in \mathbb{R}^d.$$

Proof. Note that since $X_{t_0}, \ldots, X_{t_{k-1}}$ are $\overline{\mathfrak{F}}_s^X / \mathfrak{B}_{\mathbb{R}^d}$-measurable and $f_0, \ldots, f_{k-1}$ are continuous functions on $\mathbb{R}^d$ and hence $\mathfrak{B}_{\mathbb{R}^d}/\mathfrak{B}_{\mathbb{R}}$-measurable, $f_0(X_{t_0}), \ldots, f_{k-1}(X_{t_{k-1}})$ are $\overline{\mathfrak{F}}_s^X$-measurable random variables. Similarly since X_s is $\overline{\mathfrak{F}}_s^X/\mathfrak{B}_{\mathbb{R}^d}$-measurable and φ is a continuous function on $\mathbb{R}^d$, $\varphi(X_s)$ is a $\overline{\mathfrak{F}}_s^X$-measurable random variable. Thus the product $f_0(X_{t_0}) \cdots f_{k-1}(X_{t_{k-1}}) \varphi(X_s)$ is a $\overline{\mathfrak{F}}_s^X$-measurable random variable. Therefore to prove (1), it remains to show

$$(3) \quad \int_A f_0(X_{t_0}) \cdots f_n(X_{t_n}) \, dP = \int_A f_0(X_{t_0}) \cdots f_{k-1}(X_{t_{k-1}}) \varphi(X_s) \, dP \quad \text{for } A \in \overline{\mathfrak{F}}_s^X.$$

For $0 = u_0 < \cdots < u_m$ and $y_0, \ldots, y_m \in \mathbb{R}^d$, let us define

$$(4) \quad K(u_0, \ldots, u_m; y_0, \ldots, y_m) = \left\{(2\pi)^m \prod_{j=1}^m (u_j - u_{j-1})\right\}^{-d/2} \exp\left\{-\frac{1}{2} \sum_{j=1}^m \frac{|y_j - y_{j-1}|^2}{u_j - u_{j-1}}\right\}.$$

Let $\tau^* = \{t_0, \ldots, t_{k-1}, s\}$ and $X_{\tau^*} = (X_{t_0}, \ldots, X_{t_{k-1}}, X_s)$. For the probability distribution $P_{X_{\tau^*}}$ on $(\mathbb{R}^{(k+1)d}, \mathfrak{B}_{\mathbb{R}^{(k+1)d}})$ we have for $E \in \mathfrak{B}_{\mathbb{R}^{(k+1)d}}$

$$(5) \quad P_{X_{\tau^*}}(E) = \int_E K(t_0, \ldots, t_{k-1}, s; x_0, \ldots, x_{k-1}, x)(P_{X_0} \times m_L^{kd})(d(x_0, \ldots, x_{k-1}, x)),$$

by (1) of Proposition 13.16 and (4). Let $A \in \sigma(X_{\tau^*}) = (X_{\tau^*})^{-1}(\mathfrak{B}_{\mathbb{R}^{(k+1)d}})$ so that $A = (X_{\tau^*})^{-1}(B)$ for some $B \in \mathfrak{B}_{\mathbb{R}^{(k+1)d}}$. The right side of (3) for our A is computed by the Image Probability Law, (2) and (5) as

$$(6) \quad \int_A f_0(X_{t_0}) \cdots f_{k-1}(X_{t_{k-1}}) \varphi(X_s) \, dP$$

$$= \int_B f_0(x_0) \cdots f_{k-1}(x_{k-1}) \varphi(x) P_{X_{\tau^*}}(d(x_0, \ldots, x_{k-1}, x))$$

$$= \int_B K(t_0, \ldots, t_{k-1}, s; x_0, \ldots, x_{k-1}, x) f_0(x_0) \cdots f_{k-1}(x_{k-1})$$

$$\cdot \left\{(2\pi)^{n-k+1}(t_k - s) \prod_{j=k+1}^n (t_j - t_{j-1})\right\}^{-d/2} \int_{\mathbb{R}^{(n-k+1)d}} e^{-\frac{1}{2}\left(\frac{|x_k - x|^2}{t_k - s} + \sum_{j=k+1}^n \frac{|x_j - x_{j-1}|^2}{t_j - t_{j-1}}\right)}$$

$$\cdot f_k(x_k) \cdots f_n(x_n) m_L^{(n-k+1)d}(d(x_k, \ldots, x_n))] (P_{X_{t_0}} \times m_L^{kd})(d(x_0, \ldots, x_{k-1}, x))$$

$$= \int_{B \times \mathbb{R}^{(n-k+1)d}} K(t_0, \ldots, t_{k-1}, s, t_k, \ldots, t_n; x_0, \ldots, x_{k-1}, x, x_k, \ldots, x_n)$$

$$\cdot f_0(x_0) \cdots f_n(x_n) (P_{X_{t_0}} \times m_L^{(n+1)d})(d(x_0, \ldots, x_{k-1}, x, x_k, \ldots, x_n)).$$

To evaluate the left side of (3) for our $A \in \sigma(X_{\tau^*})$, consider the probability distribution of the random vector $(X_{\tau^*}, X_{t_k}, \ldots, X_{t_n})$ on $(\mathbb{R}^{(n+2)d}, \mathfrak{B}_{\mathbb{R}^{(n+2)d}})$ which is given by (1) of Proposition 13.16 and (4) as

(7) $\quad P_{(X_{\tau^*}, X_{t_k}, \ldots, X_{t_n})}(E)$

$$= \int_E K(t_0, \ldots, t_{k-1}, s, t_k, \ldots, t_n; x_0, \ldots, x_{k-1}, x, x_k \ldots, x_n)$$
$$\cdot (P_{X_{t_0}} \times m_L^{(n+1)d})(d(x_0, \ldots, x_{k-1}, x, x_k, \ldots, x_n)) \quad \text{for } E \in \mathfrak{B}_{\mathbb{R}^{(n+2)d}}.$$

Now for our $A = X_{\tau^*}^{-1}(B)$ where $B \in \mathfrak{B}_{\mathbb{R}^{(k+1)d}}$, we have

$$\begin{aligned} X_{\tau^*}^{-1}(B) &= X_{\tau^*}^{-1}(B) \cap \Omega \cap \cdots \cap \Omega \\ &= X_{\tau^*}^{-1}(B) \cap X_{t_k}^{-1}(\mathbb{R}^d) \cap \cdots \cap X_{t_n}^{-1}(\mathbb{R}^d) \\ &= (X_{\tau^*}, X_{t_k}, \ldots, X_{t_n})^{-1}(B \times \mathbb{R}^{(n-k+1)d}). \end{aligned}$$

Thus by the Image Probability Law and (7), we have

(8) $\quad \displaystyle\int_A f_0(X_{t_0}) \cdots f_n(X_{t_n}) \, dP$

$$= \int_{B \times \mathbb{R}^{(n-k+1)d}} f_0(x_0) \cdots f_n(x_n) P_{(X_{\tau^*}, X_{t_k}, \ldots, X_{t_n})}(d(x_0, \ldots, x_{k-1}, x, x_k, \ldots, x_n))$$
$$= \int_{B \times \mathbb{R}^{(n-k+1)d}} K(t_0, \ldots, t_{k-1}, s, t_k, \ldots, t_n; x_0, \ldots, x_{k-1}, x, x_k, \ldots, x_n)$$
$$\cdot f_0(x_0) \cdots f_n(x_n)(P_{X_0} \times m_L^{(n+1)d})(d(x_0, \ldots, x_{k-1}, x, x_k, \ldots, x_n)).$$

By (6) and (8) we have the equality (3) for $A \in \sigma(X_{\tau^*})$.

Let $\mathcal{T}_s$ be the collection of all finite strictly increasing sequences τ in $[0, s]$. If $A \in \cup_{\tau \in \mathcal{T}_s} \sigma(X_\tau)$, then $A \in \sigma(X_\tau)$ for some $\tau \in \mathcal{T}_s$. Let τ' be the finite strictly increasing sequence in $[0, s]$ obtained by combining the entries of τ and $\tau^* = \{t_0, \ldots, t_{k-1}, s\}$ and let $\tau' = \{u_0, \ldots, u_{m-1}, s\}$. Clearly $A \in \sigma(X_{\tau'})$. For any u_j which is not equal to any t_j let the corresponding bounded real valued function f be identically equal to 1 on $\mathbb{R}^d$. Then by our result above, (3) holds for our A. Therefore (3) holds for every $A \in \cup_{\tau \in \mathcal{T}_s} \sigma(X_\tau)$.

According to Lemma 13.7, $\cup_{\tau \in \mathcal{T}_s} \sigma(X_\tau)$ is a π-class of subsets of Ω and we also have $\mathfrak{F}_s^X = \sigma\{X_u : u \in [0, s]\} = \sigma(\cup_{\tau \in \mathcal{T}_s} \sigma(X_\tau))$. If $f_0, \ldots, f_n$ are all nonnegative, then by Corollary 1.7 the equality (3) holds for every $A \in \mathfrak{F}_s^X$. When $f_0, \ldots, f_n$ are not necessarily nonnegative, we have $f_0 \cdots f_n = (f_0^+ - f_0^-) \cdots (f_n^+ - f_n^-)$ which is a linear combination of finite product of nonnegative bounded continuous functions. Since (3) holds for each summand in the linear combination, it holds for the linear combination by the linearity of

§13. ADAPTED BROWNIAN MOTIONS

the integrals with respect to the integrands. Thus (3) holds for $A \in \mathfrak{F}_s^X$. Since $(\Omega, \mathfrak{F}, P)$ is a complete probability space an arbitrary subset of a member of $\mathfrak{N}$ is again a subset of $\mathfrak{N}$. From this it follows that $\mathfrak{F}_s^X \cup \mathfrak{N}$ is a π-class of subsets of Ω. Since (3) holds for every $A \in \mathfrak{F}_s^X \cup \mathfrak{N}$, it holds for every $A \in \sigma(\mathfrak{F}_s^X \cup \mathfrak{N})$ by the same argument as above using Corollary 1.7. Thus (3) holds for every $A \in \overline{\mathfrak{F}}_s^X$. ∎

Lemma 13.20. *Let $X = \{X_t : t \in \mathbb{R}_+\}$ be a d-dimensional Brownian motion on a complete probability space $(\Omega, \mathfrak{F}, P)$. Let $\mathfrak{N}$ be the collection of all the null sets in $(\Omega, \mathfrak{F}, P)$ and let $\mathfrak{F}_t^X = \sigma\{X_s : s \in [0, t]\}$ and $\overline{\mathfrak{F}}_t^X = \sigma(\mathfrak{F}_t^X \cup \mathfrak{N})$ for $t \in \mathbb{R}_+$. Let $0 = t_0 < \cdots < t_n$ and let f_j, $j = 0, \ldots, n$, be bounded real valued continuous functions on $\mathbb{R}^d$. For $s \in \mathbb{R}_+$, let $\overline{\mathfrak{F}}_{s+0}^X = \bigcap_{u>s} \overline{\mathfrak{F}}_u^X$. Then*

$$E[f_0(X_{t_0}) \cdots f_n(X_{t_n}) | \overline{\mathfrak{F}}_{s+0}^X] = E[f_0(X_{t_0}) \cdots f_n(X_{t_n}) | \overline{\mathfrak{F}}_s^X],$$

that is, the two equivalence classes are identical.

Proof. Consider first the case where $t_n \leq s$. In this case, $f_0(X_{t_0}), \ldots, f_n(X_{t_n})$ are $\overline{\mathfrak{F}}_s^X$-measurable and hence also $\overline{\mathfrak{F}}_{s+0}^X$-measurable. Thus we have

$$E[f_0(X_{t_0}) \cdots f_n(X_{t_n}) | \overline{\mathfrak{F}}_{s+0}^X] = f_0(X_{t_0}) \cdots f_n(X_{t_n}) E[1 | \overline{\mathfrak{F}}_{s+0}^X],$$

and

$$E[f_0(X_{t_0}) \cdots f_n(X_{t_n}) | \overline{\mathfrak{F}}_s^X] = f_0(X_{t_0}) \cdots f_n(X_{t_n}) E[1 | \overline{\mathfrak{F}}_s^X].$$

Now $E[1 | \overline{\mathfrak{F}}_{s+0}^X]$ consists of all $\overline{\mathfrak{F}}_{s+0}^X$-measurable functions φ on Ω such that $\int_A \varphi \, dP = \int_A 1 \, dP$ for every $A \in \overline{\mathfrak{F}}_{s+0}^X$. Such a function φ is characterized by the condition that $\varphi = 1$ a.e. on $(\Omega, \overline{\mathfrak{F}}_{s+0}^X, P)$, that is, there exists a null set Λ in $(\Omega, \overline{\mathfrak{F}}_{s+0}^X, P)$ such that $\varphi = 1$ on Λ^c and φ is arbitrary on Λ since $(\Omega, \overline{\mathfrak{F}}_{s+0}^X, P)$ is a complete measure space. Similarly $E[1 | \overline{\mathfrak{F}}_s^X]$ consists of all functions φ such that $\varphi = 1$ on the complements of some null sets in $(\Omega, \overline{\mathfrak{F}}_s^X, P)$ and arbitrary on the null sets. But the collections of the null sets in these two measure spaces, $(\Omega, \overline{\mathfrak{F}}_{s+0}^X, P)$ and $(\Omega, \overline{\mathfrak{F}}_s^X, P)$, are equal to the same collection $\mathfrak{N}$. Thus $E[1 | \overline{\mathfrak{F}}_{s+0}^X] = E[1 | \overline{\mathfrak{F}}_s^X]$. This proves the lemma for the particular case.

Consider the case where $0 = t_0 < \cdots < t_{k-1} \leq s < t_k < \cdots < t_n$. Note that $\overline{\mathfrak{F}}_{s+0}^X = \bigcap_{u>s} \overline{\mathfrak{F}}_u^X = \bigcap_{m \in \mathbb{N}} \overline{\mathfrak{F}}_{s+1/m}^X$. By Theorem 8.7 for martingales with reversed time, we have

(1) $\quad E[f_0(X_{t_0}) \cdots f_n(X_{t_n}) | \overline{\mathfrak{F}}_{s+0}^X]$

$= \lim_{m \to \infty} E[f_0(X_{t_0}) \cdots f_n(X_{t_n}) | \overline{\mathfrak{F}}_{s+1/m}^X]$ a.e. on $(\Omega, \overline{\mathfrak{F}}_{s+0}^X, P),$

for arbitrary versions of the conditional expectations. Now according to Lemma 13.19

(2) $\quad \mathbf{E}[f_0(X_{t_0})\cdots f_n(X_{t_n})|\overline{\mathfrak{F}}_s^X] = f_0(X_{t_0})\cdots f_{k-1}(X_{t_{k-1}})\varphi(X_s)\quad$ a.e. on $(\Omega, \overline{\mathfrak{F}}_s^X, P)$

and similarly for $m \in \mathbb{N}$ so large that $s + 1/m < t_k$ we have

(3) $\quad \mathbf{E}[f_0(X_{t_0})\cdots f_n(X_{t_n})|\overline{\mathfrak{F}}_{s+1/m}^X]$
$\quad = f_0(X_{t_0})\cdots f_{k-1}(X_{t_{k-1}})\varphi(X_{s+1/m})\quad$ a.e. on $(\Omega, \overline{\mathfrak{F}}_{s+1/m}^X, P)$.

Note that since the filtered space $(\Omega, \mathfrak{F}, \{\overline{\mathfrak{F}}_t^X\}, P)$ is augmented, such conditions as 'a.e. on $(\Omega, \overline{\mathfrak{F}}_s^X, P)$', 'a.e. on $(\Omega, \overline{\mathfrak{F}}_{s+0}^X, P)$', and 'a.e. on $(\Omega, \overline{\mathfrak{F}}_{s+1/m}^X, P)$' are all equivalent to the condition 'a.e. on $(\Omega, \mathfrak{F}, P)$'. From (1) and (3), we have

(4) $\quad \mathbf{E}[f_0(X_{t_0})\cdots f_n(X_{t_n})|\overline{\mathfrak{F}}_{s+0}^X] = \lim_{m\to\infty} f_0(X_{t_0})\cdots f_{k-1}(X_{t_{k-1}})\varphi(X_{s+1/m})$
$\quad = f_0(X_{t_0})\cdots f_{k-1}(X_{t_{k-1}})\varphi(X_s)\quad$ a.e. on $(\Omega, \overline{\mathfrak{F}}_s^X, P)$,

where the second equality is by the continuity of X and the continuity of φ on $\mathbb{R}^d$. Thus we have $\mathbf{E}[f_0(X_{t_0})\cdots f_n(X_{t_n})|\overline{\mathfrak{F}}_{s+0}^X] = \mathbf{E}[f_0(X_{t_0})\cdots f_n(X_{t_n})|\overline{\mathfrak{F}}_s^X]$ by (4) and (2). ∎

Observation 13.21. Let G be an open set in a metric space (S, ρ). Then there exists a sequence $\{f_n : n \in \mathbb{N}\}$ of continuous functions on S such that $0 \leq f_n \leq 1$ on S for $n \in \mathbb{N}$ and $\lim_{n\to\infty} f_n = \mathbf{1}_G$ on S.

Proof. If $G = S$, then $f_n = 1$ on S for $n \in \mathbb{N}$ will do. Suppose $G \neq S$. Then G^c is a nonempty closed set. Let $\rho(x, G^c) = \inf_{y \in G^c} \rho(x, y)$ for $x \in S$. Then $\rho(x, G^c)$ is a continuous function of $x \in S$ and $\rho(x, G^c) = 0$ if and only if $x \in G^c$. For $n \in \mathbb{N}$, let f_n be defined by

$$f_n(x) = \frac{\rho(x, G^c)}{\rho(x, G^c) + n^{-1}} \quad \text{for } x \in S.$$

Then f_n is continuous on S and satisfies the condition $0 \leq f_n \leq 1$ on S. If $x \in G$, then $x \notin G^c$, $\rho(x, G^c) > 0$, and $\lim_{n\to\infty} f_n(x) = 1$. If $x \notin G$, then $x \in G^c$, $\rho(x, G^c) = 0$, $f_n(x) = 0$, and $\lim_{n\to\infty} f_n(x) = 0$. This shows that $\lim_{n\to\infty} f_n = \mathbf{1}_G$ on S. ∎

Proposition 13.22. *Let $X = \{X_t : t \in \mathbb{R}_+\}$ be a d-dimensional Brownian motion on a complete probability space $(\Omega, \mathfrak{F}, P)$. Let $\mathfrak{N}$ be the collection of all the null sets in $(\Omega, \mathfrak{F}, P)$*

§13. ADAPTED BROWNIAN MOTIONS

and let $\mathfrak{F}_t^X = \sigma\{X_s : s \in [0,t]\}$ and $\overline{\mathfrak{F}}_t^X = \sigma(\mathfrak{F}_t^X \cup \mathfrak{N})$ for $t \in \mathbb{R}_+$. Then the filtration $\{\overline{\mathfrak{F}}_t^X : t \in \mathbb{R}_+\}$ is right-continuous, that is, for every $s \in \mathbb{R}_+$, we have $\overline{\mathfrak{F}}_{s+0}^X = \overline{\mathfrak{F}}_s^X$.

Proof. Let $s \in \mathbb{R}_+$ be fixed. Let us show that for every $t \in \mathbb{R}_+$ we have

(1) $$\mathbf{E}[\mathbf{1}_A|\overline{\mathfrak{F}}_{s+0}^X] = \mathbf{E}[\mathbf{1}_A|\overline{\mathfrak{F}}_s^X] \quad \text{for every } A \in \overline{\mathfrak{F}}_t^X.$$

Now since $\overline{\mathfrak{F}}_{s+0}^X \subset \overline{\mathfrak{F}}_t^X$ for $t > s$, (1) implies

(2) $$\mathbf{E}[\mathbf{1}_A|\overline{\mathfrak{F}}_{s+0}^X] = \mathbf{E}[\mathbf{1}_A|\overline{\mathfrak{F}}_s^X] \quad \text{for every } A \in \overline{\mathfrak{F}}_{s+0}^X.$$

Since $\mathbf{1}_A \in \mathbf{E}[\mathbf{1}_A|\overline{\mathfrak{F}}_{s+0}^X]$ for A in $\overline{\mathfrak{F}}_{s+0}^X$, (2) implies $\mathbf{1}_A \in \mathbf{E}[\mathbf{1}_A|\overline{\mathfrak{F}}_s^X]$, that is, $\mathbf{1}_A$ is $\overline{\mathfrak{F}}_s^X$-measurable, in other words, $A \in \overline{\mathfrak{F}}_s^X$. Thus $\overline{\mathfrak{F}}_{s+0}^X \subset \overline{\mathfrak{F}}_s^X$ and therefore $\overline{\mathfrak{F}}_{s+0}^X = \overline{\mathfrak{F}}_s^X$. Therefore it remains to prove (1).

Now $\overline{\mathfrak{F}}_t^X = \sigma(\mathfrak{F}_t^X \cup \mathfrak{N})$ and according to Lemma 13.7 we have $\mathfrak{F}_t^X = \sigma(\cup_{\tau \in \mathcal{T}_t} \sigma(X_\tau))$ where $\mathcal{T}_t$ is the collection of all finite strictly increasing sequences τ in $[0,t]$, $X_\tau = (X_{t_1}, \ldots, X_{t_n})$ for $\tau = \{t_1, \ldots, t_n\}$, and $\cup_{\tau \in \mathcal{T}_t} \sigma(X_\tau)$ is a π-class of subsets of Ω. Let us prove (1) first for the particular case where $A \in \sigma(X_\tau)$ for some $\tau \in \mathcal{T}_t$. Now if $A \in \sigma(X_\tau) = X_\tau^{-1}(\mathfrak{B}_{\mathbb{R}^{nd}})$, then $A = X_\tau^{-1}(E)$ for some $E \in \mathfrak{B}_{\mathbb{R}^{nd}}$ and thus $\mathbf{1}_A = \mathbf{1}_{X_\tau^{-1}(E)}$. Therefore to show that (1) holds for $A \in \sigma(X_\tau)$, it suffices to show that

(3) $$\mathbf{E}[\mathbf{1}_{X_\tau^{-1}(E)}|\overline{\mathfrak{F}}_{s+0}^X] = \mathbf{E}[\mathbf{1}_{X_\tau^{-1}(E)}|\overline{\mathfrak{F}}_s^X] \quad \text{for every } E \in \mathfrak{B}_{\mathbb{R}^{nd}}.$$

Let $G_1, \ldots, G_n$ be open sets in $\mathbb{R}^d$. By Observation 13.21, there exist sequences $\{f_{j,k} : k \in \mathbb{N}\}$, $j = 1, \ldots, n$, of continuous functions on $\mathbb{R}^d$, bounded between 0 and 1 such that $\lim_{k\to\infty} f_{j,k} = \mathbf{1}_{G_j}$ for $j = 1, \ldots, n$. According to Lemma 13.20 we have

(4) $$\mathbf{E}[f_{1,k}(X_{t_1})\cdots f_{n,k}(X_{t_n})|\overline{\mathfrak{F}}_{s+0}^X] = \mathbf{E}[f_{1,k}(X_{t_1})\cdots f_{n,k}(X_{t_n})|\overline{\mathfrak{F}}_s^X].$$

By the Conditional Bounded Convergence Theorem, we have

(5) $$\lim_{k\to\infty} \mathbf{E}[f_{1,k}(X_{t_1})\cdots f_{n,k}(X_{t_n})|\overline{\mathfrak{F}}_{s+0}^X]$$
$$= \mathbf{E}[\mathbf{1}_{G_1}(X_{t_1})\cdots \mathbf{1}_{G_n}(X_{t_n})|\overline{\mathfrak{F}}_{s+0}^X] \quad \text{a.e. on } (\Omega, \overline{\mathfrak{F}}_{s+0}^X, P),$$

for arbitrary versions of the conditional expectations, and similarly

(6) $$\lim_{k\to\infty} \mathbf{E}[f_{1,k}(X_{t_1})\cdots f_{n,k}(X_{t_n})|\overline{\mathfrak{F}}_s^X]$$
$$= \mathbf{E}[\mathbf{1}_{G_1}(X_{t_1})\cdots \mathbf{1}_{G_n}(X_{t_n})|\overline{\mathfrak{F}}_s^X] \quad \text{a.e. on } (\Omega, \overline{\mathfrak{F}}_s^X, P).$$

By (4), (5) and (6), we have

(7) $\quad \mathbf{E}[1_{G_1}(X_{t_1})\cdots 1_{G_n}(X_{t_n})|\overline{\mathfrak{F}}^X_{s+0}] = \mathbf{E}[1_{G_1}(X_{t_1})\cdots 1_{G_n}(X_{t_n})|\overline{\mathfrak{F}}^X_s].$

Since $X_\tau = (X_{t_1},\ldots,X_{t_n})$, we have

$$1_{G_1}(X_{t_1})\cdots 1_{G_n}(X_{t_n}) = 1_{X^{-1}_{t_1}(G_1)}\cdots 1_{X^{-1}_{t_n}(G_n)} = 1_{X^{-1}_{t_1}(G_1)\cap\cdots\cap X^{-1}_{t_n}(G_n)}$$
$$= 1_{(X_{t_1},\ldots,X_{t_n})^{-1}(G_1\times\cdots\times G_n)} = 1_{X^{-1}_\tau(G_1\times\cdots\times G_n)}$$

so that by (7) we have

(8) $\quad \mathbf{E}[1_{X^{-1}_\tau(G_1\times\cdots\times G_n)}|\overline{\mathfrak{F}}^X_{s+0}] = \mathbf{E}[1_{X^{-1}_\tau(G_1\times\cdots\times G_n)}|\overline{\mathfrak{F}}^X_s].$

Let $\mathfrak{D}$ be the collection of all members of $\mathfrak{B}_{\mathbb{R}^{nd}}$ such that

(9) $\quad \mathbf{E}[1_{X^{-1}_\tau(D)}|\overline{\mathfrak{F}}^X_{s+0}] = \mathbf{E}[1_{X^{-1}_\tau(D)}|\overline{\mathfrak{F}}^X_s].$

It is easily verified that D is a d-class. For instance, if $D_k \in \mathfrak{D}$, $k \in \mathbb{N}$, and $D_k \uparrow$, then $\lim_{k\to\infty} D_k \in \mathfrak{D}$ by the Conditional Monotone Convergence Theorem. Let $\mathfrak{O}^{(d)}$ and $\mathfrak{O}^{(nd)}$ be the collections of all open sets in $\mathbb{R}^d$ and $\mathbb{R}^{nd}$ respectively. Now since the collection of members of $\mathfrak{B}_{\mathbb{R}^{nd}}$ of the type $G_1 \times \cdots \times G_n$ where $G_1,\ldots,G_n \in \mathfrak{O}^{(d)}$ is a π-class, and since $\mathfrak{D}$ contains this π-class according to (8), we have

$$\sigma\{G_1 \times \cdots \times G_n : G_1,\ldots,G_n \in \mathfrak{O}^{(d)}\} \subset \mathfrak{D}$$

by Theorem 1.5. Now every member of $\mathfrak{O}^{(nd)}$ is a countable union of sets of the type $G_1 \times \cdots \times G_n$ where $G_1,\ldots,G_n \in \mathfrak{O}^{(d)}$ and thus we have

$$\sigma(\mathfrak{O}^{(nd)}) = \sigma\{G_1 \times \cdots \times G_n : G_1,\ldots,G_n \in \mathfrak{O}^{(d)}\}.$$

Therefore $\mathfrak{B}_{\mathbb{R}^{nd}} = \sigma(\mathfrak{O}^{(nd)}) \subset \mathfrak{D}$. Thus (9) holds for every $E \in \mathfrak{B}_{\mathbb{R}^{nd}}$. This proves (3) and therefore (1) holds for every $A \in \sigma(X_\tau)$. From arbitrariness of $\tau \in \mathcal{T}_t$, (1) holds for every $A \in \cup_{\tau \in \mathcal{T}_t}\sigma(X_\tau)$.

It is easily verified that the collection of all members A of $\mathfrak{F}$ for which (1) holds is a d-class of subsets of Ω. Now since this d-class contains the π-class $\cup_{\tau \in \mathcal{T}_t}\sigma(X_\tau)$ as we have just shown, the d-class contains the σ-algebra generated by the π-class, namely $\mathfrak{F}^X_t$. Thus we have

(10) $\quad \mathbf{E}[1_A|\overline{\mathfrak{F}}^X_{s+0}] = \mathbf{E}[1_A|\overline{\mathfrak{F}}^X_s] \quad$ for every $A \in \mathfrak{F}^X_t$.

§13. ADAPTED BROWNIAN MOTIONS

Also

(11) $$\mathbf{E}[1_N | \overline{\mathfrak{F}}_{s+0}^X] = \mathbf{E}[1_N | \overline{\mathfrak{F}}_s^X] \quad \text{for every } N \in \mathfrak{N},$$

since the left side consists of all extended real valued functions on Ω which are equal to 1 except on a null set in $(\Omega, \overline{\mathfrak{F}}_{s+0}^X, P)$, the right side consists of all extended real valued functions on Ω which are equal to 1 except on a null set in $(\Omega, \overline{\mathfrak{F}}_s^X, P)$, and the collections of all the null sets in these two measure spaces are the same $\mathfrak{N}$. By (10) and (11), the equality (1) holds for every $A \in \mathfrak{F}_t^X \cup \mathfrak{N}$. Now by the completeness of the probability space $(\Omega, \mathfrak{F}, P)$, an arbitrary subset of a member of $\mathfrak{N}$ is again a member of $\mathfrak{N}$. This implies that $\mathfrak{F}_t^X \cup \mathfrak{N}$ is a π-class. Then since (1) holds for every A in the π-class $\mathfrak{F}_t^X \cup \mathfrak{N}$, it holds for every A in $\sigma(\mathfrak{F}_t^X \cup \mathfrak{N}) = \overline{\mathfrak{F}}_t^X$ by the same argument as above. ∎

By Remark 13.18 and Proposition 13.22 we have the following theorem.

Theorem 13.23. *Let $X = \{X_t : t \in \mathbb{R}_+\}$ be a d-dimensional Brownian motion on a complete probability space $(\Omega, \mathfrak{F}, P)$. Let $\mathfrak{N}$ be the collection of all the null sets in $(\Omega, \mathfrak{F}, P)$ and let $\mathfrak{F}_t^X = \sigma\{X_s : s \in [0, t]\}$ and $\overline{\mathfrak{F}}_t^X = \sigma(\mathfrak{F}_t^X \cup \mathfrak{N})$ for $t \in \mathbb{R}_+$. Then $(\Omega, \mathfrak{F}, \{\overline{\mathfrak{F}}_t^X\}, P)$ is a standard filtered space and X is an $\{\overline{\mathfrak{F}}_t^X\}$-adapted d-dimensional Brownian motion on it.*

We show next that conditions 2° and 3° in Definition 13.17 for an $\{\mathfrak{F}_t\}$-adapted d-dimensional Brownian motion is equivalent to a condition on the conditional expectation of the characteristic function of $X_t - X_s$ with respect to $\mathfrak{F}_s$. For this we need the following lemma.

Lemma 13.24. *Let X be a d-dimensional random vector on a probability space $(\Omega, \mathfrak{F}, P)$ and let φ_X be its characteristic function, that is, $\varphi_X(y) = \mathbf{E}[e^{i\langle y, X \rangle}]$ for $y \in \mathbb{R}^d$. Let $\mathfrak{G}$ be an arbitrary sub-σ-algebra of $\mathfrak{F}$.*
1) If $\{\mathfrak{G}, X\}$ is independent, then for every $y \in \mathbb{R}^d$ we have

(1) $$\mathbf{E}[e^{i\langle y, X \rangle} | \mathfrak{G}] = \varphi_X(y) \quad \text{a.e. on } (\Omega, \mathfrak{G}, P).$$

2) If there exists a complex valued function ψ on $\mathbb{R}^d$ such that for every $y \in \mathbb{R}^d$ we have

(2) $$\mathbf{E}[e^{i\langle y, X \rangle} | \mathfrak{G}] = \psi(y) \quad \text{a.e. on } (\Omega, \mathfrak{G}, P),$$

then $\{\mathfrak{G}, X\}$ is independent and thus $\psi = \varphi_X$.

Proof. 1) If $\{\mathfrak{G}, X\}$ is independent, then $\{\mathfrak{G}, e^{i\langle y,X\rangle}\}$ is independent for every $y \in \mathbb{R}^d$ since $e^{i\langle y,\cdot\rangle}$ is a $\mathfrak{B}_{\mathbb{R}^d}$-measurable function on $\mathbb{R}^d$. Thus

$$\mathbf{E}[e^{i\langle y,X\rangle}|\mathfrak{G}] = \mathbf{E}[e^{i\langle y,X\rangle}] = \varphi_X(y) \quad \text{a.e. on } (\Omega, \mathfrak{G}, P).$$

2) Conversely suppose (2) holds. To show the independence of $\{\mathfrak{G}, X\}$, it suffices to show the independence of $\{\mathbf{1}_G, X\}$ for every $G \in \mathfrak{G}$. According to Kac's Theorem, the d-dimensional random vector X and the random variable $\mathbf{1}_G$ constitute an independent system if and only if for every $y \in \mathbb{R}^d$ and $z \in \mathbb{R}$, we have

$$\mathbf{E}[e^{i\{\langle y,X\rangle+\langle z,\mathbf{1}_G\rangle\}}] = \mathbf{E}[e^{i\langle y,X\rangle}]\mathbf{E}[e^{i\langle z,\mathbf{1}_G\rangle}].$$

To verify this last equality note that

$$\mathbf{E}[e^{i\{\langle y,X\rangle+\langle z,\mathbf{1}_G\rangle\}}] = \mathbf{E}[\mathbf{E}[e^{i\{\langle y,X\rangle+\langle z,\mathbf{1}_G\rangle\}}|\mathfrak{G}]]$$
$$= \mathbf{E}[e^{i\langle z,\mathbf{1}_G\rangle}\mathbf{E}[e^{i\langle y,X\rangle}|\mathfrak{G}]] \quad \text{since } \mathbf{1}_G \text{ is } \mathfrak{G}\text{-measurable}$$
$$= \mathbf{E}[e^{i\langle z,\mathbf{1}_G\rangle}\psi(y)] = \psi(y)\mathbf{E}[e^{i\langle z,\mathbf{1}_G\rangle}] = \mathbf{E}[e^{i\langle y,X\rangle}]\mathbf{E}[e^{i\langle z,\mathbf{1}_G\rangle}],$$

since we have $\mathbf{E}[e^{i\langle y,X\rangle}] = \psi(y)$ by taking the expectation on both sides of (2). ∎

Proposition 13.25. *Let $X = \{X_t : t \in \mathbb{R}_+\}$ be a d-dimensional stochastic process on a filtered space $(\Omega, \mathfrak{F}, \{\mathfrak{F}_t\}, P)$. Then conditions $2°$ and $3°$ in Definition 13.17 are equivalent to the condition that for any $s, t \in \mathbb{R}_+$ such that $s < t$ and $y \in \mathbb{R}^d$ we have*

$$\mathbf{E}[e^{i\langle y,X_t-X_s\rangle}|\mathfrak{F}_s] = e^{-\frac{|y|^2}{2}(t-s)} \quad \text{a.e. on } (\Omega, \mathfrak{F}_s, P).$$

Proof. 1) Assume $2°$ and $3°$ in Definition 13.17. The independence of the system $\{\mathfrak{F}_s, X_t - X_s\}$ implies that of $\{\mathfrak{F}_s, e^{i\langle y,X_t-X_s\rangle}\}$. Thus

$$\mathbf{E}[e^{i\langle y,X_t-X_s\rangle}|\mathfrak{F}_s] = \mathbf{E}[e^{i\langle y,X_t-X_s\rangle}] = e^{-\frac{|y|^2}{2}(t-s)} \quad \text{a.e. on } (\Omega, \mathfrak{F}_s, P),$$

where the last equality is from the fact that the probability distribution $P_{X_t-X_s}$ of $X_t - X_s$ is the d-dimensional normal distribution $N_d(0, (t-s) \cdot I)$ so that by the Image Probability Law

$$\mathbf{E}[e^{i\langle y,X_t-X_s\rangle}] = \int_{\mathbb{R}^d} e^{i\langle y,x\rangle} P_{X_t-X_s}(dx)$$

and this last integral is the characteristic function of $N_d(0, (t-s) \cdot I)$ which is equal to $e^{-\frac{|y|^2}{2}(t-s)}$ for $y \in \mathbb{R}^d$ by Theorem D.5.

§13. ADAPTED BROWNIAN MOTIONS

2) Conversely assume the condition in the Proposition. Then by Lemma 13.24, the system $\{\mathfrak{F}_s, X_t - X_s\}$ is independent, and the characteristic function $\varphi_{X_t - X_s}$ of the d-dimensional random vector $X_t - X_s$ is given by

$$\varphi_{X_t - X_s}(y) = e^{-\frac{|y|^2}{2}(t-s)} \quad \text{for } y \in \mathbb{R}^d$$

so that the probability distribution of $X_t - X_s$ is the d-dimensional normal distribution $N_d(0, (t-s) \cdot I)$. ∎

As an immediate consequence of Definition 13.17 and Proposition 13.24 we have the following theorem.

Theorem 13.26. *Let $X = \{X_t : t \in \mathbb{R}_+\}$ be a d-dimensional stochastic process on a standard filtered space $(\Omega, \mathfrak{F}, \{\mathfrak{F}_t\}, P)$. Then X is an $\{\mathfrak{F}_t\}$-adapted d-dimensional Brownian motion on the filtered space if and only if*

1°. *X_t is $\mathfrak{F}_t$-measurable for every $t \in \mathbb{R}_+$,*

2°. *for every $s, t \in \mathbb{R}_+$ such that $s < t$ and $y \in \mathbb{R}^d$ we have*

$$\mathrm{E}[e^{i\langle y, X_t - X_s\rangle} | \mathfrak{F}_s] = e^{-\frac{|y|^2}{2}(t-s)} \quad \text{a.e. on } (\Omega, \mathfrak{F}_s, P),$$

3°. *every sample function of X is continuous on $\mathbb{R}_+$.*

Theorem 13.27. *An $\{\mathfrak{F}_t\}$-adapted d-dimensional process $X = \{X_t : t \in \mathbb{R}_+\}$ on a standard filtered space $(\Omega, \mathfrak{F}, \{\mathfrak{F}_t\}, P)$ is an $\{\mathfrak{F}_t\}$-adapted d-dimensional Brownian motion on the filtered space if and only if it satisfies the following two conditions.*

1°. *$\{e^{i\langle y, X_t\rangle + \frac{|y|^2}{2}t} : t \in \mathbb{R}_+\}$ is a martingale on the filtered space for every $y \in \mathbb{R}^d$,*

2°. *every sample function of X is continuous on $\mathbb{R}_+$.*

Proof. It suffices to note that condition 2° in Theorem 13.26 is equivalent to the condition that for every $s, t \in \mathbb{R}_+$ such that $s < t$ and $y \in \mathbb{R}^d$ we have

$$\mathrm{E}[e^{i\langle y, X_t\rangle + \frac{|y|^2}{2}t} | \mathfrak{F}_s] = e^{i\langle y, X_s\rangle + \frac{|y|^2}{2}s} \quad \text{a.e. on } (\Omega, \mathfrak{F}_s, P). \quad \blacksquare$$

Let $X = \{X_t : t \in \mathbb{R}_+\}$ be an $\{\mathfrak{F}_t\}$-adapted d-dimensional Brownian motion on a standard filtered space $(\Omega, \mathfrak{F}, \{\mathfrak{F}_t\}, P)$. For a fixed $t_0 \in \mathbb{R}_+$, let $\mathfrak{G}_t = \mathfrak{F}_{t_0+t}$ for $t \in \mathbb{R}_+$. Then $(\Omega, \mathfrak{F}, \{\mathfrak{G}_t\}, P)$ is a standard filtered space, and if we let $Y_t = X_{t_0+t}$ for $t \in \mathbb{R}_+$, then $Y = \{Y_t : t \in \mathbb{R}_+\}$ is a $\{\mathfrak{G}_t\}$-adapted d-dimensional process on $(\Omega, \mathfrak{F}, \{\mathfrak{G}_t\}, P)$ with initial probability distribution $P_{Y_0} = P_{X_{t_0}}$. For $s, t \in \mathbb{R}_+$ such that $s < t$, the system $\{\mathfrak{G}_t, Y_t - Y_s\} = \{\mathfrak{F}_{s+t_0}, X_{t+t_0} - X_{s+t_0}\}$ is independent. Also $P_{Y_t - Y_s} = P_{X_{t+t_0} - X_{s+t_0}} = N_d(0, (t-s) \cdot I)$. Thus Y is a $\{\mathfrak{G}_t\}$-adapted d-dimensional Brownian motion on $(\Omega, \mathfrak{F}, \{\mathfrak{G}_t\}, P)$. This shows that if X is an adapted d-dimensional Brownian motion, then at any time $t_0 \in \mathbb{R}_+$ it starts anew as an adapted d-dimensional Brownian motion Y with $P_{X_{t_0}}$ as its initial probability distribution but otherwise the probability distribution of Y does not depend on the probability distribution of X in the time interval $[0, t_0)$. This property of an adapted Brownian motion is a particular case of the following theorem.

Theorem 13.28. (Strong Markov Property) *Let $X = \{X_t : t \in \mathbb{R}_+\}$ be an $\{\mathfrak{F}_t\}$-adapted d-dimensional Brownian motion on a standard filtered space $(\Omega, \mathfrak{F}, \{\mathfrak{F}_t\}, P)$. For a finite stopping time T on the filtered space, let $\mathfrak{G}_t = \mathfrak{F}_{T+t}$ and $Y_t = X_{T+t}$ for $t \in \mathbb{R}_+$. Then $(\Omega, \mathfrak{F}, \{\mathfrak{G}_t\}, P)$ is a standard filtered space and $Y = \{Y_t : t \in \mathbb{R}_+\}$ is a $\{\mathfrak{G}_t\}$-adapted d-dimensional Brownian motion on the standard filtered space with initial distribution $P_{Y_0} = P_{X_T}$.*

Proof. Clearly $\{\mathfrak{G}_t : t \in \mathbb{R}_+\}$ is a filtration on the probability space $(\Omega, \mathfrak{F}, P)$, that is, an increasing system of sub-σ-algebras of $\mathfrak{F}$. To show that this filtration is augmented, we show that $\mathfrak{G}_0$ contains all the null sets in $(\Omega, \mathfrak{F}, P)$. Now $\mathfrak{G}_0 = \mathfrak{F}_T$ and $\mathfrak{F}_T$ consists of all $A \in \mathfrak{F}_\infty$ such that $A \cap \{T \leq t\} \in \mathfrak{F}_t$ for every $t \in \mathbb{R}_+$. Let N be an arbitrary null set in $(\Omega, \mathfrak{F}, P)$. Then $N \in \mathfrak{F}_0 \subset \mathfrak{F}_\infty$ since $\mathfrak{F}_0$ is augmented. The completeness of the measure space $(\Omega, \mathfrak{F}, P)$ implies that $N \cap \{T \leq t\}$ is a null set in it and thus $N \cap \{T \leq t\} \in \mathfrak{F}_t$ since $\mathfrak{F}_t$ is augmented for every $t \in \mathbb{R}_+$. Therefore $N \in \mathfrak{F}_T = \mathfrak{G}_0$. This shows that the filtration $\{\mathfrak{G}_t : t \in \mathbb{R}_+\}$ is augmented.

Let us show the right-continuity of the filtration $\{\mathfrak{G}_t : t \in \mathbb{R}_+\}$, that is, for every $t_0 \in \mathbb{R}_+$ we have $\cap_{u > t_0} \mathfrak{G}_u = \mathfrak{G}_{t_0}$, in other words, $\cap_{u > t_0} \mathfrak{F}_{T+u} = \mathfrak{F}_{T+t_0}$. To show this, we show that if $A \in \mathfrak{F}_{T+u}$ for all $u > t_0$, then $A \in \mathfrak{F}_{T+t_0}$. Now since the filtration $\{\mathfrak{F}_t : t \in \mathbb{R}_+\}$ is right-continuous, according to Theorem 3.4, $A \in \mathfrak{F}_{T+u}$ if and only if $A \in \mathfrak{F}_\infty$ and $A \cap \{T + u < t\} \in \mathfrak{F}_t$ for every $t \in \mathbb{R}_+$, and similarly $A \in \mathfrak{F}_{T+t_0}$ if and only if $A \in \mathfrak{F}_\infty$ and $A \cap \{T + t_0 < t\} \in \mathfrak{F}_t$ for every $t \in \mathbb{R}_+$. Note that with fixed $t \in \mathbb{R}_+$, we have $\{T + u < t\} \uparrow$ as $u \downarrow t_0$, and $\{T + u < t\} \subset \{T + t_0 < t\}$ for $u > t_0$. If $\omega \in \{T + t_0 < t\}$, then $T(\omega) + t_0 < t$ so that $T(\omega) + u < t$ and thus $\omega \in \{T + u < t\}$

§13. ADAPTED BROWNIAN MOTIONS

for some $u > t_0$. Therefore $\{T + t_0 < t\} = \cup_{u > t_0}\{T + u < t\} = \lim_{u \downarrow t_0}\{T + u < t\}$ and thus $A \cap \lim_{u \downarrow t_0}\{T + u < t\} = A \cap \{T + t_0 < t\}$ for any $A \subset \Omega$. Now let $A \in \mathfrak{F}_{T+u}$ for all $u > t_0$. Then since $A \cap \{T + u < t\} \in \mathfrak{F}_t$ for every $u > t_0$ we have $A \cap \lim_{u \downarrow t_0}\{T + u < t\} \in \mathfrak{F}_t$, that is, $A \cap \{T + t_0 < t\} \in \mathfrak{F}_t$, for every $t \in \mathbb{R}_+$, and therefore $A \in \mathfrak{F}_{T+t_0}$. This proves the right-continuity of $\{\mathfrak{G}_t : t \in \mathbb{R}_+\}$.

Let us show that Y is a $\{\mathfrak{G}_t\}$-adapted d-dimensional Brownian motion on the standard filtered space $(\Omega, \mathfrak{F}, \{\mathfrak{G}_t\}, P)$. Now since $Y_t = X_{T+t}$, $\mathfrak{G}_t = \mathfrak{F}_{T+t}$, and X_{T+t} is $\mathfrak{F}_{T+t}$-measurable, Y_t is $\mathfrak{G}_t$-measurable for every $t \in \mathbb{R}_+$, that is, Y is $\{\mathfrak{G}_t\}$-adapted. Clearly every sample function of Y is continuous. According to Theorem 13.27, it remains to show that $\{e^{i\langle y, Y_t\rangle + \frac{|y|^2}{2}t} : t \in \mathbb{R}_+\}$ is a martingale for every $y \in \mathbb{R}^d$, that is, for every $s, t \in \mathbb{R}_+$ such that $s < t$ and $y \in \mathbb{R}^d$ we have

$$\mathbf{E}[e^{i\langle y, Y_t\rangle + \frac{|y|^2}{2}t} | \mathfrak{G}_s] = e^{i\langle y, Y_s\rangle + \frac{|y|^2}{2}s} \quad \text{a.e. on } (\Omega, \mathfrak{G}_s, P),$$

in other words,

$$(1) \qquad \mathbf{E}[e^{i\langle y, X_{T+t}\rangle + \frac{|y|^2}{2}t} | \mathfrak{F}_{T+s}] = e^{i\langle y, X_{T+s}\rangle + \frac{|y|^2}{2}s} \quad \text{a.e. on } (\Omega, \mathfrak{F}_{T+s}, P).$$

The $\mathfrak{F}_{T+s}$-measurability of X_{T+s} implies that of the right side of (1). Thus it remains to verify that for every $A \in \mathfrak{F}_{T+s}$ we have

$$(2) \qquad \int_A e^{i\langle y, X_{T+t}\rangle + \frac{|y|^2}{2}t} dP = \int_A e^{i\langle y, X_{T+s}\rangle + \frac{|y|^2}{2}s} dP.$$

Now according to Remark 13.28, $\{e^{i\langle y, X_t\rangle + \frac{|y|^2}{2}t} : t \in \mathbb{R}_+\}$ is a martingale on the filtered space $(\Omega, \mathfrak{F}, \{\mathfrak{F}_t\}, P)$ for every $y \in \mathbb{R}^d$. For $n \in \mathbb{N}$, consider the stopping time $T_n = T \wedge n$. By Theorem 8.10 (Optional Sampling for Bounded Stopping Times) we have

$$\mathbf{E}[e^{i\langle y, X_{T_n+t}\rangle + \frac{|y|^2}{2}(T_n+t)} | \mathfrak{F}_{T_n+s}] = e^{i\langle y, X_{T_n+s}\rangle + \frac{|y|^2}{2}(T_n+s)} \quad \text{a.e. on } (\Omega, \mathfrak{F}_{T_n+s}, P).$$

By the $\mathfrak{F}_{T_n+s}$-measurability of $e^{\frac{|y|^2}{2}T_n}$, the last equality reduces to

$$(3) \qquad \mathbf{E}[e^{i\langle y, X_{T_n+t}\rangle + \frac{|y|^2}{2}t} | \mathfrak{F}_{T_n+s}] = e^{i\langle y, X_{T_n+s}\rangle + \frac{|y|^2}{2}s} \quad \text{a.e. on } (\Omega, \mathfrak{F}_{T_n+s}, P).$$

If $A \in \mathfrak{F}_{T+s}$, then $A \cap \{T+s \leq T_n+s\} \in \mathfrak{F}_{T_n+s}$ by Theorem 3.9. But $\{T+s \leq T_n+s\} = \{T \leq T_n\} = \{T \leq T \wedge n\} = \{T \leq n\}$. Thus $A \cap \{T \leq n\} \in \mathfrak{F}_{T_n+s}$. Therefore by (3), we have

$$(4) \qquad \int_{A \cap \{T \leq n\}} e^{i\langle y, X_{T+t}\rangle + \frac{|y|^2}{2}t} dP = \int_{A \cap \{T \leq n\}} e^{i\langle y, X_{T+s}\rangle + \frac{|y|^2}{2}s} dP.$$

Since (4) holds for every $n \in \mathbb{N}$ and since the finiteness of T implies that $\cup_{n \in \mathbb{N}} \{T \leq n\} = \Omega$ and consequently $\lim_{n \to \infty} \mathbf{1}_{A \cup \{T \leq n\}} = \mathbf{1}_A$ so that (2) follows from (4) by the Bounded Convergence Theorem. ∎

Remark 13.29. Theorem 13.28 can be extended to cover the case where the stopping time T is not finite but is finite on a set of positive probability measure. Let $X = \{X_t : t \in \mathbb{R}_+\}$ be an $\{\mathfrak{F}_t\}$-adapted d-dimensional Brownian motion on a standard filtered space $(\Omega, \mathfrak{F}, \{\mathfrak{F}_t\}, P)$. Let T be a stopping time on the filtered space such that $P\{T < \infty\} > 0$. Let $\overline{\Omega} = \{T < \infty\} \in \mathfrak{F}_\infty$, $\overline{\mathfrak{F}} = \mathfrak{F} \cap \overline{\Omega}$, $\overline{\mathfrak{F}}_t = \mathfrak{F}_t \cap \overline{\Omega}$, and $\overline{P} = P/P(\overline{\Omega})$. It is easily verified that $(\overline{\Omega}, \overline{\mathfrak{F}}, \{\overline{\mathfrak{F}}_t\}, \overline{P})$ is a standard filtered space and the restriction $\overline{X}$ of X to $\mathbb{R}_+ \times \overline{\Omega}$ is an $\{\overline{\mathfrak{F}}_t\}$-adapted d-dimensional Brownian motion on $(\overline{\Omega}, \overline{\mathfrak{F}}, \{\overline{\mathfrak{F}}_t\}, \overline{P})$. The restriction $\overline{T}$ of T to $\overline{\Omega}$ is a stopping time on the filtered space $(\overline{\Omega}, \overline{\mathfrak{F}}, \{\overline{\mathfrak{F}}_t\}, \overline{P})$. If we let $\mathfrak{G}_t = \overline{\mathfrak{F}}_{\overline{T}+t}$ and $Y_t = \overline{X}_{\overline{T}+t}$, then by Theorem 13.28, $(\overline{\Omega}, \overline{\mathfrak{F}}, \{\mathfrak{G}_t\}, \overline{P})$ is a standard filtered space and $Y = \{Y_t : t \in \mathbb{R}_+\}$ is a $\{\mathfrak{G}_t\}$-adapted d-dimensional Brownian motion on $(\overline{\Omega}, \overline{\mathfrak{F}}, \{\mathfrak{G}_t\}, \overline{P})$.

[III] 1-Dimensional Brownian Motions

A 1-dimensional Brownian motion is a particular case of d-dimensional Brownian motion we considered thus far and will be referred to simply as a Brownian motion. We shall show that if $X = \{X_t : t \in \mathbb{R}_+\}$ is an $\{\mathfrak{F}_t\}$-adapted Brownian motion on a standard filtered space $(\Omega, \mathfrak{F}, \{\mathfrak{F}_t\}, P)$ and if $X_0 = 0$ a.e. on $(\Omega, \mathfrak{F}, P)$ then $X \in \mathbf{M}_2^c(\Omega, \mathfrak{F}, \{\mathfrak{F}_t\}, P)$ and its quadratic variation process is given by $[X]_t = t$ for $t \in \mathbb{R}_t$.

Lemma 13.30. *Let $X = \{X_t : t \in \mathbb{R}_+\}$ be a Brownian motion on a probability space $(\Omega, \mathfrak{F}, P)$ with an arbitrary initial distribution P_{X_0}. Then for every $t \in \mathbb{R}_+$, we have*

(1) $$\mathbf{E}(X_t) = \mathbf{E}(X_0) = \int_\mathbb{R} x_0 P_{X_0}(dx_0),$$

provided the integral exists, and

(2) $$\mathbf{E}(X_t^2) = \mathbf{E}(X_0^2) + t = \int_\mathbb{R} x_0^2 P_{X_0}(dx_0) + t.$$

Thus X is an L_1-process if and only if $\int_\mathbb{R} x_0 P_{X_0}(dx_0) \in \mathbb{R}$, and X is an L_2-process if and only if $\int_\mathbb{R} x_0^2 P_{X_0}(dx_0) < \infty$. In particular, if P_{X_0} is a unit mass at some $a \in \mathbb{R}$, then $\mathbf{E}(X_t) = a$ and $\mathbf{E}(X_t^2) = a^2 + t$ for $t \in \mathbb{R}_+$ and X is an L_2-process.

§13. ADAPTED BROWNIAN MOTIONS

Proof. For $t > 0$, the probability distribution P_{X_t} of the random variable X_t is given by (3) of Proposition 13.16 as

$$(3) \quad P_{X_t}(E) = (2\pi t)^{-1/2} \int_{\mathbb{R}} P_{X_0}(dx_0) \int_E \exp\left\{-\frac{1}{2}\frac{|x-x_0|^2}{t-t_0}\right\} m_L(dx),$$

for $E \in \mathfrak{B}_{\mathbb{R}}$. Now from

$$(4) \quad (2\pi t)^{-1/2} \int_{\mathbb{R}} \exp\left\{-\frac{1}{2}\frac{x^2}{t}\right\} m_L(dx) = 1,$$

we have for any $x_0 \in \mathbb{R}$,

$$(5) \quad (2\pi t)^{-1/2} \int_{\mathbb{R}} x \exp\left\{-\frac{1}{2}\frac{|x-x_0|^2}{t}\right\} m_L(dx)$$

$$= (2\pi t)^{-1/2} \int_{\mathbb{R}} (x-x_0) \exp\left\{-\frac{1}{2}\frac{|x-x_0|^2}{t}\right\} m_L(dx)$$

$$+ (2\pi t)^{-1/2} \int_{\mathbb{R}} x_0 \exp\left\{-\frac{1}{2}\frac{|x-x_0|^2}{t}\right\} m_L(dx)$$

$$= x_0$$

Then by the Image Probability Law and by (3) and (5), we have

$$\mathbf{E}(X_t) = \int_{\Omega} X_t \, dP = \int_{\mathbb{R}} x \, P_{X_t}(dx)$$

$$= \int_{\mathbb{R}} P_{X_0}(dx_0) \left\{ (2\pi t)^{-1/2} \int_{\mathbb{R}} x \exp\left\{-\frac{1}{2}\frac{|x-x_0|^2}{t}\right\} m_L(dx) \right\}$$

$$= \int_{\mathbb{R}} x_0 \, P_{X_0}(dx_0),$$

proving (1). Similarly from

$$(6) \quad (2\pi t)^{-1/2} \int_{\mathbb{R}} x^2 \exp\left\{-\frac{1}{2}\frac{x^2}{t}\right\} m_L(dx) = t,$$

we have

$$(7) \quad (2\pi t)^{-1/2} \int_{\mathbb{R}} x^2 \exp\left\{-\frac{1}{2}\frac{|x-x_0|^2}{t}\right\} m_L(dx)$$

$$= (2\pi t)^{-1/2} \int_{\mathbb{R}} \{(x-x_0)^2 + 2x_0(x-x_0) + x_0^2\} \exp\left\{-\frac{1}{2}\frac{|x-x_0|^2}{t}\right\} m_L(dx)$$

$$= t + x_0^2.$$

By the Image Probability Law, (3), and (7), we have

$$\begin{aligned}\mathbf{E}(X_t^2) &= \int_\Omega X_t^2\, dP = \int_\mathbb{R} x^2\, P_{X_t}(dx) \\ &= \int_\mathbb{R} P_{X_0}(dx_0) \left\{ (2\pi t)^{-1/2} \int_\mathbb{R} x^2 \exp\left\{ -\frac{1}{2}\frac{|x-x_0|^2}{t} \right\} m_L(dx) \right\} \\ &= \int_\mathbb{R} x_0^2\, P_{X_0}(dx_0) + t,\end{aligned}$$

proving (2). ∎

Proposition 13.31. *Let* $X = \{X_t : t \in \mathbb{R}_+\}$ *be an* $\{\mathfrak{F}_t\}$*-adapted Brownian motion on a standard filtered space* $(\Omega, \mathfrak{F}, \{\mathfrak{F}_t\}, P)$. *If* X_0 *is integrable then* X *is a martingale and if* X_0 *is square-integrable then* X *is an* L_2*-martingale on the filtered space. If* $X_0 = 0$ *a.e. on* $(\Omega, \mathfrak{F}, P)$ *then* $X \in \mathbf{M}_2^c(\Omega, \mathfrak{F}, \{\mathfrak{F}_t\}, P)$ *and a quadratic variation process* $[X]$ *of* X *is given by* $[X]_t = t$ *for* $t \in \mathbb{R}_t$.

Proof. If X_0 is integrable, then by Lemma 13.30, $\mathbf{E}(X_t) = \mathbf{E}(X_0) \in \mathbb{R}$ for all $t \in \mathbb{R}_+$ and thus X is an L_1-process. To show that X is a martingale, let $s, t \in \mathbb{R}_+$ be such that $s < t$. Then

$$\mathbf{E}[X_t - X_s | \mathfrak{F}_s] = \mathbf{E}[X_t - X_s] = 0 \quad \text{a.e. on } (\Omega, \mathfrak{F}_s, P),$$

where the first equality is by the independence of $\{\mathfrak{F}_s, X_t - X_s\}$ according to 2° of Definition 13.17 and the second equality is by the fact that the probability distribution of $X_t - X_s$ is given by $N(0, t-s)$ according to 3° of Definition 13.17. If X_0 is square-integrable then X is an L_2-process by Lemma 13.30. If $X_0 = 0$ a.e. on $(\Omega, \mathfrak{F}, P)$ then $X \in \mathbf{M}_2^c(\Omega, \mathfrak{F}, \{\mathfrak{F}_t\}, P)$ by Definition 11.3. To find a quadratic variation process $[X]$, let a stochastic process $A = \{A_t : t \in \mathbb{R}_+\}$ be defined on the filtered space by setting $A(t, \omega) = t$ for $(t, \omega) \in \mathbb{R}_+ \times \Omega$. Then A is trivially a continuous increasing process on the filtered space. To show that A is a quadratic variation process of X, it remains to verify that $X^2 - A$ is a null at 0 right-continuous martingale. Clearly $X^2 - A$ is null at 0 and continuous. To show that it is a martingale, we show that for $s, t \in \mathbb{R}_+$ such that $s < t$ we have

$$\mathbf{E}[X_t^2 - A_t | \mathfrak{F}_s] = X_s^2 - A_s \quad \text{a.e. on } (\Omega, \mathfrak{F}_s, P),$$

or, equivalently,

$$\mathbf{E}[X_t^2 - X_s^2 | \mathfrak{F}_s] = t - s \quad \text{a.e. on } (\Omega, \mathfrak{F}_s, P).$$

But

$$\begin{aligned}\mathbf{E}[X_t^2 - X_s^2 | \mathfrak{F}_s] &= \mathbf{E}[\{X_t - X_s\}^2 | \mathfrak{F}_s] \\ &= \mathbf{E}[\{X_t - X_s\}^2] = t - s \quad \text{a.e. on } (\Omega, \mathfrak{F}_s, P),\end{aligned}$$

§13. ADAPTED BROWNIAN MOTIONS

where the first equality is by the martingale property of X, the second equality is by the independence of $\{\mathfrak{F}_s, \{X_t - X_s\}^2\}$ implied by that of $\{\mathfrak{F}_s, X_t - X_s\}$, and the third equality is by the fact that the probability distribution of $X_t - X_s$ is given by $N(0, t-s)$. This shows that A is a quadratic variation process of X. ∎

Proposition 13.32. *Let $X = \{X_t : t \in \mathbb{R}_+\}$ be an $\{\mathfrak{F}_t\}$-adapted d-dimensional Brownian motion on a standard filtered space $(\Omega, \mathfrak{F}, \{\mathfrak{F}_t\}, P)$. Then its components $X^{(i)}, i = 1, \ldots, d$, are $\{\mathfrak{F}_t\}$-adapted 1-dimensional Brownian motions on the standard filtered space with initial distributions $P_{X_0^{(i)}} = P_{X_0} \circ \pi_i^{-1}$ where π_i is the projection of $\mathbb{R}^d = \mathbb{R}_1 \times \cdots \times \mathbb{R}_d$ onto $\mathbb{R}_i$ for $i = 1, \ldots, d$.*

Proof. By Proposition 13.4, $X^{(i)}$ is an $\{\mathfrak{F}_t\}$-adapted process on the filtered space. Also $X^{(i)}$ is a continuous process. Thus according to Theorem 13.26, to show that $X^{(i)}$ is an $\{\mathfrak{F}_t\}$-adapted Brownian motion it remains to verify that for every $s, t \in \mathbb{R}_+$ such that $s < t$ and any $y_i \in \mathbb{R}$ we have

$$(1) \qquad \mathbf{E}[e^{i\langle y_i, X_t^{(i)} - X_s^{(i)}\rangle} | \mathfrak{F}_s] = e^{-\frac{|y_i|^2}{2}(t-s)} \quad \text{a.e. on } (\Omega, \mathfrak{F}_s, P).$$

But since X is an $\{\mathfrak{F}_t\}$-adapted d-dimensional Brownian motion, Theorem 13.26 implies that for every $y \in \mathbb{R}^d$ we have

$$(2) \qquad \mathbf{E}[e^{i\langle y, X_t - X_s\rangle} | \mathfrak{F}_s] = e^{-\frac{|y|^2}{2}(t-s)} \quad \text{a.e. on } (\Omega, \mathfrak{F}_s, P).$$

With the choice $y = (0, \ldots, 0, y_i, 0, \ldots, 0) \in \mathbb{R}^d$, (2) reduces to (1).

Regarding the initial distribution $P_{X_0^{(i)}}$ of $X^{(i)}$, we have

$$P_{X_0^{(i)}}(E) = P \circ (X_0^{(i)})^{-1}(E) = P \circ (\pi_i \circ X_0)^{-1}(E)$$
$$= P \circ X_0^{-1} \circ \pi_i^{-1}(E) = P_{X_0} \circ \pi_i^{-1}(E) \quad \text{for } E \in \mathfrak{B}_\mathbb{R}. \quad \blacksquare$$

Proposition 13.33. *Let $X = \{X_t : t \in \mathbb{R}_+\}$ be an $\{\mathfrak{F}_t\}$-adapted d-dimensional Brownian motion on a standard filtered space $(\Omega, \mathfrak{F}, \{\mathfrak{F}_t\}, P)$. If $\mathbf{E}(|X_0|^2) < \infty$, then the components $X^{(1)}, \ldots, X^{(d)}$ of X are continuous L_2-martingales on the filtered space. For every $s, t \in \mathbb{R}_+$ such that $s < t$ we have*

$$(1) \qquad \mathbf{E}[X_t^{(i)} - X_s^{(i)} | \mathfrak{F}_s] = 0 \quad \text{a.e. on } (\Omega, \mathfrak{F}_s, P),$$

and

$$(2) \qquad \mathbf{E}[\{X_t^{(i)} - X_s^{(i)}\}\{X_t^{(j)} - X_s^{(j)}\} | \mathfrak{F}_s] = \delta_{i,j}(t-s) \quad \text{a.e. on } (\Omega, \mathfrak{F}_s, P).$$

In particular when $X_0 = 0$ then $X^{(1)},\ldots,X^{(d)} \in \mathbf{M}_2^c(\Omega,\mathfrak{F},\{\mathfrak{F}_t\},P)$ and their quadratic variation processes are given by

(3) $$[X^{(i)}, X^{(j)}]_t = \delta_{i,j}t \quad for\, t \in \mathbb{R}_+.$$

Proof. By Proposition 13.32, $X^{(1)},\ldots,X^{(d)}$ are $\{\mathfrak{F}_t\}$-adapted Brownian motions on the filtered space. Since $|X_0|^2 = \sum_{i=1}^d |X_0^{(i)}|^2$, if $\mathbf{E}(|X_0|^2) < \infty$ then $\mathbf{E}(|X_0^{(i)}|^2) < \infty$ for $i=1,\ldots,d$ and thus by Proposition 13.31, $X^{(1)},\ldots,X^{(d)}$ are L_2-martingales. The equation (1) is the martingale property of $X^{(i)}$.

To prove (2), recall that since X is an $\{\mathfrak{F}_t\}$-adapted d-dimensional Brownian motion, $\{\mathfrak{F}_s, X_t - X_s\}$ is an independent system by Definition 13.17. Then since the mapping $\pi_{i,j}(x_1,\ldots,x_d) = x_i x_j$ of $\mathbb{R}^d$ into $\mathbb{R}$ is a $\mathfrak{B}_{\mathbb{R}^d}/\mathfrak{B}_{\mathbb{R}}$-measurable mapping, the system $\{\mathfrak{F}_s, \{X_t^{(i)} - X_s^{(i)}\}\{X_t^{(j)} - X_s^{(j)}\}\}$ is an independent one. Thus

$$\mathbf{E}[\{X_t^{(i)} - X_s^{(i)}\}\{X_t^{(j)} - X_s^{(j)}\}|\mathfrak{F}_s] = \mathbf{E}[\{X_t^{(i)} - X_s^{(i)}\}\{X_t^{(j)} - X_s^{(j)}\}]$$
$$= \delta_{i,j}(t-s) \quad \text{a.e. on } (\Omega,\mathfrak{F}_s,P),$$

since the probability distribution of $X_t - X_s$ is the normal distribution $N(0,(t-s)\cdot I)$ whose covariance matrix is given by $(t-s)\cdot I$ according to Theorem D.5. This proves (2).

If $X_0 = 0$, then $X_0^{(i)} = 0$ so that $X^{(i)} \in \mathbf{M}_2^c(\Omega,\mathfrak{F},\{\mathfrak{F}_t\},P)$. Let $i,j = 1,\ldots,d$ be fixed. Consider $V \in \mathbf{V}(\Omega,\mathfrak{F},\{\mathfrak{F}_t\},P)$ defined by $V_t = \delta_{i,j}t$ for $t \in \mathbb{R}_+$. Then

$$\mathbf{E}[\{X_t^{(i)}X_t^{(j)} - V_t\} - \{X_s^{(i)}X_s^{(j)} - V_s\}|\mathfrak{F}_s]$$
$$= \mathbf{E}[X_t^{(i)}X_t^{(j)} - X_s^{(i)}X_s^{(j)}|\mathfrak{F}_s] - \delta_{i,j}(t-s)$$
$$= \mathbf{E}[\{X_t^{(i)} - X_s^{(i)}\}\{X_t^{(j)} - X_s^{(j)}\}|\mathfrak{F}_s] - \delta_{i,j}(t-s)$$
$$= 0 \quad \text{a.e. on } (\Omega,\mathfrak{F}_s,P),$$

where the third equality is by the martingale property of $X^{(i)}$ and $X^{(j)}$ and the last equality is by (2). Thus $X^{(i)}X^{(j)} - V$ is a right-continuous null at 0 martingale. This proves (3). ∎

[IV] Stochastic Integrals with Respect to a Brownian Motion

If $B = \{B_t : t \in \mathbb{R}_+\}$ is an $\{\mathfrak{F}_t\}$-adapted null at 0 Brownian motion on a standard filtered space $(\Omega,\mathfrak{F},\{\mathfrak{F}_t\},P)$, then $B \in \mathbf{M}_2^c(\Omega,\mathfrak{F},\{\mathfrak{F}_t\},P)$ and its quadratic variation process is given by $[B]_t = t$ for $t \in \mathbb{R}_+$ according to Proposition 13.31. Thus for our

§13. ADAPTED BROWNIAN MOTIONS

$[B] \in \mathbf{A}(\Omega, \mathfrak{F}, \{\mathfrak{F}_t\}, P)$, the family of Lebesgue-Stieltjes measures $\{\mu_{[B]}(\cdot, \omega) : \omega \in \Omega\}$ on $(\mathbb{R}_+, \mathfrak{B}_{\mathbb{R}_+})$ determined by $[B]$ is simply

$$\mu_{[B]}(\cdot, \omega) = m_L \quad \text{for every } \omega \in \Omega,$$

where m_L is the Lebesgue measure. Since a null at 0 Brownian motion B is a particular case of martingales in $\mathbf{M}_2^c(\Omega, \mathfrak{F}, \{\mathfrak{F}_t\}, P)$, the results in §12 concerning stochastic integrals with respect to $M \in \mathbf{M}_2^c(\Omega, \mathfrak{F}, \{\mathfrak{F}_t\}, P)$ apply. We show below some implications of the special property of the quadratic variation process $[B]$ of B.

Observation 13.34. Let $p \in [1, \infty)$. Consider the collection of all measurable processes $X = \{X_t : t \in \mathbb{R}_+\}$ on a probability space $(\Omega, \mathfrak{F}, P)$ satisfying the following integrability condition

$$(1) \qquad \int_{[0,t] \times \Omega} |X(s, \omega)|^p (m_L \times P)(d(s, \omega)) < \infty \quad \text{for every } t \in \mathbb{R}_+.$$

This condition is equivalent to the condition

$$(2) \qquad \int_{[0,m] \times \Omega} |X(s, \omega)|^p (m_L \times P)(d(s, \omega)) < \infty \quad \text{for every } m \in \mathbb{N}.$$

By the Tonelli Theorem, we have

$$(3) \qquad \int_{[0,m] \times \Omega} |X(s, \omega)|^p (m_L \times P)(d(s, \omega)) = \mathbf{E}\left[\int_{[0,m]} |X(s)|^p m_L(ds)\right].$$

The condition $\mathbf{E}[\int_{[0,m]} |X(s)|^p m_L(ds)] < \infty$ implies that $\int_{[0,m]} |X(s)|^p m_L(ds) < \infty$ a.e. on $(\Omega, \mathfrak{F}, P)$. Thus for every $m \in \mathbb{N}$ there exists a null set Λ_m in $(\Omega, \mathfrak{F}, P)$ such that $\int_{[0,m]} |X(s, \omega)|^p m_L(ds) < \infty$ for $\omega \in \Lambda_m^c$. Then with the null set $\Lambda = \cup_{m \in \mathbb{N}} \Lambda_m$ we have

$$(4) \qquad \int_{[0,t]} |X(s, \omega)|^p m_L(ds) < \infty \quad \text{for every } t \in \mathbb{R}_+ \text{ when } \omega \in \Lambda^c.$$

The condition

$$(5) \qquad \int_{[0,t] \times \Omega} |X - Y|^p d(m_L \times P) = 0 \quad \text{for every } t \in \mathbb{R}_+$$

is an equivalence relation in the collection of all measurable processes on $(\Omega, \mathfrak{F}, P)$. Let $\mathbf{L}_{p,\infty}(\mathbb{R}_+ \times \Omega, m_L \times P)$ be the linear space of the equivalence classes of all measurable processes on $(\Omega, \mathfrak{F}, P)$ satisfying (1) with respect to this equivalence relation. The element

$0 \in \mathbf{L}_{p,\infty}(\mathbb{R}_+ \times \Omega, m_L \times P)$ is the equivalence class of all measurable processes X on $(\Omega, \mathfrak{F}, P)$ satisfying the condition

(6) $\quad \int_{[0,t]\times\Omega} |X(s,\omega)|^p (m_L \times P)(d(s,\omega)) = 0 \quad$ for every $t \in \mathbb{R}_+$.

A measurable process X on $(\Omega, \mathfrak{F}, P)$ satisfies condition (6) if and only if there exists a null set Λ in $(\Omega, \mathfrak{F}, P)$ such that for every $\omega \in \Lambda^c$ we have

(7) $\quad \int_{[0,t]} |X(s,\omega)|^p m_L(ds) = 0 \quad$ for every $t \in \mathbb{R}_+$.

This follows by the same argument as in Observation 11.16. Note that condition (7) is equivalent to the condition that there exists a null set Λ in $(\Omega, \mathfrak{F}, P)$ such that for every $\omega \in \Lambda^c$ we have

(8) $\quad X(\cdot, \omega) = 0 \quad$ a.e. on $(\mathbb{R}_+, \mathfrak{B}_{\mathbb{R}_+}, m_L)$.

Definition 13.35. *Let $p \in [1, \infty)$. In the linear space $\mathbf{L}_{p,\infty}(\mathbb{R}_+ \times \Omega, m_L \times P)$ of the equivalence classes of measurable processes $X = \{X_t : t \in \mathbb{R}_+\}$ on a probability space $(\Omega, \mathfrak{F}, P)$ satisfying the condition*

(1) $\quad \int_{[0,t]\times\Omega} |X(s,\omega)|^p (m_L \times P)(d(s,\omega)) < \infty \quad$ *for every $t \in \mathbb{R}_+$,*

we define

(2) $\quad \|X\|_{p,t}^{m_L \times P} = \left[\int_{[0,t]\times\Omega} |X|^p \, d(m_L \times P)\right]^{1/p} = \mathbf{E}\left[\int_{[0,t]} |X(s)|^p m_L(ds)\right]^{1/p}$

for every $t \in \mathbb{R}_+$, and define

(3) $\quad \|X\|_{p,\infty}^{m_L \times P} = \sum_{m \in \mathbb{N}} 2^{-m} \{\|X\|_{p,m}^{m_L \times P} \wedge 1\}$.

Remark 13.36. The functions $\|\cdot\|_{p,t}^{m_L \times P}$ for $t \in \mathbb{R}_+$ and $\|\cdot\|_{p,\infty}^{m_L \times P}$ on $\mathbf{L}_{p,\infty}(\mathbb{R}_+ \times \Omega, m_L \times P)$ defined above have the following properties.
1) $\|\cdot\|_{p,t}^{m_L \times P}$ is a seminorm on $\mathbf{L}_{p,\infty}(\mathbb{R}_+ \times \Omega, m_L \times P)$ for every $t \in \mathbb{R}_+$.
2) For $X, X^{(n)} \in \mathbf{L}_{p,\infty}(\mathbb{R}_+ \times \Omega, m_L \times P)$, $n \in \mathbb{N}$, we have

$$\lim_{n\to\infty} \|X^{(n)} - X\|_{p,\infty}^{m_L \times P} = 0 \Leftrightarrow \lim_{n\to\infty} \|X^{(n)} - X\|_{p,m}^{m_L \times P} = 0 \text{ for every } m \in \mathbb{N}.$$

§13. ADAPTED BROWNIAN MOTIONS

3) $\|\cdot\|_{p,\infty}$ is a quasinorm on $\mathbf{L}_{p,\infty}(\mathbb{R}_+ \times \Omega, m_L \times P)$.

Proof. These statements are proved in the same way as Remark 11.4 for the seminorm $|\cdot|_t$ and the quasinorm $|\cdot|_\infty$ on the space $\mathbf{M}_2(\Omega, \mathfrak{F}, \{\mathfrak{F}_t\}, P)$. ∎

Let the Banach space $L_p([0,m] \times \Omega, \sigma(\mathfrak{B}_{[0,m]} \times \mathfrak{F}), m_L \times P)$ for $m \in \mathbb{N}$ be abbreviated as $L_p([0,m] \times \Omega)$. The function $\|\cdot\|_{p,m}^{m_L \times P}$ is only a seminorm on the space $\mathbf{L}_{p,\infty}(\mathbb{R}_+ \times \Omega, m_L \times P)$, but it is a norm on $L_p([0,m] \times \Omega)$ and $L_p([0,m] \times \Omega)$ is complete with respect to the metric associated with this norm. We use this fact to show that the space $\mathbf{L}_{p,\infty}(\mathbb{R}_+ \times \Omega, m_L \times P)$ is complete with respect to the metric associated with the quasinorm $\|\cdot\|_{p,\infty}^{m_L \times P}$.

Theorem 13.37. *Let $p \in [1, \infty)$. The space $\mathbf{L}_{p,\infty}(\mathbb{R}_+ \times \Omega, m_L \times P)$ is a complete metric space with respect to the metric associated with the quasinorm $\|\cdot\|_{p,\infty}^{m_L \times P}$ on $\mathbf{L}_{p,\infty}(\mathbb{R}_+ \times \Omega, m_L \times P)$.*

Proof. Let $\{X^{(n)} : n \in \mathbb{N}\}$ be a Cauchy sequence in $\mathbf{L}_{p,\infty}(\mathbb{R}_+ \times \Omega, m_L \times P)$ with respect to the metric associated with the quasinorm $\|\cdot\|_{p,\infty}^{m_L \times P}$. Let $m \in \mathbb{N}$ be fixed and let $X^{(n)}_{(m)}$ be the restriction of $X^{(n)}$ to $[0,m] \times \Omega$ for every $n \in \mathbb{N}$. Now for every $\varepsilon > 0$ there exists $N \in \mathbb{N}$ such that

$$\|X^{(n)} - X^{(\ell)}\|_{p,\infty}^{m_L \times P} < 2^{-m}\varepsilon \quad \text{for } n, \ell \geq N,$$

and thus

$$2^{-m}\{\|X^{(n)}_{(m)} - X^{(\ell)}_{(m)}\|_{p,m}^{m_L \times P} \wedge 1\} < 2^{-m}\varepsilon \quad \text{for } n, \ell \geq N.$$

We may assume without loss of generality that $\varepsilon < 1$. Then we have

$$\|X^{(n)}_{(m)} - X^{(\ell)}_{(m)}\|_{p,m}^{m_L \times P} < \varepsilon \quad \text{for } n, \ell \geq N.$$

Thus $\{X^{(n)}_{(m)} : n \in \mathbb{N}\}$ is a Cauchy sequence in the Banach space $L_p([0,m] \times \Omega)$ and therefore there exists $Y_{(m)} \in L_p([0,m] \times \Omega)$ such that $\lim_{n \to \infty} \|X^{(n)}_{(m)} - Y_{(m)}\|_{p,m}^{m_L \times P} = 0$. Let us take an arbitrary real valued representative function of $Y_{(m)}$ and fix it for $m \in \mathbb{N}$.

Now for $m = 1$, $X^{(n)}_{(1)}$ converges to $Y_{(1)}$ in the L_p-norm on $L_p([0,1] \times \Omega)$ and therefore there exists a subsequence $\{n_{1,k}\}$ of $\{n\}$ such that $X^{(n_{1,k})}_{(1)}$ converges to $Y_{(1)}$ on $[0,1] \times \Omega - \Lambda_1$ where Λ_1 is a null set in $[0,1] \times \Omega$. Then since $X^{(n_{1,k})}_{(2)}$ converges to $Y_{(2)}$ in the L_p-norm on $L_p([0,2] \times \Omega)$, there exists a subsequence $\{n_{2,k}\}$ of $\{n_{1,k}\}$ such that $X^{(n_{2,k})}_{(2)}$ converges to $Y_{(2)}$ on $[0,2] \times \Omega - \Lambda_2$ where Λ_2 is a null set in $[0,2] \times \Omega$ containing Λ_1. Thus proceeding inductively we obtain a subsequence $\{n_{k,k}\}$ of $\{n\}$ such that for every $m \in \mathbb{N}$ the sequence

$X^{(n_k,k)}_{(m)}$ converges both in the L_p-norm of $L_p([0,m]\times\Omega)$ and pointwise on $[0,m]\times\Omega-\Lambda_m$ where Λ_m is a null set in $[0,m]\times\Omega$ and contains $\Lambda_1,\ldots,\Lambda_{m-1}$. Thus $Y_{(m)}=Y_{(m-1)}$ on $[0,m-1]\times\Omega-\Lambda_{(m-1)}$ for $m\geq 2$. Let $\Lambda=\cup_{m\in\mathbb{N}}\Lambda_m$. Let us define a function Y on $\mathbb{R}_+\times\Omega$ by setting

$$Y(t,\omega)=\begin{cases} Y_{(m)}(t,\omega) & \text{for } (t,\omega)\in[0,m]\times\Omega-\Lambda_m \text{ and } m\in\mathbb{N}\\ 0 & \text{for } (t,\omega)\in\Lambda.\end{cases}$$

The function Y is well defined and is a measurable process on $(\Omega,\mathfrak{F},P)$ satisfying the condition $\|Y\|^{m_L\times P}_{p,m}=\|Y_{(m)}\|^{m_L\times P}_{p,m}<\infty$ so that $Y\in\mathbf{L}_{p,\infty}(\mathbb{R}_+\times\Omega,m_L\times P)$. Also

$$\lim_{k\to\infty}\|X^{(n_k,k)}-Y\|^{m_L\times P}_{p,m}=\lim_{k\to\infty}\|X^{(n_k,k)}-Y_{(m)}\|^{m_L\times P}_{p,m}=0\quad\text{for }m\in\mathbb{N}.$$

This implies $\lim_{k\to\infty}\|X^{(n_k,k)}-Y\|^{m_L\times P}_{p,\infty}=0$ according to Remark 11.18. ∎

In Proposition 2.13 we showed that every left- or right-continuous adapted process on a filtered space is a progressively measurable process. More generally we showed in Proposition 2.23 that every well-measurable process, and in particular every predictable process, on a filtered space is a progressively measurable process.

Theorem 13.38. *Let $X=\{X_t:t\in\mathbb{R}_+\}$ be a progressively measurable process on an augmented filtered space $(\Omega,\mathfrak{F},\{\mathfrak{F}_t\},P)$. If there exists a null set Λ in $(\Omega,\mathfrak{F},P)$ such that*

(1) $$\int_{[0,t]}|X(s,\omega)|m_L(ds)<\infty\quad\text{for every }t\in\mathbb{R}_+\text{ when }\omega\in\Lambda^c,$$

then there exists a predictable process Y on the filtered space such that

(2) $$X(\cdot,\omega)=Y(\cdot,\omega)\quad\text{a.e. on }(\mathbb{R}_+,\mathfrak{B}_{\mathbb{R}_+},m_L)\text{ when }\omega\in\Lambda^c.$$

In particular if X is in $\mathbf{L}_{p,\infty}(\mathbb{R}_+\times\Omega,m_L\times P)$ for some $p\in[0,\infty)$, then (1) is satisfied and X and Y are equivalent in this space.

Proof. Since X is a progressively measurable process, it is an adapted measurable process according to Observation 2.12. Since the filtered space is augmented, $\Lambda\in\mathfrak{F}_t$ for every $t\in\mathbb{R}_+$. For every $n\in\mathbb{N}$, define a real valued function $X^{(n)}$ on $\mathbb{R}_+\times\Omega$ by setting

(3) $$X^{(n)}(t,\omega)=\begin{cases} n\int_{[t-\frac{1}{n},t]\cap\mathbb{R}_+}X(s,\omega)m_L(ds) & \text{for }(t,\omega)\in\mathbb{R}_+\times\Lambda^c\\ 0 & \text{for }(t,\omega)\in\mathbb{R}_+\times\Lambda.\end{cases}$$

§13. ADAPTED BROWNIAN MOTIONS

Let us show first that $X^{(n)}$ is an adapted process. Now since X is progressively measurable, for every $t \in \mathbb{R}_+$ the restriction of X to $[0,t] \times \Omega$ is $\sigma(\mathfrak{B}_{[0,t]} \times \mathfrak{F}_t)$-measurable. Since $\Lambda \in \mathfrak{F}_t$, $[0,t] \times \Lambda^c$ is in $\sigma(\mathfrak{B}_{[0,t]} \times \mathfrak{F}_t)$. Thus $X^{(n)}(t,\cdot)$, as the result of integrating on $[t - \frac{1}{n}, t]$ with respect to m_L, is $\mathfrak{F}_t$-measurable on Λ^c by Fubini's Theorem. On the other hand $X^{(n)}(t,\cdot) = 0$ on Λ. Thus $X^{(n)}(t,\cdot)$ is $\mathfrak{F}_t$-measurable on Ω. This shows that $X^{(n)}$ is an adapted process.

To show that $X^{(n)}$ is a continuous process, let $\omega \in \Lambda^c$ and $t_0 \in \mathbb{R}_+$ be arbitrarily fixed. To show the continuity of $X^{(n)}(\cdot,\omega)$ at t_0 we may consider only $t \in \mathbb{R}_+$ such that $|t - t_0| < 1$ for instance. Then

$$|X^{(n)}(t,\omega) - X^{(n)}(t_0,\omega)|$$
$$= n \left| \int_{[t-\frac{1}{n},t] \cap \mathbb{R}_+} X(s,\omega) m_L(ds) - \int_{[t_0-\frac{1}{n},t_0] \cap \mathbb{R}_+} X(s,\omega) m_L(ds) \right|$$
$$\leq n \int_{[0,t_0+1]} |\mathbf{1}_{[t-\frac{1}{n},t]}(s) - \mathbf{1}_{[t_0-\frac{1}{n},t_0]}(s)| |X(s,\omega)| m_L(ds)$$
$$\leq n \int_{[0,t_0+1]} |\mathbf{1}_{[t-\frac{1}{n},t] \Delta [t_0-\frac{1}{n},t_0]}(s)| |X(s,\omega)| m_L(ds).$$

Since $\lim_{t \to t_0} \mathbf{1}_{[t-\frac{1}{n},t] \Delta [t_0-\frac{1}{n},t_0]}(s) = 0$ and since $X(\cdot,\omega)$ is integrable on $[0, t_0 + 1]$, we have by the Dominated Convergence Theorem $\lim_{t \to t_0} |X^{(n)}(t,\omega) - X^{(n)}(t_0,\omega)| = 0$. This proves the continuity of $X^{(n)}(\cdot,\omega)$ at t_0 and shows that $X^{(n)}$ is a continuous process.

Now $X^{(n)}$ is an adapted continuous process and is therefore an $\mathfrak{S}/\mathfrak{B}_{\mathbb{R}}$-measurable transformation of $\mathbb{R}_+ \times \Omega$ into $\mathbb{R}$, where $\mathfrak{S}$ is the predictable σ-algebra, that is, the σ-algebra of subsets of $\mathbb{R}_+ \times \Omega$ generated by the adapted left-continuous processes on the filtered space. Let us define an extended real valued function $\overline{Y}$ on $\mathbb{R}_+ \times \Omega$ by setting

(4) $$\overline{Y}(t,\omega) = \liminf_{n \to \infty} X^{(n)}(t,\omega) \quad \text{for } (t,\omega) \in \mathbb{R}_+ \times \Omega.$$

Since $X^{(n)}$ is an $\mathfrak{S}/\mathfrak{B}_{\mathbb{R}}$-measurable mapping of $\mathbb{R}_+ \times \Omega$ into $\mathbb{R}$ for every $n \in \mathbb{N}$, $\overline{Y}$ is an $\mathfrak{S}/\mathfrak{B}_{\overline{\mathbb{R}}}$-measurable mapping of $\mathbb{R}_+ \times \Omega$ into $\overline{\mathbb{R}}$. Then $\{\overline{Y} = \pm\infty\} \in \mathfrak{S}$ so that if we define a real valued function Y on $\mathbb{R}_+ \times \Omega$ by setting

(5) $$Y(t,\omega) = \begin{cases} \overline{Y} & \text{when } \overline{Y}(t,\omega) \in \mathbb{R} \\ 0 & \text{when } \overline{Y}(t,\omega) = \pm\infty \end{cases}$$

then Y is an $\mathfrak{S}/\mathfrak{B}_{\mathbb{R}}$-measurable mapping of $\mathbb{R}_+ \times \Omega$ into $\mathbb{R}$, that is, Y is a predictable process on the filtered space.

By (1), for every $\omega \in \Lambda^c$ the sample function $X(\cdot,\omega)$ is Lebesgue integrable on every finite interval in $\mathbb{R}_+$. Then by Lebesgue's Theorem the indefinite integral of $X(\cdot,\omega)$ is differentiable with derivative equal to $X(t, \omega)$ for a.e. $t \in \mathbb{R}_+$. Thus recalling the definitions of $X^{(n)}$ and $\overline{Y}$ by (3) and (4) respectively, we have for every $\omega \in \Lambda^c$

(6) $$X(\cdot, \omega) = \overline{Y}(\cdot, \omega) = Y(\cdot, \omega) \quad \text{a.e. on } (\mathbb{R}_+, \mathfrak{B}_{\mathbb{R}_+}, m_L),$$

where the second equality is from the fact that X is real valued. This proves (2).

If X is in $\mathbf{L}_{p,\infty}(\mathbb{R}_+ \times \Omega, m_L \times P)$, then X satisfies (1) and thus by (6) for every $\omega \in \Lambda^c$ we have

$$\int_{[0,t]} |Y(s, \omega)|^p m_L(ds) = \int_{[0,t]} |X(s, \omega)|^p m_L(ds) \quad \text{for every } t \in \mathbb{R}_+,$$

and then

$$\mathbf{E}\left[\int_{[0,t]} |Y(s, \omega)|^p m_L(ds)\right] = \mathbf{E}\left[\int_{[0,t]} |X(s, \omega)|^p m_L(ds)\right] < \infty \quad \text{for every } t \in \mathbb{R}_+.$$

This shows that Y is in $\mathbf{L}_{p,\infty}(\mathbb{R}_+ \times \Omega, m_L \times P)$. From (5) we also have

$$\mathbf{E}\left[\int_{[0,t]} |Y(s, \omega) - X(s, \omega)|^p m_L(ds)\right] = 0 \quad \text{for every } t \in \mathbb{R}_+.$$

This shows according to (5) of Observation 13.34 that X and Y are equivalent processes in $\mathbf{L}_{p,\infty}(\mathbb{R}_+ \times \Omega, m_L \times P)$. ∎

For quadratic variation processes of stochastic integrals with respect to Brownian motions we have the following.

Proposition 13.39. Let $B = \{B_t : t \in \mathbb{R}_+\}$ be an $\{\mathfrak{F}_t\}$-adapted d-dimensional Brownian motion on a standard filtered space $(\Omega, \mathfrak{F}, \{\mathfrak{F}_t\}, P)$ with $B_0 = 0$ and let $B^{(i)}$, $i = 1, \ldots, d$, be its components. Let $X^{(i)}$ be a predictable process in $\mathbf{L}_{2,\infty}(\mathbb{R}_+ \times \Omega, m_L \times P)$ for $i = 1, \ldots, d$. Then for $i, j = 1, \ldots, d$, there exists a null set Λ in $(\Omega, \mathfrak{F}, P)$ such that on Λ^c we have for every $t \in \mathbb{R}_+$

(1) $$[X^{(i)} \bullet B^{(i)}, X^{(j)} \bullet B^{(j)}]_t = \delta_{i,j} \int_{[0,t]} X^{(i)}(s) X^{(j)}(s) m_L(ds),$$

and thus for any $s, t \in \mathbb{R}_+$, $s < t$, we have

(2) $$\mathbf{E}[\{(X^{(i)} \bullet B^{(i)})_t - (X^{(i)} \bullet B^{(i)})_s\}\{(X^{(j)} \bullet B^{(j)})_t - (X^{(j)} \bullet B^{(j)})_s\} | \mathfrak{F}_s]$$
$$= \mathbf{E}[(X^{(i)} \bullet B^{(i)})_t (X^{(j)} \bullet B^{(j)})_t - (X^{(i)} \bullet B^{(i)})_s (X^{(j)} \bullet B^{(j)})_s | \mathfrak{F}_s]$$
$$= \delta_{i,j} \mathbf{E}[\int_{(s,t]} X^{(i)}(u) X^{(j)}(u) m_L(du) | \mathfrak{F}_s] \quad \text{a.e. on } (\Omega, \mathfrak{F}_s, P),$$

and in particular

(3) $$\mathbb{E}[\{(X^{(i)} \bullet B^{(i)})_t - (X^{(i)} \bullet B^{(i)})_s\}\{(X^{(j)} \bullet B^{(j)})_t - (X^{(j)} \bullet B^{(j)})_s\}]$$
$$= \delta_{i,j}\mathbb{E}[\int_{(s,t]} X^{(i)}(u)X^{(j)}(u)m_L(du)].$$

Proof. According to Proposition 13.33, $[B^{(i)}, B^{(j)}]_t = \delta_{i,j}t$ for $t \in \mathbb{R}_+$. Thus the Proposition is a particular case of Theorem 12.16. ∎

Corollary 13.40. *Let B and $B^{(i)}$, $i = 1, \ldots, d$, be as in Proposition 13.39. Let $X^{(i)}$ and $Y^{(i)}$ be predictable processes in $\mathbf{L}_{2,\infty}(\mathbb{R}_+ \times \Omega, m_L \times P)$ for $i = 1, \ldots, d$. Then for $X \equiv \sum_{i=1}^d X^{(i)} \bullet B^{(i)}$ and $Y \equiv \sum_{i=1}^d Y^{(i)} \bullet B^{(i)}$ in $\mathbf{M}_2^c(\Omega, \mathfrak{F}, \{\mathfrak{F}_t\}, P)$ there exists a null set Λ in $(\Omega, \mathfrak{F}, P)$ such that on Λ^c we have for every $t \in \mathbb{R}_+$*

(1) $$[X,Y]_t = \int_{[0,t]} \sum_{i=1}^d X^{(i)}(s)Y^{(i)}(s)m_L(ds)$$

and thus for any $s, t \in \mathbb{R}_+$, $s < t$, we have

(2) $$\mathbb{E}[\{X_t - X_s\}\{Y_t - Y_s\} | \mathfrak{F}_s] = \mathbb{E}[X_tY_t - X_sY_s | \mathfrak{F}_s]$$
$$= \mathbb{E}[\int_{(s,t]} \sum_{i=1}^d X^{(i)}(u)Y^{(i)}(u)m_L(du) | \mathfrak{F}_s] \quad \text{a.e. on } (\Omega, \mathfrak{F}_s, P).$$

Proof. Since $X^{(i)} \bullet B^{(i)}$ is in $\mathbf{M}_2^c(\Omega, \mathfrak{F}, \{\mathfrak{F}_t\}, P)$, so is $X \equiv \sum_{i=1}^d X^{(i)} \bullet B^{(i)}$. Similarly for Y. Thus $[X,Y]$ is defined and according to Proposition 11.26, we have

$$[X,Y]_t = [\sum_{i=1}^d X^{(i)} \bullet B^{(i)}, \sum_{i=1}^d Y^{(i)} \bullet B^{(i)}]_t = \sum_{i,j=1}^d [X^{(i)} \bullet B^{(i)}, X^{(j)} \bullet B^{(j)}]_t$$
$$= \sum_{i,j=1}^d \delta_{i,j} \int_{[0,t]} X^{(i)}(s)Y^{(j)}(s)m_L(ds) = \int_{[0,t]} \sum_{i=1}^d X^{(i)}(s)Y^{(i)}(s)m_L(ds)$$

where the third equality is by Proposition 13.39. ∎

§14 Extensions of the Stochastic Integral

[I] Local L_2-Martingales and Their Quadratic Variation Processes

For $M \in \mathbf{M}_2(\Omega, \mathfrak{F}, \{\mathfrak{F}_t\}, P)$ and a predictable process X on the filtered space satisfying the integrability condition $X \in \mathbf{L}_{2,\infty}(\mathbb{R}_+ \times \Omega, \mu_{[M]}, P)$, that is, $\mathbb{E}[\int_{[0,t]} X^2(s)\, d[M](s)] < \infty$

for every $t \in \mathbb{R}_+$, the stochastic integral $X \bullet M$ of X with respect to M was defined in Definition 12.9 as an element in $\mathbf{M}_2(\Omega, \mathfrak{F}, \{\mathfrak{F}_t\}, P)$. We show below that if X satisfies the weaker integrability condition that $\int_{[0,t]} X^2(s)\, d[M](s) < \infty$ for every $t \in \mathbb{R}_+$ for almost every $\omega \in \Omega$, then there exists a sequence $\{X^{(n)} : n \in \mathbb{N}\}$ in $\mathbf{L}_{2,\infty}(\mathbb{R}_+ \times \Omega, \mu_{[M]}, P)$ such that $X^{(n)}$ converges pointwise on $\mathbb{R}_+ \times \Lambda^c$ to X and $X^{(n)} \bullet M$ converges pointwise on $\mathbb{R}_+ \times \Lambda^c$ where Λ is a null set in $(\Omega, \mathfrak{F}, P)$. We then extend the definition of stochastic integral with respect to M by defining the limit of the pointwise convergence of the sequence $\{X^{(n)} \bullet M : n \in \mathbb{N}\}$ in $\mathbf{M}_2(\Omega, \mathfrak{F}, \{\mathfrak{F}_t\}, P)$ as the stochastic integral of X with respect to M. This leads to the definition of local martingales.

Definition 14.1. *Let $X = \{X_t : t \in \mathbb{R}_+\}$ be an adapted process on a filtered space $(\Omega, \mathfrak{F}, \{\mathfrak{F}_t\}, P)$ and let $\{T_n : n \in \mathbb{N}\}$ be an increasing sequence of stopping times on the filtered space such that $T_n \uparrow \infty$ a.e. on $(\Omega, \mathfrak{F}, P)$. We say that X is a local martingale with respect to the sequence $\{T_n : n \in \mathbb{N}\}$ if $X^{T_n \wedge}$ is a martingale on the filtered space for every $n \in \mathbb{N}$. We say that X is a local L_2-martingale with respect to $\{T_n : n \in \mathbb{N}\}$ if $X^{T_n \wedge}$ is an L_2-martingale on the filtered space for every $n \in \mathbb{N}$.*

Thus if X is a local martingale with respect to $\{T_n : n \in \mathbb{N}\}$, then since $T_n(\omega) \uparrow \infty$ for every $\omega \in \Lambda^c$ where Λ is a null set in $(\Omega, \mathfrak{F}, P)$ we have $\lim_{n \to \infty} X^{T_n \wedge}(t, \omega) = X(t, \omega)$ for $(t, \omega) \in \mathbb{R}_+ \times \Lambda^c$. Thus X is the limit of pointwise convergence on $\mathbb{R}_+ \times \Lambda^c$ of a sequence of martingales.

Note that a martingale is always a local martingale with respect to the sequence of stopping times $\{T_n : n \in \mathbb{N}\}$ where $T_n = \infty$ for every $n \in \mathbb{N}$. Note also that the fact that X is a local martingale with respect to an increasing sequence of stopping times $\{T_n : n \in \mathbb{N}\}$ such that $T_n \uparrow \infty$ a.e. on $(\Omega, \mathfrak{F}, P)$ does not imply that it is a local martingale with respect to every other increasing sequence of stopping times tending to ∞ almost surely.

Observation 14.2. *Let $X = \{X_t : t \in \mathbb{R}_+\}$ be a local martingale on an augmented filtered space $(\Omega, \mathfrak{F}, \{\mathfrak{F}_t\}, P)$. If there exists a nonnegative integrable random variable Y on $(\Omega, \mathfrak{F}, P)$ such that $|X(t, \omega)| \leq Y(\omega)$ for $(t, \omega) \in \mathbb{R}_+ \times \Lambda^c$ where Λ is a null set in $(\Omega, \mathfrak{F}, P)$, then X is in fact a martingale. If the random variable Y^2 is integrable, then X is an L_2-martingale.*

Proof. Since X is a local martingale, there exists an increasing sequence of stopping times $\{T_n : n \in \mathbb{N}\}$ on the filtered space such that $T_n \uparrow \infty$ on Λ_0^c where Λ_0 is a null set in $(\Omega, \mathfrak{F}, P)$ and $X^{T_n \wedge}$ is a martingale on the filtered space for every $n \in \mathbb{N}$. Then for any pair

§14. EXTENSIONS OF THE STOCHASTIC INTEGRAL

$s, t \in \mathbb{R}_+$ such that $s < t$, we have according to Theorem 8.12

(1) $$\mathbb{E}[X_{T_n \wedge t} | \mathfrak{F}_s] = X_{T_n \wedge s} \quad \text{a.e. on } (\Omega, \mathfrak{F}_s, P).$$

Since $T_n \uparrow \infty$ on Λ_0^c, we have $\lim_{n \to \infty} X_{T_n \wedge t} = X_t$ on Λ_0^c. Also $|X_{T_n \wedge t}| \leq Y$ on Λ^c. Thus by the Conditional Dominated Convergence Theorem, we have

(2) $$\lim_{n \to \infty} \mathbb{E}[X_{T_n \wedge t} | \mathfrak{F}_s] = \mathbb{E}[X_t | \mathfrak{F}_s] \quad \text{a.e. on } (\Omega, \mathfrak{F}_s, P).$$

We have also $\lim_{n \to \infty} X_{T_n \wedge s} = X_s$ on Λ_0^c. Since the filtered space is augmented, the null set Λ_0 is in $\mathfrak{F}_s$. Thus

(3) $$\lim_{n \to \infty} X_{T_n \wedge s} = X_s \quad \text{a.e. on } (\Omega, \mathfrak{F}_s, P).$$

Letting $n \to \infty$ in (1) and using (2) and (3), we have $\mathbb{E}[X_t | \mathfrak{F}_s] = X_s$ a.e. on $(\Omega, \mathfrak{F}_s, P)$. This shows that X is a martingale. If Y^2 is integrable, then since $X_t^2 \leq Y^2$ on Λ^c we have $\mathbb{E}(X_t^2) \leq \mathbb{E}(Y^2) < \infty$ for every $t \in \mathbb{R}_+$ so that X is an L_2-process. ∎

Note that for a local martingale X, the right-continuity and the continuity of $X(\cdot, \omega)$ for an arbitrary $\omega \in \Omega$ is equivalent to the right-continuity and continuity of $X^{T_n \wedge}(\cdot, \omega)$ for all $n \in \mathbb{N}$.

Lemma 14.3. *Let $\{S_n : n \in \mathbb{N}\}$ and $\{T_n : n \in \mathbb{N}\}$ be two increasing sequences of stopping times on a right-continuous filtered space $(\Omega, \mathfrak{F}, \{\mathfrak{F}_t\}, P)$ such that $S_n \uparrow \infty$ and $T_n \uparrow \infty$ a.e. on $(\Omega, \mathfrak{F}, P)$. If X is a local martingale with respect to the sequence $\{S_n : n \in \mathbb{N}\}$, then it is a local martingale with respect to the sequence of stopping times $\{S_n \wedge T_n : n \in \mathbb{N}\}$. If X is a right-continuous local L_2-martingale with respect to the $\{S_n : n \in \mathbb{N}\}$, then it is a local L_2-martingale with respect to $\{S_n \wedge T_n : n \in \mathbb{N}\}$.*

Proof. Clearly $S_n \wedge T_n \uparrow \infty$ a.e. on $(\Omega, \mathfrak{F}, P)$. If X is a local martingale with respect to $\{S_n : n \in \mathbb{N}\}$, then for every $n \in \mathbb{N}$, $X^{S_n \wedge} = \{X_{S_n \wedge t} : t \in \mathbb{R}_+\}$ is a martingale. Thus $X^{S_n \wedge T_n \wedge} = \{X_{S_n \wedge T_n \wedge t} : t \in \mathbb{R}_+\}$ is a martingale by Theorem 8.12. This shows that X is a local martingale with respect to $\{S_n \wedge T_n : n \in \mathbb{N}\}$.

Suppose X is a local L_2-martingale with respect to $\{S_n : n \in \mathbb{N}\}$. Then for every $n \in \mathbb{N}$, $X^{S_n \wedge} = \{X_{S_n \wedge t} : t \in \mathbb{R}_+\}$ is an L_2-martingale so that $(X^{S_n \wedge})^2 = \{X_{S_n \wedge t}^2 : t \in \mathbb{R}_+\}$ is a submartingale on the filtered space. Since $S_n \wedge t$ and $S_n \wedge T_n \wedge t$ are stopping times on the filtered space and $S_n \wedge t \geq S_n \wedge T_n \wedge t$ for $t \in \mathbb{R}_+$, Theorem 8.10 (Optional Sampling with Bounded Stopping Times) implies that $\mathbb{E}[X_{S_n \wedge t}^2 | \mathfrak{F}_{S_n \wedge T_n \wedge t}] \geq X_{S_n \wedge T_n \wedge t}^2$

a.e. on $(\Omega, \mathfrak{F}_{S_n \wedge T_n \wedge t}, P)$ and consequently we have $\mathbf{E}[X^2_{S_n \wedge T_n \wedge t}] \leq \mathbf{E}[X^2_{S_n \wedge t}] < \infty$. This shows that $X^{S_n \wedge T_n \wedge}$ is an L_2-process and is thus an L_2-martingale. Since this holds for every $n \in \mathbb{N}$, X is a local L_2-martingale with respect to $\{S_n \wedge T_n : n \in \mathbb{N}\}$. ∎

Definition 14.4. *Let* $\mathbf{M}_2^{loc}(\Omega, \mathfrak{F}, \{\mathfrak{F}_t\}, P)$, *abbreviated as* $\mathbf{M}_2^{loc}$, *be the collection of equivalence classes of all right-continuous local L_2-martingales $X = \{X_t : t \in \mathbb{R}_+\}$ with $X_0 = 0$ almost surely on a standard filtered space* $(\Omega, \mathfrak{F}, \{\mathfrak{F}_t\}, P)$. *Let* $\mathbf{M}_2^{c,loc}(\Omega, \mathfrak{F}, \{\mathfrak{F}_t\}, P)$, *abbreviated as* $\mathbf{M}_2^{c,loc}$, *be the subcollection consisting of almost surely continuous members of* $\mathbf{M}_2^{loc}$.

Observation 14.5. The fact that $\mathbf{M}_2^{loc}(\Omega, \mathfrak{F}, \{\mathfrak{F}_t\}, P)$ is a linear space, that is, $aX + bY \in \mathbf{M}_2^{loc}$ for $X, Y \in \mathbf{M}_2^{loc}$ and $a, b \in \mathbb{R}$, can be shown as follows. Now if $X \in \mathbf{M}_2^{loc}$ the clearly $aX \in \mathbf{M}_2^{loc}$ for $a \in \mathbb{R}$. Thus it suffices to show that if $X, Y \in \mathbf{M}_2^{loc}$ then $X + Y \in \mathbf{M}_2^{loc}$. Suppose $X, Y \in \mathbf{M}_2^{loc}$. Then there exist two increasing sequences of stopping times $\{S_n : n \in \mathbb{N}\}$ and $\{T_n : n \in \mathbb{N}\}$ such that $S_n \uparrow \infty$ and $T_n \uparrow \infty$ a.e. on $(\Omega, \mathfrak{F}, P)$ and $X^{S_n \wedge}$ and $Y^{T_n \wedge}$ are L_2-martingales for every $n \in \mathbb{N}$. By Lemma 14.3, $X^{S_N \wedge T_n \wedge}$ and $Y^{S_N \wedge T_n \wedge}$ are L_2-martingales for every $n \in \mathbb{N}$. But $(X + Y)^{S_N \wedge T_n \wedge} = X^{S_N \wedge T_n \wedge} + Y^{S_N \wedge T_n \wedge}$. Thus $(X + Y)^{S_N \wedge T_n \wedge}$ is an L_2-martingales for every $n \in \mathbb{N}$. This shows that $X + Y$ is a local L_2-martingale. Since X and Y are right-continuous and null at 0, so is $X + Y$. Therefore $X + Y \in \mathbf{M}_2^{loc}$.

Definition 14.6. *Let* $\mathbf{A}^{loc}(\Omega, \mathfrak{F}, \{\mathfrak{F}_t\}, P)$ *be the collection of equivalence classes of all stochastic processes A on a standard filtered space* $(\Omega, \mathfrak{F}, \{\mathfrak{F}_t\}, P)$ *satisfying conditions* $1°$ *and* $4°$, *but not necessarily condition* $2°$ *that A be an L_1-process, of Definition 10.1. Let* $\mathbf{V}^{loc}(\Omega, \mathfrak{F}, \{\mathfrak{F}_t\}, P)$ *be the linear space of equivalence classes of all stochastic processes V on a standard filtered space* $(\Omega, \mathfrak{F}, \{\mathfrak{F}_t\}, P)$ *satisfying conditions* $1°$ *and* $4°$, *but not necessarily condition* $5°$ *that* $|V|$ *be an L_1-process, of Definition 11.10. We write* $\mathbf{A}^{c,loc}(\Omega, \mathfrak{F}, \{\mathfrak{F}_t\}, P)$ *and* $\mathbf{V}^{c,loc}(\Omega, \mathfrak{F}, \{\mathfrak{F}_t\}, P)$ *for the subcollections of* $\mathbf{A}^{loc}(\Omega, \mathfrak{F}, \{\mathfrak{F}_t\}, P)$ *and* $\mathbf{V}^{loc}(\Omega, \mathfrak{F}, \{\mathfrak{F}_t\}, P)$ *respectively consisting of almost surely continuous members.*

In what follows we write $A \in \mathbf{A}^{loc}(\Omega, \mathfrak{F}, \{\mathfrak{F}_t\}, P)$ for both an equivalence class and an arbitrary representative of the equivalence class. Similarly for $V \in \mathbf{V}^{loc}(\Omega, \mathfrak{F}, \{\mathfrak{F}_t\}, P)$.

Lemma 14.7. *If* $A', A'' \in \mathbf{A}^{loc}(\Omega, \mathfrak{F}, \{\mathfrak{F}_t\}, P)$, *then* $A' - A'' \in \mathbf{V}^{loc}(\Omega, \mathfrak{F}, \{\mathfrak{F}_t\}, P)$. *If* $V \in \mathbf{V}^{loc}(\Omega, \mathfrak{F}, \{\mathfrak{F}_t\}, P)$, *then* $V = A' - A''$ *where* $A', A'' \in \mathbf{A}^{loc}(\Omega, \mathfrak{F}, \{\mathfrak{F}_t\}, P)$.

§14. EXTENSIONS OF THE STOCHASTIC INTEGRAL

Proof. The proof parallels that of Theorem 11.12. ∎

Theorem 14.8. *For every X in $\mathbf{M}_2^{loc}(\Omega, \mathfrak{F}, \{\mathfrak{F}_t\}, P)$, there exists an equivalence class A in $\mathbf{A}^{loc}(\Omega, \mathfrak{F}, \{\mathfrak{F}_t\}, P)$ such that $X^2 - A$ is a right-continuous null at 0 local martingale. If Y is also in $\mathbf{M}_2^{loc}(\Omega, \mathfrak{F}, \{\mathfrak{F}_t\}, P)$, then there exists an equivalence class V in $\mathbf{V}^{loc}(\Omega, \mathfrak{F}, \{\mathfrak{F}_t\}, P)$ such that $XY - V$ is a right-continuous null at 0 local martingale.*

Proof. Suppose X and Y are in $\mathbf{M}_2^{loc}(\Omega, \mathfrak{F}, \{\mathfrak{F}_t\}, P)$ and are local L_2-martingales with respect to two increasing sequences of stopping times $\{S_n : n \in \mathbb{N}\}$ and $\{T_n : n \in \mathbb{N}\}$ such that $S_n \uparrow \infty$ and $T_n \uparrow \infty$ a.e. on $(\Omega, \mathfrak{F}, P)$ respectively. Let $R_n = S_n \wedge T_n$ for $n \in \mathbb{N}$. Then both X and Y are local L_2-martingales with respect to the sequence of stopping times $\{R_n; n \in \mathbb{N}\}$ according to Lemma 14.3. Let Λ_1 be a null set in $(\Omega, \mathfrak{F}, P)$ such that $R_n(\omega) \uparrow \infty$ for $\omega \in \Lambda_1^c$.

Now for every $n \in \mathbb{N}$, since $X^{R_n \wedge}$ and $Y^{R_n \wedge}$ are in $\mathbf{M}_2(\Omega, \mathfrak{F}, \{\mathfrak{F}_t\}, P)$, according to Proposition 11.22, there exists a unique natural quadratic variation process $V^{(n)}$ of $X^{R_n \wedge}$ and $Y^{R_n \wedge}$ in $\mathbf{V}(\Omega, \mathfrak{F}, \{\mathfrak{F}_t\}, P)$ such that $X^{R_n \wedge} Y^{R_n \wedge} - V^{(n)}$ is a right-continuous null at 0 martingale. Similarly there exists a unique natural $V^{(n+1)} \in \mathbf{V}(\Omega, \mathfrak{F}, \{\mathfrak{F}_t\}, P)$ such that $X^{R_{n+1} \wedge} Y^{R_{n+1} \wedge} - V^{(n+1)}$ is a right-continuous null at 0 martingale. Then $(X^{R_{n+1} \wedge} Y^{R_{n+1} \wedge} - V^{(n+1)})^{R_n \wedge}$ is a right-continuous null at 0 martingale by Theorem 8.12. But $R_n \leq R_{n+1}$ implies that

$$(X^{R_{n+1} \wedge} Y^{R_{n+1} \wedge} - V^{(n+1)})^{R_n \wedge} = X^{R_n \wedge} Y^{R_n \wedge} - (V^{(n+1)})^{R_n \wedge}.$$

Thus by the uniqueness of natural quadratic variation process of $X^{R_n \wedge}$ and $Y^{R_n \wedge}$ we have

$$V^{(n)} = (V^{(n+1)})^{R_n \wedge}.$$

In what follows we write $V^{(n)}$ for an arbitrarily fixed representative of the equivalence class. Then by the last equality there exists a null set Λ_2 in $(\Omega, \mathfrak{F}, P)$ such that

(1) $\qquad V^{(n)}(\cdot, \omega) = (V^{(n+1)})^{R_n \wedge}(\cdot, \omega) \quad$ for all $n \in \mathbb{N}$ when $\omega \in \Lambda_2^c$.

Iterating (1) and using the fact that $R_n \wedge R_{n+1} = R_n$, we have

(2) $\qquad V^{(n)}(\cdot, \omega) = (V^{(n+p)})^{R_n \wedge}(\cdot, \omega) \quad$ for $n, p \in \mathbb{N}$ when $\omega \in \Lambda_2^c$.

Since $V^{(n)} \in \mathbf{V}(\Omega, \mathfrak{F}, \{\mathfrak{F}_t\}, P)$ for every $n \in \mathbb{N}$, there exists a null set Λ_3 in $(\Omega, \mathfrak{F}, P)$ such that if $\omega \in \Lambda_3^c$ then $V^{(n)}(\cdot, \omega)$ are functions of bounded variation on every finite interval of $\mathbb{R}_+$ for all $n \in \mathbb{N}$. Let $\Lambda = \cup_{i=1}^{3} \Lambda_i$. Let $t \in \mathbb{R}_+$ be fixed. Since $R_n(\omega) \uparrow \infty$ for $\omega \in \Lambda^c$, there exists $N_\omega \in \mathbb{N}$ such that $s \leq R_n(\omega)$ for $s \in [0, t]$ when $n \geq N_\omega$. Then by (2) we have for $\omega \in \Lambda^c$,

(3) $\qquad V^{(n)}(s, \omega) = V^{(n+p)}(s, \omega) \quad$ for $s \in [0, t]$, $p \in \mathbb{N}$, $n \geq N_\omega$.

Thus $\lim_{n \to \infty} V^{(n)}(t, \omega)$ exists in $\mathbb{R}$ for every $t \in \mathbb{R}_+$ when $\omega \in \Lambda^c$. Let us define a real valued function V on $\mathbb{R}_+ \times \Omega$ by setting

(4) $\qquad V(t, \omega) = \begin{cases} \lim_{n \to \infty} V^{(n)}(t, \omega) & \text{for } (t, \omega) \in \mathbb{R}_+ \times \Lambda^c \\ 0 & \text{for } (t, \omega) \in \mathbb{R}_+ \times \Lambda. \end{cases}$

Since $V^{(n)}$ is an adapted process, $V_t^{(n)}$ is $\mathfrak{F}_t$-measurable on Ω for every $t \in \mathbb{R}_+$. Since the filtration is augmented the null set Λ is in $\mathfrak{F}_t$ for every $t \in \mathbb{R}_+$. Then V_t defined by (4) is $\mathfrak{F}_t$-measurable. Thus V is an adapted process on the filtered space. Let us show that for $\omega \in \Lambda^c$, $V(\cdot, \omega)$ is a function of bounded variation on every finite interval in $\mathbb{R}_+$. Let $t \in \mathbb{R}_+$ be fixed. Let $N \in \mathbb{N}$ be so large that $R_N(\omega) \geq t$. Then by (3) and (4) we have $V(s, \omega) = V^{(N)}(s, \omega)$ for $s \in [0, t]$. Since $V^{(N)}(\cdot, \omega)$ is a function of bounded variation on every finite interval of $\mathbb{R}_+$ and in particular on $[0, t]$, $V(\cdot, \omega)$ is a function of bounded variation on $[0, t]$. We have thus shown that if $\omega \in \Lambda^c$ then $V(\cdot, \omega)$ is a function of bounded variation on every finite interval in $\mathbb{R}_+$. This proves that the equivalence class represented by V is in $\mathbf{V}^{loc}(\Omega, \mathfrak{F}, \{\mathfrak{F}_t\}, P)$.

Finally to show that for every $n \in \mathbb{N}$, $(XY - V)^{R_n \wedge}$ is a right-continuous null at 0 martingale, note that for $(t, \omega) \in \mathbb{R}_+ \times \Lambda^c$ we have

$$\begin{aligned} (XY - V)^{R_n \wedge}(t, \omega) &= (XY)^{R_n \wedge}(t, \omega) - V^{R_n \wedge}(t, \omega) \\ &= (XY)^{R_n \wedge}(t, \omega) - (\lim_{k \to \infty} V^{(k)})^{R_n \wedge}(t, \omega) \quad \text{by (4)} \\ &= (XY)^{R_n \wedge}(t, \omega) - \lim_{k \to \infty} (V^{(k)})^{R_n \wedge}(t, \omega) \\ &= (XY)^{R_n \wedge}(t, \omega) - (V^{(n)})(t, \omega) \quad \text{by (2)}. \end{aligned}$$

Since $(XY)^{R_n \wedge} - (V^{(n)})$ is a right-continuous null at 0 martingale and since $\Lambda \in \mathfrak{F}_t$ for every $t \in \mathbb{R}_+$, the last equality implies that $(XY - V)^{R_n \wedge}$ is a right-continuous null at 0 martingale. This shows that $XY - V$ is a right-continuous null at 0 local martingale. The existence of A in $\mathbf{A}^{loc}(\Omega, \mathfrak{F}, \{\mathfrak{F}_t\}, P)$ such that $X^2 - A$ is a right-continuous null at 0 martingale is proved likewise. ∎

Corollary 14.9. *Let $X, Y \in \mathbf{M}_2^{c,loc}(\Omega, \mathfrak{F}, \{\mathfrak{F}_t\}, P)$. Then there exist an equivalence class $A \in \mathbf{A}^{c,loc}(\Omega, \mathfrak{F}, \{\mathfrak{F}_t\}, P)$ and an equivalence class $V \in \mathbf{V}^{c,loc}(\Omega, \mathfrak{F}, \{\mathfrak{F}_t\}, P)$ such that $X^2 - A$ and $XY - V$ are almost surely continuous null at 0 local martingales.*

§14. EXTENSIONS OF THE STOCHASTIC INTEGRAL

Proof. Since $X, Y \in \mathbf{M}_2^{c,loc}(\Omega, \mathfrak{F}, \{\mathfrak{F}_t\}, P)$, the processes $X^{R_n \wedge}$ and $Y^{R_n \wedge}$ in the proof of Theorem 14.8 are in $\mathbf{M}_2^c(\Omega, \mathfrak{F}, \{\mathfrak{F}_t\}, P)$. Then by Proposition 11.27, there exists $V^{(n)}$ in $\mathbf{V}^c(\Omega, \mathfrak{F}, \{\mathfrak{F}_t\}, P)$ such that $X^{R_n \wedge} Y^{R_n \wedge} - V^{(n)}$ is an almost surely continuous null at 0 martingale. The almost sure continuity of $V^{(n)}$ implies that the process V defined by (4) in the proof of Theorem 14.8 is almost surely continuous.

The existence of A in $\mathbf{A}^{c,loc}(\Omega, \mathfrak{F}, \{\mathfrak{F}_t\}, P)$ is proved likewise. ∎

Definition 14.10. Let $X, Y \in \mathbf{M}_2^{loc}(\Omega, \mathfrak{F}, \{\mathfrak{F}_t\}, P)$. An equivalence class A in $\mathbf{A}^{loc}(\Omega, \mathfrak{F}, \{\mathfrak{F}_t\}, P)$ such that $X^2 - A$ is a right-continuous null at 0 local martingale is called a quadratic variation process of X and we write $[X]$ for it. An equivalence class V in $\mathbf{V}^{loc}(\Omega, \mathfrak{F}, \{\mathfrak{F}_t\}, P)$ such that $XY - V$ is a right-continuous null at 0 local martingale is called a quadratic variation process of X and Y and we write $[X,Y]$ for it.

In what follows we write $[X]$ for both an equivalence class of processes and an arbitrary representative of an equivalence class. Similarly for $[X,Y]$. The existence of $[X]$ in $\mathbf{A}^{loc}(\Omega, \mathfrak{F}, \{\mathfrak{F}_t\}, P)$ and $[X,Y]$ in $\mathbf{V}^{loc}(\Omega, \mathfrak{F}, \{\mathfrak{F}_t\}, P)$ for X and Y in $\mathbf{M}_2^{loc}(\Omega, \mathfrak{F}, \{\mathfrak{F}_t\}, P)$ was proved in Theorem 14.8.

Theorem 14.11. Let $X, Y \in \mathbf{M}_2^{c,loc}(\Omega, \mathfrak{F}, \{\mathfrak{F}_t\}, P)$ and let $\{S_n : n \in \mathbb{N}\}$ and $\{T_n : n \in \mathbb{N}\}$ be increasing sequences of stopping times such that $S_n \uparrow \infty$, $T_n \uparrow \infty$ almost surely as $n \to \infty$, and $X^{S_n \wedge}, Y^{T_n \wedge} \in \mathbf{M}_2^c$ for $n \in \mathbb{N}$. Let $R_n = S_n \wedge T_n$ for $n \in \mathbb{N}$. Then there exists a null set Λ in $(\Omega, \mathfrak{F}, P)$ such that for every $\omega \in \Lambda^c$ we have
(1) $\lim_{n \to \infty} \mu_{[X^{R_n \wedge}, Y^{R_n \wedge}]}(E, \omega) = \mu_{[X,Y]}(E, \omega)$ for $E \in \mathfrak{B}_{[0,t]}$, $t \in \mathbb{R}_+$,
(2) $\uparrow \lim_{n \to \infty} \mu_{[X^{R_n \wedge}]}(E, \omega) = \mu_{[X]}(E, \omega)$ for $E \in \mathfrak{B}_{\mathbb{R}_+}$,
(3) $\uparrow \lim_{n \to \infty} \mu_{[Y^{R_n \wedge}]}(E, \omega) = \mu_{[Y]}(E, \omega)$ for $E \in \mathfrak{B}_{\mathbb{R}_+}$.
In particular for $t \in \mathbb{R}_+$ and $\omega \in \Lambda^c$, we have (4) $\lim_{n \to \infty} [X^{R_n \wedge}, Y^{R_n \wedge}](t, \omega) = [X, Y](t, \omega)$,
(5) $\uparrow \lim_{n \to \infty} [X^{R_n \wedge}](t, \omega) = [X](t, \omega)$,
(6) $\uparrow \lim_{n \to \infty} [Y^{R_n \wedge}](t, \omega) = [Y](t, \omega)$.

Proof. As we showed in the proof of Theorem 14.8, there exists a null set Λ_1 in $(\Omega, \mathfrak{F}, P)$ such that on $\mathbb{R}_+ \times \Lambda_1^c$ we have

(7) $\lim_{n \to \infty} [X^{R_n \wedge}, Y^{R_n \wedge}] = [X, Y]$, $\lim_{n \to \infty} [X^{R_n \wedge}] = [X]$, and $\lim_{n \to \infty} [Y^{R_n \wedge}] = [Y]$.

Since $X^{R_{n+1} \wedge} \in \mathbf{M}_2^c(\Omega, \mathfrak{F}, \{\mathfrak{F}_t\}, P)$ and $(X^{R_{n+1} \wedge})^{R_n \wedge} = X^{R_n \wedge}$, Lemma 12.23 implies that $\mu_{[X^{R_n \wedge}]}(\cdot, \omega) \leq \mu_{[X^{R_{n+1} \wedge}]}(\cdot, \omega)$ for $\omega \in \Lambda_2^c$ where Λ_2 is a null set in $(\Omega, \mathfrak{F}, P)$. Thus

$\mu_{[X^{R_n\wedge}]}(\cdot,\omega) \uparrow$ as $n \to \infty$ and in particular for any $a, b \in \mathbb{R}_+$ such that $a < b$ we have

$$\uparrow \lim_{n\to\infty} \mu_{[X^{R_n\wedge}]}((a,b],\omega) = \uparrow \lim_{n\to\infty} \{[X^{R_n\wedge}](b,\omega) - [X^{R_n\wedge}](a,\omega)\}$$
$$= [X](b,\omega) - [X](a,\omega) = \mu_{[X]}((a,b],\omega).$$

Thus $\mu_{[X^{R_n\wedge}]}((a,b],\omega) \leq \mu_{[X]}((a,b],\omega)$. From this and by the same argument for Y it follows that

(8) $\quad \mu_{[X^{R_n\wedge}]}(E,\omega) \leq \mu_{[X]}(E,\omega)$ and $\mu_{[Y^{R_n\wedge}]}(E,\omega) \leq \mu_{[X]}(E,\omega) \quad$ for $E \in \mathfrak{B}_{\mathbb{R}_+}$.

Let $\Lambda = \Lambda_1 \cup \Lambda_2$. Let $\omega\Lambda^c$ and $t \in \mathbb{R}_+$ be fixed and let $\mathfrak{G}$ be the collection of all members $E \in \mathfrak{B}_{(0,t]}$ such that $\lim_{n\to\infty} \mu_{[X^{R_n\wedge},Y^{R_n\wedge}]}(E,\omega) = \mu_{[X,Y]}(E,\omega)$. Since $R_n(\omega) \uparrow \infty$ as $n \to \infty$, for every $s \in [0,t]$ we have by (7)

$$\lim_{n\to\infty} [X^{R_n\wedge}, Y^{R_n\wedge}](s,\omega) = \lim_{n\to\infty} [X,Y]^{R_n\wedge}(s,\omega) = [X,Y](s,\omega).$$

Thus for $a, b \in (0,t]$ such that $a < b$ we have

$$\lim_{n\to\infty} \mu_{[X^{R_n\wedge},Y^{R_n\wedge}]}((a,b],\omega) = \lim_{n\to\infty} \{[X^{R_n\wedge},Y^{R_n\wedge}](b,\omega) - [X^{R_n\wedge},Y^{R_n\wedge}](a,\omega)\}$$
$$= [X,Y](b,\omega) - [X,Y](a,\omega) = \mu_{[X,Y]}((a,b],\omega).$$

Let $\mathfrak{J}$ be the semialgebra of subsets of $(0,t]$ consisting of subinterval of $(0,t]$ of the type $(a,b]$ and $\emptyset$. We have just shown that $\mathfrak{J} \subset \mathfrak{G}$. Since $\mathfrak{J}$ is also a π-class and since $\sigma(\mathfrak{J}) = \mathfrak{B}_{(0,t]}$, if we show that $\mathfrak{G}$ is a d-class then we have $\mathfrak{B}_{(0,t]} \subset \mathfrak{G}$ by Theorem 1.7.

To show that $\mathfrak{G}$ is a d-class of subsets of $(0,t]$, note first that we have shown above that $(0,t] \in \mathfrak{G}$. It follows from the definition of $\mathfrak{G}$ that if $E_1, E_2 \in \mathfrak{G}$ and $E_1 \subset E_2$ then $E_2 - E_1 \in \mathfrak{G}$. Finally let $\{E_k : k \in \mathbb{N}\}$ be an increasing sequence in $\mathfrak{G}$. We proceed to show that $E = \lim_{k\to\infty} = \cup_{k\in\mathbb{N}} $ is in $\mathfrak{G}$. Let us write $E = \cup_{k\in\mathbb{N}} F_k$ where $F_k = E_k - E_{k-1}$ for $k \in \mathbb{N}$ with $E_0 = \emptyset$. Then $\{F_k : k \in \mathbb{N}\}$ is a disjoint sequence in $\mathfrak{G}$. Let us define step functions $f_n, n \in \mathbb{N}$, f and g on $(0,\infty)$ by setting

(9) $\quad \begin{cases} f_n(x) = \mu_{[X^{R_n\wedge},Y^{R_n\wedge}]}(F_k,\omega) & \text{for } x \in (k-1,k] \text{ and } k \in \mathbb{N} \\ f(x) = \mu_{[X,Y]}(F_k,\omega) & \text{for } x \in (k-1,k] \text{ and } k \in \mathbb{N} \\ g(x) = \{\mu_{[X]}(F_k,\omega)\mu_{[Y]}(F_k,\omega)\}^{1/2} & \text{for } x \in (k-1,k] \text{ and } k \in \mathbb{N}. \end{cases}$

Since $F_k \in \mathfrak{G}$ we have $\lim_{n\to\infty} \mu_{[X^{R_n\wedge},Y^{R_n\wedge}]}(F_k,\omega) = \mu_{[X,Y]}(F_k,\omega)$ and this implies that $\lim_{n\to\infty} f_n(x) = f(x)$ for $x \in (0,\infty)$. Now

$$|\mu_{[X^{R_n\wedge},Y^{R_n\wedge}]}(F_k,\omega)| \leq \mu_{|[X^{R_n\wedge},Y^{R_n\wedge}]|}(F_k,\omega)$$
$$\leq \{\mu_{[X^{R_n\wedge}]}(F_k,\omega)\mu_{[Y^{R_n\wedge}]}(F_k,\omega)\}^{1/2} \leq \{\mu_{[X]}(F_k,\omega)\mu_{[Y]}(F_k,\omega)\}^{1/2},$$

§14. EXTENSIONS OF THE STOCHASTIC INTEGRAL

by Theorem 12.7 and (8). Thus $|f_n(x)| \leq g(x)$ for $x \in (0, \infty)$. To show that g is Lebesgue integrable on $(0, \infty)$, note that

$$\int_{(0,\infty)} g(x) m_L(dx) = \sum_{k \in \mathbb{N}} \{\mu_{[X]}(F_k, \omega) \mu_{[Y]}(F_k, \omega)\}^{1/2}$$

$$\leq \{\sum_{k \in \mathbb{N}} \mu_{[X]}(F_k, \omega)\}^{1/2} \{\sum_{k \in \mathbb{N}} \mu_{[Y]}(F_k, \omega)\}^{1/2} = \{\mu_{[X]}(E, \omega)\}^{1/2} \{\mu_{[Y]}(E, \omega)\}^{1/2}$$

$$\leq \{\mu_{[X]}((0,t], \omega)\}^{1/2} \{\mu_{[Y]}((0,t], \omega)\}^{1/2} < \infty.$$

Thus by the Dominated Convergence Theorem we have

$$\lim_{n \to \infty} \int_{(0,\infty)} f_n(x) m_L(dx) = \int_{(0,\infty)} f(x) m_L(dx),$$

in other words,

$$\lim_{n \to \infty} \sum_{k \in \mathbb{N}} \mu_{[X^{R_n \wedge}, Y^{R_n \wedge}]}(F_k, \omega) = \sum_{k \in \mathbb{N}} \mu_{[X,Y]}(F_k, \omega),$$

that is,

$$\lim_{n \to \infty} \mu_{[X^{R_n \wedge}, Y^{R_n \wedge}]}(E, \omega) = \mu_{[X,Y]}(E, \omega).$$

This shows that $E \in \mathfrak{G}$ and completes the proof that $\mathfrak{G}$ is a d-class. Therefore $\mathfrak{B}_{(0,t]} \subset \mathfrak{G}$ and thus (1) holds for every $E \in \mathfrak{B}_{(0,t]}$. By the equalities $\mu_{[X^{R_n \wedge}, Y^{R_n \wedge}]}(\{0\}, \omega) = 0$ and $\mu_{[X,Y]}(\{0\}, \omega) = 0$, (1) holds for every $E \in \mathfrak{B}_{[0,t]}$. The equalities (2) and (3) are proved in the same way as (1). Finally (4), (5), and (6) follow from (1), (2), and (3) respectively by letting $E = [0,t]$. ∎

Proposition 14.12. Let $X \in \mathbf{M}_2^{loc}(\Omega, \mathfrak{F}, \{\mathfrak{F}_t\}, P)$. If $\mathbf{E}([X]_a) < \infty$ for some $a \in \mathbb{R}_+$, then X is an L_2-martingale on $[0, a]$. Thus if X is in $\mathbf{M}_2^{loc}(\Omega, \mathfrak{F}, \{\mathfrak{F}_t\}, P)$ then X is in $\mathbf{M}_2(\Omega, \mathfrak{F}, \{\mathfrak{F}_t\}, P)$ if and only if $[X]$ is in $\mathbf{A}(\Omega, \mathfrak{F}, \{\mathfrak{F}_t\}, P)$.

Proof. Since $X \in \mathbf{M}_2^{loc}(\Omega, \mathfrak{F}, \{\mathfrak{F}_t\}, P)$, there exists an increasing sequence of stopping times $\{S_n : n \in \mathbb{N}\}$ such that $S_n \uparrow \infty$ almost surely as $n \to \infty$ and $X^{S_n \wedge}$ is in $\mathbf{M}_2(\Omega, \mathfrak{F}, \{\mathfrak{F}_t\}, P)$ for $n \in \mathbb{N}$. Now $[X] \in \mathbf{A}^{loc}(\Omega, \mathfrak{F}, \{\mathfrak{F}_t\}, P)$ and $X^2 - [X]$ is a right-continuous null at 0 local martingale. Thus there exists an increasing sequence of stopping times $\{T_n : n \in \mathbb{N}\}$ such that $T_n \uparrow \infty$ almost surely and $(X^2 - [X])^{T_n \wedge}$ is a right-continuous null at 0 martingale. Let $R_n = S_n \wedge T_n$ for $n \in \mathbb{N}$. Then there exists a null set Λ in $(\Omega, \mathfrak{F}, P)$ such that $R_n(\omega) \uparrow \infty$ for $\omega \in \Lambda^c$, $X^{R_n \wedge} \in \mathbf{M}_2(\Omega, \mathfrak{F}, \{\mathfrak{F}_t\}, P)$, and $(X^2 - [X])^{R_n \wedge}$ is a right-continuous null at 0 martingale. Thus for any $s, t \in [0, a]$ such that $s < t$ we have $\mathbf{E}(X_{R_n \wedge t}^2 - [X]_{R_n \wedge t} | \mathfrak{F}_s) = X_{R_n \wedge s}^2 - [X]_{R_n \wedge s}$ a.e. on $(\Omega, \mathfrak{F}_s, P)$. In particular for

$s = 0$, integrating the last equality and recalling that $X_0 = 0$ and $[X]_0 = 0$, we have $\mathbf{E}(X^2_{R_n \wedge t} - [X]_{R_n \wedge t}) = 0$ so that

$$\mathbf{E}(X^2_{R_n \wedge t}) = \mathbf{E}([X]_{R_n \wedge t}) \le \mathbf{E}([X]_t) \le \mathbf{E}([X]_a) < \infty.$$

Thus $\sup_{n \in \mathbb{N}} \|X_{R_n \wedge t}\|_2 < \infty$ and therefore $\{X_{R_n \wedge t} : n \in \mathbb{N}\}$ is uniformly integrable by Theorem 4.12. Now since $X^{R_n \wedge}$ is a martingale we have

(1) $\qquad \mathbf{E}(X_{R_n \wedge t} | \mathfrak{F}_s) = X_{R_n \wedge s}$ a.e. on $(\Omega, \mathfrak{F}_s, P)$.

Now $R_n(\omega) \uparrow \infty$ for $\omega \in \Lambda^c$ implies that $\lim_{n \to \infty} X_{R_n \wedge t} = X_t$ on Λ^c. This convergence and the uniform integrability of $\{X_{R_n \wedge t} : n \in \mathbb{N}\}$ imply that $\lim_{n \to \infty} \|X_{R_n \wedge t} - X_t\|_1 = 0$ by Theorem 4.16. Thus

(2) $\qquad \lim_{n \to \infty} \mathbf{E}(X_{R_n \wedge t} | \mathfrak{F}_s) = \mathbf{E}(X_t | \mathfrak{F}_s)$ a.e. on $(\Omega, \mathfrak{F}_s, P)$.

Since the filtration is augmented we have $\Lambda \in \mathfrak{F}_s$. Thus letting $n \to \infty$ in (1), we have $\mathbf{E}(X_t | \mathfrak{F}_s) = X_s$ a.e. on $(\Omega, \mathfrak{F}_s, P)$ by (2). This shows that X is a martingale on $[0, a]$. Finally since $\lim_{n \to \infty} X^2_{R_n \wedge t} = X^2_t$ on Λ^c, we have by Fatou's Lemma

$$\mathbf{E}(X^2_t) \le \liminf_{n \to \infty} \mathbf{E}(X^2_{R_n \wedge t}) \le \mathbf{E}([X]_a) < \infty.$$

This shows that X is an L_2-martingale on $[0, a]$. ∎

[II] Extensions of the Stochastic Integral to Local Martingales

Observation 14.13. For $A \in \mathbf{A}^{loc}(\Omega, \mathfrak{F}, \{\mathfrak{F}_t\}, P)$, consider the collection of all measurable processes $X = \{X_t : t \in \mathbb{R}_+\}$ on the $(\Omega, \mathfrak{F}, P)$ such that for every $\omega \in \Lambda^c$ where Λ is a null set in $(\Omega, \mathfrak{F}, P)$ depending on X we have

$$\int_{[0,t]} X^2(s, \omega) \mu_A(ds, \omega) < \infty \quad \text{for every } t \in \mathbb{R}_+.$$

For such processes X and Y the condition that there exists a null set in $(\Omega, \mathfrak{F}, P)$ depending on X and Y such that for $\omega \in \Lambda^c$ we have

$$\int_{[0,t]} |X(s, \omega) - Y(s, \omega)|^2 \mu_A(ds, \omega) = 0 \quad \text{for every } t \in \mathbb{R}_+$$

is an equivalence relation.

§14. EXTENSIONS OF THE STOCHASTIC INTEGRAL

Definition 14.14. *For $A \in \mathbf{A}^{loc}(\Omega, \mathfrak{F}, \{\mathfrak{F}_t\}, P)$, let $\mathbf{L}_{2,\infty}^{loc}(\mathbb{R}_+ \times \Omega, \mu_A, P)$ be the linear space of equivalence classes of measurable processes $X = \{X_t : t \in \mathbb{R}_+\}$ on $(\Omega, \mathfrak{F}, P)$ such that for $\omega \in \Lambda^c$ where Λ is a null set in $(\Omega, \mathfrak{F}, P)$ we have*

$$\int_{[0,t]} X^2(s,\omega)\mu_A(ds,\omega) < \infty \quad \text{for every } t \in \mathbb{R}_+.$$

Note that $\mathbf{L}_{2,\infty}(\mathbb{R}_+ \times \Omega, \mu_A, P) \subset \mathbf{L}_{2,\infty}^{loc}(\mathbb{R}_+ \times \Omega, \mu_A, P)$ for $A \in \mathbf{A}(\Omega, \mathfrak{F}, \{\mathfrak{F}_t\}, P)$ by (3) of Observation 11.16.

Lemma 14.15. *Let $M \in \mathbf{M}_2^{c,loc}(\Omega, \mathfrak{F}, \{\mathfrak{F}_t\}, P)$ so that $[M] \in \mathbf{A}^{c,loc}(\Omega, \mathfrak{F}, \{\mathfrak{F}_t\}, P)$ and let $X = \{X_t : t \in \mathbb{R}_+\}$ be a predictable process on the filtered space such that $X \in \mathbf{L}_{2,\infty}^{loc}(\mathbb{R}_+ \times \Omega, \mu_{[M]}, P)$. Let $\{a_n : n \in \mathbb{N}\}$ be a strictly increasing sequence in $\mathbb{R}_+$ such that $a_n \uparrow \infty$ as $n \to \infty$. For every $n \in \mathbb{N}$, let T_n be a nonnegative valued function on Ω defined by*

$$T_n(\omega) = \inf\{t \in \mathbb{R}_+ : \int_{[0,t]} X^2(s,\omega)\mu_{[M]}(ds,\omega) \geq a_n\} \wedge a_n \quad \text{for } \omega \in \Omega.$$

Let $\{S_n : n \in \mathbb{N}\}$ be an increasing sequence of stopping times such that $S_n \uparrow \infty$ a.e. on $(\Omega, \mathfrak{F}, P)$ as $n \to \infty$ and $M^{S_n \wedge} \in \mathbf{M}_2^c(\Omega, \mathfrak{F}, \{\mathfrak{F}_t\}, P)$ so that $[M^{S_n \wedge}] \in \mathbf{A}^c(\Omega, \mathfrak{F}, \{\mathfrak{F}_t\}, P)$. Then
1) $\{T_n : n \in \mathbb{N}\}$ is an increasing sequence of finite stopping time on the filtered space and $T_n \uparrow \infty$ a.e. on $(\Omega, \mathfrak{F}, P)$ as $n \to \infty$,
2) $X^{[T_n]}$ is a predictable process and $X^{[T_n]} \in \mathbf{L}_{2,\infty}(\mathbb{R}_+ \times \Omega, \mu_{[M^{S_k \wedge}]}, P)$ for every $k \in \mathbb{N}$.

Proof. In Proposition 3.5 we showed that the first passage time of an open set in $\mathbb{R}_+$ by an adapted right-continuous process on a right-continuous filtered space is a stopping time. Now if $Z = \{Z_t : t \in \mathbb{R}_+\}$ is an adapted continuous process on a right-continuous filtered space, then the function on the sample space defined by $\inf\{t \in \mathbb{R}_+ : Z_t \geq a_n\}$ is equal to $\inf\{t \in \mathbb{R}_+ : Z_t > a_n\}$, that is, the first passage time of the open set (a_n, ∞), and is therefore a stopping time for every $n \in \mathbb{N}$. The continuity of Z also implies that at the stopping time defined above, Z is equal to a_n. Let us construct an adapted continuous process Z on our standard filtered space $(\Omega, \mathfrak{F}, \{\mathfrak{F}_t\}, P)$ which is equivalent to the process $\{\int_{[0,t]} X^2(s)\mu_{[M]}(ds) : t \in \mathbb{R}_+\}$.

Since $X \in \mathbf{L}_{2,\infty}^{loc}(\mathbb{R}_+ \times \Omega, \mu_{[M]}, P)$, there exists a null set Λ_1 in $(\Omega, \mathfrak{F}, P)$ such that for every $\omega \in \Lambda_1^c$ we have

$$(1) \qquad \int_{[0,t]} X^2(s,\omega)\mu_{[M]}(ds,\omega) < \infty \quad \text{for every } t \in \mathbb{R}_+.$$

Since $[M]$ is an almost surely continuous process there exists a null set Λ_2 in $(\Omega, \mathfrak{F}, P)$ such that $[M](\cdot, \omega)$ is a continuous monotone increasing function on $\mathbb{R}_+$ for $\omega \in \Lambda_2^c$. Let $\Lambda = \Lambda_1 \cup \Lambda_2$ and define a real valued function Z on $\mathbb{R}_+ \times \Omega$ by setting

(2) $\quad Z(t, \omega) = \begin{cases} \int_{[0,t]} X^2(s,\omega) \mu_{[M]}(ds, \omega) & \text{for } (t,\omega) \in \mathbb{R}_+ \times \Lambda^c \\ 0 & \text{for } (t,\omega) \in \mathbb{R}_+ \times \Lambda. \end{cases}$

Now since X is a predictable process, it is a measurable process and then so is X^2. Thus $\{\int_{[0,t]} X^2(s) \mu_{[M]}(ds) : t \in \mathbb{R}_+\}$ is an adapted process by Theorem 10.11. Since the filtered space $(\Omega, \mathfrak{F}, \{\mathfrak{F}_t\}, P)$ is augmented, $\Lambda \in \mathfrak{F}_t$ for every $t \in \mathbb{R}_+$. Thus $Z(t, \cdot)$ defined by (2) is $\mathfrak{F}_t$-measurable. This shows that Z is an adapted process. For $\omega \in \Lambda^c$, $[M](\cdot, \omega)$ is a continuous function on $\mathbb{R}_+$ and therefore $\mu_{[M]}(\{s\}, \omega) = 0$ for every $s \in \mathbb{R}_+$. This implies that $\int_{[0,t]} X^2(s,\omega) \mu_{[M]}(ds,\omega)$ is a continuous function of $t \in \mathbb{R}_+$. Thus we have shown that Z is an adapted continuous process.

For every $n \in \mathbb{N}$, let R_n be a nonnegative valued function on Ω defined by

(3) $\quad R_n(\omega) = \inf\{t \in \mathbb{R}_+ : Z(t, \omega) \geq a_n\} \wedge a_n \quad \text{for } \omega \in \Omega.$

By the continuity of $Z(\cdot, \omega)$ for every $\omega \in \Omega$,

$$\inf\{t \in \mathbb{R}_+ : Z(t, \omega) \geq a_n\} = \inf\{t \in \mathbb{R}_+ : Z(t, \omega) > a_n\},$$

which is the first passage time of the open set (a_n, ∞) in $\mathbb{R}$ and hence a stopping time by Proposition 3.5. Thus R_n is a stopping time. Clearly $R_n(\omega) \uparrow$ as $n \to \infty$ for every $\omega \in \Omega$. Let us show that actually $R_n(\omega) \uparrow \infty$ for every $\omega \in \Lambda^c$. Let $\omega \in \Lambda^c$. If $Z(\cdot, \omega)$ is bounded on $\mathbb{R}_+$, say $Z(t, \omega) \leq a_m$ for some $m \in \mathbb{N}$ for all $t \in \mathbb{R}_+$, then $\inf\{t \in \mathbb{R}_+ : Z(t, \omega) > a_n\} = \inf\{\emptyset\} = \infty$ for $n > m$ and thus $R_n(\omega) = \infty \wedge a_n = a_n$ for $n > m$ and then $R_n(\omega) \uparrow \infty$ as $n \to \infty$. On the other hand if $Z(\cdot, \omega)$ is not bounded on $\mathbb{R}_+$, then $\lim_{t \to \infty} Z(t, \omega) = \infty$. If $R_n(\omega) \uparrow \infty$ does not hold, then there exists $m \in \mathbb{N}$ such that $R_n(\omega) \leq a_m$ for all $n \in \mathbb{N}$. Then for $n > m$ we have $\inf\{t \in \mathbb{R}_+ : Z(t, \omega) > a_n\} \wedge a_n \leq a_m$ and therefore $\inf\{t \in \mathbb{R}_+ : Z(t, \omega) > a_n\} \leq a_m$ for $n > m$. Thus $\lim_{t \uparrow a_m} Z(t, \omega) = \infty$. But the continuity of $Z(\cdot, \omega)$ implies $\lim_{t \uparrow a_m} Z(t, \omega) = Z(a_m, \omega)$ and then $Z(a_m, \omega) = \infty$ contradicting $Z(a_m, \omega) = \int_{[0, a_m]} X^2(s, \omega) \mu_{[M]}(ds, \omega) < \infty$. This shows that $R_n(\omega) \uparrow \infty$ for this case also.

Now by (2), $Z(\cdot, \omega) = \int_{[0, \cdot]} X^2(s, \omega) \mu_{[M]}(ds, \omega)$ for $\omega \in \Lambda^c$. Thus for R_n defined by (3) we have $R_n(\omega) = T_n(\omega)$ for $\omega \in \Lambda^c$. Since R_n is a stopping time on a standard filtered space, this implies that T_n is a stopping time on the filtered space by Lemma 3.17. Clearly $T_n(\omega) \uparrow$ for every $\omega \in \Omega$. Since $R_n(\omega) \uparrow \infty$ for $\omega \in \Lambda^c$, we have $T_n(\omega) \uparrow \infty$ for $\omega \in \Lambda^c$.

§14. EXTENSIONS OF THE STOCHASTIC INTEGRAL

Consider the truncated process $X^{[T_n]}$. Since X is a predictable process, $X^{[T_n]}$ is a predictable process by Observation 3.35. To show that $X^{[T_n]} \in \mathbf{L}_{2,\infty}(\mathbb{R}_+ \times \Omega, \mu_{[M^{S_k \wedge}]}, P)$ for any $k \in \mathbb{N}$, note first that since $X^{[T_n]}$ is a predictable process, it is a measurable process by Proposition 2.23. Next, by Observation 3.34, for every $\omega \in \Omega$ we have

$$(4) \qquad X^{[T_n]}(s,\omega) = \begin{cases} X(s,\omega) & \text{for } s \in [0, T_n(\omega)] \\ 0 & \text{for } s \in (T_n(\omega), \infty). \end{cases}$$

According to 2) of Remark 14.11 we have $\mu_{[M^{S_k \wedge}]}(\cdot,\omega) \leq \mu_{[M]}(\cdot,\omega)$ on $(\mathbb{R}_+, \mathfrak{B}_{\mathbb{R}_+})$ for $\omega \in \Lambda_k^c$ where Λ_k is a null set in $(\Omega, \mathfrak{F}, P)$. Thus for every $\omega \in (\Lambda \cup \Lambda_k)^c$ and $t \in \mathbb{R}_+$, we have by (4) and the fact that $T_n(\omega) = R_n(\omega)$

$$(5) \qquad \int_{[0,t]} (X^{[T_n]})^2(s,\omega) \mu_{[M^{S_k \wedge}]}(ds,\omega)) \leq \int_{[0,t]} (X^{[T_n]})^2(s,\omega) \mu_{[M]}(ds,\omega))$$

$$= \int_{[0, t \wedge R_n(\omega)]} X^2(s,\omega) \mu_{[M]}(ds,\omega)) = Z(t \wedge R_n(\omega), \omega) \leq a_n,$$

where the last inequality holds since by the definition of R_n by (3) if $t < R_n(\omega)$ then $Z(t \wedge R_n(\omega), \omega) = Z(t, \omega) \leq a_n$ and if $t \geq R_n(\omega)$ then $Z(t \wedge R_n(\omega), \omega) = Z(R_n(\omega), \omega) = a_n$ by the continuity of $Z(\cdot, \omega)$. Since (5) holds for $\omega \in (\Lambda \cup \Lambda_k)^c$, we have

$$\mathbf{E}\left[\int_{[0,t]} (X^{[T_n]})^2(s)\, d[M^{S_k \wedge}](ds)\right] \leq \mathbf{E}(a_n) = a_n < \infty,$$

for every $t \in \mathbb{R}_+$. This shows that $X^{[T_n]} \in \mathbf{L}_{2,\infty}(\mathbb{R}_+ \times \Omega, \mu_{[M^{S_k \wedge}]}, P)$. ∎

Theorem 14.16. *Let $M \in \mathbf{M}_2^{c,loc}(\Omega, \mathfrak{F}, \{\mathfrak{F}_t\}, P)$ and let $X = \{X_t : t \in \mathbb{R}_+\}$ be a predictable process on the filtered space such that $X \in \mathbf{L}_{2,\infty}^{loc}(\mathbb{R}_+ \times \Omega, \mu_{[M]}, P)$. Let $\{S_n : n \in \mathbb{N}\}$ be an increasing sequence of stopping times such that $S_n \uparrow \infty$ a.e. on $(\Omega, \mathfrak{F}, P)$ as $n \to \infty$ and $M^{S_n \wedge} \in \mathbf{M}_2^c(\Omega, \mathfrak{F}, \{\mathfrak{F}_t\}, P)$ for every $n \in \mathbb{N}$. Let $\{a_n : n \in \mathbb{N}\}$ be a strictly increasing sequence in $\mathbb{R}_+$ such that $a_n \uparrow \infty$ as $n \to \infty$. Let $\{T_n : n \in \mathbb{N}\}$ be an increasing sequence of stopping times defined by*

$$T_n(\omega) = \inf\{t \in \mathbb{R}_+ : \int_{[0,t]} X^2(s,\omega) \mu_{[M]}(ds,\omega) \geq a_n\} \wedge a_n \quad \text{for } \omega \in \Omega.$$

Then there exists $Y \in \mathbf{M}_2^{c,loc}(\Omega, \mathfrak{F}, \{\mathfrak{F}_t\}, P)$ such that $Y^{S_n \wedge T_n \wedge} = X^{[T_n]} \bullet M^{S_n \wedge}$ for every $n \in \mathbb{N}$ and therefore there exists a null set Λ in $(\Omega, \mathfrak{F}, P)$ such that

$$Y(t,\omega) = \lim_{n \to \infty} (X^{[T_n]} \bullet M^{S_n \wedge})(t,\omega) \quad \text{for } (t,\omega) \in \mathbb{R}_+ \times \Lambda^c.$$

Furthermore the process Y is not only unique in $\mathbf{M}_2^{c,loc}(\Omega, \mathfrak{F}, \{\mathfrak{F}_t\}, P)$ but is in fact independent of the choices of the sequences $\{a_n : n \in \mathbb{N}\}$ and $\{S_n : n \in \mathbb{N}\}$.
In particular 1) if $M \in \mathbf{M}_2^c(\Omega, \mathfrak{F}, \{\mathfrak{F}_t\}, P)$ and $X \in \mathbf{L}_{2,\infty}^{loc}(\mathbb{R}_+ \times \Omega, \mu_{[M]}, P)$, then there exists $Y \in \mathbf{M}_2^{c,loc}(\Omega, \mathfrak{F}, \{\mathfrak{F}_t\}, P)$ such that $Y^{T_n \wedge} = X^{[T_n]} \bullet M$ for every $n \in \mathbb{N}$ and there exists a null set Λ in $(\Omega, \mathfrak{F}, P)$ such that $Y(t, \omega) = \lim_{n \to \infty}(X^{[T_n]} \bullet M)(t, \omega)$ for $(t, \omega) \in \mathbb{R}_+ \times \Lambda^c$, and 2) if $M \in \mathbf{M}_2^{c,loc}(\Omega, \mathfrak{F}, \{\mathfrak{F}_t\}, P)$ and $X \in \mathbf{L}_{2,\infty}(\mathbb{R}_+ \times \Omega, \mu_{[M]}, P)$, then there exists $Y \in \mathbf{M}_2^{c,loc}(\Omega, \mathfrak{F}, \{\mathfrak{F}_t\}, P)$ such that $Y^{S_n \wedge} = X \bullet M^{S_n \wedge}$ for every $n \in \mathbb{N}$ and there exists a null set Λ in $(\Omega, \mathfrak{F}, P)$ such that $Y(t, \omega) = \lim_{n \to \infty}(X \bullet M^{S_n \wedge})(t, \omega)$ for $(t, \omega) \in \mathbb{R}_+ \times \Lambda^c$.

Proof. Since $M \in \mathbf{M}_2^{c,loc}(\Omega, \mathfrak{F}, \{\mathfrak{F}_t\}, P)$, it has an almost surely continuous quadratic variation process $[M]$ by Corollary 14.9. By Lemma 14.15, $T_n \uparrow \infty$ a.e. on $(\Omega, \mathfrak{F}, P)$ and the truncated process $X^{[T_n]}$ is a predictable process and $X^{[T_n]} \in \mathbf{L}_{2,\infty}(\mathbb{R}_+ \times \Omega, \mu_{[M]}, P)$. According to Lemma 12.23, $\mu_{[M^{S_n \wedge}]}(\cdot, \omega) \leq \mu_{[M]}(\cdot, \omega)$ on $(\mathbb{R}_+, \mathfrak{B}_{\mathbb{R}_+})$ for a.e. $\omega \in \Omega$. This implies that $X^{[T_n]} \in \mathbf{L}_{2,\infty}(\mathbb{R}_+ \times \Omega, \mu_{[M^{S_n \wedge}]}, P)$ for every $n \in \mathbb{N}$. Thus $X^{[T_n]} \bullet M^{S_n \wedge}$ is defined and is in $\mathbf{M}_2^c(\Omega, \mathfrak{F}, \{\mathfrak{F}_t\}, P)$.

Consider the subset $\{(\cdot) \leq S_n \wedge T_n\} = \{(t, \omega) \in \mathbb{R}_+ \times \Omega : t \leq S_n(\omega) \wedge T_n(\omega)\}$ of $\mathbb{R}_+ \times \Omega$. Since $S_n \wedge T_n \uparrow$ on Ω, $\{(\cdot) \leq S_n \wedge T_n\} \uparrow$ as $n \to \infty$. Let Λ be a null set in $(\Omega, \mathfrak{F}, P)$ such that $S_n(\omega) \uparrow \infty$ and $T_n(\omega) \uparrow \infty$ as $n \to \infty$ for $\omega \in \Lambda_0^c$. If $\omega \in \Lambda_0^c$ then for any $t \in \mathbb{R}_+$ we have $t \leq S_n(\omega) \wedge T_n(\omega)$ so that $(t, \omega) \in \{(\cdot) \leq S_n \wedge T_n\}$ for sufficiently large $n \in \mathbb{N}$. Thus $\mathbb{R}_+ \times \Lambda_0^c \subset \cup_{n \in \mathbb{N}}\{(\cdot) \leq S_n \wedge T_n\}$. For each $n \in \mathbb{N}$, define a real valued function $Y^{(n)}$ on $\{(\cdot) \leq S_n \wedge T_n\} \cap \{\mathbb{R}_+ \times \Lambda_0^c\}$ by setting

(1) $\quad Y^{(n)}(t, \omega) = (X^{[T_n]} \bullet M^{S_n \wedge})(t, \omega) \quad$ for $(t, \omega) \in \{(\cdot) \leq S_n \wedge T_n\} \cap \{\mathbb{R}_+ \times \Lambda_0^c\}$.

For any pair $k, n \in \mathbb{N}$ such that $k < n$, there exists according to Theorem 12.22 and Theorem 12.24 a null set $\Lambda_{k,n}$ in $(\Omega, \mathfrak{F}, P)$ such that

(2) $\quad ((X^{[T_n]})^{[T_k]} \bullet (M^{S_n \wedge})^{S_k \wedge})(t, \omega) = (X^{[T_n]} \bullet M^{S_n \wedge})(S_k(\omega) \wedge T_k(\omega) \wedge t, \omega) \quad$ for $\omega \in \Lambda_{k,n}^c$.

Let us show that

(3) $\quad Y^{(n)}(t, \omega) = Y^{(k)}(t, \omega) \quad$ for $(t, \omega) \in \{(\cdot) \leq S_k \wedge T_k\} \cap \{\mathbb{R}_+ \times (\Lambda_0 \cup \Lambda_{k,n})^c\}$.

Now for $(t, \omega) \in \{(\cdot) \leq S_k \wedge T_k\} \cap \{\mathbb{R}_+ \times (\Lambda_0 \cup \Lambda_{k,n})^c\}$, we have

$$\begin{aligned}
Y^{(k)}(t, \omega) &= (X^{[T_k]} \bullet M^{S_k \wedge})(t, \omega) \quad \text{by (1)} \\
&= ((X^{[T_n]})^{[T_k]} \bullet (M^{S_n \wedge})^{S_k \wedge})(t, \omega) \quad \text{by Observation 3.36} \\
&= (X^{[T_n]} \bullet M^{S_n \wedge})(S_k(\omega) \wedge T_k(\omega) \wedge t, \omega) \quad \text{by (2)} \\
&= (X^{[T_n]} \bullet M^{S_n \wedge})(t, \omega) \quad \text{since } t \leq S_k(\omega) \wedge T_k(\omega) \\
&= Y^{(n)}(t, \omega) \quad \text{by (1).}
\end{aligned}$$

§14. EXTENSIONS OF THE STOCHASTIC INTEGRAL

This proves (3). Now let $\Lambda = \Lambda_0 \cup (\cup_{k,n \in \mathbb{N}, k<n} \Lambda_{k,n})$. Define a real valued function Y on $\mathbb{R}_+ \times \Omega$ by setting

(4) $\quad Y(t,\omega) = \begin{cases} Y^{(n)}(t,\omega) & \text{for } (t,\omega) \in \{(\cdot) \leq S_n \wedge T_n\} \cap \{\mathbb{R}_+ \times \Lambda^c\} \text{ and } n \in \mathbb{N} \\ 0 & \text{for } (t,\omega) \in \mathbb{R}_+ \times \Lambda. \end{cases}$

Note that Y is well-defined on $\mathbb{R}_+ \times \Lambda^c$ by (3) and hence it is well-defined on $\mathbb{R}_+ \times \Omega$.

To show that Y is in $\mathbf{M}_2^{c,loc}(\Omega, \mathfrak{F}, \{\mathfrak{F}_t\}, P)$, let us show first that Y is adapted, that is, for every $s \in \mathbb{R}_+$, $Y(s, \cdot)$ is $\mathfrak{F}_s$-measurable. Now since $S_n \wedge T_n$ is a stopping time on a right-continuous filtered space, we have $\{s \leq S_n \wedge T_n\} = \{S_n \wedge T_n < s\}^c \in \mathfrak{F}_s$ by Theorem 3.4. Also since the filtered space is augmented, the null set Λ is in $\mathfrak{F}_s$. Thus $\{s \leq S_n \wedge T_n\} \cap \Lambda^c$ is in $\mathfrak{F}_s$. Since on this $\mathfrak{F}_s$-measurable set $Y(s, \cdot) = Y^{(n)}(s, \cdot) = (X^{[T_n]} \bullet M^{S_n \wedge})(s, \cdot)$ which is $\mathfrak{F}_s$-measurable, $Y(s, \cdot)$ is $\mathfrak{F}_s$-measurable on this set. Since this holds for every $n \in \mathbb{N}$ and since $\cup_{n \in \mathbb{N}}\{s \leq S_n \wedge T_n\} \cap \Lambda^c = \Lambda^c$, $Y(s, \cdot)$ is $\mathfrak{F}_s$-measurable on Λ^c. Then since $Y(s, \cdot) = 0$ on Λ, $Y(s, \cdot)$ is $\mathfrak{F}_s$-measurable on Ω. This shows that Y is adapted.

Next we show that with our increasing sequence of stopping times $\{S_n \wedge T_n : n \in \mathbb{N}\}$ such that $S_n \wedge T_n \uparrow \infty$ on Λ_0^c we have $Y^{S_n \wedge T_n \wedge} \in \mathbf{M}_2^c(\Omega, \mathfrak{F}, \{\mathfrak{F}_t\}, P)$ by showing

(5) $\quad Y^{S_n \wedge T_n \wedge} = X^{[T_n]} \bullet M^{S_n \wedge}.$

Now an arbitrary $(t,\omega) \in \mathbb{R}_+ \times \Lambda^c$ is in $\{(\cdot) \leq S_m \wedge T_m\} \cap \{\mathbb{R}_+ \times \Lambda^c\}$ for some $m \in \mathbb{N}$ and we can assume that $m \geq n$ without loss of generality. Then by (4) and (1) we have $Y(t,\omega) = (X^{[T_m]} \bullet M^{S_m \wedge})(t,\omega)$ and therefore

$$\begin{aligned} Y^{S_n \wedge T_n \wedge}(t,\omega) &= (X^{[T_m]} \bullet M^{S_m \wedge})(S_n(\omega) \wedge T_n(\omega) \wedge t, \omega) \\ &= ((X^{[T_m]})^{[T_n]} \bullet (M^{S_m})^{S_n \wedge})(t,\omega) \quad \text{by (2)} \\ &= (X^{[T_n]} \bullet M^{S_n \wedge})(t,\omega) \quad \text{since } m \geq n. \end{aligned}$$

This shows that $Y^{S_n \wedge T_n \wedge}(\cdot, \omega) = (X^{[T_n]} \bullet M^{S_n \wedge})(\cdot, \omega)$ for $\omega \in \Lambda^c$. Thus (5) holds. Therefore $Y \in \mathbf{M}_2^{c,loc}(\Omega, \mathfrak{F}, \{\mathfrak{F}_t\}, P)$. For $\omega \in \Lambda^c$ we have $S_n(\omega) \wedge T_n(\omega) \uparrow \infty$ as $n \to \infty$ so that for any $t \in \mathbb{R}_+$ we have $S_n(\omega) \wedge T_n(\omega) \wedge t = t$ for sufficiently large $n \in \mathbb{N}$ and thus $Y(t,\omega) = \lim_{n \to \infty} Y^{S_n \wedge T_n \wedge}(t,\omega)$.

To show that Y is independent of the choices of the sequences $\{a_n : n \in \mathbb{N}\}$ and $\{S_n : n \in \mathbb{N}\}$, let $\{a'_n : n \in \mathbb{N}\}$ be another strictly increasing sequence in $\mathbb{R}_+$ such that $a'_n \uparrow \infty$ and let T'_n be defined by

$$T'_n(\omega) = \inf\{t \in \mathbb{R}_+ : \int_{[0,t]} X^2(s,\omega)\mu_{[M]}(ds,\omega) \geq a'_n\} \wedge a'_n \quad \text{for } \omega \in \Omega.$$

Let $\{S'_n : n \in \mathbb{N}\}$ be the increasing sequence of stopping times such that $S'_n \uparrow \infty$ a.e. on $(\Omega, \mathfrak{F}, P)$ as $n \to \infty$ and $M^{S'_n \wedge} \in \mathbf{M}^c_2(\Omega, \mathfrak{F}, \{\mathfrak{F}_t\}, P)$ for every $n \in \mathbb{N}$. Then by what we showed above there exits $Z \in \mathbf{M}^{c,loc}_2(\Omega, \mathfrak{F}, \{\mathfrak{F}_t\}, P)$ such that

$$Z^{S'_n \wedge T'_n \wedge} = X^{[T'_n]} \bullet M^{S'_n \wedge} \quad \text{for every } n \in \mathbb{N}.$$

Let us show that there exists a null set Λ in $(\Omega, \mathfrak{F}, P)$ such that $Y(\cdot, \omega) = Z(\cdot, \omega)$ for $\omega \in \Lambda^c$. Let Λ_0 be a null set in $(\Omega, \mathfrak{F}, P)$ such that $S_n \wedge T_n \wedge S'_n \wedge T'_n \uparrow \infty$ as $n \to \infty$ on Λ_0^c. Now

$$Y^{S_n \wedge T_n \wedge S'_n \wedge T'_n \wedge} = (X^{[T_n]} \bullet M^{S_n \wedge})^{S'_n \wedge T'_n \wedge}$$
$$= (X^{[T_n]})^{[T'_n]} \bullet (M^{S_n \wedge})^{S'_n \wedge} = X^{[T_n \wedge T'_n]} \bullet M^{S_n \wedge S'_n \wedge},$$

where the second equality is by Theorem 12.22 and Theorem 12.24. Similarly

$$Z^{S_n \wedge T_n \wedge S'_n \wedge T'_n \wedge} = X^{[T_n \wedge T'_n]} \bullet M^{S_n \wedge S'_n \wedge}.$$

Thus $Y^{S_n \wedge T_n \wedge S'_n \wedge T'_n \wedge}(\cdot, \omega) = Z^{S_n \wedge T_n \wedge S'_n \wedge T'_n \wedge}(\cdot, \omega)$ for $\omega \in \Lambda_n^c$ where Λ_n is a null set in $(\Omega, \mathfrak{F}, P)$. Let $\Lambda = \cup_{n \in \mathbb{Z}_+} \Lambda_n$. Then $Y^{S_n \wedge T_n \wedge S'_n \wedge T'_n \wedge}(\cdot, \omega) = Z^{S_n \wedge T_n \wedge S'_n \wedge T'_n \wedge}(\cdot, \omega)$ for every $n \in \mathbb{N}$ when $\omega \in \Lambda^c$. Since $S_n \wedge T_n \wedge S'_n \wedge T'_n \uparrow \infty$ as $n \to \infty$ on Λ^c, by letting $n \to \infty$ in the last equality we have $Y(\cdot, \omega) = Z(\cdot, \omega)$ for $\omega \in \Lambda^c$. This proves the independence of Y from the sequences $\{a_n : n \in \mathbb{N}\}$ and $\{S_n : n \in \mathbb{N}\}$.

Finally when $M \in \mathbf{M}^c_2(\Omega, \mathfrak{F}, \{\mathfrak{F}_t\}, P)$ and $X \in \mathbf{L}^{loc}_{2,\infty}(\mathbb{R}_+ \times \Omega, \mu_{[M]}, P)$, we let $S_n(\omega) = \infty$ for all $\omega \in \Omega$ and $n \in \mathbb{N}$. Then $M^{S_n \wedge} = M$ and $Y^{S_n \wedge T_n \wedge} = Y^{T_n \wedge}$ for every $n \in \mathbb{N}$. Similarly when $M \in \mathbf{M}^{c,loc}_2(\Omega, \mathfrak{F}, \{\mathfrak{F}_t\}, P)$ and $X \in \mathbf{L}_{2,\infty}(\mathbb{R}_+ \times \Omega, \mu_{[M]}, P)$, we let $T_n(\omega) = \infty$ for all $\omega \in \Omega$ and $n \in \mathbb{N}$ so that $X^{[T_n]} = T$ and $Y^{S_n \wedge T_n \wedge} = Y^{S_n \wedge}$ for every $n \in \mathbb{N}$. ∎

Definition 14.17. *Let $M \in \mathbf{M}^{c,loc}_2(\Omega, \mathfrak{F}, \{\mathfrak{F}_t\}, P)$ and let $X = \{X_t : t \in \mathbb{R}_+\}$ be a predictable process on the filtered space such that $X \in \mathbf{L}^{loc}_{2,\infty}(\mathbb{R}_+ \times \Omega, \mu_{[M]}, P)$. Let $\{S_n : n \in \mathbb{N}\}$ be the increasing sequence of stopping times such that $S_n \uparrow \infty$ a.e. on $(\Omega, \mathfrak{F}, P)$ as $n \to \infty$ and $M^{S_n \wedge} \in \mathbf{M}^c_2(\Omega, \mathfrak{F}, \{\mathfrak{F}_t\}, P)$ for every $n \in \mathbb{N}$. Let $\{a_n : n \in \mathbb{N}\}$ be a strictly increasing sequence in $\mathbb{R}_+$ such that $a_n \uparrow \infty$ as $n \to \infty$. For every $n \in \mathbb{N}$, let T_n be defined by*

$$T_n(\omega) = \inf\{t \in \mathbb{R}_+ : \int_{[0,t]} X^2(s, \omega) \mu_{[M]}(ds, \omega) \geq a_n\} \wedge a_n \quad \text{for } \omega \in \Omega.$$

We define the stochastic integral $X \bullet M$ of X with respect to M to be the unique Y in $\mathbf{M}^{c,loc}_2(\Omega, \mathfrak{F}, \{\mathfrak{F}_t\}, P)$ such that $Y^{S_n \wedge T_n \wedge} = X^{[T_n]} \bullet M^{S_n \wedge}$ for every $n \in \mathbb{N}$ and thus $Y(t, \omega) = \lim_{n \to \infty} (X^{[T_n]} \bullet M^{S_n \wedge})(t, \omega)$ for $(t, \omega) \in \mathbb{R}_+ \times \Lambda^c$ where Λ is a null set in $(\Omega, \mathfrak{F}, P)$.

§14. EXTENSIONS OF THE STOCHASTIC INTEGRAL

As an alternate notation, we write $\int_{[0,t]} X(s)\,dM(s)$ for the random variable $(X \bullet M)_t$ for $t \in \mathbb{R}_+$.

For truncation by stopping times for the extended stochastic integral we have the following.

Theorem 14.18. *Let $M \in \mathbf{M}_2^{c,loc}(\Omega, \mathfrak{F}, \{\mathfrak{F}_t\}, P)$ and let X be a predictable process on the filtered space such that $X \in \mathbf{L}_{2,\infty}^{loc}(\mathbb{R}_+ \times \Omega, \mu_{[M]}, P)$. Then for any stopping time T on the filtered space there exists a null set Λ in $(\Omega, \mathfrak{F}, P)$ such that for every $\omega \in \Lambda^c$ we have $(X \bullet M^{T\wedge})(\cdot, \omega) = (X \bullet M)^{T\wedge}(\cdot, \omega)$.*

Proof. Let $\{S_n : n \in \mathbb{N}\}$ and $\{T_n : n \in \mathbb{N}\}$ be as in Theorem 14.16 so that there exists a null set Λ_1 in $(\Omega, \mathfrak{F}, P)$ such that

$$(1) \qquad X \bullet M = \lim_{n \to \infty} X^{[T_n]} \bullet M^{S_n \wedge} \quad \text{on } \mathbb{R}_+ \times \Lambda_1^c.$$

Now since $M^{S_n \wedge} \in \mathbf{M}_2^c(\Omega, \mathfrak{F}, \{\mathfrak{F}_t\}, P)$ we have $M^{T \wedge S_n \wedge} \in \mathbf{M}_2^c(\Omega, \mathfrak{F}, \{\mathfrak{F}_t\}, P)$ for every $n \in \mathbb{N}$ and this shows that $M^{T\wedge} \in \mathbf{M}_2^{c,loc}(\Omega, \mathfrak{F}, \{\mathfrak{F}_t\}, P)$. Since $\mu_{[M^{\wedge}]} \leq \mu_{[M]}$ according to Theorem 14.11, the fact that X is in $\mathbf{L}_{2,\infty}^{loc}(\mathbb{R}_+ \times \Omega, \mu_{[M]}, P)$ implies that X is in $\mathbf{L}_{2,\infty}^{loc}(\mathbb{R}_+ \times \Omega, \mu_{[M^{T\wedge}]}, P)$. Thus according to Theorem 14.16, $X \bullet M^{T\wedge}$ is defined and is in $\mathbf{M}_2^{c,loc}(\Omega, \mathfrak{F}, \{\mathfrak{F}_t\}, P)$. By Lemma 14.15, $X^{[T_n]} \in \mathbf{L}_{2,\infty}(\mathbb{R}_+ \times \Omega, \mu_{[M^{S_n\wedge}]}, P)$. Since $\mu_{[M^{T\wedge S_n \wedge}]} \leq \mu_{[M^{S_n\wedge}]}$ by Lemma 12.23, we have $X^{[T_n]} \in \mathbf{L}_{2,\infty}(\mathbb{R}_+ \times \Omega, \mu_{[M^{T\wedge S_n\wedge}]}, P)$. Thus $X^{[T_n]} \bullet M^{T\wedge S_n \wedge} \in \mathbf{M}_2^c(\Omega, \mathfrak{F}, \{\mathfrak{F}_t\}, P)$ and according to Theorem 14.16 there exists a null set Λ_2 in $(\Omega, \mathfrak{F}, P)$ such that

$$(2) \qquad X \bullet M^{T\wedge} = \lim_{n \to \infty} X^{[T_n]} \bullet M^{T \wedge S_n \wedge} \quad \text{on } \mathbb{R}_+ \times \Lambda_2^c.$$

By Theorem 12.22 there exists a null set Λ_3 in $(\Omega, \mathfrak{F}, P)$ such that for every $n \in \mathbb{N}$

$$(3) \qquad X^{[T_n]} \bullet M^{T \wedge S_n \wedge} = (X^{[T_n]} \bullet M^{S_n \wedge})^{\wedge T} \quad \text{on } \mathbb{R}_+ \times \Lambda_3^c.$$

Let $\Lambda = \cup_{i=1}^3 \Lambda_i$. Then $X \bullet M^{T\wedge} = \lim_{n \to \infty}(X^{[T_n]} \bullet M^{S_n \wedge})^{T\wedge} = (X \bullet M)^{T\wedge}$ on $\mathbb{R}_+ \times \Lambda^c$ by (2), (3) and (1). ∎

Theorem 14.19. *Let $M \in \mathbf{M}_2^{c,loc}(\Omega, \mathfrak{F}, \{\mathfrak{F}_t\}, P)$ and let X be a predictable process on the filtered space such that $X \in \mathbf{L}_{2,\infty}^{loc}(\mathbb{R}_+ \times \Omega, \mu_{[M]}, P)$. Let T be a stopping time defined by*

$$T(\omega) = \inf\{t \in \mathbb{R}_+ : \int_{[0,t]} X^2(s,\omega)\mu_{[M]}(ds,\omega) \geq a\} \wedge a \quad \text{for } \omega \in \Omega$$

where $a > 0$ and let S be an arbitrary stopping time. Then there exists a null set Λ in $(\Omega, \mathfrak{F}, P)$ such that for every $\omega \in \Lambda^c$ we have $(X^{[T \wedge S]} \bullet M)(\cdot, \omega) = (X \bullet M)^{T \wedge S \wedge}(\cdot, \omega)$ and in particular $(X^{[T]} \bullet M)(\cdot, \omega) = (X \bullet M)^{T \wedge}(\cdot, \omega)$.

Proof. Since $M \in \mathbf{M}_2^{c,loc}(\Omega, \mathfrak{F}, \{\mathfrak{F}_t\}, P)$ there exists an increasing sequence of stopping times $\{S_n : n \in \mathbb{N}\}$ such that $S_n \uparrow \infty$ on Λ_1^c where Λ_1 is a null set in $(\Omega, \mathfrak{F}, P)$ and $M^{S_n \wedge} \in \mathbf{M}_2^c(\Omega, \mathfrak{F}, \{\mathfrak{F}_t\}, P)$ for $n \in \mathbb{N}$. Let $\{T_n : n \in \mathbb{N}\}$ be as defined in Theorem 14.16. By Lemma 14.15, $X^{[T]}, X^{[T_n]} \in \mathbf{L}_{2,\infty}(\mathbb{R}_+ \times \Omega, \mu_{[M^{S_k \wedge}]}, P)$ for $n, k \in \mathbb{N}$. Note that for sufficiently large $n \in \mathbb{N}$ we have $T \leq T_n$ so that $T \wedge S \leq T_n$ and thus $X^{[T \wedge S]} = (X^{[T \wedge S]})^{[T_n]} = (X^{[T_n]})^{[T \wedge S]}$. By this fact and by Theorem 12.22 there exists a null set Λ_2 in $(\Omega, \mathfrak{F}, P)$ such that for sufficiently large $n \in \mathbb{N}$ we have

(1) $\quad X^{[T \wedge S]} \bullet M^{S_n \wedge} = (X^{[T_n]})^{[T \wedge S]} \bullet M^{S_n \wedge} = (X^{[T_n]} \bullet M^{S_n \wedge})^{T \wedge S \wedge} \quad$ on $\mathbb{R}_+ \times \Lambda_2^c$.

Since $X^{[T]} \in \mathbf{L}_{2,\infty}(\mathbb{R}_+ \times \Omega, \mu_{[M^{S_n \wedge}]}, P)$, we have $X^{[T]} \in \mathbf{L}_{2,\infty}^{loc}(\mathbb{R}_+ \times \Omega, \mu_{[M^{S_n \wedge}]}, P)$ and this implies that $X^{[T \wedge S]} \in \mathbf{L}_{2,\infty}^{loc}(\mathbb{R}_+ \times \Omega, \mu_{[M^{S_n \wedge}]}, P)$. Then by Theorem 14.18, there exists a null set Λ_3 in $(\Omega, \mathfrak{F}, P)$ such that for all $n \in \mathbb{N}$ we have

(2) $\quad\quad\quad\quad X^{[T \wedge S]} \bullet M^{S_n \wedge} = (X^{[T \wedge S]} \bullet M)^{S_n \wedge} \quad$ on $\mathbb{R}_+ \times \Lambda_3^c$.

On the other hand by Theorem 14.16 there exists a null set Λ_3 in $(\Omega, \mathfrak{F}, P)$ such that for all $n \in \mathbb{N}$ we have

(3) $\quad\quad\quad\quad \lim_{n \to \infty} (X^{[T_n]} \bullet M^{S_n \wedge})^{T \wedge S \wedge} = (X \bullet M)^{T \wedge S \wedge} \quad$ on $\mathbb{R}_+ \times \Lambda_4^c$.

Let $\Lambda = \cup_{i=1}^4 \Lambda_i$. Then letting $n \to \infty$ in (1) and using (2) and (3) we have $X^{[T \wedge S]} \bullet M = (X \bullet M)^{T \wedge S \wedge}$ on $\mathbb{R}_+ \times \Lambda^c$. In particular with $S = \infty$, we have $X^{[T]} \bullet M = (X \bullet M)^{T \wedge}$. ∎

For the quadratic variation process of the extended stochastic integral we have the following.

Theorem 14.20. *Let $M, N \in \mathbf{M}_2^{c,loc}(\Omega, \mathfrak{F}, \{\mathfrak{F}_t\}, P)$ and let X and Y be predictable processes on the filtered space such that $X \in \mathbf{L}_{2,\infty}^{loc}(\mathbb{R}_+ \times \Omega, \mu_{[M]}, P)$ and $Y \in \mathbf{L}_{2,\infty}^{loc}(\mathbb{R}_+ \times \Omega, \mu_{[N]}, P)$. Then there exists a null set Λ in $(\Omega, \mathfrak{F}, P)$ such that on Λ^c we have for every $t \in \mathbb{R}_+$*

(1) $\quad\quad\quad\quad [X \bullet M, Y \bullet N]_t = \int_{[0,t]} X(s) Y(s) \, d[M, N](s),$

§14. *EXTENSIONS OF THE STOCHASTIC INTEGRAL* 315

and in particular

(2) $$[X \bullet M]_t = \int_{[0,t]} X^2(s)\, d[M](s).$$

Proof. Since $M, N \in \mathbf{M}_2^{c,loc}(\Omega, \mathfrak{F}, \{\mathfrak{F}_t\}, P)$, there exists by Lemma 14.3 an increasing sequence of stopping times $\{S_n : n \in \mathbb{N}\}$ such that $S_n \uparrow \infty$ on Λ_0^c where Λ_0 is a null set in $(\Omega, \mathfrak{F}, P)$ and $M^{S_n \wedge}, N^{S_n \wedge} \in \mathbf{M}_2^c(\Omega, \mathfrak{F}, \{\mathfrak{F}_t\}, P)$ for $n \in \mathbb{N}$. Let $\{a_n : n \in \mathbb{N}\}$ be a strictly increasing sequence in $\mathbb{R}_+$ such that $a_n \uparrow \infty$ as $n \to \infty$. Let $\{T_n : n \in \mathbb{N}\}$ and $\{T_n' : n \in \mathbb{N}\}$ be two increasing sequences of stopping times defined by

$$T_n(\omega) = \inf\{t \in \mathbb{R}_+ : \int_{[0,t]} X^2(s, \omega)\mu_{[M]}(ds, \omega) \geq a_n\} \wedge a_n \quad \text{for } \omega \in \Omega,$$

and

$$T_n'(\omega) = \inf\{t \in \mathbb{R}_+ : \int_{[0,t]} Y^2(s, \omega)\mu_{[N]}(ds, \omega) \geq a_n\} \wedge a_n \quad \text{for } \omega \in \Omega.$$

By Lemma 14.15, $T_n \uparrow \infty$ and $T_n' \uparrow \infty$ on Λ_1^c where Λ_1 is a null set in $(\Omega, \mathfrak{F}, P)$ and furthermore $X^{[T_n]} \in \mathbf{L}_{2,\infty}(\mathbb{R}_+ \times \Omega, \mu_{[M^{S_n \wedge}]}, P)$ and $Y^{[T_n']} \in \mathbf{L}_{2,\infty}(\mathbb{R}_+ \times \Omega, \mu_{[N^{S_n \wedge}]}, P)$ for $n \in \mathbb{N}$. If we let $R_n = T_n \wedge T_n'$ then $\{R_n : n \in \mathbb{N}\}$ is an increasing sequence of stopping times, $R_n \uparrow \infty$ on Λ_1^c, $X^{[R_n]} \in \mathbf{L}_{2,\infty}(\mathbb{R}_+ \times \Omega, \mu_{[M^{S_n \wedge}]}, P)$, and $Y^{[R_n]} \in \mathbf{L}_{2,\infty}(\mathbb{R}_+ \times \Omega, \mu_{[N^{S_n \wedge}]}, P)$. Thus $X^{[R_n]} \bullet M^{S_n \wedge}, Y^{[R_n]} \bullet N^{S_n \wedge} \in \mathbf{M}_2^c(\Omega, \mathfrak{F}, \{\mathfrak{F}_t\}, P)$ and by Theorem 12.17 there exists a null set Λ_2 in $(\Omega, \mathfrak{F}, P)$ such that on Λ_2^c we have for every $t \in \mathbb{R}_+$ and every $n \in \mathbb{N}$

(3) $$[X^{[R_n]} \bullet M^{S_n \wedge}, Y^{[R_n]} \bullet N^{S_n \wedge}]_t = \int_{[0,t]} X^{[R_n]}(s) Y^{[R_n]}(s)\, d[M^{S_n \wedge}, N^{S_n \wedge}](s).$$

Now

(4) $$[X^{[R_n]} \bullet M^{S_n \wedge}, Y^{[R_n]} \bullet N^{S_n \wedge}]_t = [(X^{[T_n]})^{[T_n']} \bullet M^{S_n \wedge}, (Y^{[T_n']})^{[T_n]} \bullet N^{S_n \wedge}]_t$$
$$= [(X^{[T_n]} \bullet M^{S_n \wedge})^{T_n' \wedge}, (Y^{[T_n']} \bullet N^{S_n \wedge})^{T_n \wedge}]_t \quad \text{by Theorem 12.22}$$
$$= [(X \bullet M)^{T_n' \wedge S_n \wedge T_n \wedge}, (Y^{[T_n']} \bullet N)^{T_n \wedge S_n \wedge T_n' \wedge}]_t \quad \text{by Theorems 14.18 and 14.19}$$
$$= [(X \bullet M)^{R_n \wedge S_n \wedge}, (Y \bullet N)^{R_n \wedge S_n \wedge}]_t$$

on $\mathbb{R}_+ \times \Lambda_3^c$ for all $n \in \mathbb{N}$ where Λ_3 is a null set in $(\Omega, \mathfrak{F}, P)$. Now since $X \bullet M, Y \bullet N$ is in $\mathbf{M}_2^{c,loc}(\Omega, \mathfrak{F}, \{\mathfrak{F}_t\}, P)$, according to Theorem 14.11 there exists a null set Λ_4 in $(\Omega, \mathfrak{F}, P)$ such that on Λ_4^c and for every $t \in \mathbb{R}_+$ we have

(5) $$\lim_{n \to \infty} [(X \bullet M)^{R_n \wedge S_n \wedge}, (Y \bullet N)^{R_n \wedge S_n \wedge}]_t = [X \bullet M, Y \bullet N]_t$$

and

(6) $$\lim_{n\to\infty} \mu_{[M^{S_n\wedge},N^{S_n\wedge}]}(E,\omega) = \mu_{[M,N]}(E,\omega) \quad \text{for } E \in \mathfrak{B}_{[0,t]}.$$

Let $\Lambda = \cup_{i=0}^{4}\Lambda_i$. For $\omega \in \Lambda^c$ we have $(R_n \wedge S_n)(\omega) > t$ for sufficiently large n since $(R_n \wedge S_n)(\omega) \uparrow \infty$ for any $t \in \mathbb{R}_+$. Thus such n we have

$$\int_{[0,t]} X^{[R_n]}(s,\omega) Y^{[R_n]}(s,\omega)\, d[M^{S_n\wedge}, N^{S_n\wedge}](s,\omega) = \int_{[0,t]} X(s,\omega)Y(s,\omega)\, d[M,N](s,\omega).$$

Letting $n \to \infty$ in (3) and applying (5) and the last equality we have (1). We have (2) as a particular case of (1). ∎

Theorem 14.21. *Let $M \in \mathbf{M}_2^{c,loc}(\Omega, \mathfrak{F}, \{\mathfrak{F}_t\}, P)$ and let X and Y be two predictable processes on the filtered space such that X and YX are in $\mathbf{L}_{2,\infty}^{loc}(\mathbb{R}_+ \times \Omega, \mu_{[M]}, P)$. Then $X \bullet M \in \mathbf{M}^{c,loc}(\Omega, \mathfrak{F}, \{\mathfrak{F}_t\}, P)$, $Y \in \mathbf{L}_{2,\infty}^{loc}(\mathbb{R}_+ \times \Omega, \mu_{[X\bullet M]}, P)$ so that $Y \bullet (X \bullet M)$ exists in $\mathbf{M}^{c,loc}(\Omega, \mathfrak{F}, \{\mathfrak{F}_t\}, P)$ and furthermore there exists a null set Λ in $(\Omega, \mathfrak{F}, P)$ such that for every $\omega \in \Lambda^c$ we have $(Y \bullet (X \bullet M))(\cdot, \omega) = (YX \bullet M)(\cdot, \omega)$.*

Proof. Let us show that $Y \in \mathbf{L}_{2,\infty}^{loc}(\mathbb{R}_+ \times \Omega, \mu_{[X\bullet M]}, P)$. By Theorem 14.20, there exists a null set Λ_1 in $(\Omega, \mathfrak{F}, P)$ such that $[X \bullet M](t,\omega) = \int_{[0,t]} X^2(s,\omega)\, d[M](s,\omega)$ for $t \in \mathbb{R}_+$ when $\omega \in \Lambda_1^c$ so that $\mu_{[X\bullet M]}(\cdot, \omega)$ is absolutely continuous with respect to $\mu_{[M]}(\cdot, \omega)$ on $(\mathbb{R}_+, \mathfrak{B}_{\mathbb{R}_+})$ with Radon-Nikodym derivative given by $(d\mu_{[X\bullet M]}/d\mu_{[M]})(s,\omega) = X^2(s,\omega)$ for $s \in \mathbb{R}_+$. Thus by the Lebesgue-Radon-Nikodym Theorem we have

$$\int_{[0,t]} Y^2(s,\omega)\, d[X \bullet M](s,\omega) = \int_{[0,t]} Y^2(s,\omega) X^2(s,\omega)\, d[M](s,\omega) < \infty,$$

the finiteness of the last integral being from the fact that $YX \in \mathbf{L}_{2,\infty}^{loc}(\mathbb{R}_+ \times \Omega, \mu_{[M]}, P)$. This shows that $Y \in \mathbf{L}_{2,\infty}^{loc}(\mathbb{R}_+ \times \Omega, \mu_{[X\bullet M]}, P)$.

Let $\{a_n : n \in \mathbb{N}\}$ be a strictly increasing sequence in $\mathbb{R}_+$ such that $a_n \uparrow \infty$ as $n \to \infty$. Let $\{T_n : n \in \mathbb{N}\}$ and $\{T_n' : n \in \mathbb{N}\}$ be two increasing sequences of stopping times defined by

$$T_n(\omega) = \inf\{t \in \mathbb{R}_+ : \int_{[0,t]} X^2(s,\omega) \mu_{[M]}(ds,\omega) \geq a_n\} \wedge a_n \quad \text{for } \omega \in \Omega,$$

$$T_n'(\omega) = \inf\{t \in \mathbb{R}_+ : \int_{[0,t]} Y^2(s,\omega) \mu_{[X\bullet M]}(ds,\omega) \geq a_n\} \wedge a_n$$

$$= \inf\{t \in \mathbb{R}_+ : \int_{[0,t]} Y^2(s,\omega) X^2(s,\omega) \mu_{[M]}(ds,\omega) \geq a_n\} \wedge a_n \quad \text{for } \omega \in \Omega.$$

§14. EXTENSIONS OF THE STOCHASTIC INTEGRAL

Then $T_n \uparrow \infty$ and $T'_n \uparrow \infty$ almost surely. Since both M and $X \bullet M$ are in $\mathbf{M}_2^{c,loc}(\Omega, \mathfrak{F}, \{\mathfrak{F}_t\}, P)$ there exists by Lemma 14.3 an increasing sequence of stopping times $\{S_n : n \in \mathbb{N}\}$ such that $S_n \uparrow \infty$ almost surely and both $M^{S_n \wedge}$ and $(X \bullet M)^{S_n \wedge}$ are in $\mathbf{M}_2^c(\Omega, \mathfrak{F}, \{\mathfrak{F}_t\}, P)$ for every $n \in \mathbb{N}$. Thus $(X \bullet M)^{S_n \wedge T_n \wedge T'_n \wedge} \in \mathbf{M}_2^c(\Omega, \mathfrak{F}, \{\mathfrak{F}_t\}, P)$ for every $n \in \mathbb{N}$. Now

$$
\begin{aligned}
(1) \quad Y \bullet (X \bullet M) &= \lim_{n \to \infty} Y^{[T'_n]} \bullet (X \bullet M)^{S_n \wedge T_n \wedge} \quad \text{by Theorem 14.16} \\
&= \lim_{n \to \infty} Y^{[T'_n]} \bullet (X^{[T_n]} \bullet M^{S_n \wedge}) \quad \text{by Theorem 14.19 and Theorem 14.18}
\end{aligned}
$$

on $\mathbb{R}_+ \times \Lambda_2^c$ where Λ_2 is a null set in $(\Omega, \mathfrak{F}, P)$. Since X is in $\mathbf{L}_{2,\infty}^{loc}(\mathbb{R}_+ \times \Omega, \mu_{[M]}, P)$ and $\mu_{[M^{S_n \wedge}]} \leq \mu_{[M]}$, we have $X \in \mathbf{L}_{2,\infty}^{loc}(\mathbb{R}_+ \times \Omega, \mu_{[M^{S_n \wedge}]}, P)$. This implies that $X^{[T_n]}$ is in $\mathbf{L}_{2,\infty}^{loc}(\mathbb{R}_+ \times \Omega, \mu_{[M^{S_n \wedge}]}, P)$. Similarly since YX is in $\mathbf{L}_{2,\infty}^{loc}(\mathbb{R}_+ \times \Omega, \mu_{[M]}, P)$, we have YX in $\mathbf{L}_{2,\infty}^{loc}(\mathbb{R}_+ \times \Omega, \mu_{[M^{S_n \wedge}]}, P)$ and thus $Y^{[T_n]} X^{[T'_n]}$ is in $\mathbf{L}_{2,\infty}^{loc}(\mathbb{R}_+ \times \Omega, \mu_{[M^{S_n \wedge}]}, P)$. Therefore by Theorem 12.19 there exists a null set Λ_3 in $(\Omega, \mathfrak{F}, P)$ such that on $\mathbb{R}_+ \times \Lambda_3^c$ we have for all $n \in \mathbb{N}$

$$
(2) \quad Y^{[T'_n]} \bullet (X^{[T_n]} \bullet M^{S_n \wedge}) = Y^{[T_n]} X^{[T'_n]} \bullet M^{S_n \wedge} = (YX)^{[T'_n \wedge T_n]} \bullet M^{S_n \wedge}.
$$

Therefore if we let $\Lambda = \cup_{i=1}^{3} \Lambda_i$ then on $\mathbb{R}_+ \times \Lambda^c$ we have by (1) and (2)

$$
\begin{aligned}
Y \bullet (X \bullet M) &= \lim_{n \to \infty} (YX)^{[T'_n \wedge T_n]} \bullet M^{S_n \wedge} = \lim_{n \to \infty} ((YX)^{[T'_n]} \bullet M^{S_n \wedge})^{T_n \wedge} \\
&= \lim_{n \to \infty} (YX)^{[T'_n]} \bullet M^{S_n \wedge T_n \wedge} = (YX) \bullet M,
\end{aligned}
$$

where the second equality is by Theorem 12.19, the third equality is by Theorem 12.24, and the last equality is by Theorem 14.16. ∎

Corollary 14.22. *Let $M \in \mathbf{M}_2^{c,loc}(\Omega, \mathfrak{F}, \{\mathfrak{F}_t\}, P)$ and let $X^{(1)}, \ldots, X^{(n)}$ be predictable processes such that $X^{(1)}, X^{(1)} X^{(2)}, \ldots, X^{(1)} \cdots X^{(n)} \in \mathbf{L}_{2,\infty}^{loc}(\mathbb{R}_+ \times \Omega, \mu_{[M]}, P)$. Then*

$$
(X^{(n)} \bullet \cdots \bullet (X^{(2)} \bullet (X^{(1)} \bullet M)) \cdots) = (X^{(1)} \cdots X^{(n)}) \bullet M.
$$

Proof. By iterated application of Theorem 14.21, we have the Corollary. ∎

§15 Itô's Formula

[I] Continuous Local Semimartingales and Itô's Formula

Definition 15.1. *By a continuous local semimartingale we mean a stochastic process $X = \{X_t : t \in \mathbb{R}_+\}$ on a standard filtered space $(\Omega, \mathfrak{F}, \{\mathfrak{F}_t\}, P)$ such that $X = X_0 + M + V$ where $M \in \mathbf{M}_2^{c,loc}(\Omega, \mathfrak{F}, \{\mathfrak{F}_t\}, P)$, $V \in \mathbf{V}^{c,loc}(\Omega, \mathfrak{F}, \{\mathfrak{F}_t\}, P)$ and X_0 is a real valued $\mathfrak{F}_0$-measurable random variable on $(\Omega, \mathfrak{F}, P)$, The processes M and V are called the martingale part and the bounded variation part of the continuous local semimartingale X. In particular, when X_0 is integrable, M is in $\mathbf{M}_2^c(\Omega, \mathfrak{F}, \{\mathfrak{F}_t\}, P)$ and V is in $\mathbf{V}^c(\Omega, \mathfrak{F}, \{\mathfrak{F}_t\}, P)$ then X is called a continuous semimartingale. A continuous local semimartingale is also called a quasimartingale.*

Proposition 15.2. *The decomposition of a quasimartingale X in Definition 15.1 is unique, that is, if $X = X_0 + M + V$ and $X = Y_0 + N + W$ where X_0 and Y_0 are real valued $\mathfrak{F}_0$-measurable random variables on $(\Omega, \mathfrak{F}, P)$, M and N are in $\mathbf{M}_2^{c,loc}(\Omega, \mathfrak{F}, \{\mathfrak{F}_t\}, P)$, and V and W are in $\mathbf{V}^{c,loc}(\Omega, \mathfrak{F}, \{\mathfrak{F}_t\}, P)$, then there exists a null set Λ in $(\Omega, \mathfrak{F}, P)$ such that $X_0(\omega) = Y_0(\omega)$, $M(\cdot, \omega) = N(\cdot, \omega)$ and $V(\cdot, \omega) = W(\cdot, \omega)$ for $\omega \in \Lambda^c$.*

Proof. Since M, N, V, and W are almost surely continuous processes, we may assume that they are continuous without loss of generality in the proof. Now since M is in $\mathbf{M}_2^{c,loc}(\Omega, \mathfrak{F}, \{\mathfrak{F}_t\}, P)$ there exists an increasing sequence of stopping times $\{T_n : n \in \mathbb{N}\}$ such that $T_n \uparrow$ almost surely and $M^{T_n \wedge}$ is a null at 0 continuous L_2-martingale for every $n \in \mathbb{N}$. By the continuity of M, the function S_n on Ω defined by

$$S_n(\omega) = \inf\{t \in \mathbb{R}_+ : |M(t,\omega)| \geq n\} = \inf\{t \in \mathbb{R}_+ : |M(t,\omega)| > n\},$$

for $\omega \in \Omega$ and $n \in \mathbb{N}$ is the first passage time of the open set (n, ∞) in $\mathbb{R}$ and is therefore a stopping time by Proposition 3.5. Clearly $\{S_n : n \in \mathbb{N}\}$ is an increasing sequence. Since $M(\cdot, \omega)$ is a real valued continuous function on $\mathbb{R}_+$, it is bounded on $[0, t]$ for every $t \in \mathbb{R}_+$. From this it follows immediately that $S_n(\omega) \uparrow \infty$ as $n \to \infty$ for every $\omega \in \Omega$. Since V is a continuous process, so is its total variation process $|V|$. Thus if we define

$$R_n(\omega) = \inf\{t \in \mathbb{R}_+ : |V|(t,\omega) \geq n\},$$

for $\omega \in \Omega$ and $n \in \mathbb{N}$, then $\{R_n : n \in \mathbb{N}\}$ is an increasing sequence of stopping times such that $R_n(\omega) \uparrow \infty$ as $n \to \infty$ for every $\omega \in \Omega$ for the same reason as for $\{S_n : n \in \mathbb{N}\}$. Let $\{T'_n : n \in \mathbb{N}\}$, $\{S'_n : n \in \mathbb{N}\}$, and $\{R'_n : n \in \mathbb{N}\}$ be defined in the same way for N and

§15. ITÔ'S FORMULA

W in the place of M and V. Let $Q_n = T_n \wedge S_n \wedge R_n \wedge T'_n \wedge S'_n \wedge R'_n$ for $n \in \mathbb{N}$. Then $\{Q_n : n \in \mathbb{N}\}$ is an increasing sequence of stopping times such that $Q_n(\omega) \uparrow \infty$ as $n \to \infty$ for $\omega \in \Lambda_1^c$ where Λ_1 is a null set in $(\Omega, \mathfrak{F}, P)$.

Since $M^{T_n \wedge}, N^{T'_n \wedge} \in \mathbf{M}_2^c(\Omega, \mathfrak{F}, \{\mathfrak{F}_t\}, P)$, we have $M^{Q_n \wedge}, N^{Q_n \wedge} \in \mathbf{M}_2^c(\Omega, \mathfrak{F}, \{\mathfrak{F}_t\}, P)$ by Lemma 14.3. By Definition 11.10, it is clear that $\|V^{Q_n \wedge}\| = \|V\|^{Q_n \wedge}$. Since $\|V\|^{Q_n \wedge}$ is bounded by n, so is $\|V^{Q_n \wedge}\|$. Thus $\|V^{Q_n \wedge}\|$ is an L_1-process and therefore $V^{Q_n \wedge}$ is in $\mathbf{V}^c(\Omega, \mathfrak{F}, \{\mathfrak{F}_t\}, P)$. Similarly $W^{Q_n \wedge}$ is in $\mathbf{V}^c(\Omega, \mathfrak{F}, \{\mathfrak{F}_t\}, P)$.

Now since $X_0 + M + V = Y_0 + N + W$ and M, N, V, and W are null at 0, $X_0 = Y_0$, that is, there exists a null set Λ_2 such that $X_0(\omega) = Y_0(\omega)$ for $\omega \in \Lambda_2^c$. Next, from $M + V = N + W$ we have $M - N = W_V$ and then

$$(1) \qquad M^{Q_n \wedge} - N^{Q_n \wedge} = W^{Q_n \wedge} - V^{Q_n \wedge}.$$

Since $V^{Q_n \wedge}$ and $W^{Q_n \wedge}$ are in $\mathbf{V}^c(\Omega, \mathfrak{F}, \{\mathfrak{F}_t\}, P)$, so is $W^{Q_n \wedge} - V^{Q_n \wedge}$. Then there exist A and B in $\mathbf{A}^c(\Omega, \mathfrak{F}, \{\mathfrak{F}_t\}, P)$ such that $W^{Q_n \wedge} - V^{Q_n \wedge} = B - A$. Then

$$M^{Q_n \wedge} + A = N^{Q_n \wedge} + B.$$

Now $M^{Q_n \wedge}$ and $N^{Q_n \wedge}$ are continuous martingales bounded by n so that they are uniformly integrable and thus are in the class (D) by Theorem 8.22. On the other hand A and B are continuous nonnegative submartingales so that they are in the class (DL). Thus $M^{Q_n \wedge} + A$ and $N^{Q_n \wedge} + B$ are continuous submartingales of class (DL). Note also that since A and B are continuous increasing processes, they are natural by Theorem 10.18. Therefore by the uniqueness of Doob-Meyer Decomposition in Theorem 10.23, we have $M^{Q_n \wedge} = N^{Q_n \wedge}$. From this and (1), we have $V^{Q_n \wedge} = W^{Q_n \wedge}$ also. This implies that there exists a null set Λ_3 such that

$$(2) \qquad \begin{cases} M^{Q_n \wedge}(\cdot, \omega) = N^{Q_n \wedge}(\cdot, \omega) & \text{for } n \in \mathbb{N} \text{ and } \omega \in \Lambda_3^c \\ V^{Q_n \wedge}(\cdot, \omega) = W^{Q_n \wedge}(\cdot, \omega) & \text{for } n \in \mathbb{N} \text{ and } \omega \in \Lambda_3^c. \end{cases}$$

Let $\Lambda = \cup_{i=1}^3 \Lambda_i$. Since $Q_n(\omega) \uparrow \infty$ as $n \to \infty$ for $\omega \in \Lambda^c$, we have

$$\lim_{n \to \infty} M^{Q_n \wedge}(t, \omega) = \lim_{n \to \infty} M(Q_n(\omega) \wedge t, \omega) = M(t, \omega) \quad \text{for } (t, \omega) \in \mathbb{R}_+ \times \Lambda^c.$$

Similarly for $N^{T_n \wedge}, V^{T_n \wedge}$, and $W^{T_n \wedge}$. Therefore letting $n \to \infty$ in (2), we have $M(\cdot, \omega) = N(\cdot, \omega)$ and $V(\cdot, \omega) = W(\cdot, \omega)$ for $\omega \in \Lambda^c$. We have also $X_0(\omega) = Y_0(\omega)$ for $\omega \in \Lambda^c$. ∎

Let $\mathbf{C}^2(\mathbb{R})$ be the collection of all real valued continuous functions F with continuous derivatives F' and F'' on $\mathbb{R}$. Let $X = X_0 + M + V$ be a quasimartingale on a standard filtered space $(\Omega, \mathfrak{F}, \{\mathfrak{F}_t\}, P)$. Consider the real valued function $F \circ X$ on $\mathbb{R}_+ \times \Omega$. We have $(F \circ X)_t = (F \circ X)(t, \cdot) = F \circ X_t$ and therefore $F \circ X = \{(F \circ X)_t : t \in \mathbb{R}_+\} = \{F \circ X_t : t \in \mathbb{R}_+\}$. Similarly for $F' \circ X$ and $F'' \circ X$. Itô's Formula shows that $F \circ X$ is again a quasimartingale and gives its martingale part and bounded variation part in terms of M, $[M]$ and V.

Theorem 15.3. (Itô's Formula) *Let $X = X_0 + M + V$ be a quasimartingale on a standard filtered space $(\Omega, \mathfrak{F}, \{\mathfrak{F}_t\}, P)$ and let $F \in \mathbf{C}^2(\mathbb{R})$. Then for the real valued function $F \circ X$ on $\mathbb{R}_+ \times \Omega$ there exists a null set Λ in $(\Omega, \mathfrak{F}, P)$ such that on Λ^c we have for every $t \in \mathbb{R}_+$*

$$(F \circ X)_t - (F \circ X)_0$$
$$= \int_{[0,t]} (F' \circ X)(s) \, dM(s) + \int_{[0,t]} (F' \circ X)(s) \, dV(s) + \frac{1}{2} \int_{[0,t]} (F'' \circ X)(s) \, d[M](s).$$

Recall that $\int_{[0,t]} (F' \circ X)(s) \, dM(s)$ is an alternate notation for $(F' \circ X \bullet M)_t$ for $t \in \mathbb{R}_+$. Since $M \in \mathbf{M}_2^{c,loc}(\Omega, \mathfrak{F}, \{\mathfrak{F}_t\}, P)$ and $V \in \mathbf{V}^{c,loc}(\Omega, \mathfrak{F}, \{\mathfrak{F}_t\}, P)$, there exists a null set Λ in $(\Omega, \mathfrak{F}, P)$ such that $M(\cdot, \omega)$ and $V(\cdot, \omega)$ are continuous on $\mathbb{R}_+$ for $\omega \in \Lambda^c$. If we redefine M and V by setting $M(\cdot, \omega) = 0$ and $V(\cdot, \omega) = 0$ for $\omega \in \Lambda$, then since the filtered space is augmented M and V are still in $\mathbf{M}_2^{c,loc}(\Omega, \mathfrak{F}, \{\mathfrak{F}_t\}, P)$ and $\mathbf{V}^{c,loc}(\Omega, \mathfrak{F}, \{\mathfrak{F}_t\}, P)$ respectively but now every sample function of M and of V is continuous. Thus we assume that our M and V are not only almost surely continuous but are in fact continuous processes. For $[M] \in \mathbf{A}^{c,loc}(\Omega, \mathfrak{F}, \{\mathfrak{F}_t\}, P)$, we select a representative which is a continuous process.

Before we prove Itô's Formula, we show in Proposition 15.5 below that the stochastic integral $\{\int_{[0,t]} (F' \circ X)(s) \, dM(s) : t \in \mathbb{R}_+\}$ is defined and is in $\mathbf{M}_2^{c,loc}(\Omega, \mathfrak{F}, \{\mathfrak{F}_t\}, P)$ and the two processes $\{\int_{[0,t]} (F' \circ X)(s) \, dV(s) : t \in \mathbb{R}_+\}$ and $\{\int_{[0,t]} (F'' \circ X)(s) \, d[M](s) : t \in \mathbb{R}_+\}$ are defined and are in $\mathbf{V}^{c,loc}(\Omega, \mathfrak{F}, \{\mathfrak{F}_t\}, P)$. Then since $(F \circ X)_0$ is a real valued $\mathfrak{F}_0$-measurable random variable, $F \circ X$, as given by Itô's Formula, is indeed a quasimartingale.

Observation 15.4. 1) If $M \in \mathbf{M}_2^{c,loc}(\Omega, \mathfrak{F}, \{\mathfrak{F}_t\}, P)$ and $V \in \mathbf{V}^{c,loc}(\Omega, \mathfrak{F}, \{\mathfrak{F}_t\}, P)$, then M and V are right-continuous processes so that they are measurable processes by Theorem 2.10. Since M and V are also adapted processes, they are progressively measurable processes by Proposition 2.13.
2) Since F, F', and F'' are real valued continuous functions on $\mathbb{R}$, they are $\mathfrak{B}_\mathbb{R}/\mathfrak{B}_\mathbb{R}$-measurable mapping of $\mathbb{R}$ into $\mathbb{R}$. Since $X = X_0 + M + V$ is a progressively measurable process, that is, when restricted to $[0, t] \times \Omega$ the mapping X is a $\sigma(\mathfrak{B}_{[0,t]} \times \mathfrak{F})/\mathfrak{B}_\mathbb{R}$-measurable mapping of $[0, t] \times \Omega$ into $\mathbb{R}$, $F \circ X$, $F' \circ X$, and $F'' \circ X$ are progressively measurable

§15. ITÔ'S FORMULA

processes and in particular measurable processes.

3) Since $X(\cdot,\omega)$ is a real valued continuous function on $\mathbb{R}_+$ for every $\omega \in \Omega$, and since F, F', and F''' are real valued continuous functions on $\mathbb{R}$, $(F \circ X)(\cdot,\omega)$, $(F' \circ X)(\cdot,\omega)$, and $(F''' \circ X)(\cdot,\omega)$ are real valued continuous functions on $\mathbb{R}_+$. This implies that every sample function of $(F \circ X)$, $(F' \circ X)$, and $(F''' \circ X)$ is bounded on every finite interval in $\mathbb{R}_+$.

4) Since $(F' \circ X)$ and $(F''' \circ X)$ are measurable processes and since every sample function of these processes is bounded on every finite interval in $\mathbb{R}_+$, $\int_{[0,t]}(F' \circ X)(s)\,dV(s)$ and $\int_{[0,t]}(F'''\circ X)(s)\,d[M](s)$ are real valued $\mathfrak{F}_t$-measurable random variables by Theorem 10.11. Thus $\{\int_{[0,t]}(F'\circ X)(s)\,dV(s) : t \in \mathbb{R}_+\}$ and $\{\int_{[0,t]}(F'''\circ X)(s)\,d[M](s) : t \in \mathbb{R}_+\}$ are adapted processes. (Note that our $V \in \mathbf{V}^{c,loc}(\Omega,\mathfrak{F},\{\mathfrak{F}_t\},P)$ and $[M] \in \mathbf{A}^{c,loc}(\Omega,\mathfrak{F},\{\mathfrak{F}_t\},P)$ are not L_1-processes. Nevertheless Theorem 10.11 is still applicable since the fact that the process A in Theorem 10.11 is an L_1-process was not needed in the proof of Theorem 10.11.)

Proposition 15.5. *1) The two processes* $\{\int_{[0,t]}(F' \circ X)(s)\,dV(s) : t \in \mathbb{R}_+\}$ *and* $\{\int_{[0,t]}(F''' \circ X)(s)\,d[M](s) : t \in \mathbb{R}_+\}$ *are in* $\mathbf{V}^{c,loc}(\Omega,\mathfrak{F},\{\mathfrak{F}_t\},P)$.
2) The process $F' \circ X$ is in $\mathbf{L}_{2,\infty}^{loc}(\mathbb{R}_+ \times \Omega, \mu_{[M]}, P)$ *so that the stochastic integral* $\{\int_{[0,t]}(F' \circ X)(s)\,dM(s) : t \in \mathbb{R}_+\}$ *is defined and is in* $\mathbf{M}_2^{c,loc}(\Omega,\mathfrak{F},\{\mathfrak{F}_t\},P)$.

Proof. 1) Let us consider the process $\Phi = \{\Phi_t : t \in \mathbb{R}_+\}$ where

$$\Phi(t,\omega) = \int_{[0,t]} (F' \circ X)(s,\omega)\,dV(s,\omega) \quad \text{for } (t,\omega) \in \mathbb{R}_+ \times \Omega.$$

As we saw in Observation 15.4, Φ is an adapted process. Also since every sample function of $F' \circ X$ and V is continuous, $\Phi(\cdot,\omega)$ as given above is a continuous function on $\mathbb{R}_+$ for every $\omega \in \Omega$. It remains to show that $\Phi(\cdot,\omega)$ is of bounded variation on $[0,t]$ for every $t \in \mathbb{R}_+$. Let $\omega \in \Omega$ and $t \in \mathbb{R}_+$ be fixed. Since $(F' \circ X)(\cdot,\omega)$ is continuous on $\mathbb{R}_+$, we have $C \equiv \sup_{s \in [0,t]} |(F' \circ X)(s,\omega)| < \infty$. Let $\mathcal{T}$ be the collection of all finite strictly increasing sequences $\tau = \{t_0, \ldots, t_n\}$ with $t_0 = 0$ and $t_n = t$. Then

$$\begin{aligned}
\Delta_\tau &\equiv \sum_{k=1}^n |\Phi(t_k,\omega) - \Phi(t_{k-1},\omega)| \\
&= \sum_{k=1}^n \left| \int_{[0,t_k]} (F' \circ X)(s,\omega)\,dV(s,\omega) - \int_{[0,t_{k-1}]} (F' \circ X)(s,\omega)\,dV(s,\omega) \right| \\
&\leq \sum_{k=1}^n \int_{[t_{k-1},t_k]} |(F' \circ X)(s,\omega)|\,d|V|(s,\omega) \\
&\leq C|V|(t,\omega) < \infty.
\end{aligned}$$

Thus $\sup_{\tau \in T_t} \Delta_\tau < \infty$. This shows that $\Phi(\cdot, \omega)$ is of bounded variation on every finite interval in $\mathbb{R}_+$ and completes the proof that Φ is in $\mathbf{V}^{c,loc}(\Omega, \mathfrak{F}, \{\mathfrak{F}_t\}, P)$.

We show similarly that Ψ is in $\mathbf{V}^{c,loc}(\Omega, \mathfrak{F}, \{\mathfrak{F}_t\}, P)$ for the process $\Psi = \{\Psi_t : t \in \mathbb{R}_+\}$ defined by

$$\Psi(t, \omega) = \int_{[0,t]} (F'' \circ X)(s, \omega) \, d[M](s, \omega) \quad \text{for } (t, \omega) \in \mathbb{R}_+ \times \Omega.$$

2) For every $\omega \in \Omega$, $(F' \circ X)(\cdot, \omega)$ continuous and hence bounded on every finite interval in $\mathbb{R}_+$ so that for every $t \in \mathbb{R}_+$ we have $\int_{[0,t]} (F' \circ X)^2(s, \omega) \mu_{[M]}(ds, \omega) < \infty$. Thus $F' \circ X$ is in $\mathbf{L}_{2,\infty}^{loc}(\mathbb{R}_+ \times \Omega, \mu_{[M]}, P)$ and consequently the stochastic integral $\{\int_{[0,t]} (F' \circ X)(s) \, dM(s) : t \in \mathbb{R}_+\}$ is defined and is in $\mathbf{M}_2^{c,loc}(\Omega, \mathfrak{F}, \{\mathfrak{F}_t\}, P)$ by Definition 14.17. ∎

Lemma 15.6. *Let $M = \{M_t : t \in \mathbb{R}_+\}$ be a null at 0 martingale on a filtered space $(\Omega, \mathfrak{F}, \{\mathfrak{F}_t\}, P)$ such that $|M| \leq K$ on $[0, t] \times \Omega$ for some $t \in \mathbb{R}_+$ and $K \geq 0$. For $0 = t_0 < \cdots < t_n = t$, let $S = \sum_{j=1}^n |M(t_j) - M(t_{j-1})|$. Then $\mathbf{E}(S) \leq K^2$ and $\mathbf{E}(S^2) \leq (1 + K^2)K^2$.*

Proof. For brevity, let $\xi_j = M(t_j)$, $\gamma_j = |\xi_j - \xi_{j-1}|$ and $\alpha_j = \gamma_j^2 = \{\xi_j - \xi_{j-1}\}^2$ for $j = 1, \ldots, n$. Then $S = \sum_{j=1}^n \gamma_j$. By Hölder's Inequality and by the martingale property of M, we have

(1) $$\mathbf{E}(\gamma_j) \leq \mathbf{E}(\alpha_j) = \mathbf{E}[\mathbf{E}[\{M(t_j) - M(t_{j-1})\}^2 | \mathfrak{F}_{t_{j-1}}]]$$
$$= \mathbf{E}[\mathbf{E}[M(t_j)^2 - M(t_{j-1})^2 | \mathfrak{F}_{t_{j-1}}]] = \mathbf{E}[M(t_j)^2 - M(t_{j-1})^2].$$

Thus

(2) $$\mathbf{E}(S) \leq \sum_{j=1}^n \mathbf{E}[M(t_j)^2 - M(t_{j-1})^2] = \mathbf{E}[M(t)^2] \leq K^2.$$

To estimate $\mathbf{E}(S^2)$, let us write

(3) $$S^2 = \{\sum_{j=1}^n \gamma_j\}^2 = \sum_{j=1}^n \gamma_j^2 + 2 \sum_{i=1}^n \gamma_i \{\sum_{j=i+1}^n \gamma_j\}.$$

By the computations in (1) and (2), we have

(4) $$\sum_{j=1}^n \mathbf{E}(\gamma_j^2) = \sum_{j=1}^n \mathbf{E}(\alpha_j) \leq K^2.$$

§15. ITÔ'S FORMULA

By the Conditional Hölder Inequality and by (3) in the proof of Lemma 12.26 we have

$$\sum_{j=i+1}^n \mathbf{E}[\gamma_j | \mathfrak{F}_{t_i}] \leq \sum_{j=i+1}^n \mathbf{E}[\alpha_j | \mathfrak{F}_{t_i}] \leq (2K)^2 \quad \text{a.e. on } (\Omega, \mathfrak{F}_{t_i}, P)$$

and thus

(5) $$\mathbf{E}[\sum_{i=1}^n \gamma_i \{\sum_{j=i+1}^n \gamma_j\}] = \sum_{i=1}^n \mathbf{E}[\mathbf{E}[\gamma_i \{\sum_{j=i+1}^n \gamma_j\} | \mathfrak{F}_{t_i}]]$$
$$= \sum_{i=1}^n \mathbf{E}[\gamma_i \mathbf{E}[\sum_{j=i+1}^n \gamma_j | \mathfrak{F}_{t_i}]] \leq (2K)^2 \sum_{i=1}^n \mathbf{E}(\gamma_i) = (2K)^2 \mathbf{E}(S) \leq 4K^4.$$

By (3),(4), and (5), we have $\mathbf{E}(S^2) \leq (1 + 4K^2)K^2$. ∎

Proof of Theorem 15.3. Step 1. Since the process $F \circ X - (F \circ X)_0$ is continuous and since the three processes $\{\int_{[0,t]} (F' \circ X)(s) \, dV(s) : t \in \mathbb{R}_+\}$, $\{\frac{1}{2} \int_{[0,t]} (F'' \circ X)(s) \, d[M](s) : t \in \mathbb{R}_+\}$ and $\{\int_{[0,t]} (F' \circ X)(s) \, dM(s) : t \in \mathbb{R}_+\}$ are almost surely continuous according to Proposition 15.5, to show that $F \circ X - (F \circ X)_0$ is equivalent to the sum of the three processes it suffice to show according to Theorem 2.3 that for every $t \in \mathbb{R}_+$ there exists a null set Λ_t in $(\Omega, \mathfrak{F}, P)$ such that on Λ_t^c we have

(6) $(F \circ X)_t - (F \circ X)_0$
$$= \int_{[0,t]} (F' \circ X)(s) \, dM(s) + \int_{[0,t]} (F' \circ X)(s) \, dV(s) + \frac{1}{2} \int_{[0,t]} (F'' \circ X)(s) \, d[M](s).$$

Let us consider first the case where X_0, M, $[M]$ and $\|V\|$ are all bounded, that is there exists $K \geq 0$ such that

(2) $$|X_0(\omega)|, |M(t,\omega)|, [M](t,\omega), \|V\|(t,\omega) \leq K \quad \text{for } (t,\omega) \in \mathbb{R}_+ \times \Omega.$$

In this case, X_0 is an integrable random variable, M is a martingale by Observation 14.2 so that $M \in \mathbf{M}_2^c(\Omega, \mathfrak{F}, \{\mathfrak{F}_t\}, P)$, and $\|V\|$ is an L_1-process so that $V \in \mathbf{V}^c(\Omega, \mathfrak{F}, \{\mathfrak{F}_t\}, P)$. Note also that since $|V| \leq \|V\|$, we have $|X| = |X_0 + M + V| \leq 3K$ on $\mathbb{R}_+ \times \Omega$. Since F, F' and F'' are continuous on $\mathbb{R}_+$ they are bounded on the finite interval $[-3K, 3K]$. Thus there exists $C \geq 0$ such that

$$\sup_{x \in [-3K, 3K]} |F(x)|, \sup_{x \in [-3K, 3K]} |F'(x)|, \sup_{x \in [-3K, 3K]} |F''(x)| \leq C.$$

Then

(3) $$|F \circ X|, |F' \circ X|, |F'' \circ X| \leq C \quad \text{on } \mathbb{R}_+ \times \Omega.$$

The boundedness of $|F' \circ X|$ in particular implies that $|F' \circ X|$ is in $\mathbf{L}_{2,\infty}(\mathbb{R}_+ \times \Omega, \mu_{[M]}, P)$ by Observation 11.19 so that $(F' \circ X) \bullet M \in \mathbf{M}_2^c(\Omega, \mathfrak{F}, \{\mathfrak{F}_t\}, P)$ by Definition 12.9.

For $n \in \mathbb{N}$, let Δ_n be a partition of $\mathbb{R}_+$ into subintervals by a strictly increasing sequence $\{t_{n,k} : k \in \mathbb{Z}_+\}$ in $\mathbb{R}_+$ such that $t_{n,0} = 0$ and $t_{n,k} \uparrow \infty$ as $k \to \infty$ and $\lim_{n \to \infty} |\Delta_n| = 0$ where $|\Delta_n| = \sup_{k \in \mathbb{N}}(t_{n,k} - t_{n,k-1})$. Let $t \in \mathbb{R}_+$ be fixed. For each $n \in \mathbb{N}$, let $t_{n,p_n} = t$ where $t_{n,k} < t$ for $k = 0, \ldots, p_n - 1$. Let us write

(4) $$F(X(t)) - F(X(0)) = \sum_{k=1}^{p_n} \{F(X(t_{n,k})) - F(X(t_{n,k-1}))\}.$$

Now for $a, b \in \mathbb{R}$, $a \neq b$, according to Taylor's Theorem we have

$$F(b) - F(a) = F'(a)(b-a) + \frac{1}{2}F''(c)(b-a)^2 \quad \text{for some } c \in (a \wedge b, a \vee b).$$

Thus we have

(5) $$\begin{aligned} &F(X(t_{n,k}, \omega)) - F(X(t_{n,k-1}, \omega)) \\ &= F'(X(t_{n,k-1}, \omega))\{X(t_{n,k}, \omega) - X(t_{n,k-1}, \omega)\} \\ &+ \frac{1}{2}F''(\xi_{n,k}(\omega))\{X(t_{n,k}, \omega) - X(t_{n,k-1}, \omega)\}^2, \end{aligned}$$

where

(6) $$X(t_{n,k-1}, \omega) \wedge X(t_{n,k}, \omega) < \xi_{n,k}(\omega) < X(t_{n,k-1}, \omega) \vee X(t_{n,k}, \omega).$$

Now $\xi_{n,k}$ may not be a random variable, that is, it may not be $\mathfrak{F}$-measurable. However on the set $\{X(t_{n,k}) - X(t_{n,k-1}) \neq 0\} \in \mathfrak{F}_{t_{n,k}}$ we have from (5) the equality

$$\begin{aligned} F''(\xi_{n,k}) &= 2\{F(X(t_{n,k})) - F(X(t_{n,k-1}))\}\{X(t_{n,k}) - X(t_{n,k-1})\}^{-2} \\ &- 2F'(X(t_{n,k-1}))\{X(t_{n,k}) - X(t_{n,k-1})\}^{-1}, \end{aligned}$$

which is $\mathfrak{F}_{t_{n,k}}$-measurable. Let us define a function $G_{n,k}$ on Ω by setting

$$G_{n,k}(\omega) = \begin{cases} F''(\xi_{n,k}(\omega)) & \text{for } \omega \in \{X(t_{n,k}) - X(t_{n,k-1}) \neq 0\} \\ 0 & \text{otherwise.} \end{cases}$$

§15. ITÔ'S FORMULA

Then $G_{n,k}$ is an $\mathfrak{F}_{t_{n,k}}$-measurable random variable and furthermore

$$
\begin{aligned}
(7) \quad F(X(t_{n,k})) - F(X(t_{n,k-1})) &= F'(X(t_{n,k-1}))\{X(t_{n,k}) - X(t_{n,k-1})\} \\
&\quad + \frac{1}{2} G_{n,k}\{X(t_{n,k}) - X(t_{n,k-1})\}^2.
\end{aligned}
$$

Note also that by (6), the Intermediate Value Theorem applied to the continuous function $X(\cdot, \omega)$, and (3) we have

$$
(8) \qquad \sup_{\omega \in \Omega} |G_{n,k}(\omega)| \leq \sup_{\omega \in \Omega} |F''(\xi_{n,k}(\omega))| \leq C.
$$

We now have

$$
\begin{aligned}
(9) \quad F(X(t)) - F(X(0)) &= \sum_{k=1}^{p_n} F'(X(t_{n,k-1}))\{X(t_{n,k}) - X(t_{n,k-1})\} \\
&\quad + \frac{1}{2} \sum_{k=1}^{p_n} G_{n,k}\{X(t_{n,k}) - X(t_{n,k-1})\}^2.
\end{aligned}
$$

Let us write the first member on the right side of (9) as

$$
\begin{aligned}
(10) \quad & \sum_{k=1}^{p_n} F'(X(t_{n,k-1}))\{X(t_{n,k}) - X(t_{n,k-1})\} \\
&= \sum_{k=1}^{p_n} F'(X(t_{n,k-1}))\{M(t_{n,k}) - M(t_{n,k-1})\} + \sum_{k=1}^{p_n} F'(X(t_{n,k-1}))\{V(t_{n,k}) - V(t_{n,k-1})\} \\
&\equiv S_1^{(n)} + S_2^{(n)}.
\end{aligned}
$$

By Proposition 12.16, $\lim_{n \to \infty} \|S_1^{(n)} - ((F' \circ X) \bullet M)(t)\|_2 = 0$. This implies that there exist a subsequence $\{n_\ell\}$ of $\{n\}$ and a null set Λ' in $(\Omega, \mathfrak{F}, P)$ such that

$$
(11) \qquad \lim_{\ell \to \infty} S_1^{(n_\ell)}(\omega) = ((F' \circ X) \bullet M)(t, \omega) \quad \text{for } \omega \in \Lambda'^c.
$$

Regarding $S_2^{(n)}$, note that since $(F' \circ X)(\cdot, \omega)$ is a continuous function and $V(\cdot, \omega)$ is a continuous function of bounded variation on every finite interval in $\mathbb{R}_+$ for every $\omega \in \Omega$, we have

$$
(12) \qquad \lim_{n \to \infty} S_2^{(n)}(\omega) = \int_{[0,t]} (F' \circ X)(s, \omega) \, dV(s, \omega) \quad \text{for } \omega \in \Omega.
$$

Let us write the second member on the right side of (9) as

(13)
$$\frac{1}{2}\sum_{k=1}^{p_n} G_{n,k}\{X(t_{n,k}) - X(t_{n,k-1})\}^2$$

$$= \sum_{k=1}^{p_n} G_{n,k}\{M(t_{n,k}) - M(t_{n,k-1})\}\{V(t_{n,k}) - V(t_{n,k-1})\}$$

$$+ \frac{1}{2}\sum_{k=1}^{p_n} G_{n,k}\{V(t_{n,k}) - V(t_{n,k-1})\}^2 + \frac{1}{2}\sum_{k=1}^{p_n} G_{n,k}\{M(t_{n,k}) - M(t_{n,k-1})\}^2$$

$$\equiv S_3^{(n)} + S_4^{(n)} + S_5^{(n)}.$$

Now by (8) we have

$$|S_3^{(n)}| \leq C \max_{k=1,\ldots,p_n} |M(t_{n,k}) - M(t_{n,k-1})| \sum_{k=1}^{p_n} |V(t_{n,k}) - V(t_{n,k-1})|$$

$$\leq C \max_{k=1,\ldots,p_n} |M(t_{n,k}) - M(t_{n,k-1})| \|V\|_t.$$

Since every sample function of M is continuous and hence uniformly continuous on $[0, t]$, $\lim_{n\to\infty} \max_{k=1,\ldots,p_n} |M(t_{n,k}) - M(t_{n,k-1})| = 0$ on Ω. Thus we have

(14)
$$\lim_{n\to\infty} S_3^{(n)}(\omega) = 0 \quad \text{for } \omega \in \Omega.$$

Similarly we have

$$|S_4^{(n)}| \leq \frac{1}{2}C \max_{k=1,\ldots,p_n} |V(t_{n,k}) - V(t_{n,k-1})| \|V\|_t$$

so that by the uniform continuity of every sample function of V on $[0, t]$ we have

(15)
$$\lim_{n\to\infty} S_4^{(n)}(\omega) = 0 \quad \text{for } \omega \in \Omega.$$

To show $\lim_{n\to\infty} \mathbf{E}(|S_5^{(n)} - \frac{1}{2}\int_{[0,t]}(F'' \circ X)(s)\,d[M](s)|) = 0$, let

$$S_6^{(n)} = \frac{1}{2}\sum_{k=1}^{p_n}(F'' \circ X)(t_{n,k-1})\{M(t_{n,k}) - M(t_{n,k-1})\}^2,$$

and

$$S_7^{(n)} = \frac{1}{2}\sum_{k=1}^{p_n}(F'' \circ X)(t_{n,k-1})\{[M](t_{n,k}) - [M](t_{n,k-1})\}.$$

§15. ITÔ'S FORMULA

For brevity, let us write

$$\begin{cases} A_{n,k} = \{M(t_{n,k}) - M(t_{n,k-1})\}^2, & \alpha_n = \max_{k=1,\ldots,p_n} A_{n,k}, \\ B_{n,k} = \{[M](t_{n,k}) - [M](t_{n,k-1})\}, & \beta_n = \max_{k=1,\ldots,p_n} B_{n,k}. \end{cases}$$

Now

$$\begin{aligned} |S_5^{(n)} - S_6^{(n)}| &\leq \frac{1}{2} \sum_{k=1}^{p_n} |G_{n,k} - (F'' \circ X)(t_{n,k-1})|\{M(t_{n,k}) - M(t_{n,k-1})\}^2 \\ &\leq C \max_{k=1,\ldots,p_n} |M(t_{n,k}) - M(t_{n,k-1})| \sum_{k=1}^{p_n} |M(t_{n,k}) - M(t_{n,k-1})| \end{aligned}$$

so that by Hölder's Inequality and Lemma 15.6 we have

$$\mathbf{E}(|S_5^{(n)} - S_6^{(n)}|) \leq C\mathbf{E}[\alpha_n]^{\frac{1}{2}} \mathbf{E}[\{\sum_{k=1}^{p_n} |M(t_{n,k}) - M(t_{n,k-1})|\}^2]^{\frac{1}{2}} \leq C\mathbf{E}[\alpha_n]^{\frac{1}{2}} \sqrt{1 + K^2} K.$$

By the uniform continuity of the sample functions of M on $[0,t]$, we have $\lim_{n\to\infty} \alpha_n = 0$ on Ω. Since α_n is bounded by $4K^2$ we have $\lim_{n\to\infty} \mathbf{E}[\alpha_n] = 0$ by the Bounded Convergence Theorem. Therefore

$$\lim_{n\to\infty} \mathbf{E}(|S_5^{(n)} - S_6^{(n)}|) = 0. \tag{16}$$

Next we have $S_6^{(n)} - S_7^{(n)} = \frac{1}{2} \sum_{k=1}^{p_n} (F'' \circ X)(t_{n,k-1})\{A_{n,k} - B_{n,k}\}$ so that

$$\mathbf{E}(|S_6^{(n)} - S_7^{(n)}|^2) = \frac{1}{4} \sum_{k=1}^{p_n} \mathbf{E}[(F'' \circ X)(t_{n,k-1})^2 \{A_{n,k} - B_{n,k}\}^2]$$

$$+ \frac{1}{4} \sum_{j,k=1,\ldots,p_n; j\neq k} \mathbf{E}[(F'' \circ X)(t_{n,j-1})(F'' \circ X)(t_{n,k-1})\{A_{n,j} - B_{n,j}\}\{A_{n,k} - B_{n,k}\}].$$

For $j \neq k$, say $j < k$, we have

$$\mathbf{E}[(F'' \circ X)(t_{n,j-1})(F'' \circ X)(t_{n,k-1})\{A_{n,j} - B_{n,j}\}\{A_{n,k} - B_{n,k}\}|\mathfrak{F}_{t_{n,k-1}}]$$
$$= (F'' \circ X)(t_{n,j-1})(F'' \circ X)(t_{n,k-1})\{A_{n,j} - B_{n,j}\}\mathbf{E}[\{A_{n,k} - B_{n,k}\}|\mathfrak{F}_{t_{n,k-1}}],$$

a.e. on $(\Omega, \mathfrak{F}_{t_{n,k-1}}, P)$. By the definition of $[M]$ we have

$$\mathbf{E}[\{A_{n,k} - B_{n,k}\}|\mathfrak{F}_{t_{n,k-1}}]$$
$$= \mathbf{E}[\{M(t_{n,k}) - M(t_{n,k-1})\}^2 - \{[M](t_{n,k}) - [M](t_{n,k-1})\}|\mathfrak{F}_{t_{n,k-1}}] = 0,$$

a.e. on $(\Omega, \mathfrak{F}_{t_{n,k-1}}, P)$. Thus the summands in the second sum in the expression for $\mathbf{E}(|S_6^{(n)} - S_7^{(n)}|^2)$ above are all equal to 0 and therefore we have

$$\mathbf{E}(|S_6^{(n)} - S_7^{(n)}|^2) = \frac{1}{4}\sum_{k=1}^{p_n} \mathbf{E}[(F'' \circ X)(t_{n,k-1})^2\{A_{n,k} - B_{n,k}\}^2]$$

$$\leq \frac{1}{2}\sum_{k=1}^{p_n} \mathbf{E}[(F'' \circ X)(t_{n,k-1})^2\{A_{n,k}^2 + B_{n,k}^2\}]$$

$$\leq \frac{1}{2}C^2 \sum_{k=1}^{p_n} \mathbf{E}[\{M(t_{n,k}) - M(t_{n,k-1})\}^4 + \{[M](t_{n,k}) - [M](t_{n,k-1})\}^2]$$

$$\leq \frac{1}{2}C^2 \mathbf{E}[\alpha_n \sum_{k=1}^{p_n}\{M(t_{n,k}) - M(t_{n,k-1})\}^2] + \frac{1}{2}C^2 \mathbf{E}[\beta_n \sum_{k=1}^{p_n}\{[M](t_{n,k}) - [M](t_{n,k-1})\}]$$

$$\leq \frac{1}{2}C^2 \mathbf{E}[\alpha_n^2]^{1/2} \mathbf{E}[\{\sum_{k=1}^{p_n}\{M(t_{n,k}) - M(t_{n,k-1})\}^2\}^2]^{1/2} + \frac{1}{2}C^2 \mathbf{E}[\beta_n [M](t)]$$

$$\leq \frac{1}{2}C^2 \mathbf{E}[\alpha_n^2]^{1/2} \sqrt{12}K^2 + \frac{1}{2}C^2 \mathbf{E}[\beta_n [M](t)],$$

by Lemma 12.26. By the uniform continuity of every sample function of M and $[M]$ on $[0, t]$ we have $\lim_{n\to\infty} \alpha_n^2 = 0$ and $\lim_{n\to\infty} \beta_n = 0$ on Ω. Since M and $[M]$ are bounded we have $\lim_{n\to\infty} \mathbf{E}[\alpha_n^2] = 0$ and $\lim_{n\to\infty} \mathbf{E}[\beta_n[M](t)] = 0$ by the Bounded Convergence Theorem. Therefore $\lim_{n\to\infty} \mathbf{E}(|S_6^{(n)} - S_7^{(n)}|^2) = 0$ and consequently

(17) $$\lim_{n\to\infty} \mathbf{E}(|S_6^{(n)} - S_7^{(n)}|) = 0.$$

Since every sample function of $F'' \circ X$ is continuous we have

(18) $$\lim_{n\to\infty} S_7^{(n)}(\omega) = \frac{1}{2}\int_{[0,t]} (F'' \circ X)(s,\omega)\, d[M](s,\omega) \quad \text{for } \omega \in \Omega.$$

Now (16) and (17) imply that there exists a subsequence $\{n_m\}$ of $\{n_\ell\}$ and a null set Λ_t'' in $(\Omega, \mathfrak{F}, P)$ such that

(19) $$\lim_{m\to\infty} |S_5^{(n)}(\omega) - S_6^{(n)}(\omega)| = \lim_{m\to\infty} |S_6^{(n)}(\omega) - S_7^{(n)}(\omega)| = 0 \quad \text{for } \omega \in \Lambda_t''^c.$$

Let $\Lambda_t = \Lambda' \cup \Lambda_t''$. Replace n in (9) by n_m. If we let $m \to \infty$ in (9), then by (10), (11), (12), (13), (14), (15), (18), and (19) we have (1) holding on Λ_t^c. This proves (1) for the case where X_0, M, $[M]$ and $|V|$ are all bounded.

§15. ITÔ'S FORMULA

Step 2. Let us remove the assumption that X_0 is bounded maintaining the assumption that M, $[M]$, and $|V|$ are bounded. Let $X_0^{(n)} = \mathbf{1}_{[-n,n]} \circ X_0$ so that $|X_0^{(n)}| \leq n$. Let $X^{(n)} = X_0^{(n)} + M + V$ for $n \in \mathbb{N}$. For every $n \in \mathbb{N}$ according to Step 1 there exists a null set Λ_n in $(\Omega, \mathfrak{F}, P)$ such that on $\mathbb{R}_+ \times \Lambda_n^c$ we have

$$(20) \quad (F \circ X^{(n)})_t - (F \circ X^{(n)})_0 = \int_{[0,t]} (F' \circ X^{(n)})(s)\, dM(s)$$
$$+ \int_{[0,t]} (F' \circ X^{(n)})(s)\, dV(s) + \frac{1}{2} \int_{[0,t]} (F'' \circ X^{(n)})(s)\, d[M](s).$$

Since the stochastic integral does not depend on the values of the integrand process at $t = 0$ as we noted after Definition 12.3, we have

$$\int_{[0,t]} (F' \circ X^{(n)})(s)\, dM(s) = \int_{[0,t]} (F' \circ X)(s)\, dM(s).$$

For a fixed $\omega \in \Omega$, let n be so large that $|X_0(\omega)| \leq n$ so that $X_0^{(n)}(\omega) = X_0(\omega)$ and hence $X^{(n)}(\cdot, \omega) = X(\cdot, \omega)$. Then

$$\int_{[0,t]} (F' \circ X^{(n)})(s)\, dV(s) = \int_{[0,t]} (F' \circ X)(s)\, dV(s),$$

and

$$\int_{[0,t]} (F'' \circ X^{(n)})(s)\, d[M](s) = \int_{[0,t]} (F'' \circ X)(s)\, d[M](s).$$

On the other hand since $\lim_{n \to \infty} X^{(n)}(\omega) = X(\omega)$ and since F is continuous, we have $\lim_{n \to \infty} (F \circ X^{(n)})(s, \omega) = (F \circ X)(s, \omega)$ for every $s \in \mathbb{R}_+$. Let $\Lambda = \cup_{n \in \mathbb{N}} \Lambda_n$. Then letting $n \to \infty$ in (20) we have (1) holding on Λ^c.

Step 3. Let us remove the assumption that X_0, M, $[M]$, and $|V|$ are bounded. Let

$$\begin{aligned} S_{1,n} &= \inf\{t \in \mathbb{R}_+ : |M(t)| > n\} \wedge n, \\ S_{2,n} &= \inf\{t \in \mathbb{R}_+ : [M](t) > n\} \wedge n, \\ S_{3,n} &= \inf\{t \in \mathbb{R}_+ : |V|(t) > n\} \wedge n, \\ S_{4,n} &= \inf\{t \in \mathbb{R}_+ : |(F' \circ X)(t)| > n\} \wedge n, \\ T_n &= \inf\{t \in \mathbb{R}_+ : \int_{[0,t]} (F' \circ X)^2(s)\, d[M](s) > n\} \wedge n. \end{aligned}$$

Then $\{S_{i,n} : n \in \mathbb{N}\}$ for $i = 1, \ldots, 4$, and $\{T_n : n \in \mathbb{N}\}$ are increasing sequences of stopping times, and since every sample function of M, $[M]$, $|V|$ and $F' \circ X$ is continuous by our assumption, we have $S_{i,n} \uparrow \infty$ and $T_n \uparrow \infty$ on Ω as $n \to \infty$. Let $R_n = S_{1,n} \wedge \cdots \wedge S_{4,n} \wedge T_n$. Then $\{R_n : n \in \mathbb{N}\}$ is an increasing sequence of stopping times such that $R_n \uparrow \infty$ on Ω. Since the processes $M^{R_n\wedge}$, $[M]^{R_n\wedge}$, $|V|^{R_n\wedge}$ and $(F' \circ X)^{R_n\wedge}$ are bounded by n, we have $M^{R_n\wedge} \in \mathbf{M}_2^c(\Omega, \mathfrak{F}, \{\mathfrak{F}_t\}, P)$, $[M]^{R_n\wedge} \in \mathbf{A}^c(\Omega, \mathfrak{F}, \{\mathfrak{F}_t\}, P)$, $|V|^{R_n\wedge} \in \mathbf{V}^c(\Omega, \mathfrak{F}, \{\mathfrak{F}_t\}, P)$, and $(F' \circ X)^{R_n\wedge} \in \mathbf{L}_{2,\infty}(\mathbb{R}_+ \times \Omega, \mu_{[M^{R_n\wedge}]}, P)$. Note that $[M^{R_n\wedge}] = [M]^{R_n\wedge}$ by Proposition 12.25 and $|V^{R_n\wedge}| = |V|^{T_n\wedge}$. For every $n \in \mathbb{N}$, let $X^{R_n\wedge} = X_0 + M^{R_n\wedge} + V^{R_n\wedge}$. By Step 2 there exists a null set Λ in $(\Omega, \mathfrak{F}, P)$ such that on Λ^c we have for every $n \in \mathbb{N}$ and every $t \in \mathbb{R}_+$

$$(21) \quad (F \circ X^{R_n\wedge})_t - (F \circ X^{R_n\wedge})_0 = \int_{[0,t]} (F' \circ X^{R_n\wedge})(s) \, dM^{R_n\wedge}(s)$$
$$+ \int_{[0,t]} (F' \circ X^{R_n\wedge})(s) \, dV^{R_n\wedge}(s) + \frac{1}{2} \int_{[0,t]} (F'' \circ X^{R_n\wedge})(s) \, d[M^{R_n\wedge}](s).$$

Let us observe that if Y is an adapted process and S is a stopping time on a filtered space, and G is a real valued function on $\mathbb{R}$, then for the stopped process $Y^{S\wedge}$ and the truncated process $Y^{[S]}$ we have

$$(22) \quad G \circ Y^{S\wedge} = (G \circ Y)^{S\wedge}$$

and

$$(23) \quad (Y^{S\wedge})^{[S]} = Y^{[S]}$$

as can be verified easily. For $\omega \in \Lambda^c$ and $t \in \mathbb{R}_+$, for sufficiently large $n \in \mathbb{N}$ we have $R_n(\omega) \geq t$ so that $R_n(\omega) \wedge s = s$ for all $s \in [0, t]$. For such n we have by (22)

$$(24) \quad (F \circ X^{R_n\wedge})(s, \omega) = (F \circ X)^{R_n\wedge}(s, \omega) = (F \circ X)(s, \omega) \quad \text{for } s \in [0, t],$$

and similarly

$$(25) \quad \begin{cases} (F' \circ X^{R_n\wedge})(s, \omega) = (F' \circ X)(s, \omega) & \text{for } s \in [0, t] \\ (F'' \circ X^{R_n\wedge})(s, \omega) = (F'' \circ X)(s, \omega) & \text{for } s \in [0, t]. \end{cases}$$

Regarding the first term on the right side of (21) we have

$$(F' \circ X^{R_n\wedge}) \bullet M^{R_n\wedge} = (F' \circ X)^{R_n\wedge} \bullet M^{R_n\wedge} = (F' \circ X)^{R_n\wedge} \bullet M^{R_n\wedge \wedge R_n\wedge}$$
$$= ((F' \circ X)^{R_n\wedge} \bullet M^{R_n\wedge})^{R_n\wedge} = ((F' \circ X)^{R_n\wedge})^{[R_n]} \bullet M^{R_n\wedge}$$
$$= (F' \circ X)^{[R_n]} \bullet M^{R_n\wedge} = ((F' \circ X)^{[R_n]} \bullet M)^{R_n\wedge}$$

§15. ITÔ'S FORMULA

where the first equality is by (22), the third equality is by Theorem 12.24, the fourth equality is by Theorem 12.22, the fifth equality is by (23), and the last equality is by Theorem 14.18. Now $F' \circ X \in \mathbf{L}_{2,\infty}^{loc}(\mathbb{R}_+ \times \Omega, \mu_{[M]})$ by Proposition 15.5. Now by Theorem 14.19, we have

$$(F' \circ X)^{[R_n]} \bullet M = (F' \circ X)^{T_n \wedge S_{1,n} \wedge \cdots \wedge S_{4,n}} \bullet M = ((F' \circ X) \bullet M)^{R_n \wedge}.$$

Thus

$$(F' \circ X^{R_n \wedge}) \bullet M^{R_n \wedge} = ((F' \circ X) \bullet M)^{R_n \wedge}.$$

For $\omega \in \Lambda^c$, since $R_n(\omega) \wedge t = t$ for sufficiently large $n \in \mathbb{N}$, for such n we have

$$((F' \circ X^{R_n \wedge}) \bullet M^{R_n \wedge})(t, \omega) = ((F' \circ X) \bullet M)^{R_n \wedge}(t, \omega) = ((F' \circ X) \bullet M)(t, \omega).$$

Therefore we have shown that for $\omega \in \Lambda^c$ for sufficiently large $n \in \mathbb{N}$ we have

$$(26) \quad \left(\int_{[0,t]} (F' \circ X^{R_n \wedge})(s) \, dM^{R_n \wedge}(s) \right)(\omega) = \left(\int_{[0,t]} (F' \circ X)(s) \, dM(s) \right)(\omega).$$

For $s \in [0, t]$ for sufficiently large $n \in \mathbb{N}$ we have $R_n(\omega) \wedge s = s$ so that

$$V^{R_n \wedge}(s, \omega) = V(R_n(\omega) \wedge s, \omega) = V(s, \omega),$$

and similarly

$$[M^{R_n \wedge}](s, \omega) = [M]^{R_n \wedge}(s, \omega) = [M](R_n(\omega) \wedge s, \omega) = [M](s, \omega).$$

Thus by these equalities and by (25) we have for sufficiently large $n \in \mathbb{N}$

$$(27) \quad \int_{[0,t]} (F' \circ X^{R_n \wedge})(s, \omega) \, dV^{R_n \wedge}(s, \omega) = \int_{[0,t]} (F' \circ X)(s, \omega) \, dV(s, \omega),$$

and

$$(28) \quad \int_{[0,t]} (F'' \circ X^{R_n \wedge})(s, \omega) \, d[M^{R_n \wedge}](s, \omega) = \int_{[0,t]} (F'' \circ X)(s, \omega) \, d[M](s, \omega).$$

Using (24), (26), (27) and (28) in (20), we have (1) holding on Λ^c. ∎

As an application of Itô's Formula, we have the following.

Example 15.7. Let B be an $\{\mathfrak{F}_t\}$-adapted null at 0 Brownian motion on a standard filtered space $(\Omega, \mathfrak{F}, \{\mathfrak{F}_t\}, P)$. Then

$$(1) \quad \int_{[0,t]} B(s) \, dB(s) = \frac{1}{2} \{B^2(t) - t\},$$

and

(2) $$\int_{[0,t]}\left\{\int_{[0,s]} B(u)\,dB(u)\right\} dB(s) = \frac{1}{3!}\{B^3(t) - 3tB(t)\}.$$

Proof. By Proposition 13.31, $B \in \mathbf{M}_2^c(\Omega, \mathfrak{F}, \{\mathfrak{F}_t\}, P)$ with $[B]_t = t$ for $t \in \mathbb{R}_+$. For $n \in \mathbb{N}$, let $F(x) = x^n$ for $x \in \mathbb{R}$. Then $F \in \mathbf{C}^2(\mathbb{R})$ so that by Theorem 15.3 applied to $X = X_0 + M + V$ in which $X_0 = 0$, $M = B$, and $V = 0$ we have

(3) $$B^n(t) = n \int_{[0,t]} B^{n-1}(s)\,dB(s) + \frac{n(n-1)}{2} \int_{[0,t]} B^{n-2}(s) m_L(ds).$$

In particular for $n = 2$ we have

(4) $$B^2(t) = 2 \int_{[0,t]} B(s)\,dB(s) + t,$$

and therefore (1) holds. For $n = 3$, we have by (3) and (4)

(5) $$\begin{aligned} B^3(t) &= 3 \int_{[0,t]} B^2(s)\,dB(s) + 3 \int_{[0,t]} B(s) m_L(ds) \\ &= 3! \int_{[0,t]} \left\{\int_{[0,s]} B(u)\,dB(u)\right\} dB(s) \\ &\quad + 3 \left\{\int_{[0,t]} s\,dB(s) + \int_{[0,t]} B(s) m_L(ds)\right\}. \end{aligned}$$

Since the integrand in $\int_{[0,t]} s\,dB(s)$ is a continuous function, $(\int_{[0,t]} s\,dB(s))(\omega)$ is in fact equal to the Riemann-Stieltjes integral $\int_0^t s\,dB(s,\omega)$ for every $\omega \in \Omega$. Now $\int_0^t s\,dB(s,\omega) + \int_0^t B(s,\omega)\,ds = [sB(s,\omega)]_0^t = tB(t)$ by integration by parts for the Riemann-Stieltjes integral. Using this in (5), we have

$$B^3(t) = 3! \int_{[0,t]} \left\{\int_{[0,s]} B(u)\,dB(u)\right\} dB(s) + 3tB(t).$$

From this last equality we have (2). ∎

The equalities (1) and (2) in Example 15.7 make interesting contrast to the formulas $\int_0^t s\,ds = \frac{1}{2}t^2$ and $\int_0^t \{\int_0^s u\,du\}\,ds = \frac{1}{3!}t^3$ in ordinary calculus.

§15. ITÔ'S FORMULA

[II] Stochastic Integrals with Respect to Quasimartingales

Definition 15.8. *For $A \in \mathbf{A}^{loc}(\Omega, \mathfrak{F}, \{\mathfrak{F}_t\}, P)$, let $\mathbf{L}^{loc}_{1,\infty}(\mathbb{R}_+ \times \Omega, \mu_A, P)$ be the linear space of equivalence classes of measurable processes $\Phi = \{\Phi_t : t \in \mathbb{R}_+\}$ on $(\Omega, \mathfrak{F}, P)$ such that $\int_{[0,t]} |\Phi(s,\omega)| \mu_A(ds,\omega) < \infty$ for every $t \in \mathbb{R}_+$ for almost every $\omega \in \Omega$.*

Definition 15.9. *Let $\mathbf{C}(\mathbb{R}_+ \times \Omega)$ be the linear space of equivalence classes of measurable processes $\Phi = \{\Phi_t : t \in \mathbb{R}_+\}$ on a probability space $(\Omega, \mathfrak{F}, P)$ such that $\Phi(\cdot, \omega)$ is continuous on $\mathbb{R}_+$ for almost every $\omega \in \Omega$. Let $\mathbf{B}(\mathbb{R}_+ \times \Omega)$ be the linear space of equivalence classes of measurable processes Φ on $(\Omega, \mathfrak{F}, P)$ such that $\Phi(\cdot, \omega)$ is bounded on every finite interval in $\mathbb{R}_+$ for almost every $\omega \in \Omega$.*

Remark 15.10. Every quasimartingale on a standard filtered space $(\Omega, \mathfrak{F}, \{\mathfrak{F}_t\}, P)$ is in $\mathbf{C}(\mathbb{R}_+ \times \Omega)$. Note also that

$$\mathbf{C}(\mathbb{R}_+ \times \Omega) \subset \mathbf{B}(\mathbb{R}_+ \times \Omega) \subset \mathbf{L}^{loc}_{2,\infty}(\mathbb{R}_+ \times \Omega, \mu_A, P) \subset \mathbf{L}^{loc}_{1,\infty}(\mathbb{R}_+ \times \Omega, \mu_A, P)$$

for every $A \in \mathbf{A}^{loc}(\Omega, \mathfrak{F}, \{\mathfrak{F}_t\}, P)$. The last set inclusion is from the fact that for every $t \in \mathbb{R}_+$ and $\omega \in \Omega$, $\mu_A(\cdot, \omega)$ is a finite measure on $([0,t], \mathfrak{B}_{[0,t]})$ so that for every measurable process Φ on $(\Omega, \mathfrak{F}, P)$ we have

$$\int_{[0,t]} \Phi^2(s,\omega) \, dA(s,\omega) < \infty \Rightarrow \int_{[0,t]} |\Phi(s,\omega)| \, dA(s,\omega) < \infty.$$

Consider a quasimartingale $X = X_0 + M + V$ on a standard filtered space $(\Omega, \mathfrak{F}, \{\mathfrak{F}_t\}, P)$. Let Φ be a predictable process on the filtered space such that

$$\Phi \in \mathbf{L}^{loc}_{2,\infty}(\mathbb{R}_+ \times \Omega, \mu_{[M]}, P) \cap \mathbf{L}^{loc}_{1,\infty}(\mathbb{R}_+ \times \Omega, \mu_{|V|}, P).$$

Since $\Phi \in \mathbf{L}^{loc}_{2,\infty}(\mathbb{R}_+ \times \Omega, \mu_{[M]}, P)$, $\Phi \bullet M$ exists in $\mathbf{M}^{c,loc}_2(\Omega, \mathfrak{F}, \{\mathfrak{F}_t\}, P)$. The fact that Φ is in $\mathbf{L}^{loc}_{1,\infty}(\mathbb{R}_+ \times \Omega, \mu_{|V|}, P)$ implies that

$$\left| \int_{[0,t]} \Phi(s,\omega) \, dV(s,\omega) \right| \leq \int_{[0,t]} |\Phi(s,\omega)| \, d|V|(s,\omega) < \infty,$$

so that $\int_{[0,t]} \Phi(s,\omega) \, dV(s,\omega) \in \mathbb{R}$ for $(t,\omega) \in \mathbb{R}_+ \times \Omega$. The stochastic process $\{\int_{[0,t]} \Phi(s) \, dV(s) : t \in \mathbb{R}_+\}$ is adapted and its sample functions are of bounded variation on every finite interval in $\mathbb{R}_+$ by the same argument as in the proof of Proposition 15.5. Thus the process is in $\mathbf{V}^{c,loc}(\Omega, \mathfrak{F}, \{\mathfrak{F}_t\}, P)$.

Definition 15.11 *Let X be a quasimartingale given by $X = X_0 + M_X + V_X$ with $M_X \in \mathbf{M}_2^{c,loc}(\Omega, \mathfrak{F}, \{\mathfrak{F}_t\}, P)$ and $V_X \in \mathbf{V}^{c,loc}(\Omega, \mathfrak{F}, \{\mathfrak{F}_t\}, P)$. Let Φ be a predictable process on the filtered space such that*

(1) $\quad\quad\quad\quad\quad \Phi \in \mathbf{L}_{2,\infty}^{loc}(\mathbb{R}_+ \times \Omega, \mu_{[M_X]}, P) \cap \mathbf{L}_{1,\infty}^{loc}(\mathbb{R}_+ \times \Omega, \mu_{|V_X|}, P).$

By the stochastic integral of Φ with respect to X we mean the null at 0 quasimartingale $\{\int_{[0,t]} \Phi(s)\, dX(s) : t \in \mathbb{R}_+\}$ defined by

(2) $\quad\quad \int_{[0,t]} \Phi(s)\, dX(s) = \int_{[0,t]} \Phi(s)\, dM_X(s) + \int_{[0,t]} \Phi(s)\, dV_X(s) \quad \text{for } t \in \mathbb{R}_+.$

We also use the notations $\Phi \bullet X$, $\Phi \bullet M_X$, and $\Phi \bullet V_X$ for the processes $\{\int_{[0,t]} \Phi(s)\, dX(s) : t \in \mathbb{R}_+\}$, $\{\int_{[0,t]} \Phi(s)\, dM_X(s) : t \in \mathbb{R}_+\}$, and $\{\int_{[0,t]} \Phi(s)\, dV_X(s) : t \in \mathbb{R}_+\}$ respectively. Thus

$$\Phi \bullet X = \Phi \bullet M_X + \Phi \bullet V_X.$$

If we write $M_{\Phi \bullet X}$ and $V_{\Phi \bullet X}$ for the martingale part and bounded variation part of the quasimartingale $\Phi \bullet X$, then $M_{\Phi \bullet X} = \Phi \bullet M_X$ and $V_{\Phi \bullet X} = \Phi \bullet V_X$.

With a stochastic integral with respect to a quasimartingale as defined above, Theorem 15.3 can be stated as follows:
Let X be a quasimartingale on standard filtered space and let M_X be its martingale part. Then for $F \in \mathbf{C}^2(\mathbb{R})$, we have

$$(F \circ X)_t - (F \circ X)_0 = \int_{[0,t]} (F' \circ X)(s)\, dX(s) + \frac{1}{2} \int_{[0,t]} (F'' \circ X)(s)\, d[M_X](s),$$

or in the alternate notations,

$$(F \circ X)_t - (F \circ X)_0 = (F' \circ X) \bullet X + \frac{1}{2}(F'' \circ X) \bullet [M_X].$$

Contrast the first of the two expressions above with the formula $F(t) - F(0) = \int_0^t F'(s)\, ds$ for $F \in \mathbf{C}^1(\mathbb{R})$.

Theorem 15.12. *Let X and Y be two quasimartingales on a standard filtered space $(\Omega, \mathfrak{F}, \{\mathfrak{F}_t\}, P)$ given by $X = X_0 + M_X + V_X$ and $Y = Y_0 + M_Y + V_Y$ where $M_X, M_Y \in$*

§15. ITÔ'S FORMULA

$\mathbf{M}_2^{c,loc}(\Omega, \mathfrak{F}, \{\mathfrak{F}_t\}, P)$ and $V_X, V_Y \in \mathbf{V}^{c,loc}(\Omega, \mathfrak{F}, \{\mathfrak{F}_t\}, P)$. Then the stochastic integral of Y with respect to X, $Y \bullet X$, always exists as a null at 0 quasimartingale with martingale part $M_{Y \bullet X} = Y \bullet M_X$ and bounded variation part $V_{Y \bullet X} = Y \bullet V_X$.

Proof. Since Y is a quasimartingale, it is in $\mathbf{C}(\mathbb{R}_+ \times \Omega)$ which is contained in $\mathbf{L}_{2,\infty}^{loc}(\mathbb{R}_+ \times \Omega, \mu_{[M_X]}, P) \cap \mathbf{L}_{1,\infty}^{loc}(\mathbb{R}_+ \times \Omega, \mu_{|V_X|}, P)$. Thus by Definition 15.11, $Y \bullet X$ exists and $M_{Y \bullet X} = Y \bullet M_X$ and $V_{Y \bullet X} = Y \bullet V_X$. ∎

Theorem 15.13. *Let $X = X_0 + M_X + V_X$ and $Y = Y_0 + M_Y + V_Y$ be quasimartingales on a standard filtered space $(\Omega, \mathfrak{F}, \{\mathfrak{F}_t\}, P)$ and let Φ and Ψ be predictable processes on the filtered space such that*

(1) $\quad\quad \Phi, \Psi \ \in \ \mathbf{L}_{2,\infty}^{loc}(\mathbb{R}_+ \times \Omega, \mu_{[M_X]}, P) \cap \mathbf{L}_{1,\infty}^{loc}(\mathbb{R}_+ \times \Omega, \mu_{|V_X|}, P)$
$\quad\quad\quad\quad \cap \ \mathbf{L}_{2,\infty}^{loc}(\mathbb{R}_+ \times \Omega, \mu_{[M_Y]}, P) \cap \mathbf{L}_{1,\infty}^{loc}(\mathbb{R}_+ \times \Omega, \mu_{|V_Y|}, P).$

Then for $a, b, \alpha, \beta \in \mathbb{R}$, we have

(2) $\quad\quad (a\Phi + b\Psi) \bullet (\alpha X + \beta Y) = a\alpha\Phi \bullet X + a\beta\Phi \bullet Y + b\alpha\Psi \bullet X + b\beta\Psi \bullet Y.$

In particular when $\Phi, \Psi \in \mathbf{B}(\mathbb{R}_+ \times \Omega)$, (2) is satisfied.

Proof. By the linearity of stochastic integrals with respect to local L_2 martingales and Lebesgue-Stieltjes integrals we have the linearity of stochastic integrals with respect to quasimartingales. ∎

[III] Exponential Quasimartingales

If X is a quasimartingale then since the exponential function has continuous derivatives of all orders Theorem 15.3 implies that $e^X = \{e^{X(t)} : t \in \mathbb{R}_+\}$ is a quasimartingale.

Definition 15.14. *For a quasimartingale X, e^X is called the exponential quasimartingale of X.*

Theorem 15.15. *Let X be a null at 0 quasimartingale given by $X = M_X + V_X$ where $M_X \in \mathbf{M}_2^{c,loc}(\Omega, \mathfrak{F}, \{\mathfrak{F}_t\}, P)$ and $V_X \in \mathbf{V}^{c,loc}(\Omega, \mathfrak{F}, \{\mathfrak{F}_t\}, P)$. Then for the null at 0 quasimartingale defined by $Y = X - \frac{1}{2}[M_X]$, we have*

(1) $$e^{Y(t)} = 1 + \int_{[0,t]} e^{Y(s)} \, dX(s)$$

on $\mathbb{R}_+ \times \Lambda^c$ where Λ is a null set in $(\Omega, \mathfrak{F}, P)$.

Proof. Since $Y = X - \frac{1}{2}[M_X] = M_X + V_X - \frac{1}{2}[M_X]$, Y is a null at 0 quasimartingale with its martingale part M_Y and bounded variation part V_Y given by

(2) $$\begin{cases} M_Y = M_X \\ V_Y = V_X - \frac{1}{2}[M_X]. \end{cases}$$

By Theorem 15.3, we have

(3) $$e^{Y(t)} - e^{Y(0)} = \int_{[0,t]} e^{Y(s)} dM_Y(s) + \int_{[0,t]} e^{Y(s)} dV_Y(s) + \frac{1}{2} \int_{[0,t]} e^{Y(s)} d[M_Y](s).$$

Now $e^{Y(0)} = e^0 = 1$. Substituting (2) in (3) we have (1). ∎

Corollary 15.16. *Let $M \in \mathbf{M}_2^{c,loc}(\Omega, \mathfrak{F}, \{\mathfrak{F}_t\}, P)$ and let Φ be a predictable process such that $\Phi \in \mathbf{L}_{2,\infty}^{loc}(\mathbb{R}_+ \times \Omega, \mu_{[M]}, P)$. If we define a null at 0 quasimartingale by setting*

(1) $$Y(t) = \int_{[0,t]} \Phi(s) dM(s) - \frac{1}{2} \int_{[0,t]} \Phi^2(s) d[M](s) \quad \text{for } t \in \mathbb{R}_+,$$

then its exponential quasimartingale e^Y satisfies the condition

(2) $$e^{Y(t)} = 1 + \int_{[0,t]} e^{Y(s)} \Phi(s) dM(s)$$

on $\mathbb{R}_+ \times \Lambda^c$ where Λ is a null set in $(\Omega, \mathfrak{F}, P)$. In particular the exponential quasimartingale e^Z of the quasimartingale Z defined by $Z = M - \frac{1}{2}[M]$ satisfies the condition

(3) $$e^{Z(t)} = 1 + \int_{[0,t]} e^{Z(s)} dM(s)$$

on $\mathbb{R}_+ \times \Lambda^c$.

Proof. Let $X = \Phi \bullet M \in \mathbf{M}_2^{c,loc}(\Omega, \mathfrak{F}, \{\mathfrak{F}_t\}, P)$. Then by Theorem 14.20 we have $[X]_t = \int_{[0,t]} \Phi^2(s) d[M](s)$. Thus $Y = X - \frac{1}{2}[X]$. Then by Theorem 15.15 and Theorem 14.20, we have (2). ∎

Remark 15.17 The exponential function satisfies the condition $e^t - 1 = \int_0^t e^s \, ds$. According to Theorem 15.15, for a quasimartingale $X = X_0 + M_X + V_X$, the quasimartingale $Y =$

§15. ITÔ'S FORMULA

$X - \frac{1}{2}[M_X]$ satisfies the condition $e^{Y(t)} - 1 = \int_{[0,t]} e^{Y(s)} dX(s)$. Thus for stochastic integrals with respect to M the quasimartingale $e^{X - \frac{1}{2}[M_X]}$ corresponds to the exponential function in Riemann integrals.

Observation 15.18. Let us define the *Hermite polynomials* $H_n(t, x)$, $(t, x) \in (0, \infty) \times \mathbb{R}$ for $n \in \mathbb{Z}_+$ as the coefficients in the power series expansion of $\exp\{\gamma x - \frac{\gamma^2 t}{2}\}$ in $\gamma \in \mathbb{R}$. Thus

$$\exp\left\{\gamma x - \frac{\gamma^2 t}{2}\right\} = \sum_{n \in \mathbb{Z}_+} \gamma^n H_n(t, x) \quad \text{for } \gamma \in \mathbb{R}. \tag{1}$$

Differentiating the power series n times, we have

$$\begin{aligned}
H_n(t, x) &= \frac{1}{n!} \left[\frac{\partial^n}{\partial \gamma^n} \exp\left\{\gamma x - \frac{\gamma^2 t}{2}\right\}\right]_{\gamma=0} \\
&= \frac{1}{n!} \exp\left\{\frac{x^2}{2t}\right\} \left[\frac{\partial^n}{\partial \gamma^n} \exp\left\{-\frac{t}{2}\left(\gamma - \frac{x}{t}\right)^2\right\}\right]_{\gamma=0},
\end{aligned} \tag{2}$$

where the second equality is by completing square for γ in $\gamma x - \frac{\gamma^2 t}{2}$. Now with $y = \gamma - \frac{x}{t}$, we have $\frac{\partial y}{\partial \gamma} = 1$ and thus

$$\frac{\partial}{\partial \gamma} \exp\left\{-\frac{t}{2}\left(\gamma - \frac{x}{t}\right)^2\right\} = \frac{\partial}{\partial y} \exp\left\{-\frac{t}{2} y^2\right\}.$$

By iteration we have

$$\frac{\partial^n}{\partial \gamma^n} \exp\left\{-\frac{t}{2}\left(\gamma - \frac{x}{t}\right)^2\right\} = \frac{\partial^n}{\partial y^n} \exp\left\{-\frac{t}{2} y^2\right\}.$$

If we set $\gamma = 0$, then $y = -\frac{x}{t}$ and $dy = -\frac{1}{t} dx$ so that

$$\left[\frac{\partial^n}{\partial \gamma^n} \exp\left\{-\frac{t}{2}\left(\gamma - \frac{x}{t}\right)^2\right\}\right]_{\gamma=0} = (-t)^n \frac{\partial^n}{\partial x^n} \exp\left\{-\frac{x^2}{2t}\right\}.$$

Using this in (2), we obtain

$$H_n(t, x) = \frac{(-t)^n}{n!} \exp\left\{\frac{x^2}{2t}\right\} \frac{\partial^n}{\partial x^n} \exp\left\{-\frac{x^2}{2t}\right\} \quad \text{for } n \in \mathbb{Z}_+, \tag{3}$$

which is the customary definition of $H_n(t,x)$. It follows from (3) that $H_n(t,x)$ is a polynomial in the two variables t and x of mixed degree 4 with leading term $(n!)^{-1}x^n$ for $n \in \mathbb{Z}_+$. For instance we have

(4)
$$\begin{array}{ll} H_0(t,x) = 1, & H_3(t,x) = \frac{1}{6}x^3 - \frac{1}{2}xt, \\ H_1(t,x) = x, & H_4(t,x) = \frac{1}{24}x^4 - \frac{1}{4}x^2t + \frac{1}{8}t^2, \\ H_2(t,x) = \frac{1}{2}x^2 - \frac{1}{2}t, & H_5(t,x) = \frac{1}{120}x^5 - \frac{1}{12}x^3t + \frac{1}{8}xt^2. \end{array}$$

For small values of n, $H_n(t,x)$ can be obtained more easily by writing

$$\exp\left\{\gamma x - \frac{\gamma^2 t}{2}\right\} = \left\{\sum_{n \in \mathbb{Z}_+} \frac{(\gamma x)^n}{n!}\right\}\left\{\sum_{n \in \mathbb{Z}_+} \left(\frac{-\gamma^2 t}{2}\right)^n \frac{1}{n!}\right\}$$

and then by long multiplication of the two power series on the right side and by equating the coefficients of γ to $H_n(t,x)$ according to (1).

Theorem 15.19. *Let $X = M_X + V_X$ be a null at 0 quasimartingale on a standard filtered space. Then for $n \in \mathbb{N}$, we have*

$$\int_{[0,t]} dX(t_1) \int_{[0,t_1]} dX(t_2) \cdots \int_{[0,t_{n-1}]} 1 \, dX(t_n) = H_n([M_X]_t, X_t),$$

that is,

$$(\cdots((1 \bullet X) \bullet X) \bullet \cdots) \bullet X = H_n([M_X], X).$$

Proof. For $\gamma \in \mathbb{R}$, let

(1)
$$Y = \gamma X - \frac{1}{2}[\gamma M_X] = \gamma X - \frac{\gamma^2}{2}[M_X].$$

With the quasimartingale γX replacing the quasimartingale X in Theorem 15.15, we have

(2)
$$e^{Y(t)} = 1 + \gamma \int_{[0,t]} e^{Y(s)} dX(s).$$

By (1) above and (1) of Observation 15.18, we have

(3)
$$e^{Y(t)} = \exp\left\{\gamma X_t - \frac{\gamma^2}{2}[M_X]_t\right\} = \sum_{n \in \mathbb{Z}_+} \gamma^n H_n([M_X]_t, X_t).$$

§15. ITÔ'S FORMULA

Note that since $H_n(t, x)$ is a polynomial in t and x and since $[M_X]$ and X are quasimartingales, $H_n([M_X], X)$ is a quasimartingale by Theorem 15.3. For brevity, let us write L_n for $H_n([M_X], X)$. Then by (2) and (3), we have

$$\sum_{n \in \mathbb{Z}_+} \gamma^n L_n(t) = 1 + \gamma \int_{[0,t]} \sum_{n \in \mathbb{Z}_+} \gamma^n L_n(s) \, dX(s) = 1 + \sum_{n \in \mathbb{Z}_+} \gamma^{n+1} \int_{[0,t]} L_n(s) \, dX(s).$$

Equating the coefficients of γ^n of the two sides of the last equation for $n \in \mathbb{Z}_+$ we have

$$\begin{cases} L_0(t) = 1 \\ L_n(t) = \int_{[0,t]} L_{n-1}(s) \, dX(s) \quad \text{for } n \in \mathbb{N} \end{cases}$$

and therefore by iteration

$$\begin{aligned} L_n(t) &= \int_{[0,t]} L_{n-1}(t_1) \, dX(t_1) \\ &= \int_{[0,t]} \left\{ \int_{[0,t_1]} L_{n-2}(t_2) \, dX(t_2) \right\} dX(t_1) \\ &= \int_{[0,t]} \left\{ \int_{[0,t_1]} \left\{ \int_{[0,t_2]} L_{n-3}(t_3) \, dX(t_3) \right\} dX(t_2) \right\} dX(t_1) \\ &\vdots \\ &= \int_{[0,t]} \cdots \left\{ \int_{[0,t_{n-2}]} \left\{ \int_{[0,t_{n-1}]} 1 \, dX(t_n) \right\} dX(t_{n-1}) \right\} \cdots dX(t_1), \end{aligned}$$

that is,

$$L_n(t) = \int_{[0,t]} dX(t_1) \int_{[0,t_1]} dX(t_2) \cdots \int_{[0,t_{n-1}]} 1 \, dX(t_n).$$

Since $L_n = H_n([M_X], X)$ by definition this completes the proof. ∎

Example 15.20 To continue with Example 15.7, let B be an $\{\mathfrak{F}_t\}$-adapted null at 0 Brownian motion on a standard filtered space $(\Omega, \mathfrak{F}, \{\mathfrak{F}_t\}, P)$. By Theorem 15.19 and by (4) of Observation 15.18, we have

$$\int_{[0,t]} dB(t_1) \int_{[0,t_1]} dB(t_2) \int_{[0,t_2]} dB(t_3) \int_{[0,t_3]} 1 \, dB(t_4) = H_4([B]_t, B_t) = \frac{B^4(t)}{24} - \frac{tB^2(t)}{4} + \frac{t^2}{8},$$

and similarly

$$\int_{[0,t]} dB(t_1) \int_{[0,t_1]} dB(t_2) \int_{[0,t_2]} dB(t_3) \int_{[0,t_3]} dB(t_4) \int_{[0,t_4]} 1 \, dB(t_5)$$
$$= H_5([B]_t, B_t) = \frac{B^5(t)}{120} - \frac{tB^3(t)}{12} + \frac{t^2 B(t)}{8}.$$

[IV] Multidimensional Itô's Formula

Definition 15.21. *By a d-dimensional quasimartingale we mean a d-dimensional stochastic process $X = \{X_t : t \in \mathbb{R}_+\}$ on a standard filtered space $(\Omega, \mathfrak{F}, \{\mathfrak{F}_t\}, P)$ whose components $X^{(1)}, \ldots, X^{(d)}$ are quasimartingales on the filtered space, that is, $X^{(i)} = X_0^{(i)} + M^{(i)} + V^{(i)}$ where $X_0^{(i)}$ is a real valued $\mathfrak{F}_0$-measurable random variable, $M^{(i)} \in \mathbf{M}_2^{c,loc}(\Omega, \mathfrak{F}, \{\mathfrak{F}_t\}, P)$, and $V^{(i)} \in \mathbf{V}^{c,loc}(\Omega, \mathfrak{F}, \{\mathfrak{F}_t\}, P)$, for $i = 1, \ldots, d$.*

Let $\mathbf{C}^2(\mathbb{R}^d)$ be the collection of all real valued functions F on $\mathbb{R}^d$ whose first order partial derivatives F_i', $i = 1, \ldots, d$ and second order partial derivatives $F_{i,j}''$, $i, j = 1, \ldots d$, are all continuous on $\mathbb{R}^d$.

Theorem 15.22. *(Multidimensional Itô's Formula) Let $X = \{X_t : t \in \mathbb{R}_+\}$ be a d-dimensional quasimartingale on a standard filtered space $(\Omega, \mathfrak{F}, \{\mathfrak{F}_t\}, P)$ with components given by $X^{(i)} = X_0^{(i)} + M^{(i)} + V^{(i)}$ for $i = 1, \ldots, d$. Let $F \in \mathbf{C}^2(\mathbb{R}^d)$. Then $F \circ X$ is a 1-dimensional quasimartingale given by*

$$(F \circ X)_t - (F \circ X)_0 = \sum_{i=1}^d \int_{[0,t]} (F_i' \circ X)(s) \, dM^{(i)}(s)$$

$$+ \sum_{i=1}^d \int_{[0,t]} (F_i' \circ X)(s) \, dV^{(i)}(s) + \frac{1}{2} \sum_{i,j=1}^d \int_{[0,t]} (F_{i,j}'' \circ X)(s) \, d[M^{(i)}, M^{(j)}](s)$$

on $\mathbb{R}_+ \times \Lambda^c$ where Λ is a null set in $(\Omega, \mathfrak{F}, P)$.

Proof. Theorem 15.22 can be proved in the same way as Theorem 15.3 using Taylor's Theorem for $F \in \mathbf{C}^2(\mathbb{R}^d)$. that is, for any $a, b \in \mathbb{R}^d$,

$$F(b) - F(a) = \sum_{i=1}^d F_i'(a)(b_i - a_i) + \frac{1}{2} \sum_{i,j=1}^d F_{i,j=1}''(c)(b_i - a_i)(b_j - a_j),$$

where $c = (c_1, \ldots, c_d) \in \mathbb{R}^d$ with $c_i \in (a_i \wedge b_i, a_i \vee b_i)$ for $i = 1, \ldots, d$. ∎

Let us note that if we write $M_{X^{(i)}}$ for the martingale part of the quasimartingale $X^{(i)}$ in Theorem 15.22 then in the notations introduced in Definition 15.11 the multidimensional Itô's Formula takes the following form:

$$(F \circ X)_t - (F \circ X)_0 = \sum_{i=1}^d (F_i' \circ X) \bullet X^{(i)} + \frac{1}{2} \sum_{i,j=1}^d (F_{i,j}'' \circ X)(s) \bullet [M_{X^{(i)}}, M_{X^{(j)}}].$$

§15. ITÔ'S FORMULA

Theorem 15.23. Let $B = (B^{(1)}, \ldots, B^{(r)})$ be an $\{\mathfrak{F}_t\}$-adapted r-dimensional null at 0 Brownian motion on a standard filtered space $(\Omega, \mathfrak{F}, \{\mathfrak{F}_t\}, P)$. Let $Y^{(i,k)}$ be predictable processes on the filtered space such that $Y^{(i,k)} \in \mathbf{L}_{2,\infty}^{loc}(\mathbb{R}_+ \times \Omega, m_L \times P)$ for $k = 1, \ldots, r$ and $i = 1, \ldots, d$. Let $Z^{(i)}$ be a predictable process on the filtered space such that $Z^{(i)} \in \mathbf{L}_{1,\infty}^{loc}(\mathbb{R}_+ \times \Omega, m_L \times P)$ for $i = 1, \ldots, d$. Let $X_0^{(i)}$ be real valued $\mathfrak{F}_0$-measurable random variables on $(\Omega, \mathfrak{F}, P)$ for $i = 1, \ldots, d$. Let $X = (X^{(1)}, \ldots, X^{(d)})$ be a d-dimensional stochastic process on the filtered space defined by setting

$$(1) \qquad X^{(i)}(t) = X_0^{(i)} + \sum_{k=1}^{r} \int_{[0,t]} Y^{(i,k)}(s) \, dB^{(k)}(s) + \int_{[0,t]} Z^{(i)}(s) m_L(ds).$$

Then X is a d-dimensional quasimartingale and for any $F \in \mathbf{C}^2(\mathbb{R}^d)$, $F \circ X$ is a quasimartingale given by

$$(2) \quad (F \circ X)_t - (F \circ X)_0 = \sum_{i=1}^{d} \sum_{k=1}^{r} \int_{[0,t]} (F_i' \circ X)(s) Y^{(i,k)}(s) \, dB^{(k)}(s)$$
$$+ \sum_{i=1}^{d} \int_{[0,t]} (F_i' \circ X)(s) Z^{(i)}(s) m_L(ds)$$
$$+ \frac{1}{2} \sum_{i,j=1}^{d} \sum_{k=1}^{r} \int_{[0,t]} (F_{i,j}'' \circ X)(s) Y^{(i,k)}(s) Y^{(j,k)}(s) m_L(ds)$$

on $\mathbb{R}_+ \times \Lambda^c$ where Λ is a null set in $(\Omega, \mathfrak{F}, P)$.

Proof. Let us show first that $X^{(i)}$ is a quasimartingale. Now since $Y^{(i,k)} \in \mathbf{L}_{2,\infty}^{loc}(\mathbb{R}_+ \times \Omega, m_L \times P)$, the stochastic integral $Y^{(i,k)} \bullet B^{(k)}$ is defined and is in $\mathbf{M}_2^{c,loc}(\Omega, \mathfrak{F}, \{\mathfrak{F}_t\}, P)$. Then

$$(3) \qquad M^{(i)} \equiv \sum_{k=1}^{r} Y^{(i,k)} \bullet B^{(k)} \in \mathbf{M}_2^{c,loc}(\Omega, \mathfrak{F}, \{\mathfrak{F}_t\}, P) \quad \text{for } i = 1, \ldots, d.$$

For $i = 1, \ldots, d$, let us define a stochastic process $V^{(i)}$ by setting

$$V^{(i)}(t, \omega) = \int_{[0,t]} Z^{(i)}(s, \omega) m_L(ds) \quad \text{for } (t, \omega) \in \mathbb{R}_+ \times \Omega.$$

Every sample function of $V^{(i)}$ is absolutely continuous and in particular of bounded variation on every finite interval in $\mathbb{R}_+$. The continuity of the sample functions also implies that $V^{(i)}$ is a measurable process. By Theorem 10.11, $V^{(i)}$ is an adapted process. Thus

$$(4) \qquad V^{(i)} \in \mathbf{V}^{c,loc}(\Omega, \mathfrak{F}, \{\mathfrak{F}_t\}, P) \quad \text{for } i = 1, \ldots, d.$$

By (3) and (4), $X^{(i)} = X_0^{(i)} + M^{(i)} + V^{(i)}$ is a quasimartingale for $i = 1, \ldots, d$ and therefore X is a d-dimensional quasimartingale.

By Theorem 15.22, we have

(5) $$(F \circ X)_t - (F \circ X)_0 = \sum_{i=1}^{d} \int_{[0,t]} (F_i' \circ X)(s) \, dM^{(i)}(s)$$
$$+ \sum_{i=1}^{d} \int_{[0,t]} (F_i' \circ X)(s) \, dV^{(i)}(s) + \frac{1}{2} \sum_{i,j=1}^{d} \int_{[0,t]} (F_{i,j}'' \circ X)(s) \, d[M^{(i)}, M^{(j)}](s),$$

on $\mathbb{R}_+ \times \Lambda^c$ where Λ is a null set in $(\Omega, \mathfrak{F}, P)$. Now by Theorem 14.21, we have

(6) $$(F_i' \circ X) \bullet M^{(i)} = \sum_{k=1}^{r} (F_i' \circ X) \bullet (Y^{(i,k)} \bullet B^{(k)}) = \sum_{k=1}^{r} (F_i' \circ X) Y^{(i,k)} \bullet B^{(k)}.$$

Next by the Lebesgue-Radon-Nikodym Theorem we have

(7) $$\int_{[0,t]} (F_i' \circ X)(s) \, dV^{(i)}(s) = \int_{[0,t]} (F_i' \circ X)(s) Z^{(i)}(s) m_L(ds).$$

Finally

$$[M^{(i)}, M^{(j)}] = \sum_{k=1}^{r} \sum_{\ell=1}^{r} [Y^{(i,k)} \bullet B^{(k)}, Y^{(j,\ell)} \bullet B^{(\ell)}]$$
$$= \sum_{k=1}^{r} \sum_{\ell=1}^{r} \int_{[0,t]} Y^{(i,k)}(s) Y^{(j,\ell)}(s) \, d[B^{(k)}, B^{(\ell)}](s)$$
$$= \sum_{k=1}^{r} \int_{[0,t]} Y^{(i,k)}(s) Y^{(j,\ell)}(s) m_L(ds),$$

where the second equality is by Theorem 14.20 and the third equality is by Proposition 13.33. Therefore

(8) $$\int_{[0,t]} (F_{i,j}'' \circ X)(s) \, d[M^{(i)}, M^{(j)}](s) = \sum_{k=1}^{r} \int_{[0,t]} (F_{i,j}'' \circ X)(s) Y^{(i,k)}(s) Y^{(j,\ell)}(s) m_L(ds).$$

Using (6), (7), and (8) in (5) we have (2). ∎

Theorem 15.24. *Let X be a d-dimensional quasimartingale on a standard filtered space $(\Omega, \mathfrak{F}, \{\mathfrak{F}_t\}, P)$ with components given by $X^{(j)} = X_0^{(j)} + M^{(j)}$ where $X_0^{(j)}$ is an $\mathfrak{F}_0$-measurable*

§15. ITÔ'S FORMULA

real valued random variable and $M^{(j)} \in \mathbf{M}_2^{loc}(\Omega, \mathfrak{F}, \{\mathfrak{F}_t\}, P)$ *for* $j = 1, \ldots, d$. *Then* X *is a* d-*dimensional* $\{\mathfrak{F}_t\}$-*adapted Brownian motion if and only if every sample function of* X *is continuous and*

$$[M^{(j)}, M^{(k)}]_t = \delta_{j,k} t \quad \text{for } t \in \mathbb{R}_+ \text{ and } j, k = 1, \ldots, d. \tag{1}$$

Proof. If X is a d-dimensional $\{\mathfrak{F}_t\}$-adapted Brownian motion then every sample function of X is continuous and $M = (M^{(1)}, \ldots, M^{(d)})$ is a d-dimensional $\{\mathfrak{F}_t\}$-adapted null at 0 Brownian motion so that (1) holds according to Proposition 13.33.

To prove the converse, according to Proposition 13.25 it suffices to show that for every $s, t \in \mathbb{R}_+$ such that $s < t$ and $y \in \mathbb{R}^d$ we have

$$\mathbf{E}[e^{i\langle y, X_t - X_s\rangle} | \mathfrak{F}_s] = e^{-\frac{|y|^2}{2}(t-s)} \quad \text{a.e. on } (\Omega, \mathfrak{F}_s, P),$$

or equivalently

$$\mathbf{E}\left[e^{i\langle y, X_t - X_s\rangle} \mathbf{1}_A\right] = P(A) e^{-\frac{|y|^2}{2}(t-s)} \quad \text{for } A \in \mathfrak{F}_s. \tag{2}$$

Consider the function $F(x) = e^{i\langle y, x\rangle}$ for $x \in \mathbb{R}^d$. We have $F \in \mathbf{C}^2(\mathbb{R}^d)$ with $F_j'(x) = iy_j e^{i\langle y, x\rangle}$ and $F_{j,k}''(x) = -y_j y_k e^{i\langle y, x\rangle}$ for $j, k = 1, \ldots, d$. Thus by Theorem 15.22 and by (1) we have

$$e^{i\langle y, X_t\rangle} - e^{i\langle y, X_s\rangle} = i \sum_{j=1}^d y_j \int_{(s,t]} e^{i\langle y, X(u)\rangle} dM^{(j)}(u) - \frac{|y|^2}{2} \int_{(s,t]} e^{i\langle y, X(u)\rangle} m_L(du). \tag{3}$$

Let $A \in \mathfrak{F}_s$. Multiplying both sides of (3) by $e^{-i\langle y, X_s\rangle} \mathbf{1}_A$ and integrating we have

$$\begin{aligned}
\mathbf{E}\left[e^{i\langle y, X_t - X_s\rangle} \mathbf{1}_A\right] - P(A) &= i \sum_{j=1}^d y_j \mathbf{E}\left[e^{-i\langle y, X(s)\rangle} \mathbf{1}_A \int_{(s,t]} e^{i\langle y, X(u)\rangle} dM^{(j)}(u)\right] \\
&\quad - \frac{|y|^2}{2} \mathbf{E}\left[\mathbf{1}_A \int_{(s,t]} e^{i\langle y, X(u) - X(s)\rangle} m_L(du)\right].
\end{aligned} \tag{4}$$

By (1), $[M^{(j)}]_t = t$ and then $\mathbf{E}([M^{(j)}]_t) = t < \infty$ for every $t \in \mathbb{R}_+$. Thus by Theorem 12.36, $M^{(j)} \in \mathbf{M}_2^c(\Omega, \mathfrak{F}, \{\mathfrak{F}_t\}, P)$ for $j = 1, \ldots, d$. The process $e^{i\langle y, X\rangle}$ is continuous and hence predictable. It is also bounded so that it is in $\mathbf{L}_{2,\infty}(\mathbb{R}_+ \times \Omega, \mu_{[M^{(j)}]}, P)$ by Observation 11.19. Then the stochastic integral $e^{i\langle y, X\rangle} \bullet M^{(j)}$ is defined and is in $\mathbf{M}_2^c(\Omega, \mathfrak{F}, \{\mathfrak{F}_t\}, P)$. Thus we have $\mathbf{E}\left[\int_{(s,t]} e^{i\langle y, X(u)\rangle} dM^{(j)}(u) | \mathfrak{F}_s\right] = 0$ a.e. on $(\Omega, \mathfrak{F}_s, P)$. By this equality and by the fact that $e^{-i\langle y, X(s)\rangle} 1_A$ is $\mathfrak{F}_s$-measurable, the expectations in the first member on the right side of (4) are all equal to 0. On the other hand since $\int_A \left\{\int_{(s,t]} |e^{i\langle y, X(u) - X(s)\rangle}| dm_L\right\} dP \leq P(A)(t-s) < \infty$, Fubini's Theorem is applicable to the second term on the right side of (4). Thus (4) is reduced to

$$(5) \qquad \mathbf{E}\left[e^{i\langle y, X_t - X_s\rangle} 1_A\right] - P(A) = -\frac{|y|^2}{2} \int_{(s,t]} \mathbf{E}\left[e^{i\langle y, X(u) - X(s)\rangle} 1_A\right] m_L(du).$$

Now if we define a real valued function φ on $[s, \infty)$ by setting

$$(6) \qquad \varphi(t) = \mathbf{E}\left[e^{i\langle y, X_t - X_s\rangle} 1_A\right] \quad \text{for } t \in [s, \infty),$$

then by (5) we have

$$(7) \qquad \varphi(t) - P(A) = -\frac{|y|^2}{2} \int_{(s,t]} \varphi(u) m_L(du).$$

The unique solution of this integral equation is given by

$$(8) \qquad \varphi(t) = \varphi(s) e^{-\frac{|y|^2}{2}(t-s)}.$$

By (6) and (8) we have (2). ∎

As a particular case of Theorem 15.24, we have

Corollary 15.25. *Let* $X \in M_2(\Omega, \mathfrak{F}, \{\mathfrak{F}_t\}, P)$. *Then X is an $\{\mathfrak{F}_t\}$-adapted Brownian motion if and only if*

1°. *every sample function of X is continuous,*

2°. $\{X_t^2 - t : t \in \mathbb{R}_+\}$ *is a right-continuous null at 0 martingale.*

§16 Itô's Stochastic Calculus

[I] The Space of Stochastic Differentials

For a real valued function f with a continuous derivative f' on $\mathbb{R}$, we have $f(t_2) - f(t_1) = \int_{t_1}^{t_2} f'(t) \, dt$ for any $t_1, t_2 \in \mathbb{R}$ such that $t_1 < t_2$ according to the Fundamental Theorem

§16. ITÔ'S STOCHASTIC CALCULUS

of Integral Calculus. We write df for $f'\,dt$ and thus $f(t_2) - f(t_1) = \int_{t_1}^{t_2} df(t)$. Every antiderivative g of f' satisfies the condition $g(t_2) - g(t_1) = \int_{t_1}^{t_2} f'(t)\,dt = f(t_2) - f(t_1)$ for any $t_1, t_2 \in \mathbb{R}$ such that $t_1 < t_2$. Also every antiderivative g of f' has the form $g = f + C$ for some $C \in \mathbb{R}$. For a quasimartingale X on a standard filtered space $(\Omega, \mathfrak{F}, \{\mathfrak{F}_t\}, P)$ the derivative is undefined. Instead we define the stochastic differential dX of X as the collection of every quasimartingale Y on the filtered space satisfying the condition that $Y(t_2) - Y(t_1) = Y(t_2) - Y(t_1)$ any $t_1, t_2 \in \mathbb{R}_+$ such that $t_1 < t_2$ almost surely. Equivalently, dX is the collection of every the quasimartingales Y on the filtered space of the form $Y = X + C$ where C is a real valued $\mathfrak{F}_0$-measurable random variable on $(\Omega, \mathfrak{F}, P)$. Note then that $X \in dX$ and if $Y \in dX$ then $dX = dY$.

Definition 16.1. *Let* $\mathbf{Q}(\Omega, \mathfrak{F}, \{\mathfrak{F}_t\}, P)$ *be the collection of all quasimartingales X on a standard filtered space $(\Omega, \mathfrak{F}, \{\mathfrak{F}_t\}, P)$, that is, $X = X_0 + M_X + V_X$ where $M_X \in \mathbf{M}_2^{c,loc}(\Omega, \mathfrak{F}, \{\mathfrak{F}_t\}, P)$, $V_X \in \mathbf{V}^{c,loc}(\Omega, \mathfrak{F}, \{\mathfrak{F}_t\}, P)$, and $X_0 \in (\mathfrak{F}_0)$, the collection of all real valued $\mathfrak{F}_0$-measurable random variables on $(\Omega, \mathfrak{F}, P)$.*

Observation 16.2. By Proposition 15.2, the decomposition of $X \in \mathbf{Q}(\Omega, \mathfrak{F}, \{\mathfrak{F}_t\}, P)$ as a sum of the random variable $X_0 \in (\mathfrak{F}_0)$, the martingale part $M_X \in \mathbf{M}_2^{c,loc}(\Omega, \mathfrak{F}, \{\mathfrak{F}_t\}, P)$, and the bounded variation part $V_X \in \mathbf{V}^{c,loc}(\Omega, \mathfrak{F}, \{\mathfrak{F}_t\}, P)$ is unique. Let us call this decomposition the canonical decomposition of a quasimartingale. Since $\mathbf{M}_2^{c,loc}$, $\mathbf{V}^{c,loc}$, and $(\mathfrak{F}_0)$ are linear spaces, so is $\mathbf{Q}$. Furthermore for $X, Y \in \mathbf{Q}$ and $a, b \in \mathbb{R}$, if we let M_{aX+bY} and V_{aX+bY} be the martingale part and bounded variation part of the quasimartingale $aX + bY$, that is,

$$aX + bY = (aX_0 + bY_0) + (aM_X + bM_Y) + (aV_X + bV_Y),$$

then by the uniqueness of decomposition we have

$$\begin{cases} M_{aX+bY} = aM_X + bM_Y \\ V_{aX+bY} = aV_X + bV_Y. \end{cases}$$

Let $X \in \mathbf{Q}(\Omega, \mathfrak{F}, \{\mathfrak{F}_t\}, P)$ given as $X = X_0 + M_X + V_X$. For a predictable process Φ on the filtered space which is in $\mathbf{B}(\mathbb{R}_+ \times \Omega)$, we have by Remark 15.10 and Definition 15.11

$$\begin{aligned} \Phi \bullet M_X &= \{ \int_{[0,t]} \Phi(s)\,dM_X(s) : t \in \mathbb{R}_+ \} \in \mathbf{M}_2^{c,loc}, \\ \Phi \bullet V_X &= \{ \int_{[0,t]} \Phi(s)\,dV_X(s) : t \in \mathbb{R}_+ \} \in \mathbf{V}^{c,loc}, \\ \Phi \bullet X &= \Phi \bullet M_X + \Phi \bullet V_X \in \mathbf{Q}, \end{aligned}$$

and as an alternate notation for $(\Phi \bullet X)_t$ for $t \in \mathbb{R}_+$, we have

$$\int_{[0,t]} \Phi(s)\, dX(s) = \int_{[0,t]} \Phi(s)\, dM_X(s) + \int_{[0,t]} \Phi(s)\, dV_X(s).$$

If $X, Y \in \mathbf{Q}$ and if we assume that every sample function of Y is continuous then Y is a predictable process which is also in $\mathbf{B}(\mathbb{R}_+ \times \Omega)$ so that $Y \bullet X$ exists in $\mathbf{Q}$.

Definition 16.3. *We say that $X, Y \in \mathbf{Q}(\Omega, \mathfrak{F}, \{\mathfrak{F}_t\}, P)$ are equivalent as quasimartingales and write $X \overset{q}{\sim} Y$ if there exists a null set Λ in $(\Omega, \mathfrak{F}, P)$ such that for $\omega \in \Lambda^c$ we have*

$$X(t, \omega) - X(s, \omega) = Y(t, \omega) - Y(s, \omega) \quad \text{for every } s, t \in \mathbb{R}_+, \ s < t.$$

For $X \in \mathbf{Q}(\Omega, \mathfrak{F}, \{\mathfrak{F}_t\}, P)$, we write dX for the equivalence class in $\mathbf{Q}(\Omega, \mathfrak{F}, \{\mathfrak{F}_t\}, P)$ with respect to the equivalence relation $\overset{q}{\sim}$ to which X belongs and call it the stochastic differential of X. We write $\mathbf{dQ}$ for the collection of the equivalence classes in $\mathbf{Q}(\Omega, \mathfrak{F}, \{\mathfrak{F}_t\}, P)$ with respect to $\overset{q}{\sim}$.

For two stochastic processes X and Y on a probability space $(\Omega, \mathfrak{F}, P)$ we write $X = Y$ to indicate their equivalence in the sense of Definition 2.1, that is, there exists a null set Λ in $(\Omega, \mathfrak{F}, P)$ such that $X(t, \omega) = Y(t, \omega)$ for $(t, \omega) \in \mathbb{R}_+ \times \Lambda^c$. If ξ is a real valued random variable on $(\Omega, \mathfrak{F}, P)$, then we write $X = \xi$ to indicate that $X(t, \omega) = \xi(\omega)$ for $(t, \omega) \in \mathbb{R}_+ \times \Lambda^c$.

Remark 16.4. Let $X, Y \in \mathbf{Q}(\Omega, \mathfrak{F}, \{\mathfrak{F}_t\}, P)$ given by $X = X_0 + M_X + V_X$ and $Y = Y_0 + M_Y + V_Y$ respectively. Then

(1) $\qquad X \overset{q}{\sim} Y \ \Leftrightarrow \ X - Y = X_0 - Y_0,$

(2) $\qquad X \overset{q}{\sim} Y \ \Leftrightarrow \ M_X = M_Y \text{ and } V_X = V_Y,$

(3) $\qquad X \overset{q}{\sim} Y \ \Leftrightarrow \ X - Y = Z_0 \text{ for some } Z_0 \in (\mathfrak{F}_0).$

Proof. 1) (1) is immediate from Definition 16.3 by considering $s = 0$.

2) If $X \overset{q}{\sim} Y$ then by (1) we have $X - Y = X_0 - Y_0 \in (\mathfrak{F}_0)$. Since $X - Y \in \mathbf{Q}$ this implies $M_{X-Y} = 0$ and $V_{X-Y} = 0$ by Proposition 15.2. But according to Observation 16.2, $M_{X-Y} = M_X - M_Y$ and $V_{X-Y} = V_X - V_Y$. Thus $M_X = M_Y$ and $V_X = V_Y$. Conversely, if $M_X = M_Y$ and $V_X = V_Y$, then $X - Y = X_0 - Y_0$ and thus $X \overset{q}{\sim} Y$ by (1). This proves (2).

3) By (1) the implication $\Rightarrow$ in (3) holds. Conversely if $X - Y = Z_0 \in (\mathfrak{F}_0)$ then by Proposition 15.2 we have $M_{X-Y} = 0$ and $V_{X-Y} = 0$. Then by Observation 16.2, we have $M_X = M_Y$ and $V_X = V_Y$. Thus by (2) we have $X \overset{q}{\sim} Y$. This proves (3). ∎

§16. ITÔ'S STOCHASTIC CALCULUS

Observation 16.5. 1) Since $(\mathfrak{F}_0)$ contains the identically vanishing random variable 0 and $\mathbf{M}_2^{c,loc}(\Omega, \mathfrak{F}, \{\mathfrak{F}_t\}, P)$ and $\mathbf{V}^{c,loc}(\Omega, \mathfrak{F}, \{\mathfrak{F}_t\}, P)$ contain the identically vanishing stochastic process 0, $M = 0 + M + 0 \in \mathbf{Q}(\Omega, \mathfrak{F}, \{\mathfrak{F}_t\}, P)$ for any $M \in \mathbf{M}_2^{c,loc}(\Omega, \mathfrak{F}, \{\mathfrak{F}_t\}, P)$ and similarly $V = 0 + 0 + V \in \mathbf{Q}(\Omega, \mathfrak{F}, \{\mathfrak{F}_t\}, P)$ for any $V \in \mathbf{V}^{c,loc}(\Omega, \mathfrak{F}, \{\mathfrak{F}_t\}, P)$. Thus $\mathbf{M}_2^{c,loc}(\Omega, \mathfrak{F}, \{\mathfrak{F}_t\}, P)$ and $\mathbf{V}^{c,loc}(\Omega, \mathfrak{F}, \{\mathfrak{F}_t\}, P)$ are linear subspaces of $\mathbf{Q}(\Omega, \mathfrak{F}, \{\mathfrak{F}_t\}, P)$. Note also $\mathbf{M}_2^{c,loc}(\Omega, \mathfrak{F}, \{\mathfrak{F}_t\}, P) \cap \mathbf{V}^{c,loc}(\Omega, \mathfrak{F}, \{\mathfrak{F}_t\}, P) = \{0\}$ by Proposition 15.2.

2) Let $X \in \mathbf{Q}(\Omega, \mathfrak{F}, \{\mathfrak{F}_t\}, P)$ be given in its canonical decomposition as $X = X_0 + M_X + V_X$. If $X \overset{q}{\sim} M$ for some $M \in \mathbf{M}_2^{c,loc}(\Omega, \mathfrak{F}, \{\mathfrak{F}_t\}, P)$, then $M_X = M$ and $V_X = 0$ by (2) of Remark 16.4. Similarly if $X \overset{q}{\sim} V$ for some $V \in \mathbf{V}^{c,loc}(\Omega, \mathfrak{F}, \{\mathfrak{F}_t\}, P)$, then $M_X = 0$ and $V_X = V$.

3) Since $\mathbf{M}_2^{c,loc}(\Omega, \mathfrak{F}, \{\mathfrak{F}_t\}, P) \subset \mathbf{Q}(\Omega, \mathfrak{F}, \{\mathfrak{F}_t\}, P)$, for $M \in \mathbf{M}_2^{c,loc}(\Omega, \mathfrak{F}, \{\mathfrak{F}_t\}, P)$ the stochastic differential dM is defined and consists of all $X \in \mathbf{Q}(\Omega, \mathfrak{F}, \{\mathfrak{F}_t\}, P)$ such that $X = Z_0 + M$ where $Z_0 \in (\mathfrak{F}_0)$ according to (3) of Remark 16.4. Similarly for $V \in \mathbf{V}^{c,loc}(\Omega, \mathfrak{F}, \{\mathfrak{F}_t\}, P)$, dV is defined and consists of all $X \in \mathbf{Q}(\Omega, \mathfrak{F}, \{\mathfrak{F}_t\}, P)$ such that $X = Z_0 + V$ where $Z_0 \in (\mathfrak{F}_0)$.

Definition 16.6. *We define* $\mathbf{dM}_2^{c,loc}$ *as the subcollection of* $\mathbf{dQ}$ *consisting of* dM *for* $M \in \mathbf{M}_2^{c,loc}(\Omega, \mathfrak{F}, \{\mathfrak{F}_t\}, P)$. *Similarly we define* $\mathbf{dV}^{c,loc}$ *as the subcollection of* $\mathbf{dQ}$ *consisting of* dV *for* $V \in \mathbf{V}^{c,loc}(\Omega, \mathfrak{F}, \{\mathfrak{F}_t\}, P)$.

Definition 16.7. *We define four operations in* $\mathbf{dQ}$. *For* $X, Y \in \mathbf{Q}(\Omega, \mathfrak{F}, \{\mathfrak{F}_t\}, P)$ *and* $c \in \mathbb{R}$, *let addition, scalar multiplication and product be defined respectively by*

(1) $$dX + dY = d(X + Y),$$
(2) $$c\, dX = d(cX),$$
(3) $$dX \cdot dY = d[M_X, M_Y],$$

and for a predictable process $\Phi \in \mathbf{L}_{2,\infty}^{loc}(\mathbb{R}_+ \times \Omega, \mu_{[M_X]}, P) \cap \mathbf{L}_{1,\infty}^{loc}(\mathbb{R}_+ \times \Omega, \mu_{|V_X|}, P)$, *let* $\bullet$-*multiplication of* Φ *by* dX *be defined by*

(4) $$\Phi \bullet dX = d(\Phi \bullet X).$$

Sometimes $dX \cdot dY$ *is written as* $dX\, dY$ *and* $\Phi \bullet dX$ *is written as* ΦdX.

Regarding (3) and (4), note that since $[M_X, M_Y] \in \mathbf{V}^{c,loc} \subset \mathbf{Q}$, $d[M_X, M_Y]$ is defined and since $\Phi \bullet X \in \mathbf{Q}$, $d(\Phi \bullet X)$ is defined.

With addition and scalar multiplication as defined above **dQ** is a linear space over $\mathbb{R}$. We show next that **dQ** is a commutative ring with respect to addition and product. If we define addition and product in $\mathbf{B}(\mathbb{R}_+ \times \Omega)$ by pointwise addition and product of its elements on $\mathbb{R}_+ \times \Omega$, then $\mathbf{B}(\mathbb{R}_+ \times \Omega)$ is a commutative ring with identity. We show below that with respect to addition and •-multiplication, **dQ** is a commutative algebra over the ring $\mathbf{B}(\mathbb{R}_+ \times \Omega)$.

Observation 16.8. For $X, Y, Z \in \mathbf{Q}(\Omega, \mathfrak{F}, \{\mathfrak{F}_t\}, P)$, we have

(1) $$(dX + dY) \cdot dZ = dX \cdot dZ + dY \cdot dZ,$$
(2) $$dX \cdot dY = dY \cdot dX.$$

Thus **dQ** is a commutative ring with respect to addition and product in Definition 16.7.

Proof. To prove (1), note that by (1) and (3) of Definition 16.7 we have

$$(dX + dY) \cdot dZ = d(X + Y) \cdot dZ = d[M_{X+Y}, M_Z] = d[M_X + M_Y, M_Z]$$
$$= d([M_X, M_Z] + [M_Y, M_Z]) = d[M_X, M_Z] + d[M_Y, M_Z] = dX \cdot dZ + dY \cdot dZ.$$

To prove (2), note that by (3) of Definition 16.7 we have

$$dX \cdot dY = d[M_X, M_Y] = d[M_Y, M_X] = dY \cdot dX. \quad \blacksquare$$

Theorem 16.9. *Let $X, Y \in \mathbf{Q}(\Omega, \mathfrak{F}, \{\mathfrak{F}_t\}, P)$ and Φ and Ψ be predictable processes on the filtered space such that*

$$\Phi, \Psi \in \mathbf{L}_{2,\infty}^{loc}(\mathbb{R}_+ \times \Omega, \mu_{[M_X]}, P) \cap \mathbf{L}_{1,\infty}^{loc}(\mathbb{R}_+ \times \Omega, \mu_{|V_X|}, P)$$
$$\cap \; \mathbf{L}_{2,\infty}^{loc}(\mathbb{R}_+ \times \Omega, \mu_{[M_Y]}, P) \cap \mathbf{L}_{1,\infty}^{loc}(\mathbb{R}_+ \times \Omega, \mu_{|V_Y|}, P).$$

Then

(1) $$\Phi \bullet (dX + dY) = \Phi \bullet dX + \Phi \bullet dY,$$
(2) $$(\Phi + \Psi) \bullet dX = \Phi \bullet dX + \Psi \bullet dX,$$
(3) $$\Phi \bullet (dX \cdot dY) = (\Phi \bullet dX) \cdot dY.$$

*In particular, (1), (2), and (3) hold for $\Phi, \Psi \in \mathbf{B}(\mathbb{R}_+ \times \Omega)$. Thus **dQ** is a commutative algebra over $\mathbf{B}(\mathbb{R}_+ \times \Omega)$ with respect to addition and •-multiplication in Definition 16.7. Also for $\Phi, \Psi \in \mathbf{B}(\mathbb{R}_+ \times \Omega)$ we have*

(4) $$\Phi\Psi \bullet dX = \Phi \bullet (\Psi \bullet dX).$$

§16. ITÔ'S STOCHASTIC CALCULUS

Proof. 1) To prove (1), note that by (1) and (4) of Definition 16.7 and by the linearity of the stochastic integral we have

$$\Phi \bullet (dX + dY) = \Phi \bullet d(X + Y) = d(\Phi \bullet (X + Y))$$
$$= d(\Phi \bullet X + \Phi \bullet Y) = d(\Phi \bullet X) + d(\Phi \bullet Y) = \Phi \bullet dX + \Phi \bullet dY.$$

2) To prove (2), note that by (4) and (1) of Definition 16.7 and the linearity of the stochastic integral, we have

$$(\Phi+\Psi)\bullet dX = d((\Phi+\Psi)\bullet X) = d(\Phi\bullet X + \Psi\bullet X) = d(\Phi\bullet X) + d(\Psi\bullet X) = \Phi\bullet dX + \Psi\bullet dX.$$

3) To prove (4), note that by (3) and (4) of Definition 16.7, we have on one hand

$$(5) \qquad \Phi \bullet (dX \cdot dY) = \Phi \bullet d[M_X, M_Y] = d(\Phi \bullet [M_X, M_Y]),$$

and on the other hand

$$(6) \qquad (\Phi \bullet dX) \cdot dY = d(\Phi \bullet X) \cdot dY = d[M_{\Phi \bullet X}, M_Y] = d[\Phi \bullet M_X, M_Y],$$

where the last equality is from the fact that the martingale part of the stochastic integral of Φ with respect to X is the stochastic integral of Φ with respect to the martingale part of X. By Theorem 14.20, we have $[\Phi \bullet M_X, M_Y]_t = \int_{[0,t]} \Phi(s) d[M_X, M_Y](s)$ for $t \in \mathbb{R}_+$, that is, $[\Phi \bullet M_X, M_Y] = \Phi \bullet [M_X, M_Y]$ and thus $d[\Phi \bullet M_X, M_Y] = d(\Phi \bullet [M_X, M_Y])$. Using this in (6), we have (4) from (5) and (6).

4) To prove (4), let us observe first that $\Phi\Psi \bullet X = \Phi\Psi \bullet M_X + \Phi\Psi \bullet V_X$. By Theorem 14.21, we have $\Phi\Psi \bullet M_X = \Phi \bullet (\Psi \bullet M_X)$. Also for every $t \in \mathbb{R}_+$ we have

$$(\Phi\Psi \bullet V_X)_t = \int_{[0,t]} \Phi(s)\Psi(s) \, dV_X(s)$$
$$= \int_{[0,t]} \Phi(s) \, d\left\{\int_{[0,s]} \Psi(u) \, dV_X(u)\right\} = (\Phi \bullet (\Psi \bullet V_X))_t$$

by the Lebesgue-Radon-Nikodym Theorem so that $\Phi\Psi \bullet V_X = \Phi \bullet (\Psi \bullet V_X)$. Thus

$$\Phi\Psi \bullet X = \Phi \bullet (\Psi \bullet M_X) + \Phi \bullet (\Psi \bullet V_X) = \Phi \bullet (\Psi \bullet M_X + \Psi \bullet V_X) = \Phi \bullet (\Psi \bullet X).$$

By (4) of Definition 16.7 and by the last equality, we have

$$\Phi\Psi \bullet dX = d(\Phi\Psi \bullet X) = d(\Phi \bullet (\Psi \bullet X)) = \Phi \bullet d(\Psi \bullet X)) = \Phi \bullet (\Psi \bullet dX).$$

This proves (4). ∎

Corollary 16.10. *Under the same assumptions as in Theorem 16.9, we have*

(1) $$\Phi \bullet (dX \cdot dY) = (\Phi \bullet dX) \cdot dY = (\Phi \bullet dY) \cdot dX,$$

(2) $$d(\Phi \bullet X) \cdot d(\Psi \bullet Y) = \Phi\Psi \bullet (dX \cdot dY) = d(\Psi \bullet X) \cdot d(\Phi \bullet Y),$$

and for predictable processes $\Phi_1, \ldots, \Phi_n \in \mathbf{B}(\mathbb{R}_+ \times \Omega)$, *we have*

(2) $$\Phi_1 \cdots \Phi_n \bullet dX = \Phi_1 \bullet (\cdots (\Phi_{n-1} \bullet (\Phi_n \bullet dX)) \cdots).$$

Proof. The first equality is by (3) of Theorem 16.9. The remaining equalities are from the fact that the product in **dQ** is commutative and the fact that $\Phi \bullet dX, \Phi \bullet dY \in \mathbf{dQ}$.

To prove (2), note first that by (3) of Definition 16.7 and by Definition 15.11, we have

$$d(\Phi \bullet X) \cdot d(\Psi \bullet Y) = d[M_{\Phi \bullet X}, M_{\Psi \bullet Y}] = d[\Phi \bullet M_X, \Psi \bullet M_Y].$$

Now $[\Phi \bullet M_X, \Psi \bullet M_Y] = \Phi\Psi \bullet [M_X, M_Y]$ by Theorem 14.20 so that by (4) and (3) of Definition 16.7 we have

$$d[\Phi \bullet M_X, \Psi \bullet M_Y] = d(\Phi\Psi \bullet [M_X, M_Y]) = \Phi\Psi \bullet d[M_X, M_Y] = \Phi\Psi \bullet (dX \cdot dY).$$

Therefore $d(\Phi \bullet X) \cdot d(\Psi \bullet Y) = \Phi\Psi \bullet (dX \cdot dY)$. Similarly $d(\Psi \bullet X) \cdot d(\Phi \bullet Y) = \Psi\Phi \bullet (dX \cdot dY)$. This proves (2).

The equality (3) is obtained by iterated application of (4) of Theorem 16.9. Thus

$$\Phi_1 \cdots \Phi_n \bullet dX = \Phi_1 \bullet (\Phi_2 \cdots \Phi_n \bullet dX) = \Phi_1 \bullet (\Phi_2 \bullet (\Phi_3 \cdots \Phi_n \bullet dX)),$$

and so on. ∎

For the identically vanishing process 0, $d0 \in \mathbf{dQ}$ consists of all almost surely constant (in time) processes according to Definition 16.3. By (1) of Definition 16.7, $d0$ is the identity of addition in the linear space **dQ**.

Theorem 16.11. *For the product in* **dQ** *we have*

(1) $$\mathbf{dQ} \cdot \mathbf{dQ} \subset \mathbf{dV}^{c,loc},$$
(2) $$\mathbf{dV}^{c,loc} \cdot \mathbf{dQ} = \{d0\},$$
(3) $$\mathbf{dQ} \cdot \mathbf{dQ} \cdot \mathbf{dQ} = \{d0\}.$$

§16. ITÔ'S STOCHASTIC CALCULUS

Proof. To prove (1), note that if $X, Y \in \mathbf{Q}$ then by (3) of Definition 16.7 we have $dX \cdot dY = d[M_X, M_Y] \in \mathbf{dV}^{c,loc}$. To prove (2), note that if $V \in \mathbf{dV}^{c,loc}$, then $M_V = 0$ so that $[M_V, M_X] = 0$ for any $X \in \mathbf{Q}$ and thus $dV \cdot dX = d0$. Finally for $X, Y, Z \in \mathbf{Q}$ we have $dX \cdot dY \in \mathbf{dV}^{c,loc}$ by (1) and then $dX \cdot dY \cdot dZ = d0$ by (2). This proves (3). ∎

Definition 16.12. *Let $X \in \mathbf{Q}(\Omega, \mathfrak{F}, \{\mathfrak{F}_t\}, P)$. For the stochastic differential $dX \in \mathbf{dQ}$ and for $s, t \in \mathbb{R}_+$ such that $s < t$ we define*

(1) $$\int_{(s,t]} dX(u) = X(t) - X(s).$$

For $dX, dY \in \mathbf{dQ}$ and $a, b \in \mathbb{R}$, we define

(2) $$\int_{[0,t]} (a\, dX + b\, dY)(u) = a \int_{[0,t]} dX(u) + b \int_{[0,t]} dY(u).$$

Remark 16.13. 1) Recall that the stochastic differential dX consists of all quasimartingales $Y \in \mathbf{Q}(\Omega, \mathfrak{F}, \{\mathfrak{F}_t\}, P)$ such that $Y \overset{q}{\sim} X$, that is, $Y(t, \omega) - Y(s, \omega) = X(t, \omega) - X(s, \omega)$ for every $s, t \in \mathbb{R}_+$, $s < t$, and $\omega \in \Lambda^c$ where Λ is a null set in $(\Omega, \mathfrak{F}, P)$. Thus if Y is an arbitrary representative of dX, then we have $\int_{(s,t]} dX(u) = Y(t) - Y(s) = X(t) - X(s)$. This shows that $\int_{(s,t]} dX(u)$ in (1) of Definition 16.12 is well defined.

2) According to Definition 15.11, for the stochastic integral of the process 1, that is, the process $\Phi(t, \omega) = 1$ for $(t, \omega) \in \mathbb{R}_+ \times \Omega$, with respect to a quasimartingale $X \in \mathbf{Q}(\Omega, \mathfrak{F}, \{\mathfrak{F}_t\}, P)$ given by $X = X_0 + M_X + V_X$, we have

$$\int_{(s,t]} \mathbf{1}_{(s,t]}(u)\, dX(u) = \int_{(s,t]} \mathbf{1}_{(s,t]}(u)\, dM_X(u) + \int_{(s,t]} \mathbf{1}_{(s,t]}(u)\, dV_X(u)$$
$$= \{M_X(t) - M_X(s)\} + \{V_X(t) - V_X(s)\} = X(t) - X(s).$$

Thus $\int_{[0,t]} dX(u)$ in Definition 16.12 is equal to $\int_{[0,t]} 1\, dX(u)$ in Definition 15.11. In particular for the stochastic differential $d(\Phi \bullet X)$ where Φ is a predictable process in $\mathbf{L}_{2,\infty}^{loc}(\mathbb{R}_+ \times \Omega, \mu_{[M_X]}, P) \cap \mathbf{L}_{1,\infty}^{loc}(\mathbb{R}_+ \times \Omega, \mu_{|V_X|}, P)$, we have

$$\int_{(s,t]} (\Phi \bullet dX)(u) = \int_{(s,t]} d(\Phi \bullet X)(u) = (\Phi \bullet X)(t) - (\Phi \bullet X)(s) = \int_{(s,t]} \Phi(u)\, dX(u).$$

Observation 16.14. Let $X = (X^{(1)}, \ldots, X^{(d)})$ where $X^{(1)}, \ldots, X^{(d)} \in \mathbf{Q}(\Omega, \mathfrak{F}, \{\mathfrak{F}_t\}, P)$ and let $F \in \mathbf{C}^2(\mathbb{R}^d)$. According to Theorem 15.21 there exists a null set Λ in $(\Omega, \mathfrak{F}, P)$ such that on Λ^c we have for any $s,t \in \mathbb{R}_+$ such that $s < t$

$$(F \circ X)(t) - (F \circ X)(s) = \sum_{i=1}^{d}\{((F_i' \circ X) \bullet X^{(i)})(t) - ((F_i' \circ X) \bullet X^{(i)})(s)\}$$

$$= \sum_{i,j=1}^{d} \frac{1}{2}\{((F_{i,j}'' \circ X) \bullet [M_{X^{(i)}}, M_{X^{(j)}}])(t) - ((F_{i,j}'' \circ X) \bullet [M_{X^{(i)}}, M_{X^{(j)}}])(s)\}.$$

By Definition 16.3 and Theorem 16.9, we have

$$d(F \circ X) = \sum_{i=1}^{d} (F_i' \circ X) \bullet dX^{(i)} + \frac{1}{2} \sum_{i,j=1}^{d} (F_{i,j}'' \circ X) \bullet dX^{(i)} \cdot dX^{(j)}.$$

[II] Fisk-Stratonovich Integrals

Let $X, Y \in \mathbf{Q}(\Omega, \mathfrak{F}, \{\mathfrak{F}_t\}, P)$. Then $Y \in \mathbf{B}(\mathbb{R}_+ \times \Omega) \subset \mathbf{L}_{2,\infty}^{loc}(\mathbb{R}_+ \times \Omega, \mu_{[M_X]}, P) \cap \mathbf{L}_{1,\infty}^{loc}(\mathbb{R}_+ \times \Omega, \mu_{|V_X|}, P)$ so that $Y \bullet X$ exists in $\mathbf{Q}(\Omega, \mathfrak{F}, \{\mathfrak{F}_t\}, P)$ by Definition 15.11 and $d(Y \bullet X) = Y \bullet dX$ by Definition 16.7.

Lemma 16.15. *If $X, Y \in \mathbf{Q}(\Omega, \mathfrak{F}, \{\mathfrak{F}_t\}, P)$, then $XY \in \mathbf{Q}(\Omega, \mathfrak{F}, \{\mathfrak{F}_t\}, P)$ and*

$$d(XY) = X \bullet dY + Y \bullet dX + dX \cdot dY.$$

Proof. Consider the 2-dimensional quasimartingale $(X^{(1)}, X^{(1)}) = (X, Y)$ and $F \in \mathbf{C}^2(\mathbb{R}^2)$ defined by $F(x, y) = xy$ for $(x, y) \in \mathbb{R}^2$. By the multidimensional Itô's Formula we have $F \circ (X^{(1)}, X^{(2)}) \in \mathbf{Q}$, that is, $XY \in \mathbf{Q}$, and furthermore since $F_1'(x,y) = y$, $F_2'(x,y) = x$, $F_{1,1}''(x,y) = F_{2,2}''(x,y) = 0$, and $F_{1,2}''(x,y) = F_{2,1}''(x,y) = 1$ for $(x,y) \in \mathbb{R}^2$, we have according to the multidimensional Itô's Formula as given in Observation 16.14

$$d(XY) = X^{(2)} \bullet dX^{(1)} + X^{(1)} \bullet dX^{(2)} + \frac{1}{2}\{dX^{(1)} \cdot dX^{(2)} + dX^{(2)} \cdot dX^{(1)}\}$$
$$= Y \bullet dX + X \bullet dY + dX \cdot dY. \blacksquare$$

Definition 16.16. *Let $X, Y \in \mathbf{Q}(\Omega, \mathfrak{F}, \{\mathfrak{F}_t\}, P)$. The symmetric multiplication of Y by dX is defined by*

$$Y \diamond dX = Y \bullet dX + \frac{1}{2} dY \cdot dX.$$

§16. ITÔ'S STOCHASTIC CALCULUS

Note that $Y \diamond dX \in \mathbf{dQ}$.

Lemma 16.17. *Let* $X, Y \in \mathbf{Q}(\Omega, \mathfrak{F}, \{\mathfrak{F}_t\}, P)$. *Then*

$$d(XY) = X \diamond dY + Y \diamond dX.$$

Proof. By Definition 16.16, we have $Y \diamond dX = Y \bullet dX + \frac{1}{2} dY \cdot dX$ and $X \diamond dY = X \bullet dY + \frac{1}{2} dX \cdot dY$. Adding the two equalities side by side and recalling Lemma 16.15 we have

$$X \diamond dY + Y \diamond dX = X \bullet dY + Y \bullet dX + dX \cdot dY = d(XY). \quad \blacksquare$$

Theorem 16.18. *For* $X, Y, Z \in \mathbf{Q}(\Omega, \mathfrak{F}, \{\mathfrak{F}_t\}, P)$, *we have*

(1) $\quad X \diamond (dY + dZ) = X \diamond dY + X \diamond dZ,$

(2) $\quad (X + Y) \diamond dZ = X \diamond dZ + Y \diamond dZ,$

(3) $\quad XY \diamond dZ = X \diamond (Y \diamond dZ),$

(4) $\quad X \diamond (dY \cdot dZ) = X \bullet (dY \cdot dZ),$

(5) $\quad X \diamond (dY \cdot dZ) = (X \diamond dY) \cdot dZ.$

Proof. To prove (1), note that

$$X \diamond (dY + dZ) = X \diamond d(Y + Z) = X \bullet d(Y + Z) + \frac{1}{2} dX \cdot d(Y + Z)$$
$$= X \bullet dY + X \bullet dZ + \frac{1}{2}(dX \cdot dY + dX \cdot dZ) = X \diamond dY + X \diamond dZ,$$

where the first equality is by (1) of Definition 16.7, the second equality is by Definition 16.16, the third equality is by (1) of Definition 16.7 and Observation 16.8, and the last equality is by Definition 16.16. The equality (2) is proved likewise.

To prove (3), note that

(6) $\quad XY \diamond dZ = XY \bullet dZ + \frac{1}{2} d(XY) \cdot dZ$

$\quad\quad\quad\quad = XY \bullet dZ + \frac{1}{2}(X \bullet dY + Y \bullet dX + dX \cdot dY) \cdot dZ$

by Lemma 16.15. Since $Y \diamond dZ \in \mathbf{dQ}$, we have by Definition 16.16

(7) $$X \diamond (Y \diamond dZ) = X \bullet (Y \diamond dZ) + \frac{1}{2} dX \cdot (Y \diamond dZ)$$
$$= X \bullet (Y \bullet dZ + \frac{1}{2} dY \cdot dZ) + \frac{1}{2} dX \cdot (Y \bullet dZ + \frac{1}{2} dY \cdot dZ)$$
$$= XY \bullet dZ + \frac{1}{2}(X \bullet dY) \cdot dZ + \frac{1}{2}(Y \bullet dX) \cdot dZ + \frac{1}{4} dX \cdot dY \cdot dZ$$

by (4) of Theorem 16.9 and (1) of Corollary 16.10. By (6), (7), and the fact that $dX \cdot dY \cdot dX = d0$, we have (3).

To prove (4), note that since $dY \cdot dZ \in \mathbf{dQ}$ we have by Definition 16.16 and (3) of Theorem 16.11

$$X \diamond (dY \cdot dZ) = X \bullet (dY \cdot dZ) + \frac{1}{2} dX \cdot (dY \cdot dZ) = X \bullet (dY \cdot dZ).$$

To prove (5), note that

$$(X \diamond dY) \cdot dZ = (X \bullet dY + \frac{1}{2} dX \cdot dY) \cdot dZ$$
$$= (X \bullet dY) \cdot dZ = X \bullet (dY \cdot dZ) = X \diamond (dY \cdot dZ),$$

where the second equality is by (3) of Theorem 16.11, the third equality is by (3) of Theorem 16.9, and the last equality is by (4). ∎

Theorem 16.19. *Let $X = (X^{(1)}, \ldots, X^{(d)})$ where $X^{(i)} \in \mathbf{Q}(\Omega, \mathfrak{F}, \{\mathfrak{F}_t\}, P)$ for $i = 1, \ldots, d$ and let $F \in \mathbf{C}^3(\mathbb{R}^d)$. Then $F \circ X \in \mathbf{Q}(\Omega, \mathfrak{F}, \{\mathfrak{F}_t\}, P)$ and*

$$d(F \circ X) = \sum_{i=1}^{d} (F_i' \circ X) \diamond dX^{(i)}.$$

Proof. By Definition 16.16, we have

(1) $$(F_i' \circ X) \diamond dX^{(i)} = (F_i' \circ X) \bullet dX^{(i)} + \frac{1}{2} d(F_i' \circ X) \cdot dX^{(i)}.$$

Since $F_i' \in \mathbf{C}^2(\mathbb{R}^d)$, by Itô's Formula as given in Observation 16.14 we have

$$d(F_i' \circ X) = \sum_{j=1}^{d} (F_{i,j}'' \circ X) \bullet dX^{(j)} + \frac{1}{2} \sum_{j,k=1}^{d} (F_{i,j,k}''' \circ X) \bullet dX^{(j)} \cdot dX^{(k)}.$$

§16. ITÔ'S STOCHASTIC CALCULUS

Substituting this in (1) and recalling that $dX^{(i)} \cdot dX^{(j)} \cdot dX^{(k)} = d0$ by Theorem 16.11, we have

$$\sum_{i=1}^{d}(F_i' \circ X) \diamond dX^{(i)} = \sum_{i=1}^{d}(F_i' \circ X) \bullet dX^{(i)} + \frac{1}{2}\sum_{i,j=1}^{d}(F_{i,j}'' \circ X) \bullet dX^{(i)} \cdot dX^{(j)} = d(F \circ X). \blacksquare$$

Definition 16.20. Let $X, Y \in \mathbf{Q}(\Omega, \mathfrak{F}, \{\mathfrak{F}_t\}, P)$ be given by $X = X_0 + M_X + V_X$ and $Y = Y_0 + M_Y + V_Y$. The Fisk-Stratonovich integral of Y with respect to X, $Y \diamond X$, is the quasimartingale $Y \diamond X$ defined by

(1) $$Y \diamond X = Y \bullet X + \frac{1}{2}[M_Y, M_X]$$

and thus

(2) $$d(Y \diamond X) = d(Y \bullet X) + \frac{1}{2}d[M_Y, M_X] = Y \bullet dX + \frac{1}{2}dY \cdot dX = Y \diamond dX.$$

We also use the notation $\int_{[0,t]} Y \diamond dX$ for $(Y \diamond X)_t$, that is, $\int_{[0,t]} Y \diamond dX = \int_{[0,t]} d(Y \diamond X)(s)$ for $t \in \mathbb{R}_+$.

Example 16.21. Let B be an $\{\mathfrak{F}_t\}$-adapted null at 0 Brownian motion on a standard filtered space $(\Omega, \mathfrak{F}, \{\mathfrak{F}_t\}, P)$. Then

$$\int_{[0,t]} B \diamond dB = \int_{[0,t]} B(s)\, dB(s) + \frac{1}{2}\int_{[0,t]} d[B, B](s)$$

Since $\int_{[0,t]} B(s)\, dB(s) = \frac{1}{2}\{B^2(t) - t\}$ as we showed in Example 15.7 and since $[B, B]_t = t$, we have

$$\int_{[0,t]} B \diamond dB = \frac{1}{2}B^2(t),$$

which resembles $\int_0^t s\, ds = \frac{1}{2}t^2$.

Theorem 16.22. Let $X \in \mathbf{M}_2^c(\Omega, \mathfrak{F}, \{\mathfrak{F}_t\}, P)$ and let Y be a bounded continuous martingale on $(\Omega, \mathfrak{F}, \{\mathfrak{F}_t\}, P)$. For $n \in \mathbb{N}$, let Δ_n be a partition of $\mathbb{R}_+$ into subintervals by a strictly increasing sequence $\{t_{n,k} : k \in \mathbb{Z}_+\}$ in $\mathbb{R}_+$ such that $t_{n,0} = 0$ and $t_{n,k} \uparrow \infty$ as $k \to \infty$ and

$\lim_{n\to\infty} |\Delta_n| = 0$ *where* $|\Delta_n| = \sup_{k\in\mathbb{N}}(t_{n,k} - t_{n,k-1})$. *Let* $t \in \mathbb{R}_+$ *be fixed. For each* $n \in \mathbb{N}$, *let* $t_{n,p_n} = t$ *where* $t_{n,k} < t$ *for* $k = 0,\ldots,p_n - 1$. *Then*

$$\int_{[0,t]} Y \diamond dX = P \cdot \lim_{n\to\infty} \sum_{k=1}^{p_n} \frac{1}{2}\{Y(t_{n,k}) + Y(t_{n,k-1})\}\{X(t_{n,k}) - X(t_{n,k-1})\}.$$

Proof. We have

$$\sum_{k=1}^{p_n} \frac{1}{2}\{Y(t_{n,k}) + Y(t_{n,k-1})\}\{X(t_{n,k}) - X(t_{n,k-1})\}$$

$$= \sum_{k=1}^{p_n} Y(t_{n,k-1})\{X(t_{n,k}) - X(t_{n,k-1})\}$$

$$+ \frac{1}{2}\sum_{k=1}^{p_n}\{Y(t_{n,k}) - Y(t_{n,k-1})\}\{X(t_{n,k}) - X(t_{n,k-1})\}.$$

By Proposition 12.16

$$P \cdot \lim_{n\to\infty} \sum_{k=1}^{p_n} Y(t_{n,k-1})\{X(t_{n,k}) - X(t_{n,k-1})\} = (Y \bullet X)(t).$$

By Theorem 12.27

$$P \cdot \lim_{n\to\infty} \sum_{k=1}^{p_n} \{Y(t_{n,k}) - Y(t_{n,k-1})\}\{X(t_{n,k}) - X(t_{n,k-1})\} = [Y,X](t).$$

Thus

$$P \cdot \lim_{n\to\infty} \sum_{k=1}^{p_n} \frac{1}{2}\{Y(t_{n,k}) + Y(t_{n,k-1})\}\{X(t_{n,k}) - X(t_{n,k-1})\}$$

$$= (Y \bullet X)(t) + \frac{1}{2}[Y,X](t) = \int_{[0,t]} Y \diamond dX. \quad \blacksquare$$

Chapter 4

Stochastic Differential Equations

§17 The Space of Continuous Functions

[I] Function Space Representation of Continuous Processes

Let $\mathbf{W}^d$ be the collection of all $\mathbb{R}^d$-valued continuous functions w on $\mathbb{R}_+$. Our objective is to introduce a σ-algebra $\mathfrak{W}^d$ of subsets of $\mathbf{W}^d$ and an increasing system of sub-σ-algebras of $\mathfrak{W}^d$, $\{\mathfrak{W}^d_t : t \in \mathbb{R}_+\}$, with the following properties:

1°. For every $t \in \mathbb{R}_+$, the mapping q_t of $\mathbf{W}^d$ into $\mathbb{R}^d$ defined by $q_t(w) = w(t)$ for $w \in \mathbf{W}^d$ is a $\mathfrak{W}^d_t/\mathfrak{B}_{\mathbb{R}^d}$-measurable mapping so that $q = \{q_t : t \in \mathbb{R}_+\}$ is a d-dimensional $\{\mathfrak{W}^d_t\}$-adapted stochastic process on the filtered measurable space $(\mathbf{W}^d, \mathfrak{W}^d, \{\mathfrak{W}^d_t\})$.

2°. For an arbitrary continuous d-dimensional stochastic process $X = \{X_t : t \in \mathbb{R}_+\}$ on an arbitrary probability space $(\Omega, \mathfrak{F}, P)$, the mapping $\mathcal{X}$ of Ω into $\mathbf{W}^d$ defined by $\mathcal{X}(\omega) = X(\cdot, \omega)$ for $\omega \in \Omega$ is $\mathfrak{F}/\mathfrak{W}^d$-measurable.

Under the assumption of 1° and 2° let $P_\mathcal{X}$ be the probability distribution of $\mathcal{X}$ on the measurable space $(\mathbf{W}^d, \mathfrak{W}^d)$, that is, $P_\mathcal{X}(E) = P(\mathcal{X}^{-1}(E))$ for $E \in \mathfrak{W}^d$. Strictly speaking, $P_\mathcal{X}$ is the probability distribution of $\mathcal{X}$ on $(\mathbf{W}^d, \mathfrak{W}^d)$. However we will also call it the probability distribution of X on $(\mathbf{W}^d, \mathfrak{W}^d)$ and use the alternate notation P_X for it. Consider the d-dimensional $\{\mathfrak{W}^d_t\}$-adapted stochastic process $q = \{q_t : t \in \mathbb{R}_+\}$ on the filtered space $(\mathbf{W}^d, \mathfrak{W}^d, \{\mathfrak{W}^d_t\}, P_\mathcal{X})$ and the continuous d-dimensional stochastic process $X = \{X_t : t \in \mathbb{R}_+\}$ on the probability space $(\Omega, \mathfrak{F}, P)$. For any $0 \leq t_1 < \cdots < t_n$ in $\mathbb{R}_+$ the two random vectors $(q_{t_1}, \ldots, q_{t_n})$ and $(X_{t_1}, \ldots, X_{t_n})$ have identical probability distributions on $(\mathbb{R}^{nd}, \mathfrak{B}_{\mathbb{R}^{nd}})$.

Definition 17.1. *Let $\mathbf{W}^d$ be the collection of all $\mathbb{R}^d$-valued continuous functions on $\mathbb{R}_+$. For $t \in \mathbb{R}_+$, let q_t be a mapping of $\mathbf{W}^d$ into $\mathbb{R}^d$ defined by $q_t(w) = w(t)$ for $w \in \mathbf{W}^d$. Let $\tau = \{t_1, \ldots, t_n\}$ be a finite strictly increasing sequence in $\mathbb{R}_+$ and let $E_1, \ldots, E_n \in \mathfrak{B}_{\mathbb{R}^d}$. By a cylinder set in $\mathbf{W}^d$ with index τ and base $E_1 \times \cdots \times E_n$ we mean a subset Z of $\mathbf{W}^d$ defined by*

$$Z = q_{t_1}^{-1}(E_1) \cap \cdots \cap q_{t_n}^{-1}(E_n) = \{w \in \mathbf{W}^d : (w(t_1), \ldots, w(t_n)) \in E_1 \times \cdots \times E_n\}$$

Let $\mathfrak{Z}$ be the collection of all cylinder sets in $\mathbf{W}^d$, and for $t \in \mathbb{R}_+$ let $\mathfrak{Z}_{[0,t]}$ be the collection of all cylinder sets with indices $\tau = \{t_1, \ldots, t_n\}$ preceding t, that is, $t_n \leq t$. We write $\mathfrak{Z}_t$ for the collection of all cylinder sets with index $\tau = \{t\}$, that is, $\mathfrak{Z}_t = q_t^{-1}(\mathfrak{B}_{\mathbb{R}^d})$.

Lemma 17.2. *$\mathfrak{Z}$ is a semialgebra of subsets of $\mathbf{W}^d$. For every $t \in \mathbb{R}_+$, $\mathfrak{Z}_{[0,t]}$ is also a semialgebra of subsets of $\mathbf{W}^d$.*

Proof. Since $\emptyset = q_t^{-1}(\emptyset)$ and $\mathbf{W}^d = q_t^{-1}(\mathbb{R}^d)$ for an arbitrary $t \in \mathbb{R}_+$, $\emptyset$ and $\mathbf{W}^d$ are in $\mathfrak{Z}$. Clearly $\mathfrak{Z}$ is closed under intersections. Therefore it remains to show that if $Z \in \mathfrak{Z}$ then there exists a finite disjoint collection $\{Z_k : k = 0, \ldots, n\}$ in $\mathfrak{Z}$ such that $Z_0 = Z$ and $\cup_{j=0}^k Z_j \in \mathfrak{Z}$ for $k = 0, \ldots, n$ with $\cup_{j=0}^n Z_j = \mathbf{W}^d$. Let $Z = q_{t_1}^{-1}(E_1) \cap \cdots \cap q_{t_n}^{-1}(E_n)$ with $0 \leq t_1 < \cdots < t_n$ and $E_j \in \mathfrak{B}_{\mathbb{R}^d}$ for $j = 1, \ldots, n$. Let

$$\begin{cases} Z_0 = Z \\ Z_k = q_{t_k}^{-1}(E_k^c) \cap q_{t_{k+1}}^{-1}(E_{k+1}) \cap \cdots \cap q_{t_n}^{-1}(E_n) & \text{for } k = 1, \ldots, n \end{cases}$$

and in particular $Z_n = q_{t_n}^{-1}(E_n^c)$. Then $\{Z_k : k = 0, \ldots, n\}$ is a disjoint collection in $\mathfrak{Z}$. Also

$$Z_0 \cup \cdots \cup Z_{k-1} = q_{t_k}^{-1}(E_k) \cap \cdots \cap q_{t_n}^{-1}(E_n) \in \mathfrak{Z}$$

so that in particular

$$Z_0 \cup \cdots \cup Z_{n-1} \cup Z_n = q_{t_n}^{-1}(E_n) \cup q_{t_n}^{-1}(E_n^c) = q_{t_n}^{-1}(\mathbb{R}^d) = \mathbf{W}^d.$$

This completes the proof that $\mathfrak{Z}$ is a semialgebra of subsets of $\mathbf{W}^d$. The fact that $\mathfrak{Z}_{[0,t]}$ is a semialgebra of subsets of $\mathbf{W}^d$ can be shown in the same way. ∎

Since $\mathfrak{Z}$ is a semialgebra of subsets of $\mathbf{W}^d$, the algebra of subsets of $\mathbf{W}^d$ generated by $\mathfrak{Z}$ is the collection of all finite unions of members of $\mathfrak{Z}$. Since $\mathfrak{Z}_{[0,s]} \subset \mathfrak{Z}_{[0,t]} \subset \mathfrak{Z}$ we have $\sigma(\mathfrak{Z}_{[0,s]}) \subset \sigma(\mathfrak{Z}_{[0,t]}) \subset \sigma(\mathfrak{Z})$ for $0 \leq s < t < \infty$.

§17. THE SPACE OF CONTINUOUS FUNCTIONS

Definition 17.3. Let $\mathfrak{W}^d = \sigma(\mathfrak{Z})$ and $\mathfrak{W}^d_t = \sigma(\mathfrak{Z}_{[0,t]})$ for $t \in \mathbb{R}_+$. We then have a filtered measurable space $(\mathbf{W}^d, \mathfrak{W}^d, \{\mathfrak{W}^d_t\})$.

Proposition 17.4. For the filtered measurable space $(\mathbf{W}^d, \mathfrak{W}^d, \{\mathfrak{W}^d_t\})$, we have
(1) $\sigma(\cup_{t \in \mathbb{R}_+} \mathfrak{W}^d_t) = \mathfrak{W}^d$,
(2) $\sigma(\cup_{t \in \mathbb{R}_+} \mathfrak{Z}_t) = \mathfrak{W}^d$,
(3) $\sigma(\cup_{s \in [0,t]} \mathfrak{Z}_s) = \mathfrak{W}^d_t$ for every $t \in \mathbb{R}_+$.

Let $(\Omega, \mathfrak{F})$ be an arbitrary measurable space and let T be a mapping of Ω into $\mathbf{W}^d$. Then
(4) T is $\mathfrak{F}/\mathfrak{W}^d$-measurable if and only if $q_t \circ T$ is $\mathfrak{F}/\mathfrak{B}_{\mathbb{R}^d}$-measurable for every $t \in \mathbb{R}_+$,
(5) For fixed $t \in \mathbb{R}_+$, T is $\mathfrak{F}/\mathfrak{W}^d_t$-measurable if and only if $q_s \circ T$ is $\mathfrak{F}/\mathfrak{B}_{\mathbb{R}^d}$-measurable for every $s \in [0,t]$.

Proof. To prove (1), note that for $t \in \mathbb{R}_+$ we have $\mathfrak{Z}_{[0,t]} \subset \mathfrak{Z}$, $\sigma(\mathfrak{Z}_{[0,t]}) \subset \sigma(\mathfrak{Z})$, and thus $\cup_{t \in \mathbb{R}_+} \sigma(\mathfrak{Z}_{[0,t]}) \subset \sigma(\mathfrak{Z})$ so that $\sigma(\cup_{t \in \mathbb{R}_+} \sigma(\mathfrak{Z}_{[0,t]})) \subset \sigma(\mathfrak{Z})$. Conversely $\mathfrak{Z} = \cup_{t \in \mathbb{R}_+} \mathfrak{Z}_{[0,t]} \subset \cup_{t \in \mathbb{R}_+} \sigma(\mathfrak{Z}_{[0,t]})$ so that $\sigma(\mathfrak{Z}) \subset \sigma(\cup_{t \in \mathbb{R}_+} \sigma(\mathfrak{Z}_{[0,t]}))$. This shows that $\sigma(\cup_{t \in \mathbb{R}_+} \sigma(\mathfrak{Z}_{[0,t]})) = \sigma(\mathfrak{Z}) = \mathfrak{W}^d$.

To prove (2), note first that $\cup_{t \in \mathbb{R}_+} \mathfrak{Z}_t \subset \mathfrak{Z}$ so that $\sigma(\cup_{t \in \mathbb{R}_+} \mathfrak{Z}_t) \subset \sigma(\mathfrak{Z}) = \mathfrak{W}^d$. On the other hand for an arbitrary $Z \in \mathfrak{Z}$ given by $Z = q_{t_1}^{-1}(E_1) \cap \cdots \cap q_{t_n}^{-1}(E_n)$, we have $Z \in \sigma(\cup_{t \in \mathbb{R}_+} \mathfrak{Z}_t)$ and thus $\mathfrak{Z} \subset \sigma(\cup_{t \in \mathbb{R}_+} \mathfrak{Z}_t)$. Then $\mathfrak{W}^d = \sigma(\mathfrak{Z}) \subset \sigma(\cup_{t \in \mathbb{R}_+} \mathfrak{Z}_t)$. Therefore (2) holds. The equality (3) is proved likewise.

Since (4) and (5) are proved in the same way, let us prove (5) here. Suppose T is $\mathfrak{F}/\mathfrak{W}^d_t$-measurable. Let $s \in [0,t]$. Since q_s is $\mathfrak{W}^d_s/\mathfrak{B}_{\mathbb{R}^d}$-measurable and thus $\mathfrak{W}^d_t/\mathfrak{B}_{\mathbb{R}^d}$-measurable, $q_s \circ T$ is $\mathfrak{F}/\mathfrak{B}_{\mathbb{R}^d}$-measurable. Conversely suppose $q_s \circ T$ is $\mathfrak{F}/\mathfrak{B}_{\mathbb{R}^d}$-measurable for every $s \in [0,t]$. Let $Z \in \mathfrak{Z}_s$ given by $Z = q_s^{-1}(E)$ for some $E \in \mathfrak{B}_{\mathbb{R}^d}$. Then $T^{-1}(Z) = T^{-1} \circ q_s^{-1}(E) = (q_s \circ T)^{-1}(E) \subset \mathfrak{F}$ and thus $T^{-1}(\mathfrak{Z}_s) \subset \mathfrak{F}$. Then $T^{-1}(\cup_{s \in [0,t]} \mathfrak{Z}_s) \subset \mathfrak{F}$. Therefore by (3) we have $T^{-1}(\mathfrak{W}^d_t) = T^{-1}(\sigma(\cup_{s \in [0,t]} \mathfrak{Z}_s)) = \sigma(T^{-1}(\cup_{s \in [0,t]} \mathfrak{Z}_s)) \subset \mathfrak{F}$, that is, T is $\mathfrak{F}/\mathfrak{W}^d_t$-measurable. This proves (5). ∎

Observation 17.5. For a finite strictly increasing sequence $\tau = \{t_1, \ldots, t_n\}$ in $\mathbb{R}_+$, consider the mapping of $\mathbf{W}^d$ into $\mathbb{R}^{nd}$ defined by $q_\tau = (q_{t_1}, \ldots, q_{t_n})$. Then $q_\tau^{-1}(\mathfrak{B}_{\mathbb{R}^{nd}}) \subset \sigma(\mathfrak{Z}_{[0,t]})$ for any $t \geq t_n$, in other words, any cylinder set of index τ with non-factorable base $E \in \mathfrak{B}_{\mathbb{R}^{nd}}$, $Z = \{w \in \mathbf{W}^d : (w(t_1), \ldots, w(t_n)) \in E\}$, is in the σ-algebra $\sigma(\mathfrak{Z}_{[0,t]})$ with $t \geq t_n$.

Proof. By Proposition 13.2 and Theorem 1.1 we have

$$\begin{aligned} q_\tau^{-1}(\mathfrak{B}_{\mathbb{R}^{nd}}) &= q_\tau^{-1}(\sigma(\mathfrak{B}_{\mathbb{R}^d} \times \cdots \times \mathfrak{B}_{\mathbb{R}^d})) = \sigma(q_\tau^{-1}(\mathfrak{B}_{\mathbb{R}^d} \times \cdots \times \mathfrak{B}_{\mathbb{R}^d})) \\ &= \sigma(q_{t_1}^{-1}(\mathfrak{B}_{\mathbb{R}^d}) \cap \cdots \cap q_{t_n}^{-1}(\mathfrak{B}_{\mathbb{R}^d})) \subset \sigma(\mathfrak{Z}_{[0,t]}). \quad \blacksquare \end{aligned}$$

$\mathbf{W}^d$ is a linear space over $\mathbb{R}$ if we define addition and scalar multiplication by $(w_1 + w_2)(t) = w_1(t) + w_2(t)$ for $t \in \mathbb{R}_+$ and $w_1, w_2 \in \mathbf{W}^d$ and $(\lambda w)(t) = \lambda w(t)$ for $t \in \mathbb{R}_+$, $\lambda \in \mathbb{R}$, and $w \in \mathbf{W}^d$. Consider the translate of a subset E of $\mathbf{W}^d$ by an element w_0 of $\mathbf{W}^d$ defined by $E + w_0 = \{w + w_0 : w \in E\}$ and the λ-multiple of E defined by $\lambda E = \{\lambda w : w \in E\}$.

Remark 17.6. $\sigma(\mathfrak{Z})$ is closed under translations and scalar multiplications, that is, if $E \in \sigma(\mathfrak{Z})$ then $E + w_0, \lambda E \in \sigma(\mathfrak{Z})$ for any $w_0 \in \mathbf{W}^d$ an $\lambda \in \mathbb{R}$.

Proof. Let $\mathfrak{G}$ be the collection of all members E of $\sigma(\mathfrak{Z})$ such that $E + w_0$ is in $\sigma(\mathfrak{Z})$. Let us show that $\mathfrak{G}$ is a d-class of subsets of $\mathbf{W}^d$. Now $\mathbf{W}^d + w_0 = \mathbf{W}^d \in \mathfrak{Z} \subset \sigma(\mathfrak{Z})$. Thus $\mathbf{W}^d \in \mathfrak{G}$. Suppose $A, B \in \mathfrak{G}$ and $A \subset B$. Then $(B - A) + w_0 = (B + w_0) - (A + w_0) \in \sigma(\mathfrak{Z})$ since both $B + w_0$ and $A + w_0$ are in the σ-algebra $\sigma(\mathfrak{Z})$. Let $\{A_n : n \in \mathbb{N}\}$ be an increasing sequence in $\mathfrak{G}$. Then $(\cup_{n \in \mathbb{N}} A_n) + w_0 = \cup_{n \in \mathbb{N}} (A_n + w_0) \in \sigma(\mathfrak{Z})$ since $\sigma(\mathfrak{Z})$ is a σ-algebra. Thus $(\cup_{n \in \mathbb{N}} A_n) \in \mathfrak{G}$. This shows that $\mathfrak{G}$ is a d-class of subsets of $\mathbf{W}^d$.

To show $\mathfrak{Z} \subset \mathfrak{G}$, take an arbitrary member of $\mathfrak{Z}$ given by $Z = \cap_{i=1}^n q_{t_i}^{-1}(E_i)$ where $\{t_1, \ldots, t_n\}$ is a finite strictly increasing sequence in $\mathbb{R}_+$ and $E_1, \ldots, E_n \in \mathfrak{B}_{\mathbb{R}^d}$. Then $Z + w_0 = \cap_{i=1}^n (q_{t_i}^{-1}(E_i) + w_0)$. But

$$q_{t_i}^{-1}(E_i) + w_0 = \{w \in \mathbf{W}^d : w(t_i) \in E_i\} + w_0 = \{w \in \mathbf{W}^d : w(t_i) \in E_i + w_0(t_i)\} \in \mathfrak{Z}$$

since $E_i + w_0(t_i) \in \mathfrak{B}_{\mathbb{R}^d}$. Therefore $Z + w_0 \in \mathfrak{Z} \subset \sigma(\mathfrak{Z})$. This shows that $Z \in \mathfrak{G}$ and thus $\mathfrak{Z} \subset \mathfrak{G}$. Since $\mathfrak{G}$ is a d-class containing the π-class $\mathfrak{Z}$, we have $\sigma(\mathfrak{Z}) \subset \mathfrak{G}$ by Theorem 1.7. This shows that $E + w_0 \in \sigma(\mathfrak{Z})$ for every $E \in \sigma(\mathfrak{Z})$. Similarly $\lambda E \in \sigma(\mathfrak{Z})$ for every $E \in \sigma(\mathfrak{Z})$. ∎

Let us consider $\sigma\{q_t : t \in \mathbb{R}_+\}$, the σ-algebra of subsets of $\mathbf{W}^d$ generated by q_t for $t \in \mathbb{R}_+$, that is the smallest σ-algebra of subsets of $\mathbf{W}^d$ with respect to which every q_t is a measurable mapping of $\mathbf{W}^d$ into $(\mathbb{R}^d, \mathfrak{B}_{\mathbb{R}^d})$.

Proposition 17.7. *Consider the filtered measurable space* $(\mathbf{W}^d, \mathfrak{W}^d, \{\mathfrak{W}_t^d\})$ *and the mappings* $\{q_t : t \in \mathbb{R}_+\}$ *of* $\mathbf{W}^d$ *into* $\mathbb{R}^d$ *defined by* $q_t(w) = w(t)$ *for* $(t, w) \in \mathbb{R}_+ \times \mathbf{W}^d$. *Then*
1) $\sigma\{q_s : s \in [0, t]\} = \mathfrak{W}_t^d$ *for every* $t \in \mathbb{R}_+$ *and* $\sigma\{q_t : t \in \mathbb{R}_+\} = \mathfrak{W}^d$,
2) $q = \{q_t : t \in \mathbb{R}_+\}$ *is a d-dimensional $\{\mathfrak{W}_t^d\}$-adapted process on* $(\mathbf{W}^d, \mathfrak{W}^d, \{\mathfrak{W}_t^d\})$,
3) for the process q we have $q(\cdot, w) = w$ *for every* $w \in \mathbf{W}^d$, *that is, every sample point is also the corresponding sample function of the process q.*

Proof. 1) For every $t \in \mathbb{R}_+$ we have $\sigma\{q_s : s \in [0, t]\} = \sigma(\cup_{s \in [0,t]} q_s^{-1}(\mathfrak{B}_{\mathbb{R}^d}))$. Now

§17. THE SPACE OF CONTINUOUS FUNCTIONS

$\mathfrak{Z}_{[0,t]}$ is the collection of all finite intersections of members of $\cup_{s \in [0,t]} q_s^{-1}(\mathfrak{B}_{\mathbb{R}^d})$ and thus $\mathfrak{Z}_{[0,t]} \subset \sigma(\cup_{s \in [0,t]} q_s^{-1}(\mathfrak{B}_{\mathbb{R}^d}))$. Consequently $\sigma(\mathfrak{Z}_{[0,t]}) \subset \sigma(\cup_{s \in [0,t]} q_s^{-1}(\mathfrak{B}_{\mathbb{R}^d}))$. On the other hand, $q_s^{-1}(\mathfrak{B}_{\mathbb{R}^d}) \subset \mathfrak{Z}_{[0,t]}$ for every $s \in [0,t]$. This implies that $\cup_{s \in [0,t]} q_s^{-1}(\mathfrak{B}_{\mathbb{R}^d}) \subset \mathfrak{Z}_{[0,t]}$ and then $\sigma(\cup_{s \in [0,t]} q_s^{-1}(\mathfrak{B}_{\mathbb{R}^d})) \subset \sigma(\mathfrak{Z}_{[0,t]})$. Thus we have $\sigma\{q_s : s \in [0,t]\} = \sigma(\mathfrak{Z}_{[0,t]}) = \mathfrak{W}_t^d$. The equality $\sigma\{q_t : t \in \mathbb{R}_+\} = \sigma(\mathfrak{Z}) = \mathfrak{W}^d$ is proved in the same way.

2) By 1), q_t is $\mathfrak{W}_t^d/\mathfrak{B}_{\mathbb{R}^d}$-measurable for every $t \in \mathbb{R}_+$. Thus q is a $\{\mathfrak{W}_t^d\}$-adapted process on $(\mathbf{W}^d, \mathfrak{W}^d, \{\mathfrak{W}_t^d\})$.

3) We have $q(t, w) = q_t(w) = w(t)$ for $(t, w) \in \mathbb{R}_+ \times \mathbf{W}^d$. Thus $q(\cdot, w) = w$. ∎

Theorem 17.8. *Let* $X = \{X_t : t \in \mathbb{R}_+\}$ *be a continuous d-dimensional stochastic process on a probability space* $(\Omega, \mathfrak{F}, P)$. *Then the mapping* $\mathcal{X}$ *of* Ω *into* $\mathbf{W}^d$ *defined by* $\mathcal{X}(\omega) = X(\cdot, \omega)$ *for* $\omega \in \Omega$ *is* $\mathfrak{F}/\mathfrak{W}^d$-*measurable. Let* $P_\mathcal{X}$ *be the probability distribution of* $\mathcal{X}$ *on* $(\mathbf{W}^d, \mathfrak{W}^d)$. *Then the d-dimensional* $\{\mathfrak{W}_t^d\}$-*adapted process* $q = \{q_t : t \in \mathbb{R}_+\}$ *on the filtered space* $(\mathbf{W}^d, \mathfrak{W}^d, \{\mathfrak{W}_t^d\}, P_\mathcal{X})$ *and the d-dimensional process* X *on* $(\Omega, \mathfrak{F}, P)$ *have identical finite dimensional probability distributions, that is, for any finite strictly increasing sequence* $\tau = \{t_1, \ldots, t_n\}$ *in* $\mathbb{R}_+$ *the probability distribution* $(P_\mathcal{X})_{q_\tau}$ *on* $(\mathbb{R}^{nd}, \mathfrak{B}_{\mathbb{R}^{nd}})$ *of the random vector* $q_\tau = (q_{t_1}, \ldots, q_{t_n})$ *and the probability distribution* P_{X_τ} *on* $(\mathbb{R}^{nd}, \mathfrak{B}_{\mathbb{R}^{nd}})$ *of the random vector* $X_\tau = (X_{t_1}, \ldots, X_{t_n})$ *are identical. Also, for every* $E \in \mathfrak{B}_{\mathbb{R}^{nd}}$ *and the cylinder set* $Z = \{w \in \mathbf{W}^d : (w(t_1), \ldots, w(t_n)) \in E\}$ *in* $\mathbf{W}^d$ *we have* $P_{X_\tau}(E) = P_\mathcal{X}(Z)$. *If in addition* X *is adapted to a filtration* $\{\mathfrak{F}_t : t \in \mathbb{R}_+\}$ *on* $(\Omega, \mathfrak{F}, P)$, *then* $\mathcal{X}$ *is an* $\mathfrak{F}_t/\mathfrak{W}_t^d$-*measurable mapping of* Ω *into* $\mathbf{W}^d$ *for every* $t \in \mathbb{R}_+$.

Proof. Let us prove the $\mathfrak{F}/\mathfrak{W}^d$-measurability of $\mathcal{X}$, that is, $\mathcal{X}^{-1}(\mathfrak{W}^d) \subset \mathfrak{F}$. Since $\mathfrak{W}^d = \sigma(\mathfrak{Z})$, if we show that $\mathcal{X}^{-1}(\mathfrak{Z}) \subset \mathfrak{F}$ then we have $\mathcal{X}^{-1}(\mathfrak{W}^d) = \mathcal{X}^{-1}(\sigma(\mathfrak{Z})) = \sigma(\mathcal{X}^{-1}(\mathfrak{Z})) \subset \mathfrak{F}$ by Theorem 1.1 and we are done. Take an arbitrary $Z \in \mathfrak{Z}$ given by $Z = \cap_{i=1}^n q_{t_i}^{-1}(E_i)$ where $0 \leq t_1 < \cdots < t_n < \infty$ and $E_1, \ldots, E_n \in \mathfrak{B}_{\mathbb{R}^d}$. Then

$$\begin{aligned}
\mathcal{X}^{-1}(Z) &= \mathcal{X}^{-1}(\cap_{i=1}^n q_{t_i}^{-1}(E_i)) = \cap_{i=1}^n \mathcal{X}^{-1}(q_{t_i}^{-1}(E_i)) \\
&= \cap_{i=1}^n (q_{t_i} \circ \mathcal{X})^{-1}(E_i) = \cap_{i=1}^n X_{t_i}^{-1}(E_i) \in \mathfrak{F}.
\end{aligned} \tag{1}$$

This shows that $\mathcal{X}^{-1}(\mathfrak{Z}) \subset \mathfrak{F}$.

To show that $(P_\mathcal{X})_{q_\tau} = P_{X_\tau}$ on $(\mathbb{R}^{nd}, \mathfrak{B}_{\mathbb{R}^{nd}})$, let $E \in \mathfrak{B}_{\mathbb{R}^{nd}}$. Then

$$\begin{aligned}
(P_\mathcal{X})_{q_\tau}(E) &= P_\mathcal{X}(q_\tau^{-1}(E)) = P \circ \mathcal{X}^{-1} \circ q_\tau^{-1}(E) \\
&= P \circ (q_\tau \circ \mathcal{X})^{-1}(E) = P \circ X_\tau^{-1}(E) = P_{X_\tau}(E).
\end{aligned}$$

To show that $P_{X_\tau}(E) = P_\mathcal{X}(Z)$, note that

$$P_{X_\tau}(E) = P \circ X_\tau^{-1}(E) = P\{\omega \in \Omega : X_\tau(E) \in E\}$$

$$= P\{\omega \in \Omega : X(\cdot, \omega) \in Z\} = P \circ \mathcal{X}^{-1}(Z) = P_{\mathcal{X}}(Z).$$

When X is adapted to a filtration $\{\mathfrak{F}_t : t \in \mathbb{R}_+\}$, we show that $\mathcal{X}$ is $\mathfrak{F}_t/\mathfrak{W}_t^d$-measurable by starting with $Z \in \mathfrak{Z}_{[0,t]}$ in (1). ∎

Definition 17.9. *Let B be a d-dimensional null at 0 Brownian motion on a probability space $(\Omega, \mathfrak{F}, P)$. Let $\mathcal{B}$ be the mapping of Ω into $\mathbf{W}^d$ defined by $\mathcal{B}(\omega) = B(\cdot, \omega)$ for $\omega \in \Omega$. We call the probability distribution $P_{\mathcal{B}}$ of $\mathcal{B}$ on $(\mathbf{W}^d, \mathfrak{W}^d)$ the Wiener measure on $(\mathbf{W}^d, \mathfrak{W}^d)$ and write m_W^d for it. We call $(\mathbf{W}^d, \mathfrak{W}^d, m_W^d)$ the d-dimensional Wiener space.*

For a cylinder set Z in $\mathbf{W}^d$ with index $\tau = \{t_1, \ldots, t_n\}$ where $t_1 > 0$ and base $E \in \mathfrak{B}_{\mathbb{R}^{nd}}$ we have by Proposition 13.16

$$\begin{aligned} m_W^d(Z) &= P_{\mathcal{B}}(Z) \\ &= P_{\mathcal{B}}\{w \in \mathbf{W}^d : (w(t_1), \ldots, w(t_n)) \in E\} = P\{(B(t_1), \ldots, B(t_n)) \in E\} \\ &= \{(2\pi)^n \prod_{j=1}^n (t_j - t_{j-1})\}^{-d/2} \int_E \exp\left\{-\frac{1}{2}\sum_{j=1}^n \frac{|x_j - x_{j-1}|^2}{t_j - t_{j-1}}\right\} m_L^{nd}(d(x_1, \ldots, x_n)) \end{aligned}$$

where $t_0 \equiv 0$.

Theorem 17.10. *Let W be a stochastic process on $(\mathbf{W}^d, \mathfrak{W}^d, m_W^d)$ defined by setting*

$$W(t, w) = w(t) \quad \text{for } (t, w) \in \mathbb{R}_+ \times \mathbf{W}^d.$$

Then W is a d-dimensional null at 0 Brownian motion on $(\mathbf{W}^d, \mathfrak{W}^d, m_W^d)$.

Proof. Let us verify that W satisfies the conditions in Definition 13.14. Clearly every sample function of W is an $\mathbb{R}^d$-valued continuous function on $\mathbb{R}_+$. Let us show that for every $s, t \in \mathbb{R}_+$, $s < t$, the probability distribution Q on $(\mathbb{R}^d, \mathfrak{B}_{\mathbb{R}^d})$ of the random vector $W_t - W_s$ is the d-dimensional normal distribution $N_d(0, (t-s) \cdot I)$. Let $x = T(y)$ be the nonsingular linear mapping of $\mathbb{R}^{2d}$ onto $\mathbb{R}^{2d}$ defined by $x_1 = y_1$ and $x_2 = y_1 + y_2$ for $y = (y_1, y_2) \in \mathbb{R}^{2d}$. Then for any $E \in \mathfrak{B}_{\mathbb{R}^d}$ we have

$$\begin{aligned} Q(E) &= m_W^d \circ (W_t - W_s)^{-1}(E) = m_W^d \circ (W_s, W_t - W_s)^{-1}(\mathbb{R}^d \times E) \\ &= m_W^d\{w \in \mathbf{W}^d : (w(s), w(t) - w(s)) \in \mathbb{R}^d \times E\} \\ &= m_W^d\{w \in \mathbf{W}^d : (w(s), w(t)) \in T(\mathbb{R}^d \times E)\} \end{aligned}$$

§17. THE SPACE OF CONTINUOUS FUNCTIONS

$$= \{(2\pi)^2 s(t-s)\}^{-d/2} \int_{T(\mathbb{R}^d \times E)} \exp\left\{-\frac{|x_1|^2}{2s} - \frac{|x_2 - x_1|^2}{2(t-s)}\right\} m_L^{2d}(d(x_1, x_2))$$

$$= \{(2\pi)^2 s(t-s)\}^{-d/2} \int_{\mathbb{R}^d \times E} \exp\left\{-\frac{|y_1|^2}{2s} - \frac{|y_2|^2}{2(t-s)}\right\} m_L^{2d}(d(y_1, y_2))$$

$$= \{2\pi(t-s)\}^{-d/2} \int_E \exp\left\{-\frac{|y_2|^2}{2(t-s)}\right\} m_L^d(dy_2).$$

This shows that $Q = N_d(0, (t-s) \cdot I)$.

To show that W is a process with independent increments, we show that if $\{t_1, \ldots, t_n\}$ is a strictly increasing sequence in $\mathbb{R}_+$, Q is the probability distribution of the random vector $(W_{t_1}, W_{t_2} - W_{t_1}, \ldots, W_{t_n} - W_{t_{n-1}})$ on $(\mathbb{R}^{nd}, \mathfrak{B}_{\mathbb{R}^{nd}})$, and $Q_1, Q_2, \ldots, Q_n$ are the probability distributions of the random vectors $W_{t_1}, W_{t_2} - W_{t_1}, \ldots, W_{t_n} - W_{t_{n-1}}$ on $(\mathbb{R}^d, \mathfrak{B}_{\mathbb{R}^d})$ respectively, then $Q(E) = (Q_1 \times \cdots \times Q_n)(E)$ for every $E \in \mathfrak{B}_{\mathbb{R}^{nd}}$. For this it suffices according to Corollary 1.8 to show

$$Q(E_1 \times \cdots \times E_n) = Q_1(E_1) \cdots Q_n(E_n) \quad \text{for } E_j \in \mathfrak{B}_{\mathbb{R}^d}, j = 1, \ldots, n.$$

Let $x = T(y)$ be the nonsingular linear mapping of $\mathbb{R}^{nd}$ onto $\mathbb{R}^{nd}$ defined by

$$x_1 = y_1, x_2 = y_1 + y_2, \ldots, x_n = y_1 + \cdots + y_n \quad \text{for } y = (y_1, \ldots, y_n) \in \mathbb{R}^{nd}$$

so that

$$y_1 = x_1, y_2 = x_2 - x_1, \ldots, y_n = x_n - x_{n-1} \quad \text{for } x = (x_1, \ldots, x_n) \in \mathbb{R}^{nd}.$$

Let us write $E = E_1 \times \cdots \times E_n$. Then

$$Q(E_1 \times \cdots \times E_n) = m_W^d \circ (W_{t_1}, W_{t_2} - W_{t_1}, \ldots, W_{t_n} - W_{t_{n-1}})^{-1}(E)$$

$$= m_W^d\{w \in \mathbf{W}^d : (w(t_1), w(t_2) - w(t_1), \ldots, w(t_n) - w(t_{n-1})) \in E\}$$

$$= m_W^d\{w \in \mathbf{W}^d : (w(t_1), \ldots, w(t_n)) \in T(E)\}$$

$$= \{(2\pi)^n \prod_{j=1}^n (t_j - t_{j-1})\}^{-d/2} \int_{T(E)} \exp\left\{-\frac{1}{2}\sum_{j=1}^n \frac{|x_j - x_{j-1}|^2}{t_j - t_{j-1}}\right\} m_L^{nd}(d(x_1, \ldots, x_n))$$

$$= \{(2\pi)^n \prod_{j=1}^n (t_j - t_{j-1})\}^{-d/2} \int_E \exp\left\{-\frac{1}{2}\sum_{j=1}^n \frac{|y_j|^2}{t_j - t_{j-1}}\right\} m_L^{nd}(d(y_1, \ldots, y_n))$$

$$= \prod_{j=1}^n \{2\pi(t_j - t_{j-1})\}^{-d/2} \int_{E_j} \exp\left\{-\frac{1}{2}\frac{|y_j|^2}{t_j - t_{j-1}}\right\} m_L^d(dy_j)$$

$$= \prod_{j=1}^n Q_j(E_j).$$

This proves the independence of increments for W. ∎

[II] Metrization of the Space of Continuous Functions

Our next objective is to introduce a metric to $\mathbf{W}^d$ with respect to which $\mathbf{W}^d$ is a complete separable metric space and the Borel σ-algebra of this metric space is equal to the σ-algebra $\mathfrak{W}^d = \sigma(3)$.

Definition 17.11. *For every $t \in \mathbb{R}_+$, let us define a seminorm ν_t on $\mathbf{W}^d$ by setting*

$$\nu_t(w) = \max_{s \in [0,t]} |w(s)| \wedge 1 \quad \text{for } w \in \mathbf{W}^d$$

where $|\cdot|$ is the Euclidean norm on $\mathbb{R}^d$, and let

$$\rho_t(w_1, w_2) = \nu_t(w_1 - w_2) \quad \text{for } w_1, w_2 \in \mathbf{W}^d.$$

Let us define a quasinorm ν_∞ on $\mathbf{W}^d$ by

$$\nu_\infty(w) = \sum_{m \in \mathbb{N}} 2^{-m} \nu_m(w) \quad \text{for } w \in \mathbf{W}^d$$

and define a metric ρ_∞ on $\mathbf{W}^d$ by

$$\rho_\infty(w_1, w_2) = \nu_\infty(w_1 - w_2) = \sum_{m \in \mathbb{N}} 2^{-m} \rho_m(w_1, w_2) \quad \text{for } w_1, w_2 \in \mathbf{W}^d.$$

Lemma 17.12. *Let $\{w_n : n \in \mathbb{N}\}$ be a sequence and w be an element in $\mathbf{W}^d$. Then $\lim_{n \to \infty} \rho_\infty(w_n, w) = 0$ if and only if $\lim_{n \to \infty} w_n(s) = w(s)$ uniformly in $s \in [0, t]$ for every $t \in \mathbb{R}_+$.*

Proof. It suffices to show that $\lim_{n \to \infty} \rho_\infty(w_n, w) = 0$ if and only if $\lim_{n \to \infty} \rho_m(w_n, w) = 0$ for every $m \in \mathbb{N}$. Clearly $\lim_{n \to \infty} \rho_\infty(w_n, w) = 0$ implies that $\lim_{n \to \infty} \rho_m(w_n, w) = 0$ for every $m \in \mathbb{N}$. Conversely suppose $\lim_{n \to \infty} \rho_m(w_n, w) = 0$ for every $m \in \mathbb{N}$. Given $\varepsilon > 0$ let $N \in \mathbb{N}$ be so large that $\sum_{m \geq N+1} 2^{-m} < \varepsilon$. Then since $\rho_m(w_n, w) \leq 1$ for every $m, n \in \mathbb{N}$, we have $\sum_{m \geq N+1} 2^{-m} \rho_m(w_n, w) < \varepsilon$ for every $n \in \mathbb{N}$ so that $\limsup_{n \to \infty} \sum_{m \geq N+1} 2^{-m} \rho_m(w_n, w) \leq \varepsilon$.

§17. THE SPACE OF CONTINUOUS FUNCTIONS

Then

$$\limsup_{n\to\infty} \rho_\infty(w_n, w) = \limsup_{n\to\infty} \left\{ \sum_{m=1}^{N} 2^{-m} \rho_m(w_n, w) + \sum_{m\geq N+1} 2^{-m} \rho_m(w_n, w) \right\}$$
$$\leq \lim_{n\to\infty} \sum_{m=1}^{N} 2^{-m} \rho_m(w_n, w) + \limsup_{n\to\infty} \sum_{m\geq N+1} 2^{-m} \rho_m(w_n, w) \leq \varepsilon$$

since by assumption $\lim_{n\to\infty} \rho_m(w_n, w) = 0$ for every $m \in \mathbb{N}$. By the arbitrariness of $\varepsilon > 0$ we have $\limsup_{n\to\infty} \rho_\infty(w_n, w) = 0$ and thus $\lim_{n\to\infty} \rho_\infty(w_n, w) = 0$. ∎

Theorem 17.13. $(\mathbf{W}^d, \rho_\infty)$ *is a complete separable metric space.*

Proof. To show that $(\mathbf{W}^d, \rho_\infty)$ is a complete metric space, let $\{w_n : n \in \mathbb{N}\}$ be a Cauchy sequence in $(\mathbf{W}^d, \rho_\infty)$. By Lemma 17.12, it suffices to show that there exists $w \in \mathbf{W}^d$ such that $\lim_{n\to\infty} \rho_m(w_n, w) = 0$ for every $m \in \mathbb{N}$. Let $\varepsilon \in (0,1)$. Let $m \in \mathbb{N}$ be fixed. Since $\{w_n : n \in \mathbb{N}\}$ is a Cauchy sequence there exists $N \in \mathbb{N}$ such that $\rho_\infty(w_\ell, w_n) < 2^{-m}\varepsilon$ when $\ell, n \geq N$. Then $2^{-m}\rho_m(w_\ell, w_n) < 2^{-m}\varepsilon$ when $\ell, n \geq N$ and thus $\rho_m(w_\ell, w_n) < \varepsilon$ when $\ell, n \geq N$. Therefore $\max_{s\in[0,m]} |w_\ell(s) - w_n(s)| \wedge 1 < \varepsilon$, or equivalently, $\max_{s\in[0,m]} |w_\ell(s) - w_n(s)| < \varepsilon$, when $\ell, n \geq N$. This shows that the restrictions of w_n to $[0, m]$ for $n \in \mathbb{N}$ is a Cauchy sequence with respect to the metric of the uniform norm on the space of continuous $\mathbb{R}^d$-valued functions on $[0, m]$. By the completeness of this metric space there exists a continuous $\mathbb{R}^d$-valued functions $f^{(m)}$ on $[0, m]$ such that the restrictions of w_n, $n \in \mathbb{N}$, to $[0, m]$ converge uniformly on $[0, m]$ to $f^{(m)}$. For $m_1, m_2 \in \mathbb{N}$, $m_1 < m_2$, we have $f^{(m_1)} = f^{(m_2)}$ on $[0, m_1]$ by the uniqueness of the limit of the convergence. Thus there exists a continuous $\mathbb{R}^d$-valued functions f on $\mathbb{R}_+$ such that $f = f^{(m)}$ on $[0, m]$ for every $m \in \mathbb{N}$. Since $\lim_{n\to\infty} \max_{s\in[0,m]} |w_n(s) - f^{(m)}(s)| = 0$ and $f = f^{(m)}$ on $[0, m]$ for every $m \in \mathbb{N}$, we have $\lim_{n\to\infty} \max_{s\in[0,m]} |w_n(s) - f(s)| \wedge 1 = 0$, that is, $\lim_{n\to\infty} \rho_m(w_n, f) = 0$. This proves the completeness of the metric space $(\mathbf{W}^d, \rho_\infty)$.

To show the separability of the metric space $(\mathbf{W}^d, \rho_\infty)$ let us show that it has a countable dense subset. For every $k \in \mathbb{N}$ let V_k be the collection of all $\mathbb{R}^d$-valued polygonal functions v on $\mathbb{R}_+$ with vertices occurring at $\ell 2^{-k}$ where $\ell \in \mathbb{Z}_+$ and with $v(\ell 2^{-k})$ equal to rational points in $\mathbb{R}^d$. Then V_k is a countable subset of $\mathbf{W}^d$ and so is $V = \cup_{k\in\mathbb{N}} V_k$. To show that V is dense in $\mathbf{W}^d$ we show that for every $w \in \mathbf{W}^d$ and $\varepsilon > 0$ there exists some $v \in V$ such that $\rho_\infty(w, v) < \varepsilon$. Let $N \in \mathbb{N}$ be so large that $\sum_{m\geq N+1} 2^{-m} < 2^{-1}\varepsilon$. By the uniform continuity of w on $[0, N]$ there exists $v \in V$ such that $\max_{t \in [0,N]} |w(t) - v(t)| < 2^{-1}\varepsilon$. Then

$$\rho_\infty(w, v) = \sum_{m=1}^{N} 2^{-m} \{\max_{t \in [0,m]} |w(t) - v(t)| \wedge 1\} + \sum_{m \geq N+1} 2^{-m} \{\max_{t \in [0,m]} |w(t) - v(t)| \wedge 1\}$$

$$\leq \sum_{m=1}^{N} 2^{-m} 2^{-1}\varepsilon + \sum_{m \geq N+1} 2^{-m} < \varepsilon. \quad \blacksquare$$

Lemma 17.14. *Let $\mathfrak{O}_{\mathbb{R}^d}$ be the collection of all open sets in $\mathbb{R}^d$ and let $\mathfrak{O}_{\mathbf{W}^d}$ be the collection of all open sets in $(\mathbf{W}^d, \rho_\infty)$. Then $q_t^{-1}(\mathfrak{O}_{\mathbb{R}^d}) \subset \mathfrak{O}_{\mathbf{W}^d}$ for every $t \in \mathbb{R}_+$.*

Proof. Let $G \in \mathfrak{O}_{\mathbb{R}^d}$. Consider $q_t^{-1}(G)$ for an arbitrary $t \in \mathbb{R}_+$. If $G = \emptyset$, then $q_t^{-1}(G) = \emptyset \in \mathfrak{O}_{\mathbf{W}^d}$. Consider the case where $G \neq \emptyset$ and thus $q_t^{-1}(G) \neq \emptyset$. To show that $q_t^{-1}(G) \in \mathfrak{O}_{\mathbf{W}^d}$ we show that for every $w_0 \in q_t^{-1}(G)$ there exists an open sphere in $\mathbf{W}^d$ with center w_0 and radius η, $S(w_0, \eta)$, contained in $q_t^{-1}(G)$. Now since $w_0 \in q_t^{-1}(G)$, we have $w_0(t) \in G$. Then there exists $\delta \in (0, 1)$ such that the open sphere in $\mathbb{R}^d$ with center $w_0(t)$ and radius δ, $S(w_0(t), \delta)$, is contained in G. Let $N \in \mathbb{N}$ be so large that $t \in [0, N]$. We proceed to show that $S(w_0, 2^{-N}\delta) \subset q_t^{-1}(G)$. Let $w \in S(w_0, 2^{-N}\delta)$. Then $\rho_\infty(w, w_0) < 2^{-N}\delta$ and thus $2^{-N}\{\max_{s \in [0,N]} |w(s) - w_0(s)| \wedge 1\} < 2^{-N}\delta$. Since $\delta \in (0, 1)$ we have $\max_{s \in [0,N]} |w(s) - w_0(s)| < \delta$. Since $t \in [0, N]$ we have $|w(t) - w_0(t)| < \delta$. Thus $w(t) \in S(w_0(t), \delta) \subset G$. Therefore $w \in q_t^{-1}(G)$. Since w is an arbitrary element in $S(w_0, 2^{-N}\delta)$, we have $S(w_0, 2^{-N}\delta) \subset q_t^{-1}(G)$. $\blacksquare$

Theorem 17.15. *Let $\mathfrak{O}_{\mathbf{W}^d}$ be the collection of all open sets in $(\mathbf{W}^d, \rho_\infty)$. Consider the Borel σ-algebra $\mathfrak{B}_{\mathbf{W}^d} = \sigma(\mathfrak{O}_{\mathbf{W}^d})$ in $\mathbf{W}^d$. Then*
1) $\mathfrak{Z} \subset \mathfrak{B}_{\mathbf{W}^d}$,
2) $\mathfrak{O}_{\mathbf{W}^d} \subset \sigma(\mathfrak{Z})$,
3) $\sigma(\mathfrak{Z}) = \mathfrak{B}_{\mathbf{W}^d}$, that is, $\mathfrak{W}^d = \mathfrak{B}_{\mathbf{W}^d}$.

Proof. 1) Let us show that $q_t^{-1}(\mathfrak{B}_{\mathbb{R}^d}) \subset \mathfrak{B}_{\mathbf{W}^d}$ for every $t \in \mathbb{R}_+$. Let $\mathfrak{O}_{\mathbb{R}^d}$ be the collection of all open sets in $\mathbb{R}^d$. According to Lemma 17.14, we have $q_t^{-1}(\mathfrak{O}_{\mathbb{R}^d}) \subset \mathfrak{O}_{\mathbf{W}^d}$. Then by Theorem 1.1,

$$q_t^{-1}(\mathfrak{B}_{\mathbb{R}^d}) = q_t^{-1}(\sigma(\mathfrak{O}_{\mathbb{R}^d})) = \sigma(q_t^{-1}(\mathfrak{O}_{\mathbb{R}^d})) \subset \sigma(\mathfrak{O}_{\mathbf{W}^d}) = \mathfrak{B}_{\mathbf{W}^d}.$$

Since every member Z of $\mathfrak{Z}$ given by $Z = \cap_{i=1}^{n} q_{t_i}^{-1}(E_i)$ where $0 \leq t_1 < \cdots < t_n < \infty$ and $E_1, \ldots, E_n \in \mathfrak{B}_{\mathbb{R}^d}$, and since $\mathfrak{B}_{\mathbf{W}^d}$ is a σ-algebra and in particular closed under finite intersections, Z is in $\mathfrak{B}_{\mathbf{W}^d}$. This shows that $\mathfrak{Z} \subset \mathfrak{B}_{\mathbf{W}^d}$.

§17. THE SPACE OF CONTINUOUS FUNCTIONS

2) To show $\mathfrak{O}_{\mathbf{W}^d} \subset \sigma(3)$, let us show first that for every $w_0 \in \mathbf{W}^d$ and $\varepsilon > 0$ the closed sphere in $\mathbf{W}^d$ with center w_0 and radius ε, $K(w_0, \varepsilon) = \{w \in \mathbf{W}^d : \rho_\infty(w, w_0) \leq \varepsilon\}$, is in $\sigma(3)$. Now

$$\rho_\infty(w, w_0) \leq \varepsilon \Leftrightarrow \sum_{m=1}^{N} 2^{-m} \rho_m(w, w_0) \leq \varepsilon \quad \text{for every } N \in \mathbb{N}.$$

Thus

$$K(w_0, \varepsilon) = \bigcap_{N \in \mathbb{N}} \left\{ w \in \mathbf{W}^d : \sum_{m=1}^{N} 2^{-m} \rho_m(w, w_0) \leq \varepsilon \right\}.$$

Since $\sigma(3)$ is closed under countable intersections, to show that $K(w_0, \varepsilon) \in \sigma(3)$, it suffices to show that

(1) $\quad \left\{ w \in \mathbf{W}^d : \sum_{m=1}^{N} 2^{-m} \rho_m(w, w_0) \leq \varepsilon \right\} \in \sigma(3) \quad \text{for every } N \in \mathbb{N}.$

Note that

(2) $\quad \{w \in \mathbf{W}^d : \rho_m(w, w_0) \leq \varepsilon\} = \{w \in \mathbf{W}^d : \max_{s \in [0,m]} |w(s) - w_0(s)| \wedge 1 \leq \varepsilon\}.$

If $\varepsilon \in [1, \infty)$, then the last set is equal to $\mathbf{W}^d$, which is a member of 3. On the other hand if $\varepsilon \in (0, 1)$ then the set on the right side of (2) is equal to

$$\{w \in \mathbf{W}^d : \max_{s \in [0,m]} |w(s) - w_0(s)| \leq \varepsilon\} = \bigcap_{r \in [0,m] \cap Q} \{w \in \mathbf{W}^d : |w(r) - w_0(r)| \leq \varepsilon]\}$$
$$= \bigcap_{r \in [0,m] \cap Q} q_r^{-1}(K(w_0(r), \varepsilon)),$$

where Q is the collection of all nonnegative rationals and $K(w_0(r), \varepsilon)$ is the closed sphere in $\mathbb{R}^d$ with center $w_0(r)$ and radius ε, by the continuity of w_0 and w. Since $q_r^{-1}(K(w_0(r), \varepsilon))$ is a member of 3 the last countable intersection above is in $\sigma(3)$. Thus we have shown that

$$\{w \in \mathbf{W}^d : \rho_m(w, w_0) \leq \varepsilon\} \in \sigma(3) \quad \text{for any } w_0 \in \mathbf{W}^d, \varepsilon > 0, \text{ and } m \in \mathbb{N}.$$

Consequently we have

$$\{w \in \mathbf{W}^d : \rho_m(w, w_0) < \varepsilon\} = \bigcup_{\ell \in \mathbb{N}} \{w \in \mathbf{W}^d : \rho_m(w, w_0) \leq \varepsilon - \ell^{-1}\} \in \sigma(3)$$

and then

(3) $\quad \{w \in \mathbf{W}^d : 2^{-m}\rho_m(w, w_0) < \varepsilon\} = \{w \in \mathbf{W}^d : \rho_m(w, w_0) < 2^m\varepsilon\} \in \sigma(3).$

Let $(r_1, \ldots, r_N)$ be an N-tuple of positive rational numbers $r_1, \ldots, r_N$ such that $r_1 + \cdots + r_N < \varepsilon$. Let $\mathbf{Q}$ be the countable collection of all such N-tuples. Then

(4) $\quad \left\{w \in \mathbf{W}^d : \sum_{m=1}^N 2^{-m}\rho_m(w, w_0) < \varepsilon\right\}$

$\quad = \bigcup_{(r_1,\ldots,r_N) \in \mathbf{Q}} \bigcap_{m=1}^N \{w \in \mathbf{W}^d : 2^{-m}\rho_m(w, w_0) < r_m\} \in \sigma(3)$

by (3). From this we have

$\left\{w \in \mathbf{W}^d : \sum_{m=1}^N 2^{-m}\rho_m(w, w_0) \leq \varepsilon\right\}$

$= \bigcap_{\ell \in \mathbb{N}} \left\{w \in \mathbf{W}^d : \sum_{m=1}^N 2^{-m}\rho_m(w, w_0) < \varepsilon + \ell^{-1}\right\} \in \sigma(3).$

This proves (1) and therefore $K(w_0, \varepsilon) \in \sigma(3)$ for every $w_0 \in \mathbf{W}^d$ and $\varepsilon > 0$.

For an open sphere in $\mathbf{W}^d$ with center $w_0 \in \mathbf{W}^d$ and radius $\varepsilon > 0$, we have $S(w_0, \varepsilon) = \bigcup_{\ell \in \mathbb{N}} K(w_0, \varepsilon - \ell^{-1}) \in \sigma(3)$. Since $(\mathbf{W}^d, \rho_\infty)$ is a separable metric space, it has a countable dense subset D. Every open set in $\mathbf{W}^d$ can be given as the union of open spheres with centers in D and with rational radii. Thus as a countable union of open spheres, an open set in $\mathbf{W}^d$ is always in $\sigma(3)$. Therefore $\mathfrak{O}_{\mathbf{W}^d} \subset \sigma(3)$.

3) By 1) we have $\sigma(3) \subset \mathfrak{B}_{\mathbf{W}^d}$. By 2) we have $\mathfrak{B}_{\mathbf{W}^d} = \sigma(\mathfrak{O}_{\mathbf{W}^d}) \subset \sigma(3)$. Thus $\sigma(3) = \mathfrak{B}_{\mathbf{W}^d}$. ∎

§18 Definition and Function Space Representation of Solution

[I] Definition of Solutions

Consider stochastic differential equations, that is, equations involving stochastic differentials, of the type

(1) $\quad dX_t^i = \sum_{j=1}^r \alpha_j^i(t, X)\, dB^j(t) + \beta^i(t, X)\, dt \quad \text{for } i = 1, \ldots, d,$

§18. DEFINITION AND FUNCTION SPACE REPRESENTATION

where $B = (B^1, \ldots, B^r)$ is an r-dimensional $\{\mathfrak{F}_t\}$-adapted Brownian motion, $X = (X^1, \ldots, X^d)$ is a d-dimensional continuous $\{\mathfrak{F}_t\}$-adapted process on a standard filtered space $(\Omega, \mathfrak{F}, \{\mathfrak{F}_t\}, P)$, and the coefficients α_j^i and β^i for $i = 1, \ldots, d, j = 1, \ldots, r$, are real valued functions on $\mathbb{R}_+ \times \mathbf{W}^d$ satisfying certain measurability and integrability conditions to ensure the existence of the stochastic integrals $\{\int_{[0,t]} \alpha_j^i(s, X) \, dB^j(s) : t \in \mathbb{R}_+\}$ and $\{\int_{[0,t]} \beta^i(s, X) m_L(ds) : t \in \mathbb{R}_+\}$. The first and the second term on the right side of (1) are called the diffusion term and the drift term respectively of the stochastic differential equation. Note that at any $t \in \mathbb{R}_+$ the coefficients α_j^i and β^i depend on $X(\cdot, \omega)$, not just $X(t, \omega)$, for $\omega \in \Omega$. The particular case of (1) where at any $t \in \mathbb{R}_+$ the coefficients depend only on $X(t, \omega)$, that is, a stochastic differential equation of the type

$$(2) \qquad dX_t^i = \sum_{j=1}^r a_j^i(t, X(t)) \, dB^j(t) + b^i(t, X(t)) \, dt \quad \text{for } i = 1, \ldots, d,$$

is called Markovian. Here a_j^i and b^i are real valued functions on $\mathbb{R}_+ \times \mathbb{R}^d$ satisfying some measurability and integrability conditions. As a further specialized case we have

$$(3) \qquad dX_t^i = \sum_{j=1}^r f_j^i(X(t)) \, dB^j(t) + g^i(X(t)) \, dt \quad \text{for } i = 1, \ldots, d,$$

where f_j^i and g^i are real valued functions on $\mathbb{R}^d$ satisfying some measurability and integrability conditions. This type of equation is called time-homogeneous Markovian. In (2), if $a_j^i = 0$ for $i = 1, \ldots, d, j = 1, \ldots, r$, then we have

$$dX_t^i = b^i(t, X(t)) \, dt \quad \text{for } i = 1, \ldots, d,$$

which is a randomization of a dynamical system.

Definition 18.1. *Let $\mathbb{R}^d \otimes \mathbb{R}^r$ be the collection of all $d \times r$ matrices of real numbers. We identify $\mathbb{R}^d \otimes \mathbb{R}^r$ with $\mathbb{R}^{dr}$ and consider the measurable space $(\mathbb{R}^d \otimes \mathbb{R}^r, \mathfrak{B}_{\mathbb{R}^{dr}})$.*

Definition 18.2. *Let $\mathbf{M}^{d \times r}(\mathbf{W}^d, \mathfrak{W}^d, \{\mathfrak{W}_t^d\})$ be the collection of all $\mathbb{R}^d \otimes \mathbb{R}^r$-valued $\{\mathfrak{W}_t^d\}$-progressively measurable processes α on the filtered measurable space $(\mathbf{W}^d, \mathfrak{W}^d, \{\mathfrak{W}_t^d\})$, that is, every α in this collection is a mapping of $\mathbb{R}_+ \times \mathbf{W}^d$ into $\mathbb{R}^d \otimes \mathbb{R}^r$ such that for every $t \in \mathbb{R}_+$ the restriction of α to $[0, t] \times \mathbf{W}^d$ is a $\sigma(\mathfrak{B}_{[0,t]} \times \mathfrak{W}_t^d)/\mathfrak{B}_{\mathbb{R}^{dr}}$-measurable mapping of $[0, t] \times \mathbf{W}^d$ into $\mathbb{R}^d \otimes \mathbb{R}^r$.*

Remark 18.3. According to Observation 2.12, a progressively measurable process is always an adapted measurable process. Thus if $\alpha \in \mathbf{M}^{d \times r}(\mathbf{W}^d, \mathfrak{W}^d, \{\mathfrak{W}_t^d\})$ then

1°. α is a $\sigma(\mathfrak{B}_{\mathbb{R}_+} \times \mathfrak{W}^d)/\mathfrak{B}_{\mathbb{R}^{dr}}$-measurable mapping of $\mathbb{R}_+ \times \mathbf{W}^d$ into $\mathbb{R}^d \otimes \mathbb{R}^r$.

2°. $\alpha(t, \cdot)$ is a $\mathfrak{W}_t^d/\mathfrak{B}_{\mathbb{R}^{dr}}$-measurable mapping of $\mathbf{W}^d$ into $\mathbb{R}^d \otimes \mathbb{R}^r$ for every $t \in \mathbb{R}_+$.

Remark 18.4. Let a be a mapping of $\mathbb{R}_+ \times \mathbb{R}^d$ into $\mathbb{R}^d \otimes \mathbb{R}^r$ such that for every $t \in \mathbb{R}_+$ the restriction of a to $[0,t] \times \mathbb{R}^d$ is a $\sigma(\mathfrak{B}_{[0,t]} \times \mathfrak{B}_{\mathbb{R}^d})/\mathfrak{B}_{\mathbb{R}^{dr}}$-measurable mapping of $[0,t] \times \mathbb{R}^d$ into $\mathbb{R}^d \otimes \mathbb{R}^r$. If we define a mapping α of $\mathbb{R}_+ \times \mathbf{W}^d$ into $\mathbb{R}^d \otimes \mathbb{R}^r$ by setting

$$(1) \qquad \alpha(t,w) = a(t, w(t)) \quad \text{for } (t,w) \in \mathbb{R}_+ \times \mathbf{W}^d,$$

then $\alpha \in \mathbf{M}^{d \times r}(\mathbf{W}^d, \mathfrak{W}^d, \{\mathfrak{W}_t^d\})$. In particular, if f is a $\mathfrak{B}_{\mathbb{R}^d}/\mathfrak{B}_{\mathbb{R}^{dr}}$-measurable mapping of $\mathbb{R}^d$ into $\mathbb{R}^d \otimes \mathbb{R}^r$, then the mapping α of $\mathbb{R}_+ \times \mathbf{W}^d$ into $\mathbb{R}^d \otimes \mathbb{R}^r$ defined by

$$(2) \qquad \alpha(t,w) = f(w(t)) \quad \text{for } (t,w) \in \mathbb{R}_+ \times \mathbf{W}^d$$

is a member of $\mathbf{M}^{d \times r}(\mathbf{W}^d, \mathfrak{W}^d, \{\mathfrak{W}_t^d\})$.

Proof. 1) With $t \in \mathbb{R}_+$ fixed, consider the restriction of a to $[0,t] \times \mathbb{R}^d$. Since $s \mapsto s$ for $s \in [0,t]$ is a $\mathfrak{B}_{[0,t]}/\mathfrak{B}_{[0,t]}$-measurable mapping of $[0,t]$ into $[0,t]$ and $w \mapsto w(t)$ for $w \in \mathbf{W}^d$ is a $\mathfrak{W}_t^d/\mathfrak{B}_{\mathbb{R}^d}$-measurable mapping of $\mathbf{W}^d$ into $\mathbb{R}^d$, $(s,w) \mapsto (s, w(t))$ is a $\sigma(\mathfrak{B}_{[0,t]} \times \mathfrak{W}_t^d)/\sigma(\mathfrak{B}_{[0,t]} \times \mathfrak{B}_{\mathbb{R}^d})$-measurable mapping of $[0,t] \times \mathbf{W}^d$ into $[0,t] \times \mathbb{R}^d$. Then since the restriction of a to $[0,t] \times \mathbb{R}^d$ is a $\sigma(\mathfrak{B}_{[0,t]} \times \mathfrak{B}_{\mathbb{R}^d})/\mathfrak{B}_{\mathbb{R}^{dr}}$-measurable mapping of $[0,t] \times \mathbb{R}^d$ into $\mathbb{R}^d \otimes \mathbb{R}^r$, $(s,w) \mapsto a(s, w(t))$ is a $\sigma(\mathfrak{B}_{[0,t]} \times \mathfrak{W}_t^d)/\mathfrak{B}_{\mathbb{R}^{dr}}$-measurable mapping of $[0,t] \times \mathbf{W}^d$ into $\mathbb{R}^d \otimes \mathbb{R}^r$. This shows that α defined by (1) is in $\mathbf{M}^{d \times r}(\mathbf{W}^d, \mathfrak{W}^d, \{\mathfrak{W}_t^d\})$.

2) Let f be a $\mathfrak{B}_{\mathbb{R}^d}/\mathfrak{B}_{\mathbb{R}^{dr}}$-measurable mapping of $\mathbb{R}^d$ into $\mathbb{R}^d \otimes \mathbb{R}^r$. Consider a mapping a of $\mathbb{R}_+ \times \mathbb{R}^d$ into $\mathbb{R}^d \otimes \mathbb{R}^r$ defined by setting $a(t,x) = f(x)$ for $(t,x) \in \mathbb{R}_+ \times \mathbb{R}^d$. For every $t \in \mathbb{R}_+$, the restriction of a to $[0,t] \times \mathbb{R}^d$ is a $\sigma(\mathfrak{B}_{[0,t]} \times \mathfrak{B}_{\mathbb{R}^d})/\mathfrak{B}_{\mathbb{R}^{dr}}$-measurable mapping of $[0,t] \times \mathbb{R}^d$ into $\mathbb{R}^d \otimes \mathbb{R}^r$. Then for α defined by (2) we have $\alpha(t,w) = f(w(t)) = a(t, w(t))$ so that α is in $\mathbf{M}^{d \times r}(\mathbf{W}^d, \mathfrak{W}^d, \{\mathfrak{W}_t^d\})$ by 1). ∎

Lemma 18.5. *Let X be a continuous d-dimensional adapted process on a standard filtered space $(\Omega, \mathfrak{F}, \{\mathfrak{F}_t\}, P)$. Then for every $\alpha \in \mathbf{M}^{d \times r}(\mathbf{W}^d, \mathfrak{W}^d, \{\mathfrak{W}_t^d\})$ the function α_X on $\mathbb{R}_+ \times \Omega$ defined by*

$$\alpha_X(t, \omega) = \alpha(t, X(\cdot, \omega)) \quad \text{for } (t, \omega) \in \mathbb{R}_+ \times \Omega$$

is an $\{\mathfrak{F}_t\}$-progressively measurable process on $(\Omega, \mathfrak{F}, \{\mathfrak{F}_t\}, P)$. In particular α_X is an $\{\mathfrak{F}_t\}$-adapted measurable process.

Proof. Recall that according to Theorem 17.8 if X is a continuous d-dimensional stochastic process on a standard filtered space $(\Omega, \mathfrak{F}, \{\mathfrak{F}_t\}, P)$, then the mapping $\mathcal{X}$ of Ω into $\mathbf{W}^d$

§18. DEFINITION AND FUNCTION SPACE REPRESENTATION

defined by $\mathcal{X}(\omega) = X(\cdot, \omega)$ for $\omega \in \Omega$ is an $\mathfrak{F}/\mathfrak{W}^d$-measurable mapping, and if X is adapted then $\mathcal{X}$ is $\mathfrak{F}_t/\mathfrak{W}_t^d$-measurable for every $t \in \mathbb{R}_+$. Let us show that for every $t \in \mathbb{R}_+$, the restriction of α_X to $[0,t] \times \Omega$ is a $\sigma(\mathfrak{B}_{[0,t]} \times \mathfrak{F}_t)/\mathfrak{B}_{\mathbb{R}^{dr}}$-measurable mapping of $[0,t] \times \Omega$ into $\mathbb{R}^d \otimes \mathbb{R}^r$. Now since $s \mapsto s$ for $s \in [0,t]$ is a $\mathfrak{B}_{[0,t]}/\mathfrak{B}_{[0,t]}$-measurable mapping of $[0,t]$ into $[0,t]$ and $\omega \mapsto X(\cdot, \omega)$ for $\omega \in \Omega$ is an $\mathfrak{F}_t/\mathfrak{W}_t^d$-measurable mapping of Ω into $\mathbf{W}^d$, $(s, \omega) \mapsto (s, X(\cdot, \omega))$ is a $\sigma(\mathfrak{B}_{[0,t]} \times \mathfrak{F}_t)/\sigma(\mathfrak{B}_{[0,t]} \times \mathfrak{W}_t^d)$-measurable mapping of $[0,t] \times \Omega$ into $[0,t] \times \mathbf{W}^d$. But the restriction of α to $[0,t] \times \mathbf{W}^d$ is a $\sigma(\mathfrak{B}_{[0,t]} \times \mathfrak{W}_t^d)/\mathfrak{B}_{\mathbb{R}^{dr}}$-measurable mapping of $[0,t] \times \mathbf{W}^d$ into $\mathbb{R}^d \otimes \mathbb{R}^r$ by assumption. Thus $(s, \omega) \mapsto \alpha(s, X(\cdot, \omega))$ is a $\sigma(\mathfrak{B}_{[0,t]} \times \mathfrak{F}_t)/\mathfrak{B}_{\mathbb{R}^{dr}}$-measurable mapping of $[0,t] \times \Omega$ into $\mathbb{R}^d \otimes \mathbb{R}^r$. ∎

Definition 18.6. *Let* $\alpha \in \mathbf{M}^{d \times r}(\mathbf{W}^d, \mathfrak{W}^d, \{\mathfrak{W}_t^d\})$ *and* $\beta \in \mathbf{M}^{d \times 1}(\mathbf{W}^d, \mathfrak{W}^d, \{\mathfrak{W}_t^d\})$ *and consider the stochastic differential equation*

$$dX_t = \alpha(t, X) \, dB(t) + \beta(t, X) \, dt,$$

that is,

(1) $$dX_t^i = \sum_{j=1}^r \alpha_j^i(t, X) \, dB^j(t) + \beta^i(t, X) \, dt \quad \text{for } i = 1, \ldots, d.$$

We say that the stochastic differential equation has a solution if there exists a standard filtered space $(\Omega, \mathfrak{F}, \{\mathfrak{F}_t\}, P)$ *with a pair of stochastic process* (B, X) *on it satisfying the following conditions.*

1°. *B is an r-dimensional $\{\mathfrak{F}_t\}$-adapted null at 0 Brownian motion on $(\Omega, \mathfrak{F}, \{\mathfrak{F}_t\}, P)$.*

2°. *X is a continuous d-dimensional adapted process on $(\Omega, \mathfrak{F}, \{\mathfrak{F}_t\}, P)$.*

3°. *The $\{\mathfrak{F}_t\}$-progressively measurable processes $(\alpha_X)_j^i$ for $i = 1, \ldots, d$, $j = 1, \ldots, r$ defined by $(\alpha_X)_j^i(t, \omega) = \alpha_j^i(t, X(\cdot, \omega))$ for $(t, \omega) \in \mathbb{R}_+ \times \Omega$ are in $\mathbf{L}_{2,\infty}(\mathbb{R}_+ \times \Omega, m_L \times P)$, and the $\{\mathfrak{F}_t\}$-progressively measurable processes $(\beta_X)^i$ for $i = 1, \ldots, d$, defined by $(\beta_X)^i(t, \omega) = \beta^i(t, X(\cdot, \omega))$ for $(t, \omega) \in \mathbb{R}_+ \times \Omega$ are in $\mathbf{L}_{1,\infty}(\mathbb{R}_+ \times \Omega, m_L \times P)$.*

4°. *There exists a null set* $\Lambda \in (\Omega, \mathfrak{F}, P)$ *such that on Λ^c we have for every $t \in \mathbb{R}_+$*

$$X^i(t) - X^i(0) = \sum_{j=1}^r \int_{[0,t]} (\alpha_X)_j^i(s) \, dB^j(s) + \int_{[0,t]} (\beta_X)^i(s) m_L(ds)$$

for $i = 1, \ldots, d$.

In this case we say that (B,X) is a solution of the stochastic differential equation.

Note that the fact that X^i satisfies condition 4° above implies that X_i is the sum of a process in $\mathbf{M}_2^c(\Omega,\mathfrak{F},\{\mathfrak{F}_t\},P)$, a process in $\mathbf{V}^c(\Omega,\mathfrak{F},\{\mathfrak{F}_t\},P)$, and an $\mathfrak{F}_0$-measurable real valued random variable and is therefore a quasimartingale on $(\Omega,\mathfrak{F},\{\mathfrak{F}_t\},P)$. Hence the stochastic differential dX^i is defined.

Remark 18.7. If X satisfies condition 2° of Definition 18.6, then by Lemma 18.5, $(\alpha_X)_j^i$ and $(\beta_X)^i$ are $\{\mathfrak{F}_t\}$-progressively measurable processes on $(\Omega,\mathfrak{F},\{\mathfrak{F}_t\},P)$.
1) If $(\alpha_X)_j^i$ are in $\mathbf{L}_{2,\infty}(\mathbb{R}_+ \times \Omega, m_L \times P)$ as required by condition 3° of Definition 18.6, then by Theorem 13.38 there exist equivalent processes which are $\{\mathfrak{F}_t\}$-predictable. Let us understand $(\alpha_X)_j^i$ to be $\{\mathfrak{F}_t\}$-predictable versions so that the stochastic integrals $\{\int_{[0,t]}(\alpha_X)_j^i(s)\,dB^j(s) : t \in \mathbb{R}_+\}$ exist in $\mathbf{M}_2^c(\Omega,\mathfrak{F},\{\mathfrak{F}_t\},P)$. Condition 4° in Definition 18.6 refers to these stochastic integrals. Note also that by Theorem 12.8 there exist sequences $\{(a^n)_j^i : n \in \mathbb{N}\}$ in $\mathbf{L}_0(\Omega,\mathfrak{F},\{\mathfrak{F}_t\},P)$ such that $\lim_{n\to\infty}\|(a^n)_j^i - (\alpha_X)_j^i\|_{2,\infty}^{m_L \times P} = 0$.
2) If $(\beta_X)^i$ are in $\mathbf{L}_{1,\infty}(\mathbb{R}_+ \times \Omega, m_L \times P)$ as required by condition 3° of Definition 18.6, then the stochastic integrals $\{\int_{[0,t]}(\beta_X)^i(s)m_L(ds) : t \in \mathbb{R}_+\}$ exist in $\mathbf{V}^c(\Omega,\mathfrak{F},\{\mathfrak{F}_t\},P)$. By Theorem 13.38, $(\beta_X)^i$ have $\{\mathfrak{F}_t\}$-predictable versions. Let us understand $(\beta_X)^i$ to be $\{\mathfrak{F}_t\}$-predictable versions. Note also that by Theorem 12.8 there exist sequences $\{(b^n)^i : n \in \mathbb{N}\}$ in $\mathbf{L}_0(\Omega,\mathfrak{F},\{\mathfrak{F}_t\},P)$ such that $\lim_{n\to\infty}\|(b^n)^i - (\beta_X)^i\|_{1,\infty}^{m_L \times P} = 0$.

Whether or not a solution exists for the stochastic differential equation (1) in Definition 18.6 depends entirely on the given coefficients $\alpha \in \mathbf{M}^{d\times r}(\mathbf{W}^d, \mathfrak{W}^d, \{\mathfrak{W}_t^d\})$ and $\beta \in \mathbf{M}^{d\times 1}(\mathbf{W}^d, \mathfrak{W}^d, \{\mathfrak{W}_t^d\})$. We shall show this by showing that whenever the stochastic differential equation has a solution on some standard filtered space it has a solution on the function space $\mathbf{W}^{r+d}$.

[II] Function Space Representation of Solutions

Our objective here is to show that if the stochastic differential equation (1) in Definition 18.6 has a solution (B,X) on some standard filtered space $(\Omega,\mathfrak{F},\{\mathfrak{F}_t\},P)$, then it has a solution (W,Y) on a standard filtered space constructed on the function space $\mathbf{W}^{r+d}$ and furthermore (B,X) and (W,Y) have the same probability distribution on $(\mathbf{W}^{r+d}, \mathfrak{W}^{r+d})$.

Proposition 18.8. *Let $\mathbf{W}^{r+d} = \mathbf{W}^r \times \mathbf{W}^d$. Let $\mathfrak{W}^{r+d}$, $\mathfrak{W}^r$, and $\mathfrak{W}^d$ be the σ-algebras gener-*

8. DEFINITION AND FUNCTION SPACE REPRESENTATION

ated by the collections of the cylinder sets, $\mathfrak{Z}$, $\mathfrak{Z}'$, and $\mathfrak{Z}''$ in $\mathbf{W}^{r+d}$, $\mathbf{W}^r$, and $\mathbf{W}^d$ respectively. Then

$$\mathfrak{W}^{r+d} = \sigma(\mathfrak{W}^r \times \mathfrak{W}^d). \tag{1}$$

Similarly for the σ-algebras $\mathfrak{W}_t^{r+d}$, $\mathfrak{W}_t^r$, and $\mathfrak{W}_t^d$ generated by the collections of the cylinder sets with indices preceding $t \in \mathbb{R}_+$, $\mathfrak{Z}_{[0,t]}$, $\mathfrak{Z}'_{[0,t]}$, and $\mathfrak{Z}''_{[0,t]}$ in $\mathbf{W}^{r+d}$, $\mathbf{W}^r$, and $\mathbf{W}^d$, we have

$$\mathfrak{W}_t^{r+d} = \sigma(\mathfrak{W}_t^r \times \mathfrak{W}_t^d). \tag{2}$$

Proof. To prove (1), it suffices to show that

$$\sigma(\mathfrak{Z}) = \sigma(\mathfrak{Z}' \times \mathfrak{Z}'') = \sigma(\sigma(\mathfrak{Z}') \times \sigma(\mathfrak{Z}'')). \tag{3}$$

To prove the first equality in (3), note that $\mathfrak{Z}' \times \mathfrak{Z}'' \subset \mathfrak{Z}$ so that $\sigma(\mathfrak{Z}' \times \mathfrak{Z}'') \subset \sigma(\mathfrak{Z})$. To show the reverse inclusion, let $w \in \mathbf{W}^{r+d}$ be given as $w = (w', w'')$ with $w' \in \mathbf{W}^r$ and $w'' \in \mathbf{W}^d$. For $t \in \mathbb{R}_+$, let q_t, q_t', and q_t'' be mappings of $\mathbf{W}^{r+d}$, $\mathbf{W}^r$, and $\mathbf{W}^d$ into $\mathbb{R}^d$ defined by $q_t(w) = w(t)$ for $w \in \mathbf{W}^{r+d}$, $q_t'(w') = w'(t)$ for $w' \in \mathbf{W}^r$, and $q_t''(w'') = w''(t)$ for $w'' \in \mathbf{W}^d$. Then for $E' \in \mathfrak{B}_{\mathbb{R}^r}$ and $E'' \in \mathfrak{B}_{\mathbb{R}^d}$, we have $q_t^{-1}(E' \times E'') = (q_t')^{-1}(E') \times (q_t'')^{-1}(E'') \in \mathfrak{Z}' \times \mathfrak{Z}''$. Thus $q_t^{-1}(\mathfrak{B}_{\mathbb{R}^r} \times \mathfrak{B}_{\mathbb{R}^d}) \subset \mathfrak{Z}' \times \mathfrak{Z}''$ and then by Proposition 13.2 and Theorem 1.1 we have

$$q_t^{-1}(\mathfrak{B}_{\mathbb{R}^{r+d}}) = q_t^{-1}(\sigma(\mathfrak{B}_{\mathbb{R}^r} \times \mathfrak{B}_{\mathbb{R}^d})) = \sigma(q_t^{-1}(\mathfrak{B}_{\mathbb{R}^r} \times \mathfrak{B}_{\mathbb{R}^d})) \subset \sigma(\mathfrak{Z}' \times \mathfrak{Z}'').$$

Since every $Z \in \mathfrak{Z}$ is of the form $Z = \cap_{i=1}^n q_{t_i}^{-1}(E_i)$ where $E_i \in \mathfrak{B}_{\mathbb{R}^{r+d}}$ for $i = 1, \ldots, d$, we have $\mathfrak{Z} \subset \sigma(\mathfrak{Z}' \times \mathfrak{Z}'')$. Thus $\sigma(\mathfrak{Z}) \subset \sigma(\mathfrak{Z}' \times \mathfrak{Z}'')$. This proves the first equality in (3). To prove the second equality in (3), note that $\mathfrak{Z}' \times \mathfrak{Z}'' \subset \sigma(\mathfrak{Z}') \times \sigma(\mathfrak{Z}'')$ so that $\sigma(\mathfrak{Z}' \times \mathfrak{Z}'') \subset \sigma(\sigma(\mathfrak{Z}') \times \sigma(\mathfrak{Z}''))$. On the other hand by Lemma 1.3 we have $\sigma(\sigma(\mathfrak{Z}') \times \sigma(\mathfrak{Z}'')) \subset \sigma(\mathfrak{Z}' \times \mathfrak{Z}'')$. This proves the second equality in (3).

For any $t \in \mathbb{R}_+$, we have $\sigma(\mathfrak{Z}_{[0,t]}) = \sigma(\mathfrak{Z}'_{[0,t]} \times \mathfrak{Z}''_{[0,t]}) = \sigma(\sigma(\mathfrak{Z}'_t) \times \sigma(\mathfrak{Z}''_t))$ by the same argument as above in proving (3). Thus (2) holds. ∎

Observation 18.9. For $w \in \mathbf{W}^{r+d}$, let us write $w = (w', w'')$ where $w' \in \mathbf{W}^r$ and $w'' \in \mathbf{W}^d$. Define an r-dimensional continuous $\{\mathfrak{W}_t^{r+d}\}$-adapted process W and a d-dimensional continuous $\{\mathfrak{W}_t^{r+d}\}$-adapted process Y on $(\mathbf{W}^{r+d}, \mathfrak{W}^{r+d}, \{\mathfrak{W}_t^{r+d}\})$ by setting

$$\begin{cases} W(t,w) = w'(t) & \text{for } (t,w) \in \mathbb{R}_+ \times \mathbf{W}^{r+d} \\ Y(t,w) = w''(t) & \text{for } (t,w) \in \mathbb{R}_+ \times \mathbf{W}^{r+d}. \end{cases}$$

Suppose the stochastic differential equation (1) in Definition 18.6 has a solution $(\mathcal{B}, \mathcal{X})$ on a standard filtered space $(\Omega, \mathfrak{F}, \{\mathfrak{F}_t\}, P)$. According to Theorem 17.8, the mapping $(\mathcal{B}, \mathcal{X})$ of Ω into $\mathbf{W}^{r+d}$ defined by $(\mathcal{B}, \mathcal{X})(\omega) = (B(\cdot, \omega), X(\cdot, \omega))$ for $\omega \in \Omega$ is $\mathfrak{F}/\mathfrak{W}^{r+d}$-measurable and $\mathfrak{F}_t/\mathfrak{W}^{r+d}_t$-measurable for every $t \in \mathbb{R}_+$. Let $P_{(\mathcal{B},\mathcal{X})}$ be the probability distribution of $(\mathcal{B}, \mathcal{X})$ on $(\mathbf{W}^{r+d}, \mathfrak{W}^{r+d})$. We define a standard filtered space on the probability space $(\mathbf{W}^{r+d}, \mathfrak{W}^{r+d}, P_{(\mathcal{B},\mathcal{X})})$ as follows.

Definition 18.10 *Consider the probability space* $(\mathbf{W}^{r+d}, \mathfrak{W}^{r+d}, P_{(\mathcal{B},\mathcal{X})})$ *where* (B, X) *is a solution of the stochastic differential equation (1) in Definition 18.6 on some standard filtered space. Let*

1°. $\mathfrak{W}^{r+d,*}$ *be the completion of* $\mathfrak{W}^{r+d}$ *with respect to* $P_{(\mathcal{B},\mathcal{X})}$.

2°. $\mathfrak{W}^{r+d,0}_t = \sigma(\mathfrak{W}^{r+d}_t \cup \mathfrak{N})$ *where* $\mathfrak{N}$ *is the collection of all the null sets in the complete measure space* $(\mathbf{W}^{r+d}, \mathfrak{W}^{r+d,*}, P_{(\mathcal{B},\mathcal{X})})$.

3°. $\mathfrak{W}^{r+d,*}_t = \cap_{\varepsilon > 0} \mathfrak{W}^{r+d,0}_{t+\varepsilon}$.

We then have a standard filtered space $(\mathbf{W}^{r+d}, \mathfrak{W}^{r+d,*}, \{\mathfrak{W}^{r+d,*}_t\}, P_{(\mathcal{B},\mathcal{X})})$ *in which the σ-algebra, the filtration, as well as the probability measure, depend on* (B, X). *Let us call this standard filtered space the standard filtered space on* $(\mathbf{W}^{r+d}, \mathfrak{W}^{r+d})$ *generated by* $P_{(\mathcal{B},\mathcal{X})}$).

Our aim is to show that with W and Y defined in Observation 18.9, (W, Y) is a solution of the stochastic differential equation on $(\mathbf{W}^{r+d}, \mathfrak{W}^{r+d,*}, \{\mathfrak{W}^{r+d,*}_t\}, P_{(\mathcal{B},\mathcal{X})})$. This is done in Theorem 18.13 below. For this we show in Lemma 18.11 that W is a Brownian motion satisfying condition 1° of Definition 18.6. In Lemma 18.12 we show that α_Y and β_Y satisfy condition 3° of Definition 18.6.

Lemma 18.11. *The r-dimensional continuous adapted process W on the filtered space* $(\mathbf{W}^{r+d}, \mathfrak{W}^{r+d}, \{\mathfrak{W}^{r+d}_t\})$ *defined in Observation 18.9 is an r-dimensional $\{\mathfrak{W}^{r+d,*}_t\}$-adapted null at 0 Brownian motion on the standard filtered space* $(\mathbf{W}^{r+d}, \mathfrak{W}^{r+d,*}, \{\mathfrak{W}^{r+d,*}_t\}, P_{(\mathcal{B},\mathcal{X})})$.

Proof. As we noted in Observation 18.9, W is an r-dimensional continuous $\{\mathfrak{W}^{r+d}_t\}$-adapted process on $(\mathbf{W}^{r+d}, \mathfrak{W}^{r+d}, \{\mathfrak{W}^{r+d}_t\})$. Since $\mathfrak{W}^{r+d}_t$ is a sub-σ-algebra of $\mathfrak{W}^{r+d,*}_t$, W is a $\mathfrak{W}^{r+d,*}_t$-adapted process on $(\mathbf{W}^{r+d}, \mathfrak{W}^{r+d,*}, \{\mathfrak{W}^{r+d,*}_t\}, P_{(\mathcal{B},\mathcal{X})})$. To show that W is null at 0, let q_0 be the $\mathfrak{W}^{r+d}/\mathfrak{B}_{\mathbb{R}^d}$-measurable mapping of $\mathbf{W}^{r+d}$ into $\mathbb{R}^d$ defined by $q_0(w) = w(0)$ for $w \in \mathbf{W}^{r+d}$. Then

(1) $\quad P_{(\mathcal{B},\mathcal{X})}\{W_0 = 0\} = P_{(\mathcal{B},\mathcal{X})}\{(W, Y)(0) \in \{0\} \times \mathbb{R}^d\} = P\{q_0 \circ (\mathcal{B}, \mathcal{X}) \in \{0\} \times \mathbb{R}^d\}$

§18. DEFINITION AND FUNCTION SPACE REPRESENTATION

$$= P\{(B,X)(0) \in \{0\} \times \mathbb{R}^d\} = P\{B = 0\} = 1$$

by the fact that B is an r-dimensional $\{\mathfrak{F}_t\}$-adapted null at 0 Brownian motion on the standard filtered space $(\Omega, \mathfrak{F}, \{\mathfrak{F}_t\}, P)$. Thus W is null at 0. To show that W is an r-dimensional $\{\mathfrak{W}_t^{r+d,*}\}$-Brownian motion on $((\mathbf{W}^{r+d}, \mathfrak{W}^{r+d,*}, \{\mathfrak{W}_t^{r+d,*}\}, P_{(B,X)})$, it remains according to Theorem 13.26 to verify that for any $s, t \in \mathbb{R}_+$, $s < t$, and $y \in \mathbb{R}^r$ we have

(2) $\quad \mathbf{E}[e^{i\langle y, W_t - W_s \rangle} | \mathfrak{W}_s^{r+d,*}] = e^{-\frac{|y|^2}{2}(t-s)} \quad$ a.e. on $(\mathbf{W}^{r+d}, \mathfrak{W}_s^{r+d,*}, P_{(B,X)})$.

Since the right side of (2) is $\mathfrak{W}_s^{r+d,*}$-measurable, it remains to verify

(3) $\quad \int_A e^{i\langle y, W_t - W_s \rangle} dP_{(B,X)} = P_{(B,X)}(A) e^{-\frac{|y|^2}{2}(t-s)} \quad$ for $A \in \mathfrak{W}_s^{r+d,*}$,

or equivalently according to the Image Probability Law

(4) $\quad \int_{(\mathcal{B},\mathcal{X})^{-1}(A)} e^{i\langle y, B_t - B_s \rangle} dP = P((\mathcal{B},\mathcal{X})^{-1}(A)) e^{-\frac{|y|^2}{2}(t-s)} \quad$ for $A \in \mathfrak{W}_s^{r+d,*}$.

To verify (2), consider first the case where $A \in \mathfrak{W}_s^{r+d}$. Since B is an r-dimensional continuous $\{\mathfrak{F}_t\}$-adapted process, $\mathcal{B}$ is an $\mathfrak{F}_t/\mathfrak{W}_t^r$-measurable mapping of Ω into $\mathbf{W}^r$ for every $t \in \mathbb{R}_+$ by Theorem 17.8. By condition 2° of Definition 18.6, $\mathcal{X}$ is an $\mathfrak{F}_t/\mathfrak{W}_t^d$-measurable mapping of Ω into $\mathbf{W}^d$. Then since $\mathfrak{W}_t^{r+d} = \sigma(\mathfrak{W}_t^r \times \mathfrak{W}_t^d)$ by Proposition 18.8, $(\mathcal{B}, \mathcal{X})$ is an $\mathfrak{F}_t/\mathfrak{W}_t^{r+d}$-measurable mapping of Ω into $\mathbf{W}^{r+d}$ for every $t \in \mathbb{R}_+$. Thus we have $(\mathcal{B}, \mathcal{X})^{-1}(\mathfrak{W}_t^{r+d}) \subset \mathfrak{F}_s$. Since B is an r-dimensional $\{\mathfrak{F}_t\}$-adapted Brownian motion on $(\Omega, \mathfrak{F}, \{\mathfrak{F}_t\}, P)$, (4) holds by Theorem 13.26 for our $A \in \mathfrak{W}_s^{r+d}$. Then since $\mathfrak{W}_s^{r+d,0} = \sigma(\mathfrak{W}_s^{r+d} \cup \mathfrak{N})$ and since (3) holds for every $A \in \mathfrak{W}_s^{r+d}$, (3) holds for every $A \in \mathfrak{W}_s^{r+d,0}$. Thus we have shown that

(5) $\quad \mathbf{E}[e^{i\langle y, W_t - W_s \rangle} | \mathfrak{W}_s^{r+d,0}] = e^{-\frac{|y|^2}{2}(t-s)} \quad$ a.e. on $(\mathbf{W}^{r+d}, \mathfrak{W}_s^{r+d,0}, P_{(B,X)})$.

Now $\mathfrak{W}_s^{r+d,*} = \cap_{\varepsilon>0} \mathfrak{W}_{s+\varepsilon}^{r+d,0} = \cap_{n\in\mathbb{N}} \mathfrak{W}_{s+1/n}^{r+d,0}$. Let $n \in \mathbb{N}$ be so large that $s + 1/n < t$. Then

(6) $\quad \mathbf{E}[e^{i\langle y, W_t - W_{s+1/n} \rangle} | \mathfrak{W}_s^{r+d,*}] = \mathbf{E}[\mathbf{E}[e^{i\langle y, W_t - W_{s+1/n} \rangle} | \mathfrak{W}_{s+1/n}^{r+d,0}] | \mathfrak{W}_s^{r+d,*}]$

$\quad = \mathbf{E}[e^{-\frac{|y|^2}{2}(t-(s+1/n))} | \mathfrak{W}_s^{r+d,*}] = e^{-\frac{|y|^2}{2}(t-(s+1/n))} \quad$ a.e. on $(\mathbf{W}^{r+d}, \mathfrak{W}_s^{r+d,*}, P_{(B,X)})$,

where the second equality is by (5). Letting $n \to \infty$ in (6) and applying the Conditional Bounded Convergence Theorem we have (2). ∎

Lemma 18.12. *For the d-dimensional continuous adapted process Y on the filtered space $(\mathbf{W}^{r+d}, \mathfrak{W}^{r+d}, \{\mathfrak{W}_t^{r+d}\})$ defined in Observation 18.9 and for the coefficients α and β in the stochastic differential equation (1) in Definition 16.8 for which (B, X) is a solution on a standard filtered space $(\Omega, \mathfrak{F}, \{\mathfrak{F}_t\}, P)$, the $\{\mathfrak{W}_t^{r+d,*}\}$-progressively measurable processes $(\alpha_Y)^i_j$ for $i = 1, \ldots, d$, $j = 1, \ldots, r$, defined by $(\alpha_Y)^i_j(t, w) = \alpha^i_j(t, Y(\cdot, w))$ for $(t, w) \in \mathbb{R}_+ \times \mathbf{W}^{r+d}$ are in $\mathbf{L}_{2,\infty}(\mathbb{R}_+ \times \mathbf{W}^{r+d}, m_L \times P_{(B,X)})$, and the $\{\mathfrak{W}_t^{r+d,*}\}$-progressively measurable processes $(\beta_Y)^i$ for $i = 1, \ldots, d$, defined by $(\beta_Y)^i(t, w) = \beta^i(t, Y(\cdot, w))$ for $(t, w) \in \mathbb{R}_+ \times \mathbf{W}^{r+d}$ are in $\mathbf{L}_{1,\infty}(\mathbb{R}_+ \times \mathbf{W}^{r+d}, m_L \times P_{(B,X)})$.*

Proof. We noted in Observation 18.9 that Y is a d-dimensional $\{\mathfrak{W}_t^{r+d}\}$-adapted process on $(\mathbf{W}^{r+d}, \mathfrak{W}^{r+d}, \{\mathfrak{W}_t^{r+d}\})$. Thus Y is a $\{\mathfrak{W}_t^{r+d,*}\}$-adapted process on the standard filtered space $(\mathbf{W}^{r+d}, \mathfrak{W}^{r+d,*}, \{\mathfrak{W}_t^{r+d,*}\}, P_{(B,X)})$. Then the mapping of $\mathcal{Y}$ of $\mathbf{W}^{r+d}$ into $\mathbf{W}^d$ defined by $\mathcal{Y}(w) = Y(\cdot, w)$ for $w \in \mathbf{W}^{r+d}$ is $\mathfrak{W}_t^{r+d}/\mathfrak{W}_t^d$-measurable for every $t \in \mathbb{R}_+$ by Theorem 17.8. Thus $\mathcal{Y}$ is $\mathfrak{W}_t^{r+d,*}/\mathfrak{W}_t^d$-measurable for every $t \in \mathbb{R}_+$. From this we have the $\{\mathfrak{W}_t^{r+d,*}\}$-progressive measurability of $(\alpha_Y)^i_j$ for $i = 1, \ldots, d$, $j = 1, \ldots, r$ and $(\beta_Y)^i$ for $i = 1, \ldots, d$ by Lemma 18.5. To show that $(\alpha_Y)^i_j$ are in $\mathbf{L}_{2,\infty}(\mathbb{R}_+ \times \mathbf{W}^{r+d}, m_L \times P_{(B,X)})$, we show that for every $t \in \mathbb{R}_+$ we have

$$(1) \quad \int_{[0,t] \times \mathbf{W}^{r+d}} |\alpha^i_j(s, Y(\cdot, w))|^2 (m_L \times P_{(B,X)})(d(s, w)) < \infty.$$

Consider the measure space $([0, t] \times \Omega, \sigma(\mathfrak{B}_{[0,t]} \times \mathfrak{F}_t), m_L \times P)$. Let ι be the identity mapping of $[0, t]$ into $[0, t]$. Since $(\mathcal{B}, \mathcal{X})$ is an $\mathfrak{F}_t/\mathfrak{W}_t^{r+d}$-measurable mapping of Ω into $\mathbf{W}^{r+d}$ for every $t \in \mathbb{R}_+$ as we noted in the proof of Lemma 18.11, the mapping $(\iota, (\mathcal{B}, \mathcal{X}))$ of $[0, t] \times \Omega$ into $[0, t] \times \mathbf{W}^{r+d}$ is $\sigma(\mathfrak{B}_{[0,t]} \times \mathfrak{F}_t)/\sigma(\mathfrak{B}_{[0,t]} \times \mathfrak{W}_t^{r+d})$-measurable. For the probability distribution $P_{(\iota,(\mathcal{B},\mathcal{X}))}$ of $(\iota, (\mathcal{B}, \mathcal{X}))$ on $([0, t] \times \mathbf{W}^{r+d}, \sigma(\mathfrak{B}_{[0,t]} \times \mathfrak{W}_t^{r+d}))$, we have for any $E \in \mathfrak{B}_{[0,t]}$ and $F \in \mathfrak{W}_t^{r+d}$,

$$P_{(\iota,(\mathcal{B},\mathcal{X}))}(E \times F) = (m_L \times P) \circ (\iota, (\mathcal{B}, \mathcal{X}))^{-1}(E \times F)$$
$$= (m_L \times P)(\iota)^{-1}(E) \times (\mathcal{B}, \mathcal{X})^{-1}(F) = m_L(E) \cdot P \circ (\mathcal{B}, \mathcal{X})^{-1}(F)$$
$$= m_L(E) \cdot P_{(B,X)}(F).$$

This shows that the two finite measures $P_{(\iota,(\mathcal{B},\mathcal{X}))}$ and $m_L \times P_{(B,X)}$ on $\sigma(\mathfrak{B}_{[0,t]} \times \mathfrak{W}_t^{r+d}))$ are equal on the π-class $\mathfrak{B}_{[0,t]} \times \mathfrak{W}_t^{r+d}$ and therefore equal on $\sigma(\mathfrak{B}_{[0,t]} \times \mathfrak{W}_t^{r+d}))$ by Corollary 1.8. Thus for any extended real valued $\sigma(\mathfrak{B}_{[0,t]} \times \mathfrak{W}_t^{r+d})$-measurable function φ on $[0, t] \times \mathbf{W}^{r+d}$, we have by the Image Probability Law

$$(2) \quad \int_{[0,t] \times \Omega} \varphi(s, (\mathcal{B}, \mathcal{X})(\omega))(m_L \times P)(d(s, \omega)) = \int_{[0,t] \times \mathbf{W}^{r+d}} \varphi(s, w)(m_L \times P_{(B,X)})(d(s, w))$$

§18. DEFINITION AND FUNCTION SPACE REPRESENTATION   377

in the sense that the existence of one side implies that of the other and the equality of the two. Now since $w \mapsto w''$ for $w = (w', w'') \in \mathbf{W}^{r+d}$ is a $\mathfrak{W}_t^{r+d}/\mathfrak{W}_t^d$-measurable mapping of $\mathbf{W}^{r+d}$ into $\mathbf{W}^d$ for every $t \in \mathbb{R}_+$, $(s, w) \mapsto (s, w'')$ for $(s, w) \in [0, t] \times \mathbf{W}^{r+d}$ is a $\sigma(\mathfrak{B}_{[0,t]} \times \mathfrak{W}_t^{r+d})/\sigma(\mathfrak{B}_{[0,t]} \times \mathfrak{W}_t^d)$-measurable mapping of $[0, t] \times \mathbf{W}^{r+d}$ into $[0, t] \times \mathbf{W}^d$. Then since $(s, w'') \mapsto \alpha_j^i(s, w'')$ for $(s, w'') \in [0, t] \times \mathbf{W}^d$ is a $\sigma(\mathfrak{B}_{[0,t]} \times \mathfrak{W}_t^d)/\mathfrak{B}_\mathbb{R}$-measurable mapping of $[0, t] \times \mathbf{W}^d$ into $\mathbb{R}$, $(s, w) \mapsto \alpha_j^i(s, Y(\cdot, w))$ for $(s, w) \in [0, t] \times \mathbf{W}^{r+d}$ is a $\sigma(\mathfrak{B}_{[0,t]} \times \mathfrak{W}_t^{r+d})/\mathfrak{B}_\mathbb{R}$-measurable mapping of $[0, t] \times \mathbf{W}^{r+d}$ into $\mathbb{R}$. Thus applying (2), we have

$$(3) \quad \int_{[0,t] \times \Omega} |\alpha_j^i(s, X(\cdot, \omega))|^2 (m_L \times P)(d(s, \omega))$$

$$= \int_{[0,t] \times \mathbf{W}^{r+d}} |\alpha_j^i(s, Y(\cdot, w))|^2 (m_L \times P_{(B,X)})(d(s, w))$$

in the sense that the existence of one side implies that of the other and the equality of the two. But the right side of (3) exists and is finite since $(\alpha_X)_j^i$ are in $\mathbf{L}_{2,\infty}(\mathbb{R}_+ \times \Omega, m_L \times P)$. This proves (1). Thus $(\alpha_Y)_j^i$ are in $\mathbf{L}_{2,\infty}(\mathbb{R}_+ \times \mathbf{W}^{r+d}, m_L \times P_{(B,X)})$. Similarly $(\beta_Y)^i$ are in $\mathbf{L}_{1,\infty}(\mathbb{R}_+ \times \mathbf{W}^{r+d}, m_L \times P_{(B,X)})$. ∎

Theorem 18.13. *Consider the r-dimensional continuous $\{\mathfrak{W}_t^{r+d}\}$-adapted process W and the d-dimensional continuous $\{\mathfrak{W}_t^{r+d}\}$-adapted process Y on $(\mathbf{W}^{r+d}, \mathfrak{W}^{r+d}, \{\mathfrak{W}_t^{r+d}\})$ defined by $W(t, w) = w'(t)$ and $Y(t, w) = w''(t)$ for $w = (w', w'') \in \mathbf{W}^{r+d}$, $w' \in \mathbf{W}^r$, $w'' \in \mathbf{W}^d$, and $t \in \mathbb{R}_+$. Suppose the stochastic differential equation (1) in Definition 18.6 has a solution (B, X) on a standard filtered space $(\Omega, \mathfrak{F}, \{\mathfrak{F}_t\}, P)$. Then (W, Y) is a solution of the stochastic differential equation on $(\mathbf{W}^{r+d}, \mathfrak{W}^{r+d,*}, \{\mathfrak{W}_t^{r+d,*}\}, P_{(B,X)})$ in Definition 18.10. Furthermore the two processes (W, Y) and (B, X) have identical probability distributions on $(\mathbf{W}^{r+d}, \mathfrak{B}_{\mathbf{W}^{r+d}})$ and the two processes Y and X have identical probability distributions on $(\mathbf{W}^d, \mathfrak{B}_{\mathbf{W}^d})$.*

Proof. Since (B, X) is a solution, there exists a null set Λ in $(\Omega, \mathfrak{F}, P)$ such that on Λ^c and for every $t \in \mathbb{R}_+$ we have

$$X^i(t) - X^i(0) = \sum_{j=1}^r \int_{[0,t]} (\alpha_X)_j^i(s) \, dB^j(s) + \int_{[0,t]} (\beta_X)^i(s) m_L(ds) \quad \text{for } i = 1, \ldots, d,$$

or equivalently, writing $(\alpha_X)_j^i \bullet B^j$ and $(\beta_X)^i \bullet m_L$ respectively for the stochastic integrals $\{\int_{[0,t]} (\alpha_X)_j^i(s) \, dB^j(s) : t \in \mathbb{R}_+\}$ and $\{\int_{[0,t]} (\beta_X)^i(s) m_L(ds) : t \in \mathbb{R}_+\}$, we have

$$(1) \quad \Phi^i(t) \equiv \sum_{j=1}^r ((\alpha_X)_j^i \bullet B^j)(t) + ((\beta_X)^i \bullet m_L)(t) + X^i(0) - X^i(t) = 0 \quad \text{for } i = 1, \ldots, d.$$

By Lemma 18.11, W is an r-dimensional $\{\mathfrak{W}_t^{r+d,*}\}$-adapted null at 0 Brownian motion on the standard filtered space $(\mathbf{W}^{r+d}, \mathfrak{W}^{r+d,*}, \{\mathfrak{W}_t^{r+d,*}\}, P_{(B,X)})$. By Lemma 18.12, Y is a d-dimensional continuous $\{\mathfrak{W}_t^{r+d,*}\}$-adapted process on $(\mathbf{W}^{r+d}, \mathfrak{W}^{r+d,*}, \{\mathfrak{W}_t^{r+d,*}\}, P_{(B,X)})$ and furthermore the $\{\mathfrak{W}_t^*\}$-progressively measurable processes $(\alpha_Y)_j^i$ and $(\beta_Y)^i$ are respectively in $\mathbf{L}_{2,\infty}(\mathbb{R}_+ \times \mathbf{W}^{r+d}, m_L \times P_{(B,X)})$ and $\mathbf{L}_{1,\infty}(\mathbb{R}_+ \times \mathbf{W}^{r+d}, m_L \times P_{(B,X)})$. Thus the stochastic integrals $(\alpha_Y)_j^i \bullet W^j$ and $(\beta_Y)^i \bullet m_L$ exist respectively in $\mathbf{M}_2^c(\mathbf{W}^{r+d}, \mathfrak{W}^{r+d,*}, \{\mathfrak{W}_t^{r+d,*}\}, P_{(B,X)})$ and $\mathbf{V}^c(\mathbf{W}^{r+d}, \mathfrak{W}^{r+d,*}, \{\mathfrak{W}_t^{r+d,*}\}, P_{(B,X)})$. Therefore to show that (W, Y) is a solution of the stochastic differential equation on $(\mathbf{W}^{r+d}, \mathfrak{W}^{r+d,*}, \{\mathfrak{W}_t^{r+d,*}\}, P_{(B,X)})$ it remains to show that there exists a null set Λ in $(\mathbf{W}^{r+d}, \mathfrak{W}^{r+d,*}, P_{(B,X)})$ such that on Λ^c we have for every $t \in \mathbb{R}_+$

$$(2) \quad \Psi^i(t) \equiv \sum_{j=1}^r ((\alpha_Y)_j^i \bullet W^j)(t) + ((\beta_Y)^i \bullet m_L)(t) + Y^i(0) - Y^i(t) = 0 \quad \text{for } i = 1, \ldots, d.$$

Since Y, $(\alpha_Y)_j^i \bullet W^j$ and $(\beta_Y)^i \bullet m_L$ are all continuous processes, it suffices to show that for every $t \in \mathbb{R}_+$ there exists a null set Λ_t in $(\mathbf{W}^{r+d}, \mathfrak{W}^{r+d,*}, P_{(B,X)})$ such that on Λ_t^c the equality (2) holds.

Now if we show that for every $i = 1, \ldots, d$, the random variable $\Phi^i(t)$ on $(\Omega, \mathfrak{F}, P)$ and the random variable $\Psi^i(t)$ on $(\mathbf{W}^{r+d}, \mathfrak{W}^{r+d,*}, P_{(B,X)})$ have the same probability distribution on $(\mathbb{R}, \mathfrak{B}_\mathbb{R})$, that is, $P \circ (\Phi^i(t))^{-1}(E) = P_{(B,X)} \circ (\Psi^i(t))^{-1}(E)$ for every $E \in \mathfrak{B}_\mathbb{R}$, then

$$P_{(B,X)}\{\Psi^i(t) = 0\} = P_{(B,X)} \circ (\Psi^i(t))^{-1}(\{0\})$$
$$= P \circ (\Phi^i(t))^{-1}(\{0\}) = P\{\Phi^i(t) = 0\} = 0$$

by (1), that is, $\Psi^i(t) = 0$ a.e. on $(\mathbf{W}^{r+d}, \mathfrak{W}^{r+d,*}, P_{(B,X)})$ and we are done. Thus it remains to show that for every $i = 1, \ldots, d$, $\Phi^i(t)$ and $\Psi^i(t)$ have the same probability distribution on $(\mathbb{R}, \mathfrak{B}_\mathbb{R})$.

Since the $\{\mathfrak{W}_t^{r+d,*}\}$-progressively measurable processes $(\alpha_Y)_j^i$ and $(\beta_Y)^i$ are in $\mathbf{L}_{2,\infty}(\mathbb{R}_+ \times \mathbf{W}^{r+d}, m_L \times P_{(B,X)})$ and $\mathbf{L}_{1,\infty}(\mathbb{R}_+ \times \mathbf{W}^{r+d}, m_L \times P_{(B,X)})$ there exist by Remark 18.7 sequences $\{(\alpha_Y^n)_j^i : n \in \mathbb{N}\}$ and $\{(\beta_Y^n)^i : n \in \mathbb{N}\}$ in $\mathbf{L}_0(\mathbf{W}^{r+d}, \mathfrak{W}^{r+d,*}, \{\mathfrak{W}_t^{r+d,*}\}, P_{(B,X)})$ such that

$$(3) \quad \begin{cases} \lim_{n \to \infty} \|(\alpha_Y^n)_j^i - (\alpha_Y)_j^i\|_{2,\infty}^{m_L \times P_{(B,X)}} = 0, \\ \lim_{n \to \infty} \|(\beta_Y^n)^i - (\beta_Y)^i\|_{1,\infty}^{m_L \times P_{(B,X)}} = 0. \end{cases}$$

Thus by Definition 12.9, we have $\lim_{n \to \infty} |(\alpha_Y^n)_j^i \bullet W^j - (\alpha_Y)_j^i \bullet W^j|_\infty = 0$ in the space $\mathbf{M}_2^c(\mathbf{W}^{r+d}, \mathfrak{W}^{r+d,*}, \{\mathfrak{W}_t^{r+d,*}\}, P_{(B,X)})$ and then by Remark 11.5 for every $t \in \mathbb{R}_+$ we have

§18. DEFINITION AND FUNCTION SPACE REPRESENTATION

$$\lim_{n\to\infty} \left\| \sum_{j=1}^{r}(\alpha_Y^n)_j^i \bullet W^j - \sum_{j=1}^{r}(\alpha_Y)_j^i \bullet W^j \right\|_t = 0, \text{ in other words,}$$

$$\lim_{n\to\infty} \int_{\mathbf{W}^{r+d}} \left| \sum_{j=1}^{r}((\alpha_Y^n)_j^i \bullet W^j)(t) - \sum_{j=1}^{r}((\alpha_Y)_j^i \bullet W^j)(t) \right|^2 dP_{(B,X)} = 0.$$

Similarly we have

$$\lim_{n\to\infty} \int_{\mathbf{W}^{r+d}} \left| ((\beta_Y^n)^i \bullet m_L)(t) - ((\beta_Y)^i \bullet m_L)(t) \right| dP_{(B,X)} = 0.$$

Since convergence of a sequence in $\mathbf{L}_2(\mathbf{W}^{r+d}, \mathfrak{W}^{r+d}, P_{(B,X)})$ implies convergence in $\mathbf{L}_1(\mathbf{W}^{r+d}, \mathfrak{W}^{r+d}, P_{(B,X)})$ which then implies convergence in the probability measure $P_{(B,X)}$, the last two equalities imply

(4) $$P_{(B,X)} \cdot \lim_{n\to\infty} \left\{ \sum_{j=1}^{r}((\alpha_Y^n)_j^i \bullet W^j)_t + ((\beta_Y^n)^i \bullet m_L)_t + Y_0^i - Y_t^i \right\}$$

$$= \sum_{j=1}^{r}((\alpha_Y)_j^i \bullet W^j)_t + ((\beta_Y)^i \bullet m_L)_t + Y_0^i - Y_t^i.$$

Now since $(\alpha_Y^n)_j^i$ and $(\beta_Y^n)^i$ are in $\mathbf{L}_0(\mathbf{W}^{r+d}, \mathfrak{W}^{r+d,*}, \{\mathfrak{W}_t^{r+d,*}\}, P_{(B,X)})$ we have

(5) $$\begin{cases} (\alpha_Y^n)(s,w) = (f_0^n)_j^i(w)\mathbf{1}_{\{0\}}(s) + \sum_{k\in\mathbb{N}}(f_k^n)_j^i(w)\mathbf{1}_{I_{n,k}}(s), \\ (\beta_Y^n)(s,w) = (g_0^n)^i(w)\mathbf{1}_{\{0\}}(s) + \sum_{k\in\mathbb{N}}(g_k^n)^i(w)\mathbf{1}_{I_{n,k}}(s) \quad \text{for } (s,w) \in \mathbb{R}_+ \times \mathbf{W}^{r+d}, \end{cases}$$

where $I_{n,k} = (t_{n,k-1}, t_{n,k}]$ for $n \in \mathbb{N}$; $\{t_{n,k} : k \in \mathbb{Z}_+\}$ are strictly increasing sequences in $\mathbb{R}_+$ such that $t_{n,0} = 0$ and $\lim_{k\to\infty} t_{n,k} = 0$; $\{(f_k^n)_j^i : k \in \mathbb{Z}_+\}$ and $\{(g_k^n)^i : k \in \mathbb{Z}_+\}$ are bounded sequences of real valued random variables on $(\mathbf{W}^{r+d}, \mathfrak{W}^{r+d,*}, P_{(B,X)})$ such that $(f_0^n)_j^i$ and $(g_0^n)^i$ are $\mathfrak{W}_{t_{n,0}}^{r+d,*}$-measurable and $(f_k^n)_j^i$ and $(g_k^n)^i$ are $\mathfrak{W}_{t_{n,k-1}}^{r+d,*}$-measurable for $k \in \mathbb{N}$. If we replace $(f_k^n)_j^i$ and $(g_k^n)^i$ with $\mathbf{E}[(f_k^n)_j^i | \mathfrak{W}_{t_{n,k-1}}^{r+d}]$ and $\mathbf{E}[(g_k^n)^i | \mathfrak{W}_{t_{n,k-1}}^{r+d}]$ respectively in (5), then (3) and (4) still hold. Thus we may and do assume that $(f_k^n)_j^i$ and $(g_k^n)^i$ are $\mathfrak{W}_{t_{n,k-1}}^{r+d}$-measurable for $k \in \mathbb{N}$.

Consider the mapping $(\mathcal{B}, \mathcal{X})$ of Ω into $\mathbf{W}^{r+d}$ defined by $(\mathcal{B}, \mathcal{X})(\omega) = (B(\cdot,\omega), X(\cdot,\omega))$ for $\omega \in \Omega$. Since (B,X) is $\{\mathfrak{F}_t\}$-adapted, $(\mathcal{B}, \mathcal{X})$ is $\mathfrak{F}_{t_{n,k-1}}/\mathfrak{W}_{t_{n,k-1}}^{r+d}$-measurable for every

$t_{n,k-1}$ by Theorem 17.8. Thus $(f_k^n)_j^i((\mathcal{B}, \mathcal{X})(\omega))$ and $(g_k^n)^i((\mathcal{B}, \mathcal{X})(\omega))$ for $\omega \in \Omega$ are $\mathfrak{F}_{t_{n,k-1}}$-measurable. Then the random variables $(a_k^n)_j^i$ and $(b_k^n)^i$ on $(\Omega, \mathfrak{F}, P)$ defined by

(6) $$\begin{cases} (a_k^n)_j^i(\omega) = (f_k^n)_j^i((\mathcal{B}, \mathcal{X})(\omega)), \\ (b_k^n)^i(\omega) = (g_k^n)^i((\mathcal{B}, \mathcal{X})(\omega)) & \text{for } \omega \in \Omega, \end{cases}$$

are $\mathfrak{F}_{t_{n,k-1}}$-measurable. This implies that the stochastic processes $(\alpha_X^n)_j^i$ and $(\beta_X^n)^i$ on the standard filtered space $(\Omega, \mathfrak{F}, \{\mathfrak{F}_t\}, P)$ defined by

(7) $$\begin{cases} (\alpha_X^n)(s, \omega) = (a_0^n)_j^i(\omega)\mathbf{1}_{\{0\}}(s) + \sum_{k \in \mathbb{N}}(a_k^n)_j^i(\omega)\mathbf{1}_{I_{n,k}}(s), \\ (\beta_X^n)(s, \omega) = (b_0^n)^i(\omega)\mathbf{1}_{\{0\}}(s) + \sum_{k \in \mathbb{N}}(b_k^n)^i(\omega)\mathbf{1}_{I_{n,k}}(s) & \text{for } (s, \omega) \in \mathbb{R}_+ \times \Omega, \end{cases}$$

are in $L_0(\Omega, \mathfrak{F}, \{\mathfrak{F}_t\}, P)$. Let us show

(8) $$\begin{cases} \lim_{n \to \infty} \|(\alpha_X^n)_j^i - (\alpha_X)_j^i\|_{2,\infty}^{m_L \times P_{(B,X)}} = 0, \\ \lim_{n \to \infty} \|(\beta_X^n)^i - (\beta_X)^i\|_{1,\infty}^{m_L \times P_{(B,X)}} = 0. \end{cases}$$

Note that according to Remark 11.18, the first of these two expressions are equivalent to the condition that $\lim_{n \to \infty} \|(\alpha_X^n)_j^i - (\alpha_X)_j^i\|_{2,t}^{m_L \times P_{(B,X)}} = 0$ for every $t \in \mathbb{R}_+$. Now for fixed $t \in \mathbb{R}_+$, letting $t_{n,p_n} = t$ for $n \in \mathbb{N}$, we have

$$\|(\alpha_X^n)_j^i - (\alpha_X)_j^i\|_{2,t}^{m_L \times P_{(B,X)}}$$
$$= \int_{[0,t] \times \Omega} |(a_0^n)_j^i(\omega)\mathbf{1}_{\{0\}}(s) + \sum_{k=1}^{p_n}(a_k^n)_j^i(\omega)\mathbf{1}_{I_{n,k}}(s) - \alpha_j^i(s, X(\cdot, \omega))|^2 (m_L \times P)(d(s,\omega))$$
$$= \int_{[0,t] \times \mathbf{W}^{r+d}} |(f_0^n)_j^i(w)\mathbf{1}_{\{0\}}(s) + \sum_{k=1}^{p_n}(f_k^n)_j^i(w)\mathbf{1}_{I_{n,k}}(s) - \alpha_j^i(s, w'')|^2 (m_L \times P_{(B,X)})(d(s,w))$$
$$= \|(\alpha_Y^n)_j^i - (\alpha_Y)_j^i\|_{2,t}^{m_L \times P_{(B,X)}}$$

by the Image Probability Law. But according to Remark 11.18, the first equality in (3) is equivalent to having $\lim_{n \to \infty} \|(\alpha_Y^n)_j^i - (\alpha_Y)_j^i\|_{2,t}^{m_L \times P_{(B,X)}} = 0$ for every $t \in \mathbb{R}_+$. Thus the first equality in (8) holds. The second equality in (8) is proved likewise. From (8) we derive

(9) $$P \cdot \lim_{n \to \infty}\left\{\sum_{j=1}^{r}((\alpha_X^n)_j^i \bullet B^j)_t + ((\beta_X^n)^i \bullet m_L)_t + X_0^i - X_t^i\right\}$$
$$= \sum_{j=1}^{r}((\alpha_X)_j^i \bullet B^j)_t + ((\beta_X)^i \bullet m_L)_t + X_0^i - X_t^i$$

§18. DEFINITION AND FUNCTION SPACE REPRESENTATION

in the same way we derived (4) from (3).

We show next that the sum on the left side of (4) and the sum on the left side of (9) have identical probability distributions on $(\mathbb{R}, \mathfrak{B}_{\mathbb{R}})$. For brevity, let these two sums be denoted by $S^n_{(W,Y)}$ and $S^n_{(B,X)}$ respectively. Then by (5), we have

$$
\begin{aligned}
(10) \quad S^n_{(W,Y)}(w) &= \sum_{j=1}^{r}((\alpha^n_Y)^i_j \bullet W^j)_t(w) + ((\beta^n_Y)^i \bullet m_L)_t(w) + Y^i_0(w) - Y^i_t(w) \\
&= \sum_{j=1}^{r}\sum_{k=1}^{p_n}(f^n_k)^i_j(w)\{W^j_{t_{n,k}}(w) - W^j_{t_{n,k-1}}(w)\} \\
&\quad + \sum_{k=1}^{p_n}(g^n_k)^i(w)(t_{n,k} - t_{n,k-1}) + Y^i_0(w) - Y^i_t(w) \quad \text{for } w \in \mathbf{W}^{r+d},
\end{aligned}
$$

and similarly by (7) and (6) we have

$$
\begin{aligned}
(11) \quad S^n_{(B,X)}(\omega) &= \sum_{j=1}^{r}((\alpha^n_X)^i_j \bullet B^j)_t(\omega) + ((\beta^n_X)^i \bullet m_L)_t(\omega) + X^i_0(\omega) - X^i_t(\omega) \\
&= \sum_{j=1}^{r}\sum_{k=1}^{p_n}(f^n_k)^i_j((\mathcal{B},\mathcal{X})(\omega))\{B^j_{t_{n,k}}(\omega) - B^j_{t_{n,k-1}}(\omega)\} \\
&\quad + \sum_{k=1}^{p_n}(g^n_k)^i((\mathcal{B},\mathcal{X})(\omega))(t_{n,k} - t_{n,k-1}) + X^i_0(\omega) - X^i_t(\omega) \quad \text{for } \omega \in \Omega.
\end{aligned}
$$

Consider two $(2r+1)(p_n+1)+2$-dimensional random vectors $V_{(W,Y)}$ on the probability space $(\mathbf{W}^{r+d}, \mathfrak{W}^{r+d,*}, P_{(B,X)})$ and $V_{(B,X)}$ on $(\Omega, \mathfrak{F}, P)$ defined by

$$
\begin{cases}
V_{(W,Y)} = \times_{j=1}^{r}\times_{k=0}^{p_n}W^j_{t_{n,k}} \cdot \times_{j=1}^{r}\times_{k=0}^{p_n}(f^n_k)^i_j \cdot \times_{k=0}^{p_n}(g^n_k)^i \times Y^i_{t_{n,0}} \times Y^i_{t_{n,p_n}}, \\
V_{(B,X)} = \times_{j=1}^{r}\times_{k=0}^{p_n}B^j_{t_{n,k}} \cdot \times_{j=1}^{r}\times_{k=0}^{p_n}(a^n_k)^i_j \cdot \times_{k=0}^{p_n}(b^n_k)^i \times X^i_{t_{n,0}} \times X^i_{t_{n,p_n}}.
\end{cases}
$$

Let us show that their probability distributions $P_{(B,X)} \circ V^{-1}_{(W,Y)}$ and $P \circ V^{-1}_{(B,X)}$ are equal on $(\mathbb{R}^{(2r+1)(p_n+1)+2}, \mathfrak{B}_{\mathbb{R}^{(2r+1)(p_n+1)+2}})$. Now for a product set $A \in \times_{m=1}^{(2r+1)(p_n+1)+2}\mathfrak{B}_{\mathbb{R}_m}$ given by $A = \times_{j=1}^{r}(\times_{k=0}^{p_n}E^j_{n,k} \cdot \times_{k=0}^{p_n}F^j_{n,k}) \times_{k=0}^{p_n}G_{n,k} \times H_{n,0} \times H_{n,p_n}$ with $E^j_{n,k}, F^j_{n,k}, G_{n,k}, H_{n,k} \in \mathfrak{B}_{\mathbb{R}}$, we have

$$
\begin{aligned}
P_{(B,X)} \circ V^{-1}_{(W,Y)}(A) &= P_{(B,X)}(\cap_{j=1}^{r}\cap_{k=0}^{p_n}(W^j_{t_{n,k}})^{-1}(E^j_{n,k}) \cdot \cap_{j=1}^{r}\cap_{k=0}^{p_n}((f^n_k)^i_j)^{-1}(F^j_{n,k}) \\
&\quad \cdot \cap_{k=0}^{p_n}((g^n_k)^i)^{-1}(G^j_{n,k}) \cap (Y^i_{t_{n,0}})^{-1}(H_{n,0}) \cap (Y^i_{t_{n,p_n}})^{-1}(H_{n,p_n}))
\end{aligned}
$$

and

$$P \circ V_{(B,X)}^{-1}(A) = P(\cap_{j=1}^{r} \cap_{k=0}^{p_n} (B_{t_{n,k}}^j)^{-1}(E_{n,k}^j) \cdot \cap_{j=1}^{r} \cap_{k=0}^{p_n} ((a_k^n)_j^i)^{-1}(F_{n,k}^j)$$
$$\cap_{k=0}^{p_n} ((b_k^n)^i)^{-1}(G_{n,k}^j) \cap (X_{t_{n,0}}^i)^{-1}(H_{n,0}) \cap (X_{t_{n,p_n}}^i)^{-1}(H_{n,p_n})).$$

Since $B_{t_{n,k}}^j(\omega) = W_{t_{n,k}}^j((B,X)(\omega))$ and $X_{t_{n,k}}^i(\omega) = Y_{t_{n,k}}^i((B,X)(\omega))$ for $\omega \in \Omega$ we have

$$(B_{t_{n,k}}^j)^{-1}(E_{n,k}^j) = (B,X)^{-1}(W_{t_{n,k}}^j)^{-1}(E_{n,k}^j),$$

and

$$(X_{t_{n,k}}^i)^{-1}(H_{n,k}) = (B,X)^{-1}(Y_{t_{n,k}}^i)^{-1}(H_{n,k}).$$

Also from (6), we have

$$((a_k^n)_j^i)^{-1}(F_{n,k}^j) = ((f_k^n)_j^i \circ (B,X))^{-1}(F_{n,k}^j) = (B,X)^{-1} \circ ((f_k^n)_j^i)^{-1}(F_{n,k}^j)$$

and similarly

$$((b_k^n)^i)^{-1}(G_{n,k}^j) = (B,X)^{-1} \circ ((g_k^n)^i)^{-1}(G_{n,k}).$$

Substituting these equalities in the expression above for $P \circ V_{(B,X)}^{-1}(A)$ and recalling $P_{(B,X)} = P \circ (B,X)^{-1}$, we have $P_{(B,X)} \circ V_{(W,Y)}^{-1}(A) = P \circ V_{(B,X)}^{-1}(A)$ Now since $\times_{m=1}^{(2r+1)(p_n+1)+2} \mathfrak{B}_{\mathbb{R}_m}$ is a π-class and the σ-algebra generated by it is $\mathfrak{B}_{\mathbb{R}^{(2r+1)(p_n+1)+2}}$, the equality of the two probability measures $P_{(B,X)} \circ V_{(W,Y)}^{-1}$ and $P \circ V_{(B,X)}^{-1}$ on the π-class implies their equality on the σ-algebra by Corollary 1.8.

Let the elements of $\mathbb{R}^{(2r+1)(p_n+1)+2}$ be given as

$$\times_{j=1}^{r} \times_{k=0}^{p_n} x_{n,k}^j \cdot \times_{j=1}^{r} \times_{k=0}^{p_n} y_{n,k}^j \cdot \times_{k=0}^{p_n} z_{n,k} \times v_{n,0} \times v_{n,p_n}$$

with $x_{n,k}^j, y_{n,k}^j, z_{n,k}, v_{n,k} \in \mathbb{R}$. Let us define a mapping T of $\mathbb{R}^{(2r+1)(p_n+1)+2}$ into $\mathbb{R}$ by setting

$$T(\times_{j=1}^{r} \times_{k=0}^{p_n} x_{n,k}^j \cdot \times_{j=1}^{r} \times_{k=0}^{p_n} y_{n,k}^j \cdot \times_{k=0}^{p_n} z_{n,k} \times v_{n,0} \times v_{n,p_n})$$
$$= \sum_{j=1}^{r} \sum_{k=1}^{p_n} y_{n,k}^j(x_{n,k}^j - x_{n,k-1}^j) + \sum_{k=1}^{p_n} z_{n,k}(t_{n,k} - t_{n,k-1}) + v_{n,0} - v_{n,p_n}.$$

Clearly T is a $\mathfrak{B}_{\mathbb{R}^{(2r+1)(p_n+1)+2}}/\mathfrak{B}_{\mathbb{R}}$-measurable mapping of $\mathbb{R}^{(2r+1)(p_n+1)+2}$ into $\mathbb{R}$. Thus $T \circ V_{(W,Y)}$ and $T \circ V_{(B,X)}$ are real valued random variables on $(\mathbf{W}^{r+d}, \mathfrak{W}^{r+d,*}, P_{(B,X)})$ and $(\Omega, \mathfrak{F}, P)$ respectively. Furthermore by (10) and (11), we have $T \circ V_{(W,Y)} = S_{(W,Y)}^n$ and $T \circ V_{(B,X)} = S_{(B,X)}^n$. Thus for the probability distributions of $S_{(W,Y)}^n$ and $S_{(B,X)}^n$ on $(\mathbb{R}, \mathfrak{B}_{\mathbb{R}})$,

§18. DEFINITION AND FUNCTION SPACE REPRESENTATION

which we denote by $\mu_{S^n_{(W,Y)}}$ and $\mu_{S^n_{(B,X)}}$, we have $\mu_{S^n_{(W,Y)}} = P_{(B,X)} \circ (T \circ V_{(W,Y)})^{-1} = P_{(B,X)} \circ (V_{(W,Y)})^{-1} \circ T^{-1}$ and $\mu_{S^n_{(B,X)}} = P \circ (T \circ V_{(B,X)})^{-1} = P \circ (V_{(B,X)})^{-1} \circ T^{-1}$. But $P_{(B,X)} \circ (V_{(W,Y)})^{-1} = P \circ (V_{(B,X)})^{-1}$ on $\mathfrak{B}_\mathbb{R}$ as we showed above. Therefore $\mu_{S^n_{(W,Y)}} = \mu_{S^n_{(B,X)}}$ on $(\mathbb{R}, \mathfrak{B}_\mathbb{R})$.

Let μ_n, $n \in \mathbb{N}$, and μ be probability measures on a measurable space $(S, \mathfrak{B}_S)$ where S is a metric space and $\mathfrak{B}_S$ is the Borel σ-algebra of subsets of S. We say that μ_n converges weakly to μ and write $w \cdot \lim_{n \to \infty} \mu_n = \mu$ if $\lim_{n \to \infty} \int_S f \, d\mu_n = \int_S f \, d\mu$ for every real valued bounded continuous function f on S. The limit of weak convergence of probability measures is unique in the sense that if ν is a probability measure on $(S, \mathfrak{B}_S)$ and $w \cdot \lim_{n \to \infty} \mu_n = \mu$ as well as $w \cdot \lim_{n \to \infty} \mu_n = \nu$, then $\mu = \nu$ on $(S, \mathfrak{B}_S)$. If ξ_n, $n \in \mathbb{N}$, and ξ are S-valued random variables on a probability space and their probability distributions on $(S, \mathfrak{B}_S)$ are denoted by μ_n, $n \in \mathbb{N}$, and μ respectively, then we say that ξ_n converges in distribution to ξ if $w \cdot \lim_{n \to \infty} \mu_n = \mu$. If ξ_n converges in probability to ξ, then ξ_n converges in distribution to ξ. Now according to (4), the random variables $S^n_{(W,Y)}$ on $(\mathbf{W}^{r+d}, \mathfrak{W}^{r+d,*}, P_{(B,X)})$ defined by (10) converge in probability $P_{(B,X)}$ to the random variable $\Psi^i(t)$ defined by (2) and therefore their probability distributions $\mu_{S^n_{(W,Y)}}$ on $(\mathbb{R}, \mathfrak{B}_\mathbb{R})$ converge weakly to the probability distribution of $\Psi^i(t)$. On the other hand according to (9), the random variables $S^n_{(B,X)}$ on $(\Omega, \mathfrak{F}, P)$ defined by (11) converge in probability P to the random variable $\Phi^i(t)$ defined by (1) and thus their probability distributions $\mu_{S^n_{(B,X)}}$ on $(\mathbb{R}, \mathfrak{B}_\mathbb{R})$ converge weakly to the probability distribution of $\Phi^i(t)$. But $\mu_{S^n_{(W,Y)}} = \mu_{S^n_{(B,X)}}$ on $(\mathbb{R}, \mathfrak{B}_\mathbb{R})$ for every $n \in \mathbb{N}$ as we showed above. Thus by the uniqueness of limit in weak convergence of probability measures the probability distributions of $\Psi^i(t)$ and $\Phi^i(t)$ are equal on $(\mathbb{R}, \mathfrak{B}_\mathbb{R})$.

Regarding the probability distribution of the process (W, Y), note that since (W, Y) is an $r + d$-dimensional continuous process on $(\mathbf{W}^{r+d}, \mathfrak{W}^{r+d,*}, \{\mathfrak{W}^{r+d,*}_t\}, P_{(B,X)})$, its probability distribution on $(\mathbf{W}^{r+d}, \mathfrak{B}_{\mathbf{W}^{r+d}})$ is given by $(P_{(B,X)})_{(W,Y)} \equiv P_{(B,X)} \circ (\mathcal{W}, \mathcal{Y})^{-1}$ on $\mathfrak{B}_{\mathbf{W}^{r+d}}$ where $(\mathcal{W}, \mathcal{Y})$ is the mapping of $\mathbf{W}^{r+d}$ into $\mathbf{W}^{r+d}$ defined by $(\mathcal{W}, \mathcal{Y})(w) = w$ for $w \in \mathbf{W}^{r+d}$. Since $(\mathcal{W}, \mathcal{Y})$ is an identity mapping, we have $(P_{(B,X)})_{(W,Y)} = P_{(B,X)}$ on $\mathfrak{B}_{\mathbf{W}^{r+d}}$. But $P_{(B,X)}$ is the probability distribution of (B, X) on $(\mathbf{W}^{r+d}, \mathfrak{B}_{\mathbf{W}^{r+d}})$. This shows that (W, Y) and (B, X) have identical probability distributions. The probability distribution P_X of X on $(\mathbf{W}^d, \mathfrak{B}_{\mathbf{W}^d})$ is given by

$$P_X(A) = P \circ \mathcal{X}^{-1}(A) = P \circ (\mathcal{B}, \mathcal{X})^{-1}(\mathbf{W}^r \times A) = P_{(B,X)}(\mathbf{W}^r \times A)$$

for $A \in \mathfrak{B}_{\mathbf{W}^d}$. For the probability distribution on $(\mathbf{W}^d, \mathfrak{B}_{\mathbf{W}^d})$ of the process Y on the standard filtered space $(\mathbf{W}^{r+d}, \mathfrak{W}^*, \{\mathfrak{W}^*_t\}, P_{(B,X)})$, we have

$$(P_{(B,X)})_Y(A) = (P_{(B,X)})_{(W,Y)}(\mathbf{W}^r \times A) = P_{(B,X)}(\mathbf{W}^r \times A)$$

for $A \in \mathfrak{B}_{\mathbf{W}^d}$. This shows that Y and X have identical probability distributions. ∎

As a converse of Theorem 18.13 we have the following.

Theorem 18.14. *Let B be an r-dimensional $\{\mathfrak{F}_t\}$-adapted null at 0 Brownian motion and X be a d-dimensional $\{\mathfrak{F}_t\}$-adapted continuous process on a standard filtered space $(\Omega, \mathfrak{F}, \{\mathfrak{F}_t\}, P)$. Let $P_{(B,X)}$ be the probability distribution of (B,X) on $(\mathbf{W}^{r+d}, \mathfrak{W}^{r+d})$ and let $(\mathbf{W}^{r+d}, \mathfrak{W}^{r+d,*}, \{\mathfrak{W}_t^{r+d,*}\}, P_{(B,X)})$ be the standard filtered space generated by $P_{(B,X)}$ as in Definition 18.10 and let W and Y be two processes on this filtered space defined by setting*

$$\begin{cases} W(t,w) = w'(t) & \text{for } (t,w) \in \mathbb{R}_+ \times \mathbf{W}^{r+d} \\ Y(t,w) = w''(t) & \text{for } (t,w) \in \mathbb{R}_+ \times \mathbf{W}^{r+d}, \end{cases}$$

where we write $w = (w', w'')$ with $w' \in \mathbf{W}^r$ and $w'' \in \mathbf{W}^d$. If (W, Y) is a solution of the stochastic differential equation (1) in Definition 18.6 on $(\mathbf{W}^{r+d}, \mathfrak{W}^{r+d,*}, \{\mathfrak{W}_t^{r+d,*}\}, P_{(B,X)})$, then (B, X) is a solution of the stochastic differential equation on $(\Omega, \mathfrak{F}, \{\mathfrak{F}_t\}, P)$.

Proof. To show that $(\alpha_X)_j^i \in \mathbf{L}_{2,\infty}(\mathbb{R}_+ \times \Omega, m_L \times P)$, note that since (W, Y) is a solution of the stochastic differential equation, we have $(\alpha_Y)_j^i \in \mathbf{L}_{2,\infty}(\mathbb{R}_+ \times \mathbf{W}^{r+d}, m_L \times P_{(B,X)})$. Thus

$$\int_{[0,t] \times \mathbf{W}^{r+d}} |\alpha_j^i(s, Y(\cdot, w))|^2 (m_L \times P_{(B,X)})(d(s,w)) < \infty.$$

Then by the Image Probability Law as in the proof of Lemma 18.12, we have

$$\int_{[0,t] \times \Omega} |\alpha_j^i(s, X(\cdot, \omega))|^2 (m_L \times P)(d(s,\omega)) < \infty,$$

and thus $(\alpha_X)_j^i \in \mathbf{L}_{2,\infty}(\mathbb{R}_+ \times \Omega, m_L \times P)$. Similarly from $(\beta_Y)^i \in \mathbf{L}_{1,\infty}(\mathbb{R}_+ \times \mathbf{W}^{r+d}, m_L \times P_{(B,X)})$ we have $(\beta_X)^i \in \mathbf{L}_{1,\infty}(\mathbb{R}_+ \times \Omega, m_L \times P)$. Then the stochastic integrals $(\alpha_X)_j^i \bullet B^j$ and $(\beta_X)^i \bullet m_L$ exist. For $t \in \mathbb{R}_+$, let

$$\Phi^i(t) = \sum_{j=1}^r ((\alpha_X)_j^i \bullet B^j)(t) + ((\beta_X)^i \bullet m_L)(t) + X^i(0) - X^i(t) \quad \text{for } i = 1, \ldots, d,$$

and

$$\Psi^i(t) = \sum_{j=1}^r ((\alpha_Y)_j^i \bullet W^j)(t) + ((\beta_Y)^i \bullet m_L)(t) + Y^i(0) - Y^i(t) \quad \text{for } i = 1, \ldots, d.$$

The fact that $\Phi^i(t)$ and $\Psi^i(t)$ have identical probability distributions on $(\mathbb{R}, \mathfrak{B}_\mathbb{R})$ can be verified by the same argument as in the proof of Theorem 18.13. Since (W, Y) is a solution

§18. DEFINITION AND FUNCTION SPACE REPRESENTATION

of the stochastic differential equation we have $\Psi^i(t) = 0$ a.e. on $(\mathbf{W}^{r+d}, \mathfrak{W}^{r+d}, P_{(B,X)})$ and thus $\Phi^i(t) = 0$ a.e. on $(\Omega, \mathfrak{F}, P)$. This shows that (B, X) is a solution of the stochastic differential equation on $(\Omega, \mathfrak{F}, \{\mathfrak{F}_t\}, P)$. ∎

Combining Theorem 18.13 and Theorem 18.14 we have the following.

Theorem 18.15. *Let $B^{(q)}$ be an r-dimensional $\{\mathfrak{F}_t^{(q)}\}$-adapted null at 0 Brownian motion and $X^{(q)}$ be a d-dimensional $\{\mathfrak{F}_t^{(q)}\}$-adapted continuous process on a standard filtered space $(\Omega^{(q)}, \mathfrak{W}^{(q)}, \{\mathfrak{F}_t^{(q)}\}, P^{(q)})$ for $q = 1, 2$. If $(B^{(1)}, X^{(1)})$ is a solution of the stochastic differential equation (1) in Definition 18.6 on $(\Omega^{(1)}, \mathfrak{W}^{(1)}, \{\mathfrak{F}_t^{(1)}\}, P^{(1)})$ and $(B^{(1)}, X^{(1)})$ and $(B^{(2)}, X^{(2)})$ have identical probability distributions on $(\mathbf{W}^{r+d}, \mathfrak{W}^{r+d})$, then $(B^{(2)}, X^{(2)})$ is a solution of the stochastic differential equation on $(\Omega^{(2)}, \mathfrak{W}^{(2)}, \{\mathfrak{F}_t^{(2)}\}, P^{(2)})$.*

Proof. Let $P_{(B^{(q)},X^{(q)})}^{(q)}$ be the probability distribution of $(B^{(q)}, X^{(q)})$ on $(\mathbf{W}^{r+d}, \mathfrak{W}^{r+d})$. Consider for $q = 1, 2$ the standard filtered space $(\mathbf{W}^{r+d}, \mathfrak{W}^{r+d,*,(q)}, \{\mathfrak{W}_t^{r+d,*,(q)}\}, P_{(B,X)}^{(q)})$ generated by $P_{(B^{(q)},X^{(q)})}^{(q)}$. Since $P_{(B^{(1)},X^{(1)})}^{(1)} = P_{(B^{(2)},X^{(2)})}^{(2)}$ by our assumption these two standard filtered spaces are identical. Let W and Y be two processes on the standard filtered space defined by $W(t, w) = w'(t)$ and $Y(t, w) = w''(t)$ for $(t, w) \in \mathbb{R}_+ \times \mathbf{W}^{r+d}$, where $w = (w', w'')$ with $w' \in \mathbf{W}^r$ and $w'' \in \mathbf{W}^d$. Since $(B^{(1)}, X^{(1)})$ is a solution of the stochastic differential equation on $(\Omega^{(1)}, \mathfrak{W}^{(1)}, \{\mathfrak{F}_t^{(1)}\}, P^{(1)})$, (W, Y) is a solution of the stochastic differential equation on $(\mathbf{W}^{r+d}, \mathfrak{W}^{r+d,*,(q)}, \{\mathfrak{W}_t^{r+d,*,(q)}\}, P_{(B,X)}^{(q)})$ by Theorem 18.13. Then by Theorem 18.14, $(B^{(2)}, X^{(2)})$ is a solution of the stochastic differential equation on $(\Omega^{(2)}, \mathfrak{W}^{(2)}, \{\mathfrak{F}_t^{(2)}\}, P^{(2)})$. ∎

[III] Initial Value Problems

In Theorem 18.13 we showed that if the stochastic differential equation (1) of Definition 18.6 has a solution (B, X) on some standard filtered space $(\Omega, \mathfrak{F}, \{\mathfrak{F}_t\}, P)$, then (W, Y), where W and Y are defined by $W(t, w) = w'(t)$ and $Y(t, w) = w''(t)$ for $t \in \mathbb{R}_+$ and $w = (w', w'') \in \mathbf{W}^{r+d}$ with $w' \in \mathbf{W}^r$ and $w'' \in \mathbf{W}^d$, is a solution of the stochastic differential equation on $(\mathbf{W}^{r+d}, \mathfrak{W}^{r+d,*}, \{\mathfrak{W}_t^{r+d,*}\}, P_{(B,X)})$. Let μ be the probability distribution on $(\mathbb{R}^d, \mathfrak{B}_{\mathbb{R}^d})$ of the d-dimensional random vector X_0 on $(\Omega, \mathfrak{F}, P)$. We shall show that for a.e. x in $(\mathbb{R}^d, \mathfrak{B}_{\mathbb{R}^d}, \mu)$, the mapping (W, Y) is a solution on a certain standard filtered space $(\mathbf{W}^{r+d}, \mathfrak{W}^{r+d,*,x}, \{\mathfrak{W}_t^{r+d,*,x}\}, P_{(B,X)}^x)$ of the initial value problem

$$\begin{cases} dX_t = \alpha(t, X)dB(t) + \beta(t, X)dt \\ X_0 = x. \end{cases}$$

Observation 18.16. Consider the solution (W, Y) of the stochastic differential equation on $(\mathbf{W}^{r+d}, \mathfrak{W}^{r+d,*}, \{\mathfrak{W}^{r+d,*}_t\}, P_{(B,X)})$ in Theorem 18.13. Now Y is a d-dimensional $\{\mathfrak{W}^{r+d}_t\}$-adapted process on $(\mathbf{W}^{r+d}, \mathfrak{W}^{r+d}, \{\mathfrak{W}^{r+d}_t\}, P_{(B,X)})$ and Y_0 is a $\mathfrak{W}^{r+d}_0/\mathfrak{B}_{\mathbb{R}^d}$-measurable mapping of $\mathbf{W}^{r+d}$ into $\mathbb{R}^d$. Let μ be the probability distribution on $(\mathbb{R}^d, \mathfrak{B}_{\mathbb{R}^d})$ of the random vector X_0 on $(\Omega, \mathfrak{F}, P)$. Since X_0 and Y_0 have equal probability distributions, μ is also the probability distribution of Y_0 on $(\mathbb{R}^d, \mathfrak{B}_{\mathbb{R}^d})$. Consider the probability space $(\mathbf{W}^{r+d}, \mathfrak{W}^{r+d}, P_{(B,X)})$. Since $\mathbf{W}^{r+d}$ is a complete separable metric space and $\mathfrak{W}^{r+d}$ is the Borel σ-algebra of subsets of $\mathbf{W}^{r+d}$, a regular image conditional probability given Y_0 exists, that is, there exists a function $P^{\mathfrak{W}^{r+d}|Y_0}_{(B,X)}$ on $\mathfrak{W} \times \mathbb{R}^d$ satisfying the following conditions according to Definition C.11.

1°. There exists a null set N_1 in $(\mathbb{R}^d, \mathfrak{B}_{\mathbb{R}^d}, \mu)$ such that $P^{\mathfrak{W}^{r+d}|Y_0}_{(B,X)}(\cdot, x)$ is a probability measure on $\mathfrak{W}^{r+d}$ for every $x \in N_1^c$.

2°. $P^{\mathfrak{W}^{r+d}|Y_0}_{(B,X)}(A, \cdot)$ is a $\mathfrak{B}_{\mathbb{R}^d}$-measurable function on $\mathbb{R}^d$ for every $A \in \mathfrak{W}^{r+d}$.

3°. $P_{(B,X)}(A \cap Y_0^{-1}(G)) = \int_G P^{\mathfrak{W}^{r+d}|Y_0}_{(B,X)}(A, x)\mu(dx)$ for every $A \in \mathfrak{W}^{r+d}$ and $G \in \mathfrak{B}_{\mathbb{R}^d}$.

By Theorem C.14, for an extended real valued random variable ξ on $(\mathbf{W}^{r+d}, \mathfrak{W}^{r+d}, P_{(B,X)})$, we have for every $G \in \mathfrak{B}_{\mathbb{R}^d}$

4°. $$\int_{Y_0^{-1}(G)} \xi \, dP_{(B,X)} = \int_G \left[\int_{\mathbf{W}^{r+d}} \xi(w) P^{\mathfrak{W}^{r+d}|Y_0}_{(B,X)}(dw, x) \right] \mu(dx)$$

in the sense that the existence of one side implies that of the other and the equality of the two. According to Theorem C.15, there exists a null set N_2 in $(\mathbb{R}^d, \mathfrak{B}_{\mathbb{R}^d}, \mu)$ such that

5°. $\qquad P^{\mathfrak{W}^{r+d}|Y_0}_{(B,X)}(Y_0^{-1}(\{x\}), x) = 1 \quad$ for $x \in N_2^c$.

Let $N_{Y_0} = N_1 \cup N_2$. Let us define

(1) $$P^x_{(B,X)}(\cdot) = \begin{cases} P^{\mathfrak{W}^{r+d}|Y_0}_{(B,X)}(\cdot, x) & \text{for } x \in N_{Y_0}^c \\ 0 & \text{for } x \in N_{Y_0}. \end{cases}$$

Thus defined, $P^x_{(B,X)}(A)$ is a $\mathfrak{B}_{\mathbb{R}^d}$-measurable function of x on $\mathbb{R}^d$ for every $A \in \mathfrak{W}^{r+d}$, and $P^x_{(B,X)}(\cdot)$ is a measure on $\mathfrak{W}^{r+d}$ for every $x \in \mathbb{R}^d$ and in particular for $x \in N_{Y_0}^c$, $P^x_{(B,X)}(\cdot)$ is equal to the probability measure $P^{\mathfrak{W}^{r+d}|Y_0}_{(B,X)}(\cdot, x)$ on $\mathfrak{W}^{r+d}$. From 3°, 4°, and 5°, we have

(2) $$P_{(B,X)}(A \cap Y_0^{-1}(G)) = \int_G P^x_{(B,X)}(A) \mu(dx)$$

§18. DEFINITION AND FUNCTION SPACE REPRESENTATION

for every A in $\mathfrak{W}^{r+d}$ and $G \in \mathfrak{B}_{\mathbb{R}^d}$,

(3) $$\int_{Y_0^{-1}(G)} \xi \, dP_{(B,X)} = \int_G \left[\int_{\mathbf{W}^{r+d}} \xi(w) P^x_{(B,X)}(dw) \right] \mu(dx),$$

in the sense that the existence of one side implies that of the other and the equality of the two, and

(4) $$P^x_{(B,X)}(Y_0^{-1}(\{x\})) = 1 \quad \text{for } x \in N^c_{Y_0}.$$

Definition 18.17. *For each $x \in N^c_{Y_0}$, let*

1°. *$\mathfrak{W}^{r+d,*,x}$ be the completion of $\mathfrak{W}^{r+d}$ with respect to $P^x_{(B,X)}$.*

2°. *$\mathfrak{W}^{r+d,0,x}_t = \sigma(\mathfrak{W}^{r+d}_t \cup \mathfrak{N})$ where $\mathfrak{N}$ is the collection of all the null sets in the complete measure space $(\mathbf{W}^{r+d}, \mathfrak{W}^{r+d,*,x}, P^x_{(B,X)})$.*

3°. *$\mathfrak{W}^{r+d,*,x}_t = \cap_{\varepsilon>0} \mathfrak{W}^{r+d,0,x}_{t+\varepsilon}$.*

Thus $(\mathbf{W}^{r+d}, \mathfrak{W}^{r+d,*,x}, \{\mathfrak{W}^{r+d,*,x}_t\}, P^x_{(B,X)})$ is a standard filtered space for each $x \in N^c_{Y_0}$.

Observation 18.18. The Borel σ-algebra $\mathfrak{W}^d$ of subsets of $\mathbf{W}^d$ for any $d \in \mathbb{N}$ is countably determined according to Proposition C.8 since $\mathbf{W}^d$ is a separable metric space. We show here that for any $t \in \mathbb{R}_+$, the σ-algebra $\mathfrak{W}^d_t = \sigma(\mathfrak{Z}_{[0,t]})$ where $\mathfrak{Z}_{[0,t]}$ is the collection of all cylinder sets in $\mathbf{W}^d$ with indices preceding t is also countably determined.

Proof. With fixed $t \in \mathbb{R}_+$, let $\mathbf{V}^d$ be the collection of all d-dimensional continuous functions v on $[0,t]$. With respect to the metric ρ defined by

$$\rho(v_1, v_2) = \max_{s \in [0,t]} |v_1(s) - v_2(s)| \quad \text{for } v_1, v_2 \in \mathbf{V}^d,$$

$\mathbf{V}^d$ is a complete separable metric space. Let $\mathfrak{V}$ be the Borel σ-algebra of subsets of $\mathbf{V}^d$. Let $\mathfrak{C}$ be the collection of all cylinder sets in $\mathbf{V}^d$, that is, sets C of the type

$$C = \{v \in \mathbf{V}^d : (v(t_1), \ldots, v(t_n)) \in E_1 \times \cdots \times E_n\},$$

where $0 \leq t_1 < \cdots < t_n \leq t$ and $E_i \in \mathfrak{B}_{\mathbb{R}}$ for $i = 1, \ldots, n$. The fact that $\mathfrak{V} = \sigma(\mathfrak{C})$ can be shown by using some of the arguments in the proof of Theorem 17.15.

Let T be the mapping of $\mathbf{W}^d$ into $\mathbf{V}^d$ defined by letting $T(w)$ be the restriction of $w \in \mathbf{W}^d$ to $[0, t]$. For $\tau = \{t_1, \ldots, t_n\}$ where $0 \le t_1 < \cdots < t_n \le t$ and $E = E_1 \times \cdots \times E_n$ where $E_i \in \mathfrak{B}_\mathbb{R}$ for $i = 1, \ldots, n$, let

$$\begin{cases} Z_{\tau,E} = \{w \in \mathbf{W}^d : (w(t_1), \ldots, w(t_n)) \in E_1 \times \cdots \times E_n\} \in \mathfrak{Z}_{[0,t]} \\ C_{\tau,E} = \{v \in \mathbf{V}^d : (v(t_1), \ldots, v(t_n)) \in E_1 \times \cdots \times E_n\} \in \mathfrak{C}. \end{cases}$$

Clearly $T(Z_{\tau,E}) = C_{\tau,E}$ and $T^{-1}(C_{\tau,E}) = Z_{\tau,E}$. Thus T establishes a one-to-one correspondence between $\mathfrak{Z}_{[0,t]}$ and $\mathfrak{C}$. This implies $T^{-1}(\mathfrak{C}) = \mathfrak{Z}_{[0,t]}$ and then

$$T^{-1}(\mathfrak{V}) = T^{-1}(\sigma(\mathfrak{C})) = \sigma(T^{-1}(\mathfrak{C})) = \sigma(\mathfrak{Z}_{[0,t]}) = \mathfrak{W}_t^d.$$

This shows that T is a $\mathfrak{W}_t^d/\mathfrak{V}$-measurable mapping of $\mathbf{W}^d$ into $\mathbf{V}^d$. Now since $\mathbf{V}^d$ is a separable metric space, its Borel σ-algebra of subsets $\mathfrak{V}$ is countably determined according to Proposition C.8. Let $\mathfrak{D} = \{D_n : n \in \mathbb{N}\}$ be a countable collection of determining sets for $\mathfrak{V}$. Let $\mathfrak{E} = \{E_n : n \in \mathbb{N}\}$ where $E_n = T^{-1}(E_n)$ for $n \in \mathbb{N}$. Let us show that $\mathfrak{E}$ is a countable collection of determining sets for $\mathfrak{W}_t^d$. Let P_1 and P_2 be two probability measures on $(\mathbf{W}^d, \mathfrak{W}_t^d)$ and suppose $P_1 = P_2$ on $\mathfrak{E}$. Let Q_1 and Q_2 be the probability distributions of T on $(\mathbf{V}^d, \mathfrak{V})$ relative to the two probability measures P_1 and P_2 on $(\mathbf{W}^d, \mathfrak{W}_t^d)$, that is, $Q_1 = P_1 \circ T^{-1}$ and $Q_2 = P_1 \circ T^{-2}$ on $\mathfrak{V}$. Now the fact that $P_1 = P_2$ on $\mathfrak{E}$ implies that

$$Q_1(D_n) = P_1 \circ T^{-1}(D_n) = P_1(E_n) = P_2(E_n) = P_2 \circ T^{-1}(D_n) = Q_2(D_n)$$

for every $n \in \mathbb{N}$. Thus $Q_1 = Q_2$ on $\mathfrak{D}$ and therefore $Q_1 = Q_2$ on $\mathfrak{V}$ since $\mathfrak{D}$ is a countable collection of determining sets for $\mathfrak{V}$. Take an arbitrary $A \in \mathfrak{W}_t^d$. Since $\mathfrak{W}_t^d = T^{-1}(\mathfrak{V})$, there exists $B \in \mathfrak{V}$ such that $A = T^{-1}(B)$. Then

$$P_1(A) = P_1(T^{-1}(B)) = Q_1(B) = Q_2(B) = P_2(T^{-1}(B)) = P_2(A).$$

Thus $P_1 = P_2$ on $\mathfrak{W}_t^d$. This shows that $\mathfrak{E}$ is a countable collection of determining sets for $\mathfrak{W}_t^d$. ∎

Lemma 18.19. *There exists a null set N in $(\mathbb{R}^d, \mathfrak{B}_{\mathbb{R}^d}, \mu)$ such that for every $x \in N^c$, the r-dimensional continuous adapted process W on $(\mathbf{W}^{r+d}, \mathfrak{W}^{r+d}, \{\mathfrak{W}_t^{r+d}\})$ is an r-dimensional $\{\mathfrak{W}_t^{r+d,*,x}\}$-adapted null at 0 Brownian motion on the standard filtered space $(\mathbf{W}^{r+d}, \mathfrak{W}^{r+d,*,x}, \{\mathfrak{W}_t^{r+d,*,x}\}, P_{(B,X)}^x)$.*

Proof. W is an r-dimensional continuous $\{\mathfrak{W}_t^{r+d}\}$-adapted process on the filtered space $(\mathbf{W}^{r+d}, \mathfrak{W}^{r+d}, \{\mathfrak{W}_t^{r+d}\})$. For $x \in N_{Y_0}^c$, $\mathfrak{W}_t^{r+d}$ is a sub-σ-algebra of $\mathfrak{W}_t^{r+d,*,x}$ for every $t \in \mathbb{R}_+$.

§18. DEFINITION AND FUNCTION SPACE REPRESENTATION

Thus W is a $\mathfrak{W}_t^{r+d,*,x}$-adapted process on $(\mathbf{W}^{r+d}, \mathfrak{W}^{r+d,*,x}, \{\mathfrak{W}_t^{r+d,*,x}\}, P_{(B,X)}^x)$ for $x \in N_{Y_0}^c$. We show next that there exists a null set N_0 in $(\mathbb{R}^d, \mathfrak{B}_{\mathbb{R}^d}, \mu)$ such that for $x \in N_0^c$, W is a null at 0 process on $(\mathbf{W}^{r+d}, \mathfrak{W}^{r+d,*,x}, \{\mathfrak{W}_t^{r+d,*,x}\}, P_{(B,X)}^x)$. Now since $P_{(B,X)}^x$ is a $\mathfrak{B}_{\mathbb{R}^d}$-measurable function of x on $\mathbb{R}^d$, integrating with respect to μ we have

$$\text{(1)} \qquad \int_{\mathbb{R}^d} P_{(B,X)}^x \{W_0 = 0\} \mu(dx) = P_{(B,X)}(W_0^{-1}(\{0\}) \cap Y_0^{-1}(\mathbb{R}^d))$$
$$= P_{(B,X)}(W_0^{-1}(\{0\})) = 1,$$

by (2) of Observation 18.16 and (1) in the proof of Lemma 18.11. Now since $P_{(B,X)}^x \{W_0 = 0\} \in [0, 1]$ for $x \in \mathbb{R}^d$ and since $\int_{\mathbb{R}^d} \mu(dx) = 1$, (1) implies that there exists a null set N_0 in $(\mathbb{R}^d, \mathfrak{B}_{\mathbb{R}^d}, \mu)$ containing the null set $N_{Y_0}^c$ such that $P_{(B,X)}^x \{W_0 = 0\} = 1$, that is, W is null at 0 when $x \in N_0^c$.

To show that there exists a null set N in $(\mathbb{R}^d, \mathfrak{B}_{\mathbb{R}^d}, \mu)$ such that for every $x \in N^c$ the process W is an r-dimensional $\{\mathfrak{W}_t^{r+d,*,x}\}$-adapted Brownian motion on a standard filtered space $(\mathbf{W}^{r+d}, \mathfrak{W}^{r+d,*,x}, \{\mathfrak{W}_t^{r+d,*,x}\}, P_{(B,X)}^x)$, it suffices to show that there exists a null set N in $(\mathbb{R}^d, \mathfrak{B}_{\mathbb{R}^d}, \mu)$ such that when $x \in N^c$ then $(\mathbf{W}^{r+d}, \mathfrak{W}^{r+d,*,x}, \{\mathfrak{W}_t^{r+d,*,x}\}, P_{(B,X)}^x)$ is a standard filtered space and according to Theorem 13.26, for any $s, t \in \mathbb{R}_+$, $s < t$, and $y \in \mathbb{R}^r$ we have

$$\text{(2)} \qquad \mathbf{E}[e^{i\langle y, W_t - W_s\rangle} | \mathfrak{W}_s^{r+d,*,x}] = e^{-\frac{|y|^2}{2}(t-s)} \quad \text{a.e. on } (\mathbf{W}^{r+d}, \mathfrak{W}_s^{r+d,*,x}, P_{(B,X)}^x),$$

that is,

$$\text{(3)} \qquad \int_A e^{i\langle y, W_t - W_s\rangle} dP_{(B,X)}^x = P_{(B,X)}^x(A) e^{-\frac{|y|^2}{2}(t-s)} \quad \text{for } A \in \mathfrak{W}_s^{r+d,*,x},$$

Now for $A \in \mathfrak{W}_s^{r+d}$ and $G \in \mathfrak{B}_{\mathbb{R}^d}$ we have

$$\text{(4)} \qquad \int_G \left\{ \int_{\mathbf{W}^{r+d}} e^{i\langle y, W_t - W_s\rangle} \mathbf{1}_A \, dP_{(B,X)}^x \right\} \mu(dx) = \int_{Y_0^{-1}(G)} e^{i\langle y, W_t - W_s\rangle} \mathbf{1}_A \, dP_{(B,X)}$$

by (3) of Observation 18.16. By Lemma 18.11, W is an r-dimensional $\{\mathfrak{W}_t^{r+d,*}\}$-adapted Brownian motion on $(\mathbf{W}^{r+d}, \mathfrak{W}^{r+d,*}, \{\mathfrak{W}_t^{r+d,*}\}, P_{(B,X)})$. Also since $A \in \mathfrak{W}_s^{r+d}$ and since $Y_0^{-1}(G) \in \mathfrak{W}_0^{r+d}$, we have $A \cap Y_0^{-1}(G) \in \mathfrak{W}_s^{r+d} \subset \mathfrak{W}_s^{r+d,*}$. Thus by (3) in the proof of Lemma 18.11 and (2) of Observation 18.16, we have

$$\text{(5)} \qquad \int_{A \cap Y_0^{-1}(G)} e^{i\langle y, W_t - W_s\rangle} dP_{(B,X)} = P_{(B,X)}(A \cap Y_0^{-1}(G)) e^{-\frac{|y|^2}{2}(t-s)}$$
$$= e^{-\frac{|y|^2}{2}(t-s)} \int_G P_{(B,X)}^x(A) \mu(dx).$$

Combining (4) and (5) we have for $A \in \mathfrak{W}_s^{r+d}$ and $G \in \mathfrak{B}_{\mathbb{R}^d}$

$$\int_G \left\{ \int_{\mathbf{W}^{r+d}} e^{i\langle y, W_t - W_s \rangle} \mathbf{1}_A \, dP_{(B,X)}^x \right\} \mu(dx) = e^{-\frac{|y|^2}{2}(t-s)} \int_G P_{(B,X)}^x(A) \mu(dx).$$

Since this equality holds for every $G \in \mathfrak{B}_{\mathbb{R}^d}$ and the integrands are $\mathfrak{B}_{\mathbb{R}^d}$-measurable functions, there exists a null set $N_{s,t,y,A}$ in $(\mathbb{R}^d, \mathfrak{B}_{\mathbb{R}^d}, \mu)$ containing N_0 such that for $x \in N_{s,t,y,A}^c$ we have

(6) $$\int_{\mathbf{W}^{r+d}} e^{i\langle y, W_t - W_s \rangle} \mathbf{1}_A \, dP_{(B,X)}^x = e^{-\frac{|y|^2}{2}(t-s)} P_{(B,X)}^x(A).$$

Let $\mathfrak{D}$ be a countable collection of determining sets of the countably determined σ-algebra $\mathfrak{W}_s^{r+d}$. Let Q_+ be the collection of all nonnegative rational numbers and let Q^d be the collection of all rational points in $\mathbb{R}^d$. Let $N = \bigcup_{s,t \in Q_+, s < t, y \in Q^d, A \in \mathfrak{D}} N_{s,t,y,A}$. Then for $x \in N^c$, (6) holds for all $s, t \in Q_+$, $s < t$, $y \in Q^d$, and $A \in \mathfrak{D}$. By the continuity of both sides of (6) in s, t, and $y \in \mathbb{R}^d$, (6) holds for all $s, t \in \mathbb{R}_+$, $s < t$, $y \in \mathbb{R}^d$, and $A \in \mathfrak{D}$. Now with fixed s, t, and y, both sides of (6) as functions of $A \in \mathfrak{W}_s^{r+d}$ are measures on $\mathfrak{W}_s^{r+d}$. Since these two measures are equal on the countable collection $\mathfrak{D}$ of determining sets for $\mathfrak{W}_s^{r+d}$, they are equal on $\mathfrak{W}_s^{r+d}$. Thus (6) holds for all $s, t \in \mathbb{R}_+$, $s < t$, $y \in \mathbb{R}^d$, and $A \in \mathfrak{W}_s^{r+d}$. From the definition of $\mathfrak{W}_s^{r+d,0,x}$, it is clear that (6) holds for $A \in \mathfrak{W}_s^{r+d,0,x}$. Thus for every $s, t \in \mathbb{R}_+$, $s < t$ and $y \in \mathbb{R}^d$ we have

$$\mathbf{E}[e^{i\langle y, W_t - W_s \rangle} | \mathfrak{W}_s^{r+d,0,x}] = e^{-\frac{|y|^2}{2}(t-s)} \quad \text{a.e. on } (\mathbf{W}^{r+d}, \mathfrak{W}_s^{r+d,0,x}, p_{(B,X)}^x).$$

Then by same argument as in the last part of the proof of Theorem 18.11 we have (2). ∎

According to Lemma 18.11, W is an r-dimensional $\{\mathfrak{W}_t^{r+d,*}\}$-adapted null at 0 Brownian motion on $(\mathbf{W}^{r+d}, \mathfrak{W}^{r+d,*}, \{\mathfrak{W}_t^{r+d,*}\}, P_{(B,X)})$. According to Lemma 18.19 there exists a null set N in $(\mathbb{R}^d, \mathfrak{B}_{\mathbb{R}^d}, \mu)$ such that for every $x \in N^c$, W is an r-dimensional $\{\mathfrak{W}_t^{r+d,*,x}\}$-adapted null at 0 Brownian motion on $(\mathbf{W}^{r+d}, \mathfrak{W}^{r+d,*,x}, \{\mathfrak{W}_t^{r+d,*,x}\}, P_{(B,X)}^x)$. If Φ is a stochastic process on $(\mathbf{W}^{r+d}, \mathfrak{W}^{r+d})$ which is both $\{\mathfrak{W}_t^{r+d,*}\}$-predictable and $\{\mathfrak{W}_t^{r+d,*,x}\}$-predictable and in both $\mathbf{L}_{2,\infty}(\mathbb{R}_+ \times \mathbf{W}^{r+d}, m_L \times P_{(B,X)})$ and $\mathbf{L}_{2,\infty}(\mathbb{R}_+ \times \mathbf{W}^{r+d}, m_L \times P_{(B,X)}^x)$ for some $x \in N^c$, then the stochastic integral of Φ with respect to the $\{\mathfrak{W}_t^{r+d,*}\}$-adapted Brownian motion W^j exists as an element of $\mathbf{M}_2^c(\mathbf{W}^{r+d}, \mathfrak{W}^{r+d,*}, \{\mathfrak{W}_t^{r+d,*}\}, P_{(B,X)})$ and the stochastic integral of Φ with respect to the $\{\mathfrak{W}_t^{r+d,*,x}\}$-adapted Brownian motion W^j exists as an element of $\mathbf{M}_2^c(\mathbf{W}^{r+d}, \mathfrak{W}^{r+d,*,x}, \{\mathfrak{W}_t^{r+d,*,x}\}, P_{(B,X)}^x)$. To distinguish between these two stochastic integrals we use the notations $P_{(B,X)} \cdot \Phi \bullet W^j$ and $P_{(B,X)}^x \cdot \Phi \bullet W^j$ respectively. We use also the alternate notations $\{P_{(B,X)} \cdot \int_{[0,t]} \Phi(s) \, dW^j(s) : t \in \mathbb{R}_+\}$ and

§18. DEFINITION AND FUNCTION SPACE REPRESENTATION

$\{P^x_{(B,X)} \cdot \int_{[0,t]} \Phi(s)\,dW^j(s) : t \in \mathbb{R}_+\}$ respectively. We write $|\cdot|_\infty^{P_{(B,X)}}$ and $|\cdot|_\infty^{P^x_{(B,X)}}$ for the quasinorm $|\cdot|_\infty$ in Definition 11.4 for the two spaces $\mathbf{M}^c_2(\mathbf{W}^{r+d}, \mathfrak{W}^{r+d,*}, \{\mathfrak{W}^{r+d,*}_t\}, P_{(B,X)})$ and $\mathbf{M}^c_2(\mathbf{W}^{r+d}, \mathfrak{W}^{r+d,*,x}, \{\mathfrak{W}^{r+d,*,x}_t\}, P^x_{(B,X)})$ respectively. Note that if the process Φ is in both $\mathbf{L}_0(\mathbf{W}^{r+d}, \mathfrak{W}^{r+d,*}, \{\mathfrak{W}^{r+d,*}_t\}, P_{(B,X)})$ and $\mathbf{L}_0(\mathbf{W}^{r+d}, \mathfrak{W}^{r+d,*,x}, \{\mathfrak{W}^{r+d,*,x}_t\}, P^x_{(B,X)})$, then this distinction is unnecessary by Definition 12.3 of stochastic integrals of processes in $\mathbf{L}_0$.

Likewise if Φ is in both $\mathbf{L}_{1,\infty}(\mathbb{R}_+ \times \mathbf{W}^{r+d}, m_L \times P_{(B,X)})$ and $\mathbf{L}_{1,\infty}(\mathbb{R}_+ \times \mathbf{W}^{r+d}, m_L \times P^x_{(B,X)})$, then we write $P_{(B,X)} \cdot \Phi \bullet m_L$ or $\{P_{(B,X)} \cdot \int_{[0,t]} \Phi(s) m_L(ds) : t \in \mathbb{R}_+\}$ for $\Phi \bullet m_L$ in $\mathbf{V}^c(\mathbf{W}^{r+d}, \mathfrak{W}^{r+d,*}, \{\mathfrak{W}^{r+d,*}_t\}, P_{(B,X)})$ and write $P^x_{(B,X)} \cdot \Phi \bullet m_L$ or $\{P^x_{(B,X)} \cdot \int_{[0,t]} \Phi(s) m_L(ds) : t \in \mathbb{R}_+\}$ for $\Phi \bullet m_L$ in $\mathbf{V}^c(\mathbf{W}^{r+d}, \mathfrak{W}^{r+d,*,x}, \{\mathfrak{W}^{r+d,*,x}_t\}, P^x_{(B,X)})$.

Lemma 18.20. *There exists a null set N_0 in $(\mathbb{R}^d, \mathfrak{B}_{\mathbb{R}^d}, \mu)$ such that for every $x \in N^c_0$ the stochastic integral $P^x_{(B,X)} \cdot (\alpha_Y)^i_j \bullet W^j$ exists in $\mathbf{M}^c_2$ in $\mathbf{M}^c_2(\mathbf{W}^{r+d}, \mathfrak{W}^{r+d,*,x}, \{\mathfrak{W}^{r+d,*,x}_t\}, P^x_{(B,X)})$ and $P^x_{(B,X)} \cdot (\beta_Y)^i \bullet m_L$ exists in $\mathbf{V}^c(\mathbf{W}^{r+d}, \mathfrak{W}^{r+d,*,x}, \{\mathfrak{W}^{r+d,*,x}_t\}, P^x_{(B,X)})$. Furthermore there exists a null set Λ_x in $(\mathbf{W}^d, \mathfrak{W}^d, P^x_{(B,X)})$ such that*

(1) $$\begin{cases} P^x_{(B,X)} \cdot (\alpha_Y)^i_j \bullet W^j = P_{(B,X)} \cdot (\alpha_Y)^i_j \bullet W^j, \\ P^x_{(B,X)} \cdot (\beta_Y)^i \bullet m_L = P_{(B,X)} \cdot (\beta_Y)^i \bullet m_L \end{cases} \text{ on } \Lambda^c_x.$$

Proof. According to Lemma 18.19 there exists a null set N_1 in $(\mathbb{R}^d, \mathfrak{B}_{\mathbb{R}^d}, \mu)$ such that for every $x \in N^c_1$, W is an r-dimensional $\{\mathfrak{W}^{r+d,*,x}_t\}$-adapted null at 0 Brownian motion on $(\mathbf{W}^{r+d}, \mathfrak{W}^{r+d,*,x}, \{\mathfrak{W}^{r+d,*,x}_t\}, P^x_{(B,X)})$. Since Y is a d-dimensional $\{\mathfrak{W}^{r+d}_t\}$-adapted process on $(\mathbf{W}^{r+d}, \mathfrak{W}^{r+d}, \{\mathfrak{W}^{r+d}_t\})$, for every $x \in N^c_1$ it is a d-dimensional $\{\mathfrak{W}^{r+d,x}_t\}$-adapted process on $(\mathbf{W}^{r+d}, \mathfrak{W}^{r+d,*,x}, \{\mathfrak{W}^{r+d,*,x}_t\}, P^x_{(B,X)})$. The mapping $\mathcal{Y}$ of $\mathbf{W}^{r+d}$ into $\mathbf{W}^d$ defined by setting $\mathcal{Y}(w) = Y(\cdot, w)$ for $w \in \mathbf{W}^{r+d}$ is $\mathfrak{W}^{r+d}_t/\mathfrak{W}^d_t$-measurable and hence $\mathfrak{W}^{r+d,*,x}_t/\mathfrak{W}^d_t$-measurable for every $t \in \mathbb{R}_+$. Thus by Lemma 18.5, $(\alpha_Y)^i_j$ and $(\beta_Y)^i$ are $\{\mathfrak{W}^{r+d,*,x}_t\}$-progressively measurable processes on $(\mathbf{W}^{r+d}, \mathfrak{W}^{r+d,*,x}, \{\mathfrak{W}^{r+d,*,x}_t\}, P^x_{(B,X)})$ for every $x \in N^c_1$.

Now for every $m \in \mathbb{N}$, we have

$$\int_{\mathbb{R}^d} \left[\int_{[0,m] \times \mathbf{W}^{r+d}} |(\alpha_Y)^i_j|^2 d(m_L \times P^x_{(B,X)}) \right] \mu(dx)$$
$$= \int_{\mathbb{R}^d} \left[\int_{\mathbf{W}^{r+d}} \left\{ \int_{[0,m]} |(\alpha_Y)^i_j|^2 d(m_L) \right\} dP^x_{(B,X)} \right] \mu(dx)$$
$$= \int_{\mathbf{W}^{r+d}} \left[\int_{[0,m]} |(\alpha_Y)^i_j|^2 d(m_L) \right] dP_{(B,X)}$$

$$= \int_{[0,m]\times \mathbf{W}^{r+d}} |(\alpha_Y)^i_j|^2 d(m_L \times P_{(B,X)}) < \infty,$$

where the first equality is by applying the Fubini-Tonelli Theorem to the measurable process $(\alpha_Y)^i_j$, the second equality is by (3) of Observation 18.16, and the finiteness of the last integral is by (1) in the proof of Lemma 18.12. Thus there exists a null set in $(\mathbb{R}^d, \mathfrak{B}_{\mathbb{R}^d}, \mu)$, whose union with N_1 we call N_2, such that

$$\int_{[0,m]\times \mathbf{W}^{r+d}} |(\alpha_Y)^i_j|^2 d(m_L \times P^x_{(B,X)}) < \infty$$

for all $m \in \mathbb{N}$ when $x \in N_2^c$. This shows that $(\alpha_Y)^i_j$ is in $\mathbf{L}_{2,\infty}(\mathbb{R}_+ \times \mathbf{W}^{r+d}, m_L \times P^x_{(B,X)})$ for $x \in N_2^c$.

According to Lemma 18.12, $(\alpha_Y)^i_j$ is in $\mathbf{L}_{2,\infty}(\mathbb{R}_+ \times \mathbf{W}^{r+d}, m_L \times P_{(B,X)})$ so that there exists a sequence $\{(\alpha_Y^n)^i_j : n \in \mathbb{N}\}$ in $\mathbf{L}_0(\mathbf{W}^{r+d}, \mathfrak{W}^{r+d,*}, \{\mathfrak{W}^{r+d,*}_t\}, P_{(B,X)})$ such that

$$(2) \qquad \lim_{n\to\infty} \|(\alpha_Y^n)^i_j - (\alpha_Y)^i_j\|^{m_L \times P_{(B,X)}}_{2,\infty} = 0.$$

Recall that $(\alpha_Y^n)^i_j$ can be chosen from $\mathbf{L}_0(\mathbf{W}^{r+d}, \mathfrak{W}^{r+d}, \{\mathfrak{W}^{r+d}_t\}, P_{(B,X)})$ as we showed in the proof of Theorem 18.13. By Definition 12.9, for the stochastic integral $P_{(B,X)} \cdot (\alpha_Y)^i_j \bullet W^j$ of $(\alpha_Y)^i_j$ with respect to the $\{\mathfrak{W}^{r+d,*}_t\}$-adapted null at 0 Brownian motions W^j on the standard filtered space $(\mathbf{W}^{r+d}, \mathfrak{W}^{r+d,*}, \{\mathfrak{W}^{r+d,*}_t\}, P_{(B,X)})$, which is a member of $\mathbf{M}^{c}_2(\mathbf{W}^{r+d}, \mathfrak{W}^{r+d,*}, \{\mathfrak{W}^{r+d,*}_t\}, P_{(B,X)})$, we have

$$\lim_{n\to\infty} \mathbf{I}(\alpha_Y^n)^i_j \bullet W^j - P_{(B,X)} \cdot (\alpha_Y)^i_j \bullet W^j \mathbf{I}^{P_{(B,X)}}_{\infty} = 0,$$

or equivalently, $\lim_{n\to\infty} \mathbf{I}(\alpha_Y^n)^i_j \bullet W^j - P_{(B,X)} \cdot (\alpha_Y)^i_j \bullet W^j \mathbf{I}^{P_{(B,X)}}_m = 0$ for every $m \in \mathbb{N}$ by Remark 11.5, that is, $\lim_{n\to\infty} \int_{[0,m]} (\alpha_Y^n)^i_j(s) \, dW^j(s) = P_{(B,X)} \cdot \int_{[0,m]} (\alpha_Y)^i_j(s) \, dW^j(s)$ in $\mathbf{L}_2(\mathbf{W}^{r+d}, \mathfrak{W}^{r+d,*}, P_{(B,X)})$ for every $m \in \mathbb{N}$. Now convergence in L_2 implies the existence of a subsequence which converges a.e. Note that a null set in $(\mathbf{W}^{r+d}, \mathfrak{W}^{r+d,*}, P_{(B,X)})$ is always a subset of a null set in $(\mathbf{W}^{r+d}, \mathfrak{W}^{r+d}, P_{(B,X)})$ since $\mathfrak{W}^{r+d,*}$ is the completion of $\mathfrak{W}^{r+d}$ with respect to $P_{(B,X)}$. Thus there exist a null set Λ in $(\mathbf{W}^{r+d}, \mathfrak{W}^{r+d}, P_{(B,X)})$ and a collection of subsequences $\{(m,k) : k \in \mathbb{N}\}$ of $\{n\}$ for $m \in \mathbb{N}$ in which $\{(m+1,k) : k \in \mathbb{N}\}$ is a subsequence of $\{(m,k) : k \in \mathbb{N}\}$ for $m \in \mathbb{N}$ such that

$$(3) \qquad \lim_{k\to\infty} \int_{[0,m]} (\alpha_Y^{(m,k)})^i_j(s) \, dW^j(s) = P_{(B,X)} \cdot \int_{[0,m]} (\alpha_Y)^i_j(s) \, dW^j(s)$$

§18. DEFINITION AND FUNCTION SPACE REPRESENTATION

for every $m \in \mathbb{N}$ on Λ^c. On the other hand

$$
\begin{aligned}
(4) \quad & \|(\alpha_Y^{(m,k)})_j^i - (\alpha_Y)_j^i\|_{2,m}^{m_L \times P_{(B,X)}} \\
&= \int_{\mathbf{W}^{r+d}} \left\{ \int_{[0,m]} |(\alpha_Y^{(m,k)})_j^i - (\alpha_Y)_j^i|^2 dm_L \right\} dP_{(B,X)} \\
&= \int_{\mathbb{R}^d} \left[\int_{\mathbf{W}^{r+d}} \left\{ \int_{[0,m]} |(\alpha_Y^{(m,k)})_j^i - (\alpha_Y)_j^i|^2 dm_L \right\} dP_{(B,X)}^x \right] \mu(dx) \\
&= \int_{\mathbb{R}^d} \left[\int_{[0,m] \times \mathbf{W}^{r+d}} |(\alpha_Y^{(m,k)})_j^i - (\alpha_Y)_j^i|^2 d(m_L \times P_{(B,X)}^x) \right] \mu(dx),
\end{aligned}
$$

where the second equality is by (3) of Observation 18.16. Since (2) is equivalent to $\lim_{n \to \infty} \|(\alpha_Y^n)_j^i - (\alpha_Y)_j^i\|_{2,m}^{m_L \times P_{(B,X)}} = 0$ for every $m \in \mathbb{N}$ by Remark 11.18, (4) implies that there exists a null set in $(\mathbb{R}^d, \mathfrak{B}_{\mathbb{R}^d}, \mu)$, whose union with N_2 we call N_3, and a subsequence $\{(m,\ell) : \ell \in \mathbb{N}\}$ of $\{(m,k) : \ell \in \mathbb{N}\}$ for $m \in \mathbb{N}$ chosen so that $\{(m+1,\ell) : \ell \in \mathbb{N}\}$ is a subsequence of $\{(m,\ell) : \ell \in \mathbb{N}\}$ such that

$$\lim_{\ell \to \infty} \int_{[0,m] \times \mathbf{W}^{r+d}} |(\alpha_Y^{(m,\ell)})_j^i(s) - (\alpha_Y)_j^i|^2 d(m_L \times P_{(B,X)}^x) = 0$$

for every $m \in \mathbb{N}$ when $x \in N_3^c$. If we take the diagonal sequence $\{(\ell,\ell)\}$ and write $\{\ell\}$ for this subsequence of $\{n\}$, then we have

$$\lim_{\ell \to \infty} \int_{[0,m] \times \mathbf{W}^{r+d}} |(\alpha_Y^\ell)_j^i(s) - (\alpha_Y)_j^i|^2 d(m_L \times P_{(B,X)}^x) = 0,$$

that is, $\lim_{\ell \to \infty} \|(\alpha_Y^\ell)_j^i - (\alpha_Y)_j^i\|_{2,m}^{m_L \times P_{(B,X)}^x} = 0$ for every $m \in \mathbb{N}$ when $x \in N_3^c$. Thus by Remark 11.18 we have

$$(5) \quad \lim_{\ell \to \infty} \|(\alpha_Y^\ell)_j^i - (\alpha_Y)_j^i\|_{2,\infty}^{m_L \times P_{(B,X)}^x} = 0 \quad \text{for } x \in N_3^c.$$

Recall that our $(\alpha_Y^n)_j^i$ are in $\mathbf{L}_0(\mathbf{W}^{r+d}, \mathfrak{W}^{r+d}, \{\mathfrak{W}_t^{r+d}\}, P_{(B,X)})$. Since the spaces $\mathbf{L}_0$ do not actually depend on the probability measure in the underlying filtered space, $(\alpha_Y^n)_j^i$ are also in $\mathbf{L}_0(\mathbf{W}^{r+d}, \mathfrak{W}^{r+d}, \{\mathfrak{W}_t^{r+d}\}, P_{(B,X)}^x)$ for $x \in N_3^c$. Thus by Definition 12.9, (5) implies

$$\lim_{\ell \to \infty} \|(\alpha_Y^\ell)_j^i \bullet W^j - P_{(B,X)}^x \cdot (\alpha_Y)_j^i \bullet W^j\|_\infty^{P_{(B,X)}^x} = 0 \quad \text{for } x \in N_3^c,$$

and then by Remark 11.5, $\lim_{\ell \to \infty} \int_{[0,m]} (\alpha_Y^\ell)_j^i(s) \, dW^j(s) = P_{(B,X)}^x \cdot \int_{[0,m]} (\alpha_Y)_j^i(s) \, dW^j(s)$ in $L_2(\mathbf{W}^{r+d}, \mathfrak{W}^{r+d,*,x}, P_{(B,X)}^x)$ for every $m \in \mathbb{N}$ when $x \in N_3^c$. Since convergence in L_2

implies the existence of a subsequence which converges a.e. and since a null set in $(\mathbf{W}^{r+d}, \mathfrak{W}^{r+d,*,x}, P^x_{(B,X)})$ is always contained in a null set in $(\mathbf{W}^{r+d}, \mathfrak{W}^{r+d}, P^x_{(B,X)})$, for every $x \in N_3^c$ there exists a null set Λ'_x in $(\mathbf{W}^{r+d}, \mathfrak{W}^{r+d}, P^x_{(B,X)})$ and a subsequence $\{k\}$ of $\{\ell\}$ such that

$$(6) \qquad \lim_{k \to \infty} \int_{[0,m]} (\alpha_Y^k)^i_j(s)\, dW^j(s) = P^x_{(B,X)} \cdot \int_{[0,m]} (\alpha_Y)^i_j(s)\, dW^j(s)$$

for all $m \in \mathbb{N}$ on $(\Lambda'_x)^c$ when $x \in N_3^c$. From (3) we have

$$(7) \qquad \lim_{k \to \infty} \int_{[0,m]} (\alpha_Y^k)^i_j(s)\, dW^j(s) = P_{(B,X)} \cdot \int_{[0,m]} (\alpha_Y)^i_j(s)\, dW^j(s)$$

for all $m \in \mathbb{N}$ on Λ^c. Now by (2) of Observation 18.16, we have

$$(8) \qquad \int_{\mathbb{R}^d} P^x_{(B,X)}(\Lambda)\mu(dx) = P_{(B,X)}(\Lambda) = 0.$$

Thus there exists a null set N_4 in $(\mathbb{R}^d, \mathfrak{B}_{\mathbb{R}^d}, \mu)$ such that $P^x_{(B,X)}(\Lambda) = 0$ for $x \in N_4^c$. Let $N_0 = N_3 \cup N_4$. For every $x \in N_0^c$, let $\Lambda_x = \Lambda'_x \cup \Lambda$. Then for every $x \in N_0^c$, Λ_x is a null set in $(\mathbf{W}^{r+d}, \mathfrak{W}^{r+d}, P^x_{(B,X)})$ and furthermore by (6) and (7) we have

$$P^x_{(B,X)} \cdot \int_{[0,m]} (\alpha_Y)^i_j(s)\, dW^j(s) = P_{(B,X)} \cdot \int_{[0,m]} (\alpha_Y)^i_j(s)\, dW^j(s)$$

for all $m \in \mathbb{N}$ on Λ_x^c, that is, $P^x_{(B,X)} \cdot (\alpha_Y)^i_j \bullet W^j = P_{(B,X)} \cdot (\alpha_Y)^i_j \bullet W^j$ on Λ_x^c. Following similar line of argument we have $P^x_{(B,X)} \cdot (\beta_Y)^i \bullet m_L = P_{(B,X)} \cdot (\beta_Y)^i \bullet m_L$ on Λ_x^c. ∎

Theorem 18.21. *Suppose the stochastic differential equation (1) in Definition 18.6 has a solution (B, X) on some standard filtered space $(\Omega, \mathfrak{F}, \{\mathfrak{F}_t\}, P)$. Let μ be the probability distribution of X_0 on $(\mathbb{R}^d, \mathfrak{B}_{\mathbb{R}^d})$. Then there exists a null set N in $(\mathbb{R}^d, \mathfrak{B}_{\mathbb{R}^d}, \mu)$ such that for every $x \in N^c$ the pair of processes (W, Y) on $(\mathbf{W}^{r+d}, \mathfrak{W}^{r+d})$ defined by $W(t, w) = w'(t)$ and $Y(t, w) = w''(t)$ for $t \in \mathbb{R}_+$ and $w = (w', w'') \in \mathbf{W}^{r+d}$ with $w' \in \mathbf{W}^r$ and $w'' \in \mathbf{W}^d$ is a solution of the stochastic differential equation (1) in Definition 18.6 on the standard filtered space $(\mathbf{W}^{r+d}, \mathfrak{W}^{r+d,*,x}, \{\mathfrak{W}^{r+d,*,x}_t\}, P^x_{(B,X)})$ satisfying the condition $Y_0 = x$ a.e. on $(\mathbf{W}^{r+d}, \mathfrak{W}^{r+d}, P^x_{(B,X)})$.*

Proof. Let N_0 and Λ_x for $x \in N_0^c$ be as specified in Lemma 18.20. For $x \in N_0^c$, let

$$(1) \qquad \Psi_t^{x,i} = \sum_{j=1}^r (P^x_{(B,X)} \cdot (\alpha_Y)^i_j \bullet W^j)_t + (P^x_{(B,X)} \cdot (\beta_Y)^i \bullet m_L)_t + Y_0^i - Y_t^i.$$

Let

(2) $$\Psi_t^i = \sum_{j=1}^{r}(P_{(B,X)} \cdot (\alpha_Y)_j^i \bullet W^j)_t + (P_{(B,X)} \cdot (\beta_Y)^i \bullet m_L)_t + Y_0^i - Y_t^i.$$

According to Theorem 18.13, (W, Y) is a solution of the stochastic differential equation on $(\mathbf{W}^{r+d}, \mathfrak{W}^{r+d,*}, \{\mathfrak{W}_t^{r+d,*}\}, P_{(B,X)})$ and thus there exists a null set Λ in $(\mathbf{W}^{r+d}, \mathfrak{W}^{r+d}, P_{(B,X)})$ such that $\Psi_t^i = 0$ on Λ^c. Since a null set in $(\mathbf{W}^{r+d}, \mathfrak{W}^{r+d,*}, P_{(B,X)})$ is always contained in a null set in $(\mathbf{W}^{r+d}, \mathfrak{W}^{r+d}, P_{(B,X)})$, we can choose Λ to be a null set in $(\mathbf{W}^{r+d}, \mathfrak{W}^{r+d}, P_{(B,X)})$. Then as we saw in (8) in the proof of Lemma 18.20, there exists a null set N_1 in $(\mathbb{R}^d, \mathfrak{B}_{\mathbb{R}^d}, \mu)$ such that Λ is a null set in $(\mathbf{W}^d, \mathfrak{W}^d, P_{(B,X)}^x)$ for every $x \in N_1^c$. Let $N = N_0 \cup N_1$ and $\Lambda'_x = \Lambda_x \cup \Lambda$ for $x \in N^c$. Applying (1) of Lemma 18.20 to (1), we have $\Psi_t^{x,i} = \Psi_t^i = 0$ on $(\Lambda')^c$ for $x \in N^c$. This shows that (W, Y) is a solution of the stochastic differential equation on $(\mathbf{W}^{r+d}, \mathfrak{W}^{r+d,*,x}, \{\mathfrak{W}_t^{r+d,*,x}\}, P_{(B,X)}^x)$ for every $x \in N^c$. Finally according to (4) of Observation 18.16, we have $P_{(B,X)}^x(Y_0^{-1}(\{0\})) = 1$ for $x \in N^c$, that is, $Y_0 = x$ a.e. on $(\mathbf{W}^{r+d}, \mathfrak{W}^{r+d}, P_{(B,X)}^x)$. ∎

§19 Existence and Uniqueness of Solutions

[I] Uniqueness in Probability Law and Pathwise Uniqueness of Solutions

Let $(B^{(q)}, X^{(q)})$ be a solution of the stochastic differential equation (1) in Definition 18.6 on a standard filtered space $(\Omega^{(q)}, \mathfrak{F}^{(q)}, \{\mathfrak{F}^{(q)}\}, P^{(q)})$ for $q = 1, 2$. Consider the standard filtered space $(\mathbf{W}^{r+d}, \mathfrak{W}^{r+d,*,(q)}, \{\mathfrak{W}_t^{r+d,*,(q)}\}, P_{(B^{(q)}, X^{(q)})}^{(q)})$ generated by $P_{(B^{(q)}, X^{(q)})}^{(q)}$ for $q = 1, 2$ as in Definition 18.10. The mapping (W, Y) of $\mathbb{R}_+ \times \mathbf{W}^{r+d}$ into $\mathbb{R}^{r+d}$ defined by setting $(W, Y)(t, w) = w(t)$ for $(t, w) \in \mathbb{R}_+ \times \mathbf{W}^{r+d}$ represents the solution $(B^{(q)}, X^{(q)})$ in the sense that (W, Y) is a solution of the stochastic differential equation on the standard filtered space $(\mathbf{W}^{r+d}, \mathfrak{W}^{r+d,*,(q)}, \{\mathfrak{W}_t^{r+d,*,(q)}\}, P_{(B^{(q)}, X^{(q)})}^{(q)})$ and has the same probability distribution on $(\mathbf{W}^{r+d}, \mathfrak{W}^{r+d})$ as $(B^{(q)}, X^{(q)})$ for $q = 1, 2$ as we showed in Theorem 18.13. Let us write $(W^{(q)}, Y^{(q)})$ for (W, Y) as representations of $(B^{(q)}, X^{(q)})$ for $q = 1, 2$.

Definition 19.1. *We say that the solution of the stochastic differential equation (1) in Definition 18.6 is unique in the sense of probability law if* $(B^{(1)}, X^{(1)})$ *and* $(B^{(2)}, X^{(2)})$ *are two solutions on two standard filtered spaces* $(\Omega^{(1)}, \mathfrak{F}^{(1)}, \{\mathfrak{F}^{(1)}\}, P^{(1)})$ *and* $(\Omega^{(2)}, \mathfrak{F}^{(2)}, \{\mathfrak{F}^{(2)}\}, P^{(2)})$ *respectively and* $X^{(1)}$ *and* $X^{(2)}$ *have identical initial probability distributions, that is,* $X_0^{(1)}$ *and* $X_0^{(2)}$ *have identical probability distributions on* $(\mathbb{R}^d, \mathfrak{B}_{\mathbb{R}^d})$, *then* $X^{(1)}$ *and* $X^{(2)}$ *have* identical probability distributions on $(\mathbf{W}^d, \mathfrak{W}^d)$. We say that the solution is unique in the sense of probability law under deterministic initial conditions if whenever $X_0^{(1)} = x$ a.e. on $(\Omega^{(1)}, \mathfrak{F}^{(1)}, P^{(1)})$ and $X_0^{(2)} = x$ a.e. on $(\Omega^{(2)}, \mathfrak{F}^{(2)}, P^{(2)})$ for some $x \in \mathbb{R}^d$, then $X^{(1)}$ and $X^{(2)}$ have identical probability distributions on $(\mathbf{W}^d, \mathfrak{W}^d)$.

Lemma 19.2 *Uniqueness of solution in the sense of probability law for the stochastic differential equation (1) in Definition 18.6 is equivalent to uniqueness in the sense of probability law under deterministic initial conditions.*

Proof. Clearly uniqueness in the sense of probability law contains uniqueness in the sense of probability law under deterministic initial conditions as particular cases in which the initial probability distributions are unit masses.

Conversely assume that whenever $X_0^{(1)} = x$ a.e. on $(\Omega^{(1)}, \mathfrak{F}^{(1)}, P^{(1)})$ and $X_0^{(2)} = x$ a.e. on $(\Omega^{(2)}, \mathfrak{F}^{(2)}, P^{(2)})$ for some $x \in \mathbb{R}^d$, then $X^{(1)}$ and $X^{(2)}$ have identical probability distributions on $(\mathbf{W}^d, \mathfrak{W}^d)$. Suppose that $X_0^{(1)}$ and $X_0^{(2)}$ have identical probability distribution μ on $(\mathbb{R}^d, \mathfrak{B}_{\mathbb{R}^d})$. We are to show that $X^{(1)}$ and $X^{(2)}$ have the identical probability distributions on $(\mathbf{W}^d, \mathfrak{W}^d)$, that is,

(1) $$P^{(1)} \circ (\mathcal{X}^{(1)})^{-1}(A) = P^{(1)} \circ (\mathcal{X}^{(2)})^{-1}(A) \quad \text{for } A \in \mathfrak{W}^d$$

where $(\mathcal{X}^{(q)})$ is the mapping of $\Omega^{(q)}$ into $\mathbf{W}^d$ defined by $(\mathcal{X}^{(q)})(\omega) = X^{(q)}(\cdot, \omega)$ for $\omega \in \Omega^{(q)}$. For the mapping $(\mathcal{B}^{(q)}, \mathcal{X}^{(q)})(\omega) = (B^{(q)}(\cdot, \omega), X^{(q)}(\cdot, \omega))$ for $\omega \in \Omega$ into $\mathbf{W}^{r+d}$ we have

(2) $$P^{(q)} \circ (\mathcal{X}^{(q)})^{-1}(A) = P^{(q)} \circ (\mathcal{B}^{(q)}, \mathcal{X}^{(q)})^{-1}(\mathbf{W}^r \times A) \quad \text{for } A \in \mathfrak{W}^d.$$

Now since the process $(W^{(q)}, Y^{(q)})$ on $(\mathbf{W}^{r+d}, \mathfrak{W}^{r+d,*,(q)}, \{\mathfrak{W}_t^{r+d,*,(q)}\}, P_{(B^{(q)}, X^{(q)})}^{(q)})$ and the process $(B^{(q)}, X^{(q)})$ on $(\Omega^{(q)}, \mathfrak{F}^{(q)}, \{\mathfrak{F}_t^{(q)}\}, P^{(q)})$ have identical probability distributions on $(\mathbf{W}^{r+d}, \mathfrak{W}^{r+d})$ according to Theorem 18.13, we have

(3) $$P^{(q)} \circ (\mathcal{B}^{(q)}, \mathcal{X}^{(q)})^{-1}(\mathbf{W}^r \times A) = P_{(B^{(q)}, X^{(q)})}^{(q)} \circ (\mathcal{W}^{(q)}, \mathcal{Y}^{(q)})^{-1}(\mathbf{W}^r \times A)$$
$$= P_{(B^{(q)}, X^{(q)})}^{(q)}(\mathbf{W}^r \times A) \quad \text{for } A \in \mathfrak{W}^d,$$

where the last equality is by the fact that $(\mathcal{W}^{(q)}, \mathcal{Y}^{(q)})(w) = (W^{(q)}(\cdot, w), Y^{(q)}(\cdot, w)) = w$ for $w \in \mathbf{W}^{r+d}$, that is $(\mathcal{W}^{(q)}, \mathcal{Y}^{(q)})$ is an identity mapping of $\mathbf{W}^{r+d}$ into $\mathbf{W}^{r+d}$. By (2) and (3), to prove (1) it suffices to show

(4) $$P_{(B^{(1)}, X^{(1)})}^{(1)}(\mathbf{W}^r \times A) = P_{(B^{(2)}, X^{(2)})}^{(2)}(\mathbf{W}^r \times A) \quad \text{for } A \in \mathfrak{W}^d.$$

§19. EXISTENCE AND UNIQUENESS OF SOLUTIONS

Now since $(W^{(q)}, Y^{(q)})$ and $(B^{(q)}, X^{(q)})$ have identical probability distributions on $(\mathbf{W}^{r+d}, \mathfrak{W}^{r+d})$, $Y^{(q)}$ and $X^{(q)}$ have identical probability distributions on $(\mathbf{W}^d, \mathfrak{W}^d)$. This implies according to Theorem 17.8 that $Y_0^{(q)}$ and $X_0^{(q)}$ have identical probability distributions on $(\mathbb{R}^d, \mathfrak{B}_{\mathbb{R}^d})$. Therefore both $Y_0^{(1)}$ and $Y_0^{(2)}$ have μ as their probability distributions on $(\mathbb{R}^d, \mathfrak{B}_{\mathbb{R}^d})$. According to Theorem 18.21, there exists a null set N in $(\mathbb{R}^d, \mathfrak{B}_{\mathbb{R}^d}, \mu)$ such that for every $x \in N^c$, $(W^{(q)}, Y^{(q)})$ are solutions of the stochastic differential equation on $(\mathbf{W}^{r+d}, \mathfrak{W}^{r+d,*,x,(q)}, \{\mathfrak{W}_t^{r+d,*,x,(q)}\}, P_{(B^{(q)}, X^{(q)})}^{(q),x})$ satisfying the condition $Y_0^{(q)} = x$ a.e. on $(\mathbf{W}^{r+d}, \mathfrak{W}^{r+d}, P_{(B^{(q)}, X^{(q)})}^{x,(q)})$. Therefore $Y^{(1)}$ and $Y^{(2)}$ have identical probability distribution on $(\mathbf{W}^d, \mathfrak{W}^d)$. Thus for $x \in N^c$ and $A \in \mathfrak{W}^d$, we have

$$P_{(B^{(1)}, X^{(1)})}^{(1),x} \circ (\mathcal{Y}^{(1)})^{-1}(A) = P_{(B^{(2)}, X^{(2)})}^{(2),x} \circ (\mathcal{Y}^{(2)})^{-1}(A),$$

and consequently

$$P_{(B^{(1)}, X^{(1)})}^{(1),x} \circ (\mathcal{W}^{(1)}, \mathcal{Y}^{(1)})^{-1}(\mathbf{W}^r \times A) = P_{(B^{(2)}, X^{(2)})}^{(2),x} \circ (\mathcal{W}^{(2)}, \mathcal{Y}^{(2)})^{-1}(\mathbf{W}^r \times A).$$

Since $(\mathcal{W}^{(q)}, \mathcal{Y}^{(q)})$ are identity mappings, the last equality reduces to

(5) $\qquad P_{(B^{(1)}, X^{(1)})}^{(1),x}(\mathbf{W}^r \times A) = P_{(B^{(2)}, X^{(2)})}^{(2),x}(\mathbf{W}^r \times A).$

But according to (2) of Observation 18.16, for every $x \in N^c$ we have

(6) $\qquad P_{(B^{(q)}, X^{(q)})}^{(q)}(\mathbf{W}^r \times A) = \int_{\mathbb{R}^d} P_{(B^{(q)}, X^{(q)})}^{(q),x}(\mathbf{W}^r \times A) \mu(dx) \quad \text{for } A \in \mathfrak{W}^d.$

Using (5) in (6), we have (4). ∎

Definition 19.3. *We say that the solution of the stochastic differential equation (1) in Definition 18.6 is pathwise unique if whenever $(B, X^{(1)})$ and $(B, X^{(2)})$ are two solutions on the same standard filtered space $(\Omega, \mathfrak{F}, \{\mathfrak{F}_t\}, P)$ such that $X_0^{(1)} = X_0^{(2)}$ a.e. on $(\Omega, \mathfrak{F}, P)$ then there exists a null set Λ in $(\Omega, \mathfrak{F}, P)$ such that $X^{(1)}(\cdot, \omega) = X^{(2)}(\cdot, \omega)$ for $\omega \in \Lambda^c$.*
We say that the solution satisfy the pathwise uniqueness condition under deterministic initial conditions if whenever $X_0^{(1)} = X_0^{(2)} = x$ a.e. on $(\Omega, \mathfrak{F}, P)$ for some $x \in \mathbb{R}^d$ then there exists a null set Λ in $(\Omega, \mathfrak{F}, P)$ such that $X^{(1)}(\cdot, \omega) = X^{(2)}(\cdot, \omega)$ for $\omega \in \Lambda^c$.

Unlike uniqueness in the sense of probability law, pathwise uniqueness assumes that $(B, X^{(1)})$ and $(B, X^{(2)})$ are two solutions on the same standard filtered space with the same Brownian motion B. Nevertheless pathwise uniqueness implies uniqueness in the sense of probability law. This will be shown in Lemma 19.19 below. Also the equivalence of pathwise uniqueness and pathwise uniqueness under deterministic initial conditions will be shown in Theorem 20.20. Let us define two conditions on the coefficients α and β of the stochastic differential equation. The first is a Lipschitz condition which will imply the pathwise uniqueness of solutions. The second is a growth condition which together with Lipschitz condition will imply the existence of solutions.

Definition 19.4. *We define two conditions on the coefficients α and β in the stochastic differential equation (1) in Definition 18.6 as follows: There exists a measure λ on $(\mathbb{R}_+, \mathfrak{B}_{\mathbb{R}_+})$ which is finite for every compact subset of $\mathbb{R}_+$ and satisfies the condition that for every $T \in \mathbb{R}_+$ there exists $L_T \in \mathbb{R}_+$ such that for $t \in [0, T]$ and $w_1, w_2 \in \mathbf{W}^d$ we have*

$$(1) \quad |\alpha(t, w_1) - \alpha(t, w_2)|^2 + |\beta(t, w_1) - \beta(t, w_2)|^2$$
$$\leq L_T \left\{ \int_{[0,t]} |w_1(s) - w_2(s)|^2 \lambda(ds) + |w_1(t) - w_2(t)|^2 \right\},$$

and for $t \in [0, T]$ and $w \in \mathbf{W}^d$ we have

$$(2) \quad |\alpha(t, w)|^2 + |\beta(t, w)|^2 \leq L_T \left\{ \int_{[0,t]} |w(s)|^2 \lambda(ds) + |w(t)|^2 + 1 \right\}.$$

Note that in the definition above, $|\cdot|$ is the generic Euclidean norm for all finite dimensional Euclidean spaces. Thus $|\alpha(t, w)| = \{\sum_{i=1}^{d} \sum_{j=1}^{r} |\alpha_j^i(t, w)|^2\}^{1/2}$ and $|\beta(t, w)| = \{\sum_{i=1}^{d} |\beta^i(t, w)|^2\}^{1/2}$.

Lemma 19.5. *Let Z be a nonnegative left-continuous adapted process on a standard filtered space $(\Omega, \mathfrak{F}, \{\mathfrak{F}_t\}, P)$ which is increasing in the sense that $Z_s \leq Z_t$ for any $s, t \in \mathbb{R}_+$ such that $s < t$. For $N \in \mathbb{N}$ and $t \in \mathbb{R}_+$, let*

$$(1) \quad A_t^N = \{\omega \in \Omega : Z(t, \omega) \leq N\}.$$

The process I^N on $(\Omega, \mathfrak{F}, \{\mathfrak{F}_t\}, P)$ defined by

$$(2) \quad I^N(t, \omega) = \mathbf{1}_{A_t^N}(\omega) \quad \text{for } (t, \omega) \in \mathbb{R}_+ \times \Omega$$

is a $\{0, 1\}$-valued decreasing left-continuous adapted process. For every $\omega \in \Omega$, the sample function $I^N(\cdot, \omega)$, if it is not identically equal to 0 or 1, is given by

$$(3) \quad I^N(t, \omega) = \begin{cases} 1 & \text{for } t \in [0, \tau(\omega)] \\ 0 & \text{for } t \in (\tau(\omega), \infty) \end{cases}$$

§19. EXISTENCE AND UNIQUENESS OF SOLUTIONS

for some $\tau(\omega) \in \mathbb{R}_+$.

Proof. Since Z is an increasing process, A_t^N decreases as t increases. Since Z is an adapted process, $A_t^N \in \mathfrak{F}_t$ for $t \in \mathbb{R}_+$. Thus $\mathbf{1}_{A_t^N}$ is an $\mathfrak{F}_t$-measurable random variable and I^N is an adapted process. Since $A_s^N \supset A_t^N$ for $s < t$, we have $\mathbf{1}_{A_s^N} \geq \mathbf{1}_{A_t^N}$ and thus I^N is a decreasing process. To show that every sample function of I^N is left-continuous, let $\omega \in \Omega$ and $t \in \mathbb{R}_+$ be fixed. If $I^N(t,\omega) = 1$, then $\mathbf{1}_{A_t^N}(\omega) = 1$ so that $\omega \in A_t^N$. This implies that $\omega \in A_s^N$ for every $s \in [0,t]$ so that $\mathbf{1}_{A_s^N}(\omega) = 1$, that is, $I^N(s,\omega) = 1$ for $s \in [0,t]$, proving the left-continuity of $I^N(\cdot,\omega)$ at t. If on the other hand $I^N(t,\omega) = 0$, then $\mathbf{1}_{A_t^N}(\omega) = 0$, that is, $\omega \notin A_t^N$. This implies by (1) that $Z(t,\omega) > N$. Then by the left-continuity of $Z(\cdot,\omega)$ there exists $\delta > 0$ such that $Z(s,\omega) > N$ for $s \in (t-\delta,t]$. Thus $\omega \notin A_s^N$ and $\mathbf{1}_{A_s^N}(\omega) = 0$, that is, $I^N(s,\omega) = 0$ for $s \in (t-\delta,t]$, proving the left-continuity of $I^N(\cdot,\omega)$ at t. This shows the left-continuity of every sample function of I^N. Now since every sample function of I^N is a $\{0,1\}$-valued, left-continuous and decreasing function on $\mathbb{R}_+$, it is given by (3) unless it is identically equal to 0 or 1 on $\mathbb{R}_+$. ∎

Lemma 19.6. *Let J be an adapted process on a standard filtered space $(\Omega, \mathfrak{F}, \{\mathfrak{F}_t\}, P)$ such that every sample function is of the form*

$$(1) \qquad J(t,\omega) = \begin{cases} 1 & \text{for } t \in [0, \tau(\omega)] \\ 0 & \text{for } t \in (\tau(\omega), \infty) \end{cases}$$

for some $\tau(\omega) \in \mathbb{R}_+$ unless $J(\cdot, \omega)$ is identically equal to 0 or 1 on $\mathbb{R}_+$. Then

$$(2) \qquad \begin{cases} J(t,\omega)^2 = J(t,\omega) & \text{for } t \in \mathbb{R}_+ \text{ and } \omega \in \Omega \\ J(s,\omega)J(t,\omega) = J(s,\omega) & \text{for } s,t \in \mathbb{R}_+, s<t, \text{ and } \omega \in \Omega. \end{cases}$$

For $M \in \mathbf{M}_2(\Omega, \mathfrak{F}, \{\mathfrak{F}_t\}, P)$, $X \in \mathbf{L}_{1,\infty}(\mathbb{R}_+ \times \Omega, \mu_{[M]}, P)$ and $t \in \mathbb{R}_+$, we have

$$(3) \qquad J(t)|(X \bullet m_L)(t)| \leq |(JX \bullet m_L)(t)| \quad \text{on } \Omega.$$

If X is a predictable process and is in $\mathbf{L}_{2,\infty}(\mathbb{R}_+ \times \Omega, \mu_{[M]}, P)$, then there exists a null set Λ in $(\Omega, \mathfrak{F}, P)$ such that

$$(4) \qquad J(t)|(X \bullet M)(t)| \leq |(JX \bullet M)(t)| \quad \text{on } \Lambda^c.$$

Proof. The equalities in (2) are immediate form the properties of the sample functions of J. Regarding (3) and (4), let us note that since J is a left-continuous adapted process, it is a predictable process and in particular a measurable process. Thus if X is a process in $\mathbf{L}_{1,\infty}(\mathbb{R}_+ \times \Omega, \mu_{[M]}, P)$, then the measurability and the boundedness of J imply that JX too is in $\mathbf{L}_{1,\infty}(\mathbb{R}_+ \times \Omega, \mu_{[M]}, P)$. Similarly, if X is a predictable process and is in $\mathbf{L}_{2,\infty}(\mathbb{R}_+ \times \Omega, \mu_{[M]}, P)$, then so is JX.

To prove (3), note that if $J(t, \omega) = 0$, then

$$J(t,\omega) | \int_{[0,t]} X(s,\omega) m_L(ds) | \leq | \int_{[0,t]} J(s,\omega) X(s,\omega) m_L(ds) |$$

and if $J(t, \omega) = 1$, then $J(s, \omega) = 1$ for $s \in [0, t]$ so that

$$J(t,\omega) | \int_{[0,t]} X(s,\omega) m_L(ds) | \leq | \int_{[0,t]} J(s,\omega) X(s,\omega) m_L(ds) |.$$

Therefore (3) holds.

To prove (4), let us consider first the case where X is a bounded left-continuous adapted process. In this case JX too is a bounded left-continuous process. Thus by Proposition 12.16 and by the fact that convergence in probability implies the existence of a subsequence which converges almost surely, for $t \in \mathbb{R}_+$ there exists a null set Λ in $(\Omega, \mathfrak{F}, P)$ such that

(5) $$\lim_{n \to \infty} \sum_{k=1}^{p_n} X(t_{n,k-1}) \{M(t_{n,k}) - M(t_{n,k-1})\} = (X \bullet M)(t) \quad \text{on } \Lambda^c$$

and

(6) $$\lim_{n \to \infty} \sum_{k=1}^{p_n} J(t_{n,k-1}) X(t_{n,k-1}) \{M(t_{n,k}) - M(t_{n,k-1})\} = (JX \bullet M)(t) \quad \text{on } \Lambda^c$$

where $\{t_{n,k} : k \in \mathbb{Z}_+\}$ is a strictly increasing sequence in $\mathbb{R}_+$ with $t_{n,0} = 0$ and $t_{n,k} \uparrow \infty$ as $k \to \infty$ for each $n \in \mathbb{N}$ and $\lim_{n \to \infty} \sup_{k \in \mathbb{N}}(t_{n,k} - t_{n,k-1}) = 0$, $t_{n,p_n} = t$ and $t_{n,k} < t$ for $k = 0, \ldots, p_n - 1$ for $n \in \mathbb{N}$. From (5) we have

(7) $$\lim_{n \to \infty} J(t) \sum_{k=1}^{p_n} X(t_{n,k-1}) \{M(t_{n,k}) - M(t_{n,k-1})\} = J(t)(X \bullet M)(t) \quad \text{on } \Lambda^c.$$

Let $\omega \in \Lambda^c$ be fixed. If $J(\omega) = 1$, then $J(t_{n,k-1}) = 1$ for $k = 1, \ldots, p_n$ so that the left side of (7) and that of (6) are identical and thus we have $J(t,\omega)(X \bullet M)(t,\omega) = (JX \bullet M)(t,\omega)$ by (7) and (6). On the other hand if $J(t,\omega) = 0$, then $J(t,\omega)|(X \bullet M)(t,\omega)| \leq |(JX \bullet M)(t,\omega)|$. Thus (4) holds in this case.

§19. EXISTENCE AND UNIQUENESS OF SOLUTIONS

For the general case where X is a predictable process and is in $\mathbf{L}_{2,\infty}(\mathbb{R}_+ \times \Omega, \mu_{[M]}, P)$ there exists by Theorem 12.8 a sequence $\{X^{(n)} : n \in \mathbb{N}\}$ in $\mathbf{L}_0(\Omega, \mathfrak{F}, \{\mathfrak{F}_t\}, P)$ such that $\lim_{n\to\infty} \|X^{(n)} - X\|_{2,\infty}^{[M],P} = 0$ and thus $\lim_{n\to\infty} \mathbf{I} X^{(n)} \bullet M - X \bullet M \mathbf{I}_\infty = 0$ by Definition 12.9. Now $\lim_{n\to\infty} \|X^{(n)} - X\|_{2,\infty}^{[M],P} = 0$ implies that $\lim_{n\to\infty} \|JX^{(n)} - JX\|_{2,\infty}^{[M],P} = 0$ as can be verified easily. This convergence implies $\lim_{n\to\infty} \mathbf{I} JX^{(n)} \bullet M - JX \bullet M \mathbf{I}_\infty = 0$ according to Proposition 12.13. Now since $X^{(n)}$ are bounded left-continuous adapted processes, there exists according to our result above a null set Λ_0 in $(\Omega, \mathfrak{F}, P)$ for our $t \in \mathbb{R}$ and all $n \in \mathbb{N}$ such that we have

(8) $$J(t)|(X^{(n)} \bullet M)(t)| \leq |(JX^{(n)} \bullet M)(t)| \quad \text{on } \Lambda_0^c.$$

Now $\lim_{n\to\infty} \mathbf{I} X^{(n)} \bullet M - X \bullet M \mathbf{I}_\infty = 0$ implies $\lim_{n\to\infty} \mathbf{E}(|X^{(n)} \bullet M - X \bullet M|^2) = 0$ according to Remark 11.5 and thus the existence of a subsequence $\{n'\}$ of $\{n\}$ such that $\lim_{n'\to\infty} X^{(n')} \bullet M = X \bullet M$ on Λ_1^c where Λ_1 is a null set in $(\Omega, \mathfrak{F}, P)$. Similarly the convergence $\lim_{n'\to\infty} \mathbf{I} JX^{(n')} \bullet M - JX \bullet M \mathbf{I}_\infty = 0$ implies the existence of a subsequence $\{n''\}$ of $\{n'\}$ such that $\lim_{n''\to\infty} JX^{(n'')} \bullet M = JX \bullet M$ on Λ_2^c where Λ_2 is a null set in $(\Omega, \mathfrak{F}, P)$. Let $\Lambda = \cup_{i=0}^{2} \Lambda_i$. Now (8) holds for every n'' on Λ^c. Letting $n'' \to \infty$ we have (4). ∎

Lemma 19.7. *Let c be a bounded nonnegative Lebesgue measurable function on a finite interval $[0, T]$. If for some $a, b \geq 0$ we have*

(1) $$c(t) \leq a + b \int_{[0,t]} c(s) m_L(ds) \quad \text{for } t \in [0, T]$$

then

(2) $$c(t) \leq a e^{bt} \quad \text{for } t \in [0, T].$$

Proof. By substituting the inequality (1) into its right side and by repeating this substitution we obtain for every $n \in \mathbb{N}$ the inequality

$$c(t) \leq a \sum_{k=0}^{n} \frac{(bt)^k}{k!} + b^{n+1} \int_{[0,t]} \int_{[0,t_1]} \cdots \int_{[0,t_n]} c(t_{n+1}) m_L(dt_n) \cdots m_L(dt_2) m_L(dt_1).$$

This inequality can then be proved by mathematical induction on n. Let $M \in \mathbb{R}_+$ be a bound of c on $[0, T]$. Then

$$c(t) \leq a \sum_{k=0}^{n} \frac{(bt)^k}{k!} + \frac{(bt)^{n+1}}{(n+1)!} M \quad \text{for } t \in [0, T].$$

Letting $n \to \infty$, we have (2). ∎

Theorem 19.8. *If the coefficients α and β in the stochastic differential equation (1) in Definition 18.6 satisfy the Lipschitz condition (1) in Definition 19.4, the solution, if it exists, is pathwise unique.*

Proof. Suppose $(B, X^{(1)})$ and $(B, X^{(2)})$ are two solutions on the same standard filtered space $(\Omega, \mathfrak{F}, \{\mathfrak{F}_t\}, P)$ such that $X_0^{(1)} = X_0^{(2)}$ a.e. on $(\Omega, \mathfrak{F}, P)$ Assume that α and β satisfy the Lipschitz condition. We are to show that there exists a null set Λ in $(\Omega, \mathfrak{F}, P)$ such that $X^{(1)}(\cdot, \omega) = X^{(2)}(\cdot, \omega)$ for $\omega \in \Lambda^c$. Now since $X^{(1)}$ and $X^{(2)}$ are continuous processes, it suffices to show that for every $t \in \mathbb{R}_+$ there exists a null set Λ_t in $(\Omega, \mathfrak{F}, P)$ such that $X^{(1)} = X^{(2)}$ on Λ_t^c. To show this we show equivalently that for every $t \in \mathbb{R}_+$ we have

(1) $$P\{|X_t^{(1)} - X_t^{(2)}| > 0\} = 0.$$

Now since $(B, X^{(1)})$ and $(B, X^{(2)})$ are two solutions and $X_0^{(1)} = X_0^{(2)}$ a.e. on $(\Omega, \mathfrak{F}, P)$, condition 4° of Definition 16.8 implies that for $i = 1, \ldots, d$ we have

$$X_t^{(1),i} - X_t^{(2),i} = \sum_{j=1}^{r}(\{\alpha_{X^{(1)}} - \alpha_{X^{(2)}}\}_j^i \bullet B^j)_t + (\{\beta_{X^{(1)}} - \beta_{X^{(2)}}\}^i \bullet m_L)_t.$$

For brevity in notation, let us define

(2) $$\begin{cases} \Delta_\alpha(t, \omega) = \alpha_{X^{(1)}}(t, \omega) - \alpha_{X^{(2)}}(t, \omega) = \alpha(t, X^{(1)}(\cdot, \omega)) - \alpha(t, X^{(2)}(\cdot, \omega)) \\ \Delta_\beta(t, \omega) = \beta_{X^{(1)}}(t, \omega) - \beta_{X^{(2)}}(t, \omega) = \beta(t, X^{(1)}(\cdot, \omega)) - \beta(t, X^{(2)}(\cdot, \omega)) \end{cases}$$

for $(t, \omega) \in \mathbb{R}_+ \times \Omega$. We write $(\Delta_\alpha)_j^i$ and $(\Delta_\beta)^i$ for the component processes of Δ_α and Δ_β. Then

$$X_t^{(1),i} - X_t^{(2),i} = \sum_{j=1}^{r}((\Delta_\alpha)_j^i \bullet B^j)_t + ((\Delta_\beta)^i \bullet m_L)_t.$$

For any real numbers $c_1, \ldots, c_n$, we have $\{\sum_{j=1}^n c_j\}^2 \leq n \sum_{j=1}^n c_j^2$. Thus

(3) $$|X_t^{(1),i} - X_t^{(2),i}|^2 \leq (r+1)\left\{\sum_{j=1}^{r}((\Delta_\alpha)_j^i \bullet B^j)_t^2 + ((\Delta_\beta)^i \bullet m_L)_t^2\right\}.$$

Let us define a nonnegative continuous increasing process Z by setting

$$Z(t, \omega) = \sup_{s \in [0,t]} \{|X^{(1)}(s, \omega)|^2 + |X^{(2)}(s, \omega)|^2\} \quad \text{for } (t, \omega) \in \mathbb{R}_+ \times \Omega.$$

§19. EXISTENCE AND UNIQUENESS OF SOLUTIONS

By the continuity of $X^{(1)}$ and $X^{(2)}$, $\sup_{s\in[0,t]}$ is equal to $\sup_{s\in[0,t]\cap Q}$ where Q is the countable collection of the rational numbers and thus $Z(t)$ is $\mathfrak{F}$-measurable. Actually since $X^{(1)}$ and $X^{(2)}$ are adapted processes, $Z(t)$ is $\mathfrak{F}_t$-measurable and thus Z is an adapted process. The continuity of $X^{(1)}$ and $X^{(2)}$ also implies that of Z. As in Lemma 19.5, for $N \in \mathbb{N}$ and $t \in \mathbb{R}_+$ let $A_t^N = \{\omega \in \Omega : Z(t,\omega) \leq N\}$ and define a $\{0,1\}$-valued decreasing left-continuous adapted process by setting $I^N(t,\omega) = \mathbf{1}_{A_t^N}(\omega)$ for $(t,\omega) \in \mathbb{R}_+ \times \Omega$. By Lemma 19.6, for every $t \in \mathbb{R}_+$ we have

$$\begin{cases} I^N(t)|((\Delta_\alpha)_j^i \bullet B^j)(t)| \leq |(I^N(\Delta_\alpha)_j^i \bullet B^j)(t)| \\ I^N(t)|((\Delta_\beta)^i \bullet m_L)(t)| \leq |(I^N(\Delta_\beta)^i \bullet m_L)(t)|. \end{cases}$$

Multiplying both sides of (3) by $I^N(t)$, using the inequalities above and the fact that $I^N(t)^2 = I^N(t)$, we have

$$I^N(t)|X_t^{(1),i} - X_t^{(2),i}|^2 \leq (r+1)\left\{\sum_{j=1}^r (I^N(\Delta_\alpha)_j^i \bullet B^j)_t^2 + (I^N(\Delta_\beta)^i \bullet m_L)_t^2\right\}$$

and hence

(4). $\mathbf{E}[I^N(t)|X_t^{(1),i} - X_t^{(2),i}|^2] \leq (r+1)\left\{\sum_{j=1}^r \mathbf{E}[(I^N(\Delta_\alpha)_j^i \bullet B^j)_t^2] + \mathbf{E}[(I^N(\Delta_\beta)^i \bullet m_L)_t^2]\right\}$

By Observation 12.11,

$$\mathbf{E}[(I^N(\Delta_\alpha)_j^i \bullet B^j)_t^2] = \mathbf{I}\,I^N(\Delta_\alpha)_j^i \bullet B^j\mathbf{I}_t = \|I^N(\Delta_\alpha)_j^i\|_{2,t}^{m_L \times P}$$
$$= \int_{[0,t]} \mathbf{E}[I^N(s)|(\Delta_\alpha)_j^i(s)|^2] m_L(ds) \leq \int_{[0,t]} \mathbf{E}[I^N(s)|\Delta_\alpha(s)|^2] m_L(ds)$$

for $i = 1,\ldots,d$ and $j = 1,\ldots,r$. Also, applying the Schwarz Inequality we have

$$\mathbf{E}[(I^N(\Delta_\beta)^i \bullet m_L)_t^2] = \mathbf{E}[|\int_{[0,t]} I^N(s)(\Delta_\beta)^i(s) m_L(ds)|^2]$$
$$\leq \mathbf{E}[t\int_{[0,t]} I^N(s)|(\Delta_\beta)^i(s)|^2 m_L(ds)] \leq t\int_{[0,t]} \mathbf{E}[I^N(s)|(\Delta_\beta)(s)|^2] m_L(ds)$$

for $i = 1,\ldots,d$. Using these estimates in (4), we have

(5) $\mathbf{E}[I^N(t)|X_t^{(1)} - X_t^{(2)}|^2] = \sum_{j=1}^d \mathbf{E}[I^N(t)|X_t^{(1),i} - X_t^{(2),i}|^2]$

$\leq d(r+1)(r+t)\int_{[0,t]} \mathbf{E}[I^N(s)\{|\Delta_\alpha(s)|^2 + |\Delta_\beta(s)|^2\}]m_L(ds).$

Let $T \in \mathbb{R}_+$ be arbitrarily fixed. By the Lipschitz condition we have for $s \in [0,t] \subset [0,T]$ and $\omega \in \Omega$

$$\begin{aligned}&|\Delta_\alpha(s,\omega)|^2 + |\Delta_\beta(s,\omega)|^2 \\ &= |\alpha(t, X^{(1)}(\cdot,\omega)) - \alpha(t, X^{(2)}(\cdot,\omega))|^2 + |\beta(t, X^{(1)}(\cdot,\omega)) - \beta(t, X^{(2)}(\cdot,\omega))|^2 \\ &\leq L_T \left\{ \int_{[0,s]} |X^{(1)}(u,\omega) - X^{(2)}(u,\omega)|^2 \lambda(du) + |X^{(1)}(s,\omega) - X^{(2)}(s,\omega)|^2 \right\}\end{aligned}$$

Substituting this in (5) and recalling $I^N(s) \leq I^N(u)$ for $u \in [0,s]$,

(6) $\quad \mathbf{E}[I^N(t)|X_t^{(1)} - X_t^{(2)}|^2]$
$$\begin{aligned}&\leq d(r+1)(r+T)L_T \int_{[0,t]} \left\{ \int_{[0,s]} \mathbf{E}[I^N(u)|X^{(1)}(u) - X^{(2)}(u)|^2]\lambda(du) \right\} m_L(ds) \\ &+ d(r+1)(r+T)L_T \int_{[0,t]} \mathbf{E}[I^N(s)|X^{(1)}(s) - X^{(2)}(s)|^2] m_L(ds).\end{aligned}$$

Since $|X^{(1)}(u) - X^{(2)}(u)|^2 \leq 2\{|X^{(1)}(u)|^2 + |X^{(2)}(u)|^2\} \leq 2Z(u)$ and $I^N(u) = \mathbf{1}_{A_u^N}$ and $A_u^N = \{\omega \in \Omega : 2Z(u) \leq 2N\}$, we have $I^N(u)|X^{(1)}(u) - X^{(2)}(u)|^2 \leq 2N$. Thus

(7) $\quad \mathbf{E}[I^N(t)|X_t^{(1)} - X_t^{(2)}|^2] \leq d(r+1)(r+T)L_T 2N\{\lambda([0,T]) + 1\}T \quad \text{for } t \in [0,T].$

Define a function c on $[0,T]$ by setting

(8) $\quad c(t) = \sup_{s \in [0,t]} \mathbf{E}[I^N(s)|X^{(1)}(s) - X^{(2)}(s)|^2] \quad \text{for } t \in [0,T].$

By (7), c is bounded on $[0,T]$. From (6) and (8) we have

$$\begin{aligned}\mathbf{E}[I^N(t)|X_t^{(1)} - X_t^{(2)}|^2] &\leq d(r+1)(r+T)L_T \int_{[0,t]} \{c(s)\lambda([0,s]) + c(s)\} m_L(ds) \\ &\leq d(r+1)(r+T)L_T \{\lambda([0,T]) + 1\} \int_{[0,t]} c(s) m_L(ds) \quad \text{for } t \in [0,T].\end{aligned}$$

Substituting this in the right side of (8), we have

$$c(t) \leq d(r+1)(r+T)L_T\{\lambda([0,T]) + 1\} \int_{[0,t]} c(s) m_L(ds) \quad \text{for } t \in [0,T].$$

Applying Lemma 19.7 with $a = 0$ and $b = d(r+1)(r+T)L_T\{\lambda([0,T])+1\}$, we have $c(t) = 0$ for $t \in [0,T]$. Then by (8), $\sup_{s \in [0,t]} \mathbf{E}[I^N(s)|X^{(1)}(s) - X^{(2)}(s)|^2] = 0$ for $t \in [0,T]$ and in particular $\mathbf{E}[I^N(t)|X^{(1)}(t) - X^{(2)}(t)|^2] = 0$ for $t \in [0,T]$ and then for every $t \in \mathbb{R}_+$ by the

§19. EXISTENCE AND UNIQUENESS OF SOLUTIONS   405

arbitrariness of $T \in \mathbb{R}_+$. Thus for every $t \in \mathbb{R}_+$ we have $\mathbf{E}[1_{A_t^N}|X^{(1)}(t) - X^{(2)}(t)|^2] = 0$. This implies
$$\int_{A_t^N \cap \{|X^{(1)}(t) - X^{(2)}(t)| > 0\}} |X^{(1)}(t) - X^{(2)}(t)|^2 \, dP = 0.$$

Thus $P(A_t^N \cap \{|X^{(1)}(t) - X^{(2)}(t)| > 0\}) = 0$ and then

(9) $\qquad P\{|X^{(1)}(t) - X^{(2)}(t)| > 0\} \leq P((A_t^N)^c) \quad \text{for } N \in \mathbb{N}.$

Since $A_t^N \uparrow$ as $N \to \infty$, $(A_t^N)^c \downarrow$ and $P((A_t^N)^c) \downarrow$ as $N \to \infty$. To show $P((A_t^N)^c) \downarrow 0$ as $N \to \infty$, assume the contrary, that is, there exists $\varepsilon > 0$ such that $P((A_t^N)^c) \geq \varepsilon$ for all $N \in \mathbb{N}$. Then $P(\cap_{N \in \mathbb{N}}(A_t^N)^c) = \lim_{N \to \infty} P((A_t^N)^c) \geq \varepsilon$, that is,

$$P\{\omega \in \Omega : \sup_{s \in [0,t]} \{|X^{(1)}(s,\omega)|^2 + |X^{(2)}(s,\omega)|^2 > N \text{ for all } N\} \geq \varepsilon,$$

in other words,

$$P\{\omega \in \Omega : \sup_{s \in [0,t]} \{|X^{(1)}(s,\omega)|^2 + |X^{(2)}(s,\omega)|^2 = \infty\} \geq \varepsilon,$$

which is impossible since the continuity of $|X^{(1)}|^2 + |X^{(2)}|^2$ implies that for every $\omega \in \Omega$ we have $\sup_{s \in [0,t]}\{|X^{(1)}(s,\omega)|^2 + |X^{(2)}(s,\omega)|^2\} < \infty$. This shows that $P((A_t^N)^c) \downarrow 0$ as $N \to \infty$. Using this in (9), we have (1). ∎

[II] Simultaneous Representation of Two Solutions on a Function Space

Our immediate goal here is to show that any two solutions of the stochastic differential equation can be represented on one function space with a common Brownian motion on it. We need this to show that pathwise uniqueness implies uniqueness in the sense of probability law. To be specific, if $(B^{(q)}, X^{(q)})$, $q = 1, 2$, are two solutions of the stochastic differential equation (1) in Definition 18.6 on two standard filtered spaces $(\Omega^{(q)}, \mathfrak{F}^{(q)}, \{\mathfrak{F}_t^{(q)}\}, P^{(q)})$, $q = 1, 2$, then we can introduce a probability measure and a filtration to the measurable space $(\mathbf{W}^{r+2d}, \mathfrak{W}^{r+2d})$ in such a way that three processes W and $Y^{(q)}$, $q = 1, 2$, on the resulting standard filtered space defined by $W(t, w) = v(t)$, $Y^{(1)}(t, w) = v'(t)$ and $Y^{(2)}(t, w) = v''(t)$ for $(t, w) \in \mathbb{R}_+ \times \mathbf{W}^{r+2d}$ where $w = (v, v', v'')$ with $v \in \mathbf{W}^r$ and $v', v'' \in \mathbf{W}^d$ are such that $(W, Y^{(q)})$ is a solution of the stochastic differential equation for $q = 1, 2$ and $(W, Y^{(1)})$ and $(W, Y^{(2)})$ have identical probability distributions on $(\mathbf{W}^{r+d}, \mathfrak{W}^{r+d})$.

Observation 19.9. Let (B, X) be a solution of the stochastic differential equation (1) in Definition 18.6 on a standard filtered space $(\Omega, \mathfrak{F}, \{\mathfrak{F}_t\}, P)$. Consider the probability space $(\mathbf{W}^{r+d}, \mathfrak{W}^{r+d}, P_{(B,X)})$ where $P_{(B,X)}$ is the probability distribution of (B, X) on $(\mathbf{W}^{r+d}, \mathfrak{W}^{r+d})$. Let π_0 be the projection of $\mathbf{W}^{r+d}$ onto $\mathbf{W}^r$. The probability distribution $(P_{(B,X)})_{\pi_0}$ of π_0 on $(\mathbf{W}^r, \mathfrak{W}^r)$ is given by

(1) $\quad (P_{(B,X)})_{\pi_0} = P_{(B,X)} \circ \pi_0^{-1} = P \circ (\mathcal{B}, \mathcal{X})^{-1} \circ \pi_0^{-1}$
$\quad = P \circ (\pi_0 \circ (\mathcal{B}, \mathcal{X}))^{-1} = P \circ \mathcal{B}^{-1} = P_\mathcal{B} = m_W^r$

by Definition 17.9.

Definition 19.10. On the r-dimensional Wiener space $(\mathbf{W}^r, \mathfrak{W}^r, m_W^r)$, we define a filtered space $(\mathbf{W}^r, \mathfrak{W}^{r,w}, \{\mathfrak{W}_t^{r,w}\}, m_W^r)$ by letting $\mathfrak{W}^{r,w}$ be the completion of $\mathfrak{W}^r$ with respect to the Wiener measure m_W^r and letting $\mathfrak{W}_t^{r,w} = \sigma(\mathfrak{W}_t^r \cup \mathfrak{N})$ for every $t \in \mathbb{R}_+$ where $\mathfrak{N}$ is the collection of all the null sets in $(\mathbf{W}^r, \mathfrak{W}^{r,w}, m_W^r)$.

Lemma 19.11. $\{\mathfrak{W}_t^{r,w} : t \in \mathbb{R}_+\}$ is a right-continuous filtration on $(\mathbf{W}^r, \mathfrak{W}^{r,w}, m_W^r)$ and $(\mathbf{W}^r, \mathfrak{W}^{r,w}, \{\mathfrak{W}_t^{r,w}\}, m_W^r)$ is a standard filtered space.

Proof. The stochastic process W on $(\mathbf{W}^r, \mathfrak{W}^r, m_W^r)$ defined by $W(t, w) = w(t)$ for $(t, w) \in \mathbb{R}_+ \times \mathbf{W}^r$ is an r-dimensional null at 0 Brownian motion on $(\mathbf{W}^r, \mathfrak{W}^r, m_W^r)$ by Theorem 17.10. Since the completion of $(\mathbf{W}^r, \mathfrak{W}^r, m_W^r)$ to $(\mathbf{W}^r, \mathfrak{W}^{r,w}, m_W^r)$ has no effect on the probability distributions of random vectors defined on $(\mathbf{W}^r, \mathfrak{W}^r, m_W^r)$, our W remains an r-dimensional null at 0 Brownian motion on $(\mathbf{W}^r, \mathfrak{W}^{r,w}, m_W^r)$. Thus by Proposition 13.22, $\sigma\{\sigma\{W_s : s \in [0,t]\} \cup \mathfrak{N}\}$ for $t \in \mathbb{R}_+$ is a right-continuous filtration on $(\mathbf{W}^r, \mathfrak{W}^{r,w}, m_W^r)$. But according to Proposition 17.7, $\sigma\{W_s : s \in [0,t]\} = \mathfrak{W}_t^r$ for $t \in \mathbb{R}_+$. Thus $\mathfrak{W}_t^{r,w} \equiv \sigma(\mathfrak{W}_t^r \cup \mathfrak{N})$ for $t \in \mathbb{R}_+$ is a right-continuous filtration. Since $\mathfrak{W}_t^{r,w}$ is augmented for every $t \in \mathbb{R}_+$, $(\mathbf{W}^r, \mathfrak{W}^{r,w}, \{\mathfrak{W}_t^{r,w}\}, m_W^r)$ is a standard filtered space. ∎

Let $(\mathbf{W}^{r+d}, \mathfrak{W}^{r+d,*}, \{\mathfrak{W}_t^{r+d,*}\}, P_{(B,X)})$ be the standard filtered space generated by $P_{(B,X)}$ as in Definition 18.10. On the probability space $(\mathbf{W}^{r+d}, \mathfrak{W}^{r+d,*}, P_{(B,X)})$, consider a regular image conditional probability given the projection π_0 of $\mathbf{W}^{r+d}$ onto $\mathbf{W}^r$, $P_{(B,X)}^{\mathfrak{W}^{r+d,*}} |^{\pi_0}(A, v)$ for $(A, v) \in \mathfrak{W}^{r+d,*} \times \mathbf{W}^r$. (See Definiton C.11.) For brevity let us write $Q^v(A)$ for $P_{(B,X)}^{\mathfrak{W}^{r+d,*}} |^{\pi_0}(A, v)$. Then

1°. there exists a null set N in $(\mathbf{W}^r, \mathfrak{W}^{r,w}, m_W^r)$ such that Q^v is a probability measure on $(\mathbf{W}^{r+d}, \mathfrak{W}^{r+d,*})$ for $v \in N^c$,

§19. EXISTENCE AND UNIQUENESS OF SOLUTIONS

2°. $Q^{(\cdot)}(A)$ is a $\mathfrak{W}^{r,w}$-measurable function on $\mathbf{W}^r$ for every $A \in \mathfrak{W}^{r+d,*}$,
3°. for every $A \in \mathfrak{W}^{r+d,*}$ and $A_0 \in \mathfrak{W}^{r,w}$ we have

$$P_{(B,X)}(A \cap \pi_0^{-1}(A_0)) = \int_{A_0} Q^v(A) m_W^r(dv).$$

In particular if $A \in \mathfrak{W}^{r+d,*}$ is of the type $A = \mathbf{W}^r \times A_1$ with $A_1 \in \mathfrak{W}^d$, then $A \cap \pi_0^{-1}(A_0) = (\mathbf{W}^r \times A_1) \cap (A_0 \times \mathbf{W}^d) = A_0 \times A_1$. Thus by 3° we have

4°. for $A_0 \in \mathfrak{W}^{r,w}$ and $A_1 \in \mathfrak{W}^d$

$$P_{(B,X)}(A_0 \times A_1) = P_{(B,X)}((\mathbf{W}^r \times A_1) \cap (A_0 \times \mathbf{W}^d)) = \int_{A_0} Q^v(\mathbf{W}^r \times A_1) m_W^r(dv).$$

Let an element in $\mathbf{W}^{r+d}$ be arbitrarily chosen and redefine $Q^v(\cdot)$ to be a unit mass at the arbirarily chosen element for $v \in N$. With this redefinition $Q^v(\cdot)$ is a probability measure on $\mathfrak{W}^{r+d,*}$ for every $v \in \mathbf{W}^r$. Property 2° is unaffected by this redefinition.

Lemma 19.12. *For every $A_1 \in \mathfrak{W}^d$, $Q^{(\cdot)}(\mathbf{W}^r \times A_1)$ is a $\mathfrak{W}^{r,w}$-measurable function on $\mathbf{W}^r$. Furthermore if $A_1 \in \mathfrak{W}_t^d$ for some $t \in \mathbb{R}_+$, then $Q^{(\cdot)}(\mathbf{W}^r \times A_1)$ is a $\mathfrak{W}_t^{r,w}$-measurable function on $\mathbf{W}^r$.*

Proof. If $A_1 \in \mathfrak{W}^d$, then $\mathbf{W}^r \times A_1 \in \mathfrak{W}^{r+d} \subset \mathfrak{W}^{r+d,*}$ so that by 2° above $Q^{(\cdot)}(\mathbf{W}^r \times A_1)$ is a $\mathfrak{W}^{r,w}$-measurable function on $\mathbf{W}^r$.

Suppose $A_1 \in \mathfrak{W}_t^d$ for some $t \in \mathbb{R}_+$. Then $A_1 \in \mathfrak{W}^d$ and thus $Q^{(\cdot)}(\mathbf{W}^r \times A_1)$ is a $\mathfrak{W}^{r,w}$-measurable function on $\mathbf{W}^r$. The projection π_0 of $\mathbf{W}^{r+d}$ onto $\mathbf{W}^r$ is a $\mathfrak{W}^{r,w} \times \mathfrak{W}^d / \mathfrak{W}^{r,w}$-measurable mapping of $\mathbf{W}^{r+d}$ into $\mathbf{W}^r$. Thus the composite mapping $Q^{\pi_0(\cdot)}(\mathbf{W}^r \times A_1)$ is a $\mathfrak{W}^{r,w} \times \mathbf{W}^d$-measurable function on $\mathbf{W}^{r+d}$. By 4° above and (1) in Observation 19.9 and by the Image Probability Law we have

(1) $$P_{(B,X)}((\mathbf{W}^r \times A_1) \cap (A_0 \times \mathbf{W}^d)) = \int_{A_0 \times \mathbf{W}^d} Q^{\pi_0(w)}(\mathbf{W}^r \times A_1) P_{(B,X)}(dw)$$

for $A_0 \in \mathfrak{W}^{r,w}$ and $A_1 \in \mathfrak{W}_t^d$. In the probability space $(\mathbf{W}^{r+d}, \mathfrak{W}^{r+d,*}, P_{(B,X)})$ consider the conditional probability $P_{(B,X)}(\mathbf{W}^r \times A_1 | \mathfrak{W}^{r,w} \times \mathbf{W}^d)$ of the set $\mathbf{W}^r \times A_1 \in \mathbf{W}^{r+d,*}$ given the sub-σ-algebra $\mathfrak{W}^{r,w} \times \mathbf{W}^d$ of $\mathfrak{W}^{r+d,*}$. By the $\mathfrak{W}^{r,w} \times \mathbf{W}^d$-measurability of $Q^{\pi_0(\cdot)}(\mathbf{W}^r \times A_1)$ and by (1), we have

(2) $$Q^{\pi_0(\cdot)}(\mathbf{W}^r \times A_1) \in P_{(B,X)}(\mathbf{W}^r \times A_1 | \mathfrak{W}^{r,w} \times \mathbf{W}^d),$$

that is, $Q^{\pi_0(\cdot)}(\mathbf{W}^r \times A_1)$ is a version of $P_{(B,X)}(\mathbf{W}^r \times A_1 | \mathfrak{W}^{r,w} \times \mathbf{W}^d)$.

Let us show that for our $A_1 \in \mathfrak{W}_t^d$ we have

(3) $\quad P_{(B,X)}(\mathbf{W}^r \times A_1 | \mathfrak{W}_t^{r,w} \times \mathbf{W}^d) = P_{(B,X)}(\mathbf{W}^r \times A_1 | \mathfrak{W}^{r,w} \times \mathbf{W}^d).$

Now if $\mathfrak{A}_1$, $\mathfrak{A}_2$, and $\mathfrak{A}_3$ are sub-σ-algebras of $\mathfrak{F}$ in a probability space $(\Omega, \mathfrak{F}, P)$ such that $\sigma(\mathfrak{A}_1 \cup \mathfrak{A}_2)$ and $\mathfrak{A}_3$ are independent with respect to P and if $(\Omega, \sigma(\mathfrak{A}_2 \cup \mathfrak{A}_3), P)$ is a complete measure space and $\mathfrak{A}_2$ contains all the null sets in $(\Omega, \sigma(\mathfrak{A}_2 \cup \mathfrak{A}_3), P)$, then

(4) $\quad P(A | \mathfrak{A}_2) = P(A | \sigma(\mathfrak{A}_2 \cup \mathfrak{A}_3)) \quad \text{for } A \in \mathfrak{A}_1.$

(See Theorem B.34.) To apply (4) to our probability space $(\mathbf{W}^{r+d}, \mathfrak{W}^{r+d,*}, P_{(B,X)})$, let $\mathfrak{A}_1 = \mathfrak{W}_t^{r+d,*}$ and $\mathfrak{A}_2 = \mathfrak{W}_t^{r,w} \times \mathbf{W}^d \subset \mathfrak{A}_1$. The process W defined on the standard filtered space $(\mathbf{W}^{r+d}, \mathfrak{W}^{r+d,*}, \{\mathfrak{W}_t^{r+d,*}\}, P_{(B,X)})$ by setting $W(t, w) = (\pi_0 \circ w)(t)$ for $(t, w) \in \mathbb{R}_+ \times \mathbf{W}^{r+d}$ is an r-dimensional $\{\mathfrak{W}_t^{r+d,*}\}$-adapted null at 0 Brownian motion according to Lemma 18.11. Let

$$\mathfrak{A}_3 = \sigma(\sigma_{\mathbf{W}^r}\{W_{t''} - W_{t'} : t \leq t' < t'' < \infty\} \cup \mathfrak{N}) \times \mathbf{W}^d,$$

where $\mathfrak{N}$ is the collection of all the null sets in $(\mathbf{W}^r, \mathfrak{W}^{r,w}, m_W^r)$. The independence of $\mathfrak{A}_1$ and $\mathfrak{A}_3$ follows from Definition 13.17 and Theorem 13.13. Also

$$\begin{aligned}\sigma(\mathfrak{A}_2 \cup \mathfrak{A}_3) &= \sigma\left(\mathfrak{W}_t^{r,w} \cup \sigma(\sigma_{\mathbf{W}^r}\{W_{t''} - W_{t'} : t \leq t' < t'' < \infty\} \cup \mathfrak{N})\right) \times \mathbf{W}^d \\ &= \mathfrak{W}^{r,w} \times \mathbf{W}^d.\end{aligned}$$

This shows that (3) is a particular case of (4). Thus $Q^{\pi_0(\cdot)}(\mathbf{W}^r \times A_1)$ which is a version of $P_{(B,X)}(\mathbf{W}^r \times A_1 | \mathfrak{W}_t^{r,w} \times \mathbf{W}^d)$ is also a version of $P_{(B,X)}(\mathbf{W}^r \times A_1 | \mathfrak{W}_t^{r,w} \times \mathbf{W}^d)$ and is therefore $\mathfrak{W}_t^{r,w} \times \mathbf{W}^d$-measurable. Then $Q^{(\cdot)}(\mathbf{W}^r \times A_1)$ is $\mathfrak{W}_t^{r,w}$-measurable. ∎

Observation 19.13. Let $(B^{(q)}, X^{(q)})$ be a solution of the stochastic differential equation (1) in Definition 18.6 on a standard filtered space $(\Omega^{(q)}, \mathfrak{F}^{(q)}, \{\mathfrak{F}_t^{(q)}\}, P^{(q)})$ for $q = 1, 2$. Let $P_{(B^{(q)},X^{(q)})}^{(q)}$ be the probability distribution of $(B^{(q)}, X^{(q)})$ on $(\mathbf{W}^{r+d}, \mathfrak{W}^{r+d})$ for $q = 1, 2$. As in Observation 19.9, let π_0 be the projection of $\mathbf{W}^{r+d}$ onto $\mathbf{W}^r$ and consider the regular image conditional probability given π_0

(1) $\quad Q^{v,(q)}(A) \equiv \left(P_{(B^{(q)},X^{(q)})}^{(q)}\right)^{\mathfrak{W}^{r+d,*,(q)}|\pi_0}(A, v) \quad \text{for } (A, v) \in \mathfrak{W}^{r+d} \times \mathbf{W}^r.$

According to (1) of Observation 19.9, the probability distribution $(P_{(B^{(q)},X^{(q)})}^{(q)})_{\pi_0}$ of π_0 on $(\mathbf{W}^r, \mathfrak{W}^r)$ is the Wiener measure m_W^r on $(\mathbf{W}^r, \mathfrak{W}^r)$. By our convention $Q^{v,(q)}$ is a probability measure on $\mathfrak{W}^{r+d}$ for every $v \in \mathbf{W}^r$. If we let

(2) $\quad R^{v,(q)}(A_1) = Q^{v,(q)}(\mathbf{W}^r \times A_1) \quad \text{for } A_1 \in \mathfrak{W}^d,$

§19. EXISTENCE AND UNIQUENESS OF SOLUTIONS

then $R^{v,(q)}$ is a probability measure on $(\mathbf{W}^d, \mathfrak{W}^d)$ for every $v \in \mathbf{W}^r$.

Lemma 19.14. *There exists a probability measure Q on $(\mathbf{W}^{r+2d}, \mathfrak{W}^{r+2d})$ such that for every $A_0 \times A_1 \times A_2 \in \mathfrak{W}^{r,w} \times \mathfrak{W}^{d,(1)} \times \mathfrak{W}^{d,(2)}$ we have*

(1) $$Q(A_0 \times A_1 \times A_2) = \int_{A_0} R^{v,(1)}(A_1) R^{v,(2)}(A_2) m_W^r(dv).$$

Proof. For every $v \in \mathbf{W}^r$, define a set function $\mu(\cdot, v)$ on $\mathfrak{W}^{d,(1)} \times \mathfrak{W}^{d,(2)}$ by setting

(2) $\quad \mu(A_1 \times A_2, v) = R^{v,(1)}(A_1) R^{v,(2)}(A_2) \quad$ for $A_1 \times A_2 \in \mathfrak{W}^{d,(1)} \times \mathfrak{W}^{d,(2)}$.

Clearly $\mu(\emptyset, v) = 0$ and it can be verified readily that $\mu(\cdot, v)$ is a countably additive set function on the semialgebra $\mathfrak{W}^{d,(1)} \times \mathfrak{W}^{d,(2)}$ of subsets of $\mathbf{W}^{d,(1)} \times \mathbf{W}^{d,(2)}$. The finiteness of the two measures $R^{v,(q)}(\mathbf{W}^r \times \cdot)$ for $q = 1, 2$ then implies that $\mu(\cdot, v)$ can be extended uniquely to a probability measure on $(\mathbf{W}^{d,(1)} \times \mathbf{W}^{d,(2)}, \sigma(\mathfrak{W}^{d,(1)} \times \mathfrak{W}^{d,(2)}))$ for every $v \in \mathbf{W}^r$. Since $Q^{v,(q)}(A)$ is a $\mathfrak{W}^{r,w}$-measurable function of v on $\mathbf{W}^r$ for every $A \in \mathfrak{W}^{r+d}$, $R^{v,(q)}(A_1)$ defined by (2) of Observation 19.13 is a $\mathfrak{W}^{r,w}$-measurable function of v on $\mathbf{W}^r$ for every $A_1 \in \mathfrak{W}^d$. Thus $\mu(A_1 \times A_2, \cdot)$ is a $\mathfrak{W}^{r,w}$-measurable function on $\mathbf{W}^r$ for every $A_1 \times A_2 \in \mathfrak{W}^{d,(1)} \times \mathfrak{W}^{d,(2)}$. To show that $\mu(E, \cdot)$ is a $\mathfrak{W}^{r,w}$-measurable function for every $E \in \sigma(\mathfrak{W}^{d,(1)} \times \mathfrak{W}^{d,(2)})$, let $\mathfrak{G}$ be the collection of all members E of $\sigma(\mathfrak{W}^{d,(1)} \times \mathfrak{W}^{d,(2)})$ such that $\mu(E, \cdot)$ is $\mathfrak{W}^{r,w}$-measurable. Clearly $\mathbf{W}^{d,(1)} \times \mathbf{W}^{d,(2)}$ is in $\mathfrak{G}$. Let $E_n \in \mathfrak{G}$, $n \in \mathbb{N}$, and $E_n \uparrow$. The fact that $\mu(\cdot, v)$ is a measure for every $v \in \mathbf{W}^r$ implies that $\mu(\lim_{n \to \infty} E_n, v) = \lim_{n \to \infty} \mu(E_n, v)$ every $v \in \mathbf{W}^r$ and thus the $\mathfrak{W}^{r,w}$-measurability of $\mu(E_n, \cdot)$ for every $n \in \mathbb{N}$ implies the $\mathfrak{W}^{r,w}$-measurability of $\mu(\lim_{n \to \infty} E_n, v)$. Thus $\lim_{n \to \infty} E_n \in \mathfrak{G}$. Similarly if $E, F \in \mathfrak{G}$ and $E \subset F$, then $F - E \in \mathfrak{G}$. This shows that $\mathfrak{G}$ is a d-class of subsets of $\mathbf{W}^{d,(1)} \times \mathbf{W}^{d,(2)}$. As we have pointed above, $\mathfrak{G}$ contains the π-class $\mathfrak{W}^{d,(1)} \times \mathfrak{W}^{d,(2)}$. Thus it contains $\sigma(\mathfrak{W}^{d,(1)} \times \mathfrak{W}^{d,(2)})$ by Theorem 1.7. Thus $\mu(E, \cdot)$ is a $\mathfrak{W}^{r,w}$-measurable function on $\mathbf{W}^r$ for every $E \in \sigma(\mathfrak{W}^{d,(1)} \times \mathfrak{W}^{d,(2)})$.

Consider the probability space $(\mathbf{W}^r, \mathfrak{W}^{r,w}, m_W^r)$ and the family of measures $\{\mu(\cdot, v) : v \in \mathbf{W}^r\}$ on $(\mathbf{W}^{d,(1)} \times \mathbf{W}^{d,(2)}, \sigma(\mathfrak{W}^{d,(1)} \times \mathfrak{W}^{d,(2)}))$. We have shown that $\mu(E, \cdot)$ is a $\mathfrak{W}^{r,w}$-measurable function on $\mathbf{W}^r$ for every $E \in \sigma(\mathfrak{W}^{d,(1)} \times \mathfrak{W}^{d,(2)})$, that is, the family of measures is $\mathfrak{W}^{r,w}$-measurable. Thus by Theorem 10.5, a set function Q on the semialgebra $\mathfrak{W}^{r,w} \times \sigma(\mathfrak{W}^{d,(1)} \times \mathfrak{W}^{d,(2)})$ of subsets of $\mathbf{W}^r \times \mathbf{W}_1^d \times \mathbf{W}_2^d$ defined by

(3) $\quad Q(A_0 \times E) = \int_{A_0} \mu(E, v) m_W^r(dv) \quad$ for $A_0 \times E \in \mathfrak{W}^{r,w} \times \sigma(\mathfrak{W}^{d,(1)} \times \mathfrak{W}^{d,(2)})$

can be extended to be a measure on $\sigma(\mathfrak{W}^{r,w} \times \sigma(\mathfrak{W}^{d,(1)} \times \mathfrak{W}^{d,(2)}))$. Note that $\sigma(\mathfrak{W}^{d,(1)} \times \mathfrak{W}^{d,(2)}) = \mathfrak{W}^{2d}$ and $\sigma(\mathfrak{W}^{r,w} \times \sigma(\mathfrak{W}^{d,(1)} \times \mathfrak{W}^{d,(2)})) = \sigma(\mathfrak{W}^{r,w} \times \mathfrak{W}^{2d}) = \mathfrak{W}^{r+2d}$ by Proposition 18.8. Finally for $A_0 \times A_1 \times A_2 \in \mathfrak{W}^{r,w} \times \mathfrak{W}^{d,(1)} \times \mathfrak{W}^{d,(2)}$, (1) holds by (3) and (2). ∎

Definition 19.15. *Consider the probability space* $(\mathbf{W}^{r+2d}, \mathfrak{W}^{r+2d}, Q)$. *Let* $\mathfrak{W}^{r+2d,*}$ *be the completion of* $\mathfrak{W}^{r+2d}$ *with respect to* Q, *let* $\mathfrak{W}_t^{r+2d,0} = \sigma(\mathfrak{W}_t^{r+2d} \cup \mathfrak{N})$ *where* $\mathfrak{N}$ *is the collection of all the null sets in the complete measure space* $(\mathbf{W}^{r+2d}, \mathfrak{W}^{r+2d,*}, Q)$, *and let* $\mathfrak{W}_t^{r+2d,*} = \cap_{\varepsilon>0}\mathfrak{W}_{t+\varepsilon}^{r+2d,0}$. *Then* $(\mathbf{W}^{r+2d}, \mathfrak{W}^{r+2d,*}, \{\mathfrak{W}_t^{r+2d,*}\}, Q)$ *is a standard filtered space. Let* π_1 *and* π_2 *be projections of* $\mathbf{W}^{r+2d} = \mathbf{W}^r \times \mathbf{W}^{d,(1)} \times \mathbf{W}^{d,(2)}$ *onto* $\mathbf{W}_1^d$ *and* $\mathbf{W}_2^d$ *respectively. Let us define three processes* $W, Y^{(1)}$, *and* $Y^{(2)}$ *on the standard filtered space by setting*

$$\begin{cases} W(t,w) = (\pi_0(w))(t), \\ Y^{(1)}(t,w) = (\pi_1(w))(t), \\ Y^{(2)}(t,w) = (\pi_2(w))(t) \quad \text{for } (t,w) \in \mathbb{R}_+ \times \mathbf{W}^{r+2d}. \end{cases}$$

Proposition 19.16. *Let* $W, Y^{(1)}$, *and* $Y^{(2)}$ *be as in Definition 19.15. Then* W *is an r-dimensional* $\{\mathfrak{W}_t^{r+2d,*}\}$-*adapted null at 0 Brownian motion on the standard filtered space* $(\mathbf{W}^{r+2d}, \mathfrak{W}^{r+2d,*}, \{\mathfrak{W}_t^{r+2d,*}\}, Q)$, $(W, Y^{(q)})$ *and* $(B^{(q)}, X^{(q)})$ *have identical probability distributions on* $(\mathbf{W}^{r+d}, \mathfrak{W}^{r+d})$ *and* $(W, Y^{(q)})$ *is a solution of the stochastic differential equation (1) in Definition 18.6 for* $q = 1, 2$.

Proof. Clearly W is an r-dimensional continuous $\{\mathfrak{W}_t^{r+2d}\}$-adapted, and hence $\{\mathfrak{W}_t^{r+2d,*}\}$-adapted, process on $(\mathbf{W}^{r+2d}, \mathfrak{W}^{r+2d,*}, \{\mathfrak{W}_t^{r+2d,*}\}, Q)$. To show that W is null at 0, let $Z = \{v \in \mathbf{W}^r : v(0) = 0\}$. Then by (1) of Lemma 19.14, we have

$$Q\{W_0 = 0\} = Q(Z \times \mathbf{W}^{d,(1)} \times \mathbf{W}^{d,(2)})$$
$$= \int_Z R^{v,(1)}(\mathbf{W}^{d,(1)})R^{v,(2)}(\mathbf{W}^{d,(2)})m_W^r(dv) = m_W^r(Z).$$

Now the stochastic process V defined on the probability space $(\mathbf{W}^r, \mathfrak{W}^{r,w}, m_W^r)$ by $V(t,v) = v(t)$ for $(t,v) \in \mathbb{R}_+ \times \mathbf{W}^r$ is a null at 0 Brownian motion so that $m_W^r(Z) = 1$. Thus $Q\{W_0 = 0\} = 1$. This shows that W is null at 0.

To show that W is an r-dimensional $\{\mathfrak{W}_t^{r+2d,*}\}$-adapted Brownian motion, it suffices according to Theorem 13.26 to verify that for every $s, t \in \mathbb{R}_+$ such that $s < t$ and $y \in \mathbb{R}^{r+2d}$ we have

(1) $$\int_A e^{i\langle y, W_t - W_s\rangle} dQ = Q(A)e^{-\frac{|y|^2}{2}(t-s)} \quad \text{for } A \in \mathfrak{W}_s^{r+2d,*}.$$

§19. EXISTENCE AND UNIQUENESS OF SOLUTIONS

Consider first the case where $A = A_0 \times A_1 \times A_2 \in \mathfrak{W}_s^{r,w} \times \mathfrak{W}_s^{d,(1)} \times \mathfrak{W}_s^{d,(2)}$. In this case we have

$$\int_A e^{i\langle y, W_t - W_s \rangle} dQ = \int_{\mathbf{W}^{r+2d}} e^{i\langle y, W_t - W_s \rangle} \mathbf{1}_{A_0 \times A_1 \times A_2} dQ = \int_{\mathbf{W}^r} e^{i\langle y, v(t) - v(s) \rangle} \Phi(v) m_W^r(dv),$$

where

$$\Phi(v) = \mathbf{1}_{A_0} R^{v,(1)}(A_1) R^{v,(2)}(A_2) = \mathbf{1}_{A_0} Q^{v,(1)}(\mathbf{W}^r \times A_1) Q^{v,(2)}(\mathbf{W}^r \times A_2)$$

for $v \in \mathbf{W}^r$. Since $A_0 \in \mathfrak{W}_s^{r,w}$, $\mathbf{1}_{A_0}$ is a $\mathfrak{W}_s^{r,w}$-measurable function on $\mathbf{W}^r$. Since $A_1, A_2 \in \mathfrak{W}_s^d$, $Q^{(1),(\cdot)}(\mathbf{W}^r \times A_1)$ and $Q^{(2),(\cdot)}(\mathbf{W}^r \times A_2)$ are $\mathfrak{W}_s^{r,w}$-measurable functions by Lemma 19.12. Thus Φ is a $\mathfrak{W}_s^{r,w}$-measurable function on $\mathbf{W}^r$. Now if we define a process V on $(\mathbf{W}^r, \mathfrak{W}^{r,w}, m_W^r)$ by setting $V(t, v) = v(t)$ for $(t, v) \in \mathbb{R}_+ \times \mathbf{W}^r$, then V is an r-dimensional $\{\mathfrak{W}_t^{r,w}\}$-adapted Brownian motion. Thus for any $s, t \in \mathbb{R}_+$, $s < t$, the σ-algebra $\mathfrak{W}_s^{r,w}$ and the random vector $v(t) - v(s)$ are independent in the probability space $(\mathbf{W}^r, \mathfrak{W}^{r,w}, m_W^r)$ by Definition 13.17. The $\mathfrak{W}_s^r$-measurability of Φ then implies the independence of the two random variables Φ and $e^{i\langle y, v(t) - v(s) \rangle}$. Thus

$$\int_{\mathbf{W}^r} e^{i\langle y, v(t) - v(s) \rangle} \Phi(v) m_W^r(dv) = \left\{ \int_{\mathbf{W}^r} e^{i\langle y, v(t) - v(s) \rangle} m_W^r(dv) \right\} \left\{ \int_{\mathbf{W}^r} \Phi(v) m_W^r(dv) \right\}$$
$$= e^{-\frac{|y|^2}{2}(t-s)} Q(A_0 \times A_1 \times A_2) = e^{-\frac{|y|^2}{2}(t-s)} Q(A)$$

by Lemma 19.14. This verifies (1) for the case $A \in \mathfrak{W}_s^{d,(1)} \times \mathfrak{W}_s^{d,(2)}$. By applying Corollary 1.8, we then have (1) for $A \in \mathfrak{W}_s^{r+2d} = \sigma(\mathfrak{W}_s^{d,(1)} \times \mathfrak{W}_s^{d,(2)})$. By the same argument as in the proof of Lemma 18.11, we then have (1) holding for $A \in \mathfrak{W}_s^{r+2d,*}$. This completes the verification that W is W is an r-dimensional $\{\mathfrak{W}_t^{r+2d,*}\}$-adapted null at 0 Brownian motion.

Let us show next that $(W, Y^{(q)})$ and $(B^{(q)}, X^{(q)})$ have identical probability distributions on $(\mathbf{W}^{r+d}, \mathfrak{W}^{r+d})$ for each value of $q = 1, 2$. Let us consider for instance the case $q = 1$. We are to show that $P^{(1)}_{(B^{(1)}, X^{(1)})}(E) = Q \circ (W, Y^{(1)})^{-1}(E)$ for $E \in \mathfrak{W}^{r+d}$. Consider first the case where $E = A_0 \times A_1 \in \mathfrak{W}^r \times \mathfrak{W}^d$. Then

$$Q \circ (W, Y^{(1)})^{-1}(A_0 \times A_1) = Q(W^{-1}(A_0) \cap (Y^{(1)})^{-1}(A_1))$$
$$= Q(A_0 \times \mathbf{W}^d \times \mathbf{W}^d \cap \mathbf{W}^r \times A_1 \times \mathbf{W}^d) = Q(A_0 \times A_1 \times \mathbf{W}^d)$$
$$= \int_{A_0} Q^{v,(1)}(\mathbf{W}^r \times A_1) Q^{v,(2)}(\mathbf{W}^r \times \mathbf{W}^d) m_W^r(dv)$$
$$= \int_{A_0} Q^{v,(1)}(\mathbf{W}^r \times A_1) m_W^r(dv) = P^{(1)}_{(B^{(1)}, X^{(1)})}(A_0 \times A_1),$$

by 4° of Observation 19.9. This shows that $Q \circ (W, \mathcal{Y}^{(1)})^{-1}$ and $P^{(1)}_{(B^{(1)}, X^{(1)})}$ agree on $\mathfrak{W}^r \times \mathfrak{W}^d$. Then by Corollary 1.8, they agree on $\mathfrak{W}^{r+d} = \sigma(\mathfrak{W}^r \times \mathfrak{W}^d)$.

Finally $(W, Y^{(q)})$ is a solution of the stochastic differential equation since it is the image on the function space of the solution $(B^{(q)}, X^{(q)})$. ∎

Lemma 19.17. *Let μ and ν be two probability measures on a measurable space $(S, \mathfrak{B}_S)$ where S is a complete separable metric space and $\mathfrak{B}_S$ is the Borel σ-algebra of subsets of S. Let D be the diagonal in $S \times S$, that is,*

$$D = \{(s_1, s_2) \in S \times S : s_1 = s_2\}.$$

If $(\mu \times \nu)(D) = 1$, then $\mu = \nu$ on $\mathfrak{B}_S$ and furthermore there exists a unique $s_0 \in S$ such that $\mu(\{s_0\}) = \nu(\{s_0\}) = 1$.

Proof. Let $S \times S$ be topologized by the product topology and let $\mathfrak{B}_{S \times S}$ be the Borel σ-algebra of subsets of $S \times S$. Let ρ be the metric on S. Then $\rho(s_1, s_2)$ for $(s_1, s_2) \in S \times S$ is a continuous function on $S \times S$ and is thus $\mathfrak{B}_{S \times S}$-measurable. This implies that the diagonal D, which is the subset of $S \times S$ on which the $\mathfrak{B}_{S \times S}$-measurable function ρ is equal to 0, is a member of $\mathfrak{B}_{S \times S}$. Since S is a separable metric space it satisfies the second axiom of countability and therefore $\mathfrak{B}_{S \times S} = \sigma(\mathfrak{B}_S \times \mathfrak{B}_S)$ by Theorem 1.4. Thus $D \in \sigma(\mathfrak{B}_S \times \mathfrak{B}_S)$ and $(\mu \times \nu)(D)$ is defined.

Suppose $(\mu \times \nu)(D) = 1$. If $\mu \neq \nu$ on $\mathfrak{B}_S$ then there exists $A \in \mathfrak{B}_S$ such that $\mu(A) \neq \nu(A)$, say $\mu(A) > \nu(A)$. Then $\nu(A^c) > 0$ so that

$$(\mu \times \nu)(A \times A^c) = \mu(A)\nu(A^c) > 0.$$

But $(A \times A^c) \cap D = \emptyset$ and this implies $(\mu \times \nu)(A \times A^c) = 0$, contradicting the last inequality. Therefore $\mu = \nu$ on $\mathfrak{B}_S$.

If there exists $A \in \mathfrak{B}_S$ such that $\mu(A) \in (0, 1)$ then $\mu(A^c) \in (0, 1)$ also so that

$$(\mu \times \nu)(A \times A^c) = \mu(A)\nu(A^c) \in (0, 1).$$

But this contradicts the equality $(\mu \times \nu)(A \times A^c) = 0$ which is implied by $(A \times A^c) \cap D = \emptyset$. Therefore no $A \in \mathfrak{B}_S$ can have $\mu(A) \in (0, 1)$ and consequently $\mu(A) = 0$ or 1 for every $A \in \mathfrak{B}_S$.

Since a separable metric space is a Lindelöf space, for every $n \in \mathbb{N}$ there exist countably many closed spheres in S, each with radius n^{-1}, whose union is S. The μ-measure of each of these spheres is either 0 or 1. No two spheres with μ-measure 1 can be disjoint for

§19. EXISTENCE AND UNIQUENESS OF SOLUTIONS

otherwise we would have $\mu(S) \geq 2$. Let K_n be the intersection of all those closed spheres with μ-measure 1. Then $\mu(K_n) = 1$ and the radius $r(K_n) \leq n^{-1}$. Consider the sequence of the closed sets K_n, $n \in \mathbb{N}$. By the same reason as above we have $K_n \cap K_m \neq \emptyset$ for $n \neq m$. If we let $C_n = \cap_{m=1}^n K_m$, then we have a decreasing sequence of closed sets C_n, $n \in \mathbb{N}$, with $\mu(C_n) = 1$ and $r(C_n) \leq n^{-1}$ for every $n \in \mathbb{N}$. Since S is a complete metric space and $r(C_n) \downarrow 0$ as $n \to \infty$, there exists $s_0 \in S$ such that $\cap_{n \in \mathbb{N}} C_n = \{s_0\}$. Then $\mu(\{s_0\}) = \lim_{n \to \infty} \mu(C_n) = 1$. Since $\mu(S) = 1$ such $s_0 \in S$ is unique. ∎

Theorem 19.18. *Let $x \in \mathbb{R}^d$ be fixed. Suppose that the initial value problem*

$$\begin{cases} dX_t = \alpha(t, X) dB(t) + \beta(t, X) dt \\ X_0 = x \end{cases}$$

for the stochastic differential equation (1) in Definition 18.6 has a solution on some standard filtered space and suppose that the solution of the stochastic differential equation is pathwise unique under deterministic initial conditions. Then there exists a mapping F_x of $\mathbf{W}^r$ into $\mathbf{W}^d$, unique up to a null set in $(\mathbf{W}^r, \mathfrak{W}^{r,w}, m_W^r)$, such that

1°. *F_x is $\mathfrak{W}_t^{r,w}/\mathfrak{W}_t^d$-measurable for every $t \in \mathbb{R}_+$,*

2°. *if (B, X) is a solution of the initial value problem on some standard filtered space $(\Omega, \mathfrak{F}, \{\mathfrak{F}_t\}, P)$, then there exists a null set N_0 in $(\mathbf{W}^r, \mathfrak{W}^{r,w}, m_W^r)$ such that*

$$X(\cdot, \omega) = F_x[B(\cdot, \omega)] \quad \text{for } \omega \in \Omega \text{ with } B(\cdot, \omega) \in N_0^c,$$

and therefore there exists a null set Λ in $(\Omega, \mathfrak{F}, P)$ such that the equality holds for $\omega \in \Lambda^c$.

Let $(\Omega, \mathfrak{F}, P)$ be an arbitrary complete probability space on which an r-dimensional null at 0 Brownian motion B exists. Let $(\Omega, \mathfrak{F}, \{\overline{\mathfrak{F}}_t^B\}, P)$ be the standard filtered space generated by B. Let X be a continuous d-dimensional process on $(\Omega, \mathfrak{F}, \{\overline{\mathfrak{F}}_t^B\}, P)$ defined by

$$X(\cdot, \omega) = F_x[B(\cdot, \omega)] \quad \text{for } \omega \in \Omega.$$

Then (B, X) is a solution of the initial value problem on $(\Omega, \mathfrak{F}, \{\overline{\mathfrak{F}}_t^B\}, P)$.

Proof. Let $(B^{(q)}, X^{(q)})$ be a solution of the initial value problem on a standard filtered space $(\Omega^{(q)}, \mathfrak{F}^{(q)}, \{\mathfrak{F}_t^{(q)}\}, P^{(q)})$ for $q = 1, 2$. Let $P_{(B^{(q)}, X^{(q)})}^{(q)}$ be the probability distribution of

$(B^{(q)}, X^{(q)})$ on $(\mathbf{W}^{r+d}, \mathfrak{W}^{r+d})$ for $q = 1, 2$. Let $Q^{(\cdot),(q)}$ and $R^{(\cdot),(q)}$ be as in Observation 19.13 and $(\mathbf{W}^{r+2d}, \mathfrak{W}^{r+2d,*}, \{\mathfrak{W}^{r+2d,*}_t\}, Q), W, Y^{(1)}, Y^{(2)}$ be as in Definition 19.15.

According to Proposition 19.16, $(W, Y^{(q)})$ is a solution of the stochastic differential equation on $(\mathbf{W}^{r+2d}, \mathfrak{W}^{r+2d,*}, \{\mathfrak{W}^{r+2d,*}_t\}, Q)$ and its probability distribution on $(\mathbf{W}^{r+d}, \mathfrak{W}^{r+d})$ is identical to that of $(B^{(q)}, X^{(q)})$ for each value of $q = 1, 2$. Since $X^{(q)} = x$ a.e. on $(\Omega^{(q)}, \mathfrak{F}^{(q)}, P^{(q)})$ we have $Y^{(q)} = x$ a.e. on $(\mathbf{W}^{r+2d}, \mathfrak{W}^{r+2d}, Q)$ for $q = 1, 2$. Then since $(W, Y^{(1)})$ and $(W, Y^{(2)})$ are two solution of the stochastic differential equation on the same standard filtered space $(\mathbf{W}^{r+2d}, \mathfrak{W}^{r+2d,*}, \{\mathfrak{W}^{r+2d,*}_t\}, Q)$, pathwise uniqueness of solution under deterministic initial condition implies that there exists a null set N in $(\mathbf{W}^{r+2d}, \mathfrak{W}^{r+2d,*}, Q)$ such that $Y^{(1)}(\cdot, w) = Y^{(2)}(\cdot, w)$, that is, $\pi_1(w) = \pi_2(w)$, for $w \in N^c$. Thus

$$1 = Q(N^c) = Q(\{w \in \mathbf{W}^{r+2d} : \pi_1(w) = \pi_2(w)\})$$
$$= \int_{\mathbf{W}^r} \left(R^{\pi_0(w),(1)} \times R^{\pi_0(w),(2)} \right) (D_{\pi_0(w)}) m^r_W(d(\pi_0(w))),$$

where

$$D_{\pi_0(w)} = \left\{ (\pi_0(w), \pi_1(w), \pi_2(w)) \in \{\pi_0(w)\} \times \mathbf{W}^{d,(1)} \times \mathbf{W}^{d,(2)} : \pi_1(w) = \pi_2(w) \right\}.$$

Since the integrand in the last integral is nonnegative and bounded by 1, the fact that the integral is equal to 1 implies that there exists a null set N_0 in $(\mathbf{W}^r, \mathfrak{W}^{r,w}, m^r_W)$ such that for $\pi_0(w) \in N^c_0$ we have

(1) $\qquad \left(R^{\pi_0(w),(1)} \times R^{\pi_0(w),(2)} \right) (D_{\pi_0(w)}) = 1.$

If we let $N_1 = \{w \in \mathbf{W}^{r+2d} : \pi_0(w) \in N_0\}$ then

$$Q(N_1) = \int_{N_0} R^{\pi_0(w),(1)}(\mathbf{W}^{d,(1)}) R^{\pi_0(w),(2)}(\mathbf{W}^{d,(2)}) m^r_W(d(\pi_0(w))) = \int_{N_0} m^r_W(d(\pi_0(w))) = 0$$

so that N_1 is a null set in $(\mathbf{W}^{r+2d}, \mathfrak{W}^{r+2d}, Q)$ and (1) holds for $w \in N^c_1$. Thus by Lemma 19.17, for every $w \in N^c_1$ we have

(2) $\qquad \begin{cases} R^{\pi_0(w),(1)} = R^{\pi_0(w),(2)} & \text{on } (\mathbf{W}^d, \mathfrak{W}^d) \\ \pi_1(w) = \pi_2(w) \\ R^{\pi_0(w),(1)}(\{\pi_1(w)\}) = R^{\pi_0(w),(2)}(\{\pi_2(w)\}) = 1. \end{cases}$

Then for every $v \in N^c_0$ there exists a unique $\varphi(v) \in \mathbf{W}^d$ such that $R^{v,(1)}(\{\varphi(v)\}) = R^{v,(2)}(\{\varphi(v)\}) = 1$. Let us define a mapping F_x of $\mathbf{W}^r$ into $\mathbf{W}^d$ by setting

(3) $\qquad F_x(v) = \begin{cases} \varphi(v) & \text{for } v \in N^c_0 \\ 0 & \text{for } v \in N_0. \end{cases}$

§19. EXISTENCE AND UNIQUENESS OF SOLUTIONS

Since only one singleton can have probability measure 1, the last equality in (2) implies that

(4) $$F_x(\pi_0(w)) = \begin{cases} \pi_1(w) = \pi_2(w) & \text{for } w \in N_1^c \\ 0 & \text{for } w \in N_1. \end{cases}$$

Let us verify that the mapping F_x defined by (3) satisfies conditions 1° and 2°. To show that F_x is $\mathfrak{W}_t^{r,w}/\mathfrak{W}_t^d$-measurable for every $t \in \mathbb{R}_+$, note that since $\mathfrak{W}_t^{r,w}$ contains every null set in $(\mathbf{W}^r, \mathfrak{W}^{r,w}, m_W^r)$ and in particular N_0 and since $F_x(v) = 0$ for $v \in N_0$, it suffices to show that the restriction of F_x to N_0^c is $\mathfrak{W}_t^{r,w}/\mathfrak{W}_t^d$-measurable. Now for every $A_1 \in \mathfrak{W}_t^d$, we have

$$F_x^{-1}(A_1) \cap N_0^c = \{v \in N_0^c : \varphi(v) \in A_1\} = \{v \in N_0^c : R^{v,(1)}(A_1) = 1\}$$
$$= \{v \in N_0^c : Q^{v,(1)}(\mathbf{W}^r \times A_1) = 1\} \in \mathfrak{W}_t^{r,w},$$

since according to Lemma 19.12, $Q^{v,(1)}(\mathbf{W}^r \times A_1)$ is a $\mathfrak{W}_t^r$-measurable, and hence $\mathfrak{W}_t^{r,w}$-measurable, function of v.

To verify 2°, let (B, X) be a solution of the initial value problem on a standard filtered space, say our $(B^{(1)}, X^{(1)})$ on $(\Omega^{(1)}, \mathfrak{F}^{(1)}, \{\mathfrak{F}_t^{(1)}\}, P^{(1)})$. Recall that the probability distribution $P_{B^{(1)}}^{(1)}$ of $B^{(1)}$ on $(\mathbf{W}^r, \mathfrak{W}^r)$ is given by m_W^r by Definition 17.9. Let $\Lambda = (\mathcal{B}^{(1)})^{-1}(N_0)$. Then

$$P^{(1)}(\Lambda) = P^{(1)} \circ (\mathcal{B}^{(1)})^{-1}(N_0) = P_{B^{(1)}}^{(1)}(N_0) = m_W^r(N_0) = 0,$$

that is, Λ is a null set in $(\Omega^{(1)}, \mathfrak{F}^{(1)}, P^{(1)})$. For $\omega \in \Lambda^c$,

$$F_x(B^{(1)}(\cdot, \omega)) = F_x(\mathcal{B}^{(1)}(\omega)) = F_x(\pi_0 \circ (\mathcal{B}^{(1)}, \mathcal{X}^{(1)})(\omega))$$
$$= \pi_1 \circ (\mathcal{B}^{(1)}, \mathcal{X}^{(1)})(\omega) = \mathcal{X}^{(1)}(\omega) = X^{(1)}(\cdot, \omega),$$

where the third equality is by (4). Similarly for $\omega \in \Lambda$, we have $F_x(B^{(1)}(\cdot, \omega)) = 0$ by (4).

Let $(\Omega, \mathfrak{F}, P)$ be an arbitrary complete probability space on which an r-dimensional null at 0 Brownian motion B exists. Let $(\Omega, \mathfrak{F}, \{\overline{\mathfrak{F}}_t^B\}, P)$ be the standard filtered space generated by B as in Theorem 13.23, that is, $\mathfrak{F}_t^B = \sigma\{B_s : s \in [0,t]\}$ and $\overline{\mathfrak{F}}_t^B = \sigma(\mathfrak{F}_t^B \cup \mathfrak{N})$ where $\mathfrak{N}$ is the collection of all the null sets in $(\Omega, \mathfrak{F}, P)$. Then B is an r-dimensional $\{\overline{\mathfrak{F}}_t^B\}$-adapted null at 0 Brownian motion. Let us show the process X on $(\Omega, \mathfrak{F}, \{\overline{\mathfrak{F}}_t^B\}, P)$ defined by $X(\cdot, \omega) = F_x[B(\cdot, \omega)]$ for $\omega \in \Omega$ is $\overline{\mathfrak{F}}_t^B$-adapted. Now $\mathcal{B}(\omega) = B(\cdot, \omega)$ for $\omega \in \Omega$ is an $\mathfrak{F}/\mathfrak{W}^r$-measurable mapping of Ω into $\mathbf{W}^r$ and furthermore it is $\overline{\mathfrak{F}}_t^B/\mathfrak{W}_t^r$-measurable for every $t \in \mathbb{R}_+$ by Theorem 17.8. The probability distribution of $\mathcal{B}$ on $(\mathbf{W}^r, \mathfrak{W}^r)$ is given by m_W^r by Definition 17.9. Consider $\mathfrak{W}_t^{r,w}$ as in Definition 19.10. If E_0 is a null set in $(\mathbf{W}^r, \mathfrak{W}^r, m_W^r)$ then $0 = m_W^r(E_0) = P \circ \mathcal{B}^{-1}(E_0)$ so that $\mathcal{B}^{-1}(E_0)$ is a null set in $(\Omega, \mathfrak{F}, P)$.

If $E_1 \subset E_0$, then as a subset of the null set $\mathcal{B}^{-1}(E_0)$ in the complete probability space $(\Omega, \mathfrak{F}, P)$, $\mathcal{B}^{-1}(E_1)$ is a null set in $(\Omega, \mathfrak{F}, P)$ and thus $\mathcal{B}^{-1}(E_1) \in \overline{\mathfrak{F}}_t^B$ for every $t \in \mathbb{R}_+$ since $\overline{\mathfrak{F}}_t^B$ is augmented. This and the fact that $\mathcal{B}^{-1}(\mathfrak{W}_t^r) \subset \overline{\mathfrak{F}}_t^B$ implies that $\mathcal{B}^{-1}(\mathfrak{W}_t^{r,w}) \subset \overline{\mathfrak{F}}_t^B$, that is, $\mathcal{B}$ is $\overline{\mathfrak{F}}_t^B / \mathfrak{W}_t^{r,w}$-measurable for every $t \in \mathbb{R}_+$. The mapping F_x is $\mathfrak{W}_t^{r,w}/\mathfrak{W}_t^d$-measurable. The mapping q_t of $\mathbf{W}^d$ into $\mathbb{R}^d$ defined by $q_t(w) = w(t)$ for $w \in \mathbf{W}^d$ is $\mathfrak{W}_t^d/\mathfrak{B}_{\mathbb{R}^d}$-measurable as we saw in Proposition 17.7. Thus $q_t \circ F_x \circ \mathcal{B}$ is an $\overline{\mathfrak{F}}_t^B/\mathfrak{B}_{\mathbb{R}^d}$-measurable mapping of Ω into $\mathbb{R}^d$. But

$$q_t \circ F_x \circ \mathcal{B}(\omega) = q_t \circ F_x[B(\cdot, \omega)] = q_t \circ X(\cdot, \omega) = X(t, \omega) \quad \text{for } \omega \in \Omega.$$

Thus X_t is $\overline{\mathfrak{F}}_t^B/\mathfrak{B}_{\mathbb{R}^d}$-measurable, that is, X is an $\{\overline{\mathfrak{F}}_t^B\}$-adapted process.

Let us show next that the process (B, X) on $(\Omega, \mathfrak{F}, \{\overline{\mathfrak{F}}_t^B\}, P)$ and the process $(B^{(1)}, X^{(1)})$ on $(\Omega^{(1)}, \mathfrak{F}^{(1)}, \{\mathfrak{F}_t^{(1)}\}, P^{(1)})$ have identical probability distributions on $(\mathbf{W}^{r+d}, \mathfrak{W}^{r+d})$.

For the probability distribution $P_{(B,X)}$ of (B, X), writing I for the identity mapping, we have by Definition 17.9 for m_W^r

$$P_{(B,X)} = P \circ (\mathcal{B}, \mathcal{X})^{-1} = P \circ (\mathcal{B}, F_x \circ \mathcal{B})^{-1} = P \circ (I \circ \mathcal{B}, F_x \circ \mathcal{B})^{-1}$$
$$= P \circ ((I, F_x) \circ \mathcal{B})^{-1} = P \circ \mathcal{B}^{-1} \circ (I, F_x)^{-1} = m_W^r \circ (I, F_x)^{-1},$$

On the other hand the probability distribution $P^{(1)}_{(B^{(1)}, X^{(1)})}$ is given by

$$P^{(1)}_{(B^{(1)}, X^{(1)})} = P^{(1)} \circ (\mathcal{B}^{(1)}, \mathcal{X}^{(1)})^{-1} = P^{(1)} \circ (\mathcal{B}^{(1)})^{-1} \circ (I, F_x)^{-1} = m_W^r \circ (I, F_x)^{-1},$$

by 2°. Thus $P^{(1)}_{(B^{(1)}, X^{(1)})} = P_{(B,X)}$. This shows that $(B^{(1)}, X^{(1)})$ and (B, X) have identical probability distributions. Then since $(B^{(1)}, X^{(1)})$ is a solution of the stochastic differential equation on $(\Omega^{(1)}, \mathfrak{F}^{(1)}, \{\mathfrak{F}_t^{(1)}\}, P^{(1)})$, (B, X) is a solution of the stochastic differential equation on $(\Omega, \mathfrak{F}, \{\mathfrak{F}_t\}, P)$ by Theorem 18.15. Also $X_0^{(1)} = x$ a.e. on $(\Omega^{(1)}, \mathfrak{F}^{(1)}, P^{(1)})$ implies that $X_0 = x$ a.e. on $(\Omega, \mathfrak{F}, P)$. Thus (B, X) is a solution of the initial value problem on $(\Omega, \mathfrak{F}, \{\mathfrak{F}_t\}, P)$. ∎

Lemma 19.19. *Pathwise uniqueness under deterministic initial conditions of solution of the stochastic differential equation (1) in Definition 18.6 implies uniqueness of solution in the sense of probability law.*

Proof. According to Lemma 19.2, uniqueness in the sense of probability law is equivalent to uniqueness in the sense of probability law under deterministic initial conditions.

§19. EXISTENCE AND UNIQUENESS OF SOLUTIONS

Therefore it suffices to show that pathwise uniqueness under deterministic initial conditions implies uniqueness in the sense of probability law under deterministic initial conditions. Now let $(B^{(1)}, X^{(1)})$ and $(B^{(2)}, X^{(2)})$ be two solutions on two standard filtered spaces $(\Omega^{(1)}, \mathfrak{F}^{(1)}, \{\mathfrak{F}^{(1)}\}, P^{(1)})$ and $(\Omega^{(2)}, \mathfrak{F}^{(2)}, \{\mathfrak{F}^{(2)}\}, P^{(2)})$ respectively and suppose $X_0^{(1)} = x$ a.e. on $(\Omega^{(1)}, \mathfrak{F}^{(1)}, P^{(1)})$ and $X_0^{(2)} = x$ a.e. on $(\Omega^{(2)}, \mathfrak{F}^{(2)}, P^{(2)})$ for some $x \in \mathbb{R}^d$. Let F_x be the mapping of $\mathbf{W}^r$ into $\mathbf{W}^d$ defined in Theorem 19.18. Then by 2° of Theorem 18.20 there exists a null set $\Lambda^{(q)}$ in $(\Omega^{(q)}, \mathfrak{F}^{(q)}, P^{(q)})$ such that

$$X^{(q)}(\cdot, \omega) = F_x[B^{(q)}(\cdot, \omega)] \quad \text{for } \omega \in (\Lambda^{(q)})^c$$

for $q = 1, 2$. Now for the probability distribution $P_{X^{(q)}}^{(q)}$ of $X^{(q)}$ on $(\mathbf{W}^d, \mathfrak{W}^d)$ we have

$$P_{X^{(q)}}^{(q)} = P^{(q)} \circ (\mathcal{X}^{(q)})^{-1} = P^{(q)} \circ (F_x \circ \mathcal{B}^{(q)})^{-1} = P^{(q)} \circ (\mathcal{B}^{(q)})^{-1} \circ F_x^{-1} = m_W^r \circ F_x^{-1}.$$

Thus $P_{X^{(1)}}^{(1)} = P_{X^{(2)}}^{(2)}$ on $(\mathbf{W}^d, \mathfrak{W}^d)$. This proves uniqueness of solution in the sense of probability law under deterministic initial conditions. ∎

Theorem 19.20. *For the stochastic differential equation (1) in Definition 18.6, pathwise uniqueness of solution and pathwise uniqueness of solution under deterministic initial conditions are equivalent.*

Proof. Since the former contains the latter as a particular case, it suffices to show that the latter implies the former. Let us assume the latter. Let $(B, X^{(1)})$ and $(B, X^{(2)})$ be two solutions of the stochastic differential equation on a standard filtered space $(\Omega, \mathfrak{F}, \{\mathfrak{F}_t\}, P)$ such that $X_0^{(1)} = X_0^{(2)}$ a.e. on $(\Omega, \mathfrak{F}, P)$.

The representation $(W^{(q)}, Y^{(q)})$ of $(B, X^{(q)})$ on $(\mathbf{W}^{r+d}, \mathfrak{W}^{r+d,*,(q)}, \{\mathfrak{W}_t^{r+d,*,(q)}\}, P_{(B, X^{(q)})})$ is a solution of the stochastic differential equation, $(W^{(q)}, Y^{(q)})$ and $(B, X^{(q)})$ have identical probability distributions on $(\mathbf{W}^{r+d}, \mathfrak{W}^{r+d})$, and $Y^{(q)}$ and $X^{(q)}$ have identical probability distributions on $(\mathbf{W}^d, \mathfrak{W}^d)$ for each of $q = 1, 2$ according to Theorem 18.13. Since $X_0^{(1)} = X_0^{(2)}$ a.e. on $(\Omega, \mathfrak{F}, P)$, $X_0^{(1)}$ and $X_0^{(2)}$ have identical probability distributions on $(\mathbb{R}^d, \mathfrak{B}_{\mathbb{R}^d})$. Call the common probability distribution μ. Then μ is also the probability distributions of $Y_0^{(1)}$ and $Y_0^{(2)}$. According to Theorem 18.21, there exists a null set N in $(\mathbb{R}^d, \mathfrak{B}_{\mathbb{R}^d}, \mu)$ such that for every $x \in N^c$, $(W^{(q)}, Y^{(q)})$ is a solution of the stochastic differential equation on the standard filtered space $(\mathbf{W}^{r+d}, \mathfrak{W}^{r+d,*,x,(q)}, \{\mathfrak{W}_t^{r+d,*,x,(q)}\}, P_{(B, X^{(q)})}^x)$ as in Definition 18.17 satisfying the condition $Y_0^{(q)} = x$. Here as in Observation 18.16,

$$P_{(B, X^{(q)})}^x(A) = P_{(B, X^{(q)})}^{\mathfrak{W}^{r+d} | Y_0^{(q)}}(A, x) \quad \text{for } (A, x) \in \mathfrak{W}^{r+d} \times \mathbb{R}^d.$$

Now pathwise uniqueness under deterministic initial condition implies pathwise uniqueness in the sense of probability law according to Lemma 19.19. Therefore $Y_0^{(q)} = x$ implies that $Y^{(1)}$ and $Y^{(2)}$ have identical probability distributions on $(\mathbf{W}^d, \mathfrak{W}^d)$. As we noted above, the probability distribution $P_{(B,X^{(q)})}$ of $(B, X^{(q)})$ is also the probability distribution of $(W^{(q)}, Y^{(q)})$. Thus $P_{(B,X^{(q)})}(\mathbf{W}^r \times A_1)$ for $A_1 \in \mathfrak{W}^d$ is the probability distribution of $Y^{(q)}$ on $(\mathbf{W}^d, \mathfrak{W}^d)$. By (2) and (4) of Observation 18.16, we have

(1) $$P_{(B,X^{(q)})}(\mathbf{W}^r \times A_1) = \int_{\mathbb{R}^d} P^x_{(B,X^{(q)})}(\mathbf{W}^r \times A_1)\mu(dx) \quad \text{for } A_1 \in \mathfrak{W}^d$$

(2) $$P^x_{(B,X^{(q)})}\left((Y_0^{(q)})^{-1}(\{x\})\right) = 1 \quad \text{for } x \in N^c.$$

Thus for $x \in N^c$, $P^x_{(B,X^{(q)})}(\mathbf{W}^r \times A_1)$ for $A_1 \in \mathfrak{W}^d$ is the probability distribution of $Y^{(q)}$ when $Y_0^{(q)} = x$. Therefore we have

(3) $$P^x_{(B,X^{(1)})}(\mathbf{W}^r \times A_1) = P^x_{(B,X^{(2)})}(\mathbf{W}^r \times A_1) \quad \text{for } A_1 \in \mathfrak{W}^d \text{ when } x \in N^c.$$

To show that pathwise uniqueness holds, we show that there exists a null set Λ in $(\Omega, \mathfrak{F}, P)$ such that $X^{(1)}(\cdot, \omega) = X^{(2)}(\cdot, \omega)$ for $\omega \in \Lambda^c$. Suppose no such null set in $(\Omega, \mathfrak{F}, P)$ exists. The continuity of the sample functions of $X^{(1)}$ and $X^{(2)}$ then implies that there exists some $t_0 \in \mathbb{R}_+$ such that $P\{\omega \in \Omega : X_{t_0}^{(1)}(\omega) \neq X_{t_0}^{(2)}(\omega)\} > 0$. Then the probability distributions $P_{X_{t_0}^{(1)}}$ and $P_{X_{t_0}^{(2)}}$ on $(\mathbb{R}^d, \mathfrak{B}_{\mathbb{R}^d})$ are not equal. Thus there exists $E \in \mathfrak{B}_{\mathbb{R}^d}$ such that $P_{X_{t_0}^{(1)}}(E) \neq P_{X_{t_0}^{(2)}}(E)$. Since $P_{X_{t_0}^{(1)}}(E) = P_{(B_{t_0}, X_{t_0}^{(1)})}(\mathbb{R}^r \times E)$, we have

(4) $$P_{(B_{t_0}, X_{t_0}^{(1)})}(\mathbb{R}^r \times E) \neq P_{(B_{t_0}, X_{t_0}^{(2)})}(\mathbb{R}^r \times E).$$

Let q_{t_0} be the mapping of $\mathbf{W}^{r+d}$ onto $\mathbb{R}^{r+d}$ defined by $q_{t_0}(w) = w(t_0)$ for $w \in \mathbf{W}^{r+d}$. Then

$$P_{(B_{t_0}, X_{t_0}^{(1)})} = P \circ (B_{t_0}, X_{t_0}^{(1)})^{-1} = P \circ (q_{t_0} \circ (\mathcal{B}, \mathcal{X}^{(q)}))^{-1}$$
$$= P \circ (\mathcal{B}, \mathcal{X}^{(q)})^{-1} \circ q_{t_0}^{-1} = P_{(\mathcal{B}, \mathcal{X}^{(q)})} \circ q_{t_0}^{-1}$$

so that

$$P_{(B_{t_0}, X_{t_0}^{(1)})}(\mathbb{R}^r \times E) = P_{(B,X^{(q)})} \circ q_{t_0}^{-1}(\mathbb{R}^r \times E).$$

Thus from (4), we have

(5) $$P_{(B,X^{(1)})}(q_{t_0}^{-1}(\mathbb{R}^r \times E)) \neq P_{(B,X^{(2)})}(q_{t_0}^{-1}(\mathbb{R}^r \times E)).$$

Since $q_{t_0}^{-1}(\mathbb{R}^r \times E)$ is a set of the type $\mathbf{W}^r \times A_1$ with $A_1 \in \mathfrak{W}^d$, (1) and (5) imply

$$\int_{\mathbb{R}^r} P^x_{(B,X^{(1)})}(q_{t_0}^{-1}(\mathbb{R}^r \times E))\mu(dx) \neq \int_{\mathbb{R}^r} P^x_{(B,X^{(2)})}(q_{t_0}^{-1}(\mathbb{R}^r \times E))\mu(dx).$$

Thus there exists $F \in \mathfrak{B}_{\mathbb{R}^d}$ with $\mu(F) > 0$ such that for $x \in F$ we have

$$P^x_{(B,X^{(1)})}(q_{t_0}^{-1}(\mathbb{R}^r \times E)) \neq P^x_{(B,X^{(2)})}(q_{t_0}^{-1}(\mathbb{R}^r \times E)),$$

contradicting (3). This shows that pathwise uniqueness holds. ∎

§20 Strong Solutions

[I] Existence of Strong Solutions

Consider the product measure space $(\mathbb{R}^d \times \mathbf{W}^r, \sigma(\mathfrak{B}_{\mathbb{R}^d} \times \mathfrak{W}^r), \mu \times m_W^r)$ where μ is an arbitrary probability measure on $(\mathbb{R}^d, \mathfrak{B}_{\mathbb{R}^d})$. Let $\sigma(\mathfrak{B}_{\mathbb{R}^d} \times \mathfrak{W}^r)^{\mu \times m_W^r}$ be the completion of $\sigma(\mathfrak{B}_{\mathbb{R}^d} \times \mathfrak{W}^r)$ with respect to $\mu \times m_W^r$. Let $(\mathbf{W}^r, \mathfrak{W}^{r,w}, \{\mathfrak{W}_t^{r,w}\}, m_W^r)$ be the standard filtered space generated by m_W^r on $(\mathbf{W}^r, m_W^r)$ as in Definition 19.10.

Definition 20.1. *Let (B, X) be a solution of the stochastic differential equation (1) in Definition 18.6 on a standard filtered space $(\Omega, \mathfrak{F}, \{\mathfrak{F}_t\}, P)$ and let μ be the probability distribution of X_0 on $(\mathbb{R}^d, \mathfrak{B}_{\mathbb{R}^d})$. We call (B, X) a strong solution if there exists a mapping F_μ of $\mathbb{R}^d \times \mathbf{W}^r$ into $\mathbf{W}^d$ satisfying the following conditions:*

1°. F_μ *is* $\sigma(\mathfrak{B}_{\mathbb{R}^d} \times \mathfrak{W}^r)^{\mu \times m_W^r}/\mathfrak{W}^d$-*measurable,*

2°. *for every $x \in \mathbb{R}^d$, $F_\mu[x, \cdot]$ is a $\mathfrak{W}_t^{r,w}/\mathfrak{W}_t^d$-measurable mapping of $\mathbf{W}^r$ into $\mathbf{W}^d$ for every $t \in \mathbb{R}_+$,*

3°. $X(\cdot, \omega) = F_\mu[X_0(\omega), B(\cdot, \omega)]$ *for a.e.* ω *in* $(\Omega, \mathfrak{F}, P)$.

In Theorem 20.5 below we show that if the coefficients α and β in the stochastic differential equation (1) in Definition 18.6 satisfy the Lipschitz condition (1) and the growth condition (2) in Definition 19.4 then a strong solution exists on any standard filtered space $(\Omega, \mathfrak{F}, \{\mathfrak{F}_t\}, P)$ on which an r-dimensional $\{\mathfrak{F}_t\}$-adapted null at 0 Brownian motion exists.

Definition 20.2. *Let* $\mathbf{L}_{2,\infty}^{d,c}(\mathbb{R}_+ \times \Omega, \{\mathfrak{F}_t\}, m_L \times P)$, *or briefly* $\mathbf{L}_{2,\infty}^{d,c}$, *be the collection of all d-dimensional continuous $\{\mathfrak{F}_t\}$-adapted processes X on a standard filtered space $(\Omega, \mathfrak{F}, \{\mathfrak{F}_t\}, P)$ satisfying the condition that* $\sup_{s \in [0,t]} \mathbf{E}(|X(s)|^2) < \infty$ *for every* $t \in \mathbb{R}_+$.

Lemma 20.3. Let $(\Omega, \mathfrak{F}, \{\mathfrak{F}_t\}, P)$ be a standard filtered space on which an r-dimensional $\{\mathfrak{F}_t\}$-adapted null at 0 Brownian motion B exists. Assume that the coefficients α and β in the stochastic differential equation (1) in Definition 18.6 satisfy condition (2) in Definition 19.4. With fixed $x \in \mathbb{R}^d$, let us define a mapping τ of $\mathbf{L}_{2,\infty}^{d,c}$ by setting for $X \in \mathbf{L}_{2,\infty}^{d,c}$

(1) $\qquad (\tau X)(t) = x + \int_{[0,t]} \alpha(s, X) \, dB(s) + \int_{[0,t]} \beta(s, X) \, ds \quad \text{for } t \in \mathbb{R}_+.$

Then $\tau X \in \mathbf{L}_{2,\infty}^{d,c}$. If we define a sequence $\{X^{(q)} : q \in \mathbb{Z}_+\}$ in $\mathbf{L}_{2,\infty}^{d,c}$ by letting

(2) $\qquad \begin{cases} X^{(0)}(t) = x & \text{for } t \in \mathbb{R}_+ \\ X^{(q)}(t) = (\tau X^{(q-1)})(t) & \text{for } t \in \mathbb{R}_+ \text{ and } q \in \mathbb{N}, \end{cases}$

then for every $T \in \mathbb{R}_+$ there exists $K_T \in \mathbb{R}_+$ such that

(3) $\qquad \sup_{t \in [0,T]} \mathbf{E}(|X^{(q)}(t)|^2) \leq K_T \quad \text{for } q \in \mathbb{Z}_+.$

Proof. To show the existence of the stochastic integrals in (1), we show that $\alpha_j^i(\cdot, X)$ and $\beta^i(\cdot, X)$ are in $\mathbf{L}_{2,\infty}(\mathbb{R}_+ \times \Omega, m_L \times P)$ and $\mathbf{L}_{1,\infty}(\mathbb{R}_+ \times \Omega, m_L \times P)$ respectively. Now since α and β satisfy condition (2) in Definition 19.4 and since $X \in \mathbf{L}_{2,\infty}^{d,c}$, for every $T \in \mathbb{R}_+$ we have

$$\mathbf{E}\left[\int_{[0,T]} \{|\alpha(t, X)|^2 + |\beta(t, X)|^2\} m_L(dt)\right]$$

$$\leq L_T \int_{[0,T]} \mathbf{E}\left[\int_{[0,t]} |X(s)|^2 \lambda(ds) + |X(t)|^2 + 1\right] m_L(dt)$$

$$\leq L_T \left[\sup_{t \in [0,T]} \mathbf{E}(|X(t)|^2)\{\lambda([0,T]) + 1\} + 1\right] T < \infty.$$

This shows that $\alpha_j^i(\cdot, X)$ and $\beta^i(\cdot, X)$ are all in $\mathbf{L}_{2,\infty}(\mathbb{R}_+ \times \Omega, m_L \times P)$ and hence the stochastic integrals in (1) exist.

To show that τX is in $\mathbf{L}_{2,\infty}^{d,c}$, it remains to verify that for every $t \in \mathbb{R}_+$ we have

(4) $\qquad \sup_{s \in [0,t]} \mathbf{E}(|X(s)|^2) < \infty.$

Let $c_0 = \max\{|x|^2, 1\}$. For $X \in \mathbf{L}_{2,\infty}^{d,c}$ and $T \in \mathbb{R}_+$ let $A(t; X)$ be a real nonnegative valued monotone increasing function for $t \in [0, T]$ such that

(5) $\qquad \max\{\sup_{s \in [0,t]} \mathbf{E}(|X(s)|^2), 1\} \leq A(t; X) \quad \text{for } t \in [0, T].$

§20. STRONG SOLUTIONS

Since $\sup_{s\in[0,t]} \mathbf{E}(|X(s)|^2) < \infty$ for every $t \in \mathbb{R}_+$ such function $A(\cdot; X)$ always exists. Let us show that with $A(\cdot; X)$ we have for τX defined by (1) and $T \in \mathbb{R}_+$ the estimate

$$(6) \qquad \mathbf{E}(|(\tau X)(t)|^2) \leq 3\left\{c_0 + C_T \int_{[0,t]} A(s; X) m_L(ds)\right\} \quad \text{for } t \in \mathbb{R}_+,$$

where

$$(7) \qquad C_T = (1+T)L_T\{\lambda([0,T]) + 2\}.$$

Note that (6) and (7) imply $\mathbf{E}(|(\tau X)(t)|^2) \leq 3\{c_0 + C_T A(T; X)T\} < \infty$ which is (4). To prove (6), note that since $\{\sum_{j=1}^n a_j\}^2 \leq n\sum_{j=1}^n a_j^2$ for any real numbers $a_1, \ldots, a_n$, we have from (1)

$$\mathbf{E}(|(\tau X)(t)|^2) \leq 3\left\{c_0 + \mathbf{E}\left[|\int_{[0,t]} \alpha(s,X)\,dB(s)|^2\right] + \mathbf{E}\left[|\int_{[0,t]} \beta(s,X)\,ds|^2\right]\right\}.$$

Now by (3) of Proposition 13.39

$$\mathbf{E}\left[|\int_{[0,t]} \alpha(s,X)\,dB(s)|^2\right] = \int_{[0,t]} \mathbf{E}[|\alpha(s,X)|^2] m_L(ds),$$

and by the Schwarz Inequality

$$\mathbf{E}\left[|\int_{[0,t]} \beta(s,X)\,ds|^2\right] \leq t \int_{[0,t]} \mathbf{E}[|\beta(s,X)|^2] m_L(ds).$$

Thus for $t \in [0,T]$ we have by (2) of Definition 19.4 and (5)

$$\mathbf{E}(|(\tau X)(t)|^2) \leq 3\left\{c_0 + (1+T)\int_{[0,t]} \mathbf{E}\left[|\alpha(s,X)|^2 + |\beta(s,X)|^2\right] m_L(ds)\right\}$$

$$\leq 3\left\{c_0 + (1+T)L_T \int_{[0,t]} \mathbf{E}\left[\int_{[0,s]} |X(u)|^2 \lambda(du) + |X(s)|^2 + 1\right] m_L(ds)\right\}$$

$$\leq 3\left\{c_0 + (1+T)L_T\{\lambda([0,T]) + 2\}\int_{[0,t]} A(s;X) m_L(ds)\right\},$$

proving (6) and in particular (4).

Consider the sequence defined by (2). Clearly $X^{(0)} \in \mathbf{L}_{2,\infty}^{d,c}$. Then since τ is a mapping of $\mathbf{L}_{2,\infty}^{d,c}$ into $\mathbf{L}_{2,\infty}^{d,c}$ we have $X^{(q)} \in \mathbf{L}_{2,\infty}^{d,c}$ for $q \in \mathbb{N}$. To prove (3), let us show first that for every $T \in \mathbb{R}_+$ we have for $t \in [0,T]$ and $q \in \mathbb{Z}_+$ the estimate

$$(8) \qquad \mathbf{E}(|X^{(q)}(t)|^2) \leq 3c_0 \left\{\sum_{k=0}^{q-1} \frac{(3C_T t)^k}{k!} + 3^{q-1}\frac{(C_T t)^q}{q!}\right\}.$$

Now for $q = 0$, we have $\mathbf{E}(|X^{(0)}(t)|^2) = |x|^2 \leq c_0$ so that (8) holds. Suppose (8) holds for some $q \in \mathbb{Z}_+$. Let $A(t; X^{(q)})$ be defined to be equal to the right side of the inequality (8) for $t \in [0, T]$. Then $\max\{\sup_{s\in[0,t]} \mathbf{E}(|X^{(q)}(s)|^2), 1\} \leq A(t; X^{(q)})$ for $t \in [0, T]$ so that $A(\cdot; X^{(q)})$ satisfies (5). Therefore by (6) we have

$$\mathbf{E}(|X^{(q+1)}(t)|^2) = \mathbf{E}(|(\tau X^{(q)})(t)|^2) \leq 3\left\{c_0 + C_T \int_{[0,t]} A(s; X^{(q)}) m_L(ds)\right\}$$

$$= 3\left\{c_0 + 3c_0 C_T \int_{[0,t]} \left\{\sum_{k=0}^{q-1} \frac{(3C_T s)^k}{k!} + 3^{q-1}\frac{(C_T s)^q}{q!}\right\} m_L(ds)\right\}$$

$$= 3c_0 \left\{\sum_{k=0}^{q} \frac{(3C_T t)^k}{k!} + 3^q \frac{(C_T t)^{q+1}}{(q+1)!}\right\} \quad \text{for } t \in [0, T],$$

that is, (8) holds for $q + 1$. Thus by induction (8) holds for all $q \in \mathbb{Z}_+$. From (8) we then have for $t \in [0, T]$ and $q \in \mathbb{Z}_+$ the bound

$$\mathbf{E}(|X^{(q)}(t)|^2) \leq 3c_0 \sum_{k=0}^{q} \frac{(3C_T t)^k}{k!} \leq 3c_0 e^{3C_T T}.$$

With $K_T = 3c_0 e^{3C_T T}$ we have (3). ∎

Lemma 20.4. *Suppose α and β satisfy both (1) and (2) in Definition 19.4. Then for every $T \in \mathbb{R}_+$ there exists $M_T \in \mathbb{R}_+$ such that*

(1) $\quad \mathbf{E}\left[\sup_{s\in[0,t]} |(\tau X)(s) - (\tau Y)(s)|^2\right] \leq M_T \int_{[0,t]} \sup_{u\in[0,s]} \mathbf{E}(|X(u) - Y(u)|^2) m_L(ds)$

$\quad \leq M_T \int_{[0,t]} \mathbf{E}\left[\sup_{u\in[0,s]} |X(u) - Y(u)|^2\right] m_L(ds) \quad \text{for } t \in [0, T] \text{ and } X, Y \in \mathbf{L}_{2,\infty}^{d,c}.$

Proof. For $X, Y \in \mathbf{L}_{2,\infty}^{d,c}$ we have $\tau X, \tau Y \in \mathbf{L}_{2,\infty}^{d,c}$ by Lemma 20.3. By (1) of Lemma 20.3 we have

$$|(\tau X)(s) - (\tau Y)(s)|^2 \leq 2|\int_{[0,s]} \{\alpha(u, X) - \alpha(u, Y)\} dB(u)|^2$$
$$+ 2|\int_{[0,s]} \{\beta(u, X) - \beta(u, Y)\} du|^2$$

and thus

(2) $\quad \mathbf{E}\left[\sup_{s\in[0,t]} |(\tau X)(s) - (\tau Y)(s)|^2\right] \leq 2\mathbf{E}\left[\sup_{s\in[0,t]} |\int_{[0,s]} \{\alpha(u, X) - \alpha(u, Y)\} dB(s)|^2\right]$

§20. STRONG SOLUTIONS

$$+ \ 2\mathbf{E}\left[\sup_{s\in[0,t]}|\int_{[0,s]}\{\beta(u,X)-\beta(u,Y)\}\,du|^2\right].$$

Now $\mathbf{E}[\sup_{s\in[0,t]} M_t^2] \leq 4\mathbf{E}(M_t^2)$ for a continuous L_2-martingale M on $(\Omega,\mathfrak{F},\{\mathfrak{F}_t\},P)$ according to (2) of Theorem 6.16. Since

$$|\int_{[0,s]}\{\alpha(u,X)-\alpha(u,Y)\}\,dB(u)|^2 = \sum_{l=1}^{d}\sum_{j=1}^{r}|\int_{[0,s]}\{\alpha_j^i(u,X)-\alpha_j^i(u,Y)\}\,dB^i(u)|^2$$

and $\{\int_{[0,s]}\{\alpha_j^i(u,X)-\alpha_j^i(u,Y)\}\,dB^i(u) : s \in \mathbb{R}_+\}$ is a continuous L_2-martingale on $(\Omega,\mathfrak{F},\{\mathfrak{F}_t\},P)$, we have

(3)
$$\mathbf{E}\left[\sup_{s\in[0,t]}|\int_{[0,s]}\{\alpha(u,X)-\alpha(u,Y)\}\,dB(s)|^2\right]$$
$$\leq 4\mathbf{E}\left[|\int_{[0,t]}\{\alpha(u,X)-\alpha(u,Y)\}\,dB(s)|^2\right]$$
$$= 4\mathbf{E}\left[\int_{[0,t]}|\alpha(u,X)-\alpha(u,Y)|^2 m_L(du)\right].$$

Similarly by the Schwarz Inequality

$$|\int_{[0,s]}\{\beta(u,X)-\beta(u,Y)\}\,du|^2 \leq s\int_{[0,s]}|\beta(u,X)-\beta(u,Y)|^2 m_L(du)$$

so that

(4)
$$\mathbf{E}\left[\sup_{s\in[0,t]}|\int_{[0,s]}\{\beta(u,X)-\beta(u,Y)\}\,du|^2\right]$$
$$\leq t\,\mathbf{E}\left[\int_{[0,s]}|\beta(u,X)-\beta(u,Y)\}|^2 m_L(du)\right].$$

Using (3) and (4) in (2) and applying (1) of Definition 19.4 we have

$$\mathbf{E}\left[\sup_{s\in[0,t]}|(\tau X)(s)-(\tau Y)(s)|^2\right]$$
$$\leq (8+2T)\int_{[0,t]}\mathbf{E}[\{|\alpha(s,X)-\alpha(s,Y)|^2+|\beta(s,X)-\beta(s,Y)|^2\}]m_L(ds)$$
$$\leq (8+2T)\int_{[0,t]}\mathbf{E}\left[|X(u)-Y(u)|^2\lambda(du)+|X(s)-Y(s)|\right]m_L(ds)$$
$$\leq (8+2T)\int_{[0,t]}\sup_{u\in[0,s]}\mathbf{E}(|X(u)-Y(u)|^2)\{\lambda([0,s])+1\}m_L(ds)$$
$$\leq (8+2T)\{\lambda([0,s])+1\}\int_{[0,t]}\mathbf{E}(|X(u)-Y(u)|^2)m_L(ds).$$

With $M_T = (8+2T)\{\lambda([0,s])+1\}$, we have the first inequality in (1). The second inequality is immediate from $\mathbf{E}(|X(u) - Y(u)|^2) \leq \mathbf{E}[\sup_{u\in[0,s]} |X(u) - Y(u)|^2]$ for $u \in [0,s]$ and then $\sup_{u\in[0,s]}\mathbf{E}(|X(u) - Y(u)|^2) \leq \mathbf{E}[\sup_{u\in[0,s]} |X(u) - Y(u)|^2]$. ∎

Theorem 20.5. *Suppose the coefficients α and β in the stochastic differential equation (1) of Definition 18.6 satisfy conditions (1) and (2) of Definition 19.4. Let $(\Omega, \mathfrak{F}, \{\mathfrak{F}_t\}, P)$ be a standard filtered space on which an r-dimensional $\{\mathfrak{F}_t\}$-adapted null at 0 Brownian motion B exists. Then there exists a mapping F of $\mathbb{R}^d \times \mathbf{W}^r$ into $\mathbf{W}^d$ such that*
(1) F *is* $\sigma(\mathfrak{B}_{\mathbb{R}^d} \times \mathfrak{W}^r)/\mathfrak{W}^d$-*measurable*,
(2) *for every* $x \in \mathbb{R}^d$, $F[x,\cdot]$ *is a* $\mathfrak{W}_t^{r,w}/\mathfrak{W}_t^d$-*measurable mapping of* $\mathbf{W}^r$ *into* $\mathbf{W}^d$ *for every* $t \in \mathbb{R}_+$,
(3) *with an arbitrary $\mathfrak{F}_0$-measurable d-dimensional random variable Z on $(\Omega, \mathfrak{F}, P)$, if we define $X(\cdot, \omega) = F[Z(\omega), B(\cdot, \omega)]$ for $\omega \in \Omega$, then (B, X) is a solution of the stochastic differential equation satisfying $X_0 = Z$ on $(\Omega, \mathfrak{F}, \{\mathfrak{F}_t\}, P)$.*

Proof. Let $x \in \mathbb{R}^d$ and define a sequence $\{X^{(q)} : q \in \mathbb{Z}_+\}$ in $\mathbf{L}_{2,\infty}^{d,c}$ by letting

(4) $\quad \begin{cases} X^{(0)}(t) = x & \text{for } t \in \mathbb{R}_+ \\ X^{(q)}(t) = (\tau X^{(q-1)})(t) & \text{for } t \in \mathbb{R}_+ \text{ and } q \in \mathbb{N}, \end{cases}$

where τ is a mapping of $\mathbf{L}_{2,\infty}^{d,c}$ into $\mathbf{L}_{2,\infty}^{d,c}$ defined by setting

(5) $\quad (\tau X)(t) = x + \int_{[0,t]} \alpha(s, X)\, dB(s) + \int_{[0,t]} \beta(s, X)\, ds \quad \text{for } t \in \mathbb{R}_+$

for $X \in \mathbf{L}_{2,\infty}^{d,c}$. By iterated application of (1) in Lemma 20.4 we have

$$\mathbf{E}\left[\sup_{t\in[0,T]} |X^{(q+1)}(t) - X^{(q)}(t)|^2\right]$$
$$\leq M_T \int_0^T \sup_{t\in[0,t_q]} \mathbf{E}\left[|X^{(q)}(t) - X^{(q-1)}(t)|^2\right] dt_q$$
$$\leq M_T^2 \int_0^T \int_0^{t_q} \sup_{t\in[0,t_{q-1}]} \mathbf{E}\left[|X^{(q-1)}(t) - X^{(q-2)}(t)|^2\right] dt_q dt_{q-1}$$
$$\vdots$$
$$\leq M_T^q \int_0^T \int_0^{t_q} \cdots \int_0^{t_2} \sup_{t\in[0,t_1]} \mathbf{E}\left[|X^{(1)}(t) - X^{(0)}(t)|^2\right] dt_q dt_{q-1} \cdots dt_1.$$

§20. STRONG SOLUTIONS

By (3) in Lemma 20.3 we have

$$\sup_{t \in [0,t_1]} \mathbf{E}\left[|X^{(1)}(t) - X^{(0)}(t)|^2\right] \leq \sup_{t \in [0,t_1]} \mathbf{E}\left[2|X^{(1)}(t)|^2 + 2|X^{(0)}(t)|^2\right] \leq 4K_T$$

and therefore

(6)
$$\mathbf{E}\left[\sup_{t \in [0,T]} |X^{(q+1)}(t) - X^{(q)}(t)|^2\right]$$
$$\leq 4K_T M_T^q \int_0^T \int_0^{t_q} \cdots \int_0^{t_2} dt_q dt_{q-1} \cdots dt_1 = 4K_T \frac{(M_T T)^q}{q!}.$$

For $q \in \mathbb{Z}_+$ let

$$A_q = \{\omega \in \Omega : \sup_{t \in [0,T]} |X^{(q+1)}(t) - X^{(q)}(t)| > \frac{1}{2^q}\}.$$

By the Chevyshev Inequality we have

$$P(A_q) \leq (2^q)^2 4K_T \frac{(M_T T)^q}{q!} = 4K_T \frac{(4M_T T)^q}{q!}.$$

Since $\sum_{q \in \mathbb{Z}_+} (4M_T T)^q (q!)^{-1} = e^{4M_T T} < \infty$, we have $P(\liminf_{q \to \infty} A_q^c) = 1$ by the Borel-Cantelli Lemma. Thus for a.e. $\omega \in \Omega$ we have $\sup_{t \in [0,T]} |X^{(q+1)}(t,\omega) - X^{(q)}(t,\omega)| \leq 2^{-q}$ for all but finitely many $q \in \mathbb{Z}_+$. Then for an arbitrary $\varepsilon > 0$, for a.e. $\omega \in \Omega$ there exists $N(\omega) \in \mathbb{Z}_+$ such that

$$\sup_{t \in [0,T]} |X^{(m)}(t,\omega) - X^{(n)}(t,\omega)| \leq \sum_{q=n}^{m-1} \sup_{t \in [0,T]} |X^{(q+1)}(t,\omega) - X^{(q)}(t,\omega)| \leq \sum_{q=n}^{m-1} 2^{-q} < \varepsilon$$

for $m > n \geq N(\omega)$. This shows that $\{X^{(q)}(\cdot,\omega) : q \in \mathbb{Z}_+\}$ converges uniformly on $[0,T]$ for a.e. $\omega \in \Omega$. Considering $T = n$ for $n \in \mathbb{N}$, we have a null set Λ in $(\Omega, \mathfrak{F}, P)$ such that $\{X^{(q)}(\cdot,\omega) : q \in \mathbb{Z}_+\}$ converges uniformly on every finite interval in $\mathbb{R}_+$ for $\omega \in \Lambda^c$. Let us define a process X on $(\Omega, \mathfrak{F}, P)$ by

(7)
$$X(t,\omega) = \begin{cases} \lim_{q \to \infty} X^{(q)}(t,\omega) & \text{for } (t,\omega) \in \mathbb{R}_+ \times \Lambda^c \\ 0 & \text{for } (t,\omega) \in \mathbb{R}_+ \times \Lambda. \end{cases}$$

By the uniform convergence of $\{X^{(q)}(\cdot,\omega) : q \in \mathbb{Z}_+\}$ to $X(\cdot,\omega)$ on every finite interval for $\omega \in \Lambda^c$ and by the continuity of $\{X^{(q)}(\cdot,\omega) : q \in \mathbb{Z}_+\}$ on $\mathbb{R}_+$ for every $\omega \in \Omega$, X is a continuous process. The uniform convergence also implies according to Lemma 17.12 that for $\omega \in \Lambda^c$

(8) $\qquad X^{(q)}(\cdot, \omega)$ converges to $X(\cdot, \omega)$ in the metric ρ_∞ on $\mathbf{W}^d$.

By the fact that $\Lambda \in \mathfrak{F}_t$ for every $t \in \mathbb{R}_+$ since $\mathfrak{F}_t$ is augmented, and by the fact that $X^{(q)}(t, \cdot)$ is $\mathfrak{F}_t$-measurable, $X(t, \cdot)$ is $\mathfrak{F}_t$-measurable, that is, X is an $\{\mathfrak{F}_t\}$-adapted process. Regarding the convergence of $X^{(q)}$ to X we have furthermore

(9) $\qquad \lim_{q \to \infty} \left\{ \sup_{t \in [0,T]} \mathbf{E}\left[|X^{(q)}(t) - X(t)|^2\right]\right\} = 0.$

To prove (9) note that

$$\sup_{t \in [0,T]} \mathbf{E}\left[|X^{(m)}(t) - X^{(n)}(t)|^2\right]^{1/2} \le \sum_{q=n}^{m-1} \sup_{t \in [0,T]} \mathbf{E}\left[|X^{(q+1)}(t) - X^{(q)}(t)|^2\right]^{1/2}$$

$$\le \sum_{q=n}^{m-1} \mathbf{E}\left[\sup_{t \in [0,T]} |X^{(q+1)}(t) - X^{(q)}(t)|^2\right]^{1/2} \le 2\sqrt{K_T} \sum_{q=n}^{m-1} \sqrt{\frac{(M_T T)^q}{q!}}$$

by (6). Since $\sum_{q \in \mathbb{Z}_+} \sqrt{(M_T T)^q (q!)^{-1}} < \infty$, the Cauchy Criterion for uniform convergence implies that there exists $X^*(t) \in \mathbf{L}_2(\Omega, \mathfrak{F}, P)$ for $t \in [0,T]$ such that

$$\lim_{q \to \infty} \left\{ \sup_{t \in [0,T]} \mathbf{E}\left[|X^{(q)}(t) - X^*(t)|^2\right]\right\} = 0.$$

Since convergence in L_2 of $X^{(q)}(t)$ to $X^*(t)$ implies the existence of a subsequence which converges a.e. to $X^*(t)$ and since $X^{(q)}(t)$ converges a.e. to $X(t)$, we have $X^*(t) = X(t)$ a.e. on $(\Omega, \mathfrak{F}, P)$ for every $t \in [0,T]$. Thus (9) holds.

To show that $X \in \mathbf{L}_{2,\infty}^{d,c}$ note that by Fatou's Lemma and by (3) of Lemma 20.3

$$\mathbf{E}(|X_t|^2) \le \liminf_{q \to \infty} \mathbf{E}(|X_t^{(q)}|^2) \le \liminf_{q \to \infty} \sup_{t \in [0,T]} \mathbf{E}(|X_t^{(q)}|^2) \le K_T$$

so that $\sup_{t \in [0,T]} \mathbf{E}(|X_t|^2) \le K_T < \infty$. Thus $X \in \mathbf{L}_{2,\infty}^{d,c}$ and τX is defined. Let us show that $\tau X = X$ for our X defined by (7). Now for every $T \in \mathbb{R}_+$ we have

$$\sup_{t \in [0,t]} |(\tau X)(t) - X(t)|^2 \le 2 \sup_{t \in [0,t]} |(\tau X)(t) - X^{(q+1)}(t)|^2 + 2 \sup_{t \in [0,t]} |(\tau X)^{(q+1)}(t) - X(t)|^2$$

§20. STRONG SOLUTIONS

for an arbitrary $q \in \mathbb{Z}_+$. Since $\{X^{(q)}(\cdot,\omega) : q \in \mathbb{Z}_+\}$ converges uniformly on $[0,T]$ to $X(\cdot,\omega)$ for $\omega \in \Lambda^c$ we have $\lim_{q \to \infty} \sup_{t \in [0,t]} |(\tau X)^{(q+1)}(t,\omega) - X(t,\omega)|^2 = 0$ for $\omega \in \Lambda^c$. Thus

$$\sup_{t \in [0,t]} |(\tau X)(t) - X(t)|^2 \leq 2 \liminf_{q \to \infty} \left\{ \sup_{t \in [0,t]} |(\tau X)(t) - X^{(q+1)}(t)|^2 \right\}.$$

Then by Fatou's Lemma and Lemma 20.4

$$\mathbf{E}\left[\sup_{t \in [0,t]} |(\tau X)(t) - X(t)|^2 \right] \leq 2 \liminf_{q \to \infty} \mathbf{E}\left[\sup_{t \in [0,t]} |(\tau X)(t) - X^{(q+1)}(t)|^2 \right]$$

$$\leq 2M_T \liminf_{q \to \infty} \int_{[0,T]} \sup_{s \in [0,t]} \mathbf{E}[|X(s) - X^{(q)}(s)|^2] m_L(dt)$$

and therefore

(10) $\quad \mathbf{E}\left[\sup_{t \in [0,t]} |(\tau X)(t) - X(t)|^2 \right] \leq 2M_T \limsup_{q \to \infty} \int_{[0,T]} \sup_{s \in [0,t]} \mathbf{E}[|X(s) - X^{(q)}(s)|^2] m_L(dt).$

Now since $X, X^{(q)} \in \mathbf{L}_{2,\infty}^{d,c}$, we have by (3) of Lemma 20.3

$$\sup_{s \in [0,t]} \mathbf{E}[|X(s) - X^{(q)}(s)|^2] \leq 2 \left\{ \sup_{s \in [0,t]} \mathbf{E}(|X(s)|^2) + \sup_{s \in [0,t]} \mathbf{E}(|X^{(q)}(s)|^2) \right\}$$

$$\leq 2 \left\{ \sup_{s \in [0,t]} \mathbf{E}(|X(s)|^2) + K_T \right\} < \infty.$$

This shows that the integrands on the right side of (10) are bounded on $[0,T]$ uniformly in $q \in \mathbb{Z}_+$. Thus Fatou's Lemma for limit superior applies and we have

$$\mathbf{E}\left[\sup_{t \in [0,t]} |(\tau X)(t) - X(t)|^2 \right]$$

$$\leq 2M_T \int_{[0,T]} \limsup_{q \to \infty} \left\{ \sup_{s \in [0,t]} \mathbf{E}[|X(s) - X^{(q)}(s)|^2] \right\} m_L(dt) = 0$$

where the last equality is by (9). Therefore $\sup_{t \in [0,t]} |(\tau X)(t) - X(t)| = 0$ a.e. on $(\Omega, \mathfrak{F}, P)$. Thus $(\tau X)(t) = X(t)$ for $t \in [0,T]$ a.e. on $(\Omega, \mathfrak{F}, P)$. By considering $T = n$ for $n \in \mathbb{N}$, we have a null set Λ in $(\Omega, \mathfrak{F}, P)$ such that

$$(\tau X)(t,\omega) = X(t,\omega) \quad \text{for } (t,\omega) \in \mathbb{R}_+ \times \Lambda^c.$$

Recalling the definition of τX by (5) we have

(11) $$X_t = x + \int_{[0,t]} \alpha(s, X) \, dB(s) + \int_{[0,t]} \beta(s, X) \, ds \quad \text{for } t \in \mathbb{R}_+,$$

that is, X is a solution of the differential equation (1) in Definition 18.6 with $X_0 = x$.

Let us show that there exists a mapping F of $\mathbb{R}^d \times \mathbf{W}^r$ into $\mathbf{W}^d$ satisfying (1) and (2) and for X defined by (4) and (7) there exists a null set Λ_∞ in $(\Omega, \mathfrak{F}, P)$ such that we have

(12) $$X(\cdot, \omega) = F[x, B(\cdot, \omega)] \quad \text{for } \omega \in \Lambda_\infty^c.$$

For this we show first by induction that for every $q \in \mathbb{Z}_+$ there exists a mapping $F^{(q)}$ of $\mathbb{R}^d \times \mathbf{W}^r$ into $\mathbf{W}^d$ satisfying (1) and (2) and for $X^{(q)}$ defined by (4) there exists a null set Λ_q in $(\Omega, \mathfrak{F}, P)$ such that

(13) $$X^{(q)}(\cdot, \omega) = F^{(q)}[x, B(\cdot, \omega)] \quad \text{for } \omega \in \Lambda_q^c.$$

Now according to (4), $X^{(0)}(t, \omega) = x$ for $(t, \omega) \in \mathbb{R}_+ \times \Omega$. Let w_x be an element in $\mathbf{W}^d$ defined by $w_x(t) = x$ for $t \in \mathbb{R}_+$. If we define a mapping $F^{(0)}$ of $\mathbb{R}^d \times \mathbf{W}^r$ into $\mathbf{W}^d$ by setting $F^{(0)}[x, w] = w_x$ for $(x, w) \in \mathbb{R}^d \times \mathbf{W}^r$ then $F^{(0)}[x, B(\cdot, \omega)] = w_x = X^{(0)}(\cdot, \omega)$ for every $\omega \in \Omega$, verifying (13) for $F^{(0)}$. To show that $F^{(0)}$ satisfies (1), note that since $F^{(0)}$ maps the entire $\mathbb{R}^d \times \mathbf{W}^r$ into a single point $w_x \in \mathbf{W}^d$, for any $A \in \mathfrak{W}^d$ we have $(F^{(0)})^{-1}(A) = \mathbb{R}^d \times \mathbf{W}^r$ or $\emptyset$ according as $w_x \in A$ or $w_x \in A^c$. Thus $F^{(0)}$ is $\sigma(\mathfrak{B}_{\mathbb{R}^d} \times \mathfrak{W}^r)/\mathfrak{W}^d$-measurable. By the same argument $F^{(0)}$ satisfies (2).

Next suppose for some $q \in \mathbb{Z}_+$ there exists a mapping $F^{(q)}$ of $\mathbb{R}^d \times \mathbf{W}^r$ into $\mathbf{W}^d$ satisfying (1), (2), and (13). Let us show the existence of $F^{(q+1)}$. By (4) and (5) we have

$$X^{(q+1)}(t) = x + \int_{[0,t]} \alpha(s, X^{(q)}) \, dB(s) + \int_{[0,t]} \beta(s, X^{(q)}) \, ds \quad \text{for } t \in \mathbb{R}_+.$$

For brevity let us use the notations

$$\begin{cases} \alpha(\cdot, X^{(q)}) \bullet B = [\sum_{j=1}^r \alpha_j^i(\cdot, X^{(q)}) \bullet B^i : i = 1, \ldots, d] \\ \beta(\cdot, X^{(q)}) \bullet m_L = [\beta^i(\cdot, X^{(q)}) \bullet m_L : i = 1, \ldots, d] \end{cases}$$

Then

$$X^{(q+1)} = x + \alpha(\cdot, X^{(q)}) \bullet B + (\cdot, X^{(q)}) \bullet m_L.$$

Since $x = X^{(0)}(\cdot, \omega) = F^{(0)}[x, B(\cdot, \omega)]$ for $\omega \in \Omega$, if we show that there exist mappings $G^{(q)}$ and $H^{(q)}$ of $\mathbb{R}^d \times \mathbf{W}^r$ into $\mathbf{W}^d$ satisfying (1), (2) and

(14) $$\begin{cases} (\alpha(\cdot, X^{(q)}) \bullet B)(\omega) = G^{(q)}[x, B(\cdot, \omega)] & \text{for } \omega \in \Lambda_q^c \\ (\beta(\cdot, X^{(q)}) \bullet m_L)(\omega) = H^{(q)}[x, B(\cdot, \omega)] & \text{for } \omega \in \Lambda_q^c, \end{cases}$$

§20. STRONG SOLUTIONS

where Λ_q is a null set in $(\Omega, \mathfrak{F}, P)$, then by setting $F^{(q+1)} = F^{(0)} + G^{(q)} + H^{(q)}$ we have a mapping $F^{(q+1)}$ of $\mathbb{R}^d \times \mathbf{W}^r$ into $\mathbf{W}^d$ satisfying (1), (2) and

$$F^{(q+1)}[x, B(\cdot, \omega)] = x + (\alpha(\cdot, X^{(q)}) \bullet B)(\omega) + (\beta(\cdot, X^{(q)}) \bullet m_L)(\omega) = X^{(q+1)}(\cdot, \omega)$$

for $\omega \in \Lambda_q^c$, verifying (13). Thus it remains to show the existence of $G^{(q)}$ and $H^{(q)}$. Let $\Phi^{(q)}$ be a mapping of $\mathbb{R}_+ \times \Omega$ into $\mathbb{R}^{dr}$ defined by

$$(15) \qquad \Phi^{(q)}(t, \omega) = \alpha(t, X^{(q)}(\cdot, \omega)) \quad \text{for } (t, \omega) \in \mathbb{R}_+ \times \Omega.$$

Since $F^{(q)}$ satisfies (13) we have

$$(16) \qquad \Phi^{(q)}(t, \omega) = \alpha(t, F^{(q)}[x, B(\cdot, \omega)]) \quad \text{for } (t, \omega) \in \mathbb{R}_+ \times \Lambda_q^c.$$

By Lemma 18.5, $\Phi^{(q)}$ is an $\{\mathfrak{F}_t\}$-progressively measurable and in particular $\{\mathfrak{F}_t\}$-adapted measurable process on $(\Omega, \mathfrak{F}, \{\mathfrak{F}_t\}, P)$. Also since $X^{(q)} \in \mathbf{L}_{2,\infty}^{d,c}$, we have $\alpha_j^i(\cdot, X^{(q)}) \in \mathbf{L}_{2,\infty}(\mathbb{R}_+ \times \Omega, m_L \times P)$ as we saw in the proof of Lemma 20.3. Let $\varphi^{(q)}$ be a mapping of $\mathbb{R}_+ \times \mathbb{R}^d \times \mathbf{W}^r$ into $\mathbb{R}^{dr}$ defined by

$$(17) \qquad \varphi^{(q)}(t, x, w) = \alpha(t, F^{(q)}[x, w]) \quad \text{for } (t, x, w) \in \mathbb{R}_+ \times \mathbb{R}^d \times \mathbf{W}^r.$$

Let us consider first the case where the components $(\Phi^{(q)})_j^i$ of $\Phi^{(q)}$ are in $\mathbf{L}_0(\Omega, \mathfrak{F}, \{\mathfrak{F}_t\}, P)$. In this case we have for $(t, \omega) \in \mathbb{R}_+ \times \Lambda_q^c$

$$(18) \qquad \begin{aligned} (\Phi^{(q)} \bullet B)^i(t, \omega) &= \sum_{j=1}^{r} \sum_{k=1}^{m} (\Phi^{(q)})_j^i(t_k, \omega) \{B^i(t_k, \omega) - B^i(t_{k-1}, \omega)\} \\ &= \sum_{j=1}^{r} \sum_{k=1}^{m} \alpha_j^i(t_k, F^{(q)}[x, B(\cdot, \omega)]) \{B^i(t_k, \omega) - B^i(t_{k-1}, \omega)\} \end{aligned}$$

where $\{t_k : k \in \mathbb{Z}_+\}$ is a strictly increasing sequence in $\mathbb{R}_+$ with $t_0 = 0$ and $\lim_{t \to \infty} t_k = \infty$ with $t_m = t$. Define a mapping $(\varphi^{(q)} \bullet W)^i$ of $\mathbb{R}_+ \times \mathbb{R}^d \times \mathbf{W}^r$ into $\mathbb{R}$ by

$$(19) \qquad (\varphi^{(q)} \bullet W)^i(t, x, w) = \sum_{j=1}^{r} \sum_{k=1}^{m} \alpha_j^i(t_k, F^{(q)}[x, w]) \{w^i(t_k) - w^i(t_{k-1})\}$$

for $(t, x, w) \in \mathbb{R}_+ \times \mathbb{R}^d \times \mathbf{W}^r$ where w^i is the i-th component of w. Let us show first that $(\varphi^{(q)} \bullet W)^i(\cdot, x, w)$ as a mapping of $\mathbb{R}^d \times \mathbf{W}^r$ into $\mathbf{W}$ is $\sigma(\mathfrak{B}_{\mathbb{R}^d} \times \mathfrak{W}^r)/\mathfrak{W}$-measurable. According to (4) of Proposition 17.4, it suffices to show that for every $t \in \mathbb{R}_+$ the mapping

$(\varphi^{(q)} \bullet W)^i(t,\cdot,\cdot)$ of $\mathbb{R}^d \times \mathbf{W}^r$ into $\mathbb{R}$ is $\sigma(\mathfrak{B}_{\mathbb{R}^d} \times \mathfrak{W}^r)/\mathfrak{B}_{\mathbb{R}}$-measurable. Now since $F^{(q)}$ satisfies (1), it is a $\sigma(\mathfrak{B}_{\mathbb{R}^d} \times \mathfrak{W}^r)/\mathfrak{W}^d$-measurable mapping of $\mathbb{R}^d \times \mathbf{W}^r$ into $\mathbf{W}^d$. By Remark 18.3, $\alpha_j^i(t_k,\cdot)$ is a $\mathfrak{W}_t^d/\mathfrak{B}_{\mathbb{R}}$-measurable and hence $\mathfrak{W}^d/\mathfrak{B}_{\mathbb{R}}$-measurable mapping of $\mathbf{W}^d$ into $\mathbb{R}$. Thus the composite mapping $\alpha_j^i(t_k, F^{(q)}[\cdot,\cdot])$ is a $\sigma(\mathfrak{B}_{\mathbb{R}^d} \times \mathfrak{W}^r)/\mathfrak{B}_{\mathbb{R}}$-measurable mapping of $\mathbb{R}^d \times \mathbf{W}^r$ into $\mathbb{R}$. Since $w^i(t_k) - w^i(t_{k-1})$ is a $\mathfrak{W}^r/\mathfrak{B}_{\mathbb{R}}$-measurable mapping of $\mathbf{W}^r$ into $\mathbb{R}$, our $(\varphi^{(q)} \bullet W)^i(t,\cdot,\cdot)$ is a $\sigma(\mathfrak{B}_{\mathbb{R}^d} \times \mathfrak{W}^r)/\mathfrak{B}_{\mathbb{R}}$-measurable mapping of $\mathbb{R}^d \times \mathbf{W}^r$ into $\mathbb{R}$. Next let us show that with fixed $x \in \mathbb{R}^d$, $(\varphi^{(q)} \bullet W)^i(\cdot, x, w)$ as a mapping of $\mathbf{W}^r$ into $\mathbf{W}$ is $\mathfrak{W}_t^{r,w}/\mathfrak{W}_t$-measurable for every $t \in \mathbb{R}_+$. According to (5) of Proposition 17.4 it suffices to show that for every $t \in \mathbb{R}_+$, $(\varphi^{(q)} \bullet W)^i(s, x, \cdot)$ is a $\mathfrak{W}_t^{r,w}/\mathfrak{B}_{\mathbb{R}}$-measurable mapping for every $s \in [0,t]$. Now since $F^{(q)}$ satisfies (2), $F^{(q)}[x,\cdot]$ is a $\mathfrak{W}_u^{r,w}/\mathfrak{W}_u^d$-measurable mapping of $\mathbf{W}^r$ into $\mathbf{W}^d$ for every $u \in \mathbb{R}_+$. By Remark 18.3, $\alpha_j^i(t_k,\cdot)$ is a $\mathfrak{W}_{t_k}^d/\mathfrak{B}_{\mathbb{R}}$-measurable mapping of $\mathbf{W}^d$ into $\mathbb{R}$. Thus $\alpha_j^i(t_k, F^{(q)}[x,\cdot])$ is a $\mathfrak{W}_{t_k}^{r,w}/\mathfrak{B}_{\mathbb{R}}$-measurable mapping of $\mathbf{W}^r$ into $\mathbb{R}$. Also for $t_k \in [0,s]$, $w^i(t_k) - w^i(t_{k-1})$ is a $\mathfrak{W}_s^r/\mathfrak{B}_{\mathbb{R}}$-measurable of mapping of $\mathbf{W}^r$ into $\mathbb{R}$. Thus for every $s \in [0,t]$, $(\varphi^{(q)} \bullet W)^i(s,x,\cdot)$ is a $\mathfrak{W}_t^{r,w}/\mathfrak{B}_{\mathbb{R}}$-measurable mapping of $\mathbf{W}^r$ into $\mathbb{R}$. So far we have shown that $(\varphi^{(q)} \bullet W)^i(\cdot, x, w)$ is a $\sigma(\mathfrak{B}_{\mathbb{R}^d} \times \mathfrak{W}^r)/\mathfrak{W}$-measurable mapping of $\mathbb{R}^d \times \mathbf{W}^r$ into $\mathbf{W}$ and for fixed $x \in \mathbb{R}^d$, $(\varphi^{(q)} \bullet W)^i(\cdot, x, w)$ is a $\mathfrak{W}_t^{r,w}/\mathfrak{W}_t$-measurable mapping of $\mathbf{W}^r$ into $\mathbf{W}$ for every $t \in \mathbb{R}_+$. From this it follows that $(\varphi^{(q)} \bullet W)(\cdot, x, w)$ is a $\sigma(\mathfrak{B}_{\mathbb{R}^d} \times \mathfrak{W}^r)/\mathfrak{W}^d$-measurable mapping of $\mathbb{R}^d \times \mathbf{W}^r$ into $\mathbf{W}^d$ and for fixed $x \in \mathbb{R}^d$, $(\varphi^{(q)} \bullet W)(\cdot, x, w)$ is a $\mathfrak{W}_t^{r,w}/\mathfrak{W}_t^d$-measurable mapping of $\mathbf{W}^r$ into $\mathbf{W}^d$ for every $t \in \mathbb{R}_+$. Thus if we define a mapping $G^{(q)}$ of $\mathbb{R}^d \times \mathbf{W}^r$ into $\mathbf{W}^d$ by

(20) $\qquad G^{(q)}[x,w] = (\varphi^{(q)} \bullet W)(\cdot, x, w) \quad \text{for } (x,w) \in \mathbb{R}^d \times \mathbf{W}^r,$

then $G^{(q)}$ satisfies conditions (1) and (2) and also for $\omega \in \Lambda_q^c$ we have

$$G^{(q)}[x, B(\cdot,\omega)] = (\varphi^{(q)} \bullet W)(\cdot, x, B(\cdot,\omega)) = (\Phi^{(q)} \bullet B)(\cdot,\omega) = (\alpha(\cdot, X^{(q)}) \bullet B)(\omega)$$

by (20), (18), (18) and (15). This verifies (14) for the particular case where $(\Phi^{(q)})_j^i$ are in $\mathbf{L}_0(\Omega, \mathfrak{F}, \{\mathfrak{F}_t\}, P)$. Consider now the general case where $(\Phi^{(q)})_j^i$ are in $\mathbf{L}_{2,\infty}(\mathbb{R}_+ \times \Omega, m_L \times P)$. By Definition 12.9 and Remark 12.14, there exists a sequence $\Phi_n^{(q)}$, $n \in \mathbb{N}$, with components $(\Phi_n^{(q)})_j^i$ in $\mathbf{L}_0(\Omega, \mathfrak{F}, \{\mathfrak{F}_t\}, P)$ and there exists a null set $\Lambda_{q,0}$ in $(\Omega, \mathfrak{F}, P)$ such that $(\Phi_n^{(q)} \bullet B)(\cdot,\omega)$ converges to $(\Phi^{(q)} \bullet B)(\cdot,\omega)$ uniformly on every finite interval in $\mathbb{R}_+$ for $\omega \in \Lambda_{q,0}^c$. Then by Lemma 17.12 for $\omega \in \Lambda_{q,0}^c$ we have

(21) $\qquad (\Phi_n^{(q)} \bullet B)(\cdot,\omega)$ converges to $(\Phi^{(q)} \bullet B)(\cdot,\omega)$ in the metric ρ_∞ on $\mathbf{W}^d$.

Let $G_n^{(q)}$ be the mapping of $\mathbb{R}^d \times \mathbf{W}^r$ into $\mathbf{W}^d$ corresponding to $\Phi_n^{(q)}$ as defined by (20). Then $G_n^{(q)}$ satisfies (1) and (2) and

(20) $\qquad G_n^{(q)}[x, B(\cdot,\omega)] = (\Phi_n^{(q)} \bullet B)(\cdot,\omega)$

§20. STRONG SOLUTIONS

for $\omega \in \Lambda_{q,n}^c$ where $\Lambda_{q,n}$ is a null set in $(\Omega, \mathfrak{F}, P)$. Let $\Lambda_q = \cup_{n \in \mathbb{Z}_+} \Lambda_{q,n}$ and let

(23) $\quad E_q = \{(x, w) \in \mathbb{R}^d \times \mathbf{W}^r : \lim_{n \to \infty} G_n^{(q)}[x, w] \text{ exists in } \mathbf{W}^d\}.$

Since $G_n^{(q)}$ is $\sigma(\mathfrak{B}_{\mathbb{R}^d} \times \mathfrak{W}^r)/\mathfrak{W}^d$-measurable for every $n \in \mathbb{N}$, $E_q \in \sigma(\mathfrak{B}_{\mathbb{R}^d} \times \mathfrak{W}^r)$. Define a mapping $G^{(q)}$ of $\mathbb{R}^d \times \mathbf{W}^r$ into $\mathbf{W}^d$ by

(24) $\quad G^{(q)}[x, w] = \begin{cases} \lim_{n \to \infty} G_n^{(q)}[x, w] & \text{for } (x, w) \in E_q \\ 0 \in \mathbf{W}^d & \text{for } (x, w) \in E_q^c. \end{cases}$

Since $G_n^{(q)}$ satisfies (1) and (2) for every $n \in \mathbb{N}$ so does $G^{(q)}$. Let K_q be the subset of $\mathbb{R}^d \times \mathbf{W}^r$ covered by the mapping $(x, B(\cdot, \omega))$ of $\Lambda_q^c \subset \Omega$. Then for every $(x, w) \in K_q$ there exists some $\omega \in \Lambda_q^c$ such that $(x, w) = (x, B(\cdot, \omega))$ so that by (22) and (21) for $\omega \in \Lambda_q^c$ we have

$$\lim_{n \to \infty} G_n^{(q)}[x, w] = \lim_{n \to \infty} G_n^{(q)}[x, B(\cdot, \omega)] = \lim_{n \to \infty} (\Phi_n^{(q)} \bullet B)(\cdot, \omega) = (\Phi^{(q)} \bullet B)(\cdot, \omega).$$

This shows that $K_q \subset E_q$. Then by (24) we have

(25) $\quad \lim_{n \to \infty} G_n^{(q)}[x, B(\cdot, \omega)] = G^{(q)}[x, B(\cdot, \omega)] \quad \text{for } \omega \in \Lambda_q^c.$

By (15), (21), (22) and (25) we have

(26) $\quad (\alpha(\cdot, X^{(q)}) \bullet B)(\omega) = (\Phi^{(q)} \bullet B)(\cdot, \omega) = \lim_{n \to \infty} (\Phi_n^{(q)} \bullet B)(\cdot, \omega)$
$\quad = \lim_{n \to \infty} G_n^{(q)}[x, B(\cdot, \omega)] = G^{(q)}[x, B(\cdot, \omega)] \quad \text{for } \omega \in \Lambda_q^c,$

verifying (14). This proves the existence of $G^{(q)}$. The existence of $H^{(q)}$ is proved likewise. With this we have shown that the existence of $F^{(q)}$ implies that of $F^{(q+1)}$. Therefore by induction we have a sequence of mappings $\{F^{(q)} : q \in \mathbb{Z}_+\}$ where $F^{(q)}$ satisfies (1), (2), and (13).

To show the existence of a mapping F of $\mathbb{R}^d \times \mathbf{W}^r$ into $\mathbf{W}^d$ satisfying (1), (2), and (12), let

(27) $\quad E = \{(x, w) \in \mathbb{R}^d \times \mathbf{W}^r : \lim_{q \to \infty} F^{(q)}[x, w] \text{ exists in } \mathbf{W}^d\}.$

The $\sigma(\mathfrak{B}_{\mathbb{R}^d} \times \mathfrak{W}^r)$-measurability of $F^{(q)}$ for $q \in \mathbb{Z}_+$ implies that E is in $\sigma(\mathfrak{B}_{\mathbb{R}^d} \times \mathfrak{W}^r)$. Define a mapping of $\mathbb{R}^d \times \mathbf{W}^r$ into $\mathbf{W}^d$ by

(28) $\quad F[x, w] = \begin{cases} \lim_{q \to \infty} F^{(q)}[x, w] & \text{for } (x, w) \in E \\ 0 \in \mathbf{W}^d & \text{for } (x, w) \in E^c. \end{cases}$

Since $F^{(q)}$ satisfies the measurability condition (1) for $q \in \mathbb{Z}_+$, so does F. For fixed $x \in \mathbb{R}^d$, $F^{(q)}[x,\cdot]$ is $\mathfrak{W}_t^{r,w}/\mathfrak{W}_t^d$-measurable for every $t \in \mathbb{R}_+$ for $q \in \mathbb{Z}_+$ so that

$$\{w \in \mathbf{W}^r : \lim_{q \to \infty} F^{(q)}[x,w] \text{ exists in } \mathbf{W}^d\} \in \mathfrak{W}_t^{r,w}$$

for every $t \in \mathbb{R}_+$. Then by (28), $F[x,\cdot]$ is $\mathfrak{W}_t^{r,w}/\mathfrak{W}_t^d$-measurable for every $t \in \mathbb{R}_+$. This shows that F satisfies (2). To show that F satisfies (12), let $\Lambda_\infty = (\cup_{q \in \mathbb{Z}_+} \Lambda_q) \cup \Lambda$ where Λ is the null set in (8). For $\omega \in \Lambda_\infty^c$, $\lim_{q \to \infty} F^{(q)}[x,w]$ exists in $\mathbf{W}^d$ by (8) and (13). Thus by (28), (13), and (8) we have for $\omega \in \Lambda_\infty^c$

$$F[x, B(\cdot,\omega)] = \lim_{q \to \infty} F^{(q)}[x, B(\cdot,\omega)] = \lim_{q \to \infty} X^{(q)}(\cdot,\omega) = X(\cdot,\omega).$$

This proves (12).

Finally if Z is an $\mathfrak{F}_0$-measurable d-dimensional random variable on $(\Omega, \mathfrak{F}, P)$ and if we let $X(\cdot,\omega) = F[Z(\omega), B(\cdot,\omega)]$ for $\omega \in \Omega$, then (B,X) is a solution of the stochastic differential equation on $(\Omega, \mathfrak{F}, \{\mathfrak{F}_t\}, P)$ satisfying $X_0 = Z$. ∎

[II] Uniqueness of Strong Solutions

Definition 20.6. *Let* $\mathbf{S}(\mathbb{R}^d \times \mathbf{W}^r)$ *be the collection of all mappings* F *of* $\mathbb{R}^d \times \mathbf{W}^r$ *into* $\mathbf{W}^d$ *satisfying the following conditions:*

1°. *for every probability measure μ on $(\mathbb{R}^d, \mathfrak{B}_{\mathbb{R}^d})$ there exists a null set N_μ in $(\mathbb{R}^d, \mathfrak{B}_{\mathbb{R}^d}, \mu)$ such that $F[x,\cdot] = F_\mu[x,\cdot]$ a.e. on $(\mathbf{W}^r, \mathfrak{W}^{r,w}, m_W^r)$ when $x \in N_\mu^c$,*

where F_μ is a mapping of $\mathbb{R}^d \times \mathbf{W}^r$ into $\mathbf{W}^d$ such that

2°. F_μ *is* $\sigma(\mathfrak{B}_{\mathbb{R}^d} \times \mathfrak{W}^r)^{\mu \times m_W^r}/\mathfrak{W}^d$-*measurable,*

3°. *for every $x \in \mathbb{R}^d$, $F_\mu[x,\cdot]$ is a $\mathfrak{W}_t^{r,w}/\mathfrak{W}_t^d$-measurable mapping of $\mathbf{W}^r$ into $\mathbf{W}^d$ for every $t \in \mathbb{R}_+$.*

Definition 20.7. *We say that the stochastic differential equation (1) in Definition 18.6 has a unique strong solution if there exists $F \in \mathbf{S}(\mathbb{R}^d \times \mathbf{W}^r)$ such that*

1°. *if (B,X) is a solution of the stochastic differential equation on a standard filtered space $(\Omega, \mathfrak{F}, \{\mathfrak{F}_t\}, P)$ and μ is the probability distribution of X_0 on $(\mathbb{R}^d, \mathfrak{B}_{\mathbb{R}^d})$ then*

$$X(\cdot,\omega) = F_\mu[X_0(\omega), B(\cdot,\omega)] \quad \text{for a.e. } \omega \text{ in } (\Omega, \mathfrak{F}, P),$$

§20. STRONG SOLUTIONS   433

2°. *if $(\Omega, \mathfrak{F}, \{\mathfrak{F}_t\}, P)$ is a standard filtered space on which an r-dimensional $\{\mathfrak{F}_t\}$-adapted null at 0 Brownian motion B exists, Z is an arbitrary d-dimensional $\mathfrak{F}_0$-measurable random variable on $(\Omega, \mathfrak{F}, P)$ with probability distribution P_Z on $(\mathbb{R}^d, \mathfrak{B}_{\mathbb{R}^d})$, and if we define a stochastic process X by setting*

$$X(\cdot, \omega) = F_{P_Z}[Z(\omega), B(\cdot, \omega)] \quad \text{for } \omega \in \Omega,$$

then (B, X) is a solution satisfying the condition $X_0 = Z$ a.e. on $(\Omega, \mathfrak{F}, P)$.

Observation 20.8. Suppose the stochastic differential equation (1) in Definition 18.6 has a solution (B, X) on some standard filtered space $(\Omega, \mathfrak{F}, \{\mathfrak{F}_t\}, P)$. Let $P_{(B,X)}$ be the probability distribution of (B, X) on $(\mathbf{W}^{r+d}, \mathfrak{W}^{r+d})$ and μ be the probability distribution of X_0 on $(\mathbb{R}^d, \mathfrak{B}_{\mathbb{R}^d})$. Let $(\mathbf{W}^{r+d}, \mathfrak{W}^{r+d,*}, \{\mathfrak{W}^{r+d,*}_t\}, P_{(B,X)})$ be the standard filtered space generated by $P_{(B,X)}$, that is, $\mathfrak{W}^{r+d,*}$ is the completion of $\mathfrak{W}^{r+d}$ with respect to $P_{(B,X)}$, $\mathfrak{W}^{r+d,0}_t = \sigma(\mathfrak{W}^{r+d}_t \cup \mathfrak{N})$ where $\mathfrak{N}$ is the collection of all the null sets in $(\mathbf{W}^{r+d}, \mathfrak{W}^{r+d,*}, P_{(B,X)})$, and $\mathfrak{W}^{r+d,*}_t = \cap_{\varepsilon > 0} \mathfrak{W}^{r+d,0}_{t+\varepsilon}$. Let π_0 and π_1 be the projection of $\mathbf{W}^{r+d}$ onto $\mathbf{W}^r$ and $\mathbf{W}^d$ respectively and let W and Y be two processes defined on $(\mathbf{W}^{r+d}, \mathfrak{W}^{r+d,*}, \{\mathfrak{W}^{r+d,*}_t\}, P_{(B,X)})$ by setting $W(t, w) = (\pi_0(w))(t)$ and $Y(t, w) = (\pi_1(w))(t)$ for $(t, w) \in \mathbb{R}_+ \times \mathbf{W}^{r+d}$. We showed in Theorem 18.13 that (W, Y) is a solution of the stochastic differential equation on $(\mathbf{W}^{r+d}, \mathfrak{W}^{r+d,*}, \{\mathfrak{W}^{r+d,*}_t\}, P_{(B,X)})$, (W, Y) and (B, X) have the identical probability distribution $P_{(B,X)}$ on $(\mathbf{W}^{r+d}, \mathfrak{W}^{r+d})$, and Y_0 and X_0 have the identical probability distribution μ on $(\mathbb{R}^d, \mathfrak{B}_{\mathbb{R}^d})$. Let $Q = \mu \times P_{(B,X)}$ and consider the product measure space $(\mathbb{R}^d \times \mathbf{W}^{r+d}, \sigma(\mathfrak{B}_{\mathbb{R}^d} \times \mathfrak{W}^{r+d}), Q)$. Now let π_2 be the projection of $\mathbb{R}^d \times \mathbf{W}^{r+d}$ onto $\mathbb{R}^d \times \mathbf{W}^r$ and Q_{π_2} be its probability distribution on $(\mathbb{R}^d \times \mathbf{W}^r, \sigma(\mathfrak{B}_{\mathbb{R}^d} \times \mathfrak{W}^r))$. Then for $E \times A_0 \in \mathfrak{B}_{\mathbb{R}^d} \times \mathfrak{W}^r$ we have

$$Q_{\pi_2}(E \times A_0) = (\mu \times P_{(B,X)})(E \times A_0 \times \mathbf{W}^d)$$
$$= \mu(E) P_{(B,X)}(A_0 \times \mathbf{W}^d) = \mu(E) P_B(A_0) = \mu(E) m^r_W(A_0).$$

From this it follows that $Q_{\pi_2} = \mu \times m^r_W$ on $\sigma(\mathfrak{B}_{\mathbb{R}^d} \times \mathfrak{W}^{r+d})$.

Let $Q^{(x,v)}(A)$ for $A \in \sigma(\mathfrak{B}_{\mathbb{R}^d} \times \mathfrak{W}^{r+d})$ and $(x, v) \in \mathbb{R}^d \times \mathbf{W}^r$ be a regular image conditional probability of Q given π_2. Then according to Definition C.11

1°. there exists a null set N in $(\mathbb{R}^d \times \mathbf{W}^r, \sigma(\mathfrak{B}_{\mathbb{R}^d} \times \mathfrak{W}^r), \mu \times m^r_W)$ such that $Q^{(x,v)}$ is a probability measure on $(\mathbb{R}^d \times \mathbf{W}^{r+d}, \sigma(\mathfrak{B}_{\mathbb{R}^d} \times \mathfrak{W}^{r+d}))$ for $(x, v) \in N^c$,

2°. $Q^{(\cdot)}(A)$ is a $\sigma(\mathfrak{B}_{\mathbb{R}^d} \times \mathfrak{W}^r)$-measurable function on $\mathbb{R}^d \times \mathbf{W}^r$ for $A \in \sigma(\mathfrak{B}_{\mathbb{R}^d} \times \mathfrak{W}^{r+d})$,

3°. $Q(A \cap \pi_2^{-1}(G)) = \int_G Q^{(x,v)}(A)(\mu \times m_W^r)(d(x,v))$ for $A \in \sigma(\mathfrak{B}_{\mathbb{R}^d} \times \mathfrak{W}^{r+d})$ and $G \in \sigma(\mathfrak{B}_{\mathbb{R}^d} \times \mathfrak{W}^r)$.

If we define

(1) $$\overline{Q^{(x,v)}}(A_1) = Q^{(x,v)}(\mathbb{R}^d \times \mathbf{W}^r \times A_1) \quad \text{for } A_1 \in \mathfrak{W}^d$$

then $\overline{Q^{(x,v)}}$ is a probability measure on $(\mathbf{W}^d, \mathfrak{W}^d)$ for every $(x,v) \in N^c$. With $A = \mathbb{R}^d \times \mathbf{W}^r \times A_1 \in \mathfrak{B}_{\mathbb{R}^d} \times \mathfrak{W}^r \times \mathfrak{W}^d$ and $G = E \times A_0 \in \mathfrak{B}_{\mathbb{R}^d} \times \mathfrak{W}^r$, we have from 3° and (1)

(2) $$Q(E \times A_0 \times A_1) = \int_{E \times A_0} \overline{Q^{(x,v)}}(A_1)(\mu \times m_W^r)(d(x,v))$$

for $E \times A_0 \times A_1 \in \mathfrak{B}_{\mathbb{R}^d} \times \mathfrak{W}^r \times \mathfrak{W}^d$ when $(x,v) \in N^c$.

On the probability space $(\mathbb{R}^d \times \mathbf{W}^{r+d}, \sigma(\mathfrak{B}_{\mathbb{R}^d} \times \mathfrak{W}^{r+d}), Q)$ consider the projection π_3 of $\mathbb{R}^d \times \mathbf{W}^{r+d}$ onto $\mathbb{R}^d$. The probability distribution Q_{π_3} of π_3 on $(\mathbb{R}^d, \mathfrak{B}_{\mathbb{R}^d})$ is equal to μ since for every $E \in \mathfrak{B}_{\mathbb{R}^d}$ we have

$$Q_{\pi_3}(E) = Q \circ \pi_3^{-1}(E) = (\mu \times P_{(B,X)})(E \times \mathfrak{W}^{r+d}) = \mu(E) P_{(B,X)}(\mathfrak{W}^{r+d}) = \mu(E).$$

Let $Q^x(A)$ for $A \in \sigma(\mathfrak{B}_{\mathbb{R}^d} \times \mathfrak{W}^{r+d})$ and $x \in \mathbb{R}^d$ be a regular image conditional probability of Q given π_3. According to Definition C.11

4°. there exists a null set N in $(\mathbb{R}^d, \mathfrak{B}_{\mathbb{R}^d}, \mu)$ such that when $x \in N^c$, Q^x is a probability measure on $(\mathbb{R}^d \times \mathbf{W}^{r+d}, \sigma(\mathfrak{B}_{\mathbb{R}^d} \times \mathfrak{W}^{r+d}))$,

5°. $Q^{(\cdot)}(A)$ is a $\mathfrak{B}_{\mathbb{R}^d}$-measurable function on $\mathbb{R}^d$ for $A \in \sigma(\mathfrak{B}_{\mathbb{R}^d} \times \mathfrak{W}^{r+d})$,

6°. $Q(A \cap \pi_3(G)) = \int_G Q^x(A)\mu(dx)$ for $A \in \sigma(\mathfrak{B}_{\mathbb{R}^d} \times \mathfrak{W}^{r+d})$ and $G \in \mathfrak{B}_{\mathbb{R}^d}$.

If we define

(3) $$\overline{Q^x}(A) = Q^x(\mathbb{R}^d \times A) \quad \text{for } A \in \mathfrak{W}^{r+d},$$

then $\overline{Q^x}$ is a probability measure on $(\mathbf{W}^{r+d}, \mathfrak{W}^{r+d})$. With $A = \mathbb{R}^d \times A_0 \times A_1 \in \mathfrak{B}_{\mathbb{R}^d} \times \mathfrak{W}^r \times \mathfrak{W}^d$ and $G = E \in \mathfrak{B}_{\mathbb{R}^d}$, 6° and (3) give

(4) $$Q(E \times A_0 \times A_1) = \int_E \overline{Q^x}(A_0 \times A_1)\mu(dx).$$

Consider the probability space $(\mathbf{W}^{r+d}, \mathfrak{W}^{r+d}, \overline{Q^x})$ for $x \in N^c$. By Theorem 18.21, (W, Y) is a solution of the stochastic differential equation on $(\mathbf{W}^{r+d}, \mathfrak{W}^{r+d,*,x}, \{\mathfrak{W}_t^{r+d,*,x}\}, \overline{Q^x})$ satisfying the condition $Y_0 = x$ and in particular W is an r-dimensional $\{\mathfrak{W}_t^{r+d,*,x}\}$-adapted

§20. STRONG SOLUTIONS

null at 0 Brownian motion. Thus for the projection π_0 of $\mathbf{W}^{r+d}$ onto $\mathbf{W}^r$, the probability distribution $(\overline{Q^x})_{\pi_0}$ of π_0 on $(\mathbf{W}^r, \mathfrak{W}^r)$ is that of the process W which is the r-dimensional Wiener measure m_W^r by Definition 17.9. Let $(\overline{Q^x})^v(A)$ for $A \in \mathfrak{W}^{r+d}$ and $v \in \mathbf{W}^r$ be a regular image conditional probability of $\overline{Q^x}$ given π_0 when $x \in N^c$. Then

7°. there exists a null set N_x in $(\mathbf{W}^r, \mathfrak{W}^r, m_W^r)$ such that $(\overline{Q^x})^v$ is a probability measure on $(\mathbf{W}^{r+d}, \mathfrak{W}^{r+d})$ for $v \in N_x^c$,

8°. $(\overline{Q^x})^{(\cdot)}(A)$ is a $\mathfrak{W}^r$-measurable function on $\mathbf{W}^r$ for every $A \in \mathfrak{W}^{r+d}$,

9°. $(\overline{Q^x})(A \cap \pi_0^{-1}(G)) = \int_G (\overline{Q^x})^v(A) m_W^r(dv)$ for $A \in \mathfrak{W}^{r+d}$ and $G \in \mathfrak{W}^r$.

Thus if we define

$$(5) \qquad \overline{(Q^x)^v}(A_1) = (\overline{Q^x})^v(\mathbf{W}^r \times A_1) \quad \text{for } A_1 \in \mathfrak{W}^d$$

then $\overline{(Q^x)^v}$ is a probability measure on $(\mathbf{W}^d, \mathfrak{W}^d)$ for every $v \in N_x^c$. Furthermore by 9° and (5) we have for every $x \in N^c$

$$(6) \qquad \overline{Q^x}(A_0 \times A_1) = \int_{A_0} \overline{(Q^x)^v}(A_1) m_W^r(dv) \quad \text{for } A_0 \times A_1 \in \mathfrak{W}^r \times \mathfrak{W}^d.$$

Lemma 20.9. *There exists a null set N in $(\mathbb{R}^d, \mathfrak{B}_{\mathbb{R}^d}, \mu)$ and a collection $\{N_x : x \in N^c\}$ of null sets in $(\mathbf{W}^r, \mathfrak{W}^r, m_W^r)$ such that*

$$\overline{Q^{(x,v)}} = \overline{(Q^x)^v} \quad \text{on } (\mathbf{W}^d, \mathfrak{W}^d) \text{ when } v \in N_x^c \text{ and } x \in N^c.$$

Proof. Let $E \times A_0 \times A_1 \in \mathfrak{B}_{\mathbb{R}^d} \times \mathfrak{W}^r \times \mathfrak{W}^d$. Then

$$(1) \quad \int_E \left\{ \int_{A_0} \overline{Q^{(x,v)}}(A_1) m_W^r(dv) \right\} \mu(dv) = \int_{E \times A_0} \overline{Q^{(x,v)}}(A_1)(\mu \times m_W^r)(d(x,v))$$

$$= Q(E \times A_0 \times A_1) = \int_E \overline{Q^x}(A_0 \times A_1) \mu(dx) = \int_E \left\{ \int_{A_0} \overline{(Q^x)^v}(A_1) m_W^r(dv) \right\} \mu(dx),$$

where the second equality is by (2), the third equality is by (4), and the last equality is by (6) of Observation 20.8. Thus corresponding to $A_0 \times A_1 \in \mathfrak{W}^r \times \mathfrak{W}^d$, there exists a null set $N_{A_0 \times A_1}$ in $(\mathbb{R}^d, \mathfrak{B}_{\mathbb{R}^d}, \mu)$ such that for the two functions $\overline{Q^{(x,v)}}(A_1)$ and $\overline{(Q^x)^v}(A_1)$ of $x \in \mathbb{R}^d$ we have

$$\int_{A_0} \overline{Q^{(x,v)}}(A_1) m_W^r(dv) = \int_{A_0} \overline{(Q^x)^v}(A_1) m_W^r(dv) \quad \text{for } x \in N_{A_0 \times A_1}^c.$$

Consider two measures ν_1^x and ν_2^x on $(\mathbf{W}^{r+d}, \mathfrak{W}^{r+d})$ determined by

$$\begin{cases} \nu_1^x(A_0 \times A_1) = \int_{A_0} \overline{Q^{(x,v)}}(A_1) m_W^r(dv) \\ \nu_2^x(A_0 \times A_1) = \int_{A_0} \overline{(Q^x)^v}(A_1) m_W^r(dv) \end{cases}$$

for $A_0 \times A_1 \in \mathfrak{W}^r \times \mathfrak{W}^d$. Let $\mathfrak{G}$ be the collection of all $A \in \mathfrak{W}^{r+d}$ such that $\nu_1^x(A) = \nu_2^x(A)$ for a.e. x in $(\mathbb{R}^d, \mathfrak{B}_{\mathbb{R}^d}, Q_{Y_0})$. Then $\mathfrak{W}^r \times \mathfrak{W}^d \subset \mathfrak{G}$ and it is easily verified that $\mathfrak{G}$ is a d-class. Thus by Theorem 1.7 we have $\mathfrak{W}^{r+d} = \sigma(\mathfrak{W}^r \times \mathfrak{W}^d) \subset \mathfrak{G}$. By Proposition C.8, $\mathfrak{W}^{r+d}$ is a countably determined σ-algebra. Let $\mathfrak{D} = \{D_n : n \in \mathbb{N}\}$ be a countable collection of determining sets for $\mathfrak{W}^{r+d}$. Now for each $n \in \mathbb{N}$, $\nu_1^x(D_n) = \nu_2^x(D_n)$ for $x \in N_{D_n}^c$ where N_{D_n} is a null set in $(\mathbb{R}^d, \mathfrak{B}_{\mathbb{R}^d}, \mu)$. Let $N = \cup_{n \in \mathbb{N}} N_{D_n}$. Then $\nu_1^x(D_n) = \nu_2^x(D_n)$ for all $n \in \mathbb{N}$ when $x \in N^c$. This then implies that $\nu_1^x(A) = \nu_2^x(A)$ for all $A \in \mathfrak{W}^{r+d}$ when $x \in N^c$. In particular we have

$$\int_{A_0} \overline{Q^{(x,v)}}(A_1) m_W^r(dv) = \int_{A_0} \overline{(Q^x)^v}(A_1) m_W^r(dv) \quad \text{for } A_0 \times A_1 \in \mathfrak{W}^r \times \mathfrak{W}^d \text{ when } x \in N^c.$$

Thus corresponding to every $x \in N^c$ and $A_1 \in \mathfrak{W}^d$ there exists a null set N_{x,A_1} in $(\mathbf{W}^r, \mathfrak{W}^r, m_W^r)$ such that $\overline{Q^{(x,v)}}(A_1) = \overline{(Q^x)^v}(A_1)$ when $v \in N_{x,A_1}^c$. Since $\mathfrak{W}^r$ is countably determined, corresponding to our $x \in N^c$ there exists a null set N_x in $(\mathbf{W}^r, \mathfrak{W}^r, m_W^r)$ such that $\overline{Q^{(x,v)}}(A_1) = \overline{(Q^x)^v}(A_1)$ for $A_1 \in \mathfrak{W}^d$ when $v \in N_x^c$. ∎

Lemma 20.10. *Suppose the stochastic differential equation (1) in Definition 18.6 has a solution (B, X) on some standard filtered space $(\Omega, \mathfrak{F}, \{\mathfrak{F}_t\}, P)$. Let μ be the probability distribution of X_0 on $(\mathbb{R}^d, \mathfrak{B}_{\mathbb{R}^d})$. Assume further that the stochastic differential equation satisfies the pathwise uniqueness condition under deterministic initial conditions. Then there exists a null set N_μ in $(\mathbb{R}^d, \mathfrak{B}_{\mathbb{R}^d}, \mu)$ and a collection $\{N_x : x \in N_\mu^c\}$ of null sets in $(\mathbf{W}^r, \mathfrak{W}^r, m_W^r)$ such that*

$$\overline{Q^{(x,v)}} = \delta_{F_x(v)} \quad \text{on } (\mathbf{W}^d, \mathfrak{W}^d) \text{ when } v \in N_x^c \text{ and } x \in N_\mu^c,$$

where F_x is the mapping of $\mathbf{W}^r$ into $\mathbf{W}^d$ defined in Theorem 19.18, that is, the probability measure $\overline{Q^{(x,v)}}$ on $(\mathbf{W}^d, \mathfrak{W}^d)$ is the unit mass at $F_x(v) \in \mathbf{W}^d$.

Proof. According to Theorem 18.21 there exists a null set N in $(\mathbb{R}^d, \mathfrak{B}_{\mathbb{R}^d}, \mu)$ such that for every $x \in N^c$, the initial value problem

$$\begin{cases} dX_t = \alpha(t, X) dB(t) + \beta(t, X) dt \\ X_0 = x \end{cases}$$

§20. STRONG SOLUTIONS   437

has a solution on the function space $\mathbf{W}^{r+d}$. If we assume further that the stochastic differential equation satisfies the pathwise uniqueness condition under deterministic initial conditions then by Theorem 19.18 the mapping F_x of $\mathbf{W}^r$ into $\mathbf{W}^d$ exists for every $x \in N^c$. Let N_μ be the union of this null set and the null set N in Lemma 20.9. Then for every $x \in N_\mu^c$ there exists a null set N'_x in $(\mathbf{W}^r, \mathfrak{W}^r, m_W^r)$ such that $Y(\cdot, w) = F_x[W(\cdot, w)]$ for $w \in \mathbf{W}^{r+d}$ with $W(\cdot, w) \in (N'_x)^c$. In other words, for the projections π_0 and π_1 of $\mathbf{W}^{r+d}$ onto $\mathbf{W}^r$ and $\mathbf{W}^d$ we have

(1) $\qquad \pi_1(w) = F_x[\pi_0(w)] \quad \text{for } w \in \mathbf{W}^{r+d} \text{ with } \pi_0(w) \in (N'_x)^c.$

Now for $x \in N_\mu^c$, there exists a null set N''_x in $(\mathbf{W}^r, \mathfrak{W}^r, m_W^r)$ such that for $v \in (N''_x)^c$, $\overline{(Q^x)^v}$ is a regular image conditional probability of $\overline{Q^x}$ given the projection π_0 of $\mathbf{W}^{r+d}$ onto $\mathbf{W}^r$, but restricted from $\mathfrak{W}^{r+d}$ to $\mathfrak{W}^d$ by (5) of Observation 20.8. But according to (1), for every $w \in \mathbf{W}^{r+d}$ with $\pi_0(w) \in (N'_x)^c$, $\pi_1(w)$ is uniquely determined by $\pi_0(w)$ and thus $\overline{(Q^x)^v} = \delta_{F_x(v)}$. Let $N_x = N'_x \cup N''_x$. Then $\overline{(Q^x)^v} = \delta_{F_x(v)}$ for $v \in N_x^c$ and $x \in N_\mu^c$. ∎

Lemma 20.11. *Under the same assumptions as in Lemma 20.10, let*

$$\Lambda = \{(x,v) \in \mathbb{R}^d \times \mathbf{W}^r : \overline{Q^{(x,v)}} \text{ is not a unit mass on } (\mathbf{W}^d, \mathfrak{W}^d)\}.$$

Then Λ is a null set in $(\mathbb{R}^d \times \mathbf{W}^r, \sigma(\mathfrak{B}_{\mathbb{R}^d} \times \mathfrak{W}^r), \mu \times m_W^r)$.

Proof. Since $\mathbf{W}^d$ is a separable metric space, for every $n \in \mathbb{N}$ there exists a countable collection of closed spheres $\{S_{n,i} : i \in \mathbb{N}\}$ in $\mathbf{W}^d$, each with radius $r(S_{n,i}) = n^{-1}$, such that $\mathbf{W}^d = \cup_{i \in \mathbb{N}} S_{n,i}$. Let us show that for an arbitrary probability measure ν on $(\mathbf{W}^d, \mathfrak{W}^d)$ we have

(1) $\qquad \nu$ is a unit mass $\Leftrightarrow \nu(S_{n,i}) = 0$ or 1 for every n and i

Now if ν is a unit mass then clearly $\nu(S_{n,i}) = 0$ or 1 for every n and i. Conversely assume that $\nu(S_{n,i}) = 0$ or 1 for every n and i. For fixed n, no two closed spheres $S_{n,i}$ with ν measure 1 can be disjoint for otherwise we would have $\nu(\mathbf{W}^d) \geq 2$. Let K_n be the closed set which is the intersection of those closed spheres in the collection $\{S_{n,i} : i \in \mathbb{N}\}$ which have ν measure 1. Then $\nu(K_n) = 1$ and the radius $r(K_n) \leq n^{-1}$. For the collection of closed sets $\{K_n : n \in \mathbb{N}\}$ we have $K_n \cap K_m \neq \emptyset$ for otherwise we would have $\nu(\mathbf{W}^d) \geq 2$. If we let $C_n = \cap_{m=1}^n K_m$ for $n \in \mathbb{N}$, then C_n, $n \in \mathbb{N}$, is a decreasing sequence of closed sets with $\nu(C_n) = 1$ and $\delta(C_n) \leq n^{-1}$. Since $\mathbf{W}^d$ is a complete metric space and since $\delta(C_n) \downarrow 0$ as $n \to \infty$, there exists a unique $w \in \mathbf{W}^d$ such that $\cap_{n \in \mathbb{N}} C_n = \{w\}$. Then $\nu(\{w\}) = \lim_{n \to \infty} \nu(C_n) = 1$. Thus ν is a unit mass. This proves (1).

By (1) we have

$$\Lambda^c = \{(x,v) \in \mathbb{R}^d \times \mathbf{W}^r : \overline{Q^{(x,v)}} \text{ is a unit mass on } (\mathbf{W}^d, \mathfrak{W}^d)\}$$
$$= \{(x,v) \in \mathbb{R}^d \times \mathbf{W}^r : \overline{Q^{(x,v)}}(S_{n,i}) = 0 \text{ or } 1 \text{ for every } n \text{ and } i\}.$$

Since $S_{n,i} \in \mathfrak{B}_{\mathbf{W}^d} = \mathfrak{W}^d$, $\overline{Q^{(x,v)}}(S_{n,i})$ is a $\sigma(\mathfrak{B}_{\mathbb{R}^d} \times \mathfrak{W}^r)$-measurable functions of $(x,v) \in \mathbb{R}^d \times \mathbf{W}^r$ by 2° and (1) of Observation 20.8. Thus $\Lambda^c \in \sigma(\mathfrak{B}_{\mathbb{R}^d} \times \mathfrak{W}^r)$. To show that $(\mu \times m_W^r)(\Lambda) = 0$, let N_μ be a null set in $(\mathbb{R}^d, \mathfrak{B}_{\mathbb{R}^d}, \mu)$ and $\{N_x : x \in N_\mu^c\}$ be a collection of null set in $(\mathbf{W}^r, \mathfrak{W}^r, m_W^r)$ as specified in Lemma 20.10. Then for $x \in N_\mu^c$ and $v \in N_x^c$, we have $\overline{Q^{(x,v)}} = \delta_{F_x(v)}$ so that $(x,v) \in \Lambda^c$ and thus $v \in (\Lambda^c)_x$, which is the section of $\Lambda^c \subset \mathbb{R}^d \times \mathbf{W}^r$ at $x \in \mathbb{R}^d$ defined by $(\Lambda^c)_{x,\cdot} = \{v \in \mathbf{W}^r : (x,v) \in \Lambda^c\} \in \mathfrak{W}^r$. Therefore $x \in N_\mu^c$ implies $N_x^c \subset (\Lambda^c)_{x,\cdot}$, or equivalently, $\Lambda_{x,\cdot} \subset N_x$. Thus

$$(\mu \times m_W^r)(\Lambda) = \int_{\mathbb{R}^d} m_W^r(\Lambda_{x,\cdot}) \mu(dx) = \int_{N^c} m_W^r(\Lambda_{x,\cdot}) \mu(dx) \le \int_{N^c} m_W^r(N_x) \mu(dx) = 0,$$

since $m_W^r(N_x) = 0$ for $x \in N_\mu^c$. This shows that $(\mu \times m_W^r)(\Lambda) = 0$. ∎

Lemma 20.12. *Under the same assumptions on the stochastic differential equation (1) in Definition 18.6 as in Lemma 20.10, let F_μ be a mapping of $\mathbb{R}^d \times \mathbf{W}^r$ into $\mathbf{W}^d$ defined by*

(1) $\qquad\qquad \delta_{F_\mu[x,v]} = \overline{Q^{(x,v)}} \qquad \text{for } (x,v) \in (N_\mu^c \times \mathbf{W}^r) \cap \Lambda^c$

(2) $\qquad\qquad F_\mu[x,v] = 0 \in \mathbf{W}^d \qquad \text{for } (x,v) \in (N_\mu^c \times \mathbf{W}^r) \cap \Lambda$

(3) $\qquad\qquad F_\mu[x,v] = 0 \in \mathbf{W}^d \qquad \text{for } (x,v) \in N_\mu \times \mathbf{W}^r.$

Then F_μ satisfies conditions 2° and 3° of Definition 20.6, that is,

(4) *F_μ is $\sigma(\mathfrak{B}_{\mathbb{R}^d} \times \mathfrak{W}^r)^{\mu \times m_W^r}/\mathfrak{W}^d$-measurable,*

(5) *for every $x \in \mathbb{R}^d$, $F_\mu[x, \cdot]$ is a $\mathfrak{W}_t^{r,w}/\mathfrak{W}_t^d$-measurable mapping of $\mathbf{W}^r$ into $\mathbf{W}^d$ for every $t \in \mathbb{R}_+$.*

Assume further that for every $x \in \mathbb{R}_+$ the stochastic differential equation has a solution (B, X) with $X_0 = x$ on some standard filtered space so that the mapping F_x of $\mathbf{W}^r$ into $\mathbf{W}^d$ defined in Theorem 19.18 exists for every $x \in \mathbb{R}^d$. Let F be a mapping of $\mathbb{R}^d \times \mathbf{W}^r$ into $\mathbf{W}^d$ defined by

(6) $\qquad\qquad F[x,v] = F_x(v) \quad \text{for } (x,v) \in \mathbb{R}^d \times \mathbf{W}^r.$

Then F_μ and F satisfy condition 1° of Definition 20.6, that is, there exists a null set N_μ in $(\mathbb{R}^d, \mathfrak{B}_{\mathbb{R}^d}, \mu)$ such that

(7) $\qquad\qquad F[x, \cdot] = F_\mu[x, \cdot] \text{ a.e. on } (\mathbf{W}^r, \mathfrak{W}^{r,w}, m_W^r) \text{ when } x \in N_\mu^c.$

§20. STRONG SOLUTIONS

Proof. To prove (4) note that by Lemma 20.11, $\Lambda^c \in \sigma(\mathfrak{B}_{\mathbb{R}^d} \times \mathfrak{W}^r)$ and thus $(N_\mu^c \times \mathbf{W}^r) \cap \Lambda^c \in \sigma(\mathfrak{B}_{\mathbb{R}^d} \times \mathfrak{W}^r)$. Let us show that F_μ is $\sigma(\mathfrak{B}_{\mathbb{R}^d} \times \mathfrak{W}^r)/\mathfrak{W}^d$-measurable on $(N_\mu^c \times \mathbf{W}^r) \cap \Lambda^c$. Now for every $A_1 \in \mathfrak{W}^d$ we have

$$F_\mu^{-1} \cap (N_\mu^c \times \mathbf{W}^r) \cap \Lambda^c = \{(x,v) \in (N_\mu^c \times \mathbf{W}^r) \cap \Lambda^c : F_\mu(x,v) \in A_1\}$$
$$= \{(x,v) \in (N_\mu^c \times \mathbf{W}^r) \cap \Lambda^c : \overline{Q^{(x,v)}}(A_1) = 1\} \in \sigma(\mathfrak{B}_{\mathbb{R}^d} \times \mathfrak{W}^r)$$

by the definition of Λ in Lemma 20.11 and by 2° of Observation 20.8. Thus F_μ is $\sigma(\mathfrak{B}_{\mathbb{R}^d} \times \mathfrak{W}^r)/\mathfrak{W}^d$-measurable on $(N_\mu^c \times \mathbf{W}^r) \cap \Lambda^c$. Since the rest of the domain of definition of F_μ is a null set in $(\mathbb{R}^d \times \mathbf{W}^r, \sigma(\mathfrak{B}_{\mathbb{R}^d} \times \mathfrak{W}^r), \mu \times m_W^r)$, F_μ is $\sigma(\mathfrak{B}_{\mathbb{R}^d} \times \mathfrak{W}^r)^{\mu \times m_W^r}/\mathfrak{W}^d$-measurable.

To prove (5) note that for any $x \in \mathbb{R}^d$ and $A_1 \in \mathfrak{W}_t^d$, $\overline{(Q^x)}^{(\cdot)}(A_1)$ is a $\mathfrak{W}_t^{r,w}/\mathfrak{W}_t^d$-measurable mapping of $\mathbf{W}^r$ for every $t \in \mathbb{R}_+$ as can be shown by the same argument as in Lemma 19.12. Then so is $\overline{(Q^x)}^{(\cdot)}$. If $x \in N_\mu^c$, then the section $\Lambda_{x,\cdot}$ of Λ at x is a null set in $(\mathbf{W}^r, \mathfrak{W}^r, m_W^r)$ as we saw in the proof of Lemma 20.11 so that $\Lambda_{x,\cdot} \in \mathfrak{W}_t^{r,w}$ from the fact that $\mathfrak{W}_t^{r,w}$ is augmented. Therefore

$$F_\mu[x,\cdot]^{-1}(A_1) \cap (\Lambda^c)_{x,\cdot} = \{v \in (\Lambda^c)_{x,\cdot} : \overline{Q^{(x,v)}}(A_1) = 1\}$$
$$= \{v \in (\Lambda^c)_{x,\cdot} : \overline{(Q^x)^v}(A_1) = 1\} \in \mathfrak{W}_t^{r,w}$$

while $F_\mu[x,\cdot]^{-1}(A_1) \cap \Lambda_{x,\cdot} = \Lambda_{x,\cdot}$ or $\emptyset$ according as $0 \in A_1$ or $0 \in A_1^c$. Thus $F_\mu[x,\cdot]$ is $\mathfrak{W}_t^{r,w}/\mathfrak{W}_t^d$-measurable when $x \in N_\mu^c$. When $x \in N_\mu$, $F_\mu[x,\cdot] = 0$ so that $F_\mu[x,\cdot]$ is $\mathfrak{W}_t^{r,w}/\mathfrak{W}_t^d$-measurable.

To prove (7), note that

$$x \in N_\mu^c \text{ and } v \in N_x^c \Rightarrow \overline{Q^{(x,v)}} = \delta_{F_x(v)} \Rightarrow (x,v) \in \Lambda^c \Rightarrow \delta_{F_\mu[x,v]} = \overline{Q^{(x,v)}}$$

so that $F_x(v) = F_\mu[x,v]$ for $x \in N_\mu^c$ and $v \in N_x^c$. Since N_x^c is a null set in $(\mathbf{W}^r, \mathfrak{W}^r, m_W^r)$, this verifies (7). ∎

Theorem 20.13. *The stochastic differential equation (1) of Definition 18.6 has a unique strong solution if and only if*

(I). for every probability measure μ on $(\mathbb{R}^d, \mathfrak{B}_{\mathbb{R}^d})$ there exists a solution (B, X) of the stochastic differential equation on some standard filtered space $(\Omega, \mathfrak{F}, \{\mathfrak{F}_t\}, P)$ with μ as the probability distribution of X_0,

(II). the stochastic differential equation satisfies the pathwise uniqueness condition.

Proof. 1) Necessity. Suppose a unique strong solution F exists. To verify (I), let μ be an arbitrary probability measure on $(\mathbb{R}^d, \mathfrak{B}_{\mathbb{R}^d})$. Consider the product measure space $(\mathbb{R}^d \times \mathbf{W}^r, \sigma(\mathfrak{B}_{\mathbb{R}^d} \times \mathfrak{W}^r), \mu \times m_W^r)$ and consider the standard filtered space generated by $\mu \times m_W^r$, $(\mathbb{R}^d \times \mathbf{W}^r, \sigma(\mathfrak{B}_{\mathbb{R}^d} \times \mathfrak{W}^r)^{\mu \times m_W^r}, \{\sigma(\mathfrak{B}_{\mathbb{R}^d} \times \mathfrak{W}^r)_t^{\mu \times m_W^r}\}, \mu \times m_W^r)$. If we define $B(t, (x,w)) = w(t)$ for $(t, x, w) \in \mathbb{R}_+ \times \mathbb{R}^d \times \mathbf{W}^r$ then B is an r-dimensional $\{\sigma(\mathfrak{B}_{\mathbb{R}^d} \times \mathfrak{W}^r)_t^{\mu \times m_W^r}\}$-adapted null at 0 Brownian motion. Let Z be an $\mathbb{R}_d$-valued random variable on $(\mathbb{R}^d \times \mathbf{W}^r, \sigma(\mathfrak{B}_{\mathbb{R}^d} \times \mathfrak{W}^r), \mu \times m_W^r)$ with μ as its probability distribution on $(\mathbb{R}^d, \mathfrak{B}_{\mathbb{R}^d})$. If we let $X(\cdot, (x,w)) = F_\mu[Z((x,w)), B(\cdot, (x,w)]$ for $(x,w) \in \mathbb{R}^d \times \mathbf{W}^r$ then by 2° of Definition 20.7, (B, X) is a solution of the stochastic differential equation with $X_0 = Z$ and thus (I) is satisfied.

To verify (II), suppose $(B, X^{(1)})$ and $(B, X^{(2)})$ are two solutions of the stochastic differential equation on a standard filtered space $(\Omega, \mathfrak{F}, \{\mathfrak{F}_t\}, P)$ such that $X_0^{(1)} = X_0^{(2)}$ a.e. on $(\Omega, \mathfrak{F}, P)$. Letting P_Z be the probability distribution of $Z \equiv X_0^{(1)} = X_0^{(2)}$ we have by 1° of Definition 20.7 the equality $X^{(1)}(\cdot, \omega) = F_{P_Z}[Z(\omega), B(\cdot, \omega)] = X^{(2)}(\cdot, \omega)]$ for a.e. ω in $(\Omega, \mathfrak{F}, P)$. This verifies (II).

2) Sufficiency. If we assume (I) and (II) then by Lemma 20.12 mappings F_μ and F of $\mathbf{W}^r$ into $\mathbf{W}^d$ satisfying conditions 1°, 2°, and 3° of Definition 20.6 exist. It remains to verify conditions 1° and 2° of Definition 20.7.

To verify 1°, let (B, X) be a solution of the stochastic differential equation on a standard filtered space $(\Omega, \mathfrak{F}, \{\mathfrak{F}_t\}, P)$ and let μ be the probability distribution of X_0. Let (W, Y) be the representation of (B, X) on $(\mathbf{W}^{r+d}, \mathfrak{W}^{r+d,*}, \{\mathfrak{W}_t^{r+d,*}\}, P_{(B,X)})$ as in Theorem 18.13. Then $F_\mu[Y_0(w), W(\cdot, w)] = F[Y_0(w), W(\cdot, w)] = Y(\cdot, w)$ for a.e. w in $(\mathbf{W}^{r+d}, \mathfrak{W}^{r+d}, P_{(B,X)})$ by (7) of Lemma 20.12 and 2° of Theorem 19.18. Thus $F_\mu[X_0(\omega), B(\cdot, \omega)] = X(\cdot, \omega)$ for a.e. ω in $(\Omega, \mathfrak{F}, P)$.

To verify 2°, let $(\Omega, \mathfrak{F}, \{\mathfrak{F}_t\}, P)$ be a standard filtered space on which an r-dimensional $\{\mathfrak{F}_t\}$-adapted null at 0 Brownian motion B exists and let Z be an arbitrary $\mathbb{R}^d$-valued $\mathfrak{F}_0$-adapted random variable on $(\Omega, \mathfrak{F}, P)$. Let P_Z be the probability distribution of Z. By (I) there exists a solution (B', X') of the stochastic differential equation on some standard filtered space $(\Omega', \mathfrak{F}', \{\mathfrak{F}_t'\}, P')$ such that X_0' has P_Z as its probability distribution. Let (W, Y) be the function space representation of this solution on $(\mathbf{W}^{r+d}, \mathfrak{W}^{r+d,*}, \{\mathfrak{W}_t^{r+d,*}\}, P'_{(B',X')})$ as in Theorem 18.13. In particular the probability distribution of Y_0 is given by P_Z. By 1° of Definition 20.7 verified above, we have

(1) $$Y(\cdot, w) = F_{P_Z}[Y_0(w), W(\cdot, w)]$$

for a.e. w in $(\mathbf{W}^{r+d}, \mathfrak{W}^{r+d,*}, P'_{(B',X')})$. If we define

(2) $$X(\cdot, \omega) = F_{P_Z}[Z(\omega), B(\cdot, \omega)]$$

§20. STRONG SOLUTIONS   441

for $\omega \in \Omega$, then (B, X) and (W, Y) have an identical probability distribution $P'_{(B', X')}$ on $(\mathbf{W}^{r+d}, \mathfrak{W}^{r+d})$. Since (W, Y) is a solution of the stochastic differential equation, so is (B, X) by Theorem 18.15.

Let us show $X_0 = Z$. By (2) above and (7) of Lemma 20.12, we have

(3) $$X(\cdot, \omega) = F[Z(\omega), B(\cdot, \omega)]$$

for $Z(\omega) \in N^c$ where N is a null set in $(\mathbb{R}^d, \mathfrak{B}_{\mathbb{R}^d}, P_Z)$. Let $\Lambda = Z^{-1}(N)$. Then $P(\Lambda) = P \circ Z^{-1}(N) = P_Z(N) = 0$. Now (3) holds for $\omega \in \Lambda^c$. Since (B, X) is a solution of the initial value problem with the initial condition $X_0(\omega) = Z(\omega)$ for $Z(\omega) \in N^c$, we have $X_0(\omega) = Z(\omega)$ for $\omega \in \Lambda^c$ by Theorem 19.18. ∎

Appendix A
Stochastic Independence

Definition A.1. *With an arbitrary index set A, let $\mathfrak{E}_\alpha$ be a subcollection of $\mathfrak{F}$ in a probability space $(\Omega, \mathfrak{F}, P)$ for $\alpha \in A$. We say that system $\{\mathfrak{E}_\alpha : \alpha \in A\}$ is stochastically independent with respect to P, or simply independent when P is understood, if for every finite subset $\{\alpha_1, \ldots, \alpha_n\}$ of A and for arbitrary $E_{\alpha_j} \in \mathfrak{E}_{\alpha_j}$ for $j = 1, \ldots, n$ we have*

$$P(\cap_{j=1}^n E_{\alpha_j}) = \prod_{j=1}^n P(E_{\alpha_j}).$$

In particular $\{E_\alpha : \alpha \in A\}$ where $E_\alpha \in \mathfrak{F}$ is said to be independent if for every finite subset $\{\alpha_1, \ldots, \alpha_n\}$ of A we have we have $P(\cap_{j=1}^n E_{\alpha_j}) = \prod_{j=1}^n P(E_{\alpha_j})$.

Note that for a finite index set A, say $A = \{1, \ldots, n\}$ and we have $\{\mathfrak{E}_j : j = 1, \ldots, n\}$, the assumption that $P(\cap_{j=1}^n E_j) = \prod_{j=1}^n P(E_j)$ for arbitrary $E_j \in \mathfrak{E}_j$ for $j = 1, \ldots, n$ does not imply that $P(\cap_{j=1}^k E_j) = \prod_{j=1}^k P(E_j)$ for arbitrary $E_j \in \mathfrak{E}_j$ for $j = 1, \ldots, k$ for $k < n$. For instance consider the particular case $\{E_j : j = 1, \ldots, n\}$ where $E_j = E$ for $j = 1, \ldots, n-1$ for some $E \in \mathfrak{F}$ with $P(E) = 1/2$ and $E_n = \emptyset$. In this case we have $P(\cap_{j=1}^n E_j) = 0 = \prod_{j=1}^n P(E_j)$, but $P(\cap_{j=1}^{n-1} E_j) = P(E) = 1/2 \neq 1/2^{n-1} = \prod_{j=1}^{n-1} P(E_j)$. There is an exception to this when $\Omega \in \mathfrak{E}_j$, and in particular when $\mathfrak{E}_j$ is a σ-algebra, for $j = 1, \ldots, n$. This is shown in the following proposition.

Proposition A.2. *Consider a finite system $\{\mathfrak{E}_j : j = 1, \ldots, n\}$ of subcollections $\mathfrak{E}_j$ of $\mathfrak{F}$ such that $\Omega \in \mathfrak{E}_j$ for $j = 1, \ldots, n$. If*

(1) $$P(\cap_{j=1}^n E_j) = \prod_{j=1}^n P(E_j)$$

for arbitrary $E_j \in \mathfrak{E}_j$ for $j = 1, \ldots, n$, then $\{\mathfrak{E}_j : j = 1, \ldots, n\}$ is an independent system.

Proof. Take an arbitrary subset of $\{1, \ldots, n\}$, say $\{1, \ldots, k\}$ where $k < n$. Then for arbitrary $E_j \in \mathfrak{E}_j$ for $j = 1, \ldots, k$ we have

$$P(E_1 \cap \cdots \cap E_k) = P(E_1 \cap \cdots \cap E_k \cap \Omega_{k+1} \cap \cdots \cap \Omega_n)$$

$$= \left\{\prod_{j=1}^{k} P(E_j)\right\} \left\{\prod_{i=k+1}^{n} P(\Omega_i)\right\} = \left\{\prod_{j=1}^{k} P(E_j)\right\}$$

by (1). This shows that $\{\mathfrak{E}_j : j = 1, \ldots, n\}$ is an independent system. ∎

Observation A.3. As immediate consequences of Definition A.1 we have the following.
1) If $\{\mathfrak{E}_\alpha : \alpha \in A\}$ is an independent system, then for any subset A_0 of A, $\{\mathfrak{E}_\alpha : \alpha \in A_0\}$ is an independent system.
2) If $\{\mathfrak{E}_\alpha : \alpha \in A\}$ is an independent system, and $\mathfrak{C}_\alpha \subset \mathfrak{E}_\alpha$ for every $\alpha \in A$, then $\{\mathfrak{C}_\alpha : \alpha \in A\}$ is an independent system.

Theorem A.4. *Let $\{\mathfrak{E}_\alpha : \alpha \in A\}$ be an independent system of subcollections of $\mathfrak{F}$ in a probability space $(\Omega, \mathfrak{F}, P)$. If $\mathfrak{E}_\alpha$ is a π-class for every $\alpha \in A$, then $\{\sigma(\mathfrak{E}_\alpha) : \alpha \in A\}$ is an independent system of sub-σ-algebras of $\mathfrak{F}$.*

Proof. To show the independence of $\{\sigma(\mathfrak{E}_\alpha) : \alpha \in A\}$ we show that for an arbitrary finite subset $\{\alpha_1, \ldots, \alpha_n\}$ of A we have $P(\cap_{j=1}^{n} F_{\alpha_j}) = \prod_{j=1}^{n} P(F_{\alpha_j})$ for arbitrary $F_{\alpha_j} \in \sigma(\mathfrak{E}_{\alpha_j})$ for $j = 1, \ldots, n$. For brevity in notation, let us write $\{1, \ldots, n\}$ for $\{\alpha_1, \ldots, \alpha_n\}$. Thus we are to show that for arbitrary $F_j \in \sigma(\mathfrak{E}_j)$ for $j = 1, \ldots, n$ we have

(1) $$P(\cap_{j=1}^{n} F_j) = \prod_{j=1}^{n} P(F_j).$$

We prove (1) by induction. Thus let us establish first that for any $F_1 \in \sigma(\mathfrak{E}_1)$ and $E_j \in \mathfrak{E}_j$ for $j = 2, \ldots, n$ we have

(2) $$P(F_1 \cap \cap_{j=2}^{n} E_j) = P(F_1) \prod_{j=2}^{n}(E_j).$$

With fixed $E_j \in \mathfrak{E}_j$ for $j = 2, \ldots, n$, define two finite measures μ_1 and ν_1 on $\sigma(\mathfrak{E}_1)$ by setting

(3) $$\begin{cases} \mu_1(F_1) = P(F_1 \cap \cap_{j=2}^{n} E_j) & \text{for } F_1 \in \sigma(\mathfrak{E}_1) \\ \nu_1(F_1) = P(F_1) \prod_{j=2}^{n} P(E_j) & \text{for } F_1 \in \sigma(\mathfrak{E}_1). \end{cases}$$

APPENDIX A. STOCHASTIC INDEPENDENCE 445

Then for $E_j \in \sigma(\mathfrak{E}_j)$ we have $\mu_1(E_1) = P(\cap_{j=1}^n E_j) = \prod_{j=1}^n P(E_j) = \nu_1(E_1)$ where the second equality is by the independence of $\{\mathfrak{E}_\alpha : \alpha \in A\}$. Thus $\mu_1 = \nu_1$ on the π-class $\mathfrak{E}_1$. We have also $\mu_1(\Omega) = P(\Omega \cap \cap_{j=2}^n E_j) = P(\Omega) \prod_{j=2}^n P(E_j) = \nu_1(\Omega)$, where the second equality is by the independence of $\{\mathfrak{E}_\alpha : \alpha \in A\}$. Thus by Corollary 1.8 we have $\mu_1 = \nu_1$ on $\sigma(\mathfrak{E}_1)$. By (3), this implies that (2) holds. Next assume that for some $k < n$ we have

$$(4) \qquad P(\cap_{j=1}^k F_j \cap \cap_{j=k+1}^n E_j) = \prod_{j=1}^k P(F_j) \prod_{j=k+1}^n P(F_j)$$

for $F_j \in \sigma(\mathfrak{E}_j)$ for $j = 1, \ldots, k$ and $E_j \in \mathfrak{E}_j$ for $j = k+1, \ldots, n$. With fixed $F_j \in \sigma(\mathfrak{E}_j)$ for $j = 1, \ldots, k$ and $E_j \in \mathfrak{E}_j$ for $j = k+2, \ldots, n$, let us define two finite measures μ_{k+1} and ν_{k+1} on $\sigma(\mathfrak{E}_{k+1})$ by

$$(5) \qquad \begin{cases} \mu_{k+1}(F_{k+1}) = P(\cap_{j=1}^{k+1} F_j \cap \cap_{j=k+2}^n E_j) & \text{for } F_{k+1} \in \sigma(\mathfrak{E}_{k+1}) \\ \nu_{k+1}(F_{k+1}) = \prod_{j=1}^{k+1} P(F_j) \prod_{j=k+2}^n P(E_j) & \text{for } F_{k+1} \in \sigma(\mathfrak{E}_{k+1}). \end{cases}$$

Then for $E_{k+1} \in \mathfrak{E}_{k+1}$ we have

$$\mu_{k+1}(E_{k+1}) = P(\cap_{j=1}^k F_j \cap \cap_{j=k+1}^n E_j) = \prod_{j=1}^k P(F_j) \prod_{j=k+1}^n P(E_j) = \nu_{k+1}(E_{k+1})$$

where the second equality is by the induction hypothesis (4). Thus $\mu_{k+1} = \nu_{k+1}$ on $\mathfrak{E}_{k+1}$. Also

$$\mu_{k+1}(\Omega) = P(\cap_{j=1}^k F_j \cap \Omega \cap \cap_{j=k+2}^n E_j) = \prod_{j=1}^k P(E_j) P(\Omega) \prod_{j=k+2}^n P(E_j) = \nu_{k+1}(\Omega)$$

where the second equality is by (4). Therefore by Corollary 1.8 we have $\mu_{k+1} = \nu_{k+1}$ on $\sigma(\mathfrak{E}_{k+1})$. Therefore by (5) we have

$$P(\cap_{j=1}^{k+1} F_j \cap \cap_{j=k+2}^n E_j) = \prod_{j=1}^{k+1} P(F_j) \prod_{j=k+2}^n P(E_j)$$

for $F_j \in \sigma(\mathfrak{E}_j)$ for $j = 1, \ldots, k+1$ and $E_j \in \mathfrak{E}_j$ for $j = k+2, \ldots, n$. Thus (1) holds by induction on k. ∎

Corollary A.5. *Let $\{E_\alpha : \alpha \in A\}$ be an independent system of members of $\mathfrak{F}$ in a probability space $(\Omega, \mathfrak{F}, P)$. Then $\{\sigma(\{\mathfrak{E}_\alpha\}) : \alpha \in A\}$ is an independent system. For each $\alpha \in A$ let $F_\alpha = E_\alpha, E_\alpha^c, \emptyset,$ or Ω. Then $\{F_\alpha : \alpha \in A\}$ is an independent system.*

Proof. Let $\mathfrak{E}_\alpha = \{E_\alpha\}$ for every $\alpha \in A$. Then $\mathfrak{E}_\alpha$ is trivially a π-class. Thus the independence of $\{E_\alpha : \alpha \in A\}$ implies that of $\{\sigma(\mathfrak{E}_\alpha) : \alpha \in A\}$ by Theorem A.4. Now $\sigma(\mathfrak{E}_\alpha) = \{E_\alpha, E_\alpha^c, \emptyset, \Omega\}$. Thus $\{F_\alpha : \alpha \in A\}$ is an independent system by 2) of Observation A.3. ∎

Let X be a d-dimensional random vector on a probability space $(\Omega, \mathfrak{F}, P)$, that is, X is an $\mathfrak{F}/\mathfrak{B}_{\mathbb{R}^d}$-measurable mapping of Ω into $\mathbb{R}^d$. Let $\sigma(X)$ be the σ-algebra generated by X, that is, the smallest sub-σ-algebra of $\mathfrak{F}$ with respect to which X is measurable. In other words $\sigma(X) = X^{-1}(\mathfrak{B}_{\mathbb{R}^d})$.

Definition A.6. *Let $\{X_\alpha : \alpha \in A\}$ be a system of random vectors on a probability space $(\Omega, \mathfrak{F}, P)$, the dimension of X_α being $d_\alpha \in \mathbb{N}$. We say that $\{X_\alpha : \alpha \in A\}$ is an independent system if $\{\sigma(X_\alpha) : \alpha \in A\}$ is an independent system in the sense of Definition A.1.*

Remark A.7. A system $\{E_\alpha; \alpha \in A\}$ of members of $\mathfrak{F}$ in a probability space $(\Omega, \mathfrak{F}, P)$ is independent if and only if the system of random variables $\{\mathbf{1}_{E_\alpha} : \alpha \in A\}$ is independent.

Proof. By Definition A.6, $\{\mathbf{1}_{E_\alpha} : \alpha \in A\}$ is an independent system of random variables if and only if $\{\sigma(\mathbf{1}_{E_\alpha}) : \alpha \in A\}$ is an independent system of sub-σ-algebras of $\mathfrak{F}$. But $\sigma(\mathbf{1}_{E_\alpha}) = \{E_\alpha, E_\alpha^c, \emptyset, \Omega\} = \sigma(\{E_\alpha\})$ and according to Corollary A.5 the independence of $\{\sigma(\{E_\alpha\}) : \alpha \in A\}$ is equivalent to the independence of $\{E_\alpha; \alpha \in A\}$. ∎

Theorem A.8. *Let $\{X_\alpha : \alpha \in A\}$ be an independent system of random vectors on a probability space $(\Omega, \mathfrak{F}, P)$, the dimension of X being equal to $d_\alpha \in \mathbb{N}$. For each $\alpha \in A$, let T_α be a $\mathfrak{B}_{\mathbb{R}^{d_\alpha}}/\mathfrak{B}_{\mathbb{R}^{k_\alpha}}$-measurable mapping of $\mathbb{R}^{d_\alpha}$ into $\mathbb{R}^{k_\alpha}$ where $k_\alpha \in \mathbb{N}$. Then $\{T_\alpha \circ X_\alpha : \alpha \in A\}$ is an independent system of random vectors, the dimension of $T_\alpha \circ X_\alpha$ being k_α.*

Proof. Since X_α is an $\mathfrak{F}/\mathfrak{B}_{\mathbb{R}^{d_\alpha}}$-measurable mapping of Ω into $\mathbb{R}^{d_\alpha}$ and T_α is a $\mathfrak{B}_{\mathbb{R}^{d_\alpha}}/\mathfrak{B}_{\mathbb{R}^{k_\alpha}}$-measurable mapping of $\mathbb{R}^{d_\alpha}$ into $\mathbb{R}^{k_\alpha}$, $T_\alpha \circ X_\alpha$ is an $\mathfrak{F}/\mathfrak{B}_{\mathbb{R}^{k_\alpha}}$-measurable mapping of Ω into $\mathbb{R}^{k_\alpha}$, that is, a k_α-dimensional random vector on $(\Omega, \mathfrak{F}, P)$. To show that $\{T_\alpha \circ X_\alpha : \alpha \in A\}$ is an independent system of random vectors we show that $\{\sigma(T_\alpha \circ X_\alpha) : \alpha \in A\}$ is an independent system of sub-σ-algebras of $\mathfrak{F}$ according to Definition A.6. Now

$$\sigma(T_\alpha \circ X_\alpha) = (T_\alpha \circ X_\alpha)^{-1}(\mathfrak{B}_{\mathbb{R}^{k_\alpha}}) = X_\alpha^{-1}(T_\alpha^{-1}(\mathfrak{B}_{\mathbb{R}^{k_\alpha}})). \tag{1}$$

Since T_α is $\mathfrak{B}_{\mathbb{R}^{d_\alpha}}/\mathfrak{B}_{\mathbb{R}^{k_\alpha}}$-measurable, we have $T_\alpha^{-1}(\mathfrak{B}_{\mathbb{R}^k_\alpha}) \subset \mathfrak{B}_{\mathbb{R}^{d_\alpha}}$. Thus from (1) we have

(2) $$\sigma(T_\alpha \circ X_\alpha) \subset X_\alpha^{-1}(\mathfrak{B}_{\mathbb{R}^{d_\alpha}}) = \sigma(X_\alpha).$$

Now since $\{X_\alpha\} : \alpha \in A\}$ is an independent system of random vectors, $\{\sigma(X_\alpha : \alpha \in A\}$ is an independent system of sub-σ-algebras of euf. Then (2) implies that $\{\sigma(T_\alpha \circ X_\alpha) : \alpha \in A\}$ is an independent system of sub-σ-algebras of $\mathfrak{F}$ by 2) of Observation A.3. ∎

Theorem A.9. *Let X_j be a d_j-dimensional random vector on a probability space $(\Omega, \mathfrak{F}, P)$ for $j = 1, \ldots, n$. Consider the d-dimensional random vector $X = (X_1, \ldots, X_n)$ where $d = d_1 + \cdots + d_n$. Let $P_{X_1}, \ldots, P_{X_n}$, and P_X be the probability distributions of $X_1, \ldots, X_n$ and X on $(\mathbb{R}^{d_1}, \mathfrak{B}_{\mathbb{R}^{d_1}}), \ldots, (\mathbb{R}^{d_n}, \mathfrak{B}_{\mathbb{R}^{d_1}})$ and $(\mathbb{R}^d, \mathfrak{B}_{\mathbb{R}^d})$ respectively. Then*

(1) $\{X_1, \ldots, X_n\}$ *is an independent system* $\Leftrightarrow P_X = P_{X_1} \times \cdots \times P_{X_n}$ *on* $(\mathbb{R}^d, \mathfrak{B}_{\mathbb{R}^d})$.

Proof. Since $\mathfrak{B}_{\mathbb{R}^{d_1}} \times \cdots \times \mathfrak{B}_{\mathbb{R}^{d_n}}$ is a semialgebra of subsets of $\mathbb{R}^d$ and since $\sigma(\mathfrak{B}_{\mathbb{R}^{d_1}} \times \cdots \times \mathfrak{B}_{\mathbb{R}^{d_n}}) = \mathfrak{B}_{\mathbb{R}^d}$ by Proposition 13.2, we have

(2) $P_X = P_{X_1} \times \cdots \times P_{X_n}$ on $(\mathbb{R}^d, \mathfrak{B}_{\mathbb{R}^d})$

$\Leftrightarrow P_X = P_{X_1} \times \cdots \times P_{X_n}$ on $\mathfrak{B}_{\mathbb{R}^{d_1}} \times \cdots \times \mathfrak{B}_{\mathbb{R}^{d_n}}$ by Corollary 1.8

$\Leftrightarrow P_X(B_1 \times \cdots \times B_n) = \prod_{j=1}^n P_{X_j}(B_j)$ for $B_j \in \mathfrak{B}_{\mathbb{R}^{d_j}}, j = 1, \ldots, n$

$\Leftrightarrow (P \circ X^{-1})(B_1 \times \cdots \times B_n) = \prod_{j=1}^n (P \circ X_j^{-1})(B_j)$ for $B_j \in \mathfrak{B}_{\mathbb{R}^{d_j}}, j = 1, \ldots, n$

On the other hand

$\{X_1, \ldots, X_n\}$ is an independent system

$\Leftrightarrow \{\sigma(X_1), \ldots, \sigma(X_n)\}$ is an independent system

$\Leftrightarrow P(\cap_{j=1}^n E_j) = \prod_{j=1}^n P(E_j)$ for $E_j \in \sigma(X_j), j = 1, \ldots, n$ by Proposition A.2

$\Leftrightarrow P(\cap_{j=1}^n (X_j^{-1}(B_j))) = \prod_{j=1}^n P(X_j^{-1}(B_j))$ for $B_j \in \mathfrak{B}_{\mathbb{R}^{d_j}}, j = 1, \ldots, n$

since $\sigma(X_j) = X_j^{-1}(\mathfrak{B}_{\mathbb{R}^{d_j}})$. Now the fact that $\cap_{j=1}^n(X_j^{-1}(B_j)) = X^{-1}(B_1 \times \cdots \times B_n)$ implies that $P(\cap_{j=1}^n(X_j^{-1}(B_j))) = (P \circ X^{-1})(B_1 \times \cdots \times B_n)$. Thus

(3) $\{X_1, \ldots, X_n\}$ is an independent system

$$\Leftrightarrow (P \circ X^{-1})(B_1 \times \cdots \times B_n) = \prod_{j=1}^n P(X_j^{-1}(B_j)) \text{ for } B_j \in \mathfrak{B}_{\mathbb{R}^{d_j}}, j = 1, \ldots, n$$

By (2) and (3) we have (1). ∎

The characteristic function of a probability measure μ on $(\mathbb{R}^d, \mathfrak{B}_{\mathbb{R}^d})$ is a complex valued function on $\mathbb{R}^d$ defined by

$$\varphi(\eta; \mu) = \int_{\mathbb{R}^d} \exp\{i \langle \xi, \eta \rangle\} \mu(d\xi) \quad \text{for } \eta \in \mathbb{R}^d$$

where $\langle \xi, \eta \rangle = \sum_{j=1}^d \xi_j \eta_j$ for $\xi, \eta \in \mathbb{R}^d$. It is well known that the characteristic function of a probability measure on $(\mathbb{R}^d, \mathfrak{B}_{\mathbb{R}^d})$ is unique, that is, if μ and ν are two probability measures on $(\mathbb{R}^d, \mathfrak{B}_{\mathbb{R}^d})$ and $\varphi(\cdot; \mu) = \varphi(\cdot; \nu)$ on $\mathbb{R}^d$ then $\mu = \nu$ on $(\mathbb{R}^d, \mathfrak{B}_{\mathbb{R}^d})$. Let $\mu_1, \ldots, \mu_n$ be probability measures on $(\mathbb{R}^{d_1}, \mathfrak{B}_{\mathbb{R}^{d_1}}), \ldots, (\mathbb{R}^{d_n}, \mathfrak{B}_{\mathbb{R}^{d_n}})$ respectively and let $\mu = \mu_1 \times \cdots \times \mu_n$ on $(\mathbb{R}^d, \mathfrak{B}_{\mathbb{R}^d}) = (\mathbb{R}^{d_1} \times \cdots \times \mathbb{R}^{d_n}, \sigma(\mathfrak{B}_{\mathbb{R}^{d_1}} \times \cdots \times \mathfrak{B}_{\mathbb{R}^{d_n}}))$ where $d = d_1 + \cdots + d_n$. Then it follows immediately from the bilinearity of $\langle \cdot, \cdot \rangle$ and Fubini's Theorem that

$$\varphi(\eta; \mu) = \varphi(\eta_1; \mu_1) \cdots \varphi(\eta_n; \mu_n) \quad \text{for } \eta = (\eta_1, \ldots, \eta_n) \in \mathbb{R}^{d_1} \times \cdots \times \mathbb{R}^{d_n}.$$

Theorem A.10. (M. Kac) *Consider a system of random vectors $\{X_1, \ldots, X_n\}$ on a probability space $(\Omega, \mathfrak{F}, P)$, the dimension of X_j being $d_j \in \mathbb{N}$. Let P_{X_j} be the probability distribution of X_j on $(\mathbb{R}^{d_j}, \mathfrak{B}_{\mathbb{R}^{d_j}})$ for $j = 1, \ldots, n$ and let P_X be the probability distribution of the random vector $X = (X_1, \ldots, X_n)$ on $(\mathbb{R}^d, \mathfrak{B}_{\mathbb{R}^d})$ where $d = d_1 + \cdots + d_n$. Then in order that $\{X_1, \ldots, X_n\}$ be an independent system it is necessary and sufficient that*

(1) $$\varphi(\eta; P_X) = \varphi(\eta_1; P_{X_1}) \cdots \varphi(\eta_n; P_{X_n})$$

for $\eta = (\eta_1, \ldots, \eta_n) \in \mathbb{R}^{d_1} \times \cdots \times \mathbb{R}^{d_n}$, or equivalently

(2) $$\mathbf{E}[\prod_{j=1}^n \exp\{i \langle X_j, \eta_j \rangle\}] = \prod_{j=1}^n \mathbf{E}[\exp\{i \langle X_j, \eta_j \rangle\}]$$

for $\eta = (\eta_1, \ldots, \eta_n) \in \mathbb{R}^{d_1} \times \cdots \times \mathbb{R}^{d_n}$.

APPENDIX A. STOCHASTIC INDEPENDENCE

Proof. Note that (1) can be written as

(3) $$\int_{\mathbb{R}^d} \prod_{j=1}^n \exp\{i\langle \xi_j, \eta_j\rangle\} P_X(d\xi) = \prod_{j=1}^n \int_{\mathbb{R}^{d_j}} \exp\{i\langle \xi_j, \eta_j\rangle\} P_{X_j}(d\xi)$$

and (3) is equivalent to (2) by the Image Probability Law.

Now suppose $\{X_1, \ldots, X_n\}$ is an independent system. Then by Theorem A.9 we have $P_X = P_{X_1} \times \cdots \times P_{X_n}$. Fubini's Theorem then implies (3). Conversely suppose (3) holds. To shows the independence of $\{X_1, \ldots, X_n\}$ it suffices to show that $P_X = P_{X_1} \times \cdots \times P_{X_n}$ according to Theorem A.9. Let $Q = P_{X_1} \times \cdots \times P_{X_n}$. If we show that $\varphi(\eta; P_X) = \varphi(\eta; Q)$ for $\eta \in \mathbb{R}^d = \mathbb{R}^{d_1} \times \cdots \times \mathbb{R}^{d_n}$, then $P_X = Q$ and we are done. Now

$$\varphi(\eta; P_X) = \int_{\mathbb{R}^d} \exp\{i\langle \xi, \eta\rangle\} P_X(d\xi) = \prod_{j=1}^n \int_{\mathbb{R}^{d_j}} \exp\{i\langle \xi_j, \eta_j\rangle\} P_{X_j}(d\xi_j)$$

$$= \int_{\mathbb{R}^d} \exp\{i\langle \xi, \eta\rangle\} Q(d\xi) = \varphi(\eta; Q)$$

where the second equality is by (3) and the third equality is by the fact that $Q = P_{X_1} \times \cdots \times P_{X_n}$ and Fubini's Theorem. ∎

Theorem A.11. *Consider a system of random vectors on a probability space* $(\Omega, \mathfrak{F}, P)$ *given by*

(1) $$\{X_{k,q_k} : q_k = 1, \ldots, n_k; k = 1, \ldots, m\},$$

the dimension of X_{k,q_k} being $d_{k,q_k} \in \mathbb{N}$. For each $k = 1, \ldots, m$, consider the subsystem

(2) $$\{X_{k,1}, \ldots, X_{k,n_k}\}.$$

Let X_k be a $d_{k,1} + \cdots + d_{k,n_k}$-dimensional random vector defined by

$$X = (X_{k,1}, \ldots, X_{k,n_k})$$

and consider the system of random vectors

(3) $$\{X_k : k = 1, \ldots, m\}.$$

1) If the system (1) is independent, then so is the system (3).
2) Conversely if the system (3) is independent and furthermore for every $k = 1, \ldots, m$ the system (2) is independent, then the system (1) is independent.

Proof. Suppose the system (1) is independent. Then

$$E[\prod_{k=1}^{m} \exp\{i\langle X_k, \eta_k\rangle\}] = E[\exp\{i\sum_{k=1}^{m}\langle X_k, \eta_k\rangle\}]$$

$$= E[\exp\{i\sum_{k=1}^{m}\sum_{q_k=1}^{n_k}\langle X_{k,q_k}, \eta_{k,q_k}\rangle\}] = \prod_{k=1}^{m}\prod_{q_k=1}^{n_k} E[\exp\{i\langle X_{k,q_k}, \eta_{k,q_k}\rangle\}]$$

$$= \prod_{k=1}^{m} E[\exp\{i\langle X_k, \eta_k\rangle\}]$$

where the third equality is by the independence of the system (1) and Theorem A.10 and the last equality is by the independence of the system (2) as a subsystem of the independent system (1) and by Theorem A.10. According to Theorem A.10, this shows the independence of the system (3).

Conversely suppose the system (3) is independent and for every $k = 1, \ldots, m$ the system (2) is independent. Then

$$E[\prod_{k=1}^{m}\prod_{q_k=1}^{n_k} \exp\{i\langle X_{k,q_k}\eta_{k,q_k}\rangle\}] = E[\exp\{i\sum_{k=1}^{m}\sum_{q_k=1}^{n_k}\langle X_{k,q_k}\eta_{k,q_k}\rangle\}]$$

$$= E[\exp\{i\sum_{k=1}^{m}\langle X_k, \eta_k\rangle\}] = E[\prod_{k=1}^{m} \exp\{i\langle X_k, \eta_k\rangle\}]$$

$$= \prod_{k=1}^{m} E[\exp\{i\langle X_k, \eta_k\rangle\}] = \prod_{k=1}^{m}\prod_{q_k=1}^{n_k} E[\exp\{i\langle X_{k,q_k}, \eta_{k,q_k}\rangle\}]$$

where the second from last equality is by the independence of the system (3) and Theorem A.10 and the last equality is from the independence of the system (2) and Theorem A.10. This shows the independence of the system (1) according to Theorem A.10. ∎

Definition A.12. *Let $\mathfrak{E}_\alpha$ be a subcollection of $\mathfrak{F}$ for $\alpha \in A$ in a probability space $(\Omega, \mathfrak{F}, P)$ and X_β be a d_β-dimensional random vector on $(\Omega, \mathfrak{F}, P)$ for $\beta \in B$. We say that $\{\mathfrak{E}_\alpha, X_\beta : \alpha \in A, \beta \in B\}$ is an independent system if $\{\mathfrak{E}_\alpha, \sigma(X_\beta) : \alpha \in A, \beta \in B\}$ is an independent system in the sense of Definition A.1.*

Theorem A.13. *Let $\mathfrak{G}_j$ be a sub-σ-algebra of $\mathfrak{F}$ in a probability space $(\Omega, \mathfrak{F}, P)$ for $j = 1, \ldots, m$ and X_k be d_k-dimensional random vector on $(\Omega, \mathfrak{F}, P)$ for $k = 1, \ldots, n$. Then the following three conditions are equivalent:*

APPENDIX A. STOCHASTIC INDEPENDENCE

1°. $\{\mathfrak{G}_j, X_k : j = 1, \ldots, m; k = 1, \ldots, n\}$ *is an independent system in the sense of Definition A.12, that is,* $\{\mathfrak{G}_j, \sigma(X_k) : j = 1, \ldots, m; k = 1, \ldots, n\}$ *is an independent system.*

2°. $\{Z_j, X_k : j = 1, \ldots, m; k = 1, \ldots, n\}$ *is an independent system for an arbitrary real valued $\mathfrak{G}_j$-measurable random variable Z_j for $j = 1, \ldots, m$.*

3°. $\{Z_j, Y_k : j = 1, \ldots, m; k = 1, \ldots, n\}$ *is an independent system for an arbitrary real valued $\mathfrak{G}_j$-measurable random variable Z_j for $j = 1, \ldots, m$ and an arbitrary real valued $\sigma(X_k)$-measurable random variable Y_k for $k = 1, \ldots, n$.*

Proof. 1° $\Rightarrow$ 2°. Assume 1°. The $\mathfrak{G}_j$-measurability of Z_j implies $\sigma(Z_j) \subset \mathfrak{G}_j$. Then the independence of the system $\{\mathfrak{G}_j, \sigma(X_k) : j = 1, \ldots, m; k = 1, \ldots, n\}$ implies that of the system $\{\sigma(Z_j), \sigma(X_k) : j = 1, \ldots, m; k = 1, \ldots, n\}$ by 2) of Observation A.3. This shows the independence of $\{Z_j, X_k : j = 1, \ldots, m; k = 1, \ldots, n\}$.

2° $\Rightarrow$ 3°. Assume 2°. The $\sigma(X_k)$-measurability of Y_k implies $\sigma(Y_k) \subset \sigma(X_k)$. Then the independence of the system $\{\sigma(Z_j), \sigma(X_k) : j = 1, \ldots, m; k = 1, \ldots, n\}$ implies that of the system $\{\sigma(Z_j), \sigma(Y_k) : j = 1, \ldots, m; k = 1, \ldots, n\}$ by 2) of Observation A.3.

3° $\Rightarrow$ 1°. Assume 3°. Since $\Omega \in \mathfrak{G}_j$ and $\Omega \in \sigma(X_k)$, to show the independence of $\{\mathfrak{G}_j, \sigma(X_k) : j = 1, \ldots, m; k = 1, \ldots, n\}$ it suffices according to Proposition A.2 to show that for every $G_j \in \mathfrak{G}_j, j = 1, \ldots, m$, and $F_k \in \sigma(X_k), k = 1, \ldots, n$, we have

$$(1) \qquad P(\cap_{j=1}^m G_j \cap \cap_{k=1}^n F_k) = \prod_{j=1}^m P(G_j) \prod_{k=1}^n P(F_k).$$

Now $\mathbf{1}_{G_j}$ is a $\mathfrak{G}_j$-measurable random variable and $\mathbf{1}_{F_k}$ is a $\sigma(X_k)$-measurable random variable. Thus by 3°, $\{\mathbf{1}_{G_j}, \mathbf{1}_{F_k} : j = 1, \ldots, m; k = 1, \ldots, n\}$ is an independent system and thus by Definition A.6 the system $\{\sigma(\mathbf{1}_{G_j}), \sigma(\mathbf{1}_{F_k}) : j = 1, \ldots, m; k = 1, \ldots, n\}$ is independent. Then since $G_j \in \sigma(\mathbf{1}_{G_j})$ and $F_k \in \sigma(\mathbf{1}_{F_k})$, we have (1). ∎

Appendix B

Conditional Expectations

[I] Definition and Basic Equalities

Let $\mathfrak{G}$ be a sub-σ-algebra of $\mathfrak{F}$ in a probability space $(\Omega, \mathfrak{F}, P)$. Two $\mathfrak{G}$-measurable extended real valued random variables X_1 and X_2 are said to be $(\mathfrak{G}, P)$-a.e. equal if $X_1 = X_2$ on Λ^c where $\Lambda \in \mathfrak{G}$ with $P(\Lambda) = 0$. In the collection of all $\mathfrak{G}$-measurable extended real valued random variables on $(\Omega, \mathfrak{F}, P)$, $(\mathfrak{G}, P)$-a.e. equality is an equivalence relation.

Proposition B.1. *Let $\mathfrak{G}$ be a sub-σ-algebra of $\mathfrak{F}$ in a probability space $(\Omega, \mathfrak{F}, P)$. Consider the equivalence relation of $(\mathfrak{G}, P)$-a.e. equality in the collection of all $\mathfrak{G}$-measurable extended real valued random variables on $(\Omega, \mathfrak{F}, P)$. For every integrable extended real valued random variable X on $(\Omega, \mathfrak{F}, P)$ there exists a unique equivalence class consisting of all $\mathfrak{G}$-measurable extended real valued random variables Y on $(\Omega, \mathfrak{F}, P)$ such that*

$$\int_G Y \, dP = \int_G X \, dP \quad \text{for every } G \in \mathfrak{G}.$$

Proof. For an arbitrary integrable extended real valued random variable X on $(\Omega, \mathfrak{F}, P)$, let us define a set function μ on $\mathfrak{G}$ by setting $\mu(G) = \int_G X \, dP$ for $G \in \mathfrak{G}$. Then μ is a signed measure on $\mathfrak{G}$. Since $\mu(G) = 0$ for every $G \in \mathfrak{G}$ with $P(G) = 0$, μ is absolutely continuous with respect to P on $(\Omega, \mathfrak{G})$. The Radon-Nikodym derivative of μ with respect to P on $(\Omega, \mathfrak{G})$ is then an equivalence class with respect to the $(\mathfrak{G}, P)$-a.e. equality consisting of all $\mathfrak{G}$-measurable extended real valued functions Y on Ω such that $\int_G Y \, dP = \mu(G)$, that is, $\int_G Y \, dP = \int_G X \, dP$, for every $G \in \mathfrak{G}$. ∎

Definition B.2. *Let X be an integrable extended real valued random variable on a proba-*

bility space $(\Omega, \mathfrak{F}, P)$ and let $\mathfrak{G}$ be a sub-σ-algebra of $\mathfrak{F}$. By the conditional expectation of X given $\mathfrak{G}$, denoted by $\mathrm{E}(X\,|\,\mathfrak{G})$, we mean the unique equivalence class of all $\mathfrak{G}$-measurable extended real valued random variables Y on $(\Omega, \mathfrak{F}, P)$ such that

$$\int_G Y\,dP = \int_G X\,dP \quad \text{for every } G \in \mathfrak{G}.$$

The members of the equivalence class $\mathrm{E}(X\,|\,\mathfrak{G})$ are called versions of the conditional expectation of X given $\mathfrak{G}$. Thus an extended real valued random variable Y is a version of $\mathrm{E}(X\,|\,\mathfrak{G})$ if and only if it satisfies the following two conditions:

1°. Y is $\mathfrak{G}$-measurable.

2°. $\int_G Y\,dP = \int_G X\,dP$ for every $G \in \mathfrak{G}$.

For convenience the notation $\mathrm{E}(X\,|\,\mathfrak{G})$ is also used for an arbitrary version of the conditional expectation. For an arbitrary $A \in \mathfrak{F}$, the conditional probability of A given $\mathfrak{G}$, denoted by $P(A\,|\,\mathfrak{G})$, is defined as the conditional expectation of the random variable 1_A given $\mathfrak{G}$, that is, $P(A\,|\,\mathfrak{G}) = \mathrm{E}(1_A\,|\,\mathfrak{G})$.

Note that every version of Y of $\mathrm{E}(X\,|\,\mathfrak{G})$ is an integrable random variable since $\Omega \in \mathfrak{G}$ and thus $\int_\Omega Y\,dP = \int_\Omega X\,dP \in \mathbb{R}$. This implies that every version is finite a.e. on $(\Omega, \mathfrak{G}, P)$ and thus there exists a real valued version of $\mathrm{E}(X\,|\,\mathfrak{G})$.

Since the same notation $\mathrm{E}(X\,|\,\mathfrak{G})$ is used for both an equivalence class and an arbitrary representative of it, what is meant by the notation $\mathrm{E}(X\,|\,\mathfrak{G})$ should be determined from the context. For instance in expressions such as " $\int_G \mathrm{E}(X\,|\,\mathfrak{G})dP$ " and "$\mathrm{E}(X_2\,|\,\mathfrak{G}) \geq \mathrm{E}(X_1\,|\,\mathfrak{G})$ a.e. on $(\Omega, \mathfrak{G}, P)$", $\mathrm{E}(X\,|\,\mathfrak{G})$, $\mathrm{E}(X_1\,|\,\mathfrak{G})$, and $\mathrm{E}(X_2\,|\,\mathfrak{G})$ are arbitrary versions of the conditional expectations, and in expressions such as " $Z \in \mathrm{E}(X\,|\,\mathfrak{G})$" and "$\mathrm{E}(X_1\,|\,\mathfrak{G}_1) \subset \mathrm{E}(X_2\,|\,\mathfrak{G}_2)$", the notations $\mathrm{E}(X\,|\,\mathfrak{G})$, $\mathrm{E}(X_1\,|\,\mathfrak{G}_1)$, and $\mathrm{E}(X_2\,|\,\mathfrak{G}_2)$ are for equivalence classes. For a random variable Z the expression "$\mathrm{E}(X\,|\,\mathfrak{G}) = Z$" is also used to indicate that Z is a version of $\mathrm{E}(X\,|\,\mathfrak{G})$. The notation $Z\mathrm{E}(X\,|\,\mathfrak{G})$ may mean the collection of all versions multiplied by Z of an arbitrary version multiplied by Z.

Observation B.3. Let X_1 and X_2 be two integrable extended real valued random variables on a probability space $(\Omega, \mathfrak{F}, P)$ and let $\mathfrak{G}$ be a sub-σ-algebra of $\mathfrak{F}$. If there exists a $\mathfrak{G}$-measurable extended real valued random variable Y_0 on $(\Omega, \mathfrak{F}, P)$ which is a version of both $\mathrm{E}(X_1\,|\,\mathfrak{G})$ and $\mathrm{E}(X_2\,|\,\mathfrak{G})$ then $\mathrm{E}(X_1\,|\,\mathfrak{G}) = \mathrm{E}(X_2\,|\,\mathfrak{G})$, that is, if $\mathrm{E}(X_1\,|\,\mathfrak{G}) \cap \mathrm{E}(X_2\,|\,\mathfrak{G}) \neq \emptyset$ then $\mathrm{E}(X_1\,|\,\mathfrak{G}) = \mathrm{E}(X_2\,|\,\mathfrak{G})$.

APPENDIX B. CONDITIONAL EXPECTATIONS

Proof. Let Y_j be an arbitrary version of $\mathbf{E}(X_j|\mathfrak{G})$ for $j = 1, 2$. Then $Y_j = Y_0$ a.e. on $(\Omega, \mathfrak{G}, P)$ for $j = 1, 2$ so that $Y_1 = Y_2$ a.e. on $(\Omega, \mathfrak{G}, P)$. Thus the two equivalence classes with respect to $(\mathfrak{G}, P)$-a.e. equality, $\mathbf{E}(X_1|\mathfrak{G})$ and $\mathbf{E}(X_2|\mathfrak{G})$, are identical. ∎

Observation B.4. Let $\mathfrak{G}$ be a sub-σ-algebra of $\mathfrak{F}$ in a probability space $(\Omega, \mathfrak{F}, P)$. Then for any $c \in \mathbb{R}$, we have $c \in \mathbf{E}(c|\mathfrak{G})$, that is, $\{c\} \subset \mathbf{E}(c|\mathfrak{G})$, but in general $\{c\} \neq \mathbf{E}(c|\mathfrak{G})$ since $\mathbf{E}(c|\mathfrak{G})$ consists of all $\mathfrak{G}$-measurable extended real valued random variables which are $(\mathfrak{G}, P)$-a.e. equal to c on Ω.

Theorem B.5. *Let X, X_1, and X_2 be integrable extended real valued random variables on a probability space and let $\mathfrak{G}$ be a sub-σ-algebra of $\mathfrak{F}$. Then*
1) $X_1 = X_2$ *a.e. on* $(\Omega, \mathfrak{F}, P) \Rightarrow \mathbf{E}(X_1|\mathfrak{G}) = \mathbf{E}(X_2|\mathfrak{G})$,
2) X *is $\mathfrak{G}$-measurable* $\Rightarrow X \in \mathbf{E}(X|\mathfrak{G})$,
3) $X = c$ *a.e. on* $(\Omega, \mathfrak{F}, P) \Rightarrow c \in \mathbf{E}(X|\mathfrak{G})$,
4) $X \geq 0$ *a.e. on* $(\Omega, \mathfrak{F}, P) \Rightarrow \mathbf{E}(X|\mathfrak{G}) \geq 0$ *a.e. on* $(\Omega, \mathfrak{G}, P)$,
5) $\mathbf{E}(cX|\mathfrak{G}) \supset c\mathbf{E}(X|\mathfrak{G})$ *for $c \in \mathbb{R}$ and if $c \neq 0$ then* $\mathbf{E}(cX|\mathfrak{G}) = c\mathbf{E}(X|\mathfrak{G})$,
6) $\mathbf{E}(c_1X_1 + c_2X_2|\mathfrak{G}) \supset c_1\mathbf{E}(X_1|\mathfrak{G}) + c_2\mathbf{E}(X_2|\mathfrak{G})$ *for $c_1, c_2 \in \mathbb{R}$ and if $c_1c_2 \neq 0$ then* $\mathbf{E}(c_1X_1 + c_2X_2|\mathfrak{G}) = c_1\mathbf{E}(X_1|\mathfrak{G}) + c_2\mathbf{E}(X_2|\mathfrak{G})$,
7) $X_2 \geq X_1$ *a.e. on* $(\Omega, \mathfrak{F}, P) \Rightarrow \mathbf{E}(X_2|\mathfrak{G}) \geq \mathbf{E}(X_1|\mathfrak{G})$ *a.e. on* $(\Omega, \mathfrak{G}, P)$.

Proof. 1) Let $Y \in \mathbf{E}(X_1|\mathfrak{G})$. Then Y is $\mathfrak{G}$-measurable and $\int_G Y\, dP = \int_G X_1\, dP$ for every $G \in \mathfrak{G}$. Since $X_1 = X_2$ a.e. on $(\Omega, \mathfrak{F}, P)$, we have $\int_G X_1\, dP = \int_G X_2\, dP$ for every $G \in \mathfrak{G}$ and thus $\int_G Y\, dP = \int_G X_2\, dP$ for every $G \in \mathfrak{G}$. This shows that $Y \in \mathbf{E}(X_2|\mathfrak{G})$. Then by Observation B.3, we have $\mathbf{E}(X_1|\mathfrak{G}) = \mathbf{E}(X_2|\mathfrak{G})$.

2) If X is $\mathfrak{G}$-measurable, then it satisfies conditions 1° and 2° in Definition B.2 and thus $X \in \mathbf{E}(X|\mathfrak{G})$.

3) A constant c is always $\mathfrak{G}$-measurable. If $X = c$ a.e. on $(\Omega, \mathfrak{F}, P)$, then for any $G \in \mathfrak{G} \subset \mathfrak{F}$ we have $\int_G c\, dP = \int_G X\, dP$. Thus c satisfies conditions 1° and 2° in Definition B.2 and is therefore a version of $\mathbf{E}(X|\mathfrak{G})$.

4) Suppose $X \geq 0$ a.e. on $(\Omega, \mathfrak{F}, P)$. Let $Y \in \mathbf{E}(X|\mathfrak{G})$. Then $\int_G Y\, dP = \int_G X\, dP \geq 0$ for every $G \in \mathfrak{G}$ and therefore $Y \geq 0$ a.e. on $(\Omega, \mathfrak{G}, P)$.

5) Let $Y \in \mathbf{E}(X|\mathfrak{G})$. Then Y is $\mathfrak{G}$-measurable and so is cY. For every $G \in \mathfrak{G}$ we have $\int_G cY\, dP = c\int_G Y\, dP = c\int_G X\, dP = \int_G cX\, dP$ and thus $cY \in \mathbf{E}(cX|\mathfrak{G})$. This shows that $c\mathbf{E}(X|\mathfrak{G}) \subset \mathbf{E}(cX|\mathfrak{G})$ for any $c \in \mathbb{R}$. Suppose $c \neq 0$. To show that in this case we have $\mathbf{E}(cX|\mathfrak{G}) \subset c\mathbf{E}(X|\mathfrak{G})$, let $Z \in \mathbf{E}(cX|\mathfrak{G})$. Then Z is $\mathfrak{G}$-measurable and $\int_G Z\, dP = \int_G FcX\, dP = c\int_G X\, dP = c\int_G Y\, dP$ for every $G \in \mathfrak{G}$ and thus $Z = cY$ a.e.

on $(\Omega, \mathfrak{G}, P)$. This shows that $Z/c = Y$ a.e. on $(\Omega, \mathfrak{G}, P)$ and thus $Z/c \in \mathbf{E}(X|\mathfrak{G})$, that is $Z \in c\mathbf{E}(X|\mathfrak{G})$. Therefore $\mathbf{E}(cX|\mathfrak{G}) \subset c\mathbf{E}(X|\mathfrak{G})$ and thus $\mathbf{E}(cX|\mathfrak{G}) = c\mathbf{E}(X|\mathfrak{G})$ when $c \neq 0$.

6) Let $Y_1 \in \mathbf{E}(X_1|\mathfrak{G})$ and $Y_2 \in \mathbf{E}(X_2|\mathfrak{G})$. Then $c_1X_1 + c_2X_2$ is $\mathfrak{G}$-measurable and for every $G \in \mathfrak{G}$ we have

$$\int_G (c_1Y_1 + c_2Y_2)dP = c_1 \int_G Y_1\, dP + c_2 \int_G Y_2\, dP$$
$$= c_1 \int_G X_1\, dP + c_2 \int_G X_2\, dP = \int_G (c_1X_1 + c_2X_2)dP.$$

Thus $c_1Y_1 + c_2Y_2 \in \mathbf{E}(c_1X_1 + c_2X_2|\mathfrak{G})$. This shows that $c_1\mathbf{E}(X_1|\mathfrak{G}) + c_2\mathbf{E}(X_2|\mathfrak{G}) \subset \mathbf{E}(c_1X_1 + c_2X_2|\mathfrak{G})$. Suppose at least one of c_1 and c_2 is not equal to 0, say $c_2 \neq 0$. To show $\mathbf{E}(c_1X_1 + c_2X_2|\mathfrak{G}) \subset c_1\mathbf{E}(X_1|\mathfrak{G}) + c_2\mathbf{E}(X_2|\mathfrak{G})$, let $Z \in \mathbf{E}(c_1X_1 + c_2X_2|\mathfrak{G})$. Let $Y_1 \in \mathbf{E}(X_1|\mathfrak{G})$ and $Y_2 = 1/c_2\{Z - c_1Y_1\}$. Since Z and Y_1 are $\mathfrak{G}$-measurable so is Y_2. Also for every $G \in \mathfrak{G}$ we have

$$\int_G Y_2\, dP = \frac{1}{c_2}\left\{\int_G Z\, dP - c_1 \int_G Y_1\, dP\right\}$$
$$= \frac{1}{c_2} \int_G \{c_1X_1 + c_2X_2\}dP - \frac{c_1}{c_2} \int_G X_1\, dP = \int_G X_2\, dP.$$

This shows that $Y_2 \in \mathbf{E}(X_2|\mathfrak{G})$. Thus we have shown that for an arbitrary version Z of $\mathbf{E}(c_1X_1 + c_2X_2|\mathfrak{G})$ we have $Z = c_1Y_1 + c_2Y_2$ where Y_1 is a version of $\mathbf{E}(X_1|\mathfrak{G})$ and Y_2 is a version of $\mathbf{E}(X_2|\mathfrak{G})$. Thus $Z \in c_1\mathbf{E}(X_1|\mathfrak{G}) + c_2\mathbf{E}(X_2|\mathfrak{G})$ and then $\mathbf{E}(c_1X_1 + c_2X_2|\mathfrak{G}) \subset c_1\mathbf{E}(X_1|\mathfrak{G}) + c_2\mathbf{E}(X_2|\mathfrak{G})$. Consequently $\mathbf{E}(c_1X_1 + c_2X_2|\mathfrak{G}) = c_1\mathbf{E}(X_1|\mathfrak{G}) + c_2\mathbf{E}(X_2|\mathfrak{G})$.

7) If $X_2 \geq X_1$ a.e. on $(\Omega, \mathfrak{F}, P)$, then $X_2 - X_1 \geq 0$ a.e. on $(\Omega, \mathfrak{F}, P)$ so that by 2) we have $\mathbf{E}(X_2 - X_1|\mathfrak{G}) \geq 0$ a.e. on $(\Omega, \mathfrak{G}, P)$. By 6) we have $\mathbf{E}(X_2 - X_1|\mathfrak{G}) = \mathbf{E}(X_2|\mathfrak{G}) - \mathbf{E}(X_1|\mathfrak{G})$. Thus $\mathbf{E}(X_2|\mathfrak{G}) \geq \mathbf{E}(X_1|\mathfrak{G})$ a.e. on $(\Omega, \mathfrak{G}, P)$. ∎

Remark B.6. Regarding 5) of Theorem B.5 note that $c\mathbf{E}(X|\mathfrak{G}) = \mathbf{E}(cX|\mathfrak{G})$ does not hold in general when $c = 0$ since in this case we have $0 \cdot \mathbf{E}(X|\mathfrak{G}) = \{0\}$ while $\mathbf{E}(0 \cdot X|\mathfrak{G}) = \mathbf{E}(0|\mathfrak{G})$ which consists of all $\mathfrak{G}$-measurable extended real valued random variables on $(\Omega, \mathfrak{F}, P)$ which are $(\mathfrak{G}, P)$-a.e. equal to 0.

Observation B.7. Let X be an integrable extended real valued random variable on a probability space $(\Omega, \mathfrak{F}, P)$ and let $\mathfrak{G}$ be a sub-σ-algebra of $\mathfrak{F}$. Then

$$\mathbf{E}[\mathbf{E}(X|\mathfrak{G})] = \mathbf{E}(X),$$

APPENDIX B. CONDITIONAL EXPECTATIONS

that is, for every version Y of $\mathbf{E}(X|\mathfrak{G})$ we have $\mathbf{E}(Y) = \mathbf{E}(X)$.

Proof. If $Y \in \mathbf{E}(X|\mathfrak{G})$, then $\int_G Y\, dP = \int_G X\, dP$ for every $G \in \mathfrak{G}$ and in particular with $\Omega \in \mathfrak{G}$ we have $\int_\Omega Y\, dP = \int_\Omega X\, dP$, that is $\mathbf{E}(Y) = \mathbf{E}(X)$. ∎

Theorem B.8. *Let $\mathfrak{G}$ be a sub-σ-algebra of $\mathfrak{F}$ in a probability space. Let X be an integrable extended real valued random variable and Z be a $\mathfrak{G}$-measurable extended real valued random variable on $(\Omega, \mathfrak{F}, P)$ such that ZX is integrable. Then*

$$\mathbf{E}(ZX|\mathfrak{G}) = Z\mathbf{E}(X|\mathfrak{G}).$$

Proof. If we show that for an arbitrary version Y of $\mathbf{E}(X|\mathfrak{G})$, ZY is a version of $\mathbf{E}(ZX|\mathfrak{G})$ then $\mathbf{E}(ZX|\mathfrak{G}) \supset Z\mathbf{E}(X|\mathfrak{G})$. The fact that ZY is a version of $\mathbf{E}(ZX|\mathfrak{G})$ also implies that for an arbitrary version V of $\mathbf{E}(ZX|\mathfrak{G})$ we have $V = ZY$ a.e. on $(\Omega, \mathfrak{G}, P)$ and thus V is in $Z\mathbf{E}(X|\mathfrak{G})$ so that $\mathbf{E}(ZX|\mathfrak{G}) \subset Z\mathbf{E}(X|\mathfrak{G})$ and therefore $\mathbf{E}(ZX|\mathfrak{G}) = Z\mathbf{E}(X|\mathfrak{G})$. Thus it remains to show that for an arbitrary version Y of $\mathbf{E}(X|\mathfrak{G})$, ZY is a version of $\mathbf{E}(ZX|\mathfrak{G})$. Since ZY is $\mathfrak{G}$-measurable, it remains to show that

(1) $$\int_G ZY\, dP = \int_G ZX\, dP \quad \text{for every } G \in \mathfrak{G}.$$

To prove (1), consider first the case where $Z = \mathbf{1}_{G_0}$ where $G_0 \in \mathfrak{G}$. In this case we have

$$\int_G ZY\, dP = \int_G \mathbf{1}_{G_0} Y\, dP = \int_{G \cap G_0} Y\, dP$$

$$= \int_{G \cap G_0} X\, dP = \int_G \mathbf{1}_{G_0} X\, dP = \int_G ZX\, dP$$

where the third equality is by the fact that $G \cap G_0 \in \mathfrak{G}$. This verifies (1) for the particular case $Z = \mathbf{1}_{G_0}$ where $G_0 \in \mathfrak{G}$. If Z is a simple function on $(\Omega, \mathfrak{G}, P)$ then (1) holds by the linearity of the integral with respect to the integrand. Next consider the case where $Z \geq 0$ and $X \geq 0$ on Ω. Since $Z \geq 0$ on Ω there exists an increasing sequence $\{Z_n : n \in \mathbb{N}\}$ of nonnegative simple functions on $(\Omega, \mathfrak{G}, P)$ such that $Z_n \uparrow Z$ on Ω. Since Z_n is a simple function on $(\Omega, \mathfrak{G}, P)$ we have by our result above

(2) $$\int_G Z_n Y\, dP = \int_G Z_n X\, dP \quad \text{for } G \in \mathfrak{G}.$$

Since $X \geq 0$ on Ω we have $Y \geq 0$ a.e. on $(\Omega, \mathfrak{G}, P)$ by 4) of Theorem B.5. Thus $Z_n \uparrow Z$ on Ω implies $Z_n Y \uparrow ZY$ a.e. on $(\Omega, \mathfrak{G}, P)$. Since $X \geq 0$ on Ω we have also $Z_n X \uparrow ZX$

on Ω. Letting $n \to \infty$ in (2) we have (1) by the Monotone Convergence Theorem for nonnegative functions. Thus we have shown that for the case where $Z \geq 0$ and $X \geq 0$ on Ω, (1) holds and therefore we have

(3) $\qquad Z \geq 0$ and $X \geq 0$ on $\Omega \Rightarrow E(ZX|\mathfrak{G}) = ZE(X|\mathfrak{G})$.

Let us remove the condition that $Z \geq 0$ on Ω but retain the condition that $X \geq 0$ on Ω. In this case we have

$$E(ZX|\mathfrak{G}) = E(\{Z^+ - Z^-\}X|\mathfrak{G}) = E(Z^+X|\mathfrak{G}) - E(Z^-X|\mathfrak{G})$$
$$= Z^+E(X|\mathfrak{G}) - Z^-E(X|\mathfrak{G}) = ZE(X|\mathfrak{G})$$

where the third equality is by (3). Finally if we remove the condition that $X \geq 0$ on Ω, then

$$E(ZX|\mathfrak{G}) = E(Z\{X^+ - X^-\}|\mathfrak{G}) = E(ZX^+|\mathfrak{G}) - E(ZX^-|\mathfrak{G})$$
$$= ZE(X^+|\mathfrak{G}) - ZE(X^-|\mathfrak{G}) = Z\{E(X^+|\mathfrak{G}) - E(X^-|\mathfrak{G})\} = ZE(X|\mathfrak{G})$$

where the third equality is from the result above for nonnegative X and the last equality is by 6) of Theorem B.5. ∎

As a particular case of Theorem B.8, let us observe that if X and Z are two extended real valued random variables on a probability space $(\Omega, \mathfrak{F}, P)$, Z is $\mathfrak{G}$-measurable where $\mathfrak{G}$ is a sub-σ-algebra of $\mathfrak{F}$, $X \in L_p(\Omega, \mathfrak{F}, P)$ and $Z \in L_q(\Omega, \mathfrak{F}, P)$ where $p, q \in (1, \infty)$ with $\frac{1}{p} + \frac{1}{q} = 1$, then ZX is integrable so that $E(ZX|\mathfrak{G}) = ZE(X|\mathfrak{G})$.

Definition B.9. *Let X be an integrable extended real valued random variable on a probability space $(\Omega, \mathfrak{F}, P)$ and let $\mathfrak{G}$ and $\mathfrak{H}$ be sub-σ-algebras of $\mathfrak{F}$. With an arbitrary version Y of $E(X|\mathfrak{G})$, we define*

$$E[E(X|\mathfrak{G})|\mathfrak{H}] = E(Y|\mathfrak{H}).$$

An alternate notation $E(X|\mathfrak{G}|\mathfrak{H})$ is used for $E[E(X|\mathfrak{G})|\mathfrak{H}]$.

The fact that the equivalence class $E[E(X|\mathfrak{G})|\mathfrak{H}]$ of random variables does not depend on the choice of the version Y of $E(X|\mathfrak{G})$ in Definition B.9 is shown in the following proposition.

Proposition B.10. *Let X be an integrable extended real valued random variable on a probability space $(\Omega, \mathfrak{F}, P)$ and let $\mathfrak{G}$ and $\mathfrak{H}$ be sub-σ-algebras of $\mathfrak{F}$. Let Y_1 and Y_2 be two versions of $E(X|\mathfrak{G})$. Then*

$$E(Y_1|\mathfrak{H}) = E(Y_2|\mathfrak{H})$$

APPENDIX B. CONDITIONAL EXPECTATIONS 459

as equivalence classes.

Proof. For $j = 1, 2$, $\mathbf{E}(Y_j | \mathfrak{H})$ consists of all $\mathfrak{H}$-measurable extended real valued random variables Z such that $\int_H Z \, dP = \int_H Y_j \, dP$ for every $H \in \mathfrak{H}$. Now since Y_1 and Y_2 are versions of $\mathbf{E}(X | \mathfrak{G})$, we have $Y_1 = Y_2$ a.e. on $(\Omega, \mathfrak{G}, P)$ which implies that $Y_1 = Y_2$ a.e. on $(\Omega, \mathfrak{F}, P)$ so that $\int_A Y_1 \, dP = \int_A Y_2 \, dP$ for every $A \in \mathfrak{F}$ and in particular $\int_H Y_1 \, dP = \int_H Y_2 \, dP$ for every $H \in \mathfrak{H}$. If W_1 is a version of $\mathbf{E}(Y_1 | \mathfrak{H})$ and W_2 is a version of $\mathbf{E}(Y_2 | \mathfrak{H})$, then for any $H \in \mathfrak{H}$ we have $\int_H W_1 \, dP = \int_H Y_1 \, dP = \int_H Y_2 \, dP = \int_H W_2 \, dP$ and thus $W_1 = W_2$ a.e. on $(\Omega, \mathfrak{H}, P)$. From this we have $\mathbf{E}(Y_1 | \mathfrak{H}) = \mathbf{E}(Y_2 | \mathfrak{H})$. ∎

Observation B.11. For $\mathbf{E}[\mathbf{E}(X | \mathfrak{G}) | \mathfrak{H}]$ defined above, note that

$$\mathbf{E}[\mathbf{E}[\mathbf{E}(X | \mathfrak{G}) | \mathfrak{H}]] = \mathbf{E}[\mathbf{E}(X | \mathfrak{G})] = \mathbf{E}(X).$$

Note also that if $Z \in \mathbf{E}[\mathbf{E}(X | \mathfrak{G}) | \mathfrak{H}]$ and W is an $\mathfrak{H}$-measurable extended real valued function such that $W = Z$ a.e. on $(\Omega, \mathfrak{H}, P)$, then $W \in \mathbf{E}[\mathbf{E}(X | \mathfrak{G}) | \mathfrak{H}]$.

Theorem B.12. *Let X be an integrable extended real valued random variable on a probability space $(\Omega, \mathfrak{F}, P)$ and let $\mathfrak{G}_1$ be a sub-σ-algebra of $\mathfrak{F}$ and $\mathfrak{G}_2$ be a sub-σ-algebra of $\mathfrak{G}_1$. Then*

$$\mathbf{E}(X | \mathfrak{G}_1 | \mathfrak{G}_2) = \mathbf{E}(X | \mathfrak{G}_2).$$

Proof. Let Y be an arbitrary version of $\mathbf{E}(X | \mathfrak{G}_1)$ and let Z be an arbitrary version of $\mathbf{E}(Y | \mathfrak{G}_2)$. To show $\mathbf{E}(Y | \mathfrak{G}_2) = \mathbf{E}(X | \mathfrak{G}_2)$, it suffices according to Observation B.3 to show that there exists an $\mathfrak{G}_2$-measurable extended real valued random variable which is a version of both $\mathbf{E}(Y | \mathfrak{G}_2)$ and $\mathbf{E}(X | \mathfrak{G}_2)$. Since Z is a version of $\mathbf{E}(Y | \mathfrak{G}_2)$, it suffices to show that Z is also a version of $\mathbf{E}(X | \mathfrak{G}_2)$. Now Z is $\mathfrak{G}_2$-measurable. Also for every $G_2 \in \mathfrak{G}_2$ we have $\int_{G_2} Z \, dP = \int_{G_2} Y \, dP = \int_{G_2} X \, dP$ where the first equality is from the fact Z is a version of $\mathbf{E}(Y | \mathfrak{G}_2)$ and the second equality is from the fact that Y is a version of $\mathbf{E}(X | \mathfrak{G}_1)$. Thus we have shown that for an arbitrary version Y of $\mathbf{E}(X | \mathfrak{G}_1)$ we have $\mathbf{E}(Y | \mathfrak{G}_2) = \mathbf{E}(X | \mathfrak{G}_2)$. Then $\mathbf{E}[\mathbf{E}(X | \mathfrak{G}_1) | \mathfrak{G}_2] = \mathbf{E}(X | \mathfrak{G}_2)$ by Definition B.9. ∎

Theorem B.13. *Let X be an integrable extended real valued random variable on a probability space $(\Omega, \mathfrak{F}, P)$ and let $\mathfrak{G}_1$ be a sub-σ-algebra of $\mathfrak{F}$ and $\mathfrak{G}_2$ be a sub-σ-algebra of $\mathfrak{G}_1$. Then*

(1) $$\mathbf{E}(X | \mathfrak{G}_2 | \mathfrak{G}_1) \supset \mathbf{E}(X | \mathfrak{G}_2).$$

If every version of $E(X|\mathfrak{G}_2|\mathfrak{G}_1)$ *is* $\mathfrak{G}_2$-*measurable, then*

(2) $$E(X|\mathfrak{G}_2|\mathfrak{G}_1) = E(X|\mathfrak{G}_2).$$

Proof. 1) To prove (1), let Y be an arbitrary version of $E(X|\mathfrak{G}_2)$. Then Y is $\mathfrak{G}_2$-measurable and hence $\mathfrak{G}_1$-measurable. According to Observation B.11, to show that Y is a version of $E(X|\mathfrak{G}_2|\mathfrak{G}_1)$ it suffices to show that there exists a version Z of $E(X|\mathfrak{G}_2|\mathfrak{G}_1)$ such that $Z = Y$ a.e. on $(\Omega, \mathfrak{G}_1, P)$. Now let Z be an arbitrary version of $E(X|\mathfrak{G}_2|\mathfrak{G}_1)$. Then $Z \in E(W|\mathfrak{G}_1)$ for some $W \in E(X|\mathfrak{G}_2)$. Thus for an arbitrary $G_1 \in \mathfrak{G}_1$, we have $\int_{G_1} Z\, dP = \int_{G_1} W\, dP$. Since W and Y are versions of $E(X|\mathfrak{G}_2)$, we have $W = Y$ a.e. on $(\Omega, \mathfrak{G}_2, P)$ which implies $W = Y$ a.e. on $(\Omega, \mathfrak{F}, P)$ and consequently $\int_{G_1} W\, dP = \int_{G_1} Y\, dP$. Thus $\int_{G_1} Z\, dP = \int_{G_1} Y\, dP$. Since both Z and Y are $\mathfrak{G}_1$-measurable, this implies that $Z = Y$ a.e. on $E(X|\mathfrak{G}_1)$. This proves (1).

2) Suppose every version of $E(X|\mathfrak{G}_2|\mathfrak{G}_1)$ is $\mathfrak{G}_2$-measurable. To prove (2), we show that an arbitrary version of Z of $E(X|\mathfrak{G}_2|\mathfrak{G}_1)$ is a version of $E(X|\mathfrak{G}_2)$. Since Z is $\mathfrak{G}_2$-measurable, it suffices to show that there exists a version Y of $E(X|\mathfrak{G}_2)$ such that $Z = Y$ a.e. on $(\Omega, \mathfrak{G}_2, P)$ according to Observation B.11. For this it suffices to show that $\int_{G_2} Z\, dP = \int_{G_2} Y\, dP$ for every $G_2 \in \mathfrak{G}_2$. Now if $G_2 \in \mathfrak{G}_2$, then $G_2 \in \mathfrak{G}_1$ so that

$$\int_{G_2} Z\, dP = \int_{G_2} E(X|\mathfrak{G}_2)\, dP = \int_{G_2} X\, dP = \int_{G_2} Y\, dP.$$

This proves (2). ∎

According to (1) of Theorem B.13, we have $E(X|\mathfrak{G}_1|\mathfrak{G}_2) \supset E(X|\mathfrak{G}_2)$ if $\mathfrak{G}_2 \subset \mathfrak{G}_1 \subset \mathfrak{F}$. For an example of $E(X|\mathfrak{G}_1|\mathfrak{G}_2) \neq E(X|\mathfrak{G}_2)$, see Example B.14 below.

Example B.14. Let $(\Omega, \mathfrak{F}, P) = ((0,1], \mathfrak{M}_{(0,1]}, m_L)$ where $\mathfrak{M}_{(0,1]}$ is the σ-algebra of all Lebesgue measurable subsets of $(0, 1]$ and m_L is the Lebesgue measure. Let $f(x) = x$ for $x \in (0, 1]$. Consider $\mathfrak{G}_1 = \mathfrak{F} = \mathfrak{M}_{(0,1]}$ and $\mathfrak{G}_2 = \{\emptyset, (0,1]\}$. Since the only $\mathfrak{G}_2$-measurable extended real valued functions on $(0, 1]$ are the constant functions on $(0, 1]$ and since $\int_{(0,1]} c\, dm_L = \int_{(0,1]} f\, dm_L = \frac{1}{2}$ for $c \in \mathbb{R}$ implies that $c = \frac{1}{2}$, $E(f|\mathfrak{G}_2)$ consists of the constant function $\frac{1}{2}$ on $(0, 1]$. Thus $E(X|\mathfrak{G}_2|\mathfrak{G}_1)$ consists of all Lebesgue measurable extended real valued functions g on $(0, 1]$ such that $\int_{G_1} g\, dm_L = \int_{G_1} \frac{1}{2}\, dm_L = \frac{1}{2} m_L(G_1)$ for every $G_1 \in \mathfrak{M}_{(0,1]}$. The constant function $\frac{1}{2}$ is such a function but it is not the only one. Thus we have $E(X|\mathfrak{G}_1|\mathfrak{G}_2) \supset E(X|\mathfrak{G}_2)$ and $E(X|\mathfrak{G}_1|\mathfrak{G}_2) \neq E(X|\mathfrak{G}_2)$. Since $E(f|\mathfrak{G}_1)$ consists of all Lebesgue measurable extended real valued functions g on $(0, 1]$ such that

$\int_{G_1} g\, dm_L = \int_{G_1} f\, dm_L$ for every $G_1 \in \mathfrak{M}_{(0,1]}$ and since the constant function $\frac{1}{2}$ does not satisfy this condition, we have $\mathbf{E}(X\,|\,\mathfrak{G}_2) \cap \mathbf{E}(X\,|\,\mathfrak{G}_1) = \emptyset$.

The condition $\mathfrak{G}_2 \subset \mathfrak{G}_1$ implies that every version of $\mathbf{E}(X\,|\,\mathfrak{G}_2)$ is $\mathfrak{G}_1$-measurable. However in general there is no inclusion relation between $\mathbf{E}(X\,|\,\mathfrak{G}_2)$ and $\mathbf{E}(X\,|\,\mathfrak{G}_1)$ since every version Y of $\mathbf{E}(X\,|\,\mathfrak{G}_2)$ satisfies the condition $\int_G Y\, dP = \int_G X\, dP$ for $G \in \mathfrak{G}_2$ but not necessarily for $G \in \mathfrak{G}_1$. Example B.14 above presents a case where $\mathbf{E}(X\,|\,\mathfrak{G}_2) \cap \mathbf{E}(X\,|\,\mathfrak{G}_1) = \emptyset$. See Example B.15 for a case where $\mathbf{E}(X\,|\,\mathfrak{G}_2) \subset \mathbf{E}(X\,|\,\mathfrak{G}_1)$ but $\mathbf{E}(X\,|\,\mathfrak{G}_2) \neq \mathbf{E}(X\,|\,\mathfrak{G}_1)$.

Example B.15. Let $(\Omega, \mathfrak{F}, P) = ((0,1], \mathfrak{M}_{(0,1]}, m_L)$ and let f be an integrable Lebesgue measurable but not Borel measurable function on $(0,1]$. Let $\mathfrak{G}_1 = \mathfrak{F} = \mathfrak{M}_{(0,1]}$ and $\mathfrak{G}_2 = \mathfrak{B}_{(0,1]}$, the σ-algebra of Borel measurable sets in $(0,1]$. We have $\mathfrak{G}_2 \subset \mathfrak{G}_1$. Let us show that $\mathbf{E}(X\,|\,\mathfrak{G}_2) \subset \mathbf{E}(X\,|\,\mathfrak{G}_1)$ but $\mathbf{E}(X\,|\,\mathfrak{G}_2) \neq \mathbf{E}(X\,|\,\mathfrak{G}_1)$.

Now $\mathbf{E}(X\,|\,\mathfrak{G}_1)$ consists of all $\mathfrak{M}_{(0,1]}$-measurable extended real valued functions g such that

$$(1) \qquad \int_{G_1} g\, dm_L = \int_{G_1} f\, dm_L \quad \text{for } G_1 \in \mathfrak{M}_{(0,1]}$$

and $\mathbf{E}(X\,|\,\mathfrak{G}_2)$ consists of all $\mathfrak{B}_{(0,1]}$-measurable extended real valued functions h such that

$$(2) \qquad \int_{G_2} h\, dm_L = \int_{G_2} f\, dm_L \quad \text{for } G_2 \in \mathfrak{B}_{(0,1]}.$$

Let h be an arbitrary member of $\mathbf{E}(X\,|\,\mathfrak{G}_2)$. Then h is $\mathfrak{B}_{(0,1]}$-measurable and hence $\mathfrak{M}_{(0,1]}$-measurable. To show that h is a member of $\mathbf{E}(X\,|\,\mathfrak{G}_1)$, it remains to show that

$$(3) \qquad \int_{G_1} h\, dm_L = \int_{G_1} f\, dm_L \quad \text{for } G_1 \in \mathfrak{M}_{(0,1]}.$$

Let $\mathfrak{M}_\mathbb{R}$ be the σ-algebra of all Lebesgue measurable sets in $\mathbb{R}$. Recall that the measure space $(\mathbb{R}, \mathfrak{M}_\mathbb{R}, m_L)$ is the completion of the measure space $(\mathbb{R}, \mathfrak{B}_\mathbb{R}, m_L)$. If G_1 is in $\mathfrak{M}_{(0,1]}$ it is in $\mathfrak{M}_\mathbb{R}$ and thus can be given as a disjoint union $G_1 = B \cap N_0$ where $B \in \mathfrak{B}_\mathbb{R}$ and $N_0 \subset N$ where $N \in \mathfrak{B}_\mathbb{R}$ with $m_L(N) = 0$. Note that since $B \in \mathfrak{B}_\mathbb{R}$ and $B \subset (0,1]$ we have $B \in \mathfrak{B}_{(0,1]}$ and that $N_0 \in \mathfrak{M}_{(0,1]}$ with $m_L(N_0) = 0$. Thus

$$\int_{G_1} h\, dm_L = \int_B h\, dm_L + \int_{N_0} h\, dm_L = \int_B f\, dm_L + 0$$
$$= \int_B f\, dm_L + \int_{N_0} f\, dm_L = \int_{G_1} f\, dm_L,$$

probing (3). This shows that an arbitrary member h of $\mathbf{E}(X|\mathfrak{G}_2)$ is a member of $\mathbf{E}(X|\mathfrak{G}_1)$ and therefore $\mathbf{E}(X|\mathfrak{G}_2) \subset \mathbf{E}(X|\mathfrak{G}_1)$. Now since f is $\mathfrak{M}_{(0,1]}$-measurable and since $\mathfrak{G}_1 = \mathfrak{F} = \mathfrak{M}_{(0,1]}$, f is a member of $\mathbf{E}(f|\mathfrak{G}_1)$. On the other hand since f is not $\mathfrak{B}_{(0,1]}$-measurable, f is not a member of $\mathbf{E}(f|\mathfrak{G}_2)$. This shows that $\mathbf{E}(X|\mathfrak{G}_2) \neq \mathbf{E}(X|\mathfrak{G}_1)$.

Theorem B.16. *Let X be an integrable extended real valued random variable on a probability space $(\Omega, \mathfrak{F}, P)$ and let $\mathfrak{G}_1$ be a sub-σ-algebra of $\mathfrak{F}$ and $\mathfrak{G}_2$ be a sub-σ-algebra of $\mathfrak{G}_1$.*
1) $\mathbf{E}(X|\mathfrak{G}_2) \supset \mathbf{E}(X|\mathfrak{G}_1)$ if and only if every version of $\mathbf{E}(X|\mathfrak{G}_1)$ is $\mathfrak{G}_2$-measurable.
2) $\mathbf{E}(X|\mathfrak{G}_2) \subset \mathbf{E}(X|\mathfrak{G}_1)$ if and only if for every version Y of $\mathbf{E}(X|\mathfrak{G}_2)$ we have

$$(1) \qquad \int_{G_1} Y\, dP = \int_{G_1} X\, dP \quad for\ G_1 \in \mathfrak{G}_1.$$

3) If $(\Omega, \mathfrak{G}_1, P)$ is a complete measure space, $\mathfrak{G}_2$ contains all the null sets in $(\Omega, \mathfrak{G}_1, P)$, and condition (1) is satisfied, then $\mathbf{E}(X|\mathfrak{G}_2) = \mathbf{E}(X|\mathfrak{G}_1)$.

Proof. 1) If $\mathbf{E}(X|\mathfrak{G}_2) \supset \mathbf{E}(X|\mathfrak{G}_1)$, then every version of $\mathbf{E}(X|\mathfrak{G}_1)$ is $\mathfrak{G}_2$-measurable. Conversely, assume that an arbitrary version Y of $\mathbf{E}(X|\mathfrak{G}_1)$ is $\mathfrak{G}_2$-measurable. To show that Y is a version of $\mathbf{E}(X|\mathfrak{G}_2)$, it remains to verify that $\int_{G_2} Y\, dP = \int_{G_2} X\, dP$ for every $G_2 \in \mathfrak{G}_2$. Now if $G_2 \in \mathfrak{G}_2$ then $G_2 \in \mathfrak{G}_1$ so that $\int_{G_2} Y\, dP = \int_{G_2} X\, dP$ and thus Y is a version of $\mathbf{E}(X|\mathfrak{G}_2)$. Therefore $\mathbf{E}(X|\mathfrak{G}_2) \supset \mathbf{E}(X|\mathfrak{G}_1)$.

2) Since every version Y of $\mathbf{E}(X|\mathfrak{G}_2)$ is $\mathfrak{G}_2$-measurable and thus $\mathfrak{G}_1$-measurable, Y is a version of $\mathbf{E}(X|\mathfrak{G}_1)$ if and only if (1) holds. Thus $\mathbf{E}(X|\mathfrak{G}_2) \subset \mathbf{E}(X|\mathfrak{G}_1)$ if and only if (1) holds.

3) If (1) is satisfied then $\mathbf{E}(X|\mathfrak{G}_2) \subset \mathbf{E}(X|\mathfrak{G}_1)$ by 2). If we show that an arbitrary version Y_1 of $\mathbf{E}(X|\mathfrak{G}_1)$ is $\mathfrak{G}_2$-measurable, then by 1) we have $\mathbf{E}(X|\mathfrak{G}_2) \supset \mathbf{E}(X|\mathfrak{G}_1)$ and therefore $\mathbf{E}(X|\mathfrak{G}_2) = \mathbf{E}(X|\mathfrak{G}_1)$. Thus it remains to show the $\mathfrak{G}_2$-measurability of Y_1. Let Y_2 be an arbitrary version of $\mathbf{E}(X|\mathfrak{G}_2)$. Since Y_2 is a version of $\mathbf{E}(X|\mathfrak{G}_1)$ by 2), we have $Y_2 = Y_1$ a.e. on $(\Omega, \mathfrak{G}_1, P)$. Thus there exists a null set Λ_1 in $(\Omega, \mathfrak{G}_1, P)$ such that $Y_2 = Y_1$ on Λ_1^c. Since $\mathfrak{G}_2$ contains all the null sets in $(\Omega, \mathfrak{G}_1, P)$, Λ_1 is a null set in $(\Omega, \mathfrak{G}_2, P)$ and thus $\Lambda_1^c \in \mathfrak{G}_2$. The equality of Y_1 and Y_2 on Λ_1^c then implies the $\mathfrak{G}_2$-measurability of Y_1 on Λ_1^c. Since $(\Omega, \mathfrak{G}_1, P)$ is a complete measure space and since $\mathfrak{G}_2$ contains all the null sets in $(\Omega, \mathfrak{G}_1, P)$, $(\Omega, \mathfrak{G}_2, P)$ too is a complete measure space. Thus Y_1, which is $\mathfrak{G}_2$-measurable on Λ_1^c, is indeed $\mathfrak{G}_2$-measurable on Ω. ∎

[II] Conditional Convergence Theorems

Theorem B.17. (Conditional Monotone Convergence Theorem) *Let X_n, $n \in \mathbb{N}$, and X*

APPENDIX B. CONDITIONAL EXPECTATIONS

be integrable extended real valued random variables on a probability space $(\Omega, \mathfrak{F}, P)$ and let $\mathfrak{G}$ be a sub-σ-algebra of $\mathfrak{F}$. If $X_n \uparrow X$ (resp. $X_n \downarrow X$) a.e. on $(\Omega, \mathfrak{F}, P)$, then $\mathbf{E}(X_n | \mathfrak{G}) \uparrow \mathbf{E}(X | \mathfrak{G})$ (resp. $\mathbf{E}(X_n | \mathfrak{G}) \downarrow \mathbf{E}(X | \mathfrak{G})$) a.e. on $(\Omega, \mathfrak{G}, P)$.

Proof. Consider the case where $X_n \uparrow X$ a.e. on $(\Omega, \mathfrak{F}, P)$. Let $Y_n, n \in \mathbb{N}$, and Y be arbitrary versions of $\mathbf{E}(X_n | \mathfrak{G}), n \in \mathbb{N}$, and $\mathbf{E}(X | \mathfrak{G})$ respectively. Let us show that $Y_n \uparrow Y$ a.e. on $(\Omega, \mathfrak{G}, P)$. Now since $X_{n+1} \geq X_n$ a.e. on $(\Omega, \mathfrak{F}, P)$, we have $\mathbf{E}(X_{n+1} | \mathfrak{G}) \geq \mathbf{E}(X_n | \mathfrak{G})$ a.e. on $(\Omega, \mathfrak{G}, P)$ by 7) of Theorem B.5. Thus $Y_n \uparrow$ a.e. on $(\Omega, \mathfrak{G}, P)$. Since $X_n \uparrow X$ a.e. on $(\Omega, \mathfrak{F}, P)$ and since X_1 is integrable we have by the Monotone Convergence Theorem for an increasing sequence

$$(1) \qquad \lim_{n \to \infty} \int_A X_n \, dP = \int_A X \, dP \quad \text{for } A \in \mathfrak{F}$$

and similarly since $Y_n \uparrow$ a.e. on $(\Omega, \mathfrak{F}, P)$ and since Y_1 is integrable we have

$$(2) \qquad \lim_{n \to \infty} \int_G Y_n \, dP = \int_G \lim_{n \to \infty} Y_n \, dP \quad \text{for } G \in \mathfrak{G}$$

Since Y_n is a version of $\mathbf{E}(X_n | \mathfrak{G})$ we have $\int_G Y_n \, dP = \int_G X_n \, dP$. Thus from (1) and (2) we have

$$(3) \qquad \int_G \lim_{n \to \infty} Y_n \, dP = \int_G X \, dP \quad \text{for } G \in \mathfrak{G}.$$

Since Y is a version of $\mathbf{E}(X | \mathfrak{G})$ we have $\int_G Y \, dP = \int_G X \, dP$ for $G \in \mathfrak{G}$. Using this in (3) we have

$$\int_G \lim_{n \to \infty} Y_n \, dP = \int_G Y \, dP \quad \text{for } G \in \mathfrak{G}.$$

Therefore $\lim_{n \to \infty} Y_n = Y$ a.e. on $(\Omega, \mathfrak{G}, P)$. This shows that $\mathbf{E}(X_n | \mathfrak{G}) \uparrow \mathbf{E}(X | \mathfrak{G})$ for an increasing sequence. By a parallel argument we have $\mathbf{E}(X_n | \mathfrak{G}) \downarrow \mathbf{E}(X | \mathfrak{G})$ for a decreasing sequence. ∎

Theorem B.18. (Conditional Fatou's Lemma) *Let $X_n, n \in \mathbb{N}$, be integrable extended real valued random variables on a probability space $(\Omega, \mathfrak{F}, P)$ and $\mathfrak{G}$ be a sub-σ-algebra of $\mathfrak{F}$.*
1) If $\liminf_{n \to \infty} X_n$ is integrable and if there exists an integrable extended real valued random variable X such that $X_n \geq X$ a.e. on $(\Omega, \mathfrak{F}, P)$ for $n \in \mathbb{N}$, then

$$(1) \qquad \mathbf{E}(\liminf_{n \to \infty} X_n | \mathfrak{G}) \leq \liminf_{n \to \infty} \mathbf{E}(X_n | \mathfrak{G}) \quad \text{a.e. on } (\Omega, \mathfrak{G}, P).$$

2) *If* $\limsup_{n\to\infty} X_n$ *is integrable and if there exists an integrable extended real valued random variable* X *such that* $X_n \leq X$ *a.e. on* $(\Omega, \mathfrak{F}, P)$ *for* $n \in \mathbb{N}$*, then*

(2) $$\mathbf{E}(\limsup_{n\to\infty} X_n | \mathfrak{G}) \geq \limsup_{n\to\infty} \mathbf{E}(X_n | \mathfrak{G}) \quad a.e. \text{ on } (\Omega, \mathfrak{G}, P).$$

Proof. To prove (1), recall that $\liminf_{n\to\infty} X_n = \lim_{n\to\infty} \{\inf_{k\geq n} X_k\}$. Since $X \leq \inf_{k\geq n} X_k \leq X_n$ a.e. on $(\Omega, \mathfrak{F}, P)$ and since X and X_n are integrable, so is $\inf_{k\geq n} X_k$. Thus $\inf_{k\geq n} X_k$, $n \in \mathbb{N}$, is an increasing sequence of integrable random variables. Since $\inf_{k\geq n} X_k \uparrow \liminf_{n\to\infty} X_n$ on Ω as $n \to \infty$, Theorem B.17 implies that

$$\mathbf{E}(\liminf_{n\to\infty} X_n | \mathfrak{G}) = \lim_{n\to\infty} \mathbf{E}(\inf_{k\geq n} X_k | \mathfrak{G}) = \liminf_{n\to\infty} \mathbf{E}(\inf_{k\geq n} X_k | \mathfrak{G})$$

a.e. on $(\Omega, \mathfrak{G}, P)$. Since $\inf_{k\geq n} X_k \leq X_n$ on Ω, we have $\mathbf{E}(\inf_{k\geq n} X_k | \mathfrak{G}) \leq \mathbf{E}(X_n | \mathfrak{G})$ a.e. on $(\Omega, \mathfrak{G}, P)$ by 7) of Theorem B.5. Thus (1) holds. By a parallel argument we have (2). ∎

Theorem B.19. (Conditional Dominated Convergence Theorem) *Let* X_n, $n \in \mathbb{N}$, *and* X *be integrable extended real valued random variables on a probability space* $(\Omega, \mathfrak{F}, P)$ *such that* $\lim_{n\to\infty} X_n = X$ *a.e. on* $(\Omega, \mathfrak{F}, P)$ *and let* $\mathfrak{G}$ *be a sub-σ-algebra of* $\mathfrak{F}$*. If there exists an integrable extended real valued random variable* X_0 *such that* $|X_n| \leq X_0$ *a.e. on* $(\Omega, \mathfrak{F}, P)$ *for* $n \in \mathbb{N}$*, then*

$$\lim_{n\to\infty} \mathbf{E}(X_n | \mathfrak{G}) = \mathbf{E}(X | \mathfrak{G}) \quad a.e. \text{ on } (\Omega, \mathfrak{G}, P).$$

Proof. The condition $|X_n| \leq X_0$ a.e. on $(\Omega, \mathfrak{F}, P)$ for $n \in \mathbb{N}$ implies that $|\liminf_{n\to\infty} X_n| \leq X_0$ a.e. on $(\Omega, \mathfrak{F}, P)$. The integrability of X_0 then implies that of $\liminf_{n\to\infty} X_n$. We also have $-X_0 \leq X_n$ a.e. on $(\Omega, \mathfrak{F}, P)$ for $n \in \mathbb{N}$. Thus the conditions in 1) of Theorem B.18 are satisfied and consequently we have

$$\mathbf{E}(\liminf_{n\to\infty} X_n | \mathfrak{G}) \leq \liminf_{n\to\infty} \mathbf{E}(X_n | \mathfrak{G}) \quad a.e. \text{ on } (\Omega, \mathfrak{G}, P).$$

Now since $X = \liminf_{n\to\infty} X_n$ a.e. on $(\Omega, \mathfrak{F}, P)$ we have $\mathbf{E}(X | \mathfrak{G}) = \mathbf{E}(\liminf_{n\to\infty} X_n | \mathfrak{G})$ a.e. on $(\Omega, \mathfrak{G}, P)$ by 1) of Theorem B.5. Thus we have

$$\mathbf{E}(X | \mathfrak{G}) \leq \liminf_{n\to\infty} \mathbf{E}(X_n | \mathfrak{G}) \quad a.e. \text{ on } (\Omega, \mathfrak{G}, P).$$

APPENDIX B. CONDITIONAL EXPECTATIONS

Similarly by 2) of Theorem B.18, we have

$$E(X|\mathfrak{G}) \geq \limsup_{n \to \infty} E(X_n|\mathfrak{G}) \quad \text{a.e. on } (\Omega, \mathfrak{G}, P)$$

and therefore

$$E(X|\mathfrak{G}) = \lim_{n \to \infty} E(X_n|\mathfrak{G}) \quad \text{a.e. on } (\Omega, \mathfrak{G}, P). \quad \blacksquare$$

Remark B.20. Let $X_n, n \in \mathbb{N}$, and X be integrable extended real valued random variables on a probability space $(\Omega, \mathfrak{F}, P)$ and let $\mathfrak{G}$ be a sub-σ-algebra of $\mathfrak{F}$. Convergence of X_n to X on Ω alone does not imply $\lim_{n \to \infty} E(X_n|\mathfrak{G}) = E(X|\mathfrak{G})$. See the following example.

Example B.21. Let $(\Omega, \mathfrak{F}, P) = ((0, 1], \mathfrak{B}_{(0,1]}, m_L)$. For $n \in \mathbb{N}$, let X_n be defined by

$$X_n(\omega) = \begin{cases} 2^n & \text{for } \omega \in (0, \frac{1}{2^n}] \\ 0 & \text{for } \omega \in (\frac{1}{2^n}, 1] \end{cases}$$

and let X be identically equal to 0 on Ω. We have $\lim_{n \to \infty} X_n(\omega) = X(\omega)$ for every $\omega \in \Omega$. Consider a sub-σ-algebra $\mathfrak{G}$ of $\mathfrak{F}$ given by $\mathfrak{G} = \{\emptyset, (0, \frac{1}{2}], (\frac{1}{2}, 1], (0, 1]\}$. For $n \in \mathbb{N}$, let Y_n be defined by

$$Y_n(\omega) = \begin{cases} 2 & \text{for } \omega \in (0, \frac{1}{2}] \\ 0 & \text{for } \omega \in (\frac{1}{2}, 1] \end{cases}$$

and let Y be identically equal to 0 on Ω. Since Y_n is $\mathfrak{G}$-measurable and $\int_G Y_n \, dP = \int_G X_n \, dP$ for every $G \in \mathfrak{G}$, Y_n is a version of $E(X_n|\mathfrak{G})$. Furthermore since $\emptyset$ is the only null set in $(\Omega, \mathfrak{G}, P)$, any $\mathfrak{G}$-measurable extended real valued function that is equal to Y_n a.e. on $(\Omega, \mathfrak{G}, P)$ is actually equal to Y_n everywhere on Ω. Thus Y_n is the only version of $E(X_n|\mathfrak{G})$. Similarly Y is the only version of $E(X|\mathfrak{G})$. We have however $\lim_{n \to \infty} Y_n(\omega) \neq Y(\omega)$ for $\omega \in (0, \frac{1}{2}]$.

[III] Conditional Convexity Theorems

Let φ be a convex function on an open interval I in $\mathbb{R}$, that is, φ is a real valued function such that for $\xi_1, \xi_2 \in I$ and $\lambda_1, \lambda_2 \in [0, 1]$ with $\lambda_1 + \lambda_2 = 1$ we have

$$\varphi(\lambda_1 \xi_1 + \lambda_2 \xi_2) \leq \lambda_1 \varphi(\xi_1) + \lambda_2 \varphi(\xi_2).$$

Then the left derivative $(D^-\varphi)(\xi)$ and the right derivative $(D^+\varphi)(\xi)$ of φ exist at every $\xi \in I$; $(D^-\varphi)(\xi) \leq (D^+\varphi)(\xi)$ for $\xi \in I$; $(D^-\varphi)(\xi), (D^+\varphi)(\xi) \uparrow$ as $\xi \uparrow$; for any $\xi_0 \in I$

and $m \in [(D^-\varphi)(\xi_0), (D^+\varphi)(\xi_0)]$ we have $\varphi(\xi) \geq m(\xi - \xi_0) + \varphi(\xi_0)$ for $\xi \in I$; and there exists a countable collection $\{\ell_n : n \in \mathbb{N}\}$ of affine functions, that is, $\ell_n(\xi) = \alpha_n \xi + \beta_n$ with $\alpha_n, \beta_n \in \mathbb{R}$ for $\xi \in I$, such that $\varphi(\xi) = \sup_{n \in \mathbb{N}} \ell_n(\xi)$ for $\xi \in I$.

Theorem B.22. (Conditional Jensen's Inequality) *Let X be an integrable extended real valued random variable on a probability space $(\Omega, \mathfrak{F}, P)$. Let φ be a convex function on $\mathbb{R}$ such that the random variable $\varphi(X)$ is integrable. Then for an arbitrary sub-σ-algebra $\mathfrak{G}$ of $\mathfrak{F}$ we have*

$$\varphi(\mathrm{E}(X \mid \mathfrak{G})) \leq \mathrm{E}[\varphi(X) \mid \mathfrak{G}] \quad a.e. \text{ on } (\Omega, \mathfrak{G}, P).$$

Proof. Since φ is a convex function on $\mathbb{R}$ there exist $\alpha_n, \beta_n \in \mathbb{R}, n \in \mathbb{N}$, such that

(1) $$\varphi(\xi) = \sup_{n \in \mathbb{N}} \{\alpha_n \xi + \beta_n\} \quad \text{for } \xi \in \mathbb{R}.$$

Since X is integrable, $X(\omega) \in \mathbb{R}$ for a.e. ω in $(\Omega, \mathfrak{F}, P)$. Since a convex function is a continuous function, φ is Borel measurable and thus $\varphi(X)$ is a real valued random variable defined a.e. on Ω. By our assumption $\varphi(X)$ is integrable. By (1) we have

(2) $$\varphi(X(\omega)) = \sup_{n \in \mathbb{N}} \{\alpha_n X(\omega) + \beta_n\} \geq \alpha_n X(\omega) + \beta_n$$

for all $n \in \mathbb{N}$ for a.e. ω in $(\Omega, \mathfrak{F}, P)$. Applying 7) of Theorem B.5 to (2) we have

$$\mathrm{E}[\varphi(X) \mid \mathfrak{G}] \geq \alpha_n \mathrm{E}(X \mid \mathfrak{G}) + \beta_n$$

for all $n \in \mathbb{N}$ a.e. on $(\Omega, \mathfrak{G}, P)$. Recalling that every version of $\mathrm{E}(X \mid \mathfrak{G})$ is real valued a.e. on $(\Omega, \mathfrak{G}, P)$, we have by (1)

$$\mathrm{E}[\varphi(X) \mid \mathfrak{G}] \geq \sup_{n \in \mathbb{N}} \{\alpha_n \mathrm{E}(X \mid \mathfrak{G}) + \beta_n\} = \varphi(\mathrm{E}(X \mid \mathfrak{G}))$$

a.e. on $(\Omega, \mathfrak{G}, P)$. ∎

Remark B.23. Let X be an integrable extended real valued random variable on a probability space $(\Omega, \mathfrak{F}, P)$. Since

$$\xi^+ = \begin{cases} \xi & \text{for } \xi \geq 0 \\ 0 & \text{for } \xi < 0 \end{cases}$$

is a convex function on $\mathbb{R}$, Theorem B.22 implies that for an arbitrary sub-σ-algebra $\mathfrak{G}$ of $\mathfrak{F}$ we have $\mathrm{E}(X \mid \mathfrak{G})^+ \leq \mathrm{E}(X^+ \mid \mathfrak{G})$ a.e. on $(\Omega, \mathfrak{G}, P)$. Similarly $\mathrm{E}(X \mid \mathfrak{G})^- \leq \mathrm{E}(X^- \mid \mathfrak{G})$ a.e. on $(\Omega, \mathfrak{G}, P)$. Equalities in general do not hold. See the example below.

Example B.24. Let $(\Omega, \mathfrak{F}, P) = ((0, 1], \mathfrak{F}, m_L)$ where $\mathfrak{F} = \{\emptyset, (0, \frac{1}{2}], (\frac{1}{2}, 1], (0, 1]\}$ and let $\mathfrak{G} = \{\emptyset, (0, 1]\}$. Define an integrable random variable X by setting

$$X(\omega) = \begin{cases} 2 & \text{for } \omega \in (0, \frac{1}{2}] \\ -1 & \text{for } \omega \in (\frac{1}{2}, 1]. \end{cases}$$

If we define $Y(\omega) = \int_\Omega X \, dP = \frac{1}{2}$ for $\omega \in \Omega$, then as a constant function Y is $\mathfrak{G}$-measurable. Y also satisfies the condition $\int_G Y \, dP = \int_G X \, dP$ for $G = \emptyset$ and $G = \Omega$. The fact that $\emptyset$ is the only null set in $(\Omega, \mathfrak{G}, P)$ then implies that Y is the unique version of $\mathbf{E}(X \mid \mathfrak{G})$. From the nonnegativity of Y we have $\mathbf{E}(X \mid \mathfrak{G})^+ = Y^+ = Y$. If we define $Z(\omega) = \int_\Omega X^+ \, dP = 1$ for $\omega \in \Omega$ then by the same reason as above Z is the unique version of $\mathbf{E}(X^+ \mid \mathfrak{G})$. Thus we have $\mathbf{E}(X \mid \mathfrak{G})^+ < \mathbf{E}(X^+ \mid \mathfrak{G})$ on Ω.

Corollary B.25. *Let X be an integrable extended real valued random variable on a probability space $(\Omega, \mathfrak{F}, P)$ and let $\mathfrak{G}$ be a sub-σ-algebra of $\mathfrak{F}$. Then*

(1) $$|\mathbf{E}(X \mid \mathfrak{G})| \leq \mathbf{E}(|X| \mid \mathfrak{G}) \quad \text{a.e. on } (\Omega, \mathfrak{G}, P).$$

If $X \in L_p(\Omega, \mathfrak{F}, P)$ for some $p \in [1, \infty)$, then

(2) $$\mathbf{E}(|X| \mid \mathfrak{G}) \leq \mathbf{E}(|X|^p \mid \mathfrak{G})^{\frac{1}{p}} \quad \text{a.e. on } (\Omega, \mathfrak{G}, P)$$

and

(3) $$\|\mathbf{E}(X \mid \mathfrak{G})\|_p \leq \|\mathbf{E}(|X| \mid \mathfrak{G})\|_p \leq \|X\|_p.$$

Proof. Since $\xi \mapsto |\xi|$ for $\xi \in \mathbb{R}$ is a convex function on $\mathbb{R}$, (1) is a particular case of Theorem B.22.

Suppose $X \in L_p(\Omega, \mathfrak{F}, P)$ for some $p \in [1, \infty)$. Since $\xi \mapsto |\xi|^p$ for $\xi \in \mathbb{R}$ is a convex function on $\mathbb{R}$ for $p \in [1, \infty)$, we have by Theorem B.22 the inequality $|\mathbf{E}(X \mid \mathfrak{G})|^p \leq \mathbf{E}(|X|^p \mid \mathfrak{G})$ and then $|\mathbf{E}(X \mid \mathfrak{G})| \leq \mathbf{E}(|X|^p \mid \mathfrak{G})^{\frac{1}{p}}$ a.e. on $(\Omega, \mathfrak{G}, P)$. Applying the last inequality to $|X| \in L_p(\Omega, \mathfrak{F}, P)$ we have (2). By (1) and (2) we have

$$\int_\Omega |\mathbf{E}(X \mid \mathfrak{G})|^p dP \leq \int_\Omega \mathbf{E}(|X| \mid \mathfrak{G})^p \leq \int_\Omega \mathbf{E}(|X|^p \mid \mathfrak{G}) = \int_\Omega |X|^p dP.$$

Taking the p-th roots of the members in these inequalities we have (3). ∎

Theorem B.26. *Let $(\Omega, \mathfrak{F}, P)$ be a probability space, $X_n \in L_p(\Omega, \mathfrak{F}, P)$ for $n \in \mathbb{N}$ and $X \in L_p(\Omega, \mathfrak{F}, P)$ for some $p \in [1, \infty)$. If $\lim_{n \to \infty} X_n = X$ in $L_p(\Omega, \mathfrak{F}, P)$, then $\lim_{n \to \infty} \mathbf{E}(X_n | \mathfrak{G}) = \mathbf{E}(X | \mathfrak{G})$ in $L_p(\Omega, \mathfrak{F}, P)$ for an arbitrary sub-σ-algebra $\mathfrak{G}$ of $\mathfrak{F}$.*

Proof. By (3) of Corollary B.25 we have $\|\mathbf{E}(X_n | \mathfrak{G})\|_p \leq \|X_n\|_p < \infty$ so that $\mathbf{E}(X_n | \mathfrak{G}) \in L_p(\Omega, \mathfrak{F}, P)$ for $n \in \mathbb{N}$. Similarly $\mathbf{E}(X | \mathfrak{G}) \in L_p(\Omega, \mathfrak{F}, P)$. By 6) of Theorem B.5 and by (3) of Corollary B.25 we have

$$\|\mathbf{E}(X_n | \mathfrak{G}) - \mathbf{E}(X | \mathfrak{G})\|_p = \|\mathbf{E}(X_n - X | \mathfrak{G})\|_p \leq \|X_n - X\|_p$$

Thus if $\lim_{n \to \infty} \|X_n - X\|_p = 0$, then $\lim_{n \to \infty} \|\mathbf{E}(X_n | \mathfrak{G}) - \mathbf{E}(X | \mathfrak{G})\|_p = 0$. ∎

Theorem B.27. (Conditional Hölder's Inequality) *Let $\mathfrak{G}$ be a sub-σ-algebra of $\mathfrak{F}$ in a probability space $(\Omega, \mathfrak{F}, P)$. If $X \in L_p(\Omega, \mathfrak{F}, P)$ and $Y \in L_q(\Omega, \mathfrak{F}, P)$ where $p, q \in (1, \infty)$ and $\frac{1}{p} + \frac{1}{q} = 1$, then*

$$\mathbf{E}(|XY| | \mathfrak{G}) \leq \mathbf{E}(|X|^p | \mathfrak{G})^{\frac{1}{p}} \mathbf{E}(|Y|^q | \mathfrak{G})^{\frac{1}{q}} \quad \text{a.e. on } (\Omega, \mathfrak{G}, P).$$

Proof. For $n \in \mathbb{N}$, let $X_n = |X| + \frac{1}{n}$ and $Y_n = |Y| + \frac{1}{n}$. Then $X_n, Y_n \geq \frac{1}{n}$ on Ω, $X_n \in L_p(\Omega, \mathfrak{F}, P)$ and $Y_n \in L_q(\Omega, \mathfrak{F}, P)$. By 7) of Theorem B.5 we have $\alpha \equiv \mathbf{E}(|X_n|^p | \mathfrak{G})^{\frac{1}{p}} \geq \frac{1}{n}$ and $\beta \equiv \mathbf{E}(|Y_n|^q | \mathfrak{G})^{\frac{1}{q}} \geq \frac{1}{n}$ a.e. on $(\Omega, \mathfrak{G}, P)$. By the inequality $|\xi \eta| \leq p^{-1} |\xi|^p + q^{-1} |\eta|^q$ for $\xi, \eta \in \mathbb{R}$, we have

$$\frac{|X_n Y_n|}{\alpha \beta} \leq \frac{1}{p} \frac{|X_n|^p}{\alpha^p} + \frac{1}{q} \frac{|Y_n|^q}{\beta^q} \quad \text{a.e. on } (\Omega, \mathfrak{G}, P).$$

Note that since $\alpha, \beta \geq \frac{1}{n} > 0$ a.e. on $(\Omega, \mathfrak{G}, P)$, the divisions in the last inequality is possible a.e. on $(\Omega, \mathfrak{G}, P)$. By 7) and 6) of Theorem B.5,

$$\mathbf{E}\left(\frac{|X_n Y_n|}{\alpha \beta} \bigg| \mathfrak{G}\right) \leq \frac{1}{p} \mathbf{E}\left(\frac{|X_n|^p}{\alpha^p} \bigg| \mathfrak{G}\right) + \frac{1}{q} \mathbf{E}\left(\frac{|Y_n|^q}{\beta^q} \bigg| \mathfrak{G}\right) \quad \text{a.e. on } (\Omega, \mathfrak{G}, P).$$

The fact that α and β are $\mathfrak{G}$-measurable implies according to Theorem B.8

$$\frac{1}{\alpha \beta} \leq \frac{\mathbf{E}(|X_n|^p | \mathfrak{G})}{p \alpha^p} + \frac{\mathbf{E}(|Y_n|^q | \mathfrak{G})}{q \beta^q} = \frac{1}{p} + \frac{1}{q} = 1 \quad \text{a.e. on } (\Omega, \mathfrak{G}, P)$$

APPENDIX B. CONDITIONAL EXPECTATIONS

and thus

$$\mathbf{E}(|X_n Y_n| \,|\, \mathfrak{G}) \leq \alpha\beta = \mathbf{E}(|X_n|^p \,|\, \mathfrak{G})^{\frac{1}{p}} \mathbf{E}(|Y_n|^q \,|\, \mathfrak{G})^{\frac{1}{q}} \quad \text{a.e. on } (\Omega, \mathfrak{G}, P).$$

Letting $n \to \infty$ and applying Theorem B.17 (Conditional Monotone Convergence Theorem), we complete the proof. ∎

[IV] Conditioning and Independence

Theorem B.28. *Let X be an integrable extended real valued random variable on a probability space $(\Omega, \mathfrak{F}, P)$ and let $\mathfrak{G}$ be a sub-σ-algebra of $\mathfrak{F}$. If $\{X, \mathfrak{G}\}$ is an independent system, then*

$$\mathbf{E}(X \,|\, \mathfrak{G}) = \mathbf{E}(X) \quad \text{a.e. on } (\Omega, \mathfrak{G}, P).$$

Proof. Let $\{X, \mathfrak{G}\}$ be an independent system, that is, $\{\sigma(X), \mathfrak{G}\}$ is an independent system of sub-σ-algebras of $\mathfrak{F}$. We are to show that every version of $\mathbf{E}(X \,|\, \mathfrak{G})$ is equal to the constant $\mathbf{E}(X)$ a.e. on $(\Omega, \mathfrak{G}, P)$.

Let $E \in \sigma(X)$. Now the constant function $\mathbf{E}(\mathbf{1}_E)$ on Ω is $\mathfrak{G}$-measurable. Also for every $G \in \mathfrak{G}$ we have

$$\int_G \mathbf{1}_E \, dP = P(G \cap E) = P(G)P(E) = P(G)\mathbf{E}(\mathbf{1}_E) = \int_G \mathbf{E}(\mathbf{1}_E) \, dP$$

where the second equality is by the independence of $\{\sigma(X), \mathfrak{G}\}$. Thus we have shown that for an arbitrary $E \in \sigma(X)$ we have

(1) $$\mathbf{E}(\mathbf{1}_E \,|\, \mathfrak{G}) = \mathbf{E}(\mathbf{1}_E) \quad \text{a.e. on } (\Omega, \mathfrak{G}, P).$$

Let us write $X = X^+ - X^-$ for the integrable extended real valued random variable X on $(\Omega, \sigma(X), P)$. There exist two increasing sequences of nonnegative simple functions $\{\varphi_n : n \in \mathbb{N}\}$ and $\{\psi_n : n \in \mathbb{N}\}$ on $(\Omega, \sigma(X), P)$ such that $\varphi_n \uparrow X^+$ and $\psi_n \uparrow X^-$ on Ω. Let $X_n = \varphi_n - \psi_n$ for $n \in \mathbb{N}$. Then $X_n \to X$ on Ω and $|X_n| \leq \varphi_n + \psi_n \leq X^+ + X^- = |X|$. Thus by (1) and the linearity of conditional expectation, that is 6) of Theorem B.5, we have

$$\mathbf{E}(X_n \,|\, \mathfrak{G}) = \mathbf{E}(\varphi_n \,|\, \mathfrak{G}) - \mathbf{E}(\psi_n \,|\, \mathfrak{G}) = \mathbf{E}(\varphi_n) - \mathbf{E}(\psi_n) = \mathbf{E}(X_n)$$

a.e. on $(\Omega, \mathfrak{G}, P)$. Letting $n \to \infty$, applying Theorem B.19 (Conditional Dominated Convergence Theorem) on the left side and applying the Dominated Convergence Theorem on the right side we complete the proof. ∎

Remark B.29. The converse of Theorem B.28 is false, that is, $E(X|\mathfrak{G})$ a.e. on $(\Omega, \mathfrak{G}, P)$ does not imply the independence of the system $\{X, \mathfrak{G}\}$. See Example B.30 below. See also Corollary B.33.

Example B.30. Let $(\Omega, \mathfrak{F}, P) = ((0, 1], \mathfrak{B}_{(0,1]}, m_L)$. Consider $G_{1,1} = (0, \frac{1}{4}]$, $G_{1,2} = (\frac{1}{4}, \frac{1}{2}]$, $G_1 = (0, \frac{1}{2}]$, $G_2 = (\frac{1}{2}, 1]$, and let $\mathfrak{G} = \sigma\{\mathfrak{G}_1, \mathfrak{G}_2\} = \{\emptyset, (0, \frac{1}{2}], (\frac{1}{2}, 1], (0, 1]\}$. Let X be defined by
$$X(\omega) = \begin{cases} 1 & \text{for } \omega \in G_{1,1} \\ -1 & \text{for } \omega \in G_{1,2} \\ 0 & \text{for } \omega \in G_2. \end{cases}$$
Then $E(X) = 0$. Let $Y = 0$ on Ω. Then Y is $\mathfrak{G}$-measurable and for every $G \in \mathfrak{G}$ we have $\int_G Y \, dP = \int_G X \, dP$ so that Y is a version of $E(X|\mathfrak{G})$. Actually since $\emptyset$ is the only null set in $(\Omega, \mathfrak{G}, P)$, Y is the only version of $E(X|\mathfrak{G})$. To show that $\{X, \mathfrak{G}\}$ is not an independent system, that is, $\{\sigma(X), \mathfrak{G}\}$ is not an independent system, note that for $G_{1,1} \in \sigma(X) = \sigma\{G_{1,1}, G_{1,2}, G_2\}$ and $G_1 \in \mathfrak{G}$ we have $P(G_{1,1} \cap G_1) = P(G_{1,1}) = \frac{1}{4}$ while $P(G_{1,1})P(G_1) = \frac{1}{4}\frac{1}{2} = \frac{1}{8}$ so that $P(G_{1,1} \cap G_1) \neq P(G_{1,1})P(G_1)$.

Theorem B.31. *Let X be an integrable extended real valued random variable on a probability space $(\Omega, \mathfrak{F}, P)$ and let $\mathfrak{G}$ be a sub-σ-algebra of $\mathfrak{F}$. Then the following two conditions are equivalent:*

1°. $E(X|\mathfrak{G}) = E(X)$ *a.e. on* $(\Omega, \mathfrak{G}, P)$.

2°. $E(XZ) = E(X)E(Z)$ *for every bounded $\mathfrak{G}$-measurable random variable Z.*

Proof. Assume 1°. Then
$$E(XZ) = E[E(XZ|\mathfrak{G})] = E[ZE(X|\mathfrak{G})] = E(X)E(Z)$$
where the second equality is by Theorem B.8 and the third equality is by 1°. This proves 2°.

Conversely assume 2°. Then for every bounded $\mathfrak{G}$-measurable random variable Z we have

(1) $\quad E[ZE(X|\mathfrak{G})] = E[E(ZX|\mathfrak{G})] = E(ZX) = E(X)E(Z) = E[ZE(X)]$

where the first equality is by Theorem B.8 and the third equality is by 2°. Now for an arbitrary $G \in \mathfrak{G}$, $\mathbf{1}_G$ is a bounded $\mathfrak{G}$-measurable random variable so that by (1) we have

$\int_G \mathbf{E}(X|\mathfrak{G}) \, dP = \int_G \mathbf{E}(X) \, dP$. Since both $\mathbf{E}(X|\mathfrak{G})$ and the constant $\mathbf{E}(X)$ are $\mathfrak{G}$-measurable, the arbitrariness of $G \in \mathfrak{G}$ implies that $\mathbf{E}(X|\mathfrak{G}) = \mathbf{E}(X)$ a.e. on $(\Omega, \mathfrak{G}, P)$. This proves 1°. ∎

Theorem B.32. *Let $\mathfrak{G}_1$ and $\mathfrak{G}_2$ be two sub-σ-algebras of $\mathfrak{F}$. Then the following two conditions are equivalent:*

1°. $\{\mathfrak{G}_1, \mathfrak{G}_2\}$ *is an independent system.*

2°. $\mathbf{E}(X_1|\mathfrak{G}_2) = \mathbf{E}(X_1)$ *a.e. on $(\Omega, \mathfrak{G}_2, P)$ for every integrable $\mathfrak{G}_1$-measurable extended real valued random variable X_1.*

Proof. Assume 1°. Let X_1 be an integrable $\mathfrak{G}_1$-measurable extended real valued random variable. Since $\sigma(X_1) \subset \mathfrak{G}_1$, the independence of $\{\mathfrak{G}_1, \mathfrak{G}_2\}$ implies the independence of $\{\sigma(X_1), \mathfrak{G}_2\}$. Then by Theorem B.28 we have $\mathbf{E}(X_1|\mathfrak{G}_2) = \mathbf{E}(X_1)$ a.e. on $(\Omega, \mathfrak{G}_2, P)$.

Conversely assume 2°. Then for every integrable $\mathfrak{G}_1$-measurable extended real valued random variable X_1 and bounded $\mathfrak{G}_2$-measurable random variable X_2, $X_1 X_2$ is integrable and

$$\mathbf{E}(X_1 X_2) = \mathbf{E}[\mathbf{E}(X_1 X_2 | \mathfrak{G}_2)] = \mathbf{E}[X_2 \mathbf{E}(X_1|\mathfrak{G}_2)] = \mathbf{E}[X_2 \mathbf{E}(X_1)] = \mathbf{E}(X_1)\mathbf{E}(X_2)$$

where the second equality is by Theorem B.8 and the third equality is by 2°. Let $G_1 \in \mathfrak{G}_1$ and $G_2 \in \mathfrak{G}_2$. Then with $X_1 = \mathbf{1}_{G_1}$ and $X_2 = \mathbf{1}_{G_2}$ we have $P(G_1 \cap G_2) = \mathbf{E}(\mathbf{1}_{G_1} \mathbf{1}_{G_2}) = \mathbf{E}(\mathbf{1}_{G_1})\mathbf{E}(\mathbf{1}_{G_2}) = P(G_1)P(G_2)$, proving the independence of $\{\mathfrak{G}_1, \mathfrak{G}_2\}$. ∎

Corollary B.33. *Let X be an integrable extended real valued random variable on a probability space $(\Omega, \mathfrak{F}, P)$ and let $\mathfrak{G}$ be a sub-σ-algebra of $\mathfrak{F}$. Then the following two conditions are equivalent:*

1°. $\{\sigma(X), \mathfrak{G}\}$ *is an independent system.*

2°. $\mathbf{E}(X'|\mathfrak{G}) = \mathbf{E}(X')$ *a.e. on $(\Omega, \mathfrak{G}, P)$ for every integrable $\sigma(X)$-measurable extended real valued random variable X'.*

Proof. The Corollary is a particular case of Theorem B.32 in which $\mathfrak{G}_1 = \sigma(X)$ and $\mathfrak{G}_2 = \mathfrak{G}$. ∎

Theorem B.34. *Let $\mathfrak{A}_1$, $\mathfrak{A}_2$, and $\mathfrak{A}_3$ be three sub-σ-algebras of $\mathfrak{F}$ in a probability space $(\Omega, \mathfrak{F}, P)$. If $\sigma(\mathfrak{A}_1 \cup \mathfrak{A}_2)$ and $\mathfrak{A}_3$ are independent, then for every $A_1 \in \mathfrak{A}_1$ we have*

$$(1) \qquad P(A_1 | \mathfrak{A}_2) \subset P(A_1 | \sigma(\mathfrak{A}_2 \cup \mathfrak{A}_3)),$$

that is, every version of $P(A_1 | \mathfrak{A}_2)$ is a version of $P(A_1 | \sigma(\mathfrak{A}_2 \cup \mathfrak{A}_3))$. If $(\Omega, \sigma(\mathfrak{A}_2 \cup \mathfrak{A}_3), P)$ is a complete measure space and if $\mathfrak{A}_2$ contains all the null sets in $(\Omega, \sigma(\mathfrak{A}_2 \cup \mathfrak{A}_3), P)$, then

$$(2) \qquad P(A_1 | \mathfrak{A}_2) = P(A_1 | \sigma(\mathfrak{A}_2 \cup \mathfrak{A}_3)).$$

Proof. Since $\mathfrak{A}_2$ is a sub-σ-algebras of $\sigma(\mathfrak{A}_2 \cup \mathfrak{A}_3)$, to prove (1) it suffices to show according to 2) of Theorem B.16 that for every version Y of $P(A_1 | \mathfrak{A}_2)$ we have

$$(3) \qquad \int_B Y\, dP = P(A_1 \cap B) \quad \text{for } B \in \sigma(\mathfrak{A}_2 \cup \mathfrak{A}_3).$$

To prove (3), let us prove first that for $A_1 \in \mathfrak{A}_1$ and $A_3 \in \mathfrak{A}_3$ we have

$$(4) \qquad P(A_1 \cap A_3 | \mathfrak{A}_2) = P(A_1 | \mathfrak{A}_2) P(A_3 | \mathfrak{A}_2).$$

Now for $A_2 \in \mathfrak{A}_2$ we have

$$(5) \qquad \int_{A_2} P(A_1 \cap A_3 | \mathfrak{A}_2)\, dP = P(A_1 \cap A_2 \cap A_3) = P(A_1 \cap A_2) P(A_3)$$
$$= P(A_3) \int_{A_2} P(A_1 | \mathfrak{A}_2)\, dP = \int_{A_2} P(A_1 | \mathfrak{A}_2) P(A_3)\, dP,$$

where the first and the third equalities are by the definition of conditional probability, and the second equality is by the independence of $\sigma(\mathfrak{A}_1 \cup \mathfrak{A}_2)$ and $\mathfrak{A}_3$. Since $P(A_1 \cap A_3 | \mathfrak{A}_2)$ and $P(A_1 | \mathfrak{A}_2) P(A_3)$ are both $\mathfrak{A}_2$-measurable, (5) implies

$$P(A_1 \cap A_3 | \mathfrak{A}_2) = P(A_1 | \mathfrak{A}_2) P(A_3) \quad \text{a.e. on } (\Omega, \mathfrak{A}_2, P).$$

Now the independence of $\sigma(\mathfrak{A}_1 \cup \mathfrak{A}_2)$ and $\mathfrak{A}_3$ implies that of $\mathfrak{A}_2$ and $\mathfrak{A}_3$ and consequently we have $P(A_3 | \mathfrak{A}_2) = P(A_3)$ a.e. on $(\Omega, \mathfrak{A}_2, P)$ by Theorem B.28. Therefore we have $P(A_1 \cap A_3 | \mathfrak{A}_2) = P(A_1 | \mathfrak{A}_2) P(A_3 | \mathfrak{A}_2)$ a.e. on $(\Omega, \mathfrak{A}_2, P)$. This proves (4).

Next let us observe that

$$E[P(A_1 | \mathfrak{A}_2) 1_{A_3} | \mathfrak{A}_2] = P(A_1 | \mathfrak{A}_2) E(1_{A_3} | \mathfrak{A}_2) = P(A_1 \cap A_3 | \mathfrak{A}_2)$$
$$= E[P(A_1 \cap A_3 | \sigma(\mathfrak{A}_2 \cup \mathfrak{A}_3)) | \mathfrak{A}_2] = E[1_{A_3} E[1_{A_1} | \sigma(\mathfrak{A}_2 \cup \mathfrak{A}_3)] | \mathfrak{A}_2],$$

APPENDIX B. CONDITIONAL EXPECTATIONS

where the first equality is by the $\mathfrak{A}_2$-measurability of $P(A_1|\mathfrak{A}_2)$ and by Theorem B.8, the second equality is by (4), the third equality is by Theorem B.12, and the last equality is by the fact that $\mathbf{1}_{A_3}$ is $\sigma(\mathfrak{A}_2 \cup \mathfrak{A}_3)$-measurable. Thus we have

$$
\begin{aligned}
(6) \quad \int_{A_2 \cap A_3} P(A_1|\mathfrak{A}_2)\,dP &= \int_{A_2 \cap A_3} \mathbf{E}[\mathbf{1}_{A_1}|\sigma(\mathfrak{A}_2 \cup \mathfrak{A}_3)]\,dP \\
&= \int_{A_2 \cap A_3} \mathbf{1}_{A_1}\,dP = P(A_1 \cap A_2 \cap A_3),
\end{aligned}
$$

where the second equality is from the fact that $A_2 \cap A_3 \in \sigma(\mathfrak{A}_2 \cup \mathfrak{A}_3)$. Now $\mathfrak{A}_2 \cap \mathfrak{A}_3$ is a π-class of subsets of Ω and $\sigma(\mathfrak{A}_2 \cap \mathfrak{A}_3) = \sigma(\mathfrak{A}_2 \cup \mathfrak{A}_3)$. If we define two measures on $\sigma(\mathfrak{A}_2 \cup \mathfrak{A}_3)$ by setting $\mu(B) = \int_B P(A_1|\mathfrak{A}_2)\,dP$ and $\nu(B) = P(A_1 \cap B)$ for $B \in \sigma(\mathfrak{A}_2 \cup \mathfrak{A}_3)$, then according to (6) we have $\mu = \nu$ on the π-class $\mathfrak{A}_2 \cap \mathfrak{A}_3$ and thus $\mu = \nu$ on $\sigma(\mathfrak{A}_2 \cap \mathfrak{A}_3) = \sigma(\mathfrak{A}_2 \cup \mathfrak{A}_3)$ by Corollary 1.8. Therefore we have $\int_B P(A_1|\mathfrak{A}_2)\,dP = P(A_1 \cap B)$ for $B \in \sigma(\mathfrak{A}_2 \cup \mathfrak{A}_3)$. This proves (3). Thus (1) holds.

Finally, (2) follows from 3) of Theorem B.16. ∎

Appendix C
Regular Conditional Probabilities

Given a probability space $(\Omega, \mathfrak{F}, P)$. Let $\mathfrak{G}$ be a sub-σ-algebra of $\mathfrak{F}$, $A \in \mathfrak{F}$ and consider the conditional probability of A given $\mathfrak{G}$, that is, $P(A|\mathfrak{G}) = \mathbf{E}(1_A|\mathfrak{G})$. Let $\{A_n : n \in \mathbb{N}\}$ be a countable disjoint collection in $\mathfrak{F}$. By the linearity of the conditional expectation and by the Conditional Monotone Convergence Theorem, there exists a null set Λ depending on $\{A_n : n \in \mathbb{N}\}$ in $(\Omega, \mathfrak{G}, P)$ such that $P(\cup_{n \in \mathbb{N}} A_n |\mathfrak{G})(\omega) = \sum_{n \in \mathbb{N}} P(A_n|\mathfrak{G})(\omega)$ for $\omega \in \Lambda^c$. Since there may be uncountably many countable disjoint collections of members of $\mathfrak{F}$, we may have to exclude uncountably many null sets whose union may not be an $\mathfrak{F}$-measurable set. Thus a subset E of Ω such that $P(\cdot|\mathfrak{G})(\omega)$ is a measure on $\mathfrak{F}$ for $\omega \in E$ may be an empty set or it may not be an $\mathfrak{F}$-measurable set. To overcome this difficulty, let us consider regular conditional probabilities.

Definition C.1. *Let $(\Omega, \mathfrak{F}, P)$ be a probability space and let $\mathfrak{G}$ be a sub-σ-algebra of $\mathfrak{F}$. By a regular conditional probability given $\mathfrak{G}$ we mean a function $P^{\mathfrak{F}|\mathfrak{G}}$ on $\mathfrak{F} \times \Omega$ such that*

1°. *there exists a null set Λ in $(\Omega, \mathfrak{G}, P)$ such that $P^{\mathfrak{F}|\mathfrak{G}}(\cdot, \omega)$ is a probability measure on $\mathfrak{F}$ for every $\omega \in \Lambda^c$.*

2°. $P^{\mathfrak{F}|\mathfrak{G}}(A, \cdot)$ *is a $\mathfrak{G}$-measurable function on Ω for every $A \in \mathfrak{F}$.*

3°. $P(A \cap G) = \int_G P^{\mathfrak{F}|\mathfrak{G}}(A, \omega) P(d\omega)$ *for every $A \in \mathfrak{F}$ and $G \in \mathfrak{G}$.*

We say that the regular conditional probability given $\mathfrak{G}$ is unique if for any two functions p and p' on $\mathfrak{F} \times \Omega$ satisfying conditions 1°, 2°, and 3° there exists a null set Λ in $(\Omega, \mathfrak{G}, P)$ such that $p(\cdot, \omega) = p'(\cdot, \omega)$ on $\mathfrak{F}$ for $\omega \in \Lambda^c$.

We show in Theorem C.6 below that if Ω is a complete separable metric space, $\mathfrak{F}$ is the Borel σ-algebra of subsets of Ω and P is an arbitrary probability measure on $\mathfrak{F}$, then the regular conditional probability given an arbitrary sub-σ-algebra $\mathfrak{G}$ of $\mathfrak{F}$ exists uniquely. Regular conditional probability is known to exist for a *standard measurable space*, of which a complete separable metric space with its Borel σ-algebra of subsets is an example. See [26] K. P. Parthasarathy.

Observation C.2. Conditions 2° and 3° in Definition C.1 are equivalent to the single condition
$$P^{\mathfrak{F}|\mathfrak{G}}(A, \cdot) \in P(A|\mathfrak{G}) \quad \text{for every } A \in \mathfrak{F},$$
that is, for every $A \in \mathfrak{F}$, the function $P^{\mathfrak{F}|\mathfrak{G}}(A, \cdot)$ on Ω is a version of $P(A|\mathfrak{G})$.

Proof. By 2°, $P^{\mathfrak{F}|\mathfrak{G}}(A, \cdot)$ is a $\mathfrak{G}$-measurable function on Ω. By 3°, we have for every $G \in \mathfrak{G}$ the equality $\int_G P^{\mathfrak{F}|\mathfrak{G}}(A, \omega) P(d\omega) = P(A \cap G) = \int_G 1_A(\omega) P(d\omega)$. This shows that $P^{\mathfrak{F}|\mathfrak{G}}(A, \cdot)$ is a version of $P(A|\mathfrak{G})$. ∎

Proposition C.3. *If $P^{\mathfrak{F}|\mathfrak{G}}$ is a regular conditional probability given a sub-σ-algebra $\mathfrak{G}$ in a probability space $(\Omega, \mathfrak{F}, P)$, then for every integrable random variable X on the $(\Omega, \mathfrak{F}, P)$ we have*

(1) $$\mathbf{E}(X|\mathfrak{G}) = \int_\Omega X(\omega') P^{\mathfrak{F}|\mathfrak{G}}(d\omega', \cdot) \quad \text{a.e. on } (\Omega, \mathfrak{G}, P),$$

that is, the right side of (1) as a function on Ω is a version of $\mathbf{E}(X|\mathfrak{G})$. In particular for every $G \in \mathfrak{G}$, we have

(2) $$\int_G X \, dP = \int_G \left\{ \int_\Omega X(\omega') P^{\mathfrak{F}|\mathfrak{G}}(d\omega', \omega) \right\} P(d\omega).$$

Proof. Consider the case where the integrable random variable X is nonnegative. Let us show first that the right side of (1) is a $\mathfrak{G}$-measurable function on Ω. Now there exists an increasing sequence of nonnegative simple functions $\{\varphi_n : n \in \mathbb{N}\}$ on $(\Omega, \mathfrak{F})$ such that $\varphi_n \uparrow X$. Let $\varphi_n = \sum_{k=1}^{p_n} c_{n,k} 1_{A_{n,k}}$ where $A_{n,k} \in \mathfrak{F}$ and $c_{n,k} \geq 0$ for $k = 1, \ldots, p_n$ and $p_n \in \mathbb{N}$. Then
(3)
$$\int_\Omega \varphi_n(\omega') P^{\mathfrak{F}|\mathfrak{G}}(d\omega', \cdot) = \sum_{k=1}^{p_n} c_{n,k} \int_\Omega c_{n,k} 1_{A_{n,k}}(\omega') P^{\mathfrak{F}|\mathfrak{G}}(d\omega', \cdot) = \sum_{k=1}^{p_n} c_{n,k} P^{\mathfrak{F}|\mathfrak{G}}(A_{n,k}, \cdot),$$

APPENDIX C. REGULAR CONDITIONAL PROBABILITIES 477

which, being a finite sum of $\mathfrak{G}$-measurable functions according to 2° of Definition C.1, is $\mathfrak{G}$-measurable. Then by the Monotone Convergence Theorem,

$$\int_\Omega X(\omega')P^{\mathfrak{F}|\mathfrak{G}}(d\omega',\cdot) = \lim_{n\to\infty}\int_\Omega \varphi_n(\omega')P^{\mathfrak{F}|\mathfrak{G}}(d\omega',\cdot),$$

which, as the limit of a sequence of $\mathfrak{G}$-measurable functions, is $\mathfrak{G}$-measurable.

To show that $\int_\Omega X(\omega')P^{\mathfrak{F}|\mathfrak{G}}(d\omega',\cdot)$ is a version of $\mathbf{E}(X\,|\,\mathfrak{G})$, it remains to show that for every $G \in \mathfrak{G}$ we have

(4) $$\int_G \left\{\int_\Omega X(\omega')P^{\mathfrak{F}|\mathfrak{G}}(d\omega',\omega)\right\} P(d\omega) = \int_G X(\omega)P(d\omega).$$

Now by the Monotone Convergence Theorem, (3), and 3° of Definition C.1, we have

$$\int_G \left\{\int_\Omega X(\omega')P^{\mathfrak{F}|\mathfrak{G}}(d\omega',\omega)\right\} P(d\omega) = \int_G \lim_{n\to\infty}\left\{\int_\Omega \varphi_n(\omega')P^{\mathfrak{F}|\mathfrak{G}}(d\omega',\omega)\right\} P(d\omega)$$

$$= \lim_{n\to\infty}\int_G \sum_{k=1}^{p_n} c_{n,k} P^{\mathfrak{F}|\mathfrak{G}}(A_{n,k},\omega)P(d\omega) = \lim_{n\to\infty}\sum_{k=1}^{p_n} c_{n,k} P(A_{n,k}\cap G)$$

$$= \lim_{n\to\infty}\int_G \varphi(\omega)P(d\omega) = \int_G X(\omega)P(d\omega).$$

Similarly for X^-. Therefore (4) holds. This proves (1) for a nonnegative integrable random variable X. For an arbitrary integrable random variable X, let $X = X^+ - X^-$. Since $\mathbf{E}(X\,|\,\mathfrak{G}) = \mathbf{E}(X^+\,|\,\mathfrak{G}) - \mathbf{E}(X^-\,|\,\mathfrak{G})$ a.e. on $(\Omega,\mathfrak{G},P)$ and since (1) holds for $\mathbf{E}(X^+\,|\,\mathfrak{G})$ and $\mathbf{E}(X^-\,|\,\mathfrak{G})$ it holds for $\mathbf{E}(X\,|\,\mathfrak{G})$. Finally (2) is an immediate consequence of (1). ∎

Proposition C.4. *Let $\mathfrak{G}$ be a sub-σ-algebra of $\mathfrak{F}$ in a probability space $(\Omega,\mathfrak{F},P)$. Suppose a regular conditional probability $P^{\mathfrak{F}|\mathfrak{G}}$ exists. Let $(\Omega,\mathfrak{F}^*,P)$ be the completion of $(\Omega,\mathfrak{F},P)$. Then a regular conditional probability $P^{\mathfrak{F}^*|\mathfrak{G}}$ exists.*

Proof. $\mathfrak{F}^*$ consists of subsets of Ω of the form $E = A \cup N_0$ where $A \in \mathfrak{F}$ and N_0 is a subset of a null set N in $(\Omega,\mathfrak{F},P)$. Let us define a function φ on $\mathfrak{F}^* \times \Omega$ by setting

(1) $$\varphi(E,\omega) = P^{\mathfrak{F}|\mathfrak{G}}(A,\omega) \quad \text{for } (E,\omega) \in \mathfrak{F}^* \times \Omega,$$

where $A \in \mathfrak{F}$ is as specified above for the given set $E \in \mathfrak{F}^*$. To show that $\varphi(E,\cdot)$ is uniquely defined up to a null set in $(\Omega,\mathfrak{G},P)$ for each $E \in \mathfrak{F}^*$, suppose E is given as $E = A' \cup N'_0$ as well as $E = A'' \cup N''_0$ where $A', A'' \in \mathfrak{F}$, $N'_0 \subset N'$, $N''_0 \subset N''$ and N', N'' are null sets in $(\Omega, \mathfrak{F}, P)$. We assume without loss of generality that $A' \cap N_0' = \emptyset$ and $A'' \cap N_0'' = \emptyset$ and thus $A' = E - N_0'$ and $A'' = E - N_0''$. By (1) and condition 3° of Definition C.1, we have

$$\int_G P^{\mathfrak{F}|\mathfrak{G}}(A', \omega) P(d\omega) = P(A' \cap G) = P((E - N_0') \cap G)$$
$$= P((E - N_0'') \cap G) = P(A'' \cap G) = \int_G P^{\mathfrak{F}|\mathfrak{G}}(A'', \omega) P(d\omega)$$

for every $G \in \mathfrak{G}$. Thus $P^{\mathfrak{F}|\mathfrak{G}}(A', \cdot) = P^{\mathfrak{F}|\mathfrak{G}}(A'', \cdot)$ a.e. on $(\Omega, \mathfrak{G}, P)$.

Let us show that φ satisfies the conditions in Definition C.1. Now since $P^{\mathfrak{F}|\mathfrak{G}}$ satisfies condition 1° of Definition C.1 there exists a null set Λ in $(\Omega, \mathfrak{G}, P)$ such that $P^{\mathfrak{F}|\mathfrak{G}}(\cdot, \omega)$ is a probability measure on $\mathfrak{F}$ for $\omega \in \Lambda^c$. From this and from (1) it follows immediately that $\varphi(\cdot, \omega)$ is a probability measure on $\mathfrak{F}^*$ for $\omega \in \Lambda^c$. By (1), $\varphi(E, \cdot)$ is a $\mathfrak{G}$-measurable function on Ω for every $E \in \mathfrak{F}^*$. For $E \in \mathfrak{F}^*$, writing $E = A \cup N_0$ with $A \in \mathfrak{F}$ and $N_0 \subset N$ where N is a null set in $(\Omega, \mathfrak{F}, P)$, we have

$$P(E \cap G) = P(A \cap G) = \int_G P^{\mathfrak{F}|\mathfrak{G}}(A, \omega) P(d\omega) = \int_G \varphi(E, \omega) P(d\omega)$$

for every $G \in \mathfrak{G}$ by condition 3° of Definition C.1 satisfied by $P^{\mathfrak{F}|\mathfrak{G}}$. This shows that φ satisfies all the conditions in Definition C.1 and is therefore a regular conditional probability on the probability space $(\Omega, \mathfrak{F}^*, P)$ given $\mathfrak{G}$. ∎

Lemma C.5. *Let μ be a finite, nonnegative, finitely additive set function on an algebra $\mathfrak{D}$ of subsets of a Hausdorff space S. Let $\mathfrak{K}$ be the collection of all compact sets in S and let $\mathfrak{A}$ be a sub-algebra of $\mathfrak{D}$. If μ satisfies the condition that*

(1) $$\mu(A) = \sup\{\mu(K) : K \subset A, K \in \mathfrak{K} \cap \mathfrak{D}\} \quad \text{for every } A \in \mathfrak{A},$$

then μ is countably additive on $\mathfrak{A}$.

Proof. Since $\mathfrak{A}$ is a subalgebra of $\mathfrak{D}$, μ is a finite, nonnegative, finitely additive set function on $\mathfrak{A}$. A finite, nonnegative, finitely additive set function μ on an algebra is countably additive if and only if for every decreasing sequence $\{A_n : n \in \mathbb{N}\}$ in the algebra such that $A_n \downarrow \emptyset$ we have $\mu(A_n) \downarrow 0$. Thus if μ is not countably additive on $\mathfrak{A}$, then there exist $\delta > 0$ and a decreasing sequence $\{A_n : n \in \mathbb{N}\}$ in $\mathfrak{A}$ such that $A_n \downarrow \emptyset$ and $\mu(A_n) \geq \delta$ for every $n \in \mathbb{N}$. Let us show that this contradicts (1). Now according to (1) for every $n \in \mathbb{N}$ there exists $K_n \in \mathfrak{K} \cap \mathfrak{D}$ such that $K_n \subset A_n$ and $\mu(A_n - K_n) < \delta/3^n$. Then we have

$$A_n - \cap_{i=1}^n K_i \subset \cup_{i=1}^n (A_n - K_i) \subset \cup_{i=1}^n (A_i - K_i)$$

APPENDIX C. REGULAR CONDITIONAL PROBABILITIES

so that $\mu(\cap_{i=1}^n K_i) \geq \mu(A_n) - \sum_{i=1}^n \delta/3^i > \delta/2$. Then $F_n \equiv \cap_{i=1}^n K_i \neq \emptyset$. Since S is a Hausdorff space, the compact set F_n is a closed set.

Consider the decreasing sequence of closed sets $\{F_n : n \in \mathbb{N}\}$. Let us show that $\cap_{n \in \mathbb{N}} F_n \neq \emptyset$. Suppose $\cap_{n \in \mathbb{N}} F_n = \emptyset$. Then $\cup_{n \in \mathbb{N}} F_n^c = S$ so that in particular $\{F_n^c : n \in \mathbb{N}\}$ is an open covering of the compact set K_1 and thus $K_1 \subset \cup_{i=1}^N F_i^c$ for some $N \in \mathbb{N}$. Since $\{F_n^c : n \in \mathbb{N}\}$ is an increasing sequence we have $K_1 \subset F_N^c$ and thus $K_1 \cap F_N = \emptyset$. This contradicts the fact that $F_N \neq \emptyset$ and $F_N = \cap_{i=1}^N K_i \subset K_1$. Therefore $\cap_{n \in \mathbb{N}} F_n \neq \emptyset$. But $F_n \subset K_n \subset A_n$ and $\cap_{n \in \mathbb{N}} A_n = \emptyset$ so that $\cap_{n \in \mathbb{N}} F_n = \emptyset$. This is a contradiction. Therefore μ is countably additive on $\mathfrak{A}$. ∎

Let $(S, \mathfrak{B}, \mu)$ be a measure space in which S is a topological space and $\mathfrak{B}$ is an arbitrary σ-algebra of subsets of S. Let $\mathfrak{K}$ be the collection of all compact subsets of S. We say that μ is regular if for every $A \in \mathfrak{B}$ we have $\mu(A) = \sup\{\mu(K) : K \subset A, K \in \mathfrak{K} \cap \mathfrak{B}\}$. According to Ulam's Theorem if S is a complete separable metric space, $\mathfrak{B}_S$ is the Borel σ-algebra of subsets of S, and μ is a finite measure on $(S, \mathfrak{B}_S)$, then μ is regular. For a proof of Ulam's Theorem we refer to [4] R. M. Dudley.

Theorem C.6. *Let $(S, \mathfrak{B}_S, P)$ be a probability space in which S is a complete separable metric space and $\mathfrak{B}_S$ is the Borel σ-algebra of subsets of S. Then for every sub-σ-algebra $\mathfrak{G}$ of $\mathfrak{B}_S$, the regular conditional probability given $\mathfrak{G}$, $P^{\mathfrak{B}_S}|^{\mathfrak{G}}$, exists uniquely.*

Proof. 1) Let us show first that there exists a countable algebra $\mathfrak{A}$ of subsets of S such that $\mathfrak{B}_S = \sigma(\mathfrak{A})$. For an arbitrary collection $\mathfrak{C}$ of subsets of S let us write $\alpha(\mathfrak{C})$ for the algebra generated by $\mathfrak{C}$. Let $\mathfrak{O}$ be the collection of all open sets in S. Since S is a separable metric space it has a countable dense subset. Let $\mathfrak{O}_0 = \{O_n : n \in \mathbb{N}\}$ be the countable collection of all open spheres in S having centers in a countable dense subset of S and having rational radii. Every $O \in \mathfrak{O}$ is then a union of members of $\mathfrak{O}_0$. Let $\mathfrak{A} = \alpha(\mathfrak{O}_0)$. Then $\mathfrak{O} \subset \sigma(\mathfrak{A})$ so that $\mathfrak{B}_S = \sigma(\mathfrak{O}) \subset \sigma(\mathfrak{A})$. On the other hand, we have $\mathfrak{O}_0 \subset \mathfrak{O}$ and $\mathfrak{A} = \alpha(\mathfrak{O}_0) \subset \alpha(\mathfrak{O})$ so that $\sigma(\mathfrak{A}) \subset \sigma(\alpha(\mathfrak{O})) = \sigma(\mathfrak{O}) = \mathfrak{B}_S$. Thus $\mathfrak{B}_S = \sigma(\mathfrak{A})$. To show that $\mathfrak{A}$ is countable, let $\mathfrak{A}_n = \alpha(\{O_1, \ldots, O_n\})$ for $n \in \mathbb{N}$. Now $\mathfrak{A}_n$ is a finite collection for every $n \in \mathbb{N}$, $\mathfrak{A}_n \uparrow$, and $\cup_{n \in \mathbb{N}} \mathfrak{A}_n$ is an algebra containing $\mathfrak{O}_0$. Thus $\mathfrak{A} = \alpha(\mathfrak{O}_0) \subset \cup_{n \in \mathbb{N}} \mathfrak{A}_n$. On the other hand since $\mathfrak{A}_n \subset \mathfrak{A}$ for every $n \in \mathbb{N}$ we have $\cup_{n \in \mathbb{N}} \mathfrak{A}_n \subset \mathfrak{A}$. Thus $\mathfrak{A} = \cup_{n \in \mathbb{N}} \mathfrak{A}_n$. Then since $\mathfrak{A}_n$ is a finite collection for every $n \in \mathbb{N}$, $\mathfrak{A}$ is a countable collection.

2) Let $\{A_n : n \in \mathbb{N}\}$ be the members of the countable algebra $\mathfrak{A}$. Since P is a finite measure on $(S, \mathfrak{B}_S)$ it is regular. Thus for every $A_n \in \mathfrak{A} \subset \mathfrak{B}_S$, there exists a sequence of compact sets $\{K_{n,m} : m \in \mathbb{N}\}$ contained in A_n such that $P(K_{n,m}) \uparrow P(A_n)$ as $m \to \infty$.

Since a finite union of compact sets is a compact set we can choose $K_{n,m}$ so that $K_{n,m} \uparrow$ as $m \to \infty$. Let
$$\mathfrak{D} = \alpha(\{A_n, K_{n,m} : m \in \mathbb{N}, n \in \mathbb{N}\}).$$
Since $\mathfrak{D}$ is an algebra generated by a countable collection, it is a countable collection by the same argument as in 1) in showing the countability of $\mathfrak{A} = \alpha(\mathfrak{D}_0)$. Now

$1°$. $P(D|\mathfrak{G}) \geq 0$ a.e. on $(S, \mathfrak{G}, P)$ for every $D \in \mathfrak{D}$.

$2°$. $P(S|\mathfrak{G}) = 1$ a.e. on $(S, \mathfrak{G}, P)$.

$3°$. $P(\cup_{i=1}^{k} D_i | \mathfrak{G}) = \sum_{i=1}^{k} P(D_i | \mathfrak{G})$ a.e. on $(S, \mathfrak{G}, P)$ for $D_i \in \mathfrak{D}$, $i = 1, \ldots, k$, disjoint.

$4°$. $P(K_{n,m}|\mathfrak{G}) \uparrow P(A_n|\mathfrak{G})$ a.e. on $(S, \mathfrak{G}, P)$.

Let p be a function on $\mathfrak{B}_S \times S$ defined by letting $p(A, \cdot)$ be equal to an arbitrarily fixed version of $P(A|\mathfrak{G})$ for every $A \in \mathfrak{B}_S$. Now $\mathfrak{D} \subset \mathfrak{B}_S$ is a countable collection and the collection of all finite combinations of members of $\mathfrak{D}$ is countable. Therefore there exists a null set Λ_∞ in $(S, \mathfrak{G}, P)$ such that conditions $1°$, $2°$, $3°$, and $4°$ hold at x for $x \in \Lambda_\infty^c$. Thus for $x \in \Lambda_\infty^c$, $p(\cdot, x)$ is a nonnegative, finitely additive set function on the algebra $\mathfrak{D}$ with $p(S, x) = 1$ by $1°$, $2°$, and $3°$. Writing $\mathfrak{K}$ for the collection of all compact sets in S, we have by $4°$

$$p(A_n, x) = P(A_n|\mathfrak{G})(x) = \sup\{P(K|\mathfrak{G})(x) : K \subset A_n, K \in \mathfrak{K} \cap \mathfrak{D}\}$$
$$= \sup\{p(K, x) : K \subset A_n, K \in \mathfrak{K} \cap \mathfrak{D}\}.$$

Thus by Lemma C.5, $p(\cdot, x)$ is countably additive on the algebra $\mathfrak{A}$ when $x \in \Lambda_\infty^c$. Therefore when $x \in \Lambda_\infty^c$, the restriction of $p(\cdot, x)$ to $\mathfrak{A}$ can be extended uniquely to be a measure $q(\cdot, x)$ on $\mathfrak{B}_S = \sigma(\mathfrak{A})$ which is a probability measure since $p(S, x) = 1$ by $2°$. Let q be a function on $\mathfrak{B} \times S$ defined by setting $q(\cdot, x)$ to be equal to the probability measure on $\mathfrak{B}_S$ obtained by the extension of the restriction of $p(\cdot, x)$ when $x \in \Lambda_\infty^c$ and by setting $q(\cdot, x) = 0$ when $x \in \Lambda_\infty$. Thus defined, q satisfies condition $1°$ of Definition C.1. To show that q satisfies $2°$ and $3°$ of Definition C.1 we show equivalently according to Observation C.2 that $q(A, \cdot)$ is a version of $P(A|\mathfrak{G})$ for every $A \in \mathfrak{B}_S$. Let $\mathfrak{E}$ be the collection of all members A of $\mathfrak{B}_S$ such that $q(A, \cdot)$ is a version of $P(A|\mathfrak{G})$. Now if $A \in \mathfrak{A}$ then $q(A, \cdot) = p(A, \cdot)$ on Λ_∞^c and $q(A, \cdot) = 0$ on Λ_∞. But $p(A, \cdot)$ is a version of $P(A|\mathfrak{G})$. Thus $q(A, \cdot)$ is a version of $P(A|\mathfrak{G})$. This shows that $\mathfrak{A} \subset \mathfrak{E}$. Let us show that $\mathfrak{E}$ is a d-class of subsets of S. Now since $S \in \mathfrak{A}$ we have $S \in \mathfrak{E}$. Let A and B be two members of $\mathfrak{E}$ such that $A \subset B$. Then there exist two null sets Λ_A and Λ_B in $(S, \mathfrak{G}, P)$ such that $q(A, x) = P(A|\mathfrak{G})$ for $x \in \Lambda_A^c$

APPENDIX C. REGULAR CONDITIONAL PROBABILITIES

and $q(B,x) = P(B\,|\,\mathfrak{G})$ for $x \in \Lambda_B^c$. Also there exists a null set $\Lambda_{A,B}$ in $(S, \mathfrak{G}, P)$ such that $P(B-A\,|\,\mathfrak{G})(x) = P(B\,|\,\mathfrak{G})(x) - P(A\,|\,\mathfrak{G})(x)$ for $x \in \Lambda_{A,B}^c$. Let $\Lambda = \Lambda_A \cup \Lambda_B \cup \Lambda_{A,B} \cup \Lambda_\infty$. Then for $x \in \Lambda^c$ we have, recalling that $q(\cdot, x)$ is a measure on $\mathfrak{B}_S$ for $x \in \Lambda_\infty^c$,

$$q(B-A, x) = q(B, x) - q(A, x) = P(B\,|\,\mathfrak{G})(x) - P(A\,|\,\mathfrak{G})(x) = P(B-A\,|\,\mathfrak{G})(x).$$

This shows that $B - A \in \mathfrak{E}$. Next let $\{A_n : n \in \mathbb{N}\}$ be an increasing sequence in $\mathfrak{E}$. Then for every $n \in \mathbb{N}$ there exists a null set Λ_n in $(S, \mathfrak{G}, P)$ such that $q(A_n, x) = P(A_n\,|\,\mathfrak{G})(x)$ for $x \in \Lambda_n^c$. By the Conditional Monotone Convergence Theorem there exists a Λ_0 in $(S, \mathfrak{G}, P)$ such that $P(\lim_{n \to \infty} A_n\,|\,\mathfrak{G})(x) = \lim_{n \to \infty} P(A_n\,|\,\mathfrak{G})(x)$ for $x \in \Lambda_0^c$. Let $\Lambda = (\cup_{n \in \mathbb{Z}_+} \Lambda_n) \cup \Lambda_\infty$. Then for $x \in \Lambda^c$ we have

$$q(\lim_{n \to \infty} A_n, x) = \lim_{n \to \infty} q(A_n, x) = \lim_{n \to \infty} P(A_n\,|\,\mathfrak{G})(x) = P(\lim_{n \to \infty} A_n\,|\,\mathfrak{G})(x).$$

This shows that $\lim_{n \to \infty} A_n \in \mathfrak{E}$. Thus $\mathfrak{E}$ is a d-class of subsets of S. Then $\mathfrak{B}_S = \sigma(\mathfrak{A}) = d(\mathfrak{A}) \subset \mathfrak{E}$ by Theorem 1.7. Therefore $q(A, \cdot)$ is a version of $P(A\,|\,\mathfrak{G})$ for every $A \in \mathfrak{B}_S$. This shows that q is a regular conditional probability given $\mathfrak{G}$.

3) To prove the uniqueness of regular conditional probability given a sub-σ-algebra in our probability space $(S, \mathfrak{B}_S, P)$, let q_1 and q_2 be two regular conditional probability given a sub-σ-algebra $\mathfrak{G}$ of $\mathfrak{B}_S$. Then there exists a null set Λ_0 in $(S, \mathfrak{G}, P)$ such that $q_1(\cdot, x)$ and $q_2(\cdot, x)$ are probability measures on $\mathfrak{B}_S$ when $x \in \Lambda_0^c$. For every $A \in \mathfrak{B}_S$ and in particular for every A_n in the countable algebra $\mathfrak{A}$, $q_1(A_n, \cdot)$ and $q_2(A_n, \cdot)$ are versions of $P(A_n\,|\,\mathfrak{G})$ by Observation C.2 and thus there exists a null set Λ_n in $(S, \mathfrak{G}, P)$ such that $q_1(A_n, x) = q_2(A_n, x)$ when $x \in \Lambda_n^c$. Let $\Lambda = \cup_{n \in \mathbb{Z}_+} \Lambda_n$. Then for $x \in \Lambda^c$, $q_1(\cdot, x)$ and $q_2(\cdot, x)$ are probability measures on $\mathfrak{B}_S$ and are equal on $\mathfrak{A}$. For fixed $x \in \Lambda^c$, let $\mathfrak{E}$ be the collection of all members A of $\mathfrak{B}_S$ such that $q_1(A, x) = q_2(A, x)$. As we saw above, $\mathfrak{A} \subset \mathfrak{E}$. By the fact that $q_1(\cdot, x)$ and $q_2(\cdot, x)$ are measures it is easily verified that $\mathfrak{E}$ is a d-class of subsets of S. Then by Theorem 1.7, $\mathfrak{B}_S \subset \mathfrak{E}$. Therefore $q_1(A, x) = q_2(A, x)$ for $A \in \mathfrak{B}_S$ when $x \in \Lambda^c$. This completes the proof of the uniqueness of the regular conditional probability. ∎

Let $(\Omega, \mathfrak{F}, P)$ be a probability space and let $\mathfrak{G}$ be a sub-σ-algebra of $\mathfrak{F}$. Consider $P^{\mathfrak{F}\,|\,\mathfrak{G}}$, a regular conditional probability given $\mathfrak{G}$. For every $A \in \mathfrak{F}$ we have according to 3° of Definition C.1 the equality $\int_G P^{\mathfrak{F}\,|\,\mathfrak{G}}(A, \omega) P(d\omega) = P(A \cap G) = \int_G \mathbf{1}_A(\omega) P(d\omega)$ for every $G \in \mathfrak{G}$. If $A \in \mathfrak{G}$, then $\mathbf{1}_A$ is a $\mathfrak{G}$-measurable function. The last equality holding for every $G \in \mathfrak{G}$ implies that $P^{\mathfrak{F}\,|\,\mathfrak{G}}(A, \cdot) = \mathbf{1}_A$ a.e. on $(\Omega, \mathfrak{G}, P)$. Therefore there exists a null set Λ_A in $(\Omega, \mathfrak{G}, P)$ such that $P^{\mathfrak{F}\,|\,\mathfrak{G}}(A, \omega) = \mathbf{1}_A(\omega)$ for $\omega \in \Lambda_A^c$. We address the question of the existence of a null set Λ in $(\Omega, \mathfrak{G}, P)$, not depending on A, such that for every A in a sub-σ-algebra $\mathfrak{H}$ of $\mathfrak{G}$ we have $P^{\mathfrak{F}|\mathfrak{G}}(A,\omega) = \mathbf{1}_A(\omega)$ for every $A \in \mathfrak{H}$ when $\omega \in \Lambda^c$.

Definition C.7. *Let $(\Omega, \mathfrak{F})$ be a measurable space. We say that $\mathfrak{F}$ is countably determined if there exists a countable subcollection $\mathfrak{D}$ of $\mathfrak{F}$ such that whenever two probability measures P and Q are equal on $\mathfrak{D}$ then they are equal on $\mathfrak{F}$. We call $\mathfrak{D}$ a countable collection of determining sets for $\mathfrak{F}$.*

Proposition C.8. *Let S be a separable metric space. Then the Borel σ-algebra of subsets of S, $\mathfrak{B}_S$, is always countably determined.*

Proof. Let $\mathfrak{O}$ be the collection of all open sets in S. We have $\mathfrak{B}_S = \sigma(\mathfrak{O})$. Since S is a separable metric space, there is a countable subcollection $\mathfrak{O}_0$ of $\mathfrak{O}$ such that every member of $\mathfrak{O}$ is the union of some members of $\mathfrak{O}_0$. The collection $\mathfrak{D}$ of all finite unions of members of $\mathfrak{O}_0$ is a countable subcollection of $\mathfrak{B}_S$. Suppose P and Q are two probability measures on $\mathfrak{F}$ such that $P(D) = Q(D)$ for every $D \in \mathfrak{D}$. Let $O \in \mathfrak{O}$. Then $O = \cup_{n \in \mathbb{N}} O_n$ where $O_n \in \mathfrak{O}_0$ for $n \in \mathbb{N}$. Let $G_n = \cup_{k=1}^n O_k \in \mathfrak{D}$ for $n \in \mathbb{N}$. Then $G_n \uparrow$ and $O = \cup_{n \in \mathbb{N}} G_n = \lim_{n \to \infty} G_n$. Since $G_n \in \mathfrak{D}$, we have $P(G_n) = Q(G_n)$ for every $n \in \mathbb{N}$ and therefore $P(O) = \lim_{n \to \infty} P(G_n) = \lim_{n \to \infty} Q(G_n) = Q(O)$. Thus $P = Q$ on $\mathfrak{O}$. Since $\mathfrak{O}$ is a π-class, this implies that $P = Q$ on $\sigma(\mathfrak{O})$ according to Corollary 1.8. Thus the equality of P and Q on $\mathfrak{D}$ implies the equality of P and Q on $\mathfrak{B}_S$. Therefore $\mathfrak{D}$ is a countable collection of determining sets for $\mathfrak{B}_S$. ∎

Theorem C.9. *Let $(\Omega, \mathfrak{F}, P)$ be a probability space, let $\mathfrak{G}$ be a sub-σ-algebra of $\mathfrak{F}$ and suppose a regular conditional probability given $\mathfrak{G}$, $P^{\mathfrak{F}|\mathfrak{G}}$ exists. Let $\mathfrak{H}$ be a countably determined sub-σ-algebra of $\mathfrak{G}$. Then there exists a null set Λ in $(\Omega, \mathfrak{G}, P)$ such that*

$$P^{\mathfrak{F}|\mathfrak{G}}(A,\omega) = \mathbf{1}_A(\omega) \quad \text{for every } A \in \mathfrak{H} \text{ when } \omega \in \Lambda^c.$$

Proof. Let $\mathfrak{D} \subset \mathfrak{H}$ be a countable collection of determining set for $\mathfrak{H}$, that is, any two probability measures on $\mathfrak{H}$ that are equal on $\mathfrak{D}$ are equal on $\mathfrak{H}$. Let $A \in \mathfrak{D}$. Then since $A \in \mathfrak{G}$, there exists a null set Λ_A in $(\Omega, \mathfrak{G}, P)$ such that $P^{\mathfrak{F}|\mathfrak{G}}(A,\omega) = \mathbf{1}_A(\omega)$ for $\omega \in \Lambda_A^c$. Let $\Lambda_\infty = \cup_{A \in \mathfrak{D}} \Lambda_A$. Since $\mathfrak{D}$ is a countable collection, Λ is a null set in $(\Omega, \mathfrak{G}, P)$. Then

(1) $$P^{\mathfrak{F}|\mathfrak{G}}(A,\omega) = \mathbf{1}_A(\omega) \quad \text{for every } A \in \mathfrak{D} \text{ when } \omega \in \Lambda_\infty^c.$$

APPENDIX C. REGULAR CONDITIONAL PROBABILITIES 483

Now by 1° of Definition C.1 there exists a null set Λ_0 in $(\Omega, \mathfrak{G}, P)$ such that for every $\omega \in \Lambda_0^c$, $P^{\mathfrak{F}|\mathfrak{G}}(\cdot, \omega)$ is a probability measure on $\mathfrak{F}$ is hence a probability measure on the sub-σ-algebra $\mathfrak{H}$. Let $\Lambda = \Lambda_0 \cup \Lambda_\infty$. With fixed $\omega \in \Omega$, the set function $\mathbf{1}_A(\omega)$ for $A \in \mathfrak{H}$ is a probability measure on $\mathfrak{H}$ which assigns the value 1 to every $A \in \mathfrak{H}$ which contains ω and the value 0 to every $A \in \mathfrak{H}$ which does not contain ω. For $\omega \in \Lambda^c$, the two probability measures $P^{\mathfrak{F}|\mathfrak{G}}(\cdot, \omega)$ and $\mathbf{1}_{(\cdot)}(\omega)$ on $\mathfrak{H}$ are equal on $\mathfrak{D}$ according to (1). Since $\mathfrak{D}$ is a countable collection of determining sets for $\mathfrak{H}$, the two measures are equal on $\mathfrak{H}$ for $\omega \in \Lambda^c$. ∎

Corollary C.10. *Let $(\Omega, \mathfrak{F}, P)$ be a probability space, let $\mathfrak{G}$ be a sub-σ-algebra of $\mathfrak{F}$ and suppose a regular conditional probability given $\mathfrak{G}$, $P^{\mathfrak{F}|\mathfrak{G}}$, exists. Let $(S, \mathfrak{B})$ be a measurable space such that $\mathfrak{B}$ is countably determined and $\{x\} \in \mathfrak{B}$ for every $x \in S$. Let ξ be a $\mathfrak{G}/\mathfrak{B}$-measurable mapping of Ω into S. Then there exists a null set Λ in $(\Omega, \mathfrak{G}, P)$ such that $P^{\mathfrak{F}|\mathfrak{G}}(\xi^{-1}(\{\xi(\omega)\}), \omega) = 1$ for $\omega \in \Lambda^c$, that is,*

$$P^{\mathfrak{F}|\mathfrak{G}}(\{\omega' \in \Omega : \xi(\omega') = \xi(\omega)\}, \omega) = 1 \quad \text{for } \omega \in \Lambda^c.$$

Proof. Note that since ξ is $\mathfrak{G}/\mathfrak{B}$-measurable and $\{\xi(\omega)\} \in \mathfrak{B}$, we have $\xi^{-1}(\{\xi(\omega)\}) \in \mathfrak{G}$ for every $\omega \in \Omega$. Let $\mathfrak{D}$ be a countable collection of determining sets for $\mathfrak{B}$. Let $\mathfrak{H} = \xi^{-1}(\mathfrak{B})$. Since ξ is $\mathfrak{G}/\mathfrak{B}$-measurable, $\mathfrak{H}$ is a sub-σ-algebra of $\mathfrak{G}$. Let $\mathfrak{E} = \xi^{-1}(\mathfrak{D})$. To show that $\mathfrak{E}$ is a countable collection of determining sets for $\mathfrak{H}$, let μ and ν be two probability measures on $\mathfrak{H}$ which are equal on $\mathfrak{E}$. Let μ_ξ and ν_ξ be two probability measures on $\mathfrak{B}$ defined by $\mu_\xi(B) = \mu(\xi^{-1}(B))$ and $\nu_\xi(B) = \nu(\xi^{-1}(B))$ for $B \in \mathfrak{B}$. For $D \in \mathfrak{D}$, we have $\mu_\xi(D) = \mu(\xi^{-1}(D)) = \nu(\xi^{-1}(D)) = \nu_\xi(D)$ since $\mu = \nu$ on $\mathfrak{E}$. Thus $\mu_\xi = \nu_\xi$ on $\mathfrak{D}$. Since $\mathfrak{D}$ is a countable collection of determining sets for $\mathfrak{B}$, we have $\mu_\xi = \nu_\xi$ on $\mathfrak{B}$. Thus $\mu(\xi^{-1}(B)) = \nu(\xi^{-1}(B))$ for $B \in \mathfrak{B}$, that is, $\mu(H) = \nu(H)$ for $H \in \mathfrak{H}$. This shows that $\mathfrak{H}$ is countably determined. Then by Theorem C.9 there exists a null set Λ in $(\Omega, \mathfrak{G}, P)$ such that for every $A \in \mathfrak{H}$ we have $P^{\mathfrak{F}|\mathfrak{G}}(A, \omega) = \mathbf{1}_A(\omega)$ for $\omega \in \Lambda^c$. Now for every $\omega \in \Omega$ we have $\{\omega' \in \Omega : \xi(\omega') = \xi(\omega)\} = \xi^{-1}(\xi(\omega)) \in \mathfrak{H}$ since $\{\xi(\omega)\} \in \mathfrak{B}$. Thus

$$P^{\mathfrak{F}|\mathfrak{G}}(\{\omega' \in \Omega : \xi(\omega') = \xi(\omega)\}, \omega) = \mathbf{1}_{\{\omega' \in \Omega : \xi(\omega') = \xi(\omega)\}}(\omega) = 1 \quad \text{for } \omega \in \Lambda^c. \quad \blacksquare$$

Let $(\Omega, \mathfrak{F}, P)$ be a probability space and let $(S, \mathfrak{B})$ be an arbitrary measurable space. Let X be an $\mathfrak{F}/\mathfrak{B}$-measurable mapping of Ω into S. Then $X^{-1}(\mathfrak{B})$ is a sub-σ-algebra of $\mathfrak{F}$ so that a regular conditional probability given X, $P^{\mathfrak{F}|X^{-1}(\mathfrak{B})}$, is defined as a function on $\mathfrak{F} \times \Omega$. We can also define a regular image conditional probability given X as a function on $\mathfrak{F} \times S$ as follows.

Definition C.11. Let $(\Omega, \mathfrak{F}, P)$ be a probability space and $(S, \mathfrak{B})$ be a measurable space. Let X be an $\mathfrak{F}/\mathfrak{B}$-measurable mapping of Ω into S and let P_X be the probability distribution of X on $(S, \mathfrak{B})$. By a regular image conditional probability given X we mean a function $P^{\mathfrak{F}|X}$ on $\mathfrak{F} \times S$ satisfying the following conditions.

1°. There exists a null set Λ in $(S, \mathfrak{B}, P_X)$ such that $P^{\mathfrak{F}|X}(\cdot, x)$ is a probability measure on $\mathfrak{F}$ for every $x \in \Lambda^c$.

2°. $P^{\mathfrak{F}|X}(A, \cdot)$ is a $\mathfrak{B}$-measurable function on S for every $A \in \mathfrak{F}$.

3°. $P(A \cap X^{-1}(B)) = \int_B P^{\mathfrak{F}|X}(A, x) P_X(dx)$ for every $A \in \mathfrak{F}$ and $B \in \mathfrak{B}$.

We say that the regular conditional probability given X is unique if for any two functions p and p' on $\mathfrak{F} \times S$ satisfying 1°, 2°, and 3° there exists a null set Λ in $(S, \mathfrak{B}, P_X)$ such that $p(A, x) = p'(A, x)$ for every $A \in \mathfrak{F}$ when $x \in \Lambda^c$.

Remark C.12. The regular image conditional probability defined above exists uniquely when the probability space $(\Omega, \mathfrak{F}, P)$ is a complete separable metric space with the Borel σ-algebra of subsets. This can be shown by the same argument as in the proof of Theorem C.6.

Proposition C.13. Let $(\Omega, \mathfrak{F}, P)$ be a probability space, $(S, \mathfrak{B})$ be a measurable space, X be an $\mathfrak{F}/\mathfrak{B}$-measurable mapping of Ω into S, and P_X be the probability distribution of X on $(S, \mathfrak{B})$. Assume the existence of a regular image conditional probability given X, $P^{\mathfrak{F}|X}$. Let f be an extended real valued $\mathfrak{B}$-measurable function on S. Then for the extended real valued random variable $f \circ X$ on $(\Omega, \mathfrak{F}, P)$ we have for every $B \in \mathfrak{B}$

$$\int_{X^{-1}(B)} f \circ X \, dP = \int_S f(x) P^{\mathfrak{F}|X}(X^{-1}(B), x) P_X(dx),$$

in the sense that the existence of one side implies that of the other and the equality of the two.

Proof. Consider first the particular case where $f = 1_E$ where $E \in \mathfrak{B}$. Then

$$\int_{X^{-1}(B)} f \circ X \, dP = \int_{X^{-1}(B)} 1_E(X(\omega)) P(d\omega) = \int_\Omega 1_{X^{-1}(B)}(\omega) 1_{X^{-1}(E)}(\omega) P(d\omega)$$

$$= P(X^{-1}(B) \cap X^{-1}(E)) = \int_E P^{\mathfrak{F}|X}(X^{-1}(B), x) P_X(dx) \quad \text{by 3° of Definition C.11}$$

$$= \int_S 1_E(x) P^{\mathfrak{F}|X}(X^{-1}(B), x) P_X(dx) = \int_S f(x) P^{\mathfrak{F}|X}(X^{-1}(B), x) P_X(dx).$$

APPENDIX C. REGULAR CONDITIONAL PROBABILITIES

The general case follows by writing $f = f^+ - f^-$ and by the fact that there exist two increasing sequences of nonnegative simple functions on $(S, \mathfrak{B})$ which converge to f^+ and f^- respectively on S. ∎

Proposition C.14. *Let $(\Omega, \mathfrak{F}, P)$ be a probability space, $(S, \mathfrak{B})$ be a measurable space, X be an $\mathfrak{F}/\mathfrak{B}$-measurable mapping of Ω into S, and P_X be the probability distribution of X on $(S, \mathfrak{B})$. Assume the existence of a regular image conditional probability given X, $P^{\mathfrak{F}|X}$. Let ξ be an extended real valued random variable on $(\Omega, \mathfrak{F}, P)$. Then for every $B \in \mathfrak{B}$ we have*

$$\int_{X^{-1}(B)} \xi \, dP = \int_B \left[\int_\Omega \xi(\omega) P^{\mathfrak{F}|X}(d\omega, x) \right] P_X(dx),$$

in the sense that the existence of one side implies that of the other and the equality of the two.

Proof. Consider first the particular case where $\xi = 1_A$ with $A \in \mathfrak{F}$. Then

$$\int_{X^{-1}(B)} \xi \, dP = \int_{X^{-1}(B)} 1_A(\omega) P(d\omega) = P(A \cap X^{-1}(B))$$
$$= \int_B P^{\mathfrak{F}|X}(A, x) P_X(dx) = \int_B \left[\int_\Omega 1_A(\omega) P^{\mathfrak{F}|X}(d\omega, x) \right] P_X(dx)$$
$$= \int_B \left[\int_\Omega \xi(\omega) P^{\mathfrak{F}|X}(d\omega, x) \right] P_X(dx),$$

where the second equality is by 3° of Definition C.11. The general case follows by writing $\xi = \xi^+ - \xi^-$ and by the fact that there exist two increasing sequences of nonnegative simple functions on $(\Omega, \mathfrak{F})$ which converge to ξ^+ and ξ^- respectively on Ω. ∎

Theorem C.15. *Let $(\Omega, \mathfrak{F}, P)$ be a probability space, $(S, \mathfrak{B})$ be a measurable space, X be an $\mathfrak{F}/\mathfrak{B}$-measurable mapping of Ω into S, and P_X be the probability distribution of X on $(S, \mathfrak{B})$. Assume the existence of a regular image conditional probability given X, $P^{\mathfrak{F}|X}$. Assume further that $\mathfrak{B}$ is countably determined and $\{x\} \in \mathfrak{B}$ for every $x \in S$. Then there exists a null set Λ in $(S, \mathfrak{B}, P_X)$ such that*

(1) $\qquad P^{\mathfrak{F}|X}(X^{-1}(B), x) = 1_B(x)$ *for every $B \in \mathfrak{B}$ when $x \in \Lambda^c$.*

In particular we have

(2) $\qquad P^{\mathfrak{F}|X}(X^{-1}(\{x\}), x) = 1$ *for every $x \in \Lambda^c$.*

Proof. Let $\mathfrak{D} = \{D_n : n \in \mathbb{N}\}$ be a countable collection of determining sets for $\mathfrak{B}$. By choosing $A = X^{-1}(D_n)$ and $B = D_n$ in 3° of Definition C.11, we have

$$\int_{D_n} P_X(dx) = P(X^{-1}(D_n)) = \int_{D_n} P^{\mathfrak{F}|X}(X^{-1}(D_n), x) P_X(dx).$$

By 1° of Definition C.11, $P^{\mathfrak{F}|X}(X^{-1}(D_n), x) \in [0,1]$ for $x \in S$ and in particular for $x \in D_n$. Thus the last equality implies that there exists a null set Λ_n in $(S, \mathfrak{B}, P_X)$ such that $P^{\mathfrak{F}|X}(X^{-1}(D_n), x) = \mathbf{1}_{D_n}(x)$ for $x \in \Lambda_n^c$. Let $\Lambda = \cup_{n \in \mathbb{N}} \Lambda$. Then we have

(3) $\qquad P^{\mathfrak{F}|X}(X^{-1}(D_n), x) = \mathbf{1}_{D_n}(x) \quad$ for every n when $x \in \Lambda^c$.

Now for every $x \in S$, $P^{\mathfrak{F}|X}(X^{-1}(B), x)$ as a function of $B \in \mathfrak{B}$ is a probability measure on $\mathfrak{B}$ by 1° of Definition C.11. For fixed $x \in S$, $\mathbf{1}_B(x)$ as a function of $B \in \mathfrak{B}$ is also a probability measure on $\mathfrak{B}$. According to (3), when $x \in \Lambda^c$ these two probability measures are equal on $\mathfrak{D}$. Then since $\mathfrak{D}$ is a countable collection of determining sets for $\mathfrak{B}$, these two probability measures are equal on $\mathfrak{B}$ when $x \in \Lambda^c$. This proves (1). When $B = \{x\}$ for some $x \in S$, we have $\mathbf{1}_{\{x\}}(x) = 1$ and thus (2) holds. ∎

Definition C.16. *Let $(\Omega, \mathfrak{F}, P)$ be a probability space and $(S, \mathfrak{B})$ be a measurable space. Let X be an $\mathfrak{F}/\mathfrak{B}$-measurable mapping of Ω into S. Let $\mathfrak{G}$ be a sub-σ-algebra of $\mathfrak{F}$. By a regular conditional probability distribution of X given $\mathfrak{G}$ we mean a function $P^{X|\mathfrak{G}}$ on $\mathfrak{B} \times \Omega$ satisfying the following conditions.*

1°. *There exists a null set Λ in $(\Omega, \mathfrak{G}, P)$ such that $P^{X|\mathfrak{G}}(\cdot, \omega)$ is a probability measure on $\mathfrak{B}$ for every $\omega \in \Lambda^c$.*

2°. $P^{X|\mathfrak{G}}(B, \cdot)$ *is a $\mathfrak{G}$-measurable function on Ω for every $B \in \mathfrak{B}$.*

3°. $P(X^{-1}(B) \cap G) = \int_G P^{X|\mathfrak{G}}(B, \omega) P(d\omega)$ *for every $B \in \mathfrak{B}$ and $G \in \mathfrak{G}$.*

We say that the regular conditional probability distribution of X given $\mathfrak{G}$ is unique if for any two functions p and p' on $\mathfrak{B} \times \Omega$ satisfying 1°, 2°, and 3° there exists a null set Λ in $(\Omega, \mathfrak{G}, P)$ such that $p(B, \omega) = p'(B, \omega)$ for every $B \in \mathfrak{B}$ when $\omega \in \Lambda^c$.

Remark C.17. The regular conditional probability distribution defined above exists and is unique if the measurable space $(S, \mathfrak{B})$ is a complete separable metric space with the Borel σ-algebra of subsets. The proof of this parallels that of Theorem C.6.

Appendix D

Multidimensional Normal Distributions

Definition D.1. *Let Φ be a probability measure on $(\mathbb{R}^d, \mathfrak{B}_{\mathbb{R}^d})$ where $d \in \mathbb{N}$. The mean vector of Φ, namely $\mathbf{M}(\Phi) = (\mathbf{M}(\Phi)_j, j = 1, \ldots, d)$, is defined by*

$$\text{(1)} \qquad \mathbf{M}(\Phi)_j = \int_{\mathbb{R}^d} x_j \Phi(dx) \quad \text{for } j = 1, \ldots, d,$$

provided all the d integrals exist and are finite. The covariance matrix of Φ, namely $\mathbf{V}(\Phi) = (\mathbf{V}(\Phi)_{j,k}, j, k = 1, \ldots, d)$, is defined by

$$\text{(2)} \qquad \mathbf{V}(\Phi)_{j,k} = \int_{\mathbb{R}^d} \{x_j - \mathbf{M}(\Phi)_j\}\{x_k - \mathbf{M}(\Phi)_k\} \Phi(dx) \quad \text{for } j,k = 1, \ldots, d,$$

provided all the jk integrals exist and are finite. The characteristic function φ of Φ is defined by

$$\text{(3)} \qquad \varphi(y) = \int_{\mathbb{R}^d} e^{i\langle y, x \rangle} \Phi(dx) \quad \text{for } y \in \mathbb{R}^d.$$

Observation D.2. The covariance matrix $\mathbf{V}(\Phi)$ of a probability measure Φ on $(\mathbb{R}^d, \mathfrak{B}_{\mathbb{R}^d})$, if it exists, is always a nonnegative definite matrix. Indeed for every $y \in \mathbb{R}^d$ we have

$$\begin{aligned}
\langle \mathbf{V}(\Phi) y, y \rangle &= \sum_{j=1}^{d} \sum_{k=1}^{d} \mathbf{V}(\Phi)_{j,k} y_j y_k \\
&= \sum_{j=1}^{d} \sum_{k=1}^{d} \int_{\mathbb{R}^d} \{x_j - \mathbf{M}(\Phi)_j\} y_j \{x_k - \mathbf{M}(\Phi)_k\} y_k \Phi(dx) \\
&= \int_{\mathbb{R}^d} [\sum_{j=1}^{d} \{x_j - \mathbf{M}(\Phi)_j\} y_j]^2 \Phi(dx) \geq 0.
\end{aligned}$$

Under the assumption that $\int_{\mathbb{R}^d} |x_1|^{q_1} \cdots |x_d|^{q_d} \Phi(dx) < \infty$, for some $q_1, \ldots, q_d \in \mathbb{N}$, we have

(1) $\quad \left(\dfrac{\partial^{p_1 + \cdots + p_d} \varphi}{\partial y_1^{p_1} \cdots \partial y_d^{p_d}} \right)(y) = \int_{\mathbb{R}^d} (ix_1)^{p_1} \cdots (ix_d)^{p_d} e^{i\langle y, x \rangle} \Phi(dx) \quad \text{for } y \in \mathbb{R}^d,$

for $p_1 = 1, \ldots, q_1; \ldots ; p_d = 1, \ldots, q_d$, and in particular,

(2) $\quad \left(\dfrac{\partial^{p_1 + \cdots + p_d} \varphi}{\partial y_1^{p_1} \cdots \partial y_d^{p_d}} \right)(0) = i^{p_1 + \cdots + p_d} \int_{\mathbb{R}^d} x_1^{p_1} \cdots x_d^{p_d} \Phi(dx).$

Formula (1) is obtained by differentiating the integral in (3) of Definition D.1 under integral sign which is justified by the Dominated Convergence Theorem.

Definition D.3. *A probability measure Φ on $(\mathbb{R}^d, \mathcal{B}_{\mathbb{R}^d})$ is called a d-dimensional normal distribution if it is absolutely continuous with respect to the Lebesgue measure m_L^d on $(\mathbb{R}^d, \mathcal{B}_{\mathbb{R}^d})$ with the Radon-Nikodym derivative given by*

$$\frac{d\Phi}{dm_L^d}(x) = (2\pi)^{-\frac{d}{2}} (\det V)^{-\frac{1}{2}} \exp\left\{ -\frac{1}{2} \left\langle V^{-1}(x - m), x - m \right\rangle \right\} \quad \text{for } x \in \mathbb{R}^d,$$

where $m \in \mathbb{R}^d$ and V is a $d \times d$ positive definite symmetric matrix. We write $N_d(m, V)$ for this probability measure.

Observation D.4 From the well known formula

(1) $\quad \displaystyle\int_{-\infty}^{\infty} e^{-u^2} \, du = \sqrt{\pi},$

we have by a simple substitution

(2) $\quad (2\pi v)^{-\frac{1}{2}} \displaystyle\int_{-\infty}^{\infty} \exp\left\{ -\dfrac{(u - m)^2}{2v} \right\} du = 1 \quad \text{for } m \in \mathbb{R} \text{ and } v > 0.$

We obtain for $a > 0$ and b real or imaginary

(3) $\quad \displaystyle\int_{-\infty}^{\infty} e^{-(au^2 + bu)} \, du = \sqrt{\dfrac{\pi}{a}} \exp\left\{ \dfrac{b^2}{4a} \right\},$

by completing square for u and, in case b is imaginary, by contour integration. Also

(4) $\quad \displaystyle\int_{-\infty}^{\infty} |u|^p e^{-u^2} \, du < \infty \quad \text{for } p \in \mathbb{N}.$

APPENDIX D. MULTIDIMENSIONAL NORMAL DISTRIBUTIONS

Theorem D.5. *Let Φ be the d-dimensional normal distribution $N_d(m, V)$ and let φ be its characteristic function. Then*
1) $\varphi(\mathbb{R}^d) = 1$,
2) $\int_{\mathbb{R}^d} |x_1|^{p_1} \cdots |x_d|^{p_d} \Phi(dx) < \infty$ *for* $p_1, \ldots, p_d \in \mathbb{N}$,
3) $\varphi(y) = \exp\{i\langle y, m\rangle - \frac{1}{2}\langle Vy, y\rangle\}$ *for* $y \in \mathbb{R}^d$,
4) $\mathbf{M}(\Phi) = m$,
5) $\mathbf{V}(\Phi) = V$.

Proof. 1) Since V is a $d \times d$ positive definite symmetric matrix, so is V^{-1} and thus there exists a $d \times d$ nonsingular matrix C such that $C^t C = V^{-1}$ where C^t is the transpose of C. Consider a one to one mapping of $\mathbb{R}^d$ onto $\mathbb{R}^d$ defined by $z = C(x - m)$. Then

$$\langle V^{-1}(x - m), x - m\rangle = \langle C^t C(x - m), x - m\rangle = \langle z, z\rangle,$$

and the Jacobian of the inverse mapping $x = C^{-1}z + m$ is given by

$$|\det C^{-1}| = |\det C|^{-1} = (\det V)^{\frac{1}{2}}.$$

Thus we have

$$\begin{aligned}
\Phi(\mathbb{R}^d) &= (2\pi)^{-\frac{d}{2}}(\det V)^{-\frac{1}{2}} \int_{\mathbb{R}^d} \exp\left\{-\frac{1}{2}\langle V^{-1}(x-m), x-m\rangle\right\} m_L^d(dx) \\
&= (2\pi)^{-\frac{d}{2}} \int_{\mathbb{R}^d} \exp\left\{-\frac{1}{2}\langle z, z\rangle\right\} m_L^d(dz) \\
&= \prod_{j=1}^{d}(2\pi)^{-\frac{1}{2}} \int_{\mathbb{R}} \exp\left\{-\frac{z_j^2}{2}\right\} m_L(dz_j) \\
&= 1,
\end{aligned}$$

by (2) of Observation D.4.

2) Let $p_1, \ldots, p_d \in \mathbb{N}$. Then

$$\int_{\mathbb{R}^d} |x_1|^{p_1} \cdots |x_d|^{p_d} \Phi(dx) = (2\pi)^{-\frac{d}{2}} \int_{\mathbb{R}^d} |x_1|^{p_1} \cdots |x_d|^{p_d} \exp\left\{-\frac{1}{2}\langle z, z\rangle\right\} m_L^d(dz).$$

Now since $x = C^{-1}z + m$, the component x_j is a linear combination of $z_1, \ldots, z_d$ and m_j. Thus $|x_1|^{p_1} \cdots |x_d|^{p_d}$ is bounded by a polynomial in $|z_1|, \ldots, |z_d|$. But for any $q_1, \ldots, q_d \in \mathbb{N}$, we have

$$\int_{\mathbb{R}^d} |z_1|^{q_1} \cdots |z_d|^{q_d} \exp\left\{-\frac{1}{2}\langle z, z\rangle\right\} m_L^d(dz)$$
$$= \prod_{j=1}^{d} \int_{\mathbb{R}} |z_j|^{q_j} \exp\left\{-\frac{z_j^2}{2}\right\} m_L(dz_j) < \infty,$$

by (4) of Observation D.4. Therefore $\int_{\mathbb{R}^d} |x_1|^{p_1} \cdots |x_d|^{p_d} \Phi(dx) < \infty$.

3) For the characteristic function φ of Φ we have by the substitution $z = C(x - m)$

$$
\begin{aligned}
\varphi(y) &= \int_{\mathbb{R}^d} \exp\{i\langle y, x\rangle\} \Phi(dx) \\
&= (2\pi)^{-\frac{d}{2}} \int_{\mathbb{R}^d} \exp\{i\langle y, x\rangle\} \exp\left\{-\frac{1}{2}\langle z, z\rangle\right\} m_L^d(dz) \\
&= (2\pi)^{-\frac{d}{2}} \exp\{i\langle y, m\rangle\} \int_{\mathbb{R}^d} \exp\left\{-\frac{1}{2}\langle z, z\rangle + i\langle (C^{-1})^t y, z\rangle\right\} m_L^d(dz).
\end{aligned}
$$

For brevity let us write $w = (C^{-1})^t y$. Then

$$
\begin{aligned}
\varphi(y) &= (2\pi)^{-\frac{d}{2}} \exp\{i\langle y, m\rangle\} \prod_{j=1}^{d} \int_{\mathbb{R}} \exp\left\{-\frac{1}{2}z_j^2 + iw_j z_j\right\} m_L(dz_j) \\
&= (2\pi)^{-\frac{d}{2}} \exp\{i\langle y, m\rangle\} \prod_{j=1}^{d} (2\pi)^{\frac{1}{2}} \exp\left\{-\frac{1}{2}w_j^2\right\} \quad \text{by (3) of Observation D.4} \\
&= \exp\{i\langle y, m\rangle\} \exp\left\{-\frac{1}{2}\langle w, w\rangle\right\} \\
&= \exp\{i\langle y, m\rangle\} \exp\left\{-\frac{1}{2}\langle (C^{-1})^t y, (C^{-1})^t y\rangle\right\} \\
&= \exp\{i\langle y, m\rangle\} \exp\left\{-\frac{1}{2}\langle Vy, y\rangle\right\}.
\end{aligned}
$$

4) Let $m = (m_j : j = 1, \ldots, d)$ and $V = (v_{j,k} : j, k = 1, \ldots, d)$. By 3) we have

$$\varphi(y) = \exp\left\{i \sum_{p=1}^{d} m_p y_p - \frac{1}{2} \sum_{p=1}^{d} \sum_{q=1}^{d} v_{p,q} y_p y_q\right\}$$

and thus

$$\frac{\partial \varphi}{\partial y_j}(y) = \varphi(y) \left[im_j - \frac{1}{2}\{\sum_{p=1}^{d} v_{p,j} y_p + \sum_{q=1}^{d} v_{j,q} y_q\}\right].$$

According to (2) of Observation D.2, for $j = 1, \ldots, d$ we have

$$\mathbf{M}(\Phi)_j = \int_{\mathbb{R}^d} x_j \Phi(dx) = \frac{1}{i} \frac{\partial \varphi}{\partial y_j}(0) = m_j.$$

This shows that $\mathbf{M}(\Phi) = m$.

5) Since $\mathbf{M}(\Phi) = m$, we have

$$\begin{aligned}\mathbf{V}(\Phi)_{j,k} &= \int_{\mathbb{R}^d} \{x_j - m_j\}\{x_k - m_k\} \Phi(dx) \\ &= (2\pi)^{-\frac{d}{2}} (\det V)^{-\frac{1}{2}} \int_{\mathbb{R}^d} \{x_j - m_j\}\{x_k - m_k\} \\ &\quad \times \exp\left\{-\frac{1}{2}\left\langle V^{-1}(x-m), x-m\right\rangle\right\} m_L^d(dx) \\ &= (2\pi)^{-\frac{d}{2}} (\det V)^{-\frac{1}{2}} \int_{\mathbb{R}^d} x_j x_k \exp\left\{-\frac{1}{2}\left\langle V^{-1}x, x\right\rangle\right\} m_L^d(dx),\end{aligned}$$

by the translation invariance of the Lebesgue integral. Let Ψ be the d-dimensional normal distribution $N_d(0, V)$. Then $\mathbf{M}(\Psi) = 0$ by 4) and therefore

$$\begin{aligned}\mathbf{V}(\Psi)_{j,k} &= (2\pi)^{-\frac{d}{2}} (\det V)^{-\frac{1}{2}} \int_{\mathbb{R}^d} x_j x_k \exp\left\{-\frac{1}{2}\left\langle V^{-1}x, x\right\rangle\right\} m_L^d(dz) \\ &= \mathbf{V}(\Phi)_{j,k} \quad \text{for } j, k = 1, \ldots, d.\end{aligned}$$

To compute $\mathbf{V}(\Phi)_{j,k}$ we compute $\mathbf{V}(\Psi)_{j,k}$ instead. By 3), the characteristic function ψ of Ψ is given by

$$\psi(y) = \exp\left\{-\frac{1}{2}\langle Vy, y\rangle\right\} = \exp\left\{-\frac{1}{2}\sum_{p=1}^{d}\sum_{q=1}^{d} v_{p,q} y_p y_q\right\},$$

and thus

$$\frac{\partial^2 \psi}{\partial y_j \partial y_k}(y) = \psi(y)\{-v_{j,k}\} \quad \text{for } y \in \mathbb{R}^d.$$

By (2) of Observation D.2, we have

$$\mathbf{V}(\Psi)_{j,k} = \int_{\mathbb{R}^d} y_j y_k \Psi(dy) = -\frac{\partial^2 \psi}{\partial y_j \partial y_k}(0) = v_{j,k}.$$

This shows that $\mathbf{V}(\Phi) = V$. ∎

Let $X_1, \ldots, X_d$ be real valued random variables on a probability space $(\Omega, \mathfrak{F}, P)$. The joint probability distribution Φ of $X_1, \ldots, X_d$ is by definition the probability distribution of the d-dimensional random vector $X = (X_1, \ldots, X_d)$ on $(\mathbb{R}^d, \mathfrak{B}_{\mathbb{R}^d})$, that is, Φ is the probability measure on $(\mathbb{R}^d, \mathfrak{B}_{\mathbb{R}^d})$ defined by $\Phi(E) = P \circ X^{-1}(E)$ for $E \in \mathfrak{B}_{\mathbb{R}^d}$. When Φ is a d-dimensional normal distribution, we say that $X_1, \ldots, X_d$ are jointly normally distributed.

Theorem D.6. Let $X_1, \ldots, X_d$ be real valued random variables on a probability space $(\Omega, \mathfrak{F}, P)$ which are jointly normally distributed and let the probability distribution of $X = (X_1, \ldots, X_d)$ be given by $N_d(m, V)$ with mean vector $m = (m_1, \ldots, m_d)$ and covariance matrix $V = [v_{j,k} : j, k = 1, \ldots, d]$. Then
1) the probability distribution of X_j is given by $N_1(m_j, v_{j,j})$ for $j = 1, \ldots, d$,
2) $\{X_1, \ldots, X_d\}$ is an independent system if and only if the matrix V is diagonal.

Proof. 1) Let $P_{X_1}, \ldots, P_{X_d}$ and P_X be the probability distributions of $X_1, \ldots, X_d$ and X respectively. Let $\varphi_1, \ldots, \varphi_d$ and φ be the characteristic functions of $P_{X_1}, \ldots, P_{X_d}$ and P_X. Then

$$(1) \quad \varphi(y) = \int_{\mathbb{R}^d} \exp\{i\langle x, y\rangle\} P_X(dy) = \int_{\Omega} \exp\{i\langle X, y\rangle\} dP \quad \text{for } y \in \mathbb{R}^d,$$

and for every $j = 1, \ldots, d$ we have

$$(2) \quad \varphi_j(y_j) = \int_{\mathbb{R}} \exp\{i\langle x_j, y_j\rangle\} P_{X_j}(dy) = \int_{\Omega} \exp\{i\langle X_j, y_j\rangle\} dP \quad \text{for } y_j \in \mathbb{R}.$$

By selecting $y = (0, \ldots, 0, y_j, 0, \ldots, 0)$ in (1), we have

$$(3) \quad \varphi(0, \ldots, 0, y_j, 0, \ldots, 0) = \int_{\Omega} \exp\{i\langle X_j, y_j\rangle\} dP = \varphi_j(y_j).$$

Now if the probability distribution of $X = (X_1, \ldots, X_d)$ is given by $N_d(m, V)$, then by Theorem D.5 we have $\varphi(y) = \exp\{i\langle y, m\rangle - \frac{1}{2}\langle Vy, y\rangle\}$ for $y \in \mathbb{R}^d$ and therefore according to (3) we have $\varphi_j(y_j) = \exp\{i\langle y_j, m_j\rangle - \frac{1}{2} v_{j,j} y_j^2\}$ for $y_j \in \mathbb{R}$. This shows that $P_{X_j} = N_1(m_j, v_{j,j})$ for $j = 1, \ldots, d$.

2) Let $\{X_1, \ldots, X_d\}$ be an arbitrary collection of real valued random variables on a probability space $(\Omega, \mathfrak{F}, P)$ and let $\varphi_1, \ldots, \varphi_d$ and φ be the characteristic functions of the probability distributions $P_{X_1}, \ldots, P_{X_d}$ and P_X where $X = (X_1, \ldots, X_d)$. According to Kac's Theorem, $\{X_1, \ldots, X_d\}$ is an independent system if and only if

$$(4) \quad \varphi(y) = \varphi_1(y_1) \cdots \varphi_d(y_d) \quad \text{for } y = (y_1, \ldots, y_d) \in \mathbb{R}^d.$$

Suppose $X_1, \ldots, X_d$ are jointly normally distributed and the probability distribution of X is given by $N_d(m, V)$. Then according to 1), the probability distribution of X_j is given by $N_1(m_j, v_{j,j})$ so that $\varphi_j(y_j) = \exp\{i\langle y_j, m_j\rangle - \frac{1}{2} v_{j,j} y_j^2\}$ for $y_j \in \mathbb{R}$ for $j = 1, \ldots, d$.

Now if V is diagonal, then

$$\varphi(y) = \exp\{i\langle m, y\rangle - \frac{1}{2}\langle Vy, y\rangle\} = \prod_{j=1}^{d} \exp\{im_j y_j - \frac{1}{2} v_{j,j} y_j^2\} = \prod_{j=1}^{d} \varphi_j(y_j).$$

APPENDIX D. MULTIDIMENSIONAL NORMAL DISTRIBUTIONS

Thus by Kac's Theorem, $\{X_1, \ldots, X_d\}$ is an independent system.

Conversely if $\{X_1, \ldots, X_d\}$ is an independent system, then by Kac's Theorem we have

$$\exp\{i\langle m, y\rangle - \frac{1}{2}\langle Vy, y\rangle\} = \prod_{j=1}^{d} \exp\{im_j y_j - \frac{1}{2}v_{j,j}y_j^2\} \quad \text{for } y = (y_1, \ldots, y_d) \in \mathbb{R}^d.$$

Then $\exp\{-\frac{1}{2}\langle Vy, y\rangle\} = \exp\{-\frac{1}{2}\sum_{j=1}^{d} v_{j,j}y_j^2\}$ by equating the real parts on the two sides of the last equality and thus $\langle Vy, y\rangle = \sum_{j=1}^{d} v_{j,j}y_j^2$. This shows that V is diagonal. ∎

Bibliography

[1] Chung, K. L., *A Course in Probability Theory*, 2nd ed., Academic Press, New York, 1974.

[2] Chung, K. L., and Williams, R. J., *Introduction to Stochastic Integration*, 2nd ed., Birkhäuser, Boston, 1990.

[3] Doob, J. L., *Stochastic Processes*, John Wiley and Sons, New York, 1953.

[4] Dudley, R. M., *Real Analysis and Probability*, Wadsworth & Brooks/Cole, 1989.

[5] Fisk, D. L., Quasi-martingales and stochastic integrals, Tech. Rep. 1, Dept. Math. Michigan State University, 1963.

[6] Halmos, P., *Measure Theory*, Van Nostrand, New York, 1950.

[7] Ikeda, N., and Watanabe, S., *Stochastic Differential Equations and Diffusion Processes*, 2nd ed., North-Holland-Kodansha, New York, 1989.

[8] Itô, K., Stochastic integral, Proc. Imp. Acad. Tokyo, **20** (1944), 519-524.

[9] Itô, K., On a stochastic integral equation, Proc. Imp. Acad. Tokyo, **22** (1946), 32-35.

[10] Itô, K., *On stochastic differential equations*, Mem. Am. Math. Soc., **4** (1951).

[11] Itô, K., On a formula concerning stochastic differentials, Nagoya Math. J., **3** (1951), 55-65.

[12] Itô, K., *Theory of Probability*, Iwanami, Tokyo, 1953 (in Japanese).

[13] Itô, K., *Lectures on Stochastic Processes*, Tata Institute of Fundamental Research, Bombay, 1960.

[14] Itô, K., *Stochastic Processes,* Lecture Notes Series, **16**, Aarhus Univ., 1969.

[15] Itô, K., and Watanabe, S., Introduction to stochastic differential equations, Proc. Intern. Symp. SDE Kyoto 1976 (ed. by K. Itô), Kinokuniya, Tokyo, 1978.

[16] Johnson, G., and Helms, L. L., Class (D) supermartingales, Bull. Am. Math. Soc., **69** (1963), 59-62.

[17] Kallianpur, G., *Stochastic Filtering Theory,* Springer-Verlag, New York, 1980.

[18] Kolmogoroff, A. N., Über die analytischen Methoden in der Wahrscheinlichkeitsrechnung, Math. Ann., **104** (1931), 415-458.

[19] Kunita, H., and Watanabe, S., On square integrable martingales, Nagoya Math. J., **30** (1967), 209-245.

[20] McKean, H. P., *Stochastic Integrals,* Academic Press, New York, 1969.

[21] Meyer, P. A., *Probability and Potentials,* Blaisdel, Waltham, Massachusetts, 1966.

[22] Meyer, P. A., Intégrales stochastiques I-IV, Séminaire de Prob. (Univ. de Strasbourg) I, Lecture Notes in Math., **39**, 72-162, Springer Verlag, Berlin, 1967.

[23] Meyer, P. A., Martingales and stochastic integrals I, Lecture Notes in Math. **284**, Springer Verlag, Berlin, 1972.

[24] Meyer, P. A., Un cours sur les intégrales stochastiques, Séminaire de Prob. (Univ. de Strasbourg) X, Lecture Notes in Math., **511**, 245-400, Springer Verlag, Berlin, 1976.

[25] Neveu, J., *Bases Mathématiques du Calcul des Probabilités*, Mason et Cie, Paris, 1964.

[26] Parthasarathy, K. P., *Probability Measures on Metric Spaces,* Academic Press, New York, 1967.

[27] Rao, K. M., On decomposition theorems of Meyer, Math. Scand., **24** (1969), 66-78.

[28] Watanabe, S., *Stochastic Differential Equations,* Sangyō Tosho, Tokyo, 1975 (in Japanese).

[29] Williams, D., *Probability and Martingales,* Cambridge University Press, New York, 1991.

[30] Yamada, T., and Watanabe, S., On the uniqueness of solutions of stochastic differential equations, J. Math. Kyoto Univ., **11** (1971), 155-167.

[31] Yeh, J., *Stochastic Processes and the Wiener Integral,* Marcel Dekker, New York, 1973.

Index

A
augmented,
    filtered space, 149
    filtration, 149
    σ-algebra, 11

B
Borel σ-algebra, 1
Brownian motion,
    1-dimensional, 271
    adapted, 273
    multidimensional, 286
    quadratic variation process, 288
    strong Markov property, 284
bounded variation,
    locally, 207
    a. s. locally, 207

C
class (D), 141
class (DL), 141
conditional dominated convergence theorem, 464
conditional expectation, 454
conditional Fatou's lemma, 463
conditional Hölder's inequality, 468
conditional Jensen's inequality, 466
conditional monotone convergence theorem, 462
conditional probability, 454

    regular, 475
    regular image, 483
countably generated σ-algebra, 10
cylinder set, 358

D
d-class, 4
determining sets of σ-algebra, 481
deterministic initial condition, 396, 397
discontinuity,
    time of, 217
discrete time,
    a.s. increasing process, 79
    increasing process, 79
    L_2-martingale, 119, 120, 121
    predictable process, 79
    stopping time, 39
Doob decomposition theorem, 79
    discrete time L_2-martingale, 121
Doob-Kolmogorov inequality,
    97, 100, 101
Doob-Meyer decomposition theorem, 182
downcrossig, 102, 108
    number of, 102, 108
downcrossing inequalities,
    submartingale, 104, 107, 108
    supermartingale, 106, 107, 109

E
exponential quasimartingale, 335

INDEX

F

filtered space, 14, 274
    augmented, 149
    right-continuous, 14
    right-continuous modification, 149
    standard, 197
filtration, 14
    augmented, 149
    generated by a process, 14, 274
    right-continuous, 14
    right-continuous modification, 149
final element, 73, 116
    existence, 116
    uniformly integrable submartingale, 123
Fisk-Stratonovich integral, 355
Fubini's theorem extended, 164

H

Hermite polynomial, 337

I

increasing process, 158
    almost surely, 158
    natural, 173
    with discrete time, 79
independence, 443
    collections of sets, 443
    events, 443
    π-classes, 444
    random vectors, 446
independent increments, 268
Itô's formula,
    1-dimensional, 320
    multidimensional, 340

K

Kac's theorem, 448

L

L_p-bounded, 71
Lebesgue-Stieltjes measure, 157
    signed, 205
Lipschitz condition, 398, 402

M

martingale,
    definition, 71
    local, 298
    local L_2-, 298
    property, 71
    regular, 187
    reversed time, 124
martingale convergence theorem,
    continuous time, 115
    discrete time, 112
    submartingale with reversed time, 125
    uniformly integrable submartingale, 122
martingale transform, 81
maximal and minimal inequalities, 93, 98, 99
    continuous case, 99
    discrete case, 98
    finite case, 93
measure,
    on a semialgebra, 155
    on an algebra, 155
measurable process, 12
monotone class theorem,
    for functions, 7, 8
    for sets, 5
•-multiplication, 348

N

natural increasing process, 173
normal distribution, 488
null at 0, 71
π-class, 4

O
optional sampling theorem,
    with bounded stopping times, 90, 131
    with unbounded stopping times, 138, 140
optional stopping theorem, 87, 89
outer measure, 156

P
pathwise uniqueness, 397
predictable,
    process, 21
    process with discrete time, 79
    rectangle, 20
    σ-algebra, 19
progressively measurable process, 18

Q
quadratic variation process, 214
    discrete time L_2-martingale, 121
    Brownian motion, 288
quasimartingale, 318
    bounded variation part, 318
    equivalent, 346
    exponential, 335
    martingale part, 318
quasinorm, 50

R
random variable, 12
    at a stopping time, 141
random vector, 262
regular conditional probability, 475

regular image conditional probability, 483
regular submartingale, 186
reversed time, 124
right-continuous modification, 149

S
semialgebra, 155
semimartingale, 318
seminorm, 199
separable σ-algebra, 10
σ-algebra,
    at stopping time, 26
    augmented, 11
    Borel, 1
    countably determined, 481
    countably generated, 10
    predictable, 19
    separable, 10
    well-measurable, 19
signed Lebesgue-Stieltjes measure, 205
standard filtered space, 197
stochastic differential, 346
stochastic differential equation, 368
    definition of solution, 371
    deterministic initial condition, 396
    initial value problem, 385
    strong solution, 419
stochastic independence, 443
stochastic integral,
    of a bounded left-continuous adapted simple process, 223
    of a predictable process, 238
stochastic process, 12
    adapted, 14
    a.s. continuous, 12
    a.s. left-continuous, 12
    a.s. right-continuous, 12

bounded, 71
continuous, 12
equivalent, 12
left-continuous, 12
left-continuous simple, 15
L_p, 71
L_p-bounded, 71
nonnegative, 71
right-continuous, 12
right-continuous simple, 15
measurable, 12
predictable, 21
progressively measurable, 18
simple, 15
stopped, 45
truncated, 47
uniformly integrable, 71
well-measurable, 21
stopped process, 45
stopped submartingale, 132, 134
stopping time, 25
    discrete time, 39
    random variable at, 141
    σ-algebra at, 26
submartingale,
    definition, 71
    property, 71
    regular, 187
supermartingale,
    definition, 71
    property, 71
symmetric multiplication, 352

T
time of discontinuity, 217
total variation function, 206
total variation measure, 206

truncated process, 47

U
Ulam's theorem, 479
uniformly integrable, 54
    random variables at stopping times, 141
uniqueness of solution,
    in probability law, 395
    pathwise, 397
upcrossig, 101, 108
    number of, 101, 108
upcrossing inequalities,
    submartingale, 104, 107, 108
    supermartingale, 106, 107, 109

W
well-measurable,
    process, 21
    σ-algebra, 19
Wiener measure, 362
Wiener space, 362